Processes in Human Evolution

Processes in Human Evolution

The journey from early hominins to Neanderthals and modern humans

Francisco J. Ayala

and

Camilo J. Cela-Conde

UNIVERSITY PRESS

Great Clarendon Street, Oxford, OX2 6DP,
United Kingdom

Oxford University Press is a department of the University of Oxford.
It furthers the University's objective of excellence in research, scholarship,
and education by publishing worldwide. Oxford is a registered trade mark of
Oxford University Press in the UK and in certain other countries

© Francisco J. Ayala and Camilo J. Cela-Conde 2017

The moral rights of the authors have been asserted

First Edition published in 2007

Second Edition published in 2017

Published in the United States of America by Oxford University Press
198 Madison Avenue, New York, NY 10016, United States of America

British Library Cataloguing in Publication Data
Data available

Library of Congress Control Number: 2016934457

ISBN 978–0–19–873990–6 (hbk.)
ISBN 978–0–19–873991–3 (pbk.)

DOI 10.1093/acprof:oso/9780198739906.001.0001

Printed and bound by
CPI Group (UK) Ltd, Croydon, CR0 4YY

Preface

Our earlier book, *Human Evolution. Trails from the Past*, was published in 2007 by Oxford University Press, which is also publishing the present book. The necessity to update it has increasingly become obvious. The numerous human fossils discovered during the last decade, as well as advances in genetics, paleontology, ecology, archaeology, geography, and climate science, have considerably impacted changes in the feasible interpretations of human phylogeny. Eventually, it became obvious to us that more than simply updating the 2007 book, it was necessary to write a completely new and different book.

Processes in Human Evolution differs from the previous book, not only for the large amount of new information, but also because of a distinctive approach. In contrast to *Trails from the Past*, we now follow a more vigorous systematic formulation. We first introduce the research tools appropriate for investigating human evolution and describe the diverse hominin fossils from the Miocene and the Pliocene. We offer more thorough—and ambitious—interpretative methodologies, integrating as far as possible the paleontological, archeological, genetic, geography, taphonomic, environmental, and populational information. We explore fundamental concepts, such as "genus" and "species."

In spite of the different focus, some parts of *Processes* remain similar to *Trails*. We do not explicitly point out which components come from the previous book; it would have been tedious and distracting. The objective is, obviously, to investigate the evolution of the human lineage. The transition of a decade requires, in virtually every case, new observations and interpretations.

An important difference is that in *Processes* we do not explore at any length the derived traits that yield the splendid wealth of linguistic, moral, and esthetic features that distinguish modern humans from all other primates. The study of the evolutionary processes that lead to *Homo sapiens* has now required many more pages than were needed in *Trails from the Past*. Consequently, we have left out the evolutionary origin of language, morality, and other derived traits. This is an important difference between *Processes in Human Evolution* and the previous book.

Contents

1 Evolution, genetics, and systematics   **1**

1.1 The theory of evolution   1
1.2 Population and evolutionary genetics   6
1.3 Taxonomy   29
1.4 The study of the fossil record   40

2 Taxonomy   **45**

2.1 Taxonomic considerations of the tribe Hominini   45
2.2 Traits of the human lineage   54
2.3 Beyond morphology: fossil footprints   65
2.4 Bipedalism and adaptation   71

3 The hominin lineage   **78**

3.1 The origin of hominins   78
3.2 A scenario for the human evolution   83
3.3 The Rift Valley site   91
3.4. South Africa sites   107
3.5 Sites to the north of the Rift Valley   120

4 Miocene and Lower Pliocene hominins   **124**

4.1 Miocene hominins   124
4.2 The role of locomotion in the divergence of hominoid lineages   134
4.3 Change in the Lower Pliocene: genus *Australopithecus*   140
4.4 Australopithecus afarensis   142
4.5 *Australopithecus anamensis*   150
4.6 Miocene human genera   153
4.7 First phylogenetic changes in the tribe Hominini   154
4.8 Phylogenetic relationships of the Miocene and Lower Pliocene hominins   157

5 Middle and Upper Pliocene hominins   **160**

5.1 What can be included in *Australopithecus*?   160
5.2 Australopithecines found outside the Rift: South Africa   161
5.3 Australopithecines found outside the Rift: Chad   166
5.4 The diversification of *Australopithecus* in the Rift Valley during Middle
     and Upper Pleistocene   167

5.5 Adaptation; an Upper Pliocene difference   169
5.6 Dental enamel and diet   169
5.7 The genus *Paranthropus*   172
5.8 Consistency of the evolutionary scheme for the "gracile" and "robust"
     australopithecines in the Middle and Upper Pliocene   182

6 The emergence of the genus *Homo*   **192**

6.1 Homo habilis   192
6.2 The taxon *Homo rudolfensis*   203
6.3 Homo gautengensis   210
6.4 Homo naledi   211
6.5 Homo georgicus   215
6.6 The transition to *Homo*   220
6.7 Monophyly of the first *Homo*   222
6.8 The geographical issue—dispersal of ancient hominins in Africa   225

7 Lithic traditions: tool-making   **229**

7.1 Pre-cultural uses of tools   230
7.2 Taphonomic indications of culture   233
7.3 Mode 1: Oldowan culture   234
7.4 The transition Mode 1 (Oldowan) to Mode 2 (Acheulean)   239
7.5 Beyond tools: the use of fire   250
7.6 The transition Mode 2 (Acheulean) to Mode 3 (Mousterian)   253
7.7 The African Middle Stone Age   261

8 Middle and Lower Pleistocene: the *Homo* radiation   **266**

8.1 Is *Homo erectus* a well-defined species?   266
8.2 The first exit out of Africa   269
8.3 *Homo erectus* characterization   273
8.4 African specimens (*Homo ergaster*)   278
8.5 Asian specimens of *Homo erectus*   282
8.6 The colonization of Europe   304
8.7 An evolutionary model for the hominins of the Lower and Middle
     Pleistocene   335

9 Hominin transition to Upper Pliocene   **345**

9.1 European archaic *Homo sapiens*   346
9.2 African archaic *Homo sapiens*   357
9.3 African hominins from the Mindel–Riss interglacial period   357
9.4 Asian archaic *Homo sapiens*   368
9.5 Are the transitional species between *Homo ergaster* and *Homo sapiens*
     necessary?   377
9.6 *Homo floresiensis*   379

10 Species of the Upper Pleistocene **395**

    10.1 *Homo neanderthalensis* 395
    10.2 *Homo sapiens* 415

11 Neanderthals and modern humans: similarities and differences **441**

    11.1 Genetic distance between *Homo neanderthalensis* and *Homo sapiens* 441
    11.2 Brain distance between Neanderthals and modern humans 468
    11.3 Cognitive distance between Neanderthals and modern humans 474

Glossary 495
References 503
Index 553

CHAPTER 1

Evolution, genetics, and systematics

1.1 The theory of evolution

All organisms are related by descent from common ancestors. Humans and other mammals descended from shrew-like creatures that lived more than 150 million years ago; mammals, birds, reptiles, amphibians, and fishes share as ancestors aquatic worms that lived 600 million years ago; and all plants and animals derive from bacteria-like microorganisms that originated more than 3 billion years ago. Biological evolution is a process of descent with modification. The process consists of two components: lineages of organisms change through the generations (*anagenesis* or phyletic evolution); diversity arises because the lineages that descend from common ancestors diverge through time (*cladogenesis* or lineage splitting, the process by which new species arise).

1.1.1 Charles Darwin

The founder of the modern theory of evolution was Charles Darwin (1809–82), the son and grandson of physicians. He enrolled as a medical student at the University of Edinburgh. After two years, however, he left Edinburgh and moved to the University of Cambridge to pursue his studies and prepare to become a clergyman. Darwin was not an exceptional student, but he was deeply interested in natural history. On December 27, 1831, a few months after his graduation from Cambridge, he sailed as a naturalist aboard the *HMS Beagle* on a round-the-world trip that lasted until October 1836. Darwin was often able to disembark for extended trips ashore to collect natural specimens. The discovery of fossil bones from large extinct mammals in Argentina and the observation of numerous species of finches in the Galápagos Islands were among the events credited with stimulating Darwin's interest in how species originate.

The observations he made in the Galápagos Islands may have been the most influential on Darwin's thinking. The islands, on the equator 1,000 km (600 miles) off the west coast of South America, had been named Galápagos (the Spanish word for tortoises) by the Spanish discoverers because of the abundance of giant tortoises, different on different islands, and different from those known anywhere else in the world. The tortoises sluggishly clanked their way around, feeding on the vegetation and seeking the few pools of fresh water. They would have been vulnerable to predators, but these were conspicuously absent on the islands. In the Galápagos, Darwin found large lizards, feeding unlike any others of their kind, on seaweed, and mockingbirds, quite different from those found on the South American mainland. Well known is that he found several kinds of finches, varying from island to island in various features, notably their distinctive beaks, adapted to disparate feeding habits: crushing nuts, probing for insects, grasping worms, etc.

In addition to *On the Origin of Species by Means of Natural Selection* (1859), Darwin published many other books, notably *The Descent of Man and Selection in Relation to Sex* (1871), which extends the theory of natural selection to human evolution.

Darwin's theory of natural selection is summarized in *On the Origin of Species* (pp. 80–81) as follows:

Can it, then, be thought improbable, seeing that variations useful to man have undoubtedly occurred, that other variations useful in some way to each being in the great and complex battle of life, should sometimes occur in the course of thousands of generations? If such do occur, can we doubt (remembering that more individuals are born than can possibly survive) that individuals having any advantage, however slight, over others, would have the best chance of surviving and of procreating their kind? On the other hand, we may feel sure that any variation in the least degree injurious would be rigidly destroyed. This preservation of favourable variations and the rejection of injurious variations, I call Natural Selection.

The argument consists of three parts: (1) hereditary variations occur, some more favorable than others to the organisms; (2) more organisms are produced than can possibly survive and reproduce; (3) organisms

Processes in Human Evolution. Francisco J. Ayala and Camilo J. Cela-Conde, Oxford University Press (2017).
© Francisco J. Ayala and Camilo J. Cela-Conde.
DOI 10.1093/acprof:oso/9780198739906.001.0001

Box 1.1 First scientific proposals on evolution

Human cultures have advanced explanations for the origin of the world and of human beings and other creatures. Traditional Judaism and Christianity explain the origin of living beings and their adaptations to life in their environments—legs and wings, gills and lungs, leaves and flowers—as the handiwork of the Creator.

The first broad theory of evolution was proposed by the French naturalist Jean-Baptist de Monet, chevalier de Lamarck (1744–1829). In his *Philosophie Zoologique* (1809, "Zoological Philosophy"), Lamarck held the enlightened view, shared by the intellectuals of his age, that living organisms represent a progression, with humans as the highest form. Lamarck's theory of evolution asserts that organisms evolve through eons of time from lower to higher forms, a process still going on, always culminating in human beings. The remote ancestors of humans were worms and other inferior creatures, which gradually evolved into more and more advanced organisms, ultimately humans.

Lamarck's evolution theory was metaphysical rather than scientific. Lamarck postulated that life possesses an innate tendency to improve over time, so that progression from lower to higher organisms would continually occur, and always follow the same path of transformation from lower organisms to increasingly higher and more complex organisms. A somewhat similar evolution theory was formulated one century later by another Frenchman, the philosopher Henri Bergson (1859–1940) in his *L'Evolution Créatrice* (1907, "Creative Evolution").

Erasmus Darwin (1731–1802), a physician and poet, and the grandfather of Charles Darwin, proposed, in poetic rather than scientific language, a theory of the transmutation of life forms through eons of time (*Zoonomia, or the Laws of Organic Life*, 1794–96). More significant for Charles Darwin was the influence of his older contemporary and friend, the eminent geologist Sir Charles Lyell (1797–1875). In his *Principles of Geology* (1830–33), Lyell proposed that the earth's physical features were the outcome of major geological processes acting over immense periods of time, incomparably greater than the few thousand years since Creation generally assumed at the time.

Natural selection was proposed by Darwin primarily to account for the adaptive organization of living beings; it is a process that promotes and maintains adaptation. Evolutionary change through time and evolutionary diversification (multiplication of species) are not directly promoted by natural selection, but they often ensue as by-products of natural selection, as it fosters adaptation to different environments.

Wallace's independent discovery of natural selection is remarkable (see Box 1.2). But Wallace's interest and motivation was not the explanation of design but how to account for the evolution of species, as indicated in his paper's title: "On the tendency of varieties to depart indefinitely from the original type." Wallace thought that evolution proceeds indefinitely and is progressive. Darwin, on the contrary, did not accept

---

Box 1.2 Wallace and Darwin

Alfred Russell Wallace (1823–1913) is famously given credit for discovering, independently from Darwin, natural selection as the process accounting for the evolution of species. On June 18, 1858, Darwin wrote to Charles Lyell that he had received by mail a short essay from Wallace such that "if Wallace had my [manuscript] sketch written in [1844] he could not have made a better abstract." Darwin was thunderstruck. In 1858, Wallace had come upon the idea of natural selection as the explanation for evolutionary change and he wanted to know Darwin's opinion about this hypothesis, since Wallace, as well as many others, knew that Darwin had been working on the subject for years, had shared his ideas with other scientists, and was considered by them as the eminent expert on issues concerning biological evolution. Darwin hesitated as to how to proceed about Wallace's letter. He wanted to credit Wallace's discovery of natural selection, but he did not want altogether to give up his own earlier independent discovery. Eventually, two of Darwin's friends, the geologist Sir Charles Lyell and the botanist Joseph Hooker, proposed, with Darwin's consent, that Wallace's letter and two earlier writings of Darwin would be presented at a meeting of the Linnaean Society of London. On July 1, 1858, three papers were read by the society's undersecretary, George Busk, in the order of their date of composition: two short essays that Darwin had written in 1844 and 1857, and Wallace's essay, "On the tendency of varieties to depart indefinitely from the original type." The meeting was attended by some thirty people, who did not include Darwin or Wallace.

---

with more favorable variations will survive and reproduce more successfully. Two consequences follow: (1) organisms are adapted to the environments where they live because of the successful reproduction of favorable variations; (2) evolutionary change will occur over time.

that evolution would necessarily represent progress or advancement. Nor did he believe that evolution would always result in morphological change over time; rather, he knew of the existence of "living fossils," organisms that had remained unchanged for millions of years. For example, "some of the most ancient Silurian animals, as the Nautilus, Lingula, etc., do not differ much from living species."

1.1.2 The origin of species

The publication of *On the Origin of Species by Means of Natural Selection* in 1859 took the British scientific community of the mid-nineteenth century by storm. It also caused considerable public excitement. Scientists, politicians, clergymen, and notables of all kinds read and discussed the book, defending or deriding Darwin's ideas. The most visible actor in the controversies immediately following publication was the English biologist T. H. Huxley, known as "Darwin's bulldog," who defended the theory of evolution with articulate and sometimes mordant words, on public occasions as well as in numerous writings.

One important reason why Darwin's theory of evolution by natural selection encountered resistance among Darwin's contemporaries and beyond was the lack of an adequate theory of inheritance that would account for the preservation, through the generations, of the variations on which natural selection was supposed to act. Contemporary theories of "blending inheritance" proposed that offspring merely struck an average between the characteristics of their parents. Darwin's own theory of "pangenesis" proposed that each organ and tissue of an organism throws off tiny contributions of itself that are collected in the sex organs and determine the configuration of the offspring. These theories of blending inheritance could not account for the conservation of variations, because differences between variant offspring would be halved each generation, as a parent with a favorable variant crossed with a parent with the preexisting variant, rapidly reducing the favorable variation to the average of the preexisting characteristics.

1.1.3 Gregor Mendel

The missing link in Darwin's argument was provided by Mendelian genetics. About the time the *On the Origin of Species* was published, the Augustinian monk Gregor Mendel was starting a long series of experiments with peas in the garden of his monastery in Brünn, Austria–Hungary (now Brno, Czech Republic). Mendel's (1866) paper formulated the fundamental principles of the theory of heredity that is still current. His theory accounts for biological inheritance through particulate factors (now known as genes) inherited one from each parent, which do not mix or blend but segregate in the formation of the sex cells or gametes.

Darwinism in the latter part of the nineteenth century faced an alternative evolutionary theory known as neo-Lamarckism. This hypothesis shared with Lamarck's the importance of use and disuse in the development and obliteration of organs, and it added the notion that the environment acts directly on organic structures, which explained their adaptation to the way of life and environment of the organism. Adherents of this theory discarded natural selection as an explanation for adaptation to the environment.

Mendel's discoveries remained unknown to Darwin, however, and, indeed, they did not become generally known until 1900, when they were simultaneously rediscovered by a number of scientists on the Continent. Meanwhile, prominent among the defenders of natural selection was the German biologist August Weismann, who in the 1880s published his germ-plasm theory. He distinguished two substances that make up an organism: the soma, which comprises most body parts and organs, and the germ plasm, which contains the cells that give rise to the gametes and hence to progeny. Early in the development of an embryo, the germ plasm becomes segregated from the somatic cells that give rise to the rest of the body. This notion of a radical

---

Box 1.3 Herbert Spencer

A younger English contemporary of Darwin, with considerable influence during the latter part of the nineteenth and in the early twentieth century, was Herbert Spencer. A philosopher rather than a biologist, he became an energetic proponent of evolutionary ideas, popularized a number of slogans, such as "survival of the fittest" (which was taken up by Darwin in later editions of *On the Origin of Species*), and engaged in social and metaphysical speculations (which Darwin thoroughly disliked). His ideas considerably damaged proper understanding and acceptance of the theory of evolution by natural selection. Most pernicious was the crude extension by Spencer and others of the notion of the "struggle for existence" to human economic and social life that became known as Social Darwinism.

separation between germ plasm and soma—that is, between the reproductive tissues and all other body tissues—prompted Weismann to assert that inheritance of acquired characteristics was impossible, and it opened the way for his championship of natural selection as the only major process that would account for biological evolution. Weismann's ideas became known after 1896 as neo-Darwinism.

Mendel's theory of heredity was rediscovered in 1900 by the Dutch botanist Hugo de Vries, the German Carl Correns, and others. Mendel's theory provided a suitable mechanism for the natural selection of hereditary traits. But a controversy arose between two extreme ways of thinking. On one side were those who thought that the kind of characters transmitted by Mendelian heredity were not significant for natural selection (because they concerned very small, "continuous," variations among individuals) and those who thought Mendelian heredity was all, or most, that there was in evolution, with natural selection relegated to a minor role, or no role at all.

De Vries himself proposed a new theory of evolution known as mutationism, which essentially did away with natural selection as a major evolutionary process (see Box 1.4).

Many naturalists and some mathematicians, particularly in Britain but also on the Continent, rejected de Vries' mutationism, and even Mendelian heredity, as irrelevant to natural selection, because mutations produced only large, even monstrous morphological variations, while natural selection depends on minor variations impacting, most of all, lifespan and fertility. Among these scientists, called biometricians, was the English statistician Karl Pearson, who defended Darwinian natural selection as the major cause of evolution through the cumulative effects of small, continuous, individual variations (which the biometricians assumed passed from one generation to the next without being limited by Mendel's laws of inheritance).

1.1.4 The synthetic theory of evolution

The controversy between mutationists and biometricians approached a resolution in the 1920s and 1930s through the theoretical work of geneticists, such as R. A. Fisher and J. B. S. Haldane in Britain and Sewall Wright in the United States These scientists used mathematical arguments to show (1) that continuous variation (in such characteristics as body size, number of progeny, and the like) could be explained by Mendel's laws, and (2) that natural selection acting cumulatively on small variations could yield major evolutionary changes in form and function.

The synthesis of Darwin's theory of natural selection and Mendelian genetics became generally accepted by biologists only in the mid-twentieth century, after the publication of several important books by biologists who provided observations and experimental results that supported the formulations of the mathematical theorists. One important publication, in 1937, was *Genetics and the Origin of Species* by Theodosius Dobzhansky, a Russian-born American naturalist and experimental geneticist. Dobzhansky's book advanced a reasonably comprehensive account of the evolutionary process in genetic terms, laced with experimental evidence supporting the theoretical argument. *Genetics and the Origin of Species* had an enormous impact on naturalists and experimental

---

Box 1.4 The meaning of variation

According to de Vries (and other geneticists, such as William Bateson in England), two kinds of variation take place in organisms. One is the "ordinary" variability observed among individuals of a species, such as small differences in color, shape, and size. This variability would have no lasting consequence in evolution because, according to de Vries, it could not "lead to a transgression of the species border even under conditions of the most stringent and continued selection." The variation that is significant for evolution are the changes brought about by mutations, spontaneous alterations of genes that result in large modifications of the organism, which may give rise to new species: "The new species thus originates suddenly, it is produced by the existing one without any visible preparation and without transition."

Box 1.5 Impact of Darwin's theory

The theoretical framework for the integration of genetics into Darwin's theory of natural selection had a limited impact on contemporary biologists because it was formulated in a mathematical language that most biologists could not understand and was presented with little empirical corroboration.

biologists, who rapidly embraced the new understanding of the evolutionary process as one of genetic change in populations. Other significant contributions were: *Systematics and the Origin of Species from the Viewpoint of a Zoologist* (1942) by the German-born American zoologist Ernst Mayr; *Evolution: the Modern Synthesis* (1942) by the English zoologist Julian Huxley; *Tempo and Mode in Evolution* (1944) by the American paleontologist George Gaylord Simpson; *Variation and Evolution in Plants* (1950) by the American botanist George Ledyard Stebbins. The "synthetic theory of evolution," as it became known, elaborated by these scientists contributed to a burst of evolutionary studies in the traditional biological and paleontological disciplines and stimulated the development of new disciplines, such as population and evolutionary genetics, evolutionary ecology, and paleobiology (Figure 1.1).

1.1.5 Molecular evolution

In the second half of the twentieth century, population genetics and evolutionary genetics became very active disciplines, which eventually incorporated molecular biology, a new discipline that emerged from the 1953 discovery by James Watson and Francis Crick of the molecular structure of DNA, the hereditary chemical contained in the chromosomes of every cell nucleus. The genetic information is encoded within the sequence of nucleotides that make up the chainlike DNA molecules. This information determines the sequence of amino acid building blocks of protein molecules, which include structural proteins, as well as the numerous enzymes that carry out the organism's fundamental life processes. Genetic information could now be investigated by examining the sequences of amino acids in the proteins, and eventually the sequences of the nucleotides that make up the DNA.

In the mid-1960s, laboratory techniques such as electrophoresis and selective assay of enzymes became available for the rapid and inexpensive study of differences among enzymes and other proteins. These techniques made possible the pursuit of evolutionary issues, such as quantifying genetic variation in natural populations (which variation sets bounds on the evolutionary potential of a population) and determining the amount of genetic change that occurs during the formation of new species. Comparisons of the amino acid sequences of corresponding proteins in different species provided precise measures of the divergence among species evolved from common ancestors, a considerable improvement over the typically qualitative evaluations obtained by comparative anatomy and other evolutionary subdisciplines.

The laboratory techniques of DNA cloning and sequencing have provided a new and powerful means of investigating evolution at the molecular level. The fruits of this technology began to accumulate during the 1980s following the development of automated DNA-sequencing machines and the invention of the polymerase chain reaction (PCR), a simple and inexpensive technique that obtains, in a few hours, billions or trillions of copies of a specific DNA sequence or gene. Major research efforts, such as the Human Genome Project, further improved the technology for obtaining long DNA sequences rapidly and

Figure 1.1 G. Ledyard Stebbins, George Gaylord Simpson, and Theodosius Dobzhansky (left to right), three main authors of the modern theory of evolution, in 1970, at a conference in the University of California, Davis, organized by Francisco J. Ayala.

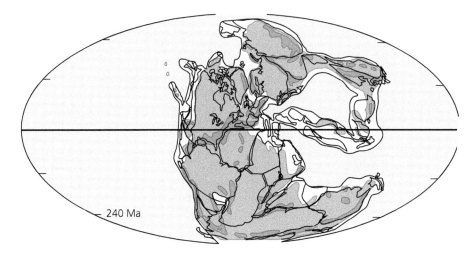

Figure 1.2 Pangea. About 240 million years ago, in the Lower Triassic, most of the continents' land was aggregated into a single mass.

inexpensively. By the first few years of the twenty-first century, the full DNA sequence—i.e., the full genetic complement or genome—had been obtained for more than 20 higher organisms, including human beings. A draft of the chimpanzee genome was published in 2005, followed by the orangutan genome in 2011, and the gorilla genome in 2012. Rapid advances have also occurred in the study of evolutionary developmental biology, which has become known as "evo-devo."

In the second half of the twentieth century, the earth sciences also experienced a conceptual revolution of great consequence for the study of evolution. The theory of plate tectonics, which was formulated in the late 1960s, revealed that the configuration and position of the continents and oceans are dynamic features of the earth. Oceans grow and shrink, while continents break into fragments or coalesce into larger masses, altering the face of the planet and causing major climatic changes along the way. The consequences for the evolutionary history of life are enormous. Thus, biogeography, the evolutionary study of plant and animal geographic distribution, has been revolutionized by the knowledge, for example, that Africa and South America were part of a single landmass some 200 million years ago and that the Indian subcontinent was not connected with Asia until geologically recent times (Figure 1.2).

New methods for dating fossils, rocks, and other materials have made it possible to determine, with much greater precision than ever before, the age of the geological periods and of the fossils themselves. This

has greatly contributed to advances in paleontology, and to the emergence of the new field of paleobiology. Increased interest and investment have favored, in particular, paleoanthropology, which has experienced a notable acceleration in the rate of discovery and investigation of hominid remains and their associated faunas and habitats. We will review these discoveries throughout this book.

Finally, ecology, the study of the interactions of organisms with their environments, has evolved from descriptive studies—"natural history"—into a vigorous biological discipline with a strong mathematical component, both in the development of theoretical models and in the collection and analysis of quantitative data. Evolutionary ecology has become a very active field of evolutionary studies. Major advances have also occurred in evolutionary ethology, the study of the evolution of animal behavior. Sociobiology, the evolutionary study of social behavior, is perhaps the most active subfield of ethology and surely the most controversial, because it seeks to explain human behavior and human societies similarly as animal social behavior, as largely determined by their genetic make-up.

1.2 Population and evolutionary genetics

1.2.1 Evolution by natural selection

Biological evolution is the process of change and diversification of living things over time, and it affects all aspects of their lives—morphology (form and structure), physiology, behavior, and ecology.[1] Underlying these changes are genetic changes.

In genetic terms, the process of evolution consists of changes through time in the genetic make-up of populations. Evolution can be seen as a two-step process: first, hereditary variation arises; second, selection is made of those genetic variants that will be passed on most effectively to the following generations. The origin of hereditary variation also entails two mechanisms: the spontaneous mutation of one variant into another; and the sexual process that recombines those variants to form a multitude of new arrangements of the variations. Selection, the second step of the evolution process, occurs because the variants that arise by mutation and recombination are not transmitted equally from one generation to another. Some may appear more frequently in the progeny because they are favorable to the organisms carrying them, which thereby leave more progeny. Other factors affect the transmission frequency of hereditary variations, particularly chance, a process called genetic drift.

Epigenetic processes yielding a stable and heritable phenotype complete the evolutionary landscape. In humans, as in other multicellular organisms, epigenetic processes impact cells and genotypes, starting in the early stages of embryonic development. The genotype of the fertilized egg is fully transmitted to the descendant cells, yet different cells develop differently: some become muscle, others skin, some develop into a kidney, others into a liver, and so on.

Darwin's argument of evolution by natural selection starts with the existence of hereditary variation. Experience with animal and plant breeding had demonstrated to Darwin that variations that are "useful to man" can be found in organisms. So, he reasoned, variations must occur in nature that are favorable or useful in some way to the organism itself in the struggle for existence. Favorable variations are ones that increase chances for survival and procreation. Those advantageous variations are preserved and multiplied from generation to generation at the expense of

less-advantageous ones. This is the process known as natural selection. The outcome of the process is an organism that is well adapted to its environment, and evolution often occurs as a consequence.

Natural selection, then, can be defined as the differential reproduction of alternative hereditary variants, determined by the fact that some variants increase the likelihood that the organisms having them will survive and reproduce more successfully than will organisms carrying alternative variants. Selection may occur as a result of differences in survival, in fertility, in rate of development, in mating success, or in any other aspect of the life cycle. All of these differences can be incorporated under the phrase "differential reproduction" because all result in natural selection to the extent that they affect the number of progeny an organism leaves.

Darwin maintained that competition for limited resources results in the survival of the most effective competitors. Nevertheless, natural selection may occur not only as a result of competition, but also as a result of some aspect of the physical environment, such as inclement weather or the action of predators. Moreover, natural selection would occur even if all the members of a population died at the same age, simply because some of them would have produced more offspring than others. Natural selection is quantified by a measure called Darwinian fitness or relative fitness. The fitness of a trait in this sense is the relative probability that a hereditary characteristic will be reproduced; that is, the degree of fitness is a measure of the reproductive efficiency of the characteristic. Fitness is also predicated of a particular set of hereditary characteristics, and even of the whole genome, always measured by their reproductive efficiency compared to their alternatives.

1.2.2 Deoxyribonucleic acid: DNA

Two related polynucleotides in organisms are: DNA (deoxyribonucleic acid) and RNA (ribonucleic acid). The hereditary chemical in most organisms is DNA; in some viruses, such as HIV, it is RNA. But RNA fulfills important functions in all organisms, such as being the messenger (mRNA) conveying the information encoded in DNA from the nucleus into the body of the cell, where it directs protein synthesis, as well as the transfer (tRNA), which brings the individual amino acids that are successively added to protein (polypeptide) chains following the instructions conveyed by mRNA. There also are other kinds of RNA molecules, such as microRNA (miRNA)—very short molecules,

---

[1] We introduce in Section 1.2 some fundamental concepts of genetics, particularly those that are relevant for understanding evolution. Students in general, but particularly those who have had a college-level course in genetics, may want to skip the subchapter, or refer to it only when seeking understanding of some particular issues. There are numerous genetics textbooks, as well as texts focused on population and evolutionary genetics, where the concepts introduced in this chapter are developed in greater detail.

typically consisting of 22 nucleotides, which are directly transcribed from the DNA, and perform important functions in gene regulation, which include early development in mammals. Hundreds of genes encoding miRNAs have been identified in animals and many more are predicted (Berezikov et al., 2006); in plants the number of known miRNAs is smaller (Mallory & Vaucheret, 2006). In animals, including humans, it is estimated that the expression of more than one-third of all genes is controlled by miRNAs.

Nucleic acids are long polymers of a basic unit, the nucleotide. A nucleotide is composed of three distinct chemical parts joined by covalent bolds. One part is a pentose sugar: deoxyribose in DNA and ribose in RNA. The second part is a nitrogenous base, which in DNA can be either a purine: adenine (A) or guamine (G), or a pyrimidine: cytosine (C) or thymine (T). RNA contains the same bases as DNA, except that it contains uracil (U), rather than T. The third part of the nucleotide is a phosphate group, which forms the joint between successive nucleotides by phosphodiester bridges between the 5'-carbon of one sugar moiety and the 3'-carbon of another. The 5'-3' links establish directionality in the nucleic acids (Figure 1.3).

DNA molecules consist of two chains of nucleotides paired in a double helix. The pairing is effected by hydrogen bonds between the nucleotides of the two strands, so that the pairing is always between A and T or between G and C (Figure 1.4). The genetic information is conveyed by linear sequences of these letters, similarly as semantic information is conveyed by sequences of the 26 letters of the English alphabet.

During replication, the two strands of the DNA double helix separate and each becomes a template for a complementary strand. Because of the strict rules of pairing, the two daughter molecules are identical to the mother molecule and to each other. This identity accounts for the fidelity of biological heredity.

The DNA of eukaryotic organisms is organized into chromosomes, which consist of several kinds of histone proteins associated with the DNA. The chromosomes occur in pairs, one inherited form each parent. The number of chromosomes, characteristic of each species, varies broadly from only one pair, as in some parasitic nematodes, to more than 100, as in some species of butterflies, and to more than 600, as in some ferns. Humans have 23 pairs of chromosomes. Other primates have 24 pairs—two of their chromosomes fused into one, chromosome 2, in our hominin ancestors. In all primates the two chromosomes of a certain pair are identical in females (XX) but not in males (XY).

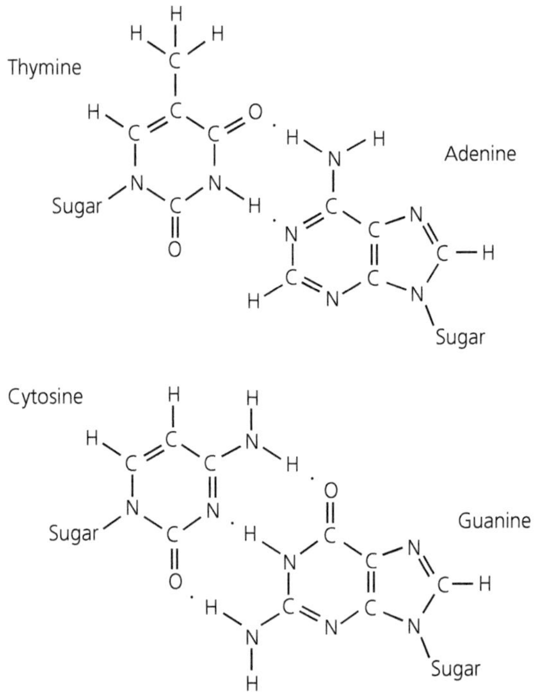

Figure 1.3 The four nitrogen bases of DNA: adenine (A), cytosine (C), guanine (G), and thymine (T). In the double helix, the bases of the two complementary strands are held together by hydrogen bonds, two between A and T, three between C and G.

A gene is a DNA segment that becomes *transcribed* into mRNA, which in turn becomes *translated* into a *polypeptide*, that is, a protein or part of a protein. (Some proteins consist of several polypeptides; e.g., hemoglobin A, the most common in adult humans, consists of four polypeptides, two of each of two different kinds, called α and β.) The number of protein-encoding genes is about 30,000 in primates and other mammals, 13,000 in *Drosophila* fruit flies, and 5,000 in yeast. Some plants, such as *Arabidopsis*, seem to have nearly as many genes as mammals. Most of the DNA of eukaryotes, which does not embrace genes, is often called *junk DNA* and a good part of it consists of sequences of various lengths, some quite small but repeated many thousands or even millions of times, such as the *Alu* sequences of the human genome. Much of the junk DNA may not be functional at all, but some sequences, such as those encoding the microRNAs, play a role in regulating the transcription or translation of other DNA sequences.

The coding part of a gene often occurs in parts (*exons*) that are separated by segments of non-coding DNA, called *introns*. Typically, a gene is preceded by

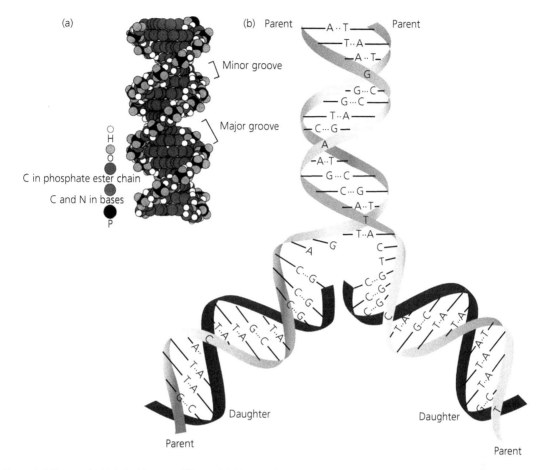

Figure 1.4 The DNA double helix. (a) A space-filling model. (b) Mode of replication: the two strands separate and each one serves as a template for the synthesis of a complementary strand, so that the two daughter double helices are identical to each other and to the original molecule.

untranscribed DNA sequences, usually short, that regulate its transcription. The rules that determine the translation of mRNA into proteins are known as the *genetic code*. Particular combinations of three consecutive nucleotides (*codons* or triplets) specify particular amino acids, out of the 20 that make up proteins (Figure 1.5). Tryptophan and methionine are specified each by only one codon; all others are specified by several, from two to six, which are said to be synonymous. Three codons are *stop* signals that indicate termination of the translation process.

Chemical reactions in organisms must occur in an orderly manner; organisms, therefore, have ways of switching genes on and off, since different sets of genes are active in different cells. Typically, a gene is turned on and off by a system of several switches acting on short DNA sequences adjacent to the coding part of the gene. There are switches acting on

a given gene activated or deactivated by feedback loops that involve molecules synthesized by other genes, as well as molecules present in the cell's environment. A variety of gene control mechanisms have been discovered, first in bacteria and other microorganisms.

The investigation of gene control mechanisms in mammals (and other complex organisms) became possible in the mid-1970s with the development of recombinant DNA techniques. This technology made it feasible to isolate single genes (and other DNA sequences) and to clone them, in billions of identical copies, in order to obtain the quantities necessary for ascertaining their nucleotide sequence. In mammals, insects, and other complex organisms, there are control circuits and master switches (such as the so-called *homeobox* genes) that operate at higher levels than the control mechanisms that activate and deactivate

Second position

First position		U	C	A	G	Third position
U	UUU	Phe	UCU (Ser)	UAU (Tyr)	UGU (Cys)	U
U	UUC	Phe	UCC (Ser)	UAC (Tyr)	UGC (Cys)	C
U	UUA	Leu	UCA (Ser)	UAA Stop	UGA Stop	A
U	UUG	Leu	UCG (Ser)	UAG Stop	UGG Trp	G
C	CUU	Leu	CCU (Pro)	CAU (His)	CGU (Arg)	U
C	CUC	Leu	CCC (Pro)	CAC (His)	CGC (Arg)	C
C	CUA	Leu	CCA (Pro)	CAA (Gln)	CGA (Arg)	A
C	CUG	Leu	CCG (Pro)	CAG (Gln)	CGG (Arg)	G
A	AUU	Ile	ACU (Thr)	AAU (Asn)	AGU (Ser)	U
A	AUC	Ile	ACC (Thr)	AAC (Asn)	AGC (Ser)	C
A	AUA	Ile	ACA (Thr)	AAA (Lys)	AGA (Arg)	A
A	AUG	Met	ACG (Thr)	AAG (Lys)	AGG (Arg)	G
G	GUU	Val	GCU (Ala)	GAU (Asp)	GGU (Gly)	U
G	GUC	Val	GCC (Ala)	GAC (Asp)	GGC (Gly)	C
G	GUA	Val	GCA (Ala)	GAA (Glu)	GGA (Gly)	A
G	GUG	Val	GCG (Ala)	GAG (Glu)	GGG (Gly)	G

Figure 1.5 The genetic code. Each set of three consecutive letters ("codon") in the DNA determines one amino acid in the encoded protein. DNA codes for RNA ("transcription"), which codes for amino acids ("translation"). RNA uses uracil (U) rather than thymine (T). The 20 amino acids making up proteins (with their three-letter and one-letter representations) are as follows: alanine (Ala, A), arginine (Arg, R), asparagine (Asn, N), aspartic acid (Asp, D), cysteine (Cys, C), glycine (Gly, G), glutamic acid (Glu, E), glutamine (Gln, Q), histidine (His, H), isoleucine (Ile, I), leucine (Leu, L), lysine (Lys, K), methionine (Met, M), phenylalanine (Phe, F), proline (Pro, P), serine (Ser, S), threonine (Thr, T), tyrosine (Tyr, Y), Tryptophan (Trp, W), and valine (Val, V).

individual genes. These higher level switches act on sets rather than individual genes. The details of how these sets are controlled, how many control systems there are, and how they interact remain largely to be elucidated, although great advances in evo-devo (short for "evolution and development"; see Section 1.2.8) have been made in recent years.

1.2.3 Mutation

The nucleotide sequence of the DNA is, as a rule, faithfully reproduced during replication. But heredity is not a perfectly conservative process—otherwise, evolution could not have taken place. Occasionally "mistakes," or mutations, occur in the DNA molecule during replication, so that daughter cells differ from the parent cells in the sequence or in the amount of DNA. A mutation first appears in a single cell of an organism, but it is passed on to all cells descended from the first. Mutations occur in all sorts of cells, but the mutations that count in evolution are those that occur in the sex cells (eggs and sperm), or in cells from which the sex cells derive, because these are the cells that produce the offspring.

Mutations can be classified into two categories—gene or point mutations, which affect only a few nucleotides within a gene, and chromosomal mutations, which either change the number of chromosomes or change the number or arrangement of genes on a chromosome.

A gene mutation may be either a substitution of one or a few nucleotides for others or an insertion or deletion of one or a few pairs of nucleotides. Substitutions in the nucleotide sequence of a structural gene may result in changes in the amino acid sequence of the protein, although this is not always the case. Consider the triplet AUA, which codes for the amino acid isoleucine. If the last A is replaced by C, the triplet still codes for isoleucine, but if it is replaced by G, it codes for methionine instead (see Figure 1.6.).

A nucleotide substitution in the DNA that results in an amino acid substitution in the corresponding protein may or may not severely affect the biological function of the protein. Some nucleotide substitutions change a codon for an amino acid into a signal ("stop" codon) that terminates translation. Those mutations are likely to have harmful effects. If, for instance, the second U in the triplet UUA, which codes for leucine, is replaced by A, the triplet becomes UAA, a stop codon; the result is that the triplets following this codon in the DNA sequence are not translated into amino acids.

Additions or deletions of nucleotides within the DNA sequence of a structural gene often result in a greatly altered sequence of amino acids in the coded protein. The addition or deletion of one or two nucleotides

shifts the "reading frame" of the nucleotide sequence all along the way from the point of the insertion or deletion to the end of the molecule. To illustrate, assume that the DNA segment . . . CATCATCATCATCAT . . . is read in groups of three as . . . CAT-CAT-CAT-CAT-CAT. . . . If a nucleotide base—say, T—is inserted after the first C of the segment, the segment will then be read as . . . CTA-TCA-TCA-TCA-TCA. . . . From the point of the insertion onward, the sequence of encoded amino acids is altered. If, however, a total of three nucleotides is either added or deleted, the original reading frame will be maintained in the rest of the sequence. Additions or deletions of nucleotides in numbers other than three or multiples of three are called frameshift mutations.

1.2.4 Effects of mutation

Newly arisen mutations are more likely to be harmful than beneficial to their carriers, because mutations are random events with respect to adaptation—that is, their occurrence is independent of any possible consequences. The allelic variants present in an existing population have already been subject to natural selection. Most are present in the population because they improve the adaptation of their carriers; their alternative alleles have been eliminated or kept at low frequencies by natural selection. A newly arisen mutation is likely to have been preceded by an identical mutation in the previous history of a population. If the previous mutation no longer exists in the population, or it exists at very low frequency, it is a sign that the new mutation is not likely to be beneficial to the organism and is likely also to be eliminated.

Occasionally, however, a new mutation may increase the organism's adaptation. The probability of such an event's happening is greater when organisms colonize a new territory or when environmental changes confront a population with new challenges. In these cases the established adaptation of a population is less than optimal and there is greater opportunity for new mutations to be better adaptive. The consequences of mutations depend on the environment. Increased melanin pigmentation may be advantageous to inhabitants of tropical Africa, where dark skin protects them from the sun's ultraviolet radiation, but it is not beneficial in Scandinavia, where the intensity of sunlight is low and light skin facilitates the synthesis of vitamin D in the deeper layers of the dermis.

Mutation rates have been measured in a great variety of organisms, mostly for mutants that exhibit conspicuous effects. Mutation rates are generally lower

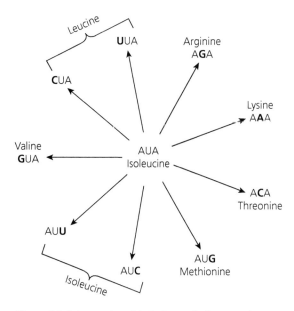

Figure 1.6 Point mutations. Substitutions at the first, second, or third position in the messenger-RNA codon for the amino acid isoleucine can give rise to nine new codons that code for six different amino acids. The effects of a mutation depend on what change takes place: arginine, threonine, and lysine have chemical properties that differ sharply from those of isoleucine.

in bacteria and other microorganisms than in more complex species. In humans and other multicellular organisms, the rate for any given mutation typically ranges from about 1 per 100,000 to 1 per 100,000,000 gametes. There is considerable variation from gene to gene, as well as from organism to organism. Moreover, there are different ways of measuring mutation rates; for example, rates with respect to changes in any given nucleotide of the DNA sequence of a gene or with respect to any change in any given gene (which encompasses hundreds or thousands of DNA nucleotides). Also, rates are quite different for gene mutations in the strict sense and for reorganizations, duplications, and deletions of sets of genes.

Although mutation rates are low, new mutants appear continuously in nature, because there are many individuals in every species and many gene loci in every individual. The process of mutation provides the organisms of each generation with many new genetic variations. Thus, it is not surprising to see that, when new environmental challenges arise, species often are able to adapt to them.

Consider the resistance of disease-causing bacteria and parasites to antibiotics and other drugs. When an individual receives an antibiotic that specifically kills the bacteria causing the disease—say, tuberculosis—the immense majority of the bacteria die, but one in a million may have a mutation that provides resistance to the antibiotic. These resistant bacteria will survive and multiply, and the antibiotic will no longer cure the disease. This is the reason why modern medicine treats bacterial diseases with cocktails of antibiotics. If the incidence of a mutation conferring resistance for a given antibiotic is one in a million, the incidence of one bacterium carrying three mutations, each conferring resistance to one of three antibiotics, is one in a million million million; such bacteria are far less likely to exist in any infected individual.

1.2.5 Chromosomal mutations

Changes in the number, size, or organization of chromosomes within a species are termed chromosomal mutations, chromosomal abnormalities, or chromosomal aberrations. Changes in number may occur by the fusion of two chromosomes into one, by fission of one chromosome into two, or by addition or subtraction of one or more whole chromosomes or sets of chromosomes. (The condition in which an organism acquires one or more additional sets of chromosomes is called polyploidy.) Changes in the structure of chromosomes may occur by inversion, when a chromosomal segment

rotates 180 degrees within the same location; by duplication, when a segment is added; by deletion, when a segment is lost; or by translocation, when a segment changes from one location to another in the same or a different chromosome (Figure 1.7). These are the processes by which chromosomes evolve.

Inversions, translocations, fusions, and fissions do not change the amount of DNA. The importance of these mutations in evolution is that they change the linkage relationships between genes. Genes that were closely linked to each other become separated and vice versa; this can affect their expression because genes are often transcribed sequentially, two or more at a time. Human chromosomes differ from those of chimps and other apes in number: they have 24, while we have 23 pairs as a consequence of the fusion of two of their chromosomes into one. In addition, inversions and translocations that distinguish human from ape chromosomes have been identified in several chromosomes.

1.2.6 Genetic variation in populations

The sum total of all genes and combinations of genes that occur in a population of organisms of the same species is called the gene pool of the population. This can be described for individual genes or sets of genes by giving the frequencies of the alternative genetic constitutions; different forms of the same gene are called alleles. Consider, for example, a particular gene, such as the one determining the M-N blood groups in humans. One allele codes for the M blood group, while the other allele codes for the N blood group. The M-N gene pool of a particular population is specified by giving the frequencies of the alleles M and N. Thus, in the United States, the M allele occurs in people of European descent with a frequency of 0.539 and the N allele with a frequency of 0.461. In other populations these frequencies are different; for instance, the frequency of the M allele is 0.917 in Navajo Indians and 0.178 in Australian Aboriginals (Table 1.1).

The genetic variation present in a population is sorted out in new ways in each generation by the process of sexual reproduction, which recombines the chromosomes inherited from the two parents during the formation of the gametes that produce the following generation. But heredity by itself does not change gene frequencies. This principle is stated by the Hardy–Weinberg law, which describes the genetic equilibrium in a population by means of an algebraic equation. It states that genotypes, the genetic constitution of individual organisms, exist in certain frequencies that are a

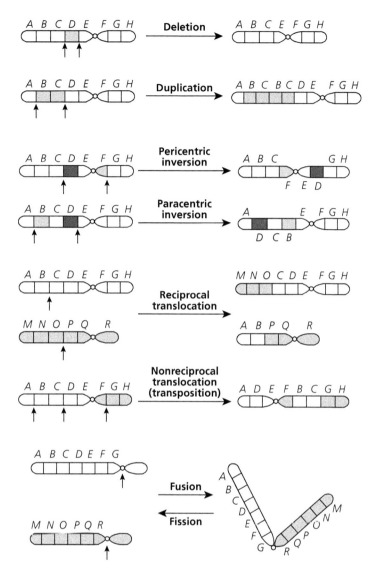

Figure 1.7 Chromosomal mutations. A deletion has a chromosome segment missing. A duplication has a chromosome segment represented twice. Inversions and translocations are chromosomal mutations that change the locations of genes in the chromosomes. Centric fusions are the joining of two chromosomes at the centromere to become one single chromosome. Centric fissions, or dissociations, are the reciprocal of fusions: one chromosome splits into two chromosomes.

simple function of the allelic frequencies—namely, the square expansion of the sum of the allelic frequencies (Figure 1.8).

If there are two alleles, A and a, at a gene locus, three genotypes will be possible: *AA, Aa,* and *aa.* If the frequencies of the alleles A and a are p and q, respectively, the equilibrium frequencies of the three genotypes will be given by:

$$(p + q)^2 = p^2 + 2pq + q^2$$

for *AA, Aa,* and *aa,* respectively. The genotype equilibrium frequencies for any number of alleles are derived in the same way.

The genetic equilibrium frequencies determined by the Hardy–Weinberg law assume that there is random mating—that is, the probability of a particular kind of mating is the same as the combined frequency of the genotypes of the two mating individuals. Random mating can occur with respect to most gene loci, even though mates may be chosen according to particular

Table 1.1 Genotypic and allelic frequencies for the M-N blood groups in three human populations

Population	Blood-group individuals			Total	Genotypic frequency			Allelic frequency	
	M	MN	N		$L^M L^M$	$L^M L^N$	$L^N L^N$	L^M	L^N
Australian Aborigines	22	216	492	730	0.030	0.296	0.674	0.178	0.822
Navajo Indians	305	52	4	361	0.845	0.144	0.011	0.917	0.083
White North Americans	1787	3039	1303	6129	0.292	0.496	0.213	0.539	0.461

Individuals with blood group M are homozygotes with genotype $L^M L^M$; those with blood group MN are heterozyotes, $L^M L^N$; those with blood group N are homozyotes, $L^N L^N$. The allelic frequency of L^M is the frequency of $L^M L^M$ plus half the frequency of the heterozyotes $L^M L^N$; for example, $0.030 + 0.148 = 0.178$. Similarly the frequency of L^N is the frequency of $L^N L^N$ plus half the frequency of $L^M L^N$.

characteristics. People, for example, choose their spouses according to all sorts of preferences concerning looks, personality, and the like. But concerning the majority of genes, people's marriages are essentially random. People are unlikely to choose their mating partners according to their M-N blood group genotypes or according to the genotype they have with respect to a particular enzyme.

Assortative, or selective, mating takes place when the choice of mates is not random. Marriages in the United States, for example, are assortative with respect to many social factors, so that members of any one social group tend to marry members of their own group more often, and people from a different group less often, than would be expected from random mating. Consider the sensitive social issue of interracial marriage in a hypothetical community in which 80% of the population is white and 20% is black. With random mating, 32% ($2 \times 0.80 \times 0.20 = 0.32$) of all marriages would be interracial, whereas only 4% ($0.20 \times 0.20 = 0.04$) would be marriages between two blacks. These statistical expectations depart from typical observations, even in modern society, as a result of persistent social customs.

The Hardy–Weinberg equilibrium expectations also assume that gene frequencies remain constant from generation to generation—that there is no gene mutation or natural selection and that populations are very large. But these assumptions are not correct. Organisms are subject to mutation, selection, and other processes that change gene frequencies, but the effects of these processes can be calculated by using the Hardy–Weinberg law as the starting point.

1.2.7 Processes of genetic change

The allelic variations that make evolution possible are generated by the process of mutation, but new mutations change gene frequencies very slowly, because mutation rates are low. If mutation were the only genetic process of evolution, this would occur very slowly. Moreover, organisms would become dysfunctional over time, because most mutations are harmful rather than beneficial.

Gene flow, or gene migration, takes place when individuals migrate from one population to another and interbreed with its members. Gene frequencies are not changed for the species as a whole, but they change locally whenever different populations that have different allele frequencies exchange genes by migration or intermarriage. In general, the greater the difference in allele frequencies between the resident and the migrant individuals, and the larger the number of migrants, the greater effect the migrants have in changing the genetic constitution of the resident population.

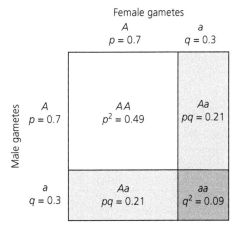

Figure 1.8 The Hardy–Weinberg law. Representation of the relationship between allele and genotype frequencies.

Gene frequencies can change from one generation to another by a process of pure chance known as genetic drift. This occurs because the number of individuals in any population is finite, and thus the frequency of a gene may change in the following generation by accidents of sampling, just as it is possible to get more or fewer than 50 "heads" in 100 throws of a coin, simply by chance.

The magnitude of the gene frequency changes due to genetic drift is inversely related to the size of the population—the larger the number of reproducing individuals, the smaller the effects of genetic drift. The reason is similar to what happens with a coin toss. If you toss a coin 10 times, you may obtain only three heads (0.30 frequency) with a probability that is not very small. But if you toss a coin 1,000 times, it is extremely unlikely that you'll get heads with a frequency of only 0.30 (300 heads) or less. The effects of genetic drift in changing gene frequencies from one generation to the next are quite small in most natural populations, which generally consist of thousands of reproducing individuals. The effects over many generations are more important.

Genetic drift can have important evolutionary consequences when a new population becomes established by only a few individuals—a phenomenon known as the founder principle. The allelic frequencies present in these few colonizers are likely to differ at many loci from those in the population they left, and those differences have a lasting impact on the evolution of the new population. The colonization of the continents of the world, starting from Africa and between continents, by *Homo erectus*, as well as by modern *Homo sapiens*, as well as the colonization of different regions of the same continent, was likely carried out at various times by relatively few individuals, which may have differed genetically, by chance, from the original population. For example, the absence of the B blood group among Native Americans is likely due to the chance absence of this relatively rare blood group among the original American colonizers.

1.2.8 Epigenetic processes

We will take epigenetics as: "DNA sequence-independent changes in chromosomal function that yield a stable and heritable phenotype" (Rissman & Adli, 2014). The phenotype may refer to the whole organism, or to only particular cells or tissues. Three conditions must be fulfilled for an epigenetic process, currently called "transgenerational epigenetic inheritance" or TEI: (1) changes in chromosomal function

expressed in a different phenotype (different from parents' phenotype); (2) independent from DNA sequences (neither coded in the genome, nor linked to parents' germ cells alterations—see Choi & Mango, 2014, p. 1440); (3) which is heritable in a stable way (persisting for at least three generations). As Choi and Mango say, "Both invertebrates and vertebrates exhibit such inheritance, and a range of environmental factors can act as a trigger." The challenge is to ascertain "what molecular mechanisms account for inheritance of TEI phenotypes" (Choi & Mango, 2014, p. 1440).

Mechanisms leading to TEI have been related to (see Rissman & Adli, 2014):

- DNA methylation (a biochemical process in which methyl chemical groups are covalently attached to cytosine residues by DNA methyltransferase enzymes);
- histone modifications (posttranslational modifications in histone proteins, mainly studied in organisms, such as *C. elegans*, lacking DNA methylation);
- ncRNAs (non-coding RNA that alters chromatin structure and DNA accessibility).

The need for this kind of mechanisms seems clear. Though all cells share a common nuclear genome, the necessity for genetic expression, from DNA to proteins, varies in the different tissues. Thus, a great part of the genome must be silenced, not leading to codify any protein or enzyme. This silencing strategy is reached during the ontogenetic development, i.e., by means of epigenetic episodes. Several factors, including environmental conditions and parental care, seem to affect epigenetics. If the alterations provoked are inherited, we are facing a TEI phenotype. Epigenetics have been used sometimes as an argument against the neo-Darwinist approach to evolution. The fact that non-coding changes could be inherited has been claimed as something like a Lamarckian approach. However, no mysterious process that might be achieved by an organisms' alleged "will" exists. DNA silencing obviously depends on genetic information appearing by means of well-known molecular mechanisms. All TEI phenotypes are the result of genetic, inheritable material that, under external—environmental, at least relative to the cell or tissue—pressures, becomes modified. TEI phenotypes are inherited because parents not only transmit DNA to the progeny, but the mother contributes a whole set of cellular organs, as well as several kind of ncRNA, included in the ovule.

1.2.9 Natural selection, again

Mutation, gene flow, and genetic drift change gene frequencies without regard for the consequences that such changes may have for the ability of organisms to survive and reproduce; they are random processes with respect to adaptation. If these were the only processes of evolutionary change, the organization of living things would gradually disintegrate. The effects of such processes alone would be analogous to those of a mechanic who changed parts in an automobile engine at random, with no regard for the role of the parts in the engine.

Natural selection keeps the disorganizing effects of mutation and other processes in check because it multiplies beneficial mutations and eliminates harmful ones. Natural selection accounts not only for the preservation and improvement of the organization of living beings, but also for their diversity. In different localities or in different circumstances, natural selection favors different traits, precisely those that make the organisms well adapted to their particular circumstances and ways of life.

The effects of natural selection are measured with a parameter called "fitness." Fitness can be expressed as an absolute or as a relative value. Consider a population consisting at a certain locus of three genotypes: A^1A^1, A^1A^2, and A^2A^2. Assume that on the average each A^1A^1 and each A^1A^2 individual produces one offspring but that each A^2A^2 individual produces two. One could use the average number of progeny left by each genotype as a measure of that genotype's absolute fitness over the generations. (This, of course, would require knowing how many of the progeny survive to adulthood and reproduce.) It is, however, mathematically more convenient to use relative fitness values (typically represented with the letter w). Evolutionists usually assign the value 1 to the genotype with the highest reproductive efficiency and calculate the other relative fitness values proportionally. For the example just used, the relative fitness of the A^2A^2 genotype would be w = 1 and that of each of the other two genotypes would be w = 0.5. A parameter related to fitness is the selection coefficient, often represented by the letter s, which is defined as s = 1 − w. The selection coefficient is a measure of the reduction in fitness of a genotype. The selection coefficients in the example are s = 0 for A^2A^2 and s = 0.5 for each A^1A^1 and A^1A^2.

Selection may favor one homozygote over the other and over the heterozygote, or may favor the heterozygote over both homozygotes. A particularly interesting example of heterozygote superiority among humans is provided by the gene responsible for sickle cell anemia in places where malaria is rife. Human hemoglobin in adults is for the most part hemoglobin A, a four-component molecule consisting of two α and two β hemoglobin chains. The gene Hb^A codes for the normal β hemoglobin chain, which consists of 146 amino acids. A mutant allele of this gene, Hb^S, causes the β chain to have, in the sixth position, the amino acid valine instead of glutamic acid (Figure 1.9). This seemingly minor substitution modifies the properties of hemoglobin so that homozygotes with the mutant allele, Hb^SHb^S, suffer from a severe form of anemia that in most cases leads to death before the age of reproduction.

The Hb^S allele occurs in some African populations with a high frequency. This seems puzzling because of the severity of the anemia. The strong natural selection against the Hb^SHb^S homozygotes should have eliminated the defective allele. But the Hb^S allele occurs at high frequency precisely in regions of the world where a particularly severe form of malaria, which is caused by the parasite *Plasmodium falciparum*, is endemic (Figure 1.10). It was hypothesized that the heterozygotes, Hb^AHb^S, were resistant to malaria, whereas the homozygotes Hb^AHb^A were not. In malaria-infested regions then the heterozygotes survived better than either of the homozygotes, which were more likely to die from either malaria (Hb^AHb^A homozygotes) or anemia (Hb^SHb^S homozygotes). This hypothesis has been confirmed in various ways. Most significant is that most hospital patients suffering from severe or fatal forms of malaria are homozygotes Hb^AHb^A. In a study of 100 children who died from malaria, only 1 was found to

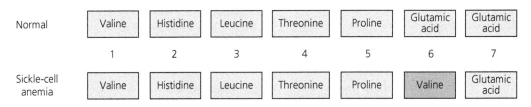

Figure 1.9 The first seven amino acids of the β chain of human hemoglobin; the β chain consists of 146 amino acids. A substitution of valine for glutamic acid at the sixth position is responsible for the severe disease known as sickle-cell anemia.

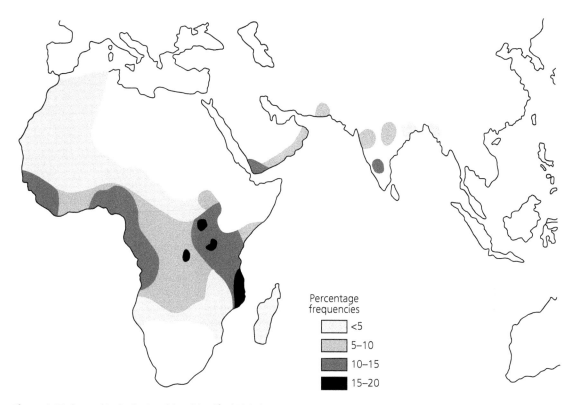

Figure 1.10 Geographic distribution of the allele *Hb^S*, which in homozygous conditions is responsible for sickle-cell anemia. The frequency of *Hb^S* is high in those regions of the world where falciparum malaria is endemic, because *Hb^A Hb^S* individuals, heterozygous for the *Hb^S* and the "normal" allele, are highly resistant to malarial infection.

be a heterozygote, whereas 22 were expected to be so according to the frequency of the HbS allele in the population.

The malaria example illustrates the general principle that the fitness of genotypes depends on the environment. Thus, dark skin is favored in the tropics where the incidence of UV-radiation from the sun is high and may cause melanoma and other cancers. At high latitudes, lighter skin may be favored because of the low-level UV-radiation, which is required for synthesizing vitamin D in the lower layers of the dermis and is less likely to cause melanoma and other UV-radiation induced diseases.

1.2.10 Modes of selection

The population density of organisms and the frequency of genotypes may impact the fitness of genotypes. Insects, for example, experience enormous yearly oscillations in density. Some genotypes may possess high fitness in the spring, when the population is rapidly expanding, because such genotypes yield more prolific individuals. Other genotypes may be favored during the summer, when populations are dense, because these genotypes make for better competitors, ones more successful at securing limited food resources.

The fitness of genotypes can also vary according to their relative numbers. Particularly interesting is the situation in which genotypic fitnesses are inversely related to their frequencies, a common situation, known as "frequency-dependent selection," that preserves genetic polymorphism in populations. Assume that two genotypes, A and B, have fitnesses related to their frequencies in such a way that the fitness of either genotype increases when its frequency decreases and vice versa. When A is rare, its fitness is high, and therefore A increases in frequency. As it becomes more and more common, however, the fitness of A gradually decreases, so that its increase in frequency eventually comes to a halt. A stable polymorphism occurs at the frequency where the two genotypes, A and B, have identical fitnesses.

Frequency-dependent selection may arise because the environment is heterogeneous in such a way that different genotypes better exploit different subenvironments. When a genotype is rare, the subenvironments that it exploits better will be relatively abundant. But as the genotype becomes common, its favored subenvironment becomes saturated. Sexual preferences also may lead to frequency-dependent selection. It has been demonstrated in some insects, birds, mammals, and other organisms that the mates preferred often are those that are rare. People also seem to experience this rare-mate advantage—blonds may seem attractively exotic to brunettes, or brunettes to blonds.

Natural selection can be explored by examining its effects on the phenotypes of individuals in a population. Distribution scales of phenotypic traits such as height, weight, number of progeny, or longevity typically show greater numbers of individuals with intermediate values and fewer and fewer toward the extremes—this is the so-called normal distribution. By reference to this distribution, we may distinguish three modes of natural selection: stabilizing, directional, and diversifying (Figure 1.11).

When individuals with intermediate phenotypes are favored and extreme phenotypes are selected against, the selection is said to be stabilizing. The range and distribution of phenotypes then remains approximately the same from one generation to another. An example of selection favoring intermediate phenotypes is mortality among newborn infants, which is highest when they are either very small or very large.

Directional selection occurs when the distribution of phenotypes in a population changes systematically in a particular direction. The physical and biological aspects of the environment are continuously changing and, over long periods of time, the changes may be substantial. The climate and even the configuration of the land or waters vary incessantly. Changes also take place in the biotic conditions—that is, in the other organisms present, whether predators, prey, parasites, or competitors. Genetic changes occur as a consequence, because the genotypic fitnesses may shift so that different sets of alleles are favored.

Over geologic time, directional selection leads to major changes in morphology and ways of life. Evolutionary changes that persist in a more or less continuous fashion over long periods of time are known as evolutionary trends. Directional evolutionary changes increased the cranial capacity of the human lineage from the small brain of *Australopithecus*—human ancestors of several million years ago—which was about 400 cc in volume, to a brain more than three times as large in modern humans. Directional selection—particularly, long-term evolutionary trends—often does not occur in a continuous or sustained manner, but rather in spurts. Surely the increase in brain size from *Australopithecus* to *Homo sapiens* did not occur at a constant rate of, say, so-many cc per thousand years.

Two or more divergent phenotypes in an environment may be favored simultaneously by diversifying selection. No natural environment is homogeneous;

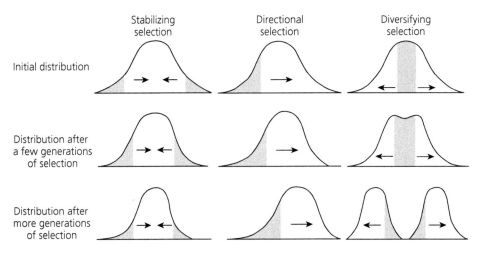

Figure 1.11 Three types of natural selection showing the effects of each on the distribution of phenotypes within a population. The shaded areas represent the phenotypes against which selection acts. Stabilizing selection acts against phenotypes at both extremes of the distribution, favoring the multiplication of intermediate phenotypes. Directional selection acts against only one extreme of phenotypes, causing a shift in distribution toward the other extreme. Diversifying selection acts against intermediate phenotypes, creating a split in distribution toward each extreme.

rather, the environment of any plant or animal population is a mosaic consisting of more or less dissimilar subenvironments. There is heterogeneity with respect to climate, food resources, and living space. Also, the heterogeneity may be temporal, with change occurring over time, as well as spatial. Species cope with environmental heterogeneity in diverse ways. One strategy is genetic monomorphism—the selection of a generalist genotype that is well adapted to all the subenvironments encountered by the species. Another strategy is genetic polymorphism—the selection of a diversified gene pool that yields different genotypes, each adapted to a specific subenvironment.

1.2.11 Sexual selection

Sexual selection is a special form of natural selection. Other things being equal, organisms more proficient in securing mates have higher fitness. There are two general circumstances leading to sexual selection. One is the preference shown by one sex (often the females) for individuals of the other sex that exhibit certain traits. The other is increased strength (usually among the males) that yields greater success in securing mates. Sexual selection explains, for example, the presence of exorbitant antlers in male deer and the spectacular plumage of male peacocks. These traits would seem disadvantageous because of increased energy costs or exposure to predators, but have evolved by natural selection because they help to secure mates.

The presence of a particular trait among the members of one sex can make them somehow more attractive to the opposite sex. This type of "sex appeal" has been experimentally demonstrated in all sorts of animals, from vinegar flies to pigeons, mice, dogs, and rhesus monkeys. Sexual selection can also come about because a trait—the size of the antlers of a stag, for example—increases prowess in competition with members of the same sex. Stags, rams, and bulls use antlers or horns in contests of strength; a winning male usually secures more female mates. Therefore, sexual selection may lead to increased size and aggressiveness in males. Male baboons are more than twice as large as females, and the behavior of the docile females contrasts with that of the aggressive males. A similar dimorphism occurs in the northern sea lion, *Eumetopias jubata*, where males weigh about 1,000 kg (2,200 pounds), about three times as much as females. The males fight fiercely in their competition for females; large, battle-scarred males occupy their own rocky islets, each holding a harem of as many as 20 females.

1.2.12 Kin selection

The apparent altruistic behavior of many animals is, like some manifestations of sexual selection, a trait that at first seems incompatible with the theory of natural selection. Altruism is a form of behavior that benefits other individuals at the expense of the one that performs the action; the fitness of the altruist is diminished by its behavior, whereas individuals that act selfishly benefit from it at no cost to themselves. Accordingly, it might be expected that natural selection would foster the development of selfish behavior and eliminate altruism. This conclusion is not so compelling when it is noticed that the beneficiaries of altruistic behavior are usually relatives. They share part of their genes, including genes that promote altruistic behavior. Altruism may evolve by kin selection, which is simply a type of natural selection in which relatives (and therefore genes in common) are taken into consideration when evaluating an individual's fitness. The fitness of a gene or genotype that takes into account its presence in relatives is known as "inclusive fitness."

Natural selection favors genes that increase the reproductive success of their carriers, but it is not necessary that all individuals that share a given genotype have higher reproductive success. It suffices that carriers of the genotype reproduce more successfully on the average than those possessing alternative genotypes. Parental care is, therefore, a form of altruism readily explained by kin selection. Parents spend energy caring for their progeny because it increases the reproductive success of the parents' genes.

Kin selection extends beyond the relationship between parents and their offspring. It facilitates the development of altruistic behavior when the energy invested, or the risk incurred, by an individual is compensated in excess by the benefits ensuing to relatives. The closer the relationship between the beneficiaries and the altruist, and the greater the number of beneficiaries, the higher the risks and efforts warranted in the altruist. Individuals that live together in a herd or troop usually are related and often behave toward each other in protective or helping ways. Adult zebras, for instance, will turn toward an attacking predator to protect the young in the herd rather than fleeing to protect themselves.

An extreme form of kin selection occurs in some species of bees, wasps, ants, and other social insects. We may use as an example the stingless bees, with hundreds of species in the tropics. These bees live in colonies, typically with a single queen, and hundreds or thousands of workers, which are morphologically different from the queen. The female workers build the hive, care for the young, and gather food, but they are sterile; the queen alone produces progeny. It would seem that the workers' behavior would in no way be promoted or maintained by natural selection. Any genes causing such behavior would seem likely to be eliminated from the population, because individuals exhibiting the behavior favor not their own reproductive success, but that of the queen.

The expectations change, however, when we take into account the genetic make-up of these social insects, in which the females are diploid (have two sets of chromosomes), but the males are haploid (have only one set of chromosomes). This genetic structure is called "haplodiploidy" (Figure 1.12). Queens produce some eggs that remain unfertilized and develop into males, or drones, and are haploid, having a mother but no father. Their main role is to engage in the nuptial flight during which one of them fertilizes a new queen. Other eggs laid by queen bees are fertilized and develop into diploid females, the large majority of which are workers. In many species of social insects, the queen typically mates with a single male once during her lifetime; the male's sperm is stored in the queen's spermatheca, from which it is gradually released as she lays fertilized eggs. All the queen's female progeny, therefore, have the same father, so that workers are more closely related to one another and to any new sister then they are to the mother queen. The female workers receive one-half of their genes from the mother and one-half from the father, but they share

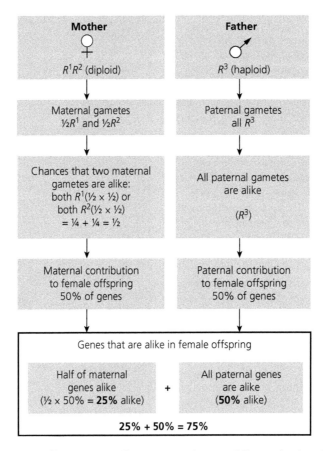

Figure 1.12 Haplodiploid reproduction of hymenopterans with respect to any given gene, R. The gene has three allele forms, R^1 and R^2 in the diploid mother and R^3 in the haploid father. The probability that two daughters will both inherit R^1 is ¼ and that both will inherit R^2 is also ¼. Therefore, the probability that both daughters will inherit the same allele from the queen is ½. They will both inherit R^3 from the father. Each daughter inherits half of her genes from the mother and half from the father. Thus, the probability that the two sisters will have inherited the same genes from both parents are ¼ for the mother genes (½ of the genes with a ½ probability of being identical) and ½ for the father genes (½ of the genes with probability 1) or ¾ for both genes. Mother and daughters share only ½ of their genes.

among themselves three-quarters of their genes. This is because the half of the set from the father is the same in every worker, given that the father had only one set of genes rather than two (the male developed from an unfertilized egg, so all his sperm carry the same set of genes). The other half of the workers' genes come from the mother and, on the average, half of them are identical in any two sisters. Consequently, with three-quarters of her genes present in her sisters (while only half of her genes would be passed on to a daughter), a worker's genes are transmitted one-and-a-half times more effectively when she raises a sister (whether another worker or a new queen), than if she were to produce a daughter of her own. With such genetic population structure, natural selection will maximize the number of sterile female workers and minimize the number of reproductive females—which is accomplished by having only one queen.

1.2.13 Reciprocal altruism and group selection

Altruism also occurs among unrelated individuals when the behavior is reciprocal and the altruist's costs are smaller than the benefits to the recipient. This reciprocal altruism is found in the mutual grooming of chimpanzees and other primates as they clean each other of lice and other pests.

Altruistic behavior may also evolve by so-called group selection, when populations ("groups") with certain attributes (such as altruistic behaviors) will persist and multiply better than populations lacking such attributes. But group selection can occur only under restrictive conditions. Within a population or group, an altruistic genotype will have lower fitness than a selfish genotype, because altruistic individuals incur a cost, from which selfish individuals benefit. Therefore, altruistic genotypes will tend to be eliminated from the population. But populations made up of selfish genotypes may become extinct more readily (for example, by over-exploiting food resources) than populations with altruistic genotypes. Altruism may evolve in a species if the rate of extinction of selfish populations is large compared to the rate at which selfish genotypes increase in frequency within populations. Evolutionists have shown that these restrictive conditions occur in nature.

1.2.14 Species and speciation

Species come about as the result of gradual change prompted by natural selection. Environments differ from place to place and change in time. Natural selection favors different characteristics in different situations. The accumulation of differences between populations exposed to different environments may eventually yield different species.

External similarity is the common basis for identifying individuals as being members of the same species. Nevertheless, there is more to a species than outward appearance. A bulldog, a terrier, and a golden retriever are very different in appearance, but they are all dogs because they can interbreed. People can also interbreed with one another, and so can cats with other cats, but people cannot interbreed with dogs or cats, nor can these with each other. Although species are usually identified by appearance, there is something basic, of great biological significance, behind similarity of appearance—individuals of a species are able to interbreed with one another but not with members of other species. Among sexual organisms, species are groups of interbreeding natural populations that are reproductively isolated from other such groups.

The ability to interbreed is of great evolutionary importance, because it determines that species are independent evolutionary units. Genetic changes originate in single individuals; they can spread by natural selection to all members of the species but not to individuals of other species. Individuals of a species share a common gene pool that is not shared by individuals of other species. Different species have independently evolving gene pools because they are reproductively isolated.

Although the criterion for deciding whether individuals belong to the same species (i.e., reproduction isolation) is clear, there may be ambiguity in practice for two reasons. One is lack of knowledge—it may not be known for certain whether individuals living in different sites belong to the same species, because it is not known whether they can naturally interbreed. The other reason for ambiguity is rooted in the nature of evolution as a gradual process. Two geographically separate populations that at one time were members of the same species may gradually diverge into two different species. Since the process is gradual, there is no particular point at which it is possible to say that the two populations have become two different species; that is, there is one particular generation in which reproductive isolation is present, but it was not present in the previous generation.

A similar kind of ambiguity obtains when we compare ancestral and descendant populations living at different times. There is no way to test if today's humans could interbreed with those who lived thousands of years ago. It seems reasonable that living people would be able to interbreed with people who lived a few generations earlier and look more or less like other people now living. But what about ancestors who lived thousands of generations earlier? There is no precise time at which *H. erectus* became *H. sapiens*, but it would not be appropriate to classify remote human ancestors and modern humans in the same species just because the changes from one generation to the next surely were small. It is useful to distinguish between two groups that look different and lived at different times by means of different species' names, just as it is useful to give different names to childhood and adulthood, even though no single moment can separate one from the other. Biologists distinguish species in organisms that lived at different times by means of a commonsense morphological criterion: If two organisms differ from each other in form and structure about as much as do two living individuals belonging to two different species, they are classified in separate species and given different names. Species that may be related as ancestral and descendant are called chronospecies. This is a matter to which we'll return later in this book because it is quite relevant in the study of fossils.

Given that species are groups of populations reproductively isolated from one another, asking about the origin of species is equivalent to asking how reproductive isolation arises between populations. This may occur as an incidental consequence of genetic divergence between populations that are geographically separated from one another. But reproductive isolation may be directly promoted by natural selection when populations are somewhat diverged, or adapted to different features of the environment, so that hybrids have low fitness. In the extreme, this occurs when hybrids are inviable or sterile. When hybrids have lower fitness than nonhybrids, genes will be favored by natural selection that reduce the probability of hybridization, and eventually complete reproductive isolation may ensue.

Geographic separation may result in complete reproductive isolation if it persists long enough. Consider, for example, the evolution of many endemic species of plants and animals in the Hawaiian archipelago. The ancestors of these species arrived on these islands several million years ago. There they evolved as they became adapted to the environmental conditions and colonizing opportunities present. Reproductive isolation between the populations evolving in Hawaii and the populations on continents was not as such directly promoted by natural selection; their geographic

remoteness forestalled any opportunities for hybridizing. Nevertheless, reproductive isolation became complete in many cases as a result of gradual genetic divergence over thousands of generations.

Box 1.6 Species' endemism in remote archipelagos

Species are called "endemic" when they have evolved in the place where they live and are not found in any other locality naturally; that is, unless they have been introduced by humans. Endemism is particularly apparent when colonizers reach geographically remote areas, such as islands, where they find few or no competitors and have an opportunity to diverge as they become adapted to the new environment.

Many examples of endemism are found in archipelagoes removed from the mainland. The Galápagos Islands are about 1,000 km (600 miles) off the west coast of South America. When Charles Darwin arrived there in 1835 during his voyage on the *HMS Beagle*, he discovered many species not found anywhere else in the world—for example, several species of finches, of which 14 are now known to exist (called Darwin's finches). These passerine birds have adapted to a diversity of habitats and diets, some feeding mostly on plants, others exclusively on insects. The various shapes of their bills are clearly adapted to probing, grasping, biting, or crushing—the diverse ways in which the different Galápagos species obtain their food. The explanation for such diversity is that the ancestor of Galápagos finches arrived in the islands before other kinds of birds and encountered an abundance of unoccupied ecological niches. Its descendants underwent adaptive radiation, evolving a variety of finch species with ways of life capable of exploiting opportunities that on various continents are already exploited by other species.

The Hawaiian archipelago also provides striking examples of endemism. Its several volcanic islands, ranging from about 1 million to more than 10 million years in age, are far from any continent or even other large islands. In their relatively small total land area, an astounding number of plant and animal species exist. Most of the species have evolved on the islands, among them about two dozen species (about one-third of them now extinct) of honeycreepers, birds of the family Drepanididae, all derived from a single immigrant form. In fact, all but one of Hawaii's 71 native bird species are endemic. More than 90% of the native species of flowering plants, land mollusks, and insects are also endemic, as are two-thirds of the 168 species of ferns.

1.2.15 Homology, analogy, and convergent evolution

Different species may exhibit features that are similar in appearance, in structure, or in function—the legs of dogs resemble the legs of leopards; bats and birds use wings for flying. Resemblances may be due to inheritance from a common ancestor or may have evolved independently as adaptations to similar functions. Correspondence of features in different organisms that is due to inheritance from a common ancestor is called homology. The forelimbs of humans, whales, dogs, and bats are homologous. The skeletons of these limbs are all constructed of bones arranged according to the same pattern because they derive from a common reptilian ancestor with similarly arranged forelimbs.

Correspondence of features due to similarity of function, but not related to common descent, is termed *analogy*. The wings of birds and of flies are analogous. These wings are not modified versions of a wing present in a common ancestor; rather they have evolved independently as adaptations to a common function, flying. Some features may be partially homologous and partially analogous; for example, the wings of bats and birds. Their skeletal structure is homologous, due to common descent from the forelimb of a reptilian ancestor; but the modifications for flying are different and independently evolved, and in this respect they are analogous.

Features that become more rather than less similar through independent evolution are said to be convergent. Convergence is often associated with similarity of function, as in the evolution of wings in birds, bats, and flies. The shark (a fish) and the dolphin (a mammal) are much alike in external morphology; their similarities are due to convergence, since they have evolved independently as adaptations to aquatic life.

Later we will return to the distinctions between homologous, analogous, and convergent features, because they play a critical role in a prevailing theory of systematics known as cladistics. We will now turn, however, to the methods for reconstructing and representing evolutionary history, both anagenesis and cladogenesis, which are often represented as evolutionary trees.

1.2.16 Evolutionary trees

The evolution of all living organisms, or of a subset of them, can be represented as a tree, with branches that divide into two or more as time progresses, which represent the splitting of species (or higher taxonomic groups). Such trees are called phylogenies. Their branches represent evolving lineages, some of which eventually die out, while others persist in themselves or in their derived lineages down to the present time. Evolutionists are interested in the history of life and hence in the topology, or configuration, of phylogenies, which represent the splitting of taxa through time. They are concerned as well with the nature of the anagenetic changes within lineages (in morphology, function, behavior, genetic make-up, etc.) and with the timing of both anagenetic and cladogenetic events.

Evolutionary trees are hypotheses or models that seek to reconstruct the evolutionary history of taxa—i.e., species or other groups of organisms, such as genera, families, or orders. The branching relationships of the trees reflect the relative relationships of ancestry, or cladogenesis. Thus, in Figure 1.13, humans and rhesus monkeys are seen to be more closely related to each other than either is to the horse. Stated another way, this tree shows that the most recent common ancestor to all three species lived in a more remote past than the most recent common ancestor to humans and monkeys.

Evolutionary trees may also indicate the changes that have occurred along each lineage, or anagenesis. Thus, in the evolution of cytochrome-c since the last common ancestor of humans and rhesus monkeys, one amino acid changed in the lineage going to humans but none in the lineage going to rhesus monkeys. In cladistic representations (see Section 1.3), decisive anagenetic changes that account for the configuration of the tree are marked by notches or otherwise along the branch leading to a particular taxon.

There exist several methods for constructing evolutionary trees. Some were developed for interpreting morphological data, others for interpreting molecular data; some can be used with either kind of data. The main methods currently in use are called distance, parsimony, and maximum likelihood.

Distance methods are primarily used with molecular data, but also with morphological information. A "distance" is the number of differences between two taxa. The differences are measured with respect to certain traits (such as morphological features) or to certain macromolecules (the sequence of amino acids in proteins or the sequence of nucleotides in DNA or RNA). The tree illustrated in Figure 1.13 was obtained by taking into account the distance, or number of amino acid differences, between three organisms with respect to a particular protein (cytochrome-c). Table 1.2 shows the (minimum) number of nucleotide differences in the genes of 20 species that account for the amino acid

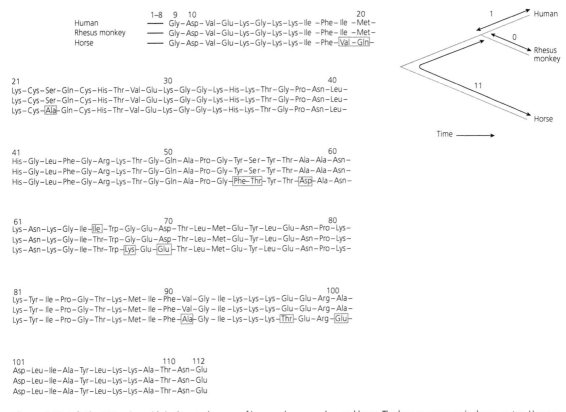

Figure 1.13 Left: The 104 amino acids in the cytochrome-c of human, rhesus monkey, and horse. The human sequence is shown on top. Humans differ from monkeys by 1, from horse by 12 amino acids; monkey and horse differ by 11 amino acids. Right: The phylogeny of human, rhesus monkey, and horse, based on their cytochrome-c. The one difference between human and monkey, at site 66, is due to a change in the human lineage, since monkey and horse are identical at this site.

differences in their cytochrome-c. An evolutionary tree based on the data in that table, showing the numbers of nucleotide changes in each branch, is illustrated in Figure 1.14.

Morphological data also can be used for constructing distance trees. The first step is to obtain a distance matrix, such as that making up Table 1.2, but one based on a set of morphological comparisons between species or other taxa. For example, in some insects one can measure body length, wing length, wing width, number and length of wing veins, or another trait.

A most common procedure to transform a distance matrix into a phylogeny is called cluster analysis. The distance matrix is scanned for the smallest distance element, and the two taxa involved (say, A and B) are joined at an internal node or branching point. The matrix is scanned again for the next smallest distance, and the two new taxa (say, C and D) are clustered. The procedure is continued until all taxa have been joined. When a distance involves a taxon that is already part of

a previous cluster (say, E and A), the average distance is obtained between the new taxon and the preexisting cluster (say, the average distance between E to A and E to B). This simple procedure, which can be used with morphological as well as molecular data, assumes that the rate of evolution is uniform along all branches.

Some distance methods relax the condition of uniform rate and allow for unequal rates of evolution along the branches. One of the most extensively used methods of this kind is called neighbor-joining. The method starts, as before, by identifying the smallest distance in the matrix and linking the two taxa involved. The next step is to remove these two taxa and calculate a new matrix in which their distances to other taxa are replaced by the distance between the node linking the two taxa and all other taxa. The smallest distance in this new matrix is used for making the next connection, which will be between two other taxa or between the previous node and another taxon. The procedure is repeated until all taxa have been

Table 1.2 Minimum number of nucleotide differences in the genes coding for cytochrome-c in 20 species

Species	1	2	3	4	5	6	7	8	9	10	11	12	13	14	15	16	17	18	19	20
1. Human	–	1	13	17*	16	13	12	12	17	16	18	18	19	20	31	33	36	63	56	66
2. Monkey		–	12	16*	15	12	11	13	16	15	17	17	18	21	32	32	35	62	57	65
3. Dog			–	10	8	4	6	7	12	12	14	14	13	30	29	24	28	64	61	66
4. Horse				–	1	5	11	11	16	16	16	17	16	32	27	24	33	64	60	68
5. Donkey					–	4	10	12	15	15	15	16	15	31	26	25	32	64	59	67
6. Pig						–	6	7	13	13	13	14	13	30	25	26	31	64	59	67
7. Rabbit							–	7	10	8	11	11	11	25	26	23	29	62	59	67
8. Kangaroo								–	14	14	15	13	14	30	27	26	31	66	58	68
9. Duck									–	3	3	3	7	24	26	25	29	61	62	66
10. Pigeon										–	4	4	8	24	27	26	30	59	62	66
11. Chicken											–	2	8	28	26	26	31	61	62	66
12. Penguin												–	8	28	27	28	30	62	61	65
13. Turtle													–	30	27	30	33	65	64	67
14. Rattlesnake														–	38	40	41	61	61	69
15. Tuna															–	34	41	72	66	69
16. Screwworm fly																–	16	58	63	65
17. Moth																	–	59	60	61
18. Neurospora																		–	57	61
19. Saccharomyces																			–	41
20. Candida																				–

Source: from Fitch and Margoliash (1967).
*These numbers are for nucleotide differences, which are greater than the amino acid differences shown in Figure 1.13.

connected with one another by intervening nodes. Maximum parsimony methods seek to reconstruct the tree that requires the fewest number of changes (i.e., it is the most parsimonious) summed along all branches. This is a reasonable assumption, because it usually will be the most likely. But evolution may not necessarily have occurred following a minimum path, because the same change may have occurred independently along different branches, and some differences may have involved intermediate steps that are not apparent in the organisms now living.

Not all evolutionary changes, even those that involve a single step, may be equally probable. For example, among the four nucleotide bases in DNA, cytosine (C) and thymine (T) are members of a family of related molecules called pyrimidines; likewise, adenine (A) and guanine (G) belong to a family of molecules called purines. A change within a DNA sequence from one pyrimidine to another (C $\rightleftarrows$ T) or from one purine to

another (A $\rightleftarrows$ G), called a transition, is more likely to occur than a change from a purine to a pyrimidine or the converse (G or A $\rightleftarrows$ C or T), called a transversion. Parsimony methods take into account different probabilities of occurrence if they are known.

Maximum parsimony methods are related to cladistics (see Section 1.3), a very formalistic theory of taxonomic classification, extensively used with morphological and paleontological data. The critical feature in cladistics is the identification of derived shared traits, called synapomorphic traits. A synapomorphic trait is shared by some taxa but not others because the former inherited it from a common ancestor that acquired the trait after its lineage separated from the lineages going to other taxa. In the evolution of carnivores, for example, domestic cats, tigers, and leopards are clustered together because of their possessing retractable claws, a trait acquired after their common ancestor branched off from the lineage leading to dogs,

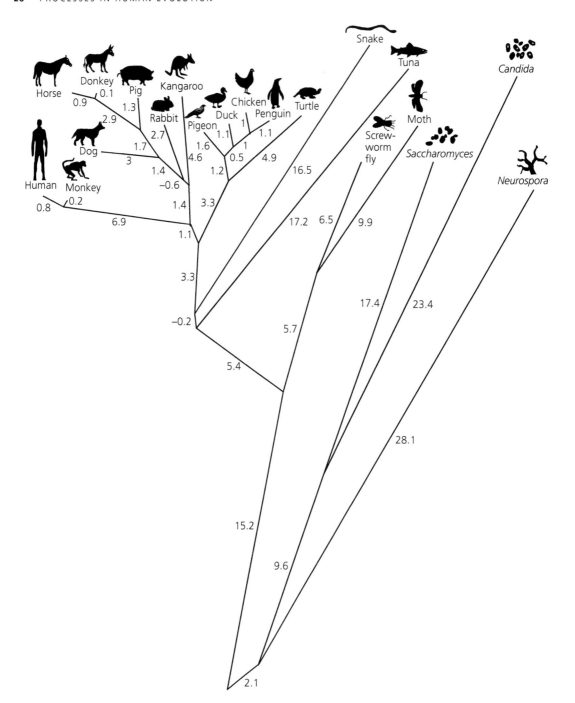

Figure 1.14 Evolutionary history of 20 species, based on the cytochrome-c amino acid sequence. The common ancestor (at the bottom) of yeast and humans lived more than 1 billion years ago. The numbers represent the (statistically) estimated number of nucleotide substitutions along each branch.

wolves, and coyotes. It is important to ascertain that the shared traits are homologous rather than analogous. For example, mammals and birds, but not lizards, have a four-chambered heart. Yet birds are more closely related to lizards than to mammals; the four-chambered heart evolved independently in the bird and mammal lineages, by parallel (or convergent) evolution. Maximum likelihood methods seek to identify the most likely tree, given the available data. They require that an evolutionary model be identified, which would make it possible to estimate the probability of each possible individual change. For example, transitions are more likely than transversions among DNA nucleotides, but a particular probability must be assigned to each. All possible trees are considered. The probabilities for each individual change are multiplied for each tree. The best tree is the one with the highest probability (or maximum likelihood) among all possible trees.

Maximum likelihood methods are computationally expensive when the number of taxa is large, because the number of possible trees (for each of which the probability must be calculated) grows factorially with the number of taxa. With 10 taxa, there are about 3.6 million possible trees; with 20 taxa, the number of possible trees is about 2 followed by 18 zeros (2×10^{18}). Even with powerful computers, maximum likelihood methods can be prohibitive if the number of taxa is large. Heuristic methods exist in which only a subsample of all possible trees is examined and thus an exhaustive search is avoided.

The statistical degree of confidence of a tree can be estimated for distance and maximum likelihood trees. The most common method is called bootstrapping. It consists of taking samples of the data by removing at least one data point at random and then constructing a tree for the new dataset. This random sampling process is repeated hundreds or thousands of times. The bootstrap value for each node is defined by the percentage of cases in which all species derived from that node appear together in the trees. Bootstrap values above 90% are regarded as statistically strongly reliable; those below 70% are considered unreliable.

1.2.17 Gene duplication

Similarity between features due to common descent is called homology and the traits are called homologous, as said earlier. Two kinds of homologous traits can be distinguished, orthologous and paralogous, a distinction that is particularly helpful with respect to genes and other genetic features. Orthologous genes are descendants of an ancestral gene that was present in the ancestral species from which the species in question have evolved. The evolution of orthologous genes, therefore, reflects the evolution of the species in which they are found. The cytochrome-c molecules of the 20 organisms shown in Figure 1.14 are orthologous, because they derive from a single ancestral gene present in a species ancestral to all 20 organisms.

Paralogous genes are descendants of a duplicated ancestral gene. Paralogous genes, therefore, evolve within the same species (as well as in different species). The genes coding for the α, β, γ, and δ hemoglobin chains in humans are paralogous. The evolution of paralogous genes reflects differences that have accumulated since the genes duplicated. Homologies between paralogous genes serve to establish gene phylogenies, i.e., the evolutionary history of duplicated genes within a given lineage.

Figure 1.15 is a phylogeny of the gene duplications giving rise to the myoglobin and hemoglobin genes found in modern humans. Hemoglobin molecules are tetramers, consisting of two polypeptides of one kind and two of another kind. In embryonic hemoglobin E, one of the two kinds of polypeptide is designated ε; in fetal hemoglobin F, it is γ; in adult hemoglobin A, it is β; and in adult hemoglobin A2, it is δ. (Hemoglobin A makes up about 98% of human adult hemoglobin, and hemoglobin A2 about 2%.) The other kind of polypeptide in embryonic hemoglobin is ζ; in both fetal and adult hemoglobin, it is α. There are yet additional complexities. Two γ genes exist (known as Gγ and Aγ), as do two α genes (α1 and α2). Furthermore, there are two β pseudogenes ($\psi\beta$1 and $\psi\beta$2) and two α pseudogenes ($\psi\alpha$1 and $\psi\alpha$2), as well as a ζ pseudogene. These pseudogenes are very similar in nucleotide sequence to the corresponding functional genes, but they include terminating codons and other mutations that make it impossible for them to yield functional hemoglobins. The similarity in the nucleotide sequence of the polypeptide genes, and pseudogenes, of both the α and β gene families indicates that they are all paralogous, arisen through various duplications and subsequent evolution from a gene ancestral to all.

1.2.18 The molecular clock of evolution

In paleontology, the time sequence of fossils is determined by the age of rocks in which they are embedded, as well as by other methods described in the chapters that follow. If the age of the rocks or of the fossils is determined, the evolutionary history of the organisms can be timed. Studies of molecular evolution rates

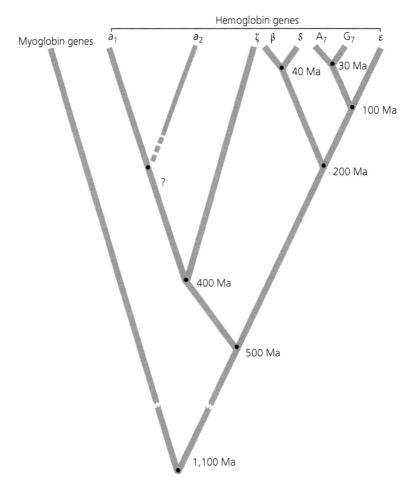

Figure 1.15 Phylogeny of the globin genes. The dots indicate points at which ancestral genes duplicated, giving rise to new gene lineages. The approximate times when these duplications occurred are indicated in million years ago (Ma).

have led to the proposition that DNA and proteins may serve as evolutionary clocks.

It was first observed in the 1960s that the number of amino acid differences between homologous proteins of any two given species seemed to be nearly proportional to the time of their divergence from a common ancestor. If the rate of evolution of a protein or gene were approximately the same in the evolutionary lineages leading to different species, proteins and DNA sequences would provide a molecular clock of evolution. The sequences could then be used to reconstruct not only the sequence of branching events of a phylogeny, but also to determine the time when the various events occurred.

Consider, for example, Figure 1.14. If the substitution of nucleotides in the gene coding for cytochrome-c occurred at a constant rate through time, we could determine the time elapsed along any branch of the phylogeny simply by examining the number of nucleotide substitutions along that branch. We would need only to calibrate the clock by reference to an outside source, such as the fossil record, that would provide the actual geologic time elapsed in at least one specific lineage or since one branching point. For example, if the time of divergence between insects and vertebrates is determined to have occurred 700 million years ago, other times of divergence can be determined by proportion of the number of amino acid changes.

The molecular evolutionary clock is not expected to be a metronomic clock, like a watch or other timepieces that measure time exactly, but a stochastic (probabilistic) clock, like radioactive decay. In a stochastic clock the probability of a certain amount of change is constant (for example, a given quantity of atoms of radium-226 is expected, through decay, to be reduced by half in 1,620 years, its "half-life"), although some variation occurs in the actual amount of change. Over fairly long periods of time a stochastic clock is quite accurate. The enormous potential of the molecular evolutionary clock lies in the fact that each gene or protein is a separate clock. Each clock "ticks" at a different rate—the rate of evolution characteristic of a particular gene or protein—but each of the thousands and thousands of genes or proteins provides an independent measure of the same evolutionary events.

Evolutionists have found that the amount of variation observed in the evolution of DNA and proteins is greater than is expected from a stochastic clock—in other words, the clock is "overdispersed," or somewhat erratic. The discrepancies in evolutionary rates along different lineages are not excessively large, however. So it is possible, in principle, to time phylogenetic events with considerable accuracy, but more genes or proteins (about two to four times as many) must be examined than would be required if the clock were stochastically constant in order to achieve a desired degree of accuracy. The average rates obtained for several proteins taken together become a fairly precise clock, particularly when many species are studied.

This conclusion is illustrated in Figure 1.16, which plots the cumulative number of nucleotide changes in seven proteins against the dates of divergence of 17 species of mammals (16 pairings) as determined from the fossil record. The overall rate of nucleotide substitution is fairly uniform. Some primate species (represented by the points below the line at the lower left of the figure) appear to have evolved at a slower rate than the average for the rest of the species. This anomaly is not unusual because the more recent the divergence of any two species, the more likely it is that the changes observed will depart from the average evolutionary rate. As the length of time increases, periods of rapid and slow evolution in any lineage tend to cancel one another out.

In the reconstruction of evolutionary history, molecular evolutionary studies have three notable advantages over paleontology, comparative anatomy, and the other classical disciplines; namely, universality, multiplicity, and variation. There is universality, because comparisons can be made between very different sorts of organisms. There is very little that comparative anatomy can say when, for example, organisms as diverse as yeasts, pine trees, and human beings are compared, but there are numerous DNA and protein sequences that can be compared in all three. A second advantage is multiplicity. Each organism possesses thousands of genes and proteins, which all reflect the same evolutionary history. If the investigation of one

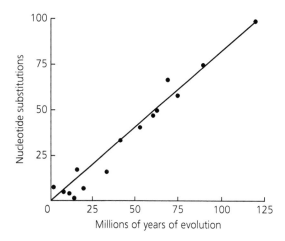

Figure 1.16 The molecular clock of evolution. The number of nucleotide substitutions for seven proteins in 17 species of mammals have been estimated for each comparison between pairs of species whose ancestors diverged at the time indicated in the abscissa. Each dot represents the number of substitutions for the seven proteins added up. The line has been drawn from the origin to the outermost point and corresponds to a rate of 0.41 nucleotide substitutions per million years for all seven proteins combined. The proteins are cytochrome-c, fibrinopeptides A and B, hemoglobins α and β, myoglobin, and insulin c-peptide.

particular gene or protein does not satisfactorily resolve the evolutionary relationship of a set of species, additional genes and proteins can be investigated until the matter has been settled.

The variation advantage of molecular evolution derives from the widely different rates of evolution of different sets of genes, which opens up the opportunity for investigating different genes in order to achieve different degrees of resolution in the tree of evolution (see Figure 1.17). Evolutionists rely on slowly evolving genes (such as cytochrome-c) for reconstructing remote evolutionary events, but increasingly faster evolving genes (hemoglobin and fibrinopeptides) for reconstructing the evolutionary history of more recently diverged organisms.

1.3 Taxonomy

1.3.1 The classification of living beings

Taxonomy is the discipline that deals with the classification of organisms on the basis of their similarities and differences. The foundations of the modern system of classification of organisms were formulated in the eighteenth century by the Swede Carolus Linnaeus

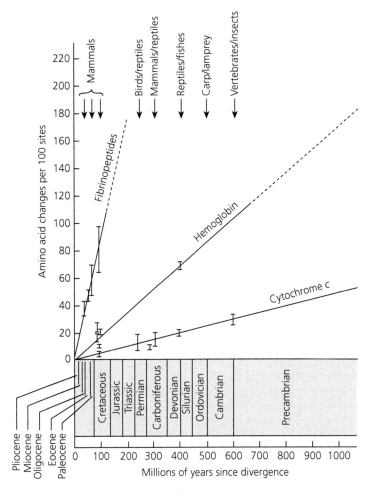

Figure 1.17 Three proteins with different rates of evolution. Cytochrome-c evolves slowly, fibrinopeptides evolve fast, and hemoglobin at an intermediate rate. The lines for each protein represent its average rate of evolution. The vertical lines encompass the variation observed.

(1707–78) in his book *Systema Naturae*. The tenth and definitive edition, published in 1758, is the historical reference for current taxonomy.

Linnaeus achieved his taxonomic hierarchy by grouping organisms according to degree of similarity. He established seven categories made up of groups of increasing inclusiveness: species are grouped into genera, genera into families, families into orders, orders into classes, classes into phyla and, finally, phyla into kingdoms. Intermediate categories were added later.

A taxon (plural: taxa) is a grouping within which organisms are classified. Thus, *Homo sapiens* is the taxon corresponding to modern humans, regarding their genus and species. But if we move up the classification hierarchy, *Homo sapiens* belongs to the tribe

Box 1.7 Hominin tribe

For reasons given in Chapter 2, we include modern humans and their direct and collateral ancestors that are not ancestors of any ape, in the tribe Hominini. Thus, we will refer to them as "hominins." Tribe is a category below family, but above genus.

"Hominini," the family "Hominidae," the order "Primates," the class "Mammals," the phylum "Chordates," and the kingdom "Metazoan" (see Table 1.3).

Linnaeus gave no scientific justification for his system of classification other than similarity. Toward

Table 1.3 Classification of three animals: human, lion, and a certain kind of mosquito. Categories are used to classify organisms (species, genus, family, order, class, phylum and kingdom). They are like drawers in which organisms are placed, so that smaller drawers are included into larger drawers. The labels that we placed in each drawer are called taxa. *Homo sapiens* refers to a "drawer" at the species level; Hominidae refers to a drawer at the family level

Category	Human	Lion	Mosquito
Kingdom	Metazoa	Metazoa	Metazoa
Phylum	Chordata	Chordata	Arthropoda
Class	Mammalia	Mammalia	Insecta
Order	Primata	Carnivora	Diptera
Family	Hominidae	Felidae	Culicidae
Genus	*Homo*	*Felis*	*Culex*
Species	*Homo sapiens*	*Felis leo*	*Cculex pipiens*

the end of the eighteenth century and beginnings of the nineteenth, the French biologist Jean Baptiste de Lamarck (1809) devoted much of his work to the systematic classification of organisms, and suggested an explanation for the resemblance-based hierarchy: degree of similarity was a consequence of evolution, a gradual transition from some kinds of organisms to others. Lamarck's (1809) evolutionary theory had little influence among contemporary or later biologists, because it was metaphysical rather than biological. Lamarck's theory of evolution postulates that all organisms have an innate tendency toward improvement over time, which will continue forever and follow again and again the same path. Our ancestors of eons of time ago were worms and today's worms will have humans as their descendants eons of time hence. Although Lamarck's evolution theory was wrong, his intuitions were correct in seeking an explanation of similarity in the degree of evolutionary relationship. A scientific understanding of the similarity relations among organisms came from Darwin's theory of evolution by natural selection.

Modern evolutionary theory offers a causal explanation for the similarities among living beings. Organisms evolve by means of a process of descent with modification. Changes, and thus differences, gradually accumulate over the generations. So, if the last common ancestor of two species is recent, they will have accumulated few differences. This is the same as saying that similarities in form and function reflect phylogenetic proximity. It follows that phylogenetic affinities can be inferred from the degrees of similarity. This principle currently is the scientific foundation for the reconstruction of phylogenetic relationships based on comparative analyses of living organisms through anatomical, taxonomical, embryological, molecular, and biogeographical studies.

1.3.2 Homology and analogy, again

The reconstruction of phylogeny faces several problems, in addition to occasional incompleteness of information. One important but well-understood difficulty comes from distinction between similarities that have a phylogenetic origin and those that have come about independently as a result of adaptation to similar environments or ways of life, known, respectively, as "homology" and "analogy."

The concept of homology was defined in 1843 by the biologist Richard Owen (independently of evolutionary theory) as "the same organ in different animals under every variety of form and function" (Owen, 1843, p. 37). Nowadays, homology is explained in phylogenetic terms. Two characters (such as human arms and dog forelimbs) are homologous when the resemblance between them reflects the presence of the same features (the various bones and muscles and their configuration and relative position) in a common ancestor from which the two current species inherited them. Analogy applies to similarities that originated independently in different lineages because they serve similar functions. For instance, the wings of bats, birds, and butterflies are analogous. These structures were not inherited from a common ancestor with wings, but evolved separately in each lineage as an adaptation to flight.

The degree of detail in the resemblance provides a practical way to distinguish between homology and

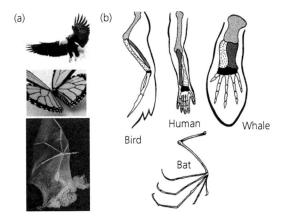

Figure 1.18 Analogy and homology. (a) The wings of different animals carry out the same function of flying, but do so through different structures separately fixed in different evolutionary lineages: they are analogous. (b) The forelimb bones of mammals are very similar although some are terrestrial, others aquatic and yet others fly; they have similar bones organized in similar ways: they are homologous.

analogy. Homology involves detailed similarity (as is the case with each of the bones and muscles of human arms and dog forelegs). Analogy involves similarities in the global configuration (the wings of butterflies, as those of eagles and bats, are wide, thin surfaces) but not in the details of structure and organization (Figure 1.18).

If we analyze the bones of the forelimbs, those of different mammals are similar in their hands, with several fingers (as in the case of humans, whales, and bats), although they perform different functions. They are homologous traits because their structural features were inherited from a common ancestor. The parts of an organ in different animals may be analogous in some respect but homologous in other respects. In Figure 1.18(a), the wings of bats and birds are analogous with respect to their function of flying, because wings evolved independently. But they are homologous with respect to their underlying bone structure. Birds and mammals inherited it from the common reptilian ancestor, Figure 1.18(b).

The example of Figure 1.18 includes two different types of convergence. "Analogy" refers to the independent but convergent modifications that lead to two similar structures from previous, non-homologous features. The wings of insects and birds are strictly "analogous." "Homoiology," on the other hand, refers to the convergent modification that happened independently in two separate lineages concerning structures that originally were homologous. The wings of birds

and bats are "homoiologous." They are homologous because they derive from the forelimbs of their common vertebrate ancestors, but their evolution as wings happened independently. Hereafter we will group both types of convergence within the concepts of "analogous trait," "analogy," or "homoplasy" in a broad sense.

1.3.3 Different types of species

Grouping organisms into different categories is primarily a nominalist activity, with only taxonomic interest: ordering the diversity of living things into sets for easier operation. Nominalist rankings are performed also outside the world of life. They are also useful, for example, to classify ceramics, books, or cars, which is achieved simply by pointing to particular features that are taken under consideration. When Linnaeus (1758) defined the order of primates as animals with two pectoral breasts and four parallel upper incisors, he was doing just that. It suffices to provide the distinctive features that justify grouping certain animals as primates (although we do not currently include among the primates, animals such as bats, which Linnaeus did).

A nominalist taxonomy that lacks an evolutionary criterion for placing specimens in a particular taxon, encounters ambiguities and uncertainties.

The concept of species is, with no doubt, one of the strongest heuristic tools of the life sciences. However, its widespread use in many disciplines, from microbiology to paleontology, which have different needs, observational techniques, and empirical evidences, leads to a considerable dispersion regarding the kind of "species" concept used. Ramon Rosselló-Mora (2003) has pointed out that there are over 22 different concepts of species, but if we are willing to restrict ourselves, we may select two of them, which have universal application: the "evolutionary" species concept and the "phenetic" or "polyphyletic" species concept (see Box 1.8).

A "universal application species" is of little use in each particular specialty of the sciences of life. If we talk about paleontology and, in particular, paleoanthropology—putting aside the demands derived from the study of microorganisms and plants— what we need is a species' concept useful to explain how the evolution of hominins or, at most, primates, took place.

The two most common seem to conflict: (1) the biological species' concept (BSC), formulated by Theodosius Dobzhansky (1935, 1937) and extensively promoted by Ernst Mayr (1942, 1963, 1970); and (2) the evolutionary species' concept (ESC), proposed in its best-known formulation by Edward Wiley (1978) after

---

Box 1.8 Phenetics

In addition to traditional criteria, accumulated through the experience of evolutionists during the nineteenth and early twentieth centuries, two new theories of classification known as "phenetics," or numerical taxonomy, and "cladistics" emerged in the decade of the 1950s. Phenetics proceeds by formulating numerical algorithms, known as phenograms, in which each character can take one of two states: "present" or "absent" (they can be morphological characters, such as the thumb, or an amino acid in a particular protein, such as valine at position six in hemoglobin β, or any other trait). Each character receives a zero (if it is absent) or a one (if it is present) for each species (or higher-ranking categories, such as genera, families or classes). The degree of phenetic affinity among different taxa is determined by the number of ones in the strings of zeroes and ones. This measure does not necessarily reflect evolutionary affinity: it only indicates the extent to which two organisms are similar in form. Indeed, phenetics seeks to avoid any theoretical underpinnings (such as evolution). It does not seek to address the reasons behind the resemblances.

Phenetics has advanced the concept of OTU (Operational Taxonomic Unit) defined according to the strings of zeroes and ones based on morphological comparisons, as described here, but with little success. Cladistics is currently the most extensively used method in paleontological taxonomy.

Box 1.9 Some definitions

By "anagenetic" we understand a lineage that is evolving without branching. Isolation is so significant when talking about species that some philosophers of science—David Hull (1977) and Michael Ghiselin (1987) among them—have argued that the species must have the ontological consideration of "individuals," like organisms. They are born, they change, and they are extinguished within a given time. Each species is distinguishable from the surrounding species and cannot be reduced to them. From this point of view, although the species' concept can be used for taxonomic purposes, grouping species is not the same procedure as grouping books or automobiles. Thus, along an anagenetic process of speciation, several chronospecies could be considered as far as we understand that a mother-organism gives rise to a son-organism. The notion of species as individuals has been the subject of interesting and, at times, acrimonious debate. One difficulty is that populations become reproductively isolated gradually; some populations are only partially reproductively isolated.

By "allopatry" we understand a speciation process in which reproductive isolation is favored by geographical separation of two populations during the time required for the appearance of such isolation mechanisms.

By "sympatry" we understand a speciation process in which reproductive isolation occurs between two populations that occupy the same space during a common time. Isolating mechanisms maybe due, for example, to different specializations in obtaining resources which then could influence mating behavior (see Barluenga et al., 2006).

---

reconsidering Simpson's (1951) previous work. Their most common definitions are:

- BSC—"A species is a group of interbreeding natural populations that is reproductively isolated from other such groups" (Mayr, 1969, p. 26).
- ESC—"A species is a lineage of ancestral descendant populations which maintains its identity from other such lineages and which has its own evolutionary tendencies and historical fate" (Wiley, 1978, p.19).

The main difference existing between these two concepts is the introduction of the variable "time" in the ESC. The ESC allows chronospecies—like anagenetic (see Box 1.9) segments within a lineage—while the BSC can only hypothetically consider such segments, assuming that if two non-coeval, separated-by-time chronospecies were contemporaries, they would comply with the conditions of reproductive isolation.

The problem that we face when trying to reconcile BSC and ESC is not ontological; it is epistemological.

Natural selection only works if diversity exists in both intraspecific (initially the most important) and interspecific (of great importance when an environmental change and/or a population dispersion occurs) cases. Species' diversity is generated by differential accumulation of changes between two populations, be they allopatric or sympatric (see Box 1.9). Changes are not shared between those two populations, provided they do not cross. And this can happen: (1) because of the emergence of mechanisms of reproductive isolation, which are the foundation of the BSC, or (2) by the factual separation—due to geographic isolation, typical of allopatric species, or by means of temporal isolation existing between two populations belonging to different epochs, which is taken into account by the ESC. In the second case, whether these separated populations belong or not to the same biological species is actually irrelevant. They will not hybridize.

Time is a critical variable for the ESC, but it does not have great relevance for the BSC. To discuss how BSC and ESC could be reconciled is worthless. They will be fully compatible if we understand that what is important is whether a separation of populations actually occurred, for whatever reason.

Paleoanthropology has devoted thousands of pages to discuss whether or not two taxa are reducible to the same species. This question often appears when a new species is proposed (e.g., *Australopithecus afarensis* vs. *Australopithecus africanus; Homo habilis* vs. *Australopithecus africanus*), but also arises through works of taxonomic revision (e.g., *Homo habilis* vs. *H. rudolfensis; Homo erectus* vs. *Homo ergaster*). In many cases, the arguments ignore the essential question of how important is the phylogenetic process existing beyond nominalist approaches. In this respect it is necessary to understand that things do not change at all if two taxa actually proposed as different species belong to two populations with effective reproductive isolation (thus legitimately entering in the BSC), or if these are just two populations kept separated due to other causes (true species only in terms of the ESC) (Cela-Conde & Nadal, 2012).

How can we know whether two fossil populations remained separate when they were alive? If features of a population P1 are not observed in P2, it is assumed that P1 and P2 did not hybridize, i.e., that they were kept separate. This is the best criterion available to describe their "own evolutionary tendencies and historical fate."

Once we identify a population that remained separate—being, thus, considered a true

ESC-species—its identity with respect to other such groups must be determined. In other words, it is a must to decide if an S2 species can be reduced to an S1 previously proposed. Using the biological species' concept, the answer would be achieved (theoretically speaking) through a cross-breeding experiment. If the crossing does not give offspring, or the giving is not fertile, then S1 and S2 are different species. Obviously, with fossil taxa considered under the ESC this resource does not apply. What is crucial then is to establish that two reputed fossil-ESCs belonging to the same epoch remained separated from each other. As we said in the previous paragraph, it is very common to use morphological measurements as the taxonomic criterion for this purpose. However, Michael Plavcan and Dana Cope (2001) examined the criteria for recognition of species through metric characteristics, indicating the absence of precise means to move from morphological characters to species. In the case of fossil taxa, a major difficulty arises: that of the amount of variability eventually existing within a single species. If it is hypothetically considered to be quite high, then it may be that two specimens, different in morphological terms, should be placed into the same species. Unfortunately, as argued by Plavcan and Cope (2001), there is no magic number of morphological distinct traits able to determine when a sample consists of one species or two. Let's consider some methods aiming to go beyond simple morphological comparisons.

1.3.4 Advanced morphometric techniques

Leslie Aiello and colleagues (2000), among others, have studied the application of advanced statistical techniques to morphometry (exact randomization) in order to determine what acceptable limits of variability exist in fossil species. These techniques are based on the morphological diversity existing within a given species of current apes. This diversity is then compared with the differences found between distinct fossil taxa. Provided the method reaches accurate conclusions, it would be possible to use it as a criterion for deciding whether a fossil specimen should be assigned to some already defined taxon, or if it is necessary to create a new species.

The exact randomization procedure has been successfully used repeatedly (see Box 1.11). For example, Katerina Harvati, Stephen Frost, and Kieran McNulty (2004), by means of randomization, compared the morphological distance between Neanderthals and modern humans, and concluded that they are two different species. But two difficulties inherent in this

---

Box 1.10 The Tree of Life

As stated by Joel Velasco (2008), "the Tree of Life, which represents phylogenetic history, is independent of our choice of species concept."

That's true, but so is that the Tree of Life is inaccessible directly and, therefore, the representation we make of it is not independent of the concepts used to build the scheme. Moreover, fossil populations were once living populations subjected to biological laws. The outline of any tree of life, although aiming to tell a story, must be respectful of the demands of the mechanisms of reproduction and heredity.

Box 1.11 Exact randomization

Exact randomization is one of the several advanced morphometric techniques existing. For an approach to other morphometric techniques, see Zelditch et al. (2004).

method appear. First, as indicated by Aiello et al. (2000), the conclusions change if we are facing a very, or just somewhat, polymorphic species. Also, this technique requires having a nearly complete description of the test samples. The species defined by means of a few features are not comparable by randomization procedure.

A qualitative test for species' identification is the detection of apomorphies, plesiomorphies, and homoplasies between two or more specimens. This method applies to incomplete samples. Thus, it is not exceptional to name species on the grounds of a few remains, provided conspicuous apomorphies can be detected.

How one can identify the apomorphic, plesiomorphic, or homoplasic kind of trait in fossil specimens leads to a matter of debate. David Begun (2007) stated that the only safe way to do it consists of having previously determined the phylogeny of specimens whose features we want to compare. A comprehensive way to reach these phylogenies is through the processual approach. It consists of establishing for the several taxa considered their adaptation, selection, exaptation—traits evolved under certain conditions that become functionally different—development, and random-change mechanisms. But, needless to say, such a procedure requires much more information than is normally provided by the limited fossil evidence, often lacking adequate taphonomy. Moreover, determining phylogenies is precisely the objective to be achieved through examination of apomorphies, plesiomorphies, and homoplasies. Are we falling into circular reasoning?

One possible way to avoid circularity is to change the zoom level, determining the phylogeny of a genus before detailing the condition of the traits of any specimen eventually belonging to it. The processual approach is comparatively easier to apply when we talk about genera. For example, immunological analyses allowed the establishment of distances between different hominoids, therefore reaching the most likely phylogeny (Goodman et al., 1960; Goodman, 1962, 1963; see Chapter 2). In actual fact, examples given by Begun (2007) refer to genera of both current and fossil hominoids.

Having a general reference to the phylogeny of the genus, it is sometimes possible to detect homologies and analogies relating to species and, from there, to establish both lineages and ranges of variation. An example is the evolutionary sense of the features of hands and feet in the genus *Ardipithecus* (Lovejoy, Latimer et al., 2009; Lovejoy, Simpson et al., 2009; White, Asfaw et al., 2009) compared with African apes' knuckle-walking. This evolutionary scenario can be achieved because the hominoids' phylogeny is previously known.

1.3.5 The molecular identification of species

Advances in molecular genetics allow for an objective method for identifying species: by means of genetic distance between any two taxa. It can only be done in a straightforward way, regarding fossils, if we have recovered their genetic material, but hypothetical extrapolations can be proposed. Though, paradoxically, molecular species' identification by means of extrapolations sometime seems to lead to contradictory results.

Darren Curnoe and Alan Thorne (2003) applied the genetic distance between humans and chimpanzees to calculate how many species fit in the human lineage, provided that they maintain similar molecular distances as current *Pan* and *Homo*. Their findings argued that *Homo sapiens* is the only species present in the last 1.7 Ma. Curnoe's subsequent work nuanced this extreme result. Initially, by means of the genetic distances' extrapolation method, Curnoe (2008) established the phylogenetic tree shown in Figure 1.19. Species identified as such largely coincide with the genera admitted in this book. Later, Curnoe (2010) adopted the traditional stance of admitting *H. erectus* as a valid species.

These results contradict those coming from studies of mitochondrial DNA, which Darren Curnoe and Alan Thorne (2003) criticized, rejecting them as valid evidence. Therefore, there is no single and consistent "molecular species' identification," and, thus, the direct contrast between genomes could not be carried out; consequently, the extrapolation of current molecular distances to identify fossil species is questionable.

To establish lineages in which the current species could be placed is the purpose of any phylogenetic approach. However, reconstruction of lineages is not straightforward. The best available solution for classifying fossil taxa is cladistics, thus far. As we shall see in Section 1.3.6, it is not free of difficulties.

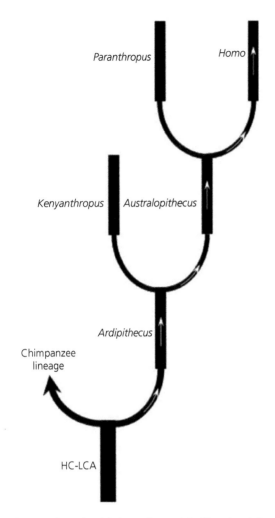

Figure 1.19 Species of the human lineage derived from the existing molecular distance between current chimpanzees and humans (Curnoe, 2008).

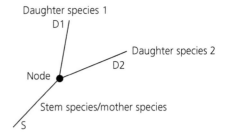

Figure 1.20 Cladistic phylogenies. Hennigian cladistics defines speciation as the split (node) of a stem or mother species (which thereby becomes extinct) into two sister species.

1.3.6 Cladistics

Cladistics is a method developed by the German zoologist, Willi Hennig (1950, 1966), as a useful tool of classification in paleontology. It starts from the requisite that species (or other taxa) be classified according to their phylogenetic relationship, rather than based on their degree of morphological or phenetic similarity. The graphical representation of phylogenetic relationships is a cladogram: a branching diagram, where one branch splits into two whenever one species (or other taxon) splits into two species (or other taxa).

Consider a certain species, which we will call a "stem species." Because it is a species, it constitutes an isolated reproductive unit. Its reproductive characteristics, according to Hennig—who called them "tokogenetic relations"—will end, and the stem species disappear, when it becomes replaced by two new descendant groups, which he called "daughter species." The set comprising the stem species and the two daughter species—sister species to one another—constitutes a clade (see Figure 1.20). The representation of the split or speciation moment is called a "node."

Cladistics distinguishes between primitive (or ancestral) characters, known as "plesiomorphic," and derived or "apomorphic" characters. When an apomorphic character is present in two or more descendant taxa, it is called "synapomorphic" (meaning "jointly derived"); if an apomorphic trait is present in only one of the descendant taxa, the trait is "autapomorphic" ("autonomously derived"). Primitive characters in any lineage are those that were already present in the ancestors. Derived characters are those that have just appeared in the lineage.

By definition, a cladistic episode requires the presence of at least one apomorphy (derived trait) characteristic of each of the sister species appearing after the node. These species can conserve, of course, primitive traits inherited from the stem species, which will be identical in the daughter species (plesiomorphic or primitive traits).

Similarities based on primitive (plesiomorphic) characters are not useful for determining relationships among descendant taxa. Characters present in only one descendant taxon (autapomorphic) are also useless for determining phylogenetic relationships. Only shared derived characters (synapomorphies) are useful to determine phylogenetic relations. For example, mammary glands and hair are found in mammals, which group them together and separates them from birds, reptiles, and fishes. From this perspective, mammary glands are a synapomorphy shared by all mammals and differentiating them from other vertebrates. However, if we want to classify different mammals, mammary glands are plesiomorphic, a primitive character that all current mammals have inherited from the first mammals. The lack of placenta is a plesiomorphy of birds, reptiles, and fishes (that is to say, a character they inherited from a common ancestor), which does not tell us anything about the phylogenetic relations among these three groups of organisms.

Thus, phylogenies can be inferred simply by comparing derived and primitive traits of current species with those of fossil species, or of fossil species living at different times. The identification of speciation events through time consists, by definition, of the identification of clades and nodes. A cladistic event requires the presence of at least one apomorphy, a characteristic derived trait in each of the sister species that appear at the node.

The species' concepts of Hennig and Mayr consider reproductive isolation as the main trait that characterizes species. Mayr wrote that "the essence of the biological species' concept is discontinuity due to reproductive isolation" (Mayr, 1957, p. 376). Hennig's species' concept refers to "reproductively isolated natural populations" (Meier & Willmann, 2000; see De Queiroz & Donoghue, 1988).

The important difference between the biological (Mayr's) and cladistic (Hennig's) species' concepts is the inclusion of the time dimension in the latter. Hennig's objective was the reconstruction of phylogenies. Time is crucial in cladistics, but it is relative time, not absolute time. When a stem species, S, splits into two sister species, D1 and D2, absolute time is present in a trivial way, in that S must precede D1 and D2 (Figure 1.20). But the key concern is relative time: after the speciation episode, each sister species occupies an isolated temporal niche. Once the two daughter species have appeared, the evolutionary events affecting species D1 have nothing to do with those involving species D2. They live two separate "specific times." For instance, D1 could originate a large clade, with many new speciation episodes, while D2 could remain as a single species for a long time. Mayr's species' concept also allows this possibility, but the difference lays in the relevance that cladistics awards to the question of what is and is not a speciation event in any given clade.

1.3.7 The problem of phyletic lineages and transformed cladistics

In the original Hennigian formulation, a speciation event involves the appearance of two sister species

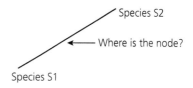

Figure 1.21 Phyletic process (speciation without ramification). The node is not identifiable.

and the extinction of the stem species. The stem species cannot survive the speciation event and the ancestral lineage of the sister species cannot include two different stem species (Meier & Willmann, 2000). In other words, anagenetic speciation (the transformation of one species into another through time without the split of a stem species into two daughter species) is not allowed. Anagenetic speciation may occur in nature, but it is irrelevant for clade reconstruction and, therefore, it is ignored in cladistics (Figure 1.21). The rationale for the decision is that there is no criterion that would establish the precise boundary at which one species becomes another; indeed, the process is gradual, as we know. But the split of a stem species into two descendant species ("cladogeny") can be unambiguously identified at the node.

In order to become a new species, S2 must achieve complete reproductive isolation from S1. This involves the appearance of mechanisms (ecological, genetic, or otherwise) that separate populations that live at the same time. Without such a temporal coincidence, the reproductive isolation concept makes little sense. To say that 1-million-year-old *Homo erectus* was reproductively isolated from any *Australopithecus* species, which lived several million years earlier, makes no sense, cladistics says. Reproductive isolation requires that S2 becomes a new species while S1 still exists. If S1 disappears when S2 appears, as in anagenetic speciation, then we have abandoned the theoretical concept of speciation through reproductive isolation episodes. The concept of anagenetic speciation depends, rather, on operational prescriptions applied to the fossil record.

Nevertheless, it is common in paleontology to name species along a phyletic lineage. The term "chronospecies" is applied to groups of organisms living in different time periods that appear to be ancestors and descendants, when these groups are morphologically as different from each other as contemporary organisms classified in different species. For instance, modern horses, *Equus*, and their 50-million-year-old ancestors, *Hyracotherium*, receive different names because, from a morphological point of view, they are at least as different from one another as either one is from, say, modern zebras. The concept of chronospecies allows for the recognition of phyletic evolution when cladogenetic events are unknown.

The classification of ancestors and descendants into different chronospecies is appropriate when the temporal sequence of known fossils is fragmentary. The absence of transitional fossils facilitates classification into different species, those fossils that are quite distinct and separated by many years of evolution. If the fossil record were sufficiently complete through time, the situation would be different. If we documented small sequential changes in a long phyletic sequence, we might consider the extreme members of the sequence as different chronospecies, but there would not be a particular point in time at which one species would have become another. Of course, the fossil record is rarely sufficiently complete to display this situation.

It would be possible to identify a chronospecies S2, daughter of S1 in a phyletic lineage, if we found contemporary species. This situation could be considered exemplar of a variant of the formation of two sister species in a node, in which the mother species takes the place of one of the daughter species (Figure 1.22a). However, original Hennigian cladistics does not allow the simultaneous presence of stem and daughter species.

Transformed cladistics (Platnick, 1979) tried to overcome this obstacle (Figure 1.22b) by allowing a daughter population to be considered a different species if it has at least one apomorphy (derived character) that distinguishes it from the mother species. But, how to represent the process? The solution adopted by transformed cladistics is to represent the mother species S1 and daughter species S2 as sister species in the cladogram.

An important consequence of placing the mother and daughter species as sister species is the transformation of the original sense of Hennig's stem species.

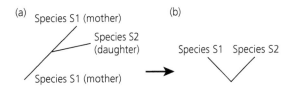

(a) Species S1 (mother)

Species S2 (daughter)

Species S1 (mother)

(b) Species S1   Species S2

Figure 1.22 A phyletic episode with the persistence of the mother species (a) in traditional cladistics (which does not allow such speciation) and (b) in transformed cladistics.

Schaeffer, Hecht, and Eldredge (1972) had already suggested, before the proposal of reformed cladistics, that all taxa, fossil or living, might be placed as terminal taxa in a cladogram. As a consequence, only hypothetical ancestors can be placed in the nodes. Once a fossil taxon is correctly identified, it must be placed as a terminal taxon. Hence, stem species disappear as parts of branches or as nodes. Not only do their representations disappear, but also the concept itself. Thus, cladograms lose their meaning as a representation of the evolution process (in the style of phylogenetic trees) and are reduced to representations of the way lineages are divided by means of sister species.

A price paid for this transformation of cladistics is the loss of the temporal dimension. Cladograms would no longer represent ancestry relations. Furthermore, speciation processes through time cannot be established by means of cladistics. According to Delson, Eldredge, and Tattersall (1977), the concept of "sister species" is a methodological instrument that must be applied even if (1) the taxa under consideration are two species that hold an ancestor–descendant relation and thus are not true "sister species"; (2) and the taxa under consideration have close relatives that are as yet unknown. According to Delson et al. (1977), a cladogram constructed in such a fashion does not allow a decision as to whether the branches stemming from a node represent sister or mother/daughter species. Ancestry relations disappear as objects of scientific inquiry given that, within cladistics, the hypothesis that a taxon is the ancestor of another cannot be tested (Nelson, 1973; Cracraft, 1974; Delson, 1997). The same idea is expressed by Siddall (1998), who states that seeking to describe evolutionary relations by searching for ancestors in the fossil record is a resurgence of the cult of the golden calf.

1.3.8 Beyond species

We have elaborated that the species' concept is fundamental in order to understand organisms and their evolution by natural selection. The Linnaean taxonomy includes other classification categories in addition to species. The category "genus" lies immediately above. A genus includes closely related species (although some genera may include a single species).

The species' concept is not only a taxonomic category, but actually refers to groups of organisms that are importantly related with one another by relations of mating and parentage. What about the concept of genus? Is it a completely artificial construct or does it have significance beyond its condition as a taxonomic artifice? If the genus category is purely an artificial construct, clustering organisms in genera would be completely arbitrary, though not insignificant. Grouping beings in certain genera would make a difference, just as we might classify books by subject. But different ways to construct genera could be suggested. An alternative is to consider that the category of genus (as well as the more inclusive ranks, such as family, order, and so on) has certain distinctive traits precisely because it refers to biological attributes. If this were the case, it would be inadequate to create a taxon belonging to a genus, or another higher category, if it distorts the distinctive sets of attributes that characterize the taxon. Cladistics emphasizes an essential feature of organisms: they evolve forming lineages. Figure 1.23 represents the lineages corresponding to seven species (A–G) that appeared by means of six speciation events (nodes 1–6). Each node is the source of two sister taxa. In this way, species A is B's sister group (and vice versa), while C + D + E + F + G is the sister group of A + B.

A genus is a set of species. If a genus is purely a taxonomic artifice, we could define a genus that would include species A, C, and E, for instance. But if taxonomy should respect evolutionary processes, then not just genera, but any category must only include taxa that constitute complete parts of the cladogram. This is the same as saying that genera (and families, and so on) are evolutionary lineages with real existence. They reflect the way in which phylogeny occurred. This is why we include bats, lions, and dolphins in the taxon "mammals." Our classification does not cluster bats, eagles, and butterflies, although they all fly. Bats, eagles, and butterflies do not constitute a lineage, just because they have wings.

"Monophyletic groups" include whole lineages: they reflect the process of evolution. In Figure 1.23,

A + B and C + D + E + F + G are monophyletic groups. The set E + F is also monophyletic because it is a group including all taxa stemming from a particular node. "Paraphyletic groups" are those that leave out some taxa pertaining to the lineage. In Figure 1.23, a group including C + D + E + G would be paraphyletic because it does not include taxon F. "Polyphyletic groups" include lineages that have not arisen from the same node, excluding intermediate ones. Thus the grouping A + B + E + F is polyphyletic.

Let's now turn to the question of rank. The rank of the genus category is superior to that of species, because a genus is a set of species; a group of genera is a "family" (if we do not consider intermediate categories, such as "tribe"); and so on. In Figure 1.23, the set A + B constitutes a genus because it is a group of species. For the same reason, E + F must be considered another genus. But now we find ourselves with a problem. Node 5 gives rise to two sister groups, E + F on the one hand and G on the other. We said that G is a species, and E + F a genus. How is it possible that the sister group (E + F) of a specific taxon (G) belongs to a higher category than the latter? A possible solution is to award sister groups the same category. Thus, although G is a species, it must be classified also as a genus, just as E + F, even if it is a genus with a single species.

Problems do not disappear with this taxonomic maneuver, however. If each new node requires elevating the rank of the categories, we will soon run into difficulties in the case of lineages with numerous branches. This is the reason why new intermediate categories are introduced (tribe, subfamily, superfamily, infraorder, and so on)—though such a proliferation could become excessive. Basing taxonomy on numbers rather than names (such as phenetics does, for example) would resolve the problem, but this is not a common practice. Thus, in order to avoid an excessive number of categories, it is advisable not to apply the strict rank equivalence of sister groups. We will now consider a practical case that refers to the group formed by the great apes and humans.

1.3.9 The adaptive concept of genus

As we have seen, it is not easy to come up with objective criteria to decide whether several species should be grouped into one or several genera. Hybridization between two organisms indicates that they belong to the same species, but there is not an equivalent test to verify whether they belong to the same genus. The international code of taxonomy

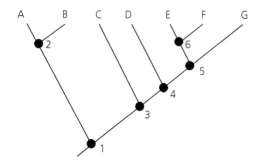

Figure 1.23 Cladogram representing seven species originated by means of six nodes.

does not provide objective classification criteria regarding categories above the species level. Authors suggest particular classifications hoping that the scientific community will accept them. But with respect to the genus category, it is often helpful to follow Ernst Mayr's (1950) proposal that a genus refers to a particular way of adaptation to specific conditions. A new genus, according to Mayr's proposal, refers to a new kind of organism that adapts to its ecosystem in a different way from other organisms included in other genera.

It is not easy to determine how fossil specimens adapted to their environment. However, certain inferences can be made from morphological traits. For instance, the presence of thick molar enamel indicates a diet that included hard materials. The robusticity of the masticatory apparatus points in the same direction. The mode of locomotion can be inferred from the analyses of forelimbs and hind limbs. A large brain in relation to body size is associated with the ability to construct and carve complex tools and instruments.

The adaptive concept of genus must be used with caution—it is not a method to decide how different lineages evolved. Cladistic analyses of apomorphies are much better suited to that end. But once the distribution of lineages is known with a certain degree of confidence, it is useful to consider adaptive specializations, because they are helpful in order to avoid naming a new genus almost every time a new fossil is discovered, as has happened in the reconstruction of human phylogeny. In the following chapters, as we review the evolution of Miocene and Pliocene hominins, we will see how convenient it is to avoid the excessive multiplication of genera; although a radical reduction of genera is not a good solution either. Mayr's adaptive criterion represents a step forward in the search for phylogenies and taxonomies unbiased by the classifier's preconceptions, even though there are serious difficulties in discovering adaptive strategies from the fossil and archaeological records.

1.4 The study of the fossil record

The fossil record is the main source of information available to understand how evolution has occurred in the human lineage. Fossils indicate which different specimens natural selection allowed to survive. And the possible age of the discovered specimens is crucial to explain the evolutionary process of any lineage. Hence, the need to find out the age of fossils, or at least that of the terrain in which they were found.

1.4.1 Stratigraphy

Fossilization, i.e., the replacement of organic materials with minerals that retain at least some of the organisms' original form, occurs primarily in sedimentary soils. These are created by accumulation of deposits that, with the passage of time and the pressure of new materials upon them, are converted into rocks—strata—which at the beginning are more or less horizontal. Deposits containing fossils are displayed as a succession of strata, successive layers of sediment. By the logic of the process, the deeper a stratum is, the older it will be (Box 1.12).

Stratigraphy is the technique that determines the correct sequence of strata. This is the fundamental task necessary to begin the study of any site and requires specifying the coordinates (horizontal and vertical dimensions) of each point of interest of the site (see Figure 1.24).

If the complete stratigraphic sequence corresponding to the time we are considering is available, this will indicate the age of the fossils, i.e., which are older and which are more recent. It is a relative age, until we know at what time the stratum containing each specimen was formed. There are several ways to identify the age of a stratum, as we will see. Please note that the notion we have of the correspondence "deeper stratum = older stratum," faces the possibility of alterations in the original sequence of deposition due to geological disturbances that may have occurred. Orogenic movements cause the ground to fold and, thus, strata alter their previous formation order by the presence of faults, thrusts, and even inversions. The reconstruction of the geological history is, therefore, necessary to reconstruct the original order of material

---

Box 1.12 Members and formations

Strata or beds are geological units that are called "members." The whole set of different members with the same geological history is called a "formation." Fossiliferous zones generally contain different formations. When we speak of "site" and "location" we are using colloquial terms that may refer to either formations or members, unless accurate identification is given. Thus, it may be said that the site of Hadar (Ethiopia) or, to be specific, that of the Sidi Hakoma member of the Hadar formation, has provided the first fossils of the species *Australopithecus afarensis* (for all sites and species referred here, see the chapters that follow).

Figure 1.24 Stratigraphy of the Aramis region. Two dated volcanic horizons constrain the main hominin-bearing stratigraphic interval in the Aramis region. The top frame shows these tephra in situ near the eastern end of the 9-km outcrop. The dark stripe in the background is the riverine forest of the modern Awash River running from right to left, south to north, through the Middle Awash study area of the Afar Rift (Figure from White, Asfaw, Beyene, et al., 2009, modified).

deposition. It helps a lot in this process to identify the absolute age, not only the relative age, of each stratum. There are different methods of absolute dating; we will refer to some of them.

1.4.2 Faunistic comparison (biochronology)

If we know the age of some of the fauna, that age can also be attributed to the stratum where it was discovered. The risk of circularity, however, may be high. This happens when we deduce the age of the fossils from the stratum, and the age of the stratum from the fossils it contains.

1.4.3 Radiometric methods: radioactive isotopes

When there are volcanoes near the sites we want to study, their past eruptions may have thrown ashes over the sedimentary land, forming tuffs. These tuffs can be dated directly, exploiting the fact that some of their material is radioactive. Radioactive atoms are characterized by instability, emitting particles (electrons, neutrons, protons, etc.) and transforming into other isotopes, a process called decay or disintegration. It is not possible to know when a particle will be emitted, but it can be established statistically the amount of time that must elapse for half of a given quantity of radioactive material (such as potassium 40)

to decay into stable material (such as argon 40). If a site has a sample containing measurable amounts of ^{40}K and ^{40}Ar, it is possible to deduce the time that has passed since the sample consisted entirely of ^{40}K, i.e., since the volcanic ashes were deposited.

Fossils are found, as we have said, in sedimentary, not in volcanic, terrain. But, the age of a stratum situated between two tuffs will range between the age of the two.

One of the radiometric methods commonly used in human paleontology is the already mentioned potassium/argon ($^{40}K/^{40}Ar$; potassium ^{40}K becomes argon ^{40}Ar gas, with a half-life of 1,260 million years Ma). If it can be determined in a sample how much is left of the original ^{40}K, then its age can be calculated. The way to do this is to heat the sample until it is completely melted, then measure the amount of argon released in the process. However, it is necessary to determine that the sample has not been contaminated by argon present in the atmosphere.

The application of the potassium/argon method is extensive. The age of the earth is estimated at 4,500 Ma by the $^{40}K/^{40}Ar$ method. The lower limits of the technique are around 1 Ma.

An improved variant of the potassium/argon method is $^{40}Ar/^{39}Ar$ laser fusion, a step forward from the previous method. It involves heating several samples at different prefixed temperatures. A determinable amount of ^{40}Ar is released. Each sample is then subjected to neutron bombardment inside a reactor. ^{39}K is the most common potassium isotope and it is not radioactive, but the bombardment results in the conversion of potassium ^{39}K to argon ^{39}Ar, which is also measurable. By comparing the ^{40}Ar released to the ^{39}Ar obtained in the nuclear reactor, it is possible to estimate the age corresponding to each of the samples of different age—the "age spectrum." The analysis of the age spectrum allows calculation of the age of the terrain.

Sites with an abundance of volcanoes have provided very precise dating by both methods. This is the case of the Rift Valley, where the site of Olduvai is located. Olduvai was the first site in which radiometric techniques were used to date their beds (Leakey et al., 1961).

Other similar techniques are U/Pb and U/Th, successfully applied in sites such as Sterkfontein (South Africa). The U/Pb method (decay of uranium ^{238}U into lead ^{204}Pb, with a half-life of 4,200 Ma), is useful for dating materials of more than 200,000 years when a sufficient concentration of uranium is present.

The technique of carbon-14 (^{14}C) measures the present amount of this isotope, obtained from stable carbon ^{12}C by the action of cosmic rays in the upper layers of the atmosphere. Carbon-14 descends and is fixed by plants, which, during their life, maintain a fixed percentage between the two isotopes. From the plants, ^{14}C passes to the herbivores that feed on them, and then goes up the trophic scale to the carnivores, so that every organism has the radioactive isotope. When the organism dies, no more contributions occur, and ^{14}C starts to disintegrate. If the proportion that remains is measured, one will obtain the absolute age directly from the fossil itself, not from the surrounding grounds. But the half-life of the ^{14}C isotope is very short, 5,730 years. Only very recent fossils can be dated this way. Although technical advances have improved this method, there is no reliable dating beyond 60,000.

1.4.4 Radiometric techniques: cosmogenic nuclides

Similar to the ^{14}C method is the cosmogenic nuclides' technique, which measures the accumulation of nuclides in materials subjected to cosmic radiation, i.e., it obtains an estimate of the time the materials have been exposed to it. Beryllium and aluminum isotopes are usually used.

The technique has been described by Timothy Partridge et al. (2003). Upon entering the atmosphere, cosmic rays not deflected by the Earth's magnetic field form cosmogenic isotopes, such as beryllium ^{10}Be, which becomes, with a half-life of 1.5 Ma, the stable isotope ^9Be. The $^{10}Be/^9Be$ method allows dating, in particular, of lacustrine sediments, but has the disadvantage that both isotopes, radioactive and stable, are initially indistinguishable from each other. So, it is first necessary to conduct the task of establishing the amounts of ^{10}Be. Nevertheless, the method has been used in sites as difficult to date as, for example, Sterkfontein (South Africa) or Toros-Menalla (Chad).

Radioactive decay of ^{26}Al (with a half-life of 1.02 ± 0.02 Ma) and ^{10}Be (half-life of 1.93 ± 0.10 Ma) produces quartz. These two radioactive cosmogonical nuclides occur near the surface, in known proportions, due to the nucleons and muons of secondary cosmic rays. Quartz grains near the surface accumulate a set of radioactive nuclides whose concentrations depend on the time the mineral has been exposed to cosmic rays, which in turn depends on the rate of erosion of the rock that contains it. If the surface quartz is buried suddenly—for example, by deposition in a cavern—then ^{26}Al and ^{10}Be production decreases or stops abruptly. Because ^{26}Al decomposes more rapidly than ^{10}Be, the rate $^{26}Al/^{10}Be$ decreases exponentially with the time of burial, offering a means of dating buried sediments. The age of a sediment is calculated by taking into account the concentration ratios of $^{26}Al/^{10}Be$ before and after the time of burial.

1.4.5 Radiometric techniques: fission-track and electron spin

The fission-track method analyzes the trails left in volcanic glass—or in ceramics when utensils of this kind are available—from the fission of ^{238}U atoms, a radioactive material with a mean lifetime of 4,500 Ma. The energy released by the fission of a nucleus leaves a mark on the glass, and the number of tracks is a function of the time elapsed since its formation.

But if the glass is heated above 60°C, tracks begin to disappear and at 120°C they are erased. Thus, the method can determine both the date of the formation of the crystal containing ^{238}U, as well as the moment when it was heated, which happens with the production of ceramics.

The fission-track technique is often used to determine the reliability of radiometric ages obtained by other methods. The KBS tuff study, by Anthony Hurford and collaborators, was the first to date recently formed zircon crystals using this technique (Hurford et al., 1976). A re-analysis done by Andrew Gleadow

(1980) found some problems in the method, and allowed the development of a new technology to work with zircon crystals, which have low-density tracks.

The method of electron spin resonance (ESR) determines the time since the last exposure to radiation from cosmic rays by measuring its effect—usually in the enamel hydroxyapatite of fossil teeth. To avoid errors, it is necessary to estimate the effects of radioactive isotopes of the site, and subtract them to get the radiation from cosmic rays. The thermoluminescence method is similar, although it usually is applied to stalactites in caves or to burned materials. The gamma ray spectroscopy technique also measures the accumulated radiation in fossils, but with the advantage over the mass spectroscopy method, that it does not require the destruction of the dated material. Gamma ray spectroscopy was used to obtain the age of Ngandong's human fossils (Java).

1.4.6 Paleomagnetism

The earth's magnetic field, with its North and South Poles located as they are today, has suffered inversions periodically throughout the history of the earth.

Within a short period in geological terms, the North and South Poles have exchanged their polarities. Examination of the orientation of magnetic minerals included in volcanic or sedimentary rocks can establish whether a particular site has a "normal" polarity (the one that exists today) or "reverse" (with the South Pole located at the actual northern end, and vice versa). The paleomagnetic column indicates the chrons sequence with changing polarities (see Figure 1.25). It is, therefore, a relative dating method, not an absolute one, whose reliability depends on the proper identification of the general age of the site. Since the main magnetic polarity reversal occurrences are known, they can be a very important aid when dating a site for complementing radiometric methods, or even as a better indicator in sites without nearby volcanic activity. But a risk of error exists in the absence of an absolute calibration. An example is associated with the column in Figure 1.25. When a stratigraphic sequence is located whose order is "− + −," with relatively long stretches of reversed polarity and a short, intermediate stretch of normal polarity, the latter can be attributed both to the subchron Jaramillo as to the Olduvai.

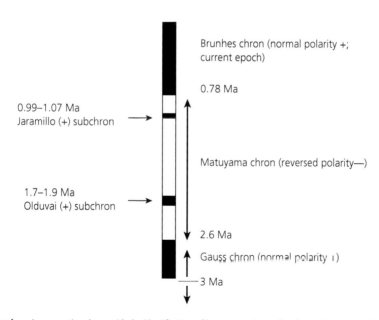

Figure 1.25 Example of a paleomagnetic column with the identification of long-range chrons, Brunhes—Matuyama—Gauss (until subchron Kaena, 3.05 Ma), and short subchrons, Jaramillo and Olduvai. In black, "normal" polarity chrons (corresponding to current times). In white, "reverse" polarity chrons (reversed North and South magnetic poles). The complete sequence for the Upper Cretaceous and Cenozoic was established by Cande and Kent (1995). Reprinted from Semendeferi, K. & Damasio, H. (2000). The brain and its main anatomical subdivisions in living hominoids using magnetic resonance imaging. *Journal of Human Evolution*, 38, 317–332, with permission from Elsevier.

Table 1.4 International Stratigraphic Chart, to which we have added their correspondence to previous denominations (https://engineering.purdue.edu/Stratigraphy/charts/chart.html)

Era	Period	Epoch	Age	Previous Denomination	Beginning (Ma)
Cenozoic	Quaternary	Holocene		Holocene	0.0118
		Pleistocene	Upper	Upper Pleistocene	0.126
			Ionian	Middle Pleistocene	0.781
			Calabrian	Lower Pleistocene	1.806
			Gelasian	Upper Pliocene	2.588
	Neogene	Pliocene	Piacenzian	Middle Pliocene	3.600
			Zanclean	Lower Pliocene	5.332
		Miocene	Messinian	Upper	7.245
			Toronian	Miocene	11.606
			Serravallian	Middle	13.820
			Langhian	Miocene	15.970
			Burdigalian	Lower	20.430
			Aquitanian	Miocene	23.030

Box 1.13 Ages and their dating

The Gelasian Age was included in the Pleistocene in order to make all recent glaciations belong to the same Age. The reason we keep the old denominations is because the vast majority of references used in this book follow the classical division of ages. However, we will use for each Age the dating indicated by the International Stratigraphic Chart. In an informal way, we refer to "Plio-Pleistocene" as the moment of transition between the Pliocene and Pleistocene (Gelasian to Calabrian Ages).

1.4.7 Geological stages

The International Commission on Stratigraphy has established the standardization of geological stages by a common reference, the International Stratigraphic Chart (ISC) whose content regarding the Cenozoic Era is shown in Table 1.4.

Despite being an international convention, we will retain the classic denominations regarding the Pliocene and Pleistocene. Thus, we will retain the Gelasian Age (between 2,588 and 1,806 Ma) in the Pliocene, under the name of Upper Pliocene, placing the beginning of the Pleistocene 1,806 Ma (Box 1.13).

If Figure 1.25 and Table 1.4 are compared, there is no strict correspondence between geological ages and paleomagnetic chrons, except when the limit between the Brunhes and Matuyama chrons is taken as the beginning of the Upper Pleistocene (Tarantian Age). The Olduvai subchron includes the transition from Pliocene to Pleistocene, although it does not coincide exactly with it. The limit between the Gauss and Gilbert chrons (not shown in the figure) coincides fairly closely with the transition from Lower to Middle Pliocene.

CHAPTER 2

Taxonomy

2.1 Taxonomic considerations of the tribe Hominini

According to the traditional classification, the order Primates includes several suborders (Figure 2.1). A suborder is an intermediate category between "order" and "family." Below this suborder category we find infraorder and, one step lower, superfamily. The suborder of anthropoids (Anthropoidea) includes two infraorders: catarrhines (African, European, and Asian monkeys) and platyrrhines (American monkeys), which diverged after continental drift separated South America from Africa. Catarrhines are divided into two superfamilies: cercopithecoids, or Old World monkeys, and hominoids (apes and humans).

The evolutionist G. G. Simpson (1931, 1945) in his classification of mammals distinguished humans from apes at the family level—Hominidae and Pongidae—but he classified australopithecines (which nowadays are considered hominins) with pongids. Except for this, Simpson's classification of primates was widely accepted for three decades.

Simpson's classification of hominoids, as including apes and humans, requires that all existing apes, their direct ancestors, and their descendants be included in a single family, Pongidae, reserving another family, Hominidae, for the human lineage. Is such a separation justified? Simpson's 1945 classification proposal was based on morphological similarities: leaving gibbons aside, it seems orangutans, gorillas, and chimpanzees are more similar among themselves than any of them is to humans. If morphological similarity reflects evolutionary relatedness, then the traditional classification implies that the hominid branch was the first to separate from the pongids, which later split into all existing ape genera and species.

2.1.1 Immunological methods

In the 1960s, immunological methods contradicted such inferences (Box 2.1). With analyses of proteins in the blood serum of hominoids, Morris Goodman, a molecular geneticist at Wayne State University in Detroit, determined that humans, chimpanzees, and gorillas are closer to each other than any of them is to orangutans (Goodman, 1962, 1963; Goodman et al., 1960).

According to Goodman's results, the evolution of hominoids proceeded very differently from what Simpson's taxonomy implied. The lineage leading to orangutans was the first to split, then the lineage of gorillas split from the others, and, finally, the chimpanzees and human lineages diverged from each other. Accordingly, the classification of apes in the family Pongidae and humans in a separate family, Hominidae, is not appropriate. Such a classification includes a paraphyletic group and artificially separates the taxon *Homo* from the common lineage that it evolved from (Figure 2.2).

According to molecular findings, a correct classification should place orangutans in a taxon of the same category as the set gorillas + chimpanzees + humans. The former would constitute the family Pongidae and the latter the family Hominidae, if we want to keep the division in two families. But before we discuss this issue any further, we'll turn to the age of separation of the different lineages.

2.1.2 Age of hominoid lineages

Morris Goodman's results on the sequence of separation of hominoid lineages had profound taxonomic consequences, but provided no information about the timing of the divergences. Although order of sequence and timing are related, determining the time of divergence between two lineages requires determining the rate of molecular evolution of the trait under consideration. We need something like a clock that would allow us to determine how much time had elapsed for each degree of immunological differentiation between proteins. Vincent Sarich and Alan Wilson (1967a, 1967b) argued that the immunological differentiation between chimpanzees and humans indicated that the

Processes in Human Evolution. Francisco J. Ayala and Camilo J. Cela-Conde, Oxford University Press (2017).
© Francisco J. Ayala and Camilo J. Cela-Conde.
DOI 10.1093/acprof:oso/9780198739906.001.0001

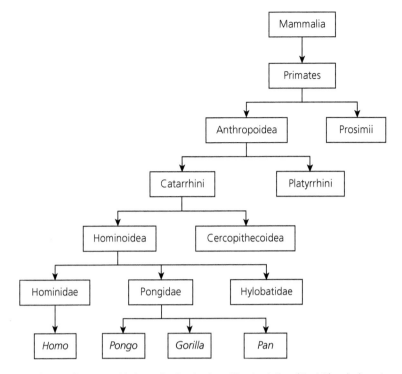

Figure 2.1 Traditional classification of primates with three suborders (Anthropoidea, Prosimii, and Tarsioidea; the latter is not included in figure). In this traditional classification, the family Hominidae includes human beings, while the family Pongidae includes the great apes.

Box 2.1 The study of immune systems

The activity of antibodies during the invasion of the organism by foreign proteins allows determination of the relatedness of immune systems. After an injection with human blood, a rabbit generates specific antigens against human proteins. If we then apply those antigens to the blood serum of other animals, we can deduce the evolutionary closeness between humans and those other animals, by the strength of their immune reaction. Vincent Sarich and Allan Wilson (1967a) confirmed the results obtained by Goodman, who established that chimpanzees and humans are closer to each other than either of them is to gorillas.

two lineages separated between 4 and 5 million years ago. According to this calculation, no Miocene fossil could be a direct ancestor of humans, given that the two lineages diverged later. Goodman (1976, for example), on the contrary, argued that the rate of molecular evolution was slow in the hominoids and that the

divergence of the hominoid lineages occurred earlier than estimated by Wilson and Sarich.

Goodman's and Sarich, and Wilson's work was based on immunological methods. Greater resolution can be achieved by other studies, such as obtaining the amino acid sequence of proteins or the nucleotide sequence of the DNA, which were not readily do-able at the time. An intermediate degree of resolution could be achieved by DNA–DNA hybridization. The two strands of the DNA helix are separated by heating and the rate of reannealing between strands from different sources are compared: human with human DNA, human with chimp DNA, and so on. This method was pursued, among others, by Charles Sibley and Jon Ahlquist (1984; see also Sibley et al., 1990).

In order to estimate time of divergence on the basis of molecular differentiation, this must occur at a constant rate. This is the "molecular clock" hypothesis, which we discussed in Chapter 1, Section 1.2. As pointed out there, the molecular clock is not expected to time events precisely, but rather it would be a stochastic clock, in which events occur with a constant probability, such as in radioactive decay. Numerous investigations have shown by now that the molecular

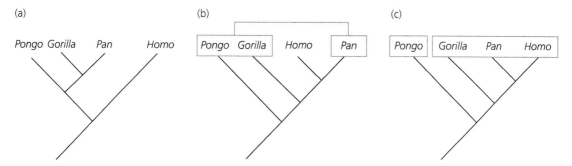

Figure 2.2 Paraphyletic and monophyletic hominoid groups. (a) Cladogram derived from traditional classification into pongids and humans. (b) Cladogram deduced from morphological and functional similarities. When grouping apes to the exclusion of humans, the former become a paraphyletic group (it leaves out a taxon of the considered lineage). (c) The correct grouping, by means of monophyletic groups, includes all the members of each lineage.

Box 2.2 Consideration of grade

Using grades for a taxonomy not narrowly tailored to the evolutionary process is a very intuitive device to distinguish closely related taxa. It is based on the similarity of phenotypic traits, rather than on genetic proximity. For example, chimpanzees and gorillas share a style of locomotion, resting the soles of their feet and hand knuckles on the ground, which is different from humans who are bipeds. Nothing is more logical than to take advantage of this noticeable difference to include gorillas and chimpanzees in the same grade, apart from humans. But the use of grades does not reflect the way lineages separated during phylogenesis.

---

clock is more erratic than expected from a stochastic clock. Nevertheless, because so many different genes and other DNA sequences, as well as proteins, can be studied, molecular investigations have provided very valuable information about the time of evolutionary events.

An issue that may be problematic with the molecular clock is that it needs to be calibrated by reference to some evolutionary event that has been dated with paleontological information. This calibration determines the rate at which a particular molecular clock (gene or protein) "ticks."

Some authors, such as Phillip Tobias, a paleoanthropologist at Witswatersrand University in South Africa, have pointed out that the molecular clock hypothesis involves a circular argument (Box 2.3). We determine

dates corresponding to the fossil record on the basis of molecular time rates obtained from the fossil record itself (Tobias, 1991d). But this misrepresents the method, which uses fairly well-ascertained fossil dates in order to determine the rate of evolution of, say, a particular protein and then uses this rate to estimate the time of divergence for lineages with an uncertain fossil record. In any case, molecular evolution dates are subject to the two problems mentioned: the assumption that the rate is constant and the determination of the rate. Thus, it is not surprising that Tobias (1986) carried out a comparison among different dates obtained for the human and chimpanzee divergence, using different kinds of calibration and observed dates that varied from 2.3 to 9.2 Ma ago.

Tobias (1991d) has pointed out that, in addition to the issues of calibration and rate constancy, molecular investigations of phylogeny face other methodological problems. Such problems are inherent to all systems for calculating phylogenetic distances. First, there is a tacit assumption that the resulting date for the divergence of two human and chimpanzee DNA sequences, for instance, reflects how long ago the two lineages themselves diverged. This excludes the possibility of mosaic molecular evolution, which might preserve certain primitive molecular features of the molecules in one or both lineages. Thus, estimates of the age of divergence between humans and great apes may be different for different proteins or DNA sequences. Second, we are ignorant of the extent to which convergent adaptation—that is to say, the appearance of analogous traits—may be expressed at the molecular level. Both of these problems are real but, as Tobias (1991d) himself has admitted, the multiplication of molecular studies carried out with different techniques is likely to

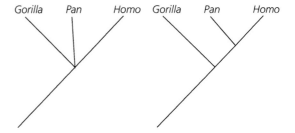

Figure 2.3 Phylogenetic relation between the genera *Gorilla*, *Pan*, and *Homo*. To the left, simultaneous appearance of the three lineages (trichotomy). To the right, initial differentiation of gorilla.

yield converging estimates that approximate true time values.

Obtaining DNA sequences has now become a readily available and relatively inexpensive process. The sequencing of the human and chimpanzee genomes has provided valuable information as a reference for phylogenetic or taxonomic investigation. The comparison of DNA sequences of genes, as well as non-coding sequences, has become the prevailing molecular method for systematics and phylogeny. Nevertheless, for historical completeness we will review earlier studies.

The early immunological and blood serum protein studies were complemented with those comparing chromosomes. The first study of this kind was carried out by the Italian anthropologist Brunetto Chiarelli (1962). Thereafter, many authors have related primate evolution with chromosomal modifications. Jean Chaline et al. (1991) investigated the branching sequence of gorillas, chimpanzees, and humans by identifying seven chromosomal mutations that differ between the African great apes and humans. More recently, Chaline

and colleagues (1991, 1996) have combined their chromosome structure investigations with the available evidence obtained by molecular and immunological methods. They have concluded that the gorilla, chimpanzee, and human lineages separated almost simultaneously, an event best represented by a so-called trichotomy (Figure 2.3). The difficulty of unraveling the divergence sequence of the three lineages had been pointed out earlier by Goodman (1975), as well as Bruce and Ayala (1979), and Smouse and Li (1987). Andrews (1992) accepted the trichotomy scenario in an influential article devoted to the reinterpretation of the status of hominoids. An attempt to resolve the issue of the classification of the gorilla, chimpanzee, and human lineages was carried out by Colin Groves and James Paterson (1991) by means of the parsimony-maximizing cladistic computer program PHYLIP. Their results pointed to a *Pan–Homo* or *Pan–Gorilla* clade, depending on the characters selected for the comparison. The trichotomy *Gorilla–Pan–Homo* has been recently defended by Amos Deinard and Kenneth Kidd (1999), based on the study of the evolution of the intergenic region *HoxB6*. But there is opposite evidence as well. Morris Goodman's team have reinterpreted beta-globin genetic sequences as favoring separate lineages for gorillas and chimpanzees + humans (Bailey et al., 1992).

It is not easy to reach a general consensus based on different molecular methods, but a revision of the available data regarding DNA sequences by Maryellen Ruvolo (1997) supports an initial separation between the gorilla clade and the one formed by chimpanzees and humans. On the whole, the molecular evidence favors the chimpanzees as the sister group of the lineage leading to modern humans: chimpanzees are our closest living relatives (Goodman et al., 1998).

2.1.3 The *Homo*/*Pan* divergence

Although both hybridization and chromosomal comparison techniques provide useful information for establishing molecular proximity between two species, the direct sequencing of nucleic acids has the last word. DNA hybridization studies (Sibley & Ahlquist, 1984) had shown that chimpanzees and humans share close to 98–99% of their genomes' DNA. The direct sequencing of the human chromosome 21 (Hattori et al., 2000) and its orthologue 22 in chimpanzees (Watanabe et al., 2004) allowed the detailed comparison of their genomes, confirming their genetic proximity. Excluding deletions and insertions, the differences between the two species amounted to only 1.44% of the nucleotides. It became obvious that the genomes of humans and chimpanzees are extremely similar in their DNA sequence; how similar, has become recently known with the publication of the draft genome sequence of the chimpanzee and its preliminary comparison to the human genome.

The Human Genome Project of the United States was initiated in 1989 and funded through two agencies, the National Institutes of Health (NIH) and the Department of Energy (DOE), with eventual participation of scientists outside the United States. A draft of the genome sequence was completed ahead of schedule in 2001. In 2003, the Human Genome Project was finished, but the analysis of the DNA sequences, chromosome by chromosome, continued over the following years. Results of these detailed analyses were published on June 1, 2006, by the Nature Publishing Group, in a special supplement entitled *Nature Collections: Human Genome*.

The draft DNA sequence of the chimpanzee genome was published on September 1, 2005, by the Chimpanzee Sequencing and Analysis Consortium, "Initial sequence of the chimpanzee genome and comparison with the human genome" in *Nature* (2005), embedded within a series of articles and commentaries (*The Chimpanzee Genome*, 2005). The last paper in the collection presents the first fossil chimpanzee ever discovered (McBrearty & Jablonski, 2005).

In the genome regions shared by humans and chimpanzees, the two species are 99% identical. These differences may seem very small or quite large, depending on how one chooses to look at them: 1% of the total appears to be very little, but it amounts to a difference of 30 million DNA nucleotides out of the 3 billion in each genome; 29% of the enzymes and other proteins encoded by the genes are identical in both species. Out of the one-hundred-to-several-hundred amino acids that make up each protein, the 71% of non-identical proteins differ between humans and chimps by only two amino acids, on the average. If one takes into account DNA stretches found in one species but not the other, the two genomes are about 96% identical, rather than nearly 99% identical as in the case of DNA sequences shared by both species. That is, a large amount of genetic material, about 3% or some 90 million DNA nucleotides, have been inserted or deleted since humans and chimps initiated their separate evolutionary ways, about 6–8 Ma ago. Most of this DNA does not contain genes coding for proteins, although it may include toolkit genes and switch genes that impact developmental processes, as the rest of the non-coding DNA surely does.

Comparison of the two genomes provides insights into the rate of evolution of particular genes in the two species. One significant finding is that genes active in the brain have changed more in the human lineage than in the chimp lineage (Khaitovich et al., 2005). Also significant is that the fastest evolving human genes are those coding for "transcription factors." These are "switch" proteins, which control the expression of other genes, i.e., they determine when other genes are turned on and off. On the whole, 585 genes have been identified as evolving faster in humans than in chimps, including genes involved in resistance to malaria and tuberculosis. (It might be mentioned that malaria is a severe disease for humans but not for chimps.) There are several regions of the human genome that contain beneficial genes that have rapidly evolved within the past 250,000 years. One region contains the *FOXP2* gene, involved in the evolution of speech.

Other regions that show a higher rate of evolution in humans than in chimpanzees and other animals include 49 segments, dubbed "human accelerated regions" or HARs. The greatest observed difference occurs in *HAR1F*, an RNA gene that "is expressed specifically in Cajal–Retzius neurons in the developing human neocortex from 7 to 19 gestational weeks, a crucial period for cortical neuron specification and migration" (Pollard et al., 2006, p. 167; see also Smith, 2006).

All this knowledge (and much more of the same kind that will be forthcoming) is of great interest, but what we so far know advances but very little our understanding of what genetic changes make us distinctively human. Extended comparisons of the human and chimpanzee genomes and experimental exploration of the functions associated with significant genes will surely advance further our understanding, over the next decade or two, of what it is that makes us distinctively human; what is it that differentiates *H.*

sapiens from our closest living species, chimpanzees and bonobos. This will surely provide some light of how and when these differences may have come about during hominid evolution.

2.1.4 Phenotypic differences

Some comparative studies of chimpanzees and humans lead to a surprising conclusion at first: even though differences between genomes are minute, it is not the case with the resulting proteins. For example, most proteins encoded by chromosome 21/22 are different in chimpanzees and humans. The computation performed by Glazko et al. (2005) amounted to a difference of 80%, which leads, naturally, to the great disparity between *Homo* and *Pan* phenotypes.

How can this fact be interpreted in evolutionary terms? As Ruvolo (2004) said in this regard, when the genomes of more distant species, such as humans and mice, are compared, the basic premise is "if it's conserved, it must be functionally important." By contrast, in closely related species, such as humans and chimpanzees, the basic premise is "if it's different, it might be important in explaining species differences." Let's proceed, then, with an attempt to understand the reasons for the differences in this case.

The discrepancy between infinitesimal molecular distance and large protein distance is explained by gene expression, by the way in which information contained in the genome leads to a protein. It is clear that genome expressions are not equivalent in chimpanzees and humans. But the genome itself also contains some notable differences that are not expressed in genetic distance when compared sequence-by-sequence. The human genome is comparatively larger than that of chimpanzees and other primates like lemurs because of a greater number of insertions. To give an example, the human *Alu* gene has gotten twice as many insertions as its corresponding gene in chimpanzees during the separate evolution of the two lineages (Hedges et al., 2004). Chromosomes of each lineage have also suffered relocations (rearrangements), which are related to their encoding different proteins (Navarro & Barton, 2003), although a later, larger study denied that relationship (Vallender & Lahn, 2004). In any case, the basic issue in evolutionary terms is, as mentioned, to identify which expression regions of genes carry functional differences. That is, what particular differences are involved in producing a human or a chimpanzee, starting from fairly similar DNA.

Much of the difference between the two sibling groups, from locomotion to the way they communicate,

---

> **Box 2.4 Genetic expression in tissues**
>
> If functional explanations are still far off, we can already identify some differences in gene expression in tissues at a molecular level. Naturally, the portrait is biased, because tissues are usually the result of the expression of different genes. Even so, differential patterns of gene expression have been identified in the brain of humans and chimpanzees (Marqués-Bonet et al., 2004), and to a greater degree in the liver (Hsieh et al., 2003).

is functional; ultimately to some extent dependent on those protein functions to which Ruvolo (2004) referred when talking about the evolutionary significance of disparities obtained, either by molecular transformations, positive selection of protein changes, or differential gene expression (Box 2.4). But, identifying genetic differences that translate into functional differences in either lineage is, for the moment, beyond our capacity. The available panorama is very poor; for instance, only one gene related to language production, *FOXP2*, is known, and its peculiarities appear to be the result of a mutation that occurred after the human lineage separated from the chimpanzee (Enard et al., 2002). However, *FOXP2* is related to motor control disorders that hinder language, not to the act of speaking (Cela-Conde et al., 2008). Perhaps the strongest evidence for a genetic, anatomical, and functional correlate is mutations suffered in the Myosin gene of *Homo* (Stedman et al., 2004), which would have allowed the development of large crania in our genus. The possible role of this gene has also been discussed by Perry et al. (2005).

2.1.5 What is *Homo* from a taxonomic point of view?

There are large functional and anatomical differences between African apes, including chimpanzees, and humans. If we are to respect phylogenetic lineages, how should the molecular similarities and phenotypic differences be reflected in the classification of hominoids?

The most commonly held point of view deduced from molecular studies proposes an evolutionary sequence that involves an initial separation of orangutans, a second separation of gorillas, and, finally, the divergence between chimpanzees and humans. Goodman (1962, 1963; Goodman et al., 1960) pointed out that this phylogeny brings into question Simpson's

Table 2.1 Changing views of the genus *Homo* and of the family Hominidae

Genus *Homo*	Family Hominidae
Traditional view:	Traditional view:
Humans and their direct and collateral ancestors not shared with australopithecines	Humans and their direct and collateral ancestors not shared with any ape
Goodman et al. (1998):	Goodman (1963):
Previous + chimpanzees	Previous + chimpanzees + gorillas
Watson et al. (2001): Previous + gorillas	Schwartz et al. (1978) and Groves (1986): Previous + orangutans
	Szalay and Delson (1979), Bailey et al. (1992), and Goodman et al. (1994): Previous + lesser apes

Box 2.5 The last common ancestor

Goodman and collaborators (1998) argue that chimpanzees and humans should be classified in the same genus because the time elapsed since their last common ancestor (LCA) is about 6 million years. This argument, however, does not require a category as low as subgenus (Cela-Conde, 2001). Taxonomic practice shows many examples, in all sorts of organisms, where the category of genus, or even higher, has been allocated to species that diverged no more than 6 million years ago; for example, Elisabeth Vrba (1984) classified two bovid African lineages, Alcelaphini and Aepycerotini, which also have a 6-million-year-old LCA, as separate tribes. The first lineage includes 27 species and the second only two, a situation somewhat similar to that of humans and chimpanzees (numerous species and several genera are generally recognized in the human lineage, while only two species are known in the chimpanzee lineage).

traditional classification of the hominoids. If chimpanzees and gorillas are placed in a single genus, we would have a paraphyletic group. This could be avoided if humans were also included in the genus with the African apes.

A solution proposed by Goodman (1963) would be broadening the hominid family to include gorillas (*Gorilla*) and chimpanzees (*Pan*) in addition to the human genus (*Homo*). Goodman's proposal had considerable resonance among primatologists, but the decision to increase the scope of the family Hominidae turned out to be a slippery slope. For instance, Schwartz et al. (1978) and Groves (1986) also placed the genus *Pongo* (orangutans) in the family Hominidae. Szalay and Delson (1979) went a step further by also including lesser apes (like gibbons, *Hylobates*) in it. Morris Goodman and others later agreed with this suggestion (Bailey et al., 1992; Goodman et al., 1994). As a consequence, the taxon that includes humans and their exclusive direct ancestors, that is to say, hominids, was transferred from the family category to the tribe category, Hominini (Schwartz et al., 1978; Groves, 1986), and, later, to the genus *Homo* (Goodman et al., 1994). This genus, which in human paleontology is usually used to group human ancestors that lived during the Upper Pliocene and Pleistocene (such as *Homo habilis*, *Homo erectus*, *Homo neanderthalensis*, and *Homo sapiens*, among others), would also include chimpanzees, according to Goodman et al. (1998). Thus, all human beings and their numerous direct and collateral ancestors would be reduced to a subgenus. An even more extreme proposal was put forward by Watson et al.

(2001), presented during the World Congress of Human Paleontology in Sun City, South Africa: to include gorillas as well within the genus *Homo* (*H. gorilla*).

In view of the numerous and diverse lineages that, as the following chapters will illustrate, appear in the human clade from the Miocene to the Pleistocene, it does not seem reasonable to include them all within a limited corset of a subgenus. More controversial yet (Cela-Conde, 1998) is the classification of such functionally different organisms as chimpanzees, gorillas, and humans in a single subgenus. If we accept Mayr's adaptive criterion for characterizing a genus (see Chapter 1), it seems clear, as it will become apparent in later chapters, that there are at least five different genera just in the human lineage.

It is not necessary to carry the taxonomic implications of the molecular evidence as far as it has been done by Goodman and others. A monophyletically based taxonomy can be achieved by granting the same consideration to the human clade as to the chimpanzee clade. Chimpanzees and humans are sister groups, which could be considered as subgenera. But they could also be awarded a higher ranking category: genus, tribe, subfamily, or even family, which would be more compatible with the spirit of Simpson's traditional classification.

It is a lineage's diversity, not its antiquity, which must determine its taxonomic level. The existence of numerous species may justify placing them into more than one genus. Several genera may justify different tribes, and so on for higher categories. The genetic proximity of chimpanzees and humans, and the very worthy attempt to avoid ideological biases that would consider our species as a superior category, which have led to the inclusion of African great apes in the family Hominidae, or even in the genus *Homo*, deserve to be praised. However, taxonomic decisions should be strictly guided by systematic criteria.

2.1.6 Controversies of morphological comparison

Morphological similarity should not be ignored when establishing taxonomies, but how to evaluate genetic similarity in order to determine taxonomic classification is a difficult issue. Peter Andrews and Lawrence Martin (1987), on the basis of all the then available morphological and molecular evidence, arrived at a cladogram of hominoid phylogenetic relations that differs little from the one resulting from exclusively molecular evidence. These authors pointed out that "the only real surprises are that the clear shared derived similarities at the molecular level among African apes and humans are not strongly reflected in the morphological analysis and that morphological similarities between African apes are not reflected at the molecular level" (p. 113). Actually, this should not have been a surprise. Rather, it is a good example of the problems we have been tackling. There is little doubt at present regarding the phylogenetic sequence of the appearance of hominoid lineages: the orangutan clade diverging from gorillas, chimpanzees, and hominids, and later gorillas separating from a common clade that includes chimpanzees and humans. But molecular similarities need not precisely translate into morphological similarities, because these also depend on factors not simply apparent by observing differences in DNA or protein sequences.

A degree of correlation between molecular proximity and morphological similarity has been pointed out in several studies. Colin Groves (1986) included chimpanzees and human beings in the same clade on the basis of morphological similarities. A similar conclusion was reached by Sally Gibbs and collaborators (2000) by means of cladograms based on soft tissue traits of living hominoids. Milford Wolpoff (1982) has also affirmed that, in morphological terms, gorillas, chimpanzees, and humans are more similar to one another than any of them is to orangutans (although he was arguing in a different context). However, Jeffrey

Schwartz (1984) carried out a detailed morphological comparative study that contradicted Wolpoff's conclusion. According to Robert Martin (1990), the morphological similarity among the great apes is due to their slow divergence from their last common ancestor. Consequently, Martin (1990) considered the possibility of grouping the apes in the same paraphyletic taxon. These diverging proposals are a consequence of the different weight given to phylogeny relative to morphological divergence. As Tobias (1991d, p. 14) has said: "there are precise definitions available [. . .] regarding morphological traits of hominids and apes; they are, in essence, the complex of anatomical and functional traits that most effectively differentiate humans from apes."

The primatologist Russell Ciochon (1983) also opted in favor of a human clade separate from an African ape clade, based on a very complete list of morphological traits that define 11 morphotypes within hominoids. But arguments in favor of such a separation are functional, as well as morphological, even though there is a correspondence between them. Bipedal gait is impossible without changes in the foot, hip, extremities, and the cranial base. In fact, paleontologists infer bipedalism based on these morphological traits. Moreover, traits do not change in isolation; usually several traits change in a coordinated fashion in order to achieve a new adaptation. Le Gros Clark (1955) and Tobias (1985b) have advanced the concept of "total morphological pattern" for taxonomically characterizing a specimen, moving away from any practice that seeks to determine adaptation and evolutionary pattern by evaluating simple traits in isolation. Nevertheless, emphasis on the relevance of a notorious and relatively isolated character (such as bipedal gait) may be more reasonable than an alternative procedure that simply quantifies the number of shared (or different) traits.

The issue at hand is not only related with the weight given to molecular data. Supporters of the close taxonomic classification of gorillas, chimpanzees, and humans do not ignore that human derived traits (from bipedalism to language) are very relevant for the adaptation of our species. What then is the base for the widespread trend among molecular primatologists to include such adaptively, morphologically, and functionally diverse beings as African apes and humans in the same family? A relevant consideration is the urgent need to avoid anthropocentrism. Often in the past an anthropocentric bias has imposed mistaken concepts, such as a hierarchical relation among living beings, with human beings in the role of "masters of nature." Nevertheless, the discrepancies concerning classification between molecular and other primatologists are largely due to the overwhelming weight attributed by some to molecular evidence, which ignores the difference between genetic distances and the determination of phylogenetic trees.

2.1.7 A monophyletic solution that respects functional aspects

Despite arguments in favor of maintaining Simpson's (1945) taxonomy and separating the family Pongidae (great apes) from the family Hominidae (humans), the results of molecular studies are too consistent and extensive to ignore them. As we have already noted, this does not necessarily lead to such a reductive classification as the one at the right-hand bottom in Table 2.1, with lower apes included in the Hominidae. The molecular evidence does not, by itself, determine the distribution of evolutionary lineages, but is relevant to formulate cladistic interpretations of how to incorporate new lineages. Cladistics argues that each speciation process involves the disappearance of the original taxon and the necessary appearance of two new taxa. The crucial issue here, as suggested by Martin (1990), is that cladistic principles do not impose the taxonomic categories to which the diverging clades belong. Cladistics requires that the two taxa that appear at a node—the idealized moment of their divergence—belong to the same category. Thus, if we assume that the taxon chimpanzees + humans separated from the taxon gorillas, the requirement is that we grant the taxon including chimpanzees + humans the same category as gorillas.

Here we will adopt a taxonomic classification that respects the increasing molecular evidence. We will largely follow Bernard Wood and Brian Richmond (2000) (Table 2.2). The family Hominidae embraces the set of great apes and humans. Orangutans constitute the subfamily Ponginae and gorillas the subfamily Gorillinae, while chimpanzees and humans form the subfamily Homininae. Within the latter, chimpanzees belong to the tribe Panini and humans to the tribe Hominini. The human lineage has the category of tribe: Hominini (informal name, "hominins"). Two subtribes are included in it. One is Australopithecina (informal name "australopiths"), which encompasses four genera: *Orrorin, Ardipithecus, Australopithecus,* and *Paranthropus.* The other subtribe is Hominina (informal name, "hominans"), with one single genus, *Homo.*

The choice of a colloquial name for *Paranthropus* deserves to be explained. It derives from the fact that for decades the oldest South African hominins were

Table 2.2 The suprageneric taxonomy of great apes and humans

Family	Subfamily	Tribe	Subtribe	Current species
Hominidae	Ponginae	Pongini	Pongina	Orangutans
	Gorillinae	Gorillini	Gorillina	Gorillas
	Homininae	Panini	Panina	Common chimpanzees; bonobos
		Hominini (hominins)	Ardipithecina (ardipiths) Australopithecina (australopiths) Paranthropitecina (robust australopiths) Hominina (hominans or, for short, homo)	Humans

In brackets, popular names. Adapted from Wood and Richmond (2000), adding the subtribes Ardipithecina and Paranthropitecina.

classified by distinguishing between "gracile" and "robust." The latter correspond to the members of the genus *Paranthropus.*

2.1.8 Hominini genera

The 4th edition of the *International Code of Zoological Nomenclature* (ICZN) states that "the Code refrains from infringing upon taxonomic judgments, which must not be made subject to regulation or restraint" (Introduction). The taxonomic freedom granted by the ICZN sometimes leads to arbitrary situations. It is easy to find in specialized literature a multitude of taxa referred to as Hominini. With regard to Neanderthal specimens, for example, six different genera have been proposed. Such scatter is obviously absurd, making it necessary to follow a more reasonable criterion. We propose to accept Ernst Mayr's idea of "genus" as indicative of a certain type of organism adapted to specific conditions. Each genus, according to this criterion, implies that we are dealing with a different type of hominin; that is, an organism that does something different, irreducible to other genera previously proposed.

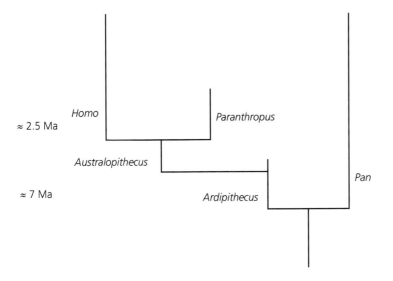

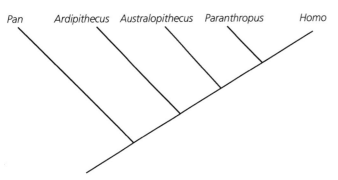

Figure 2.4 Above: Phylogenetic tree of the Hominini tribe genera. Below: Cladogram of the Hominini tribe genera.

Applying Mayr's criterion (1950) about the adaptive meaning of genus, we will see throughout this book that members of the tribe Hominini have followed one of the following four strategies:

1. The first organisms that adopt bipedal posture when moving across the ground. They retain much of the primitive mastication features.
2. Hominins that exploit locomotion to occupy areas of open savanna as the African weather cools and forests shrink, adapting their dentition to the new conditions.
3. Hominins that c. 2.5–3.5 Ma specialize in a savanna vegetarian diet and develop larger masticatory apparatus.
4. Hominins that, on dates close to 2.5–3.5 Ma, maintain a relatively gracile masticatory apparatus. The genus includes also the successors of the lineage that eventually develop larger brains and build

instruments of great precision. Our species, *Homo sapiens*, belongs to this last group.

According to Mayr (1950), every one of these different groups belongs to a single genus. Although we are anticipating arguments that will justify maintaining that taxonomic formula, we will argue that such genera would be, respectively, *Ardipithecus*, *Australopithecus*, *Paranthropus*, and *Homo*. The phylogenetic tree relating them is shown in Figure 2.4.

2.2 Traits of the human lineage

Humans are, technically speaking, apes. Common usage, when referring to "Asian apes" and "African apes," incorrectly leaves out *H. sapiens*. This popular bias has an explanation: the evolution of our derived traits involved the loss of those features we associate with our closest relatives, the apomorphies that define what an "ape" is (Table 2.3). Such derived traits, which

Table 2.3 Apomorphies identified by Carroll (2003) in his study of the genetic basis of the physical and behavioral traits that distinguish humans from other primates

Body shape and thorax
Cranial features (brain case and face)
Brain size
Brain morphology
Limb length
Long ontogeny and lifespan
Small canine teeth
Skull balanced upright on vertebral column
Reduced hair cover
Elongated thumb and shortened fingers
Dimensions of the pelvis
Presence of a chin
S-shaped spine
Language
Advanced tool making

Some apomorphies listed by Carroll are functional, like language. Others that are anatomical are not shown in fossils, like hair or the brain's topology. However, body shape, brain size, relative length of limbs, and vertically placed cranium above the vertebral column are morphological traits that clearly set us apart from any ape.

separate us from apes, are what we might call the general characteristics of the Hominini tribe or, technically speaking, the synapomorphies (shared derived traits) present in the whole human lineage.

The anatomy and behavior of chimpanzees and modern humans are very different and some differences are quite notorious, such as language, brain size, bipedal gait, and culture. One of the main features of human language is its dual patterning (we combine basic sounds, phonemes, to form words, and words to form sentences), which is absent in any ape communication system. Our brain is much larger than that of the higher apes, relative to body size, and much more complex. We usually walk on our two feet, while orangutans use brachiation to travel, and gorillas and chimpanzees are rather special quadrupeds that lean on the knuckles of their hands. The sophistication of our cultural traditions is a very notorious divergent trait.

There are other differences that may not quite jump out so conspicuously. We lack body hair or have very little. Our face is not so prognathus. Gestation is shorter than in the case of higher apes. Ovulation in human females is cryptic and they are continuously sexually receptive without specific periods when females enter into heat. Tobias (1994) lists 24 morphological traits and 5 physiological ones that differentiate humans from higher apes. One additional feature, difficult to ignore, is population size. Today, orangutans, gorillas, and chimpanzees are reduced to relatively

small populations living in receding tropical forests in Africa and southeastern Asia. There are more than 6 billion humans distributed across the planet. Numbers are not an unmixed blessing, but they are a measure of adaptive success.

Some distinctive human traits have appeared recently; if we look back in time they disappear from our lineage. Ten thousand years ago neither writing nor agriculture existed. Fifty thousand years ago there were no people in America. These are negligible time intervals relative to the 7 million years that have passed by since the divergence of the evolutionary branches leading to the African great apes and humans, or the 6 million years since the fossil *Orrorin tugenensis*, described further on (Chapter 4, Section 4.1.3), was buried in the ground.

Thus, current human features are generally not very helpful to reach conclusions about our initial apomorphies. What we are looking for are derived ancient traits that can be considered synapomorphic, shared by every hominin that ever existed. These would define the earliest member of our lineage as adaptively distinct. Leaving aside the necessarily dark period surrounding the exact moment when the lineages split, are there any such traits? Can we find a trait that will allow us to determine whether a given fossil specimen is a hominin?

2.2.1 Larger brains

If someone unfamiliar with anthropology and paleontology had to choose the most "human" apomorphy among those listed by Carroll (Table 2.3), he or she would probably choose some feature related to the mind. It could be the size of the cranium, the brain's topology, or some functional trait, like language or complex tool-making. But out of these characteristics, only the size of the cranium can be directly detected in the fossil record. Human and ape cranial volumes are very different (Figure 2.5). The average cranial capacity of modern humans is 1,350 cc, while that of chimpanzees, with a comparable body size to ours, is 450 cc. Does this mean that a large cranium evolved at the beginning of the human lineage and, thus, constitutes a hominin synapomorphy?

The notion that a large brain is an essential trait identifying the appearance of the human lineage is very old, but not enough to find it among Darwin's hypotheses about this matter. In chapter II of his *Descent of Man* (1871) he speculated about the mental powers of humans compared with those of lower animals. The earliest ancestors in the human lineage would have

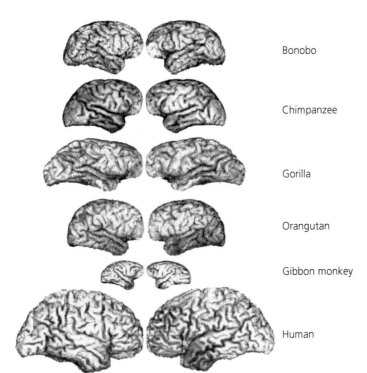

Bonobo

Chimpanzee

Gorilla

Orangutan

Gibbon monkey

Human

Figure 2.5 Hominoid brains at the same scale. A larger brain is a very notable distinctive feature of our species (Semendeferi & Damasio, 2000).

inherited, naturally, some traits from their own ancestors, such as large canines. In Darwin's opinion, the gradual reduction of the size of canines in the human lineage was a consequence of the appearance of culture: they became smaller because of their disuse in favor of tools and weapons. The manipulation of tools required a bipedal posture, or at least the former was facilitated by the latter. Because the reduction in canine size was accompanied by the reduction of the muscles that move the jaw, the cranium was able to grow, and with it, the brain and mental faculties that, of course, improved culture (*Descent of Man*, pp. 435–436).

The Darwinian notion of hominin evolutionary change may be interpreted as a closed feedback loop. Culture required bipedalism and, at the same time, reinforced it. The reduction in canine size was a consequence of the use of weapons; but that reduction facilitated brain size, through the restructuring of the cranium; further, mental development allowed devising, making, and using better weapons. Brain increase improved bipedal balance and permitted the development of language. Language facilitated the transmission of culture and collective hunting; using meat as food allowed further reductions in dentition size. This is a feedback model: each factor depends on the others

and, at the same time, promotes them. The process involves a functional and anatomical integration in which several coordinated factors participate. One of them must have appeared before the others, serving as an initial thrust for the loop to start running. Which might it be?

Darwin envisioned the following chain of events: descent from the trees, bipedalism, brain-size increase, language, and appearance of culture (with all its components, both intellectual and technological). Some of these elements can be traced in the fossil record, but not others. Phenomena such as the development of moral sense, which Darwin believed was extremely important, are not associated with fossil remains. Language does not fossilize either. But the cranium and bones of the hip and lower limbs leave fossil trails that can provide firm evidence regarding whether it was our bipedal posture or our large brain that developed first.

During the early twentieth century there were defenders of two opposite hypotheses. Arthur Keith was one of the most notorious advocates for bipedalism as the initial trait, while Grafton Elliot Smith argued that a large encephalization appeared first. The swords were drawn when the Piltdown fossil specimen appeared on the scene. The specimen had a large cranium, the size of the cranium of a modern human, combined with a very primitive mandible, resembling that of an orangutan or gorilla. The different fragments that formed the specimen were found in 1912 by Charles Dawson, an amateur archaeologist in the English town of Piltdown. The story of its discovery and the controversy it sparked has been told many times (e.g., Reader, 1981; Lewin, 1987; Spencer, 1990; Walsh, 1996; Weiner & Stringer, 2003). The article that *The London Illustrated News* devoted, in September 1913, to the finding renders a very good picture of the challenged posed by the specimen's interpretations.

The Piltdown fossil exhibited some unconvincing traits, such as the awkward connection between the cranium and mandible—raising the suspicion that they belonged to different specimens. Many paleontologists were indeed suspicious. Its discoverers did not allow its examination, alluding to the fragility of the original fossil. It was necessary to use copies made with a mold. The suspicions turned out to be well founded. In 1953 (thanks to Joseph Weiner, Kenneth Oakley, and Wilfred Le Gros Clark) it was confirmed that the Piltdown fossil was a fraud. Someone had filed an orangutan's mandible and canines to reduce them and fit them, quite sloppily, to a human cranium. The main suspect of the fraud is Martin Hinton, curator of the Natural History Museum, London. Hinton was the owner of a trunk found in the museum's attic in 1996, with bones manipulated in a similar way to those constituting the Piltdown specimen (Gee, 1996).

2.2.2 Brain vs. bipedalism

Before the Piltdown deception, there already was evidence contrary to the early evolution of a large brain. Remains of fossil beings that were very similar to us were known since the beginning of the nineteenth century, before the controversy between evolutionists and anti-evolutionists reached the virulence sparked by Darwin's work. The discovery of a very famous specimen in 1856, the Neander Valley cranium (Germany), which would christen the Neanderthals, occurred several years before the publication of Darwin's *Origin of Species*. But the first "modern" discoveries, that is to say, interpreted in terms of evolutionary ideas, were made after 1887, subsequent to the arrival of the Dutch physician Eugène Dubois in Indonesia. Dubois searched for fossils that could prove Darwin was right. At the Javanese site of Trinil, Dubois discovered, in 1891, remains that completely transcended the realm of scientists and became universally known. The specimen includes a primitive and small cranium (with a capacity of about 850 cc) found beside a femur that was very similar to that of modern humans. The name given to the taxon, *Pithecanthropus erectus* (Dubois, 1894), means upright ape-man, conveying the idea that it was an intermediate being between humans and apes (*pithecus* for ape, and *anthropus* for human); and that it had a posture distinctively upright (*erectus*) (Figure 2.6 and Box 2.7).

Subsequent discoveries have required revision of Dubois' interpretations. The Trinil fossil is not an "intermediate" form between humans and apes, but a fairly advanced hominin. The upright posture was not a new apomorphy; rather it was already present in its ancestors. But Dubois' phylogenetic interpretation was correct: the ancestors of current humans had fixed a bipedalism similar to our own before the brain reached its current size.

It is generally accepted nowadays that bipedalism is a hominin synapomorphy—an apomorphy shared by all the members of the lineage. Any specimen close to the divergence between the chimpanzee and human lineages is attributed to the latter if it is bipedal. Most of the modifications to the trunk, limbs, hip, and the insertion of the vertebral column in the skull are related to bipedalism, which distinguishes our species from the apes. But before we analyze the morphological correlates of bipedal locomotion, we will consider the Laetoli (Tanzania) fossil footprints that show that more than 3 million years ago there were creatures that walked upright.

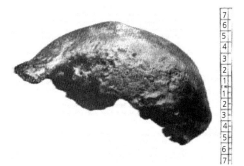 

Figure 2.6 Skullcap (left) and femur (right) of the Trinil specimen, *Pithecanthropus erectus* (now classified as *Homo erectus*), discovered by Eugène Dubois. (Photo courtesy and copyright of the National Museum of Natural History, Leiden, the Netherlands.)

Box 2.7 The meaning of *Pithecanthropus*

Eugène Dubois (1935) gave to the fossil a meaning very different from the one we give it today. For Dubois, *Pithecanthropus* was a gigantic genus of the gibbon's type, although superior because of their large brains and ability to assume an erect posture. The fact that Lydekker had discovered *Paleopithecus sivalensis*, ancestor of the gibbon, in Siwaliks (Pakistan) in 1879, must have influenced this diagnosis.

Box 2.8 Bipedalism in other animals

Birds exhibit an authentic and permanent bipedalism. The weight is balanced in front and behind the legs' vertical axis, assuring that the center of gravity falls within the legs. This form of bipedalism evolved from early reptilian tripedalism, which involved leaning the tail on the ground; it has led to a vertebral column with a pronounced S shape. The case of humans is completely different. The vertebral column is almost completely straight, and the vertical axis of the center of gravity, which practically coincides with it, passes through the articulations of the lower limbs. The skeletal and muscular anatomical modifications required by human bipedal posture are quite conspicuous. In addition to the shape and position of the vertebral column, the foramen magnum is displaced toward the inferior part of the head, the bones of the lower limbs are elongated and those of the upper limbs are shortened, and there are changes in the shape of the hip, the structure of the foot, and in flexor and extensor muscles. The modifications also affect the shape and mobility of the articulations. During walking, for instance, human bipedalism turns into a successive monopedalism that requires placing each foot on the center of gravity's vertical and leaning the distal part of the femur inward. This does not happen in quadrupeds, which can always walk on two extremities at a time

---

2.2.3 Different forms of bipedalism

Although we have spoken of bipedalism in general terms, as if it were a well-characterized feature, the fact is that there are many different types. For a quadruped to adopt and maintain a bipedal posture it must solve the problems of balance and lifting of the body (Kummer 1991). This can be done in two ways: fast and slow. Getting up by means of the thrust of acceleration requires no specific anatomical prerequisite, just having enough muscular strength. However, standing up slowly requires keeping the center of gravity within the support area, which generally is the soles of the feet, or the soles of the feet and the hind limbs. Balance can be achieved, as chimpanzees do, by means of very long upper limbs and a pronounced angulation of the lower limbs' articulations. In a bipedal posture, the center of gravity is located in a clearly ventral point and, thus, the feet must also be placed in that position. Chimpanzees and other animals that adopt an upright posture in this way keep their balance owing to intense action of ventral and dorsal muscles. This mechanism consumes great amounts of energy and does not allow the bipedal posture to be maintained for very long.

An upright posture can also be achieved starting from a sitting position, with the main supporting area constituted by the lower limbs and the pelvis, as many small mammals do. The energetic consumption is very low, but it does not allow traveling. Great apes adopt this kind of "bipedalism" for activities that take quite a time, such as eating or sleeping.

An overview of the different ways of placing the body in an upright position allows a better understanding of how human bipedalism could have evolved (Box 2.8). The second upright posture we mentioned, the one related with sitting, is common among all primates—arboreal or terrestrial—and appeared very early in their evolution. It is a first stage in the evolution of bipedalism (Tobias, 1982a, 1982b). From there on, primate locomotion diverged into many different solutions, from climbers to leapers to arboreal quadrupeds and to those that use a more or less complete brachiation (Napier, 1963; Napier & Walker, 1967). The anatomical modifications that take place during phylogenesis reflect, of course, the kind of locomotion of a given species (see Fleagle, 1992).

2.2.4 Changes related to bipedalism

If complete sequences of fossil specimens were available within the chimpanzee and hominin lineages, it would be possible to determine how and when bipedalism evolved. But this is not the case. Remains are very rare and partial, especially the oldest, and there are no informative remains of chimpanzee fossil ancestors. The best we can do is to compare the traits of different hominins with living apes and humans.

Some features, such as the insertion of the vertebral column in the cranium, the foramen magnum (Box 2.9), provide evidence regarding the bipedal habit, but the two most conspicuous morphological traits associated with bipedal locomotion are the shape of the hip—including the femur's insertion in it—and the shape of the limbs, especially the feet.

FM status, either as an integrated functional structure or a double one (Box 2.9), is a feature that serves effectively as an indication of locomotion. A ventral FM indicates a vertical spine, while the insertion of the spine in higher areas is contrary to the upright position. However, the capacity of FM to indicate apomorphies in lineages is not limited to the division between apes and humans.

The cranium base is formed by the basioccipital, sphenoid, and temporal bones. Lisa Nevell and Bernard Wood (2008) studied the state of different traits—temporomandibular and mastoid joints; petrous and squamous portions of temporal bone; outer cranial base flexion; and occipital bone—in 14 taxa of fossil hominins and other hominoids, including modern humans (Box 2.10). They performed a cladistic analysis based on previous work by Strait and Grine (2004). The results showed that the cranial base features used in their analysis largely separate the different lineages in a manner fully compatible with most taxonomic schemes available for hominins.

However, with respect to the locomotor apparatus, the most prominent morphological features imposed by bipedalism are the hip shape—including the insertion of the femur—and the form of extremities, especially the foot. Let us proceed with them.

Compared to other apes, human posture is quite orthograde, and our locomotion is achieved by using only the support of the lower extremities. Bipedalism

---

Box 2.9 The foramen magnum (FM)

Anthropological studies consider the foramen magnum as a single integrated trait whose transformation provides valuable data about both ontogenetic and phylogenetic development. However, Gary Richards and Rebecca Jabbour (2011) hypothesized that the FM is formed by two functional matrices: one ventral—related to the locomotor apparatus—and another dorsal—related to neurological functions and fluid flow dynamics. Richards and Jabbour (2011) used for their analysis a large sample of modern human crania at various stages of ontogenetic development, from 7-month fetuses to adults; they considered: (1) the ontogeny of bone size and shape for the two units, ventral and dorsal, of the foramen magnum; (2) the role of synchondrosis—skull bone sutures—in pattern and rate of observed growth; and (3) the relationship between size of FM and cranium size, shape, and growth. The hypothesis was supported by the results of their analysis.

Box 2.10 Taxa considered by Newell and Wood

The Nevell and Wood study (2008) considered the following taxa: *Ardipithecus ramidus, Australopithecus anamensis, Kenyanthropus platyops, Australopithecus garhi, Sahelanthropus tchadensis, Australopithecus afarensis, Australopithecus africanus, Paranthropus aethiopicus, Paranthropus boisei, Paranthropus robustus, Homo habilis, Homo rudolfensis,* and *Homo ergaster* (genera and species that we will see in later chapters), along with the current *H. sapiens, Pan troglodytes, Gorilla gorilla, Pongo pygmaeus, Hylobates lar,* and *Hylobates hoolock.* Taxa without basicranial representation in the hypodigm were discarded. The most significant deficiency of their study was that neither Asian *Homo erectus* nor Neanderthals were included, although there are available basicrania.

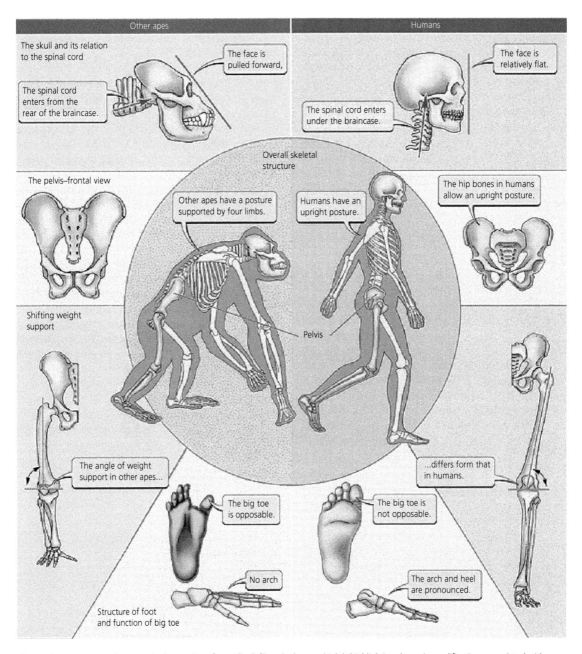

The skull and its relation to the spinal cord

The face is pulled forward,

The spinal cord enters from the rear of the braincase.

The face is relatively flat.

The spinal cord enters under the braincase.

The pelvis–frontal view

Overall skeletal structure

Other apes have a posture supported by four limbs.

Humans have an upright posture.

The hip bones in humans allow an upright posture.

Shifting weight support

Pelvis

The angle of weight support in other apes...

...differs form that in humans.

The big toe is opposable.

The big toe is not opposable.

No arch

The arch and heel are pronounced.

Structure of foot and function of big toe

Figure 2.7 Comparison between the locomotion of a gorilla (left) and a human (right), highlighting the main modifications associated with bipedalism.

involves anatomical differences related to most of the skeleton: head, torso, hips, and both upper and lower limbs (Figure 2.7).

As Poirier (1987) has pointed out, no postcranial anatomical element shows greater differences between current apes and human beings than the pelvis. He concludes that the main factor leading to the divergence between the lineages was the modification of the locomotor apparatus. Accordingly, the shape of the pelvis is a good indication of the taxonomic status of a given fossil. There are biomechanical reasons to argue, according to Yvette Deloison (1995), that the rotation of the body's axis during the transition toward an upright posture necessarily affected the structure of the pelvis. The transformations that led to bipedal locomotion very possibly began with this anatomical element.

The pelvis is a ring-shaped osseous structure constituted by two lateral parts, which articulate frontally, and a posterior part, the sacrum, composed of several welded vertebrae. Each lateral section, colloquially known as a hip, is also the result of the welding of three bones: ilium, ischium, and pubis. The differences between the shape of the hips of current chimpanzees and humans are related to the function performed by the pubic skeleton and the musculature necessary to carry out different kinds of locomotion. Most differences between the hips of apes and humans are located in the upper part, the ilium, while the lower part is quite similar between these taxa (although the ischium is shorter in our species). These similarities

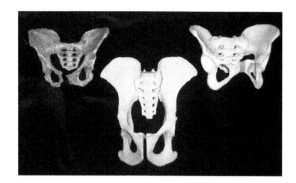

Figure 2.8 Pelvis of a chimpanzee (center), a modern human (right), and an australopith (Sts 14, *A. africanus*) (left), at different scales to facilitate morphological comparison. Matthew Murdock, *Journal of Creation*, 20(2), 104–112, 2006.

Box 2.11 Ilium compared

The ilium of African great apes is longer and thinner than that of humans. The widening and shortening in our species is the result of the adaptation of muscular insertions to allow hominins to keep their balance in an upright posture while using a bipedal locomotion.

Box 2.12 Genetics of bipedalism

What genetic changes might have led to the emergence of a biped pelvis? Evelyn Bowers (2006) indicated as the most probable those related with developmental genes, and among them, those of *HOX*. An alteration in the control region of some distal genes *HOX D* could have been the cause of the sudden emergence of bipedalism, shifting the limits of the lumbar and sacral vertebrae. As stated by this author, apes have three lumbar vertebrae instead of the six of early hominins, and 48 chromosomes instead of the 46 of current humans. Chromosome 2, which carries *HOX D*, originated by the fusion of two chromosomes that remain separated in apes. Bowers (2006) argued the hypothesis that changes related to the fusion could have been what altered the *HOX D* gene, allowing a smaller sacrum to be placed in a lower position.

Table 2.4 The oldest fossil specimens of hip bones. The description and interpretation of the different taxa and sites to which they are related will be discussed in later chapters

ARA-VP-6/500	Aramis (Ethiopia)	*Ar. Ramidus*
A.L. 288-1	Hadar (Ethiopia)	*A. afarensis*
KSD-VP-1/1	Woranso-Mille (Ethiopia)	*A. afarensis*
SK 50	Swartkrans (South Africa)	*A. africanus*
Sts 14	Sterkfontein (South Africa)	*A. africanus*
Sts 65		
Stw 431		
MLD7	Makapansgat (South Africa)	*P. boisei*
MLD252		
TM 1517		
MH1	Malapa (South Africa)	*A. sediba*
MH2		

and differences have led to a general agreement that the functional evolution of the hip, and consequently locomotion, took place through changes in the ilium (Schultz, 1930; Napier, 1967). Accordingly, the ilium is the most relevant feature for comparing the locomotor habits of different specimens (Box 2.11).

The oldest fossil pelvises—prior to *Homo erectus*—were found in South Africa and East Africa (Table 2.4). The pelvis of these early hominins is intermediate between chimpanzees and humans, but it is closer to the latter (Figure 2.8). Thus, it can be concluded that around 3 Ma there were already hominins whose locomotion was relatively similar to our own (Box 2.12).

2.2.5 Limb morphology

Human hands, compared with those of African apes, have a more robust and longer thumb, as well as shorter lateral fingers. These features, when they can be identified in the fossil record, are used to establish to what extent a specimen is biped or quadruped. Upper limbs similar to modern humans indicate that they were not used for locomotion; they were released from that specific function and, therefore, belong to a being that moved using only its feet.

John Napier's (1980) study of the human hand has shown that, among all primates, only hominins have "true hands." The morphology of the hands is an indication of the degree of bipedalism. As Jouffroy (1991) has noted, the hands of the first hominins that developed incomplete bipedalism, exhibit biomechanical features associated with their locomotor activities. Bipedalism was possibly acquired before the structure of the hand's carpal bones underwent significant modifications.

The studies of Susman and Stern (1979) and Jouffroy (1991) on the OH 7 specimen from Olduvai (Tanzania) reveal that its anatomy is associated with certain locomotor functions, but the numerous authors that have compared the morphology of different fossil hands have reached disparate conclusions (Bush, 1980, 1982; Stern & Susman, 1983; Susman et al., 1984). The same features have been interpreted as similarities between australopith specimens and modern humans, and as similarities between the former and current apes. The lower limbs provide the best evidence regarding bipedal locomotor function. However, William Jungers (1994) and Tim White (1994; White et al., 1993) disagree about the locomotor significance of the very robust MAK-VP-1/3 humerus, found at the Maka site (Ethiopia). And, finally, the comparison of two australopith scapula, one juvenile and one adult, from the sites of Dikika and Woranso-Mille (Ethiopia, in both cases) (Green & Alemseged, 2012), has allowed consideration of growth patterns related to a trait whose development differs in apes and humans.

We need to understand that along the human lineage phylogeny, hands and feet had some initial locomotor functions that were modified later. As upper extremities specialized in various functions involved in manipulating objects, and lower extremities specialized in other functions related with bipedal motion, one could think that their evolution would have been independent from the beginning. However, Campbell Rolian, Daniel Lieberman, and Benedikt Hallgrimsson (2010) have given evidence indicating the opposite: hands

---

Box 2.13 Selective pressures

An interesting aspect of the work of Rolian et al. (2010) suggests that selective pressure was first more intense in relation with the shape of the foot. This would have forced a subsequent coevolution of the hand that, according to the authors, fixed the characters that later facilitated the handling and manufacture of stone tools.

---

and feet are related homologous structures, which actually had almost identical development trends, increasing the probability that they coevolved.

Despite the nuances mentioned, it is obvious that the best evidence of bipedalism is found in the lower limbs. The fibula and tibia set and the foot bones may provide definitive evidence for ascertaining locomotor biomechanics. Whereas climbing capacity is associated with great joint mobility, bipedalism requires solid articulations able to resist the weight distribution during upright posture. But authors differ in their interpretation of the available evidence. The OH 6 specimen from Olduvai (Tanzania) is precisely formed by a tibia–fibula set. Peter Davis (1964) characterized certain traits of the fibula and the distal region of the tibia as practically identical to those of modern humans, although there are some differences in the proximal area of the tibia. After comparing the OH 6 and KNM-ER-741 specimens (Koobi Fora, Kenya), Owen Lovejoy (1975) went even further: "the australopithecine tibia approximates the modern human pattern with such fidelity that no locomotor or mechanical differences are implied by the morphology of these bones" (p. 313). With respect to the A.L. 288–1 specimen from Hadar (Ethiopia), Bruce Latimer et al. (1987) and Jack Stern and Randall Susman (1991) reached contrary conclusions concerning the tibia's flexibility, surely because of their different understanding of the type of locomotion of the specimen. According to Latimer and collaborators (1987) A. afarensis was fully bipedal, while Stern and Susman (1991) believe they were partially arboreal.

The different functional interpretations of the bipedalism of some A. afarensis specimens may be due, according to Carol Ward (2002), to two reasons: "First, there are divergent perspectives on how to interpret primitive characters [. . .]. Second, researchers are asking fundamentally different questions about the fossils. Some are interested in reconstructing the history

---

Box 2.14 Biomechanical analysis

In the discussion of human evolution in the Pliocene, we will add another perspective to clarify the degree of bipedalism: biomechanical analysis. We will see some detailed studies comparing locomotor patterns in *Australopithecus afarensis* and *Homo habilis* based on statistical analysis of available skeletons of both taxa. The viewpoints of different authors on the phylogenesis of bipedal locomotion have been summarized by Will Harcourt-Smith and Leslie Aiello (2004).

Box 2.15 Little Foot

The StW 573 bones, Little Foot, were discovered 15 years before Clarke and Tobias interpreted them as belonging to a hominin. Phillip Tobias (1997) has recounted the finding in a story that serves as an excellent testimony of the diverse demands sometimes faced by a paleontologist: "On February 28th, 1980, one of our field assistants, David Molepole, extracted a very small left astragalus, or ankle, followed the same day by the navicular, the ship-shaped bone which articulates immediately after the astragalus. As usual, each bone was carefully marked with the source of the material and the date of the extraction. The third bone, the left first cuneiform, was extracted on February 29th (it was a leap year). After the following weekend, Tuesday March 4th, Molepole extracted a fourth bone, the proximal half of the left big toe's metatarsal. Probably due to the small size of the bones, the team must have assumed that they belonged to a baboon or a monkey. As I had just been named Dean of the Medicine Faculty of the Witwatersrand University, and I had little time to work on the fossils, the four footbones were placed in a box with other small pieces of postcranial bones of primate and carnivore extremities. Alan Hughes, who was in charge of the Sterkfontein excavation from 1966 a to 1990, must have included those specimens among the 63 postcranial bones of baboons, monkeys and carnivores annotated in the Annual Report of the Palaeo-anthropology Research Unit of the year 1980 (September 1979 to September 1980). The number of limb bones rose to 190 in September 1981. So the four carefully labeled footbones remained in that box for fourteen years" (Tobias, 1997, p. 5). The second part of the story, the examination of the box containing the foot bones, their identification, and the surprising discovery of the rest of the StW 573 skeleton, has been told by Ron Clarke (1998).

---

of selection that shaped *A. afarensis*, while others are interested in reconstructing *A. afarensis* behavior." Ward concludes: "Evidence from features affected by individual behaviors during ontogeny shows that *A. afarensis* individuals were habitually traveling bipedally, but evidence presented for arboreal behavior so far is not conclusive" (p. 185). If so, bipedalism very similar to ours would be already present at least 3 Ma.

This is how Will Harcourt-Smith and Leslie Aiello (2004) see the matter: "The central point is that contemporary fossil taxa may well have been mosaic in their adaptations, but, critically, may have been mosaic in different ways to each other [. . .]. Further analyses of other skeletal elements are needed to reinforce this interpretation. If correct, this would imply that there was more locomotor diversity in the fossil record than has been suggested, and raises questions over whether there was a single origin for bipedalism or not. At the very least, if bipedalism appeared only once in the hominin radiation and is therefore monophyletic, such evidence would suggest that there were multiple evolutionary pathways responding to that selection pressure" (p. 413).

Is the bipedalism of Pliocene and Miocene hominins very similar to ours, or not? Evidence about the morphology of early hominin feet was enriched by the discovery of articulated foot bones at the Sterkfontein site (South Africa). The specimen known as Little Foot (StW 573) provided relatively solid proof that the locomotion of the first hominins cannot be characterized as functionally developed bipedalism, equivalent to that of current humans (Box 2.15).

Little Foot includes a left talus, a left navicular that articulates with the head of the talus, a medial left cuneiform that articulates with the distal surface of the navicular, and the proximal half of the first metatarsal of the left foot's big toe, articulated with the cuneiform (Clarke & Tobias, 1995) (Figure 2.9). The fact that these bones articulate together is precisely what makes possible the understanding of the biomechanical function performed by the foot to which they belonged.

The morphology of Little Foot reveals a mixture of apelike and humanlike traits. Functionally speaking, the heel suggests that the foot belonged to a biped, which supported the body's weight on its lower limbs. But its bipedalism was not completely developed; the anterior region of Little Foot retains the prehensile capacity required for arboreal activity. From back to

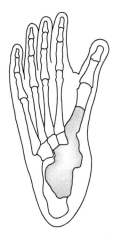

Figure 2.9 "Little Foot" (Clarke and Tobias, 1995) from 3 to 3.5 Ma sediments at Sterkfontein (South Africa). (Drawing by Ron Clarke; from Tobias, 1997.)

Box 2.16 Hadar controversies

Some discrepancies between authors regarding the bipedalism of Ethiopian australopiths would disappear if it were accepted that there are two species at Hadar. One would be represented by the A.L. 333 series and the other by specimens like A.L. 288–1. But this can hardly be taken as an argument in favor of an early functionally complete bipedalism.

---

front, the anatomy of the StW 573 articulated bones shows the intermediate condition between total bipedalism and arboreal activity. The talus is the closest bone to human morphology, though it is smaller (even accounting for the body size differences between Little Foot and modern humans). The navicular and the medial cuneiform exhibit intermediate morphology. The metatarsal and its articulation with the medial cuneiform evince an apelike character. Based on the articulation's shape, it seems that the foot's big toe adopted a diagonal position and was medially separated from the foot's axis (as it happens chimpanzees), different from the parallel position in relation to the foot's axis of the big toe of modern humans.

The mobility of the big toe is an important trait (similar in significance to the freedom of movement of the hand's thumb), determining the ability to climb trees. Tobias' (1997) conclusions are that while the talus had begun the way toward the shape associated with habitual bipedalism, part of the navicular, the medial cuneiform, and the base of the first metatarsal seem not to have done so: rather, they retained apelike traits. Little Foot represents an intermediate stage in the evolutionary conversion from a foot adapted to arboreal life, with a diverging and prehensile big toe, to an extremity adequate for regular bipedalism. "It seems that the astragalus and the proximal ankle joint adopted the human form quite early on, whereas the anterior part of the foot retained its primitive state for a long time" (Tobias 1997, p. 13).

In East Africa, Hadar (Ethiopia) has provided australopith specimens which convey information about feet morphology. The 333 series, discovered in 1975 in member DD of the site, is composed of two calcaneus bones (A.L. 333–8 and A.L. 333–55) and a cuneiform (A.L. 333–28), the proximal part of a first metatarsal (A.L. 333–54), and a partial foot (A.L. 333–115) with 13 bones, including phalanxes and metatarsals, with the head of the first metatarsal of the big toe and the first distal phalanx of the same finger, which, obviously, articulates with the previous one. The age of member DD is between 3.22 and 3.18 Ma (Walter & Aronson, 1993). Certain traits of the 333 series suggest a foot morphology intermediate between human and chimpanzee (Box 2.16). Yvette Deloison (1991) argued that the various positions of the big toe, the convexity of the calcaneus, the flattened and long phalanxes of the fingers, and the mobile articulations are features related to prehensile ability and, thus, suggest the possibility of an arboreal behavior. Randall Susman and colleagues (1984) reached a similar conclusion after a comparative examination of the metatarsal heads of gorillas, bonobos, A.L. 333–115, and modern humans.

The OH 8 specimen from Olduvai (Tanzania) is another fossil that affords information about foot traits. It is between 1 and 2 million years old, younger than Little Foot and the 333 series. Hence, it reflects the direction of the evolution of locomotion. OH 8, found in Bed I of the site and estimated at 1.85–1.71 Ma, contains (in addition to a clavicle, part of a hand, and a partial molar) an almost complete foot. Louis Leakey, Phillip Tobias, and John Napier (1964) believed it possessed most of the specializations associated with modern humans' plantigrade propulsive foot. However, the morphology of OH 8 was differently interpreted later. Some authors, such as Charles Oxnard and F. Peter Lisowski (1980), argued that its function seems to have been mainly associated with arboreal behavior. Even though this individual could probably also walk upright, this kind of locomotion would be far from human bipedalism and closer to that of gorillas and chimpanzees, as suggested mainly by the shape of the transversal arch. Other authors have described it as intermediate between complete bipedalism and arboreal life (Lewis, 1972, 1980), or as an unquestionably bipedal being but without yet reaching the posture of *Homo sapiens* (Day & Napier, 1964; Day & Wood, 1968). Some researchers who have delved deepest in the study of the evolution of bipedalism (White & Suwa, 1987; Susman & Stern, 1991; Deloison, 1995) see in the foot of OH 8 a very similar morphology to current humans. The fact that Randall Susman and Jack Stern, on one hand, and Yvette Deloison on the other, firm advocates of the gradual evolution of bipedalism, consider that the foot of *Homo habilis* is functionally modern, speaks in favor of this thesis.

Figure 2.10 The 3.5 Ma Laetoli fossil footsteps (Tanzania). (Photograph by Martha Demas.)

2.3 Beyond morphology: fossil footprints

Apart from morphological considerations, there is direct evidence of the existence of bipedalism in ancient times due to "fossil" footprints. Such evidence is very rare but we do have some of these exceptional findings.

2.3.1 Laetoli footprints

Laetoli (Tanzania) is a place of enormous paleontological value. Probably the footprints found there have given to the place its great renown (Figure 2.10). As Mary Leakey (1981) indicated, Laetoli has been identified as a promising site since 1935, and in 1975 was dated by dental and mandibular hominin specimens, "but it was not until 1976 that A. Hill [. . .] noticed the first fossilized tracks on the surface of a fine-grained tuff" (Leakey, 1981, p. 95).

Laetoli footprints comprise tracks of many mammals, including primates. However, three tracks on sites A and G have been attributed to hominins (Box 2.17).

This is the first direct proof of early bipedalism, more than 3.5 Ma. The footprints were described by Mary Leakey and Richard Hay (1979), and subjected to different interpretations (reviewed in White and Suwa, 1987).

Box 2.17 Fossil footprints?

Strictly speaking, fossils are only petrified human tissues, osseous in the vast majority of cases. But in a broad sense, any trace of morphology or behavior of an extinct organism: coprolites—petrified excrements—footprints, nests, and, in the case of humans, stone tools and objects related to the process of carving, as well as drawings, paintings, and carvings, can be considered as a "fossil."

The fossil footprints originated from steps on a freshly deposited layer of volcanic ash. The Ngorongoro, Lemagrut, Oldearn, and Sadiman volcanoes are near Laetoli. In their study of the footprints, Mary Leakey and Richard Hay (1979) did not opt for any of these volcanoes as the ash source, although the Sadiman volcano was commonly considered as the source until Anatoly Zaitsev and collaborators (2011) conducted a geochemical analysis, determining that Sadiman volcanic materials do not correspond with the nephelinite in melanite that appears at Laetoli.

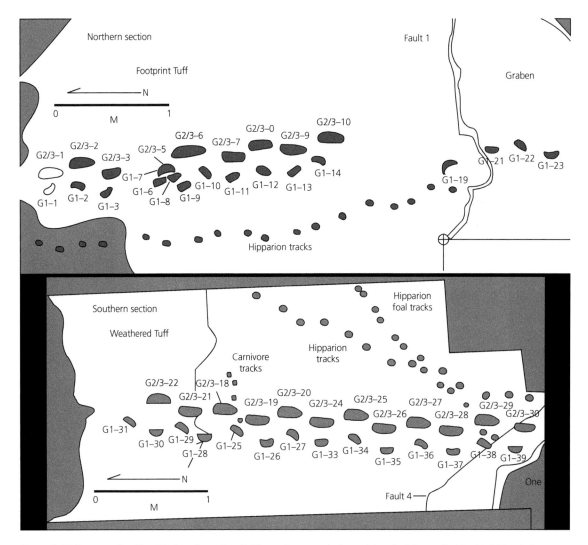

Figure 2.11 The two series of hominin footprints at Laetoli. (Pictures from www.indiana.edu/_ensiweb/lessons/footstep.html ©Laurie Grace, Scientific American.)

Ash deposits are, generally, quite ephemeral. The conservation of fossil footprints is due to the concurrence of several fortunate, unusual circumstances:

- Soft enough texture to allow the impression of the footprints.
- Adequate degree of humidity for an animal trail to be clear.
- High compression to allow large animal prints to have well-defined vertical edges.
- Rapid deposit of new and dense ashes capable of cementing and protecting the prints.
- The possibility of easily removing the covering material, though in some instances it had disappeared on its own (White & Suwa, 1987).

These circumstances came together at Laetoli, allowing the preservation of dozens of thousands of fossilized animal tracks in ash deposits. One of the three air-fall tuffs, known as the Footprint Tuff, contains 16 outcrops—sites A through P—in which the protective upper layer has disappeared exposing the underlying fossil footprints.

The footprints that most concern us are formed by two parallel series of tracks running for 25 m from south to north produced by clearly bipedal animals (Figure 2.11). One of the series (G1) corresponds to the steps of a small primate that seems to have stopped at a certain point and turned around before continuing (Leakey & Hay, 1979). The other is more difficult to

---

Box 2.18 Imagining the Laetoli footprint makers

The overlaying G2–G3 Laetoli tracks can inspire the imagination beyond reasonable limits. They have led some to envision the existence 3.5 Ma of such a human child game as stepping in someone else's footprints. Tracks G1 and G2 are about 25 cm apart, too close for both individuals to walk side by side without touching each other. Either one was walking in front of the other, or they were hugging while walking. This last image—the male embraces his mate while they stroll along the savanna—appears on the cover of *The Fossil Trail* (Tattersall, 1995). Tattersall is against indulging in empty speculations, and that book is an excellent demonstration. Tattersall does not mention the G3 series, and describes G1 and G2 as the footprints of two individuals that walked besides each other, without searching for further interpretations.

Box 2.19 Experimental studies on bipedalism

The experimental study of bipedal gait enables us to analyze how footprints are produced in different conditions and compare the results with fossil marks. There is a significant tradition of such studies (see Schmitt, 2003). It should be mentioned in reference to some of these studies that, by geometric morphometrical analysis, the Laetoli footprints were compared with those produced by humans and other animals by Christine Berge et al. (2006), confirming the idea of their human origin. An experiment conducted by Dave Raichlen and colleagues (2010) proposed to analyze, in comparable conditions, the biomechanical gait of eight participants walking barefoot on a sand bed of a texture similar to that of Laetoli, in order to study the footprints left behind. Subjects were asked to walk normally and, in a second phase of the test, to walk while bending their knees in a simian way (Raichlen et al., 2010). Only the first condition left a track comparable to the Laetoli footprint trail. The conclusion obtained by Raichlen et al. (2010) pointed out the upright gait mechanics indicated by Laetoli evidence, i.e., a bipedalism fixed in the human lineage at a much earlier time than the emergence of genus *Homo*.

---

interpret. It was described by Leakey and Hay (1979) as prints made by a larger primate (G2), but was later described as a double trail: a larger-sized biped (G2) would have left a trail of steps in which something smaller (G3) would have stepped (an interpretation that is not universally accepted). "Large" and "small" are relative terms; if we consider the relationship between the length of the foot and height of modern humans, then the smallest primate (G1) would be about 1.20 m tall and the medium-sized one (G3), which followed the footsteps of the largest, would measure close to 1.40 m. The height of the largest primate (G2) cannot be calculated because its footsteps are partially hidden by those of the medium sized one (Hay & Leakey, 1982) (Box 2.18).

2.3.2 What kind of bipedalism do Laetoli footprints indicate?

Leakey and Hay (1979) hardly addressed functional aspects in their original description. The tracks at sites A and G were considered "presumably hominin," that is all. The authors noted that the longitudinal arches of those feet were well developed, resembling those of modern humans, and that the big toe was parallel to the others. But these indications are enough to conclude that the way the foot was placed to cause these prints was similar to the way current humans do when walking, thus suggesting well-developed bipedalism.

The conclusion that such early hominins were already capable of a functionally developed bipedalism was explicitly put forward by Day and Wickens (1980), and has been suggested by many others (e.g., White, 1980a; Lovejoy, 1981; Tuttle, 1981; Robbins, 1987; White & Suwa, 1987; Tuttle et al., 1991) (Box 2.19).

Russell Tuttle, David Webb, and Nicole Tuttle (1991) have argued that the footprints found at Laetoli site G were made by feet more similar to those of *Homo sapiens* than to those of australopiths. Regarding the bipedal footprints at site A, these authors, after studying 16 circus bears, concluded that bipedal bears leave very similar footprints to the trail found at site A, although they note that there is no specimen attributable to bears in the limited collection of Laetoli carnivores. They settled the issue with a "pathetic poem" (their own expression):

Fuzzy Wuzzy was a bear/Fuzzy footprints were found there/ Was Fuzzy at Laetoli?/Or, wasn't he?

The evidence of a clearly bipedal trail at Laetoli has often been used as irrefutable proof of the existence,

3.5 Ma, of a bipedalism very similar to our own. If this is correct, a functionally complete bipedalism would be a derived trait that characterizes all the members of the tribe Hominini, distinguishing them from our closest relatives, the African apes.

However, this is far from certain. The photograph of the Laetoli footprints in the *Cambridge Encyclopedia of Human Evolution* (p. 325) is accompanied by a categorical statement: "The discovery of the footprints confirmed that early hominins walked upright on two legs in a characteristically human fashion and that their footbones were arranged like a modern human's, with no gap between the big toe (toe 1) and the other toes." Yet the author of the accompanying article, Richard Potts (1992), as well as Bernard Wood (1992c), in the article dedicated to australopiths in the same book, caution that there are reasonable doubts about the interpretation of the Laetoli footprints and favor partial bipedalism—a somewhat dubious concept to which we will return later.

Monographs devoted to footprints also reflect a diversity of interpretations. Day and Wickens (1980, pp. 386–387) wrote: "The pattern of weight and force transference throughout the foot, well known in modern man, also seems to be very similar in the fossil footprints and indicates that even at this early stage of hominid evolution bipedalism had reached an advanced and specialized stage." But some illustrations that accompany the text reveal certain traits (narrow print of the heel, absence of the medial prominence in the base of the big toe, and its orientation) that are reminiscent of the shape of a chimpanzee's print. Susman and collaborators (1984) concluded that the footprint molds are not similar to the prints left by the feet of humans. However, White and Suwa (1987), after suggesting the Laetoli footprints might have been made by *Australopithecus afarensis*, rejected Stern and Susman's (1983) idea that *A. afarensis* was intermediate between African apes and humans from a locomotor point of view, but accepted that the maker of the Laetoli footprints might differ in certain locomotor aspects from modern humans (Box 2.20).

It is difficult to arrive at an interpretation of the Laetoli footprints that would integrate the diverse points of view. Susman and colleagues (1984), who were among the first to interpret the Laetoli footprints as associated with a partial bipedalism, suggest that their scientific value is limited to the demonstration that they were produced by individuals that walked on their two lower limbs, which is not disputed by anyone. However, when the fieldwork was finished in 1979 and, with the intention of conserving the footprints, the research team covered them with sand first and lava pebbles afterward (Agnew & Demas, 1998), they did not realize that there were acacia seeds mixed in the sand. These germinated, growing into trees that menaced the destruction of the trails with their roots. A program for the conservation of the Laetoli footprints was initiated in 1994. It involved removing the trees, covering the trails again with sand that had been protected with herbicide, and extending a mantle of gravels and plastic materials crowned with lava pebbles (Agnew & Demas, 1998). Given the deterioration of the footprints, Susman and colleagues (1984) conclude that the bipedalism of early hominins should be determined by the direct study of available fossil specimens, without the need to depend on the Laetoli footprints.

2.3.3 Turkana footprints

Laetoli is not the only location with fossil footprints that can be attributed to hominins. In spite of being uncommon evidence, there are hominin footprints in places as diverse as South Africa (Nahoon Point, Mountain, 1966; Langebaan Lagoon, Roberts & Berger, 1997), Italy (Roccamonfina volcanic complex, Mietto et al., 2003), Greece (Theopetra Cave, Facorellis et al., 2001), Tibet (Quesang, Zhang & Li, 2002), and Australia (Willandra Lakes, Webb et al., 2006). All these are Pleistocene sites, but, besides Laetoli footprints, another trail of Pliocene footprints provides clues about the evolution of bipedalism: Turkana.

Box 2.20 An "intermediate form"

The thorough study carried out by Deloison (1991) agrees with Susman et al. (1984). After studying the best-defined Laetoli footprint, G1/34, Deloison compares its contour from a picture taken of a cast of the print with the prints left by the right feet of a human and a chimpanzee, arriving at the conclusion that the Laetoli footprint shows an intermediate shape between the other two. In fact, the contour is closer to that of chimpanzees' feet. Deloison pointed out, moreover, that some features of the Laetoli footprints support the prehensile functionality of the foot. In her opinion, the maker of the prints used that prehensile capability to keep its balance on the humid floor, curving the foot and separating its big toe.

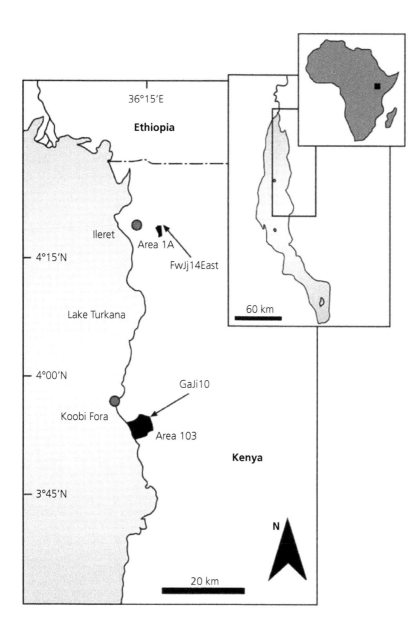

Figure 2.12 Site of Koobi Fora locations, area 103, and Ileret, area 1A (marked by the two arrows), where Turkana footprints were found. Laetoli, not indicated on the map, is located further south in Tanzania. (Map of Bennett et al., 2009; supporting online material.) From: Bennett, M. R., Harris, J. W. K., Richmond, B. G., Braun, D. R., Mbua, E., Kiura, P.,. . . Gonzalez, S. (2009). Early Hominin Foot Morphology Based on 1.5-Million-Year-Old Footprints from Ileret, Kenya. *Science*, 323(5918), 1197–1201. Reprinted with permission from AAAS.

In the same year in which the Tanzania footprints were discovered, 1978, Anna Behrensmeyer and Léo Laporte found similar tracks at the site GaJi10 of area 103, Koobi Fora formation, located on the east shore of Lake Turkana (Kenya) (Figure 2.12).

Behrensmeyer and Laporte (1981) identified several overlapping horizons containing hominin footprints, to which they attributed an age of 1.5–1.6 Ma. The footsteps occurred on fine-grained silt attributed to a lacustrine environment, an assumption that is supported by the presence of hippopotamus' and waders'

footprints. The study of these materials included the task of analyzing the size of the footprint makers.

Molds of latex and fiberglass were taken from the footprints, which are preserved in the National Museum of Kenya, a fortunate precaution, because Matthew Bennet and colleagues later restarted excavation of the site and found that some of the tracks had deteriorated.

In July 2008, the team of the Koobi Fora Research and Training Program, directed by Matthew Bennett, resumed the work by Kay Behrensmeyer and

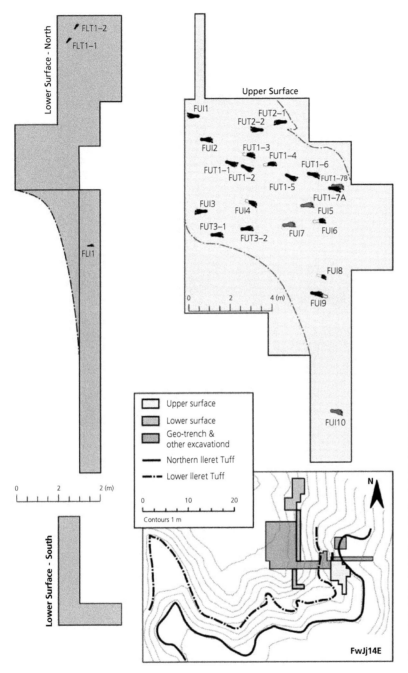

Figure 2.13 Left: Ileret fossil footprints. In light gray: The upper surface where the majority of the identified tracks are present (series FUT FUI). In dark gray: The lower surface with the rest of the tracks (FLT series). The black lines indicate volcanic tuffs. (Map created on the occasion of the reopening of the 1980 excavation conducted by Bennett et al., 2009; supporting online material.) From: Bennett, M. R., Harris, J. W. K., Richmond, B. G., Braun, D. R., Mbua, E., Kiura, P.,. . . Gonzalez, S. (2009). Early Hominin Foot Morphology Based on 1.5-Million-Year-Old Footprints from Ileret, Kenya. *Science*, 323(5918), 1197–1201. Reprinted with permission from AAAS.

Leo Laporte in Koobi Fora, expanding by 1 m the excavation area (Bennett et al., 2009). The results of the radiometric analysis of the material from the Karari Grey tuff, on site GaJi10, corresponded to an age of 1.428±0.01Ma, a reason why Bennett et al. (2009) attributed 1.43 Ma to the footprints, a slightly younger age than that assumed by Behrensmeyer and Laporte (1981).

The main objective of Bennett's team was to study another series of footprints located in the Koobi Fora formation, in Ileret (Kenya), between 2005 and 2008 (Figure 2.13).

Ileret footprints are impressed upon two horizons—upper surface in light gray and lower surface in medium gray in Figure 2.13—with 5 m of separation, Ileret locality, Okote member of Koobi Fora formation, Kenya. The new footprints were described in 2009, expanding the available information and allowing comparisons with previous materials from Koobi Fora (Bennett et al., 2009).

The age of site FwJj14E, obtained by correlation and comparison of trace elements with the volcanic tuffs of Lower Ileret and Elomaling'a, was established as 1.53 Ma. It is closer in age, therefore, to GaJi10. Although the Ileret tracks are much less clear than those of Laetoli, they were analyzed by geometric morphometrics, comparing them with footprints of current subjects from the area, according to the method of Berge et al. (2006). The comparison also took into account Neolithic fossil footprints from England.

The results of biomechanical studies confirmed the hypothesis of Bennett et al. (2009), who attributed the Ileret footprints to hominins—in particular to *H. ergaster*/*H. erectus*. Going beyond that is difficult; as Robin Crompton and Todd Pataky (2009) have indicated, other issues related to the fossil footprints of Laetoli, Koobi Fora, and Ileret, e.g., the type of gait—braking, propulsion—the degree of functionality compared to modern human bipedalism, and even the possible divergence of large toe axis compared to the other toes, cannot be resolved with the evidence existing today.

2.4 Bipedalism and adaptation

Given that bipedalism is a synapomorphy shared among different hominin genera for 7 million years, it must have had an undoubtable adaptive advantage. What was the advantage? What was it about a permanent upright posture that improved resources? As we have seen, Darwin suggested a hypothesis that related bipedalism, free hands, and tool use to the extent that their combination would amount to a single complex phenomenon with morphological and functional aspects. The *"hit'em where it hurts"* hypothesis, as Tuttle and colleagues (1990) called it, is undeniably attractive. Darwin's initial suggestions regarding this issue were preserved in later models that relate the appearance of savannas with bipedalism and tool-making. The number of published articles devoted to bipedalism and its role during human evolution possibly outnumber those dedicated to any other hominin functional feature. But, as Tobias (1965) noted, bipedalism is not a requisite for making or using tools. Chimpanzees use instruments quite ably, and they do so sitting up. The essential element in the relation between posture and the use of cultural elements is upright posture, not bipedalism. But there is more. Bipedalism appeared in human evolution long before culture.

If bipedalism is not explained by the manufacture and use of instruments, what drove its appearance? There are two separate issues underlying the search for hypotheses to explain the adaptive advantage of bipedalism. First is the motives behind the appearance of the first bipedal behaviors in a tropical forest environment. The second issue concerns the benefits of bipedalism as an adaptation to the savanna. These two questions must not be confounded: bipedal behavior existed long before savannas were extensive in the Rift Valley. The two questions are often confounded by seeking a "general explanation of bipedalism."

Yves Coppens (1983a) suggested the progressive reduction of the tropical forest thickness as a possible explanation for the gradual evolution of bipedalism. If the distance among the trees gradually increased, it would become necessary to travel longer distances on the ground to go from one to another. At the same time it would be imperative to retain the locomotor means for climbing. Distinct functional responses appeared in the different lineages leading to current primates: knuckle-walking bipedalism in the ancestors of gorillas and chimpanzees, and an incipient bipedalism in the first hominins (Coppens, 1983a, 1991; Senut, 1991).

The gradual substitution of forests for open savanna spaces would be an increasing selective pressure toward more complete bipedalism, functionally speaking. The final result of this process was two evolutionary lineages of bipedal primates based on different adaptive strategies, close to 3.5 Ma. One million years later this divergence would increase with the decrease in temperatures and the appearance of extremely robust australopiths and the genus *Homo*.

Box 2.22 Hypothetical models of the origin of bipedalism

Tuttle et al. (1990) gave colorful names to hypotheses that explain the adaptive advantages of bipedalism in pre-cultural conditions (we have retained the original names so that they can be noted):

Schlepp": food transportation, caring for offspring. Involves the presence of a kind of home-base.
Peek-a-boo": vigilant behavior, standing up over the savanna's long grasses.
Trench coat": phallic exhibition in males to attract females.
Tag along": following herds of herbivores during their migrations through the savanna.
Hot to trot": a way to lose heat when exposed to the solar radiation in the open savanna.
Two feet are better than four": bipedalism has a favorable energetic balance for long treks.

The hypotheses of Tuttle et al. (1990) refer to the adaptive advantages of bipedalism on the savanna, not in the forest. Therefore, they are not useful for explaining the entire evolution of upright gait, unless we agree that this evolution took place only during the last 2.5 million years. It seems clear that none of these explanations, except perhaps the "*trench coat*," make sense during the time when our ancestors were creatures with rudimentary bipedalism living in a tropical forest environment.

Box 2.23 Speed and efficiency

A rapid, bipedal locomotion would have been more effective for both hunting and scavenging in open savannas, when it was necessary to travel great distances (Bramble & Lieberman, 2004). But Esteban Sarmiento (Sarmiento, 1998; Sarmiento & Marcus, 2000) has argued the opposite idea: the australopiths of Hadar (Ethiopia) could have adopted quadrupedism when they needed to move quickly or to travel a great distance.

2.4.1 Energy efficiency of bipedalism

The explanation given by Coppens and Senut has a considerable advantage: simplicity. Brigitte Senut noted that the locomotor hypothesis of the origin of bipedalism has been among the least favored. This hypothesis suggests that hominins had become bipedal for reasons strictly associated with locomotion itself (Senut, 1991); that is to say, the need for traveling on the ground of open forests. Senut explored eight hypothetical ways in which bipedalism could have originated from the locomotion of other primates, but ended up developing with greater detail the explanation favored by Coppens (1983a, 1983b) (Box 2.22).

Once we accept bipedalism to be much older than the savanna expansion, it is necessary to clarify in what sense bipedalism would be useful for the adaptation to open spaces that followed the abandonment of forests. The best answer comes from energy efficiency analysis.

The importance of fast bipedalism in the evolution of the genus *Homo* has been brought to light in Denis Bramble and Daniel Lieberman's (2004) study of the role of running. It is evident that current humans are not among the fastest animals in the savanna, nor were our hominin ancestors. However, running is related not only to speed itself. After comparing the metabolic costs of running and walking, Bramble and Lieberman (2004) conclude that several anatomical traits of the genus *Homo*—including narrow pelvis, long legs, short neck of the femur, and big toe—improved the energetic balance of fast bipedalism, running, because of enhanced features of fast marching: balance, thermoregulation, shock absorption, stress reduction, stabilization of the head and trunk, energy storage, and so on. The most important characteristic of running would be related with energy balance factors and not pure speed. This kind of locomotion would have been efficient for hunting and scavenging in open savannas when long distances had to be covered (Box 2.23).

Patricia Kramer and Gerry Eck (2000) investigated *A. afarensis* locomotion by mechanical analysis of the specimen 288–1 Hadar (Ethiopia) morphology. In order to find the mechanical power that A.L. 288–1 (a female) would have required, the osteological anatomy of the specimen, combined with the movement profile of modern humans, was applied to the dynamic model of a biped walking at different speeds. These parameters were compared to a composite that represented a modern woman. The results obtained by Kramer and Eck (2000) indicated that when locomoting at walking speed, A.L. 288–1 would have spent less energy than a current woman. However, in the transition from walking to running, the energy advantage of A.L. 288–1 is lost; the australopith used more energy than the modern woman. Kramer and Eck (2000) concluded that the daily walking range of A.L. 288–1 would be substantially less than that of current humans. They argued that the locomotor anatomy of *A. afarensis* could have been optimized by natural selection for a particular ecological niche—low-speed foraging—and it could not be deemed as transitional in respect to other types of gait.

The review of the available evidence about the long-distance running capability in humans was conducted by Dennis Bramble and Daniel Lieberman (2004), and indicated that traits related to running capabilities leave traces in the skeleton.

These running capability features were then studied in the fossil record to discover their earliest appearance. The results obtained support the idea that the capability for enduring long-distance running is a derived trait of the genus *Homo*—which is compatible, of course, with the previous study of Kramer and Eck (2000). A later study by Daniel Lieberman and collaborators, using an electromyographic and kinematic analysis, supported the hypothesis that the elongation of the *gluteus maximus* was an important factor in achieving the advantage of running during human evolution (Lieberman et al., 2006).

The work of Weijei Wang and colleagues (2004) also sought to evaluate the necessary strength for bipedal locomotion from the skeletal muscle insertion. The comparison was made between A. L. 288–1, *A. afarensis*, and KNM-WT 15000, *H. erectus* sensu lato (see Section 8.3.1 for this last specimen), and a combination of both of them and current humans. Two alternative models were considered: (1) the erect gait of our own species and (2) the simian erect gait with knees bent.

The results obtained by Wang et al. (2004) indicated that, in terms of energy consumption per unit mass when walking, both A.L. 288–1 and KNM-WT 15000 would be within the human range. But if the parameter of covered distance is introduced, AL 288–1 needs more muscle power for long stretches. The conclusion reached by the authors is that the body proportions of KNM-WT 15000 indicate an evolution that tends, with respect to bipedalism, to perform more effectively during long walks or runs at higher speeds. The theses of Kramer and Eck (2000), and Bramble and Lieberman (2004) are supported by this evidence.

The study of Patricia Kramer and Adam Sylvester (2009) on the relationship between the size and shape of the body, and speed and energy efficiency, resulted in an important caveat: size—in particular, the leg length—is associated with speed and, therefore, with the range of daily movement. In the absence of confirmed morphological differences, size by itself does not relate to energy efficiency. Consequently, the difference in size between bipedal fossils must be interpreted

as an indication of the various conditions encountered in the environment; it does not mean the existence of different levels of exploiting locomotion. Kramer and Sylvester (2009) argued that smaller size hominins, such as *Homo floresiensis* and *Australopithecus*, were small because they did not need to be big to walk fast and travel far.

A new twist about the possibilities of analysis of the evolution of the locomotor system of hominins and higher apes appeared with the detailed study of specimens of *Ardipithecus ramidus*, an older member of the human lineage found in Ethiopia (Africa), whose characteristics we will discuss in Chapter 4, where we will return to this subject. For now, let us follow the evolution of bipedalism with the clues provided by the postcranial traits—hips and limbs—of the Pliocene *Australopithecus* specimens. If they are similar but not identical to the features of current humans, what would be the phylogenetic significance of these differences?

2.4.2 "Partial" vs. "complete" bipedalism

Hominin bipedalism is currently widely considered as a homologous trait, shared by the whole lineage. It is thought to have developed in several stages from the incipient bipedalism of early australopiths to the complete bipedalism of the specimen found in Java by Dubois, *Homo erectus*. But this is not the only possible interpretation. There are authors who reject the idea that there were different stages in the evolution of bipedalism along the hominin lineage. For instance, the comparative examination of the tibia of australopith specimens from Olduvai (Tanzania), Koobi Fora (Kenya), and Hadar (Ethiopia) led Owen Lovejoy, renowned specialist in hominin locomotor patterns, to the conclusion that the bipedal locomotion of early hominins was as developed as our own (Lovejoy, 1975; Latimer et al., 1987). The study of australopith specimens from South Africa also indicated, according to Lovejoy (1975), that there is no morphological reason to consider that their locomotion was "intermediate" between that of African apes and modern humans. The morphology of the pelvis of those early hominins is very similar to that of living current humans, according to this author. Their ilium is equivalent to human beings (this, by the way, had already been noted since the discovery of the first exemplars—Dart, 1949a—and generally admitted since then). The differences observed in their ischium probably have no functional consequences. And the pubis, in any case, has little bearing on the question of locomotion (Box 2.24).

Box 2.24 Australopiths' bipedalism

According to Owen Lovejoy, despite what some isolated traits might suggest, the overall biomechanical pattern of australopith postcranial anatomy supports the notion that the only difference between *Australopithecus* and *Homo sapiens* is advantageous to the former. All the necessary adaptations for bipedal locomotion, in Lovejoy's opinion, were already present in those early hominins, though in a different way in males and females.

Box 2.25 The meaning of "orthogenesis"

It makes no sense to argue that a species develops a partial organ as an intermediate step toward full versions of the same organ—the process of linear progression in stages known as "orthogenesis." How could future events affect what happens in the present? Assuming that such is the case, we will keep the expressions "partial" and "total" bipedalism in quotes, so as to preserve expressions frequently used in anthropological literature.

The idea that the very wide pelvis of australopiths would have been favorable for bipedal locomotion has been rejected, however, by Christine Berge (1991) after the examination of the A.L. 288–1 specimen from Hadar (Ethiopia). At the level of the iliac crests and the pelvic cavity, the pelvis of A.L. 288–1 is much wider than that of modern humans. In Berge's biomechanical reconstruction, the long neck of the femur, acting as a lever arm, does not constitute an advantage, as Lovejoy surmised; rather, it introduces balancing problems. The vertical of the center of gravity would fall, in the case of *A. afarensis*, far from the knee articulation when leaning on one foot while traveling, leading to a greater instability of the lower limb (Berge, 1991). As a consequence, the kind of bipedal locomotion exhibited by *A. afarensis* would have required a higher degree of hip rotation to place the leaning knee within the body's vertical axis. In her morphometric study of the mobility of the hip of *A. afarensis*, and in order to obviate the difference in height between Lucy and current humans, Berge carried out the comparison with the pelvis of a pigmy woman 137 cm tall.

With respect to the possible reconstruction of the gluteal insertion in the hip of A.L. 288–1, Christine Berge (1991) also pointed out a circumstance worthy of mention. It is not known how this insertion took place, but the two possible alternatives are those of a "human" type, with gluteus maximus inserted into the ilium, or of an "ape" type, in which the muscle would be attached mostly to the ischium. When establishing the reconstruction of the internal rotation movements of the thigh, Berge (1991) argued that the morphology of the hip, attached to the reconstruction of a "human" gluteus maximus insertion, would prevent A.L. 288–1 from performing the movements necessary for bipedalism, which could only occur through an "ape" insertion of the gluteus. This point is especially important,

since the role of the gluteus maximus in the acquisition of bipedalism had been considered as an alternative by Randall Washburn, for whom the passage from quadruped to bipedal locomotion begins precisely with the "human" change of the gluteal, and by John Napier, who thought that that change did not occur until later stages of bipedalism, and its function was only to provide balance when running or climbing hills, not when walking (see Napier, 1967). Berge (1991) agreed with Napier, and indicated that the inferred locomotion of the hip A.L. 288–1 implies the presence of a behavior that is partly arboreal.

In order to accept that australopith bipedalism was different from that of current humans, it is not necessary to consider it as the beginning of our locomotion, even disregarding the evidence of a possible "more human" gait of *A. afarensis* (Ward et al., 2012), to which we shall refer in Chapter 3. Following the criterion of genus used in this book, each taxon has a particular type of adaptation to its environment, which cannot be understood as an intermediate step toward future adaptive formulas. So, to be cautious, it is best to understand the bipedalism of *Australopithecus* on its own terms, without relating to their morphological characteristics any other variables than the environmental circumstances, the evolutionary past, and the adaptive present that correspond to this genus.

2.4.3 Walking and running

If we understand that early hominins had a functionally complete bipedal locomotion that in some ways is different from ours, then it is possible to speak of two different adaptive strategies related to bipedalism:

- A "slow-speed" bipedalism, characteristic of australopiths, shows an excellent energy efficiency in low-speed foraging, but would be ineffective when running at higher speeds. This is the morphology and function that corresponds to the adaptive strategy of a type of being that lives in the tropical forest and usually walks short distances for obtaining food.

- A "fast-speed" bipedalism, suitable for running, with greater energy efficiency when traveling long distances. This is the type of locomotion of *Homo erectus* and subsequent taxa of the genus. With respect to *Homo habilis*, the conclusions are not definitive.

Despite the anatomical and bioenergy analysis mentioned, Fred Spoor, Bernard Wood, and Frans Zonneveld (1994) proposed a new way of studying the evolution of bipedalism: a comparative examination of the vestibular apparatus (internal structure of the ear that is part of the motion perception control system). Using high-resolution computed tomography, Spoor and collaborators (1994) analyzed the bone morphology of the ear's semicircular ducts in 31 current species of primates, including ours and other higher apes, as well as other fossil specimens. With the evidence obtained, Spoor and colleagues concluded that *Homo erectus* is the first species showing undoubtedly modern human morphology. *Australopithecus* and *Paranthropus* have semicircular duct dimensions similar to current higher apes. Spoor et al. (1994) conclude that *Homo erectus* would have necessarily been completely bipedal, while *Australopithecus africanus* would have shown a locomotor repertoire that included optional bipedalism and tree climbing. Among the australopiths, bipedalism would be more a matter of posture,

---

Box 2.26 Bipedal postures in the human lineage

Studies of Susman and Stern (1991), Spoor et al. (1994), and numerous other authors indicate that posture and balance are different in australopiths and members of the genus *Homo*. But, to what extent can these differences be compared? The analysis of Carol Ward (2002) on posture and locomotion of Hadar (Ethiopia) specimens sheds doubt on the extent to which similarities and differences can be established with respect to the genus *Homo*. Ward, consequently, suggested the necessity for a better characterization of the polarity of primitive and derived features, before drawing conclusions about the *Australopithecus'* gait.

---

lacking more complex movements, such as running or jumping (Box 2.26).

The basic depiction of bipedalism is already outlined. In the African tropical forest environment, around 7 million years ago, the hominin lineage appeared, characterized by an apomorphy: bipedal locomotion. Over time that lineage diversified and spread to colonize the planet. Even the unique synapomorphy of hominin, the bipedalism, changed somewhat its function when adjusting to running in the open savanna. The distinct clades of the lineage developed adaptive specializations and one of them, the genus *Homo*, succeeded in thriving until today.

In the following chapters we will examine the steps of that evolutionary path. But, it should be anticipated that the allocation of the hominin character to every hominoid fossil that proves to be biped implies conceding bipedalism to be the essential key for the condition of being "human." Are we exaggerating the role of locomotion? Besides the morphological characteristics leading to a particular taxonomical criterion, has bipedalism any additional consequence that might be crucial in crossing the frontier into humanity?

2.4.4 The childbirth problem

Humans are indeed very different from our sibling group, the chimpanzees. We speak a language with a dual patterning (we link phonemes to words and words into sentences). We have a brain much bigger in size in relation to our body, and more complex. The technical level of our cultural solutions is also a very obvious divergent trait. We lack body hair, or have very little. Our face is not as prognathus as in apes. Human gestation is shorter than that of higher apes. There are no specific seasons when human females are sexually receptive. The universe of ethical and esthetic values and the complexity of our social life reach levels unimaginable to any ape.

At first it seems that none of these distinctive human characteristics have something to do with bipedalism. But, let us look at the pelvis. Pelvic functions are not exclusively related to locomotion: in females these bones must allow the complete development of the fetus and its birth (Figure 2.14). Locomotion and bearing offspring call for conflicting pelvic designs. The widening of the ilium, required for the acquisition of bipedal posture, tends, in females, to close the canal through which the head and body of the fetus has to pass during birth. The appearance of bipedalism and the increase in cranial size represent a paradox: natural selection has yielded traits that make healthy births

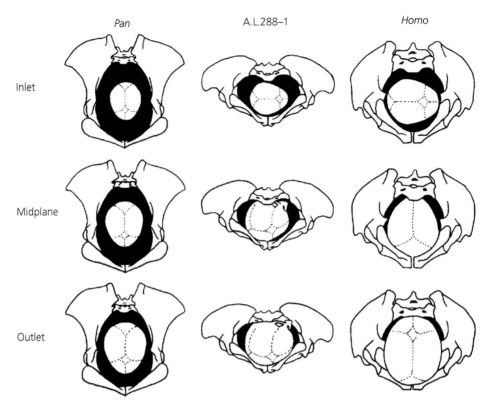

Figure 2.14 The passage of the fetus through the birth canal in a female chimpanzee (left), in *Australopithecus* (A.L. 2888–1 center), and in a modern human (right), according to Owen Lovejoy's hypotheses. The maintenance of the transversal position of the *Australopithecus* fetus has been proposed by Michael Day (1992). (Drawing by Tague & Lovejoy, 1986.) Reprinted from Tague, R. G. & Lovejoy, C. O. (1986). The Obstetrics Pelvis of AL 288–1 (Lucy). *Journal of Human Evolution*, 15, 237–255, with permission from Elsevier

difficult. But nature has alleviated this difficulty by selecting soft fetal cranial bones, birth at a very immature stage with a long period of exterogestation. In addition, human male and female pelvic anatomy is different: the design of women's pelvic structures is less strained by our bipedal posture.

The problem of the contradiction between a suitable hip for bipedal posture and an appropriate one to allow the oversized head of the fetus to pass through would not be present in the case of australopiths, according to the hypothesis of Owen Lovejoy (Lovejoy, 1975; Tague & Lovejoy, 1986). As far as we know, the size of their cranium was similar to that of current chimpanzees.

With respect to the pelvis, although a highly variable feature with many factors involved in its function and few existing samples, Lovejoy maintained the hypothesis that female australopiths would be less strained by the contradiction between bipedalism and childbirth.

However, the comparative analysis conducted by Jeremy DeSilva (2011), to establish the point in hominin evolution when infant–mother mass ratio (IMMR) equaled that of modern humans, obtained a result that differs from the hypothesis of Lovejoy (1975). In hominins of 4.4 Ma (*Ardipithecus ramidus*), the IMMR would be similar to present-day African apes—indicating that a low IMMR is a primitive trait—while *Australopithecus* females would already bear offspring that were significantly larger (DeSilva, 2011).

According to DeSilva (2011), the increasing size of the neonate could have limited the arboreal behavior of *Australopithecus* females, becoming a selective pressure toward alloparental care—care of offspring by group members who are not the parents—earlier than previously thought. This a characteristic that links premature childbirth conditions imposed by bipedalism to social behavior.

2.4.5 The rotational pattern

In modern women the entrance to the birth canal is wider in the transversal than in the anteroposterior direction. But the axes of the human head have opposite dimensions, the largest being from the front to the back. The head of a baby about to be born must, therefore, rotate 90 degrees to fit the pelvic canal dimensions of the mother. That rotation is not required in any of the great apes.

The study of pelvises from the fossil record suggests that the baby head rotation is a late feature in human evolution. The available pelvises of *A. africanus* (Sts 14) and *A. afarensis* (AL 288–1) indicate that the australopiths did not perform this rotational pattern during childbirth (Tague & Lovejoy, 1986; Rosenberg & Trevathan, 2001), and this circumstance continued among *Homo* members of the Lower Pleistocene (Ruff, 1995). Juan Luis Arsuaga and colleagues (1999) suggested that Pelvis 1 of Sima de los Huesos (Atapuerca) provides clues that a rotation similar to the modern one would have already happened in these specimens, which lived before the Neanderthals, if in fact they were not members of this taxon.

That viewpoint was criticized by Timothy Weaver and Jean-Jacques Hublin (2009) when, in order to clarify the situation regarding Neanderthal childbirth, they made a virtual reconstruction of the Tabun pelvis. They drew the conclusion that Neanderthal mothers would face the same difficulties during childbirth that women experience today. But, according to Weaver and Hublin (2009), the Neanderthal pelvic canal retained certain primitive patterns. The alternatives to gain sufficient width in the pelvic canal are:

(1) Female pubic bones similar in length to those of men, but more open in the coronal direction.
(2) Pubic bones of larger anteroposterior size.

For reasons that probably had to do with the adaptation to cold climates, the pubic bones of Neanderthal women have followed the trend (1) (Box 2.27),

---

> ### Box 2.27 The birth canal in Neanderthals
>
> Robert Franciscus (2009), in a comment on the article by Weaver and Hublin (2009), argued that the outline of different evolutionary patterns for the birth canal in Neanderthals is reasonable, but questioned the suggested adaptation to cold weather. The pelvis BSN49/P27 (Simpson et al., 2008) found in Gona, a warm place, would not suggest an adaptation to cold, and casts doubt on this explanation. However, in the opinion of Franciscus (2009), this pelvis supports the idea of a width between the iliac crests, maintained along the entire evolutionary path that led to Neanderthals.

---

according to Weaver and Hublin (2009), while modern humans have fixed the autapomorphy (2).

Beyond the problems inherent in childbirth, being born with an immature skull and, therefore, with an incomplete brain, requires a long period of exterogestation or secondary altriciality—the stage of growth that takes place outside the womb. This fact raises a number of issues related to our human way of being, all derived from our premature birth and the long period of dependence on the group, imposed by bipedalism. Communication, diversity of cultural traditions, the role of ethical and esthetic values, and, in general, the degree of socialization reached very high levels in the hominins.

What can be said about how these features evolved and the species to which they can be attributed, are issues that will be studied in the ensuing chapters. The human universe derives from the requirements imposed by bipedalism forcing anatomical changes in the set of lower limbs and pelvis. In a way, we must admit that we are human because we are bipeds, far beyond the mere taxonomic classification.

CHAPTER 3

The hominin lineage

3.1 The origin of hominins

3.1.1 Synthesis of Miocene hominoid evolution

The book *Function, Phylogeny and Fossils*, edited by David R. Begun, Carol V. Ward, and Michael D. Rose, provided in 1997 an up-to-date overview of hominoid adaptive processes. In the preface, the authors wrote that the numerous fossil ape discoveries during the late-twentieth century allowed a certain consensus about their phylogeny during the Miocene. But Laura MacLatchy (1998) pointed out in her review of the book, that the claim is rather optimistic. The only consensus that can be inferred from that book concerns the enormous functional (and taxonomic) diversity of this period's hominoids.

Certain issues are well established. First, there is agreement regarding when and where hominoids appeared: the Middle Miocene in Africa. It also seems clear that apes evolved in arboreal environments, contrary to the tendency of cercopithecoids, their sister group, to colonize the savanna. A relationship has been established between this adaptive option and certain dental and locomotor changes, which had probably already appeared in Middle Miocene specimens. Thereafter, the paucity of remains from the African continent hinders any attempt to understand how the direct ancestors of current apes appeared.

The remaining problems are not only taxonomic—to determine the taxa for each lineage—but also systematic in a general sense, concerning the difficulty of explaining how the apomorphies that appeared during the Middle Miocene, and possibly before, were transmitted to current Asian and African apes. All the suggestions made face the inconvenience of the extreme variability of the traits. Regarding locomotion, the tendency toward an orthograde posture is constant during the Miocene. But the specific type of locomotion associated with a more upright posture is not the same in all Middle Miocene or current hominoids. Regarding dental traits, the general tendency leads to low crowns on wider molars. But dental enamel thickness is also diverse in Miocene and current apes (Table 3.1).

Taking this into account, can a relationship be established among climate, diet, and hominoid evolution?

3.1.2 The ecological script of Miocene hominoid evolution

Hominoids appeared at the beginning of the Miocene—or even toward the end of the Oligocene—in tropical jungle environments in Africa. Those early hominoids can be described as very similar to current Old World monkeys, cercopithecoids (mangabeys, geladas, baboons) and colobus. This is how they have been perceived since the 1950s, for example, by Wilfrid Le Gros Clark and Louis Leakey (1951). According to Peter Andrews (1981), from a functional and ecological point of view, the first Miocene hominoids were equivalent to monkeys.

The planet's cooling around 14 Ma produced a certain degradation of East African tropical forests and the so-called proto-savanna expanded. This environment consists of forests with patches of open vegetation, similar to that associated with the appearance of *Kenyapithecus* at Fort Ternan. The climatic changes led in Europe to a dry and cold period, but by the end of the Middle Miocene, there was an increase in temperature and recovery of European and Asian forest lands (Agustí, 2000). Hominoids (*Sivapithecus* in Pakistan, *Dryopithecus* in Europe) radiated and expanded, taking advantage of the new ecological niche afforded by the floor of those forests. Jordi Agustí (2000; Agustí et al., 1996) believes that the thin enamel of *Dryopithecus* was the result of a frugivorous diet, but the phylogenetic development that led to that trait is not easily explained. It can be considered either an apomorphy—a derived trait fixed by *Dryopithecus*—or a plesiomorphy inherited from *Proconsul*. Neither option changes the fact that the African Middle Miocene ancestors of European *Dryopithecus* remain unknown.

Processes in Human Evolution. Francisco J. Ayala and Camilo J. Cela-Conde, Oxford University Press (2017).
© Francisco J. Ayala and Camilo J. Cela-Conde.
DOI 10.1093/acprof:oso/9780198739906.001.0001

The transition from the Middle to the Upper Miocene coincided with a new decrease of temperature, some 11 Ma. With the accumulation of ice on the continents and the descent of sea level, new bridges appeared, favoring faunal exchanges. Although hominoids diversified considerably, the replacements that could be expected did not happen. Agustí (2000) has explained this fact, which affects mammals as a whole. A million years later, around 9.6–8 Ma, there was a profound crisis (the Vallesian crisis, which takes its name from the studies carried out by Agustí and Moyà-Solà at Vallès-Penedès), which led to a faunal impoverishment that also affected hominoids. The essential factor that triggered the Vallesian crisis was the substitution of subtropical forests for European temperate forests, with deciduous trees. This change might have led to the extinction of hominoids in Europe. The African and Asian specimens survived and led, in time, to the ape lineages we know today.

The association of hominoid evolutionary diversification with climatic change at the end of the Miocene is both reasonable and interesting. The issue is whether it is sufficiently grounded on available evidence to be considered anything more than a mere hypothetical model.

3.1.3 Locomotion and dentition: difficulties for an integrative model

Current apes are not homogeneous in their posture or diet-related dental features (Table 3.1). They share certain general features, such as orthograde posture and a varied diet, but living apes are notoriously diverse.

Table 3.1 Locomotion and enamel of different current and Middle Miocene hominoid genera

Genus	Period	Enamel	Locomotion
Hylobates	Current	Thick	Brachiation
Dryopithecus	Middle/late Miocene	Thin	Brachiation
Kenyapithecus (K. wickeri)	Middle Miocene	Thick	Climbing
Kenyapithecus (K. africanus)	Middle Miocene	Thick	Quadrupedalism
Pongo	Current	Thick	Brachiation
Gorilla	Current	Thin	Knuckle-walking
Pan	Current	Thin	Knuckle-walking
Homo	Current	Thick	Bipedalism

Orangutans use brachiation; chimpanzees and gorillas use knuckle-walking. Chimpanzees and gorillas have thin molar enamel, but orangutans—like humans—have thicker enamel (see Box 3.1).

The work of Andrews (1981) concerning the diets of Miocene monkeys and apes has explained the paradox of the persistence of thin dental enamel in gorillas after their change from a frugivorous to a leaf-eating diet. Andrews (1981) underlined the fact that the diet of a terrestrial quadruped is actually very varied (seeds, roots, rhizomes, bulbs, fruits, and soft leaves). Thick molar enamel is not required, even for diets based on materials that are harder than fruits, as is the case with folivorous diets. In gorillas, it is the shape of the molars, with pointy cusps and long, cutting surfaces, which allows their folivorous diet. Siamangs, which are also leaf-eaters, have similar molars. Andrews

---

Box 3.1 Enamel and locomotion

The identification between thick enamel and terrestrial quadrupedal locomotion is based on Jolly's (1970) "seed-eaters" hypothesis. Clifford Jolly established a relation between the reduction of the canines, development of potent molars, feeding on seeds, bipedalism, and even development of language and social groups with a dominant male. He did not, however, mention lithic instruments as significant elements. In Milford Wolpoff's (1982) view this was not an accident. This model of a predominantly herbivorous diet moved away from the original idea in which Darwin related the use of tools and weapons with hunting behavior. But, in any case, the relation between molar enamel and the diet of an inhabitant of tropical forest floors is not easily supported. David Gantt's (1979) and Richard Kay's (1981) studies argue against any correlation between the trait of dental enamel and terrestrial locomotion.

Brigitte Senut (1991) has advised of the risk of talking about brachiation when we refer to the orangutan. True brachiators are gibbons, which use only their arms to move from branch to branch. Orangutans and, to a lesser degree, chimpanzees and gorillas, use suspension and not brachiation. Nevertheless, as it is common to use "brachiation" for suspensory locomotion, we will use that term as well. Russell Tuttle and John Basmajian (1974) preferred to refer to an "arboreal behavior"—which includes brachiation and climbing—as the trait from which the changes that would eventually lead to different types of locomotion in humans and apes are derived.

noted that those two species are the largest of their respective families, although he did not determine whether the large size is due to the folivorous diet or the other way around.

If we accept Andrews' (1981) conclusions, African great apes could have developed terrestrial quadrupedalism while retaining the trait of thin dental enamel, generally considered primitive (Andrews & Cronin, 1982). However, some authors, such as Lawrence Martin (1985) and Marc Verhaegen (1996), argue that the primitive trait is thick enamel.

A cautious approach is to establish taxonomies and evolutionary relationships based on several morphological characters, rather than just one. It does not make sense to draw phylogenies based solely on the trait of enamel thickness. This trait has an undisputable importance when classifying particular specimens, especially in instances in which there are only mandibular fragments available for identifying a species or even a genus. But progress in fossil extraction techniques has made more postcranial remains available. Moreover, advances in paleoecology have provided very relevant information. Three kinds of evidence have usually been taken into account in order to determine

taxonomic and phylogenetic relationships: dentition, locomotion, and ecological context. Explanations for the appearance of a new adaptive lineage should be grounded on a comprehensive study of, at least, these three features.

If we do so, could we obtain a cladogram of Miocene hominoid lineages? There are many proposals but they are always subject, inevitably, to the difficulties of cladistic methodology—very sensitive, as we know, to taxa selection and the allocation of the different characters to the chosen taxa.

The detailed study of John Finarelli and William Clyde (2004) obtained four most parsimonious cladograms from 18 Miocene hominoid taxa (including apes). Based on these four most parsimonious cladograms, Finarelli and Clyde (2004) prepared a strict consensus cladogram, in which the topological features common to the four previous cladograms were highlighted (Figure 3.1).

The result shows the presence of two distinct clades of hominoids that emerged from *Proconsul*, the sister group of all other apes: a group of "archaic appearance" and another of "modern appearance," according to the terminology used by Finarelli and Clyde (2004).

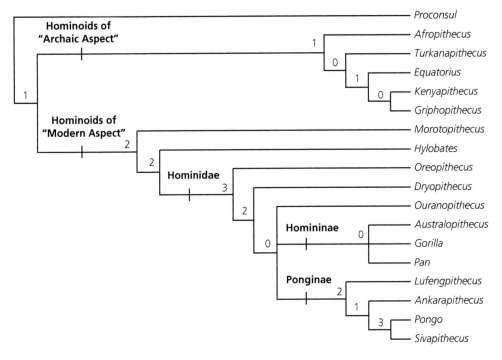

Figure 3.1 Strict consensus cladogram of the morphologically most parsimonious cladograms. Finarelli, J. A. & Clyde, W. C. (2004). Reassessing hominoid phylogeny: evaluating congruence in the morphological and temporal data. *Paleobiology*, 30, 614–651.

Box 3.2 Stratocladistics

Stratocladistics evaluates opposite phylogenetic hypotheses using a parsimony criterion that incorporates morphological characters as well as the temporal sequence of the fossil taxa stratigraphic record. Morphological characters are scored in the usual cladistics way, penalizing the resulting homoplasies in each cladogram (the more homoplasies, the less parsimonious the results). The time sequence is scored by assigning a negative point to each inconsistency between the results of applying the phylogenetic hypothesis and the stratigraphic record evidence (the more negative points, the less parsimonious the sequence). In this way it can be avoided that the cladograms, for reasons of morphological consideration, yield phylogenies that are very unlikely given the known sequence of the appearance of the taxa (see Finarelli & Clyde, 2004). It will not escape the reader's attention, that stratocladistics reincorporates the temporal dimension included in Hennig's original proposal, which was lost in reformed cladistics (see Chapter 1). Stratocladistics produces a graph that might be called a phylogenetic tree, but with greater confidence derived from the parsimony criterion than the phylogenetic trees based on pure common sense.

But the inclusion within the clade of archaic-looking apes of taxa with features to some extent derived, such as *Kenyapithecus* (or its equivalent *Equatorius*, which the authors divided into two distinct genera), raises some concerns.

The problem is, as mentioned, inherent to cladistic methods. Finarelli and Clyde (2004) went further, carrying out a morphological and stratocladistic (see Box 3.2) study—which adds time to the presence, or not, of each trait—in order to obtain a phylogenetic tree subjected to the parsimony criterion.

The most parsimonious stratocladogram—or phylogenetic tree—obtained by Finarelli and Clyde (2004) can be seen in Figure 3.2.

As can be seen, the tree follows the sequence of Miocene hominoids, but is interrupted at the taxon *Kenyapithecus*. This is due to difficulties associated with this genus and its phylogenetic interpretation. A greater problem is that the tree in Figure 3.2 positions *Pan* and *Gorilla* as sister groups, relative to *Australopithecus*, contrary to the evidence obtained by immunological

methods (see Chapter 2). As we saw there, such a solution is not acceptable: correct lineages group together humans and chimpanzees, separated from the gorillas.

To illustrate this, and following the simplified model of Kieran McNulty (2010), in Figure 3.3 we present a phylogenetic tree not subjected to any algorithmic procedure or parsimony criterion—therefore, less reliable—but which reflects in general terms the evolutionary sequence of the Miocene ape taxa.

The lineages obtained either by traditional cladistic or stratocladistic methods correspond in all cases to traits analyzed in Miocene apes. The description of a very complete hominin fossil, *Ardipithecus ramidus* (White, Asfaw et al., 2009), has allowed evaluation of the separation process of humans and African apes according to the polarity—apomorphic or plesiomorphic characters—in traits of the locomotor system. In Chapter 4, when dealing with Miocene hominins, we will return to this subject.

3.1.4 *East Side Story*: the separation of African apes and hominins

Where and how did the evolution leading to gorillas, chimpanzees, and humans take place? Jean Chaline and colleagues (1996) have noted that we still ignore how the divergence process among the three lineages occurred, but they believe that it was associated with the Upper-Miocene climatic change, which took place between 6 and 5.3 Ma. Chaline and colleagues propose a correspondence between evolutionary processes and ecological adaptation, similar to the one put forward by Yves Coppens (1994). The determining factor may have been geographical separation:

- pre-gorillas to the north of the barrier constituted by the Zaire River;
- pre-australopithecines in the eastern area of the African Rift;
- pre-chimpanzees in the center and west of Africa.

Under this scenario, the derived traits of each branch were fixed allopatrically, separately in each region (Figure 3.4). The allopatric speciation model would explain the absence of chimpanzee ancestors in hominid sites. However, the discovery of a Middle Pleistocene fossil chimpanzee specimen within the Rift Valley, in Tugen Hills (Kenya), has challenged this model of allopatric speciation, as we will see in a subsequent chapter.

The idea of linking hominoid evolutionary divergence and Upper Miocene climatic changes is both reasonable and interesting. It is another thing to say that this idea is sufficiently supported by available

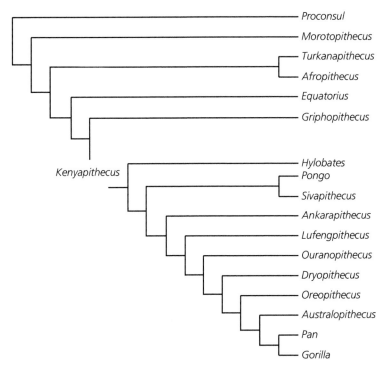

Figure 3.2 The most parsimonious phylogenetic tree recovered in the stratocladistic analysis. Finarelli, J. A. & Clyde, W. C. (2004). Reassessing hominoid phylogeny: evaluating congruence in the morphological and temporal data. *Paleobiology*, 30, 614–651.

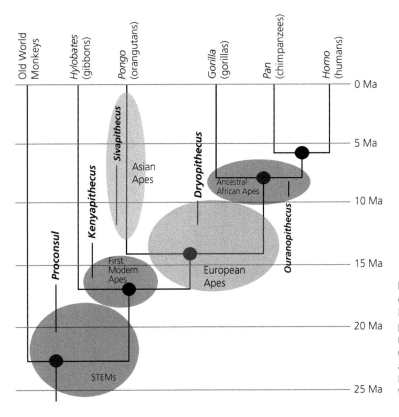

Figure 3.3 Phylogenetic tree of the different Miocene hominoid genera. Vertical lines do not indicate a timeline, but the placement of each genus higher or lower reflects their relative age. McNulty, K. P. (2010). Apes and Tricksters: The Evolution and Diversification of Humans' Closest Relatives. *Evo Edu Outreach*, 3, 322–332. With permission of Springer.

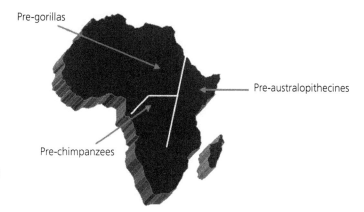

Figure 3.4 The hypothesis of the allopatric separation of chimpanzee, gorilla, and australopithecine ancestors attributed to geographical barriers (Rift Valley and Zaire River) (after Coppens 1994).

evidence to go beyond a hypothetical model. Isotopic analysis of paleoenvironments in the Kenya Rift Valley by John Kingston, Bruno Marino, and Andrew Hill (1994) led to the conclusion that savannas did not prevail at all in these places during the last 15 million years. Therefore, no significant change in this environment occurred during this period (whatever the global climate vagaries around the planet). If so, this means that neither during the Middle nor the Upper Miocene was there an overall expansion of open savanna in East Africa. Thus, there was not in this region a very different ecological situation from that in the tropical forests of North and West Africa.

The Miocene climatic conditions probably led to very diverse habitats in all sub-Saharan Africa. The divergence of hominoids is not readily related to a certain climatic episode, unless the geographical location under consideration can be determined with great precision. In later chapters dealing with the paleoclimatology of early hominid sites, we will review some of the relevant arguments. But, because we know nothing about the direct ancestors of chimpanzees and gorillas, or about the places in which they might have lived, the plausibility of an allopatric speciation process remains hypothetical.

3.2 A scenario for the human evolution

In Chapter 2 we considered hominin evolution, establishing four different types of adaptive strategy, which correspond to four different genera in agreement with the genus as an adaptive criterion of Ernst Mayr (1950).

However, the discovery of additional hominin fossils has led to the proposal of many more genera, and of many species within each. Table 3.2 provides a summary of the most common currently accepted genera and species.

It is generally accepted that tribe Hominini emerged in the Upper

Miocene in Africa; in particular, there are two enclaves that have provided the oldest known specimens: Chad, Central Africa, and the Rift Valley, in East Africa (Figure 3.5).

Thousands of miles to the south, South Africa is another key area where a further stage of hominin evolution took place. South African sites are not as old as those in East and Central Africa, but having been previously excavated, they provided the first ideas about the phylogeny of our lineage. The possibility of drawing comparisons among South African specimens, from Tchad and the Rift Valley, is an advantage when attempting to understand hominization, although occasionally these comparisons raise new questions. The appearance of very ancient hominin remains in such remote areas of Africa shows that from the very beginning members of the human lineage traveled great distances.

3.2.1 Population dispersals

The dispersal of hominins, which occurred since the emergence of the tribe in the Upper Miocene, became far-reaching at the end of the Pliocene when hominins went out of Africa to occupy Asia and, subsequently, Europe. What were the events leading to such migrations?

Jan van der Made and Ana Mateos (2010) have answered this question, considering the excess of births that occur in every healthy population with respect to the number required to reproduce its initial size.

Table 3.2 Genera and species belonging to the tribe Hominini suggested by different authors

Genus	Species	Age (million years)
Sahelanthropus	Sahelanthropus tchadensis	7
Orrorin	Orrorin tugenensis	6
Ardipithecus	Ardipithecus kadabba	5.8
	Ardipithecus ramidus	4.4
Australopithecus	Australopithecus anamensis	4
	Australopithecus africanus	3.5
	Australopithecus bahrelgazhali	3.5?
	Australopithecus garhi	2.5
Paranthropus	Paranthropus africanus	3.5?
	Paranthropus aethiopicus	2.5
	Paranthropus robustus	2.0
	Paranthropus boisei	1.7
Kenyanthropus	Kenyanthropus platyops	3.5
	Kenyanthropus rudolfensis	2.5
Homo	Homo habilis	2.5
	Homo ergaster	1.8
	Homo georgicus	1.8
	Homo erectus	1.6?
	Homo floresiensis	?
	Homo antecessor	0.8
	Homo heidelbergensis	0.4
	Homo neanderthalensis	0.3
	Homo sapiens	0.2

NB: The table does not include all species that have been named.

This is, actually, the same starting point used by Darwin to propose the mechanism of natural selection. More individuals are born than those who manage to reproduce; some of them survive and others do not, so natural selection is based on the idea of the survival of the fittest. In population terms this does not mean that the population maintains its number; rather, in fact, a continuous increase usually occurs until it reaches the so-called carrying capacity of the ecosystem, the maximum number of individuals that the available resources of a certain location can sustain (Box 3.6). As Edward Wilson (1975) stated, in a population whose birth rate exceeds its death rate, growth occurs that first appears as an exponential curve. However, as population numbers increase, members will experience various difficulties—endocrine exhaustion,

nesting problems, shortage of food resources, emergence of diseases linked to overcrowding, among others—leading to an equilibrium between birth (tn) and death (tm) rates (Figure 3.6).

A birth rate in excess of the death rate can return the number of individuals close to K (Box 3.3) after a catastrophic event has reduced the population. But under normal conditions, an excess of births means that as the population approaches K, significant migratory pressure builds, particularly on occupied territorial borders.

The main factors within a specific territory that determine its carrying capacity are climatic. Jan van der Made (1992) focused on temperature and humidity as parameters that tend to establish areas—most of them in the form of East–West belts—which adjust the distribution of animals. Each species' niche is very sensitive to small changes in temperature and humidity, and lead to rapid and extensive dispersions and contractions. For van der Made (1992), high dispersion ratios in mammals—on the order of 10 to 25 km/year—could even be the norm in species with low reproductive rates.

William Leonard and Marcia Robertson (2000) have proposed an ecomorphological model for interpreting dispersals from the areas of occupation (home range) of current human hunter-gatherer groups and chimpanzees (Box 3.4). Their results suggest that the area of occupation is between 15 and 100 times larger for human hunter-gatherers: 23 hectares for chimpanzees compared to 2,600 for hunter-gatherers (Leonard & Robertson, 2000). This fact indicates that humans engaged in foraging and hunting over a much larger area than our closest relatives, which in turn implies a tendency toward dispersal and search for new territories.

The capacity to occupy new spaces is expressed by the diffusion coefficient (D), which links the invaded area (z), the time required to occupy it (t), and the population growth rate (r). The equation relating these parameters (Anton Leonard & Robertson, 2002) is:

$$D^{1/2} = z/(t) \, (2r1/2).$$

None of the variables required to evaluate the efficiency of colonizing new areas can be accurately measured in the fossil record. The occupied area, as well as the time needed for colonization, depends on finding a few fossils, which illustrate like "sporadic flashes" a process whose image remains necessarily incomplete. And, of course, the population growth rate of any fossil taxon is completely beyond our reach. The study by Susan Antón, William Leonard, and Marcia Robertson (2002) about the dispersal process of *Homo erectus*—a process that we will see in detail in the next

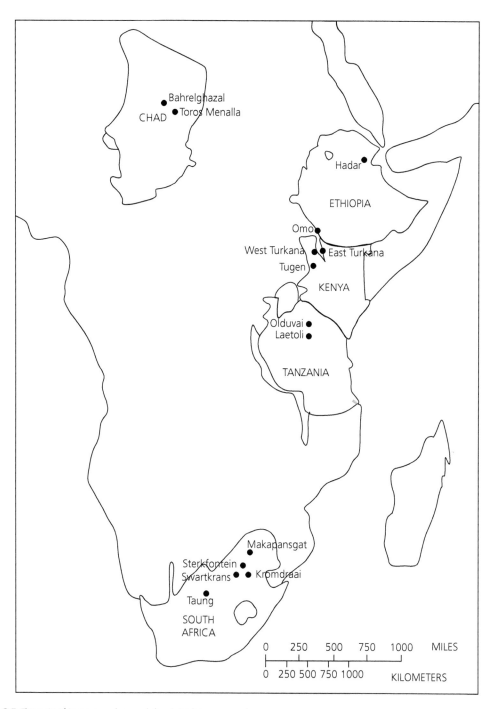

Figure 3.5 The main African areas where early hominins have appeared.

chapters—obtained the parameters by comparing them with the corresponding nonhuman primate fossils, which carried out known dispersals, and using as a reference the current closest primate to each of the fossil species.

The population expansion from the Miocene to the Pleistocene doesn't depend only on the intrinsic characteristics of the species; there are external parameters that affect the entire fauna. Hominin migration processes that led to the occupation of Asia, and later

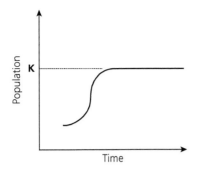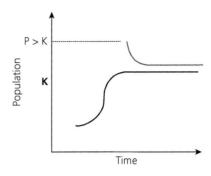

Figure 3.6 Left: Logistic curve showing incremental growth of population size until reaching the carrying capacity K. Right: Decrease of population size when this has exceeded K. The upper line represents a population size above K, which will quickly decrease until K is reached.

Box 3.3 Carrying capacity

The population at the point of equilibrium is called carrying capacity, normally represented by the letter K, and is specific for any given species, as well as for any particular ecosystem they occupy. Generally, all populations reach the equilibrium point through a logistic curve, with a first stage of rapid growth—tn > tm—until rates become equal and K is reached. If the population exceeds K—because its components increase beyond that point or because resources in the ecosystem decrease—then birth and death rates reverse—tm > tn—resulting quickly in a new K level.

Box 3.4 Home range

The home-range concept was proposed by Glynn Isaac as a *home base* or *central base foraging model*, based precisely on the contrast between current populations of hunter-gatherers and living nonhuman primates (Isaac, 1978). A critique and review of the model was done by James Steele and John Gowlett (1994), paying attention to issues such as intrasexual competition and sex "dispersal" as a mechanism to avoid inbreeding problems in the gene pool. Although these are very interesting questions, here we limit the concept of "dispersal" to the migration processes that involve the movement of a significant part of the population, leaving aside those cases of individuals changing groups for reproductive reasons.

Europe, have generally been seen as another episode in the dispersal of large mammals like *Equus, Hippopotamus, Theropithecus, Xenocyon*, or *Megantereon*. A great migration took all of them out of Africa around 1.8 Ma, and could have been the cause of the hominins' exit from the continent (Turner, 1992; Martínez Navarro & Palmqvist, 1995; Brantingham, 1998; Strait & Wood, 1999; O'Regan et al., 2011). In any case, we might want to ask, What is the reason for population movements that affect wildlife as a whole?

3.2.2 Climate and biogeography

The carrying capacity of an ecosystem depends closely on climatic and biogeographic conditions: both global and local tectonic changes, a tendency in the long run toward planetary cooling, and sporadic episodes of climate change with warm and humid stages alternating with cold and arid periods (Trauth et al., 2007). All these factors are related, because the climate is dependent on geographical conditions. But

this interrelationship is dynamic: landscapes change over time.

Specific periods in the Rift have been studied by Martin Trauth and collaborators (2007) analyzing lake sediments, which can establish the succession of humid and arid periods. This method is particularly interesting in narrowing down the climatic conditions affecting hominin dispersals, for at least two reasons: because lake basins coincide in many cases with fossil origin sites and because it is at the times when extension of lakes is greater that migration opportunities appear, at least before the episode of climate change ≈2.5 Ma; however, the concept of "lake" should be clarified. There are significant ecosystem differences when we refer to small, shallow alkaline lakes—of longer temporal presence—or to deeper and wider lakes. The latter coincide with periods of humid and warm climate. Table 3.3 indicates, again by means of a temporal sequence, the most significant localities in Ethiopia (Afar), Kenya (Turkana, Rift Central), and Tanzania

Table 3.3 Rift localities with deeper—in bold—and shallower lacustrine episodes

Ma	Afar and Ethiopian Rift	Turkana and Omo	Central Kenya Rift and Tanzania	South Kenya and Tanzania
1.1–0.9			Olorgesailiie	**Kariandusi** Natron
2.0–1.7	Konso Dik Hill Yager Dal	**Koobi Fora** Shungura	Olduvai	**Gicheru**
2.7–2.5		**Gadeb**		Chemeron
3.2–2.95	**Hadar**	Allia		
3.4–3.3	**Hadar**	Lokochot		
4.17–3.9	**Bodo**	Kanapoi Allia		
4.7–4.3	Gona			

Data from Trauth et al. (2007).

where the hominins' presence is associated with lacustrine episodes—deeper lakes appear in bold.

Figure 3.7 includes a summary of the study of Trauth et al. (2007), which related, by a temporal sequence, global climate transitions, the presence of deep and wide lakes in East Africa, and major events in the hominin evolution.

There are several conclusions to be drawn from the work of Trauth et al. (2007). First, global climate changes coincide with times in which hominin evolution goes through significant episodes. The global changes indicated by Trauth et al. (2007) correspond to the great glaciation c. 2.5 Ma, the establishment 2.0 Ma ago of the "Walker Circulation"—seasonality that affects, for example, the monsoonal regime—and the "Revolution of Middle Pleistocene"—beginning of the great glaciations—in the last million years.

The rate of climatic change caused by glaciation, far from the area covered by ice, could be explained by changes in ocean currents. This hypothesis, generated by a computer simulation model, has been suggested by Andrey Ganopolski and Stephan Rahmstorf (2001).

The changes indicated coincide with three moments in human evolution, which we will discuss later in detail: the emergence of the genera *Homo* and *Paranthropus*, coinciding with the opening of large extensions of open savanna in Africa (c. 2.5 Ma); the hominin exit out of Africa (c. 1.8 Ma); and the extinction of human genera other than *Homo*, together with the expansion of the latter through Eurasia.

3.2.3 The hypothesis of adaptation to the open savanna

As we have just indicated, the cladistics separation of *Paranthropus* and *Homo* in the Pliocene coincides with a period of global cooling on the planet, which in Africa transformed extensions of previously tropical forests into savannas and open prairies. This coincidence between open spaces and human evolution was explained decades ago by the hypothesis of adaptation to the savanna, supported among others by authors such as Elizabeth Vrba (1974, 1980, 1984,1985) and Phillip Tobias (1985a, 1991b). The hypothesis maintains that there was no mere correlation: Pliocene environmental change was the trigger that led hominins to split into two lineages. Many of the mammals of the African savanna, which appeared in the identifiable boundary between the Miocene and Pliocene, would have emerged in response to the great open spaces. They form the so-called Villafranchian fauna (Box 3.6). Hominins would be one of the many mammals that took advantage of the new open spaces. So, Elisabeth Vrba (1985) considered them both "founders" of the African savanna biota and an endemic part of its own.

The climatic transition can be dated in South Africa around 2.5–2.0 Ma. With regard to East Africa, there is a change of vegetation indicated by palynology (study of pollen) in the sedimentary area of Omo from the Shungura formation (Ethiopia). In this case, the range can be measured more accurately: between 2.52 and 2.4 Ma (Brown, McDougall et al., 1985). Anna Behrensmeyer and collaborators (1997) documented that in the Lake Turkana basin, between 58 and 77% of mammal species were replaced between 3.0 and 1.8 Ma. The most significant changes in the fauna occurred after 2.5 Ma. René Bobe and Anna Behrensmeyer (2004) confirmed, with studies of the Omo fauna, the great transformation of 2.5 Ma.

Elizabeth Vrba (1974) has suggested how the evolution of early hominins in the savanna would have occurred, relating this phenomenon to evidence from the evolution and adaptation of bovids (family Bovidae). The bovids are the most numerous large mammals among fossils that appear in African sites from the Miocene, Pliocene, and Pleistocene; thus, their presence over time is well documented. In addition, bovids and hominins share a number of notable features: large bodies, mobility, herbivorous diet (at least in large part), and the condition of being endemic members of the savanna. These correlations imply that the rhythms of speciation, morphological change, and extinction coincide between hominins and antelopes (Vrba, 1974, 1984).

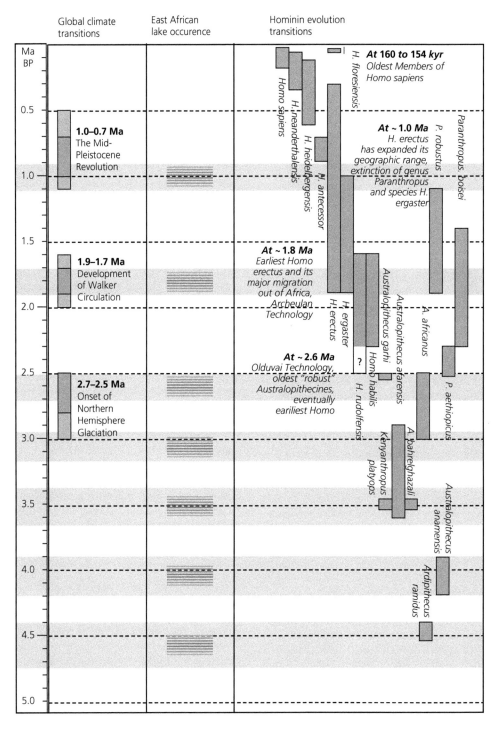

Figure 3.7 Correlation between global climate changes, presence of large lakes in East Africa, and episodes of hominin evolution. Reprinted from Trauth, M. H., Maslin, M. A., Deino, A. L., Strecker, M. R., Bergner, A. G. N., & Dühnforth, M. (2007). High- and low-latitude forcing of Plio-Pleistocene East African climate and human evolution. *Journal of Human Evolution*, 53(5), 475–486, with permission from Elsevier.

Box 3.5 The emergence of the savannas

The Pliocene climate change episode in the Rift, along with tectonic movements, gave the chain of faults its current form. The ridge that closes the Rift on the west side, full of clouds coming from the Atlantic Ocean, is a barrier to atmospheric depressions and marks as well a boundary between the tropical forest and the drier eastern lands. As a consequence of both phenomena, the drop in temperature and the climatic shield, the Rift was covered with the savannas that we see today.

Box 3.6 Villafranchian fauna

The Villafranchian fauna was described in terms of its constituent European animals, although it also appears in other continents. It persisted throughout the Upper Pliocene and the Lower Pleistocene, thus spanning the transition from one age to the other. Among the mammals of the Villafranchian fauna are the genera *Ursus* (bears), *Elephas* (elephants), *Hypperion* (horses), the large savanna ungulates and their predators, such as *Crocuta* (hyenas) and *Smilodon* or *Homotherium* (saber toothed tigers).

According to Elisabeth Vrba, this is not a mere statistical correlation. This is the result of an extensive ecological analogy. Thus, the study of bovine evolutionary patterns (easier to consider due to the abundance of fossils) could be used to generate hypotheses about hominin evolutionary patterns. A study on bovid distribution in 16 different areas of sub-Saharan Africa, conducted by a correspondence analysis, allowed Vrba to establish certain adaptive patterns by grouping the bovid taxa by their presence in the biota. A consistent association appeared between the set of two bovid tribes, Antilopini and Alcelaphini, and open savannas (Vrba, 1985; the original study is of 1980). From that established association, it is possible to use the fossil record to determine what kind of vegetation existed in a particular time and place, simply by confirming the presence of "Antilopini + Alcelaphini." The next step taken by Vrba (1982) was to confirm the presence of the set of these two bovid tribes in different hominin sites, which took place in the South African locations: Makapansgat, Sterkfontein, Taung, Swartkrans, and Kromdraai. Leaving aside the Kromdraai site, with a small presence of bovids, the results indicate a significant difference between Makapansgat members 3 and Sterkfontein 4, both with a low presence of "Antilopini + Alcelaphini," and Sterkfontein members 5 and Swartkrans 1, where these are abundant. For the Vrba theory to be consistent, vegetation on the first two sites mentioned would have been typical of a tropical forest, while in the last two sites the vegetation would have already transformed into open savanna.

With regard to Africa, the climatic zones we see now correspond roughly with those existing since the end of the Miocene, although changes in their extension were subject to the planetary heating and cooling processes we have mentioned. Different authors have studied the situation corresponding to the Pliocene and Pleistocene; the following synthesis comes from Peter deMenocal (2004) and Jan van der Made (2011):

An arid belt stretches from the Sahara to Central Asia (Gobi), extending southward into East Africa. This belt includes large areas of open savanna and prairies, especially since the great glaciation that occurred during the Pliocene, c. 2.5 Ma. However, open habitats have existed since the Lower Miocene in places like Turkey.

North and south of the arid belt appear humid areas of dense forest, although local conditions can lead to the presence of a habitat mosaic, which includes open and arid environments mixed with the dense and humid.

Considering together the African wildlife and paleoclimatic records, three intervals appear—2.9 to 2.4, 1.8 to 1.6, and 1.2 to 0.8 Ma—in which the climate undergoes change toward more open and varied habitats. In the long term, the fossil record indicates a greater abundance of taxa adapted to arid climate.

Figure 3.8 provides a graph relating climatic change cycles to cultural traditions, hominin speciation processes, open forest to savanna transformations, and glacial and interglacial periods.

The panorama that appears can be summarized as a long-term trend toward more arid conditions—evidenced by the presence in the fossil record of a fauna with a greater abundance of taxa adapted to drier climate—although as part of a succession of very changeable cycles. Periods of ≈20,000 years of warm and humid climate vs. cold and arid that occur between 4 and 3 Ma became 40,000 years in the range between 3 and 1 Ma, and 100,000 years thereafter. In turn, the climatic alternatives became more extreme, with greater differences between glacial and interglacial conditions.

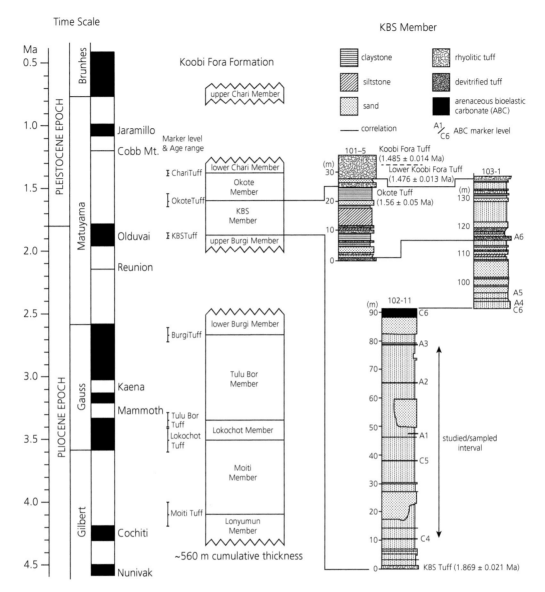

Figure 3.8 Diagram of the climatic and phylogenetic events accompanying hominin evolution in the last 4.5 Ma. From left to right, columns indicate the elapsed time between climatic alternatives of dry and humid conditions, cultural sequence, hominins' taxa, dense forests to open prairie transformations, and the succession of glacial/interglacial conditions indicated by the ratio of oxygen isotopes (see Chapter 8). Reprinted from deMenocal, P. B. (2004). African climate change and faunal evolution during the Pliocene-Pleistocene. *Earth and Planetary Science Letters (Frontiers)*, 220, 3–24, with permission from Elsevier.

Hominin dispersals are particular cases within a dynamic landscape, in which some periods were easier for migration, while in others the arid zone from the Sahara to the Gobi was an obstacle to population movements. That obstacle, as well as those represented by seas and high mountain ranges, established

"corridors," which hominins entered at different times of their evolution.

We will examine the particular geoclimatic and ecological conditions when addressing each of the major dispersions. Identifying these will force us to describe, first, the setting of their evolution: those fossil sites where specimens have been found that are able to reveal the phylogenetic events of the Miocene and Pliocene hominins.

3.3 The Rift Valley site

The Rift Valley is a long and narrow fracture depression, surrounded by faults and the presence of numerous volcanoes, extending along c. 5,000 km from South Turkey to Malawi, as well as from East Africa to Mozambique. It was formed about 30 Ma, in the Oligocene, because of the separation movement of the African and Arabian tectonic plates. From a geological perspective, the Jordan Valley and Syria are part of the Rift.

The Rift of East Africa comprises a discontinuous succession of valleys, running for 3,000 km from the region of Afar (Ethiopia) in the North, to the South of Malawi (Figure 3.9).

3.3.1 Hadar region; Hadar formation

From north to south (Figure 3.9), the first fossil area of importance is Hadar. The Hadar formation is located in the Afar Triangle, a vast desert area around the Awash River, about 300 km northeast of Addis Ababa, the Ethiopian capital. The site area that interests us encompasses more than 60 km², where, between 1972 and 1977, intensive paleontological work was carried out by the International Afar Research Expedition organized by Maurice Taieb. The paleontological research provided numerous remains of Pliocene mammals (elephants, pigs, cercopithecoid monkeys; up to 6,000 specimens of 73 species), in excellent states of preservation (Johanson & Taieb, 1976; Taieb et al., 1976; Johanson & White, 1979). However, the dating of the remains was somewhat problematic due to the geological history of the terrain. Hadar is an area with numerous crisscrossed ravines, faults, and thrusts, so that strata correlations are not obvious.

In order to give as clear a picture as possible, we will only explain the chronology of the three members of the Hadar formation in which there are hominin remains above the barren basal member. These are Kada Hadar (KH), Denen Dora (DD), and Sidi Hakoma (SH), listed by increasing age (Figure 3.10).

All the three members, KH, DH, and SH, have about 180–280 m of sedimentary deposits, depending on the zones. Sediments are of lacustrine origin, riparian—lake margin—and fluvial, intermixed with volcanic tuffs (Taieb et al., 1976). The hominin remains are concentrated in three main groups: one in the lower zone of the Sidi Hakoma member, one in the transition zone between Denen Dora and Kada Hadar, and the third in the upper zone of the latter member.

Aronson and collaborators (1977) dated the members of the Hadar formation above the basal member between 2.6 and 3.3 Ma, using geochronological, paleomagnetical, and biostratigraphical evidence. Robert Walter and James Aronson (1982, 1993) conducted subsequent revisions but, as these authors indicated, only BKT-2 deposits, in the upper-half of KH and stratigraphically above all levels containing remains

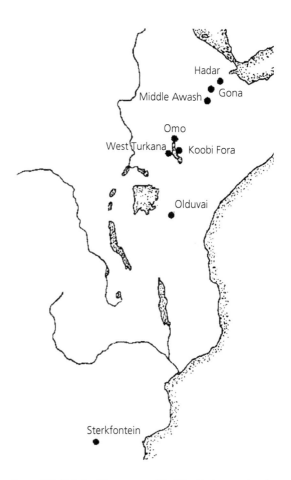

Figure 3.9 Main fossiliferous areas of the Rift with sites mentioned in this book. The position of the Sterkfontein's site (South Africa) is also included on the map in order to illustrate the distance between areas that have supplied hominin fossils in the Rift and South Africa. Plummer, T. (2004). Flaked Stones and Old Bones: Biological and Cultural Evolution at the Dawn of Technology. *Yearbook of Physical Anthropology*, 47, 118–164.

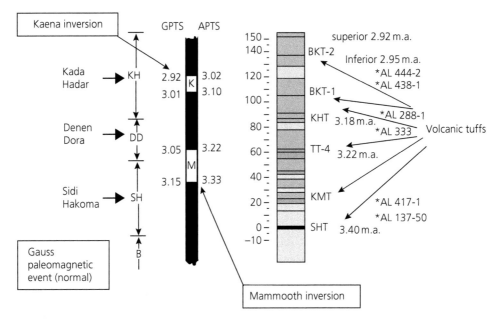

Figure 3.10 Kada Hadar, KH; Denen Dora, DD; and Sidi Hakoma, SH: members of the Hadar formation (Walter & Aronson, 1993, as amended).

of *Australopithecus*, have been dated by the conventional system K/Ar (potassium/argon) between 2.8 and 3.1 Ma. Most of the other volcanic terrains present problems for dating them. However, thanks to the use of the technique of ^{40}Ar/^{39}Ar, Walter and Aronson (1993) were able for the first time to establish directly the age of the Sidi Hakoma tuff (SHT), at 3.4 Ma. The review conducted by William Kimbel et al. (1996), to establish the age of BKT-3 deposits in the upper end of KH, confirmed those dates, providing a very complete stratigraphy of the three members of the site containing hominin remains. Thanks to this accumulation of research, the Hadar findings can be dated quite accurately.

Specimens of the Hadar formation will be described in Chapter 4.

3.3.2 Other localities of the Hadar region

The town of Dikika belongs partially to the basal member of the Hadar formation (therefore, older than 3.4 Ma) (Alemseged et al., 2005); the other part corresponds to the Sidi Hakoma member (Alemseged et al., 2006). Remains of hominins and bones with marks of having been manipulated with stones have been found in Dikika. The terrain stratigraphic study in which the latter were found suggested an age older than 3.39 Ma (McPherron et al., 2010).

The Woranso-Mille site is one of the fossiliferous areas that are part of the WORMILPRP (Woranso-Mille Paleontological Research Project). It is located in the western part of the central Afar depression (Ethiopia) and the terrains have been attributed to the Hadar formation, although there are suggestions that the site may belong to a new, unnamed formation (Deino et al., 2010).

Five different locations in Woranso-Mille (Figure 3.11) have provided fossils of hominins. The first four, Am-Ado (AMA), Aralee Issie (ARI), Mesgid Dora (MSD), and Makah Mera (MKM), form sedimentary deposits between the Kilaytoli tuff and the WM-W-1 basalt layer, dated by ^{40}Ar/^{39}Ar radiometry at 3.570±0.014 and 3.82±0.18 Ma (Haile-Selassie, Saylor et al., 2010). In all these locations, dental material has also been found. In the Korsi Dora location (KSD), inside a block, KSD-VP-1, isolated by a fault that prevented its relation to the rest of the area, a partial skeleton was discovered. By paleomagnetism and ^{40}Ar/^{39}Ar radiometry, the block KSD-VP-1 was assigned an age of 3.60/3.58 Ma (Haile-Selassie, Latimer et al., 2010).

A partial foot with an estimated age c. 3.4 Ma comes from the Burtele (BRT) location, further south of the others mentioned above (Haile-Selassie et al., 2012). Several masticatory elements with c. 3.4 Ma too come from the Waytaleyta (WYT) location, close to Burtele (Haile-Selassie et al., 2015).

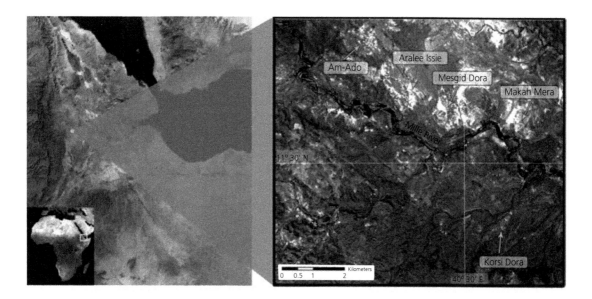

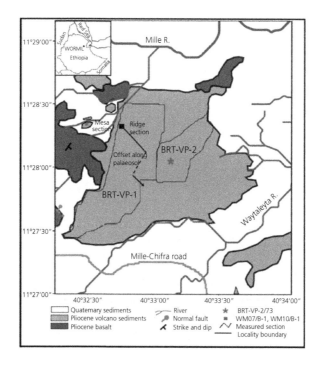

Figure 3.11 Woranso-Mille locations (Ethiopia) with fossils of hominins. Above left: Woranso-Mille area in Ethiopia. Above right: Am-Ado, Aralee Issie, Mesgid Dora, Dora Makah Mera, and Korsi locations (Haile-Selassie, Latimer et al., 2010, supplemental information). Below: Burtele location, which includes the BRT-VP-2 area where the BRT-VP-2/73 foot was found (Haile-Selassie et al., 2012). Waytaleyta location is not displayed, but it is placed about 1 km south of Burtele. Top: Haile-Selassie, Y., Latimer, B. M., Alene, M., Deino, A. L., Gibert, L., Melillo, S. M., . . . Lovejoy, C. O. (2010). An early Australopithecus afarensis postcranium from Woranso-Mille, Ethiopia. *Proceedings of the National Academy of Sciences*, 107(27), 12121–12126. Bottom: Reprinted by permission from Macmillan Publishers Ltd: Haile-Selassie, Y., Saylor, B. Z., Deino, A., Levin, N. E., Alene, M., & Latimer, B. M. (2012). A new hominin foot from Ethiopia shows multiple Pliocene bipedal adaptations. *Nature*, 483(7391), 565–569.

The Woranso-Mille specimens will be described in Chapter 4.

3.3.3 Middle Awash region

Located in the southern part of the triangle that forms the Afar Rift, that is, between Hadar—more to the North—and Omo—to the South—is the region of Middle Awash (Ethiopia), a wide area that by the beginning of the twenty-first century had provided more than 10,000 fossils, including hominins as remarkable as *Ardipithecus*, which we saw in Chapter 2. All listed formations have been dated by $^{40}Ar/^{39}Ar$ methods and paleomagnetism (Figure 3.12).

The hominins examined in this book came from the following formations, which are part of the Middle Awash:

* Adu Asa formation, with the members:
  Digiba Dora
  Dora Adu
  Asa Koma
  Alayla
  Saitune Dora

Among these, Asa Koma has supplied the oldest hominin fossil (5.57–5.54 Ma, *Ardipithecus kadabba*; see Chapter 4) (WoldeGabriel et al., 2001).

* Sagantole formation, with the members:
  Kuseralee—Amba East site—of the Lower Sagantole formation (>5.2 Ma, *Ardipithecus kadabba*; see Chapter 4) (WoldeGabriel et al., 2001).
  Aramis (4.419–4.416 Ma, *Ardipithecus ramidus*; see Chapter 4) (WoldeGabriel et al., 2009).
  Adgantole (4.2–4.1 Ma, *Australopithecus anamensis*; see Chapter 4) (White, WoldeGabriel et al., 2006).

* Matabaietu formation: Maka site—(3.40–3.39 Ma, *Australopithecus afarensis*; see Chapter 4) (White et al., 1993).

* Bouri formation, with the members:
  Hatayae ("Hata") (2.5 Ma, *Australopithecus garhi*; see Chapter 4) (Asfaw et al., 1999; de Heinzelin et al., 1999).
  Dakanihylo ("Daka") (base of the member, 1.042±0.009 Ma, $^{40}Ar/^{40}Ar$; the whole member has an inverted magnetic polarity, which means

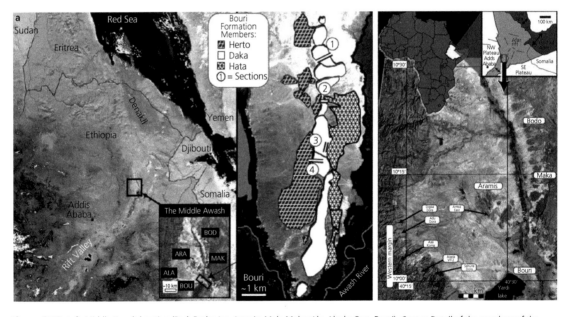

Figure 3.12 Left: Middle Awash location (Bod, Bodo; Ara, Aramis; Mak, Maka; Ala, Alayla; Bou, Bouri). Center: Detail of the members of the Bouri formation, Middle Awash (Ethiopia). The numbers 1, 2, 3, and 4 in circles correspond to stratigraphic columns indicated by the authors in the interpretation paper, not included here (Asfaw et al., 2002). Right: Most important deposits of Middle Awash (Ethiopia): Amba East, Sagantole formation, Digiba Dora, Dora Adu, Asa Koma, Alayla and Saitune Dora, Adu Asa formation. Aramis, Sagantole formation. Hata, Bouri formation. Maka, Bouri formation. Maka, Matabaietu formation. Bodo, Weahietu formation. Map from WoldeGabriel et al. (2001). Reprinted by permission from Macmillan Publishers Ltd: Asfaw, B., Glibert, W. H., Beyene, Y., Hart, W. K., Renne, P. R., WoldeGabriel, G., . . . White, T. D. (2002). Remains of Homo erectus from Bouri, Middle Awash, Ethiopia. *Nature*, 416, 317–320.

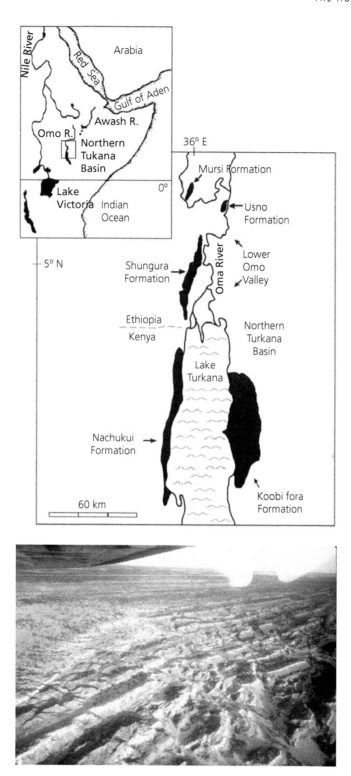

Figure 3.13 Above: Map of the Omo Valley sedimentary area (Bobe & Behrensmeyer, 2004). Below: Aerial view of the Shungura formation (Omo Valley, Ethiopia) (image source in http://www.indiana.edu/~origins/images/ShunguraFM.jpg).

a minimum of ≈0.8 Ma, *Homo ergaster*; see Chapter 8) (Asfaw et al., 2002).

- Weahietu formation:
  Bodo member (≈0.6 Ma, *H. ergaster*/"archaic" *sapiens*; see Chapter 9) (Clark et al., 1994).

Correlation between strata of two different formations is always difficult. Even with the fortunate circumstance that there are common tuffs—as in the case of Tulu Bor, in the Turkana region, Section 3.3.7—the comparative stratigraphy may become impossible to obtain. In spite of this, a fossiliferous region located south of the Afar depression has been used as a reference, due to its geological structure, to correlate members from other locations.

3.3.4 Omo region

The sedimentary area of Omo is located at the southern end of the Omo River Valley, in Ethiopia, just north of Lake Turkana, forming an extension of its geologic basement (Figure 3.13). The first vertebrate fossil of Omo was found in 1902, although it was the expedition of Camille Arambourg 1932–33 that collected a huge amount of fossil remains: up to four tons (Coppens, 1980).

The sedimentary deposits of the lower Omo Valley form a tectonic depression extending northward to the base of Lake Turkana. Up to eight different formations are shown in the region, of which three, Shungura, Usno, and Kibish, have yielded hominin remains, the most notable belonging to the taxon *P. boisei* (≈2.2 Ma; see Chapter 6), derived from the first formation.

The Shungura formation appears on the right bank of the Omo River, in an area of about 200 km², and has an impressive height of over 850 m. It is ideal for establishing chronostratigraphies because of the perfect continuity of sedimentation and the presence of numerous volcanic tuffs (Coppens, 1978a, 1978b). Starting from the basal member, B, C, D, E, F, G, H, I, J, K, and L members are discernable from oldest to youngest, providing an excellent sedimentary sequence from a bit more than 3 Ma to just under 1 Ma (Figure 3.14). Therefore, the Shungura formation serves as a horizon of reference for the correlation of other deposits nearby. Considering that the member of Lothagam Hill, near the left bank of Lake Turkana, contains a succession of strata ranging from the oldest found in Omo to c. 5.5 Ma (Patterson et al., 1970), Pliocene stratigraphy in the area is quite complete with respect to a long period of time.

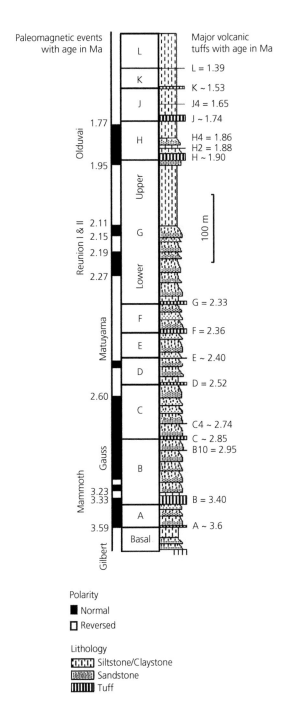

Figure 3.14 Stratigraphy of Shungura formation. Reprinted from Publication title, Bobe, R. & Behrensmeyer, A. K. (2004). The expansion of grassland ecosystems in Africa in relation to mammalian evolution and the origin of the genus Homo. *Palaeogeography, Palaeoclimatology, Palaeoecology*, 207, 399–420, with permission from Elsevier.

Even with comparable geological references, site correlation is often difficult to establish. But the presence of a great fossiliferous abundance at Shungura allowed Yves Coppens to define a number of areas based on association with large mammals, which facilitates comparative studies of faunas (Coppens, 1972; Coppens & Howell, 1976). Omo zoning establishes three different faunal associations. The first, Omo 1, extends from the basal member to the top of member C (3.2 to 2.4 Ma). The second, Omo 2, goes from the base of member C to the G (2.6 to 1.8 Ma). The third, Omo 3, goes from the base member G to the top of the formation (2.0 to 1.0 Ma) (Coppens, 1978b).

Thanks to successive reinterpretations of the volcanic tuff ages and the use of paleomagnetism, it has been possible to carry out a very detailed correlation between the bases of Turkana and Omo. Today, it is reasonable to accept a sequence of the whole area of East Africa that is compatible with biostratigraphic and magnetostratigraphic data, though, of course, with some uncertainty about the boundaries between the different members of the various formations (Howell et al., 1987; Feibel et al., 1989). We will refer to one of them, Wehaietu—from which specimens attributed to African *Homo erectus* came—in Chapter 8.

3.3.5 Other Ethiopian regions: Gadeb formation

The Gadeb formation is located in the Southeast Ethiopian Plateau, at an altitude of 2,300–2,400 m. Belonging to the Gadeb paleolake basin, its sediments contain diatomite, silts, and clays, with sands and ashes intercalated. Immediately above is the Adaba formation, of volcanic origin with ignimbrite intrusions that crown the lacustrine sequence dated at 2.35 Ma, and above that of the Mio Goro formation. Mio Goro has a pumice base of 1.45 Ma (both prior datations by K/Ar method) (Eberz et al., 1988) and continues with new sedimentary deposits of fluvial origin. Gadeb has provided tools from modes 1 and 2, and evidence of the use of fire (see Chapter 7). The archaeological remains belong entirely to fluvial deposits of the Mio Goro formation and, except the Gadeb 2A location, its age ranges between 1.45 Ma (pumice intrusions age) and 0.7 Ma (dated by paleomagnetism) (de la Torre, 2011). Gadeb 2A, the oldest deposit containing lithic tools, is embedded in a flow of sludge with volcanic intrusions dated at 1.48 Ma (de la Torre, 2011).

3.3.6 Other Ethiopian regions: Konso formation

In Ethiopia, located to the East of the Omo region and part of the basin of the Turkana bed, is the Konso formation, with more than 200 m of sediments and in which Surobo, Turoha, Kayle, and Karat members are distinguishable (Katoh et al., 2000) (Figure 3.15). Up to 21 fossil locations, whose age was established by radiometry, stratigraphy, and correlation, provided human specimens.

The localities with hominins findings are KGA10, Kayle member (Suwa et al., 1997), *P. boisei* (see Chapter 6); KGA4, 7, 8, 10, 11, and 12, Kayle and Karat members, *Homo ergaster* (Suwa et al., 2007) (see Chapter 8). The dates for these members have been established by the ^{40}Ar/^{39}Ar method as follows:

- Kayle—*Kayle Tuff* 2 located in the middle part of the member, dated at 1.72±0.03 Ma.
- Karat—BWT (Bright White Tuff) at the middle level of Karat member, has an age of 1.40±0.02 Ma, equivalent to *Chari Tuff* of Koobi Fora (Katoh et al., 2000).

3.3.7 Lake Turkana region

The two shorelines of Lake Turkana have fossiliferous areas of great interest in establishing the patterns of human evolution during the Plio-Pleistocene. These belong to the Koobi Fora formation—East Rudolf, in the colonial era—on the east bank, and the West Turkana area, with the Nachukui formation, on the west bank (Figure 3.16).

No place in East Africa has provided so many hominins as Koobi Fora. During the 1970s, up to 5,000 specimens were found there, some of fairly well-preserved skulls, and since then discoveries have been continuous—the last of them, for the moment, facial materials and mandibular KNM-ER 62000, 60000, and 62003 (Leakey et al., 2012). Such paleontological treasure was linked from the beginning to a member of the Leakey family: Richard Leakey, son of Mary and Louis.

The incorporation of Lake Turkana deposits into the field of human paleontology was the result of chance, according to Richard Leakey (1981b). Flying over the area during a trip to Nairobi, which was diverted due to a storm, Leakey glimpsed the presence of sandstone that was likely to contain fossils, where it was earlier believed there was only volcanic terrain. With a grant from *National Geographic*, and under the auspices of the National Museum of Kenya, Leakey began to carry out his own expeditions to the area in 1968. Very soon discoveries multiplied. The Koobi Fora peninsula, on the east lake shore (Figure 3.16), was particularly fertile but with some drawbacks. In Koobi Fora, the severity of the climate, desertic with sporadic and tumultuous floods, causes a spontaneous emergence of fossils, which unfortunately will quickly disappear if not collected immediately.

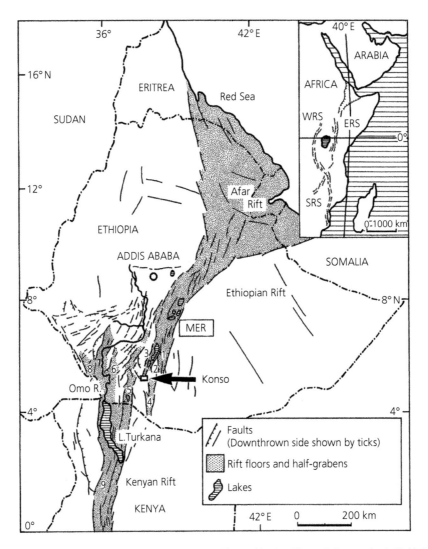

Figure 3.15 Left: Location of Konso formation (Konso-Gardula). Reprinted from Publication title, Katoh, S., Nagaoka, S., WoldeGabriel, G., Renne, P., Snow, M. G., Beyene, Y., & Suwa, G. (2000). Chronostratigraphy and correlation of the Plio-Pleistocene tephra layers of the Konso Formation, southern Main Ethiopian Rift, Ethiopia. *Quaternary Science Reviews*, 19(13), 1305–1317, with permission from Elsevier.

3.3.8 Turkana: Koobi Fora formation

Koobi Fora formation consists of eight members that took the name from the tuff on which they are found (see Box 3.7). Among them we are particularly interested in the Okote, KBS, and Upper Burgi members (Figure 3.17), where the hominin specimens considered in this book came from:

- Okote member (<1.6 Ma, *P. boisei*, see Chapter 6).
- KBS member (c. 1.6/1.7 Ma, *P. boisei*, see Chapter 6; *H. habilis*, see Chapter 6; *H. ergaster*, see Chapter 8).
- Upper Burgi member (c. 1.9 Ma, *P. boisei*, see Chapter 6; *H. habilis*, *H. rudolfensis*, see Chapter 8).

The age of Koobi Fora KBS tuff—which separates the members KBS and Upper Burgi—has been the subject of one of the most prolonged and intense debates of paleoanthropology, since in the terrains located immediately below the tuff, a specimen appeared, KNM-ER 1470, on whose accepted age the model of human evolution prevailing in the 1970s and 1980s largely depended.

In 1969, Kay Behrensmeyer discovered a thin coat of volcanic ash with inset stone tools and hippopotamus fossilized bones, which received the name *Kay Behrensmeyer Site* (KBS). The taphonomic interpretation suggested that a group of hominins had butchered the

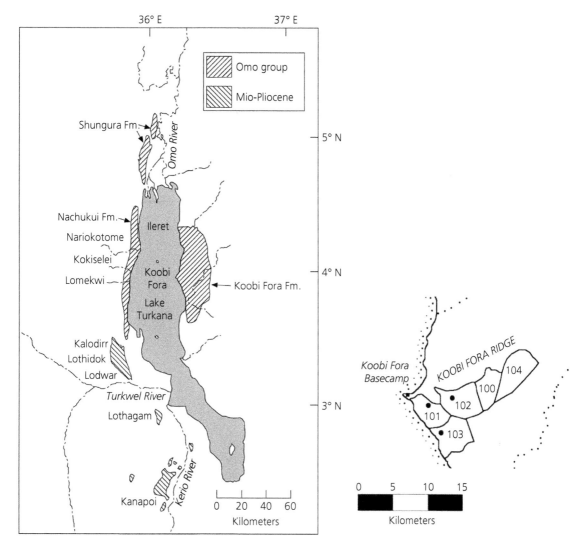

Figure 3.16 Left: Lake Turkana basin with the locations of Koobi Fora and Nachukui formations (indicated by an arrow). The Shungura formation of Omo is located to the north of the lake (Brown's map, Brown & Walker, 2001). Right: Northeastern Turkana basin map indicating the location of areas 100–104 of the Koobi Fora fossil collection. Black circles in the areas indicate 101–5, 102–11, and 103–5 locations, which together form the section-type of KBS member (Lepre & Kent, 2010). Reprinted from Publication title, (left) Brown, B., Brown, F. H., & Walker, A. (2001). New Hominids from the Lake Turkana Basin, Kenya. *Journal of Human Evolution*, 41(1), 29–44, and (right) Lepre, C. J. & Kent, D. V. (2010). New magnetostratigraphy for the Olduvai Subchron in the Koobi Fora Formation, northwest Kenya, with implications for early Homo. *Earth and Planetary Science Letters*, 290, 290, 362–374, with permission from Elsevier.

animal in situ. Obviously, neither the tools nor the fossils could be part of the ash. The coat of ash contained interspersed layers of different deposits, both volcanic and sedimentary. Therefore, it is not unusual that there were difficulties to accurately date the terrain. (An excellent summary of these difficulties, which we have followed partly here, is in MacRae, 1998.)

The first dating of the KBS sample was 200 million years—a clear indication of contamination with ancient sediments. A second dating by Frank John Fitch and Miller (1970) resulted in 2.61±0.26 Ma. That date, later reduced to 2.42 Ma (Fitch et al., 1976), became KBS tuff "official age," although more than 41 subsequent measurements produced very

Box 3.7 Koobi Fora fossils

The discovery of fossils in the terrain surface of Koobi Fora, extracted from the sediment by water currents, may raise doubts about which deposits they come from. There are not reliable stratigraphic references. However, sediment arrangement in the area is almost horizontal and the erosion gorges in which fossils appeared are not very deep. So, even assuming that they come from terrains a bit distant from the place where they were found, dating errors are likely to be small. The closed terrains normally belong to the same sedimentary strata (Walker, 1981).

- Nariokotome (c. 1.6 Ma, *Homo ergaster*, see Chapter 8) (Brown, Harris et al., 1985).
- Kaitio—sites: Kokiselei 4 (1.65 Ma, tools Mode II, see Chapter 7) (Lepre et al., 2011); Kokiselei 5 (c. 1.7 Ma, tools of transition Mode I–Mode II, see Chapter 7) (Lepre et al., 2011).
- Kalochoro—site of Lokalelei (2.34 Ma, tools Mode I, see Chapter 7) (Tiercelin et al., 2010).
- Lomekwi—sites: Lomekwi Drainage (2.50/2.45 Ma, *P. aethiopicus*, see Chapter 6) (Walker et al., 1986); Lomekwi Drainage, LO4 location (c. 3.35 Ma, *A. afarensis*, see Chapter 4) (Brown et al., 2001).
- NK 3 (1.8 Ma, *H. ergaster*, see Chapter 8) (Brown et al., 2001).
- Kataboi (c. 3.5 Ma, *A. platyops*, see Chapter 4) (Leakey et al., 2001).

sparse data, between 223 and 0.91 Ma (Lewin, 1987) (Box 3.8).

Doubts about the radiometric results were clarified when a correlate was established by a trace element (Box 3.9) comparative study between Koobi Fora KBS tuff and Shungura H2 tuff—accurately dated at 1.8 Ma (Cerling et al., 1979). Independent research conducted by Ian McDougall and collaborators obtained older ages: 1.89±0.01 by the K/Ar method (McDougall et al., 1980) and 1.88±0.02 by the ^{40}Ar/^{39}Ar method (McDougall, 1985). Ian McDougall and Francis Brown (2006) later carried out a detailed analysis of 15 volcanic tephra located between KBS tuff and Chari, which gave ages of 1.869±0.021 to KBS and 1.383±0.028 to Chari. Therefore, we will use 1.8 Ma as the most likely age of KBS tuff.

The most recent stratigraphic analysis of the various members of the Koobi Fora formation is that of Christopher Lepre and Dennis Kent (2010), who conducted a paleomagnetic column from 50 samples obtained in area 102, section type of the middle and lower parts of KBS. The results are shown in Figure 3.17.

3.3.9 Turkana: West Turkana

On the western shore of Lake Turkana, opposite to Koobi Fora, extends the fossil area of West Turkana, which has provided both tools and very robust, old specimens. The main formation is Nachukui, with eight members in a range from 4.3 Ma to around 0.7 Ma (Figure 3.18).

The excavation work began in 1987 by the West Turkana Archaeological Project. Nachukui members containing hominins examined in this book are:

- Natoo—FxJh 5 site (1.8/1.0 Ma, *P. boisei*, see Chapter 6) (Brown et al., 2001).

3.3.10 Turkana: Kanapoi and Allia Bay

Located on Lake Turkana, Kanapoi sites—western shore—and Allia Bay—eastern shore—are linked by the presence in both of *Australopithecus anamensis* specimens (see Chapter 4).

The sedimentary sequence of Kanapoi covers an interval ranging from 4.17 Ma to about 3.4 Ma, with three main paleosol intervals: basal, of fluvial type; middle, of lacustrine clay; and upper, fluvial again. Hominins come mostly from the basal interval, although they are also in the middle interval, and their dating has become very accurate. Some specimens are placed between the *Upper pumiceous tuff* and the *Lower pumiceous tuff*, i.e., in the date range of 4.12±0.02 and 4.17±0.02 Ma (Leakey et al., 1998). The minimum timescale for the remaining specimens was fixed by being located—with the exception of one—below the Kanapoi tuff. The Kanapoi tuff, belonging to the upper part of the middle interval, was dated at 4.07±0.02 Ma by the ^{40}Ar/^{39}Ar method. Thus, the age range corresponding to Kanapoi hominins corresponds to 4.17/4.07 Ma.

Allia Bay sedimentary terrains in which hominins were found are younger than Kanapoi. Most of the Allia Bay specimens come from 261–1 site, just below Moiti tuff (Ward et al., 2001). Moiti tuff was dated in this region, although in a different location, at 3.94±0.03 Ma (Leakey et al., 1995). VT-1 tuff in Maka (Ethiopia), by correlation with that of Moiti, resulted in 3.89±0.02 Ma (White et al., 1993). The dating of the sedimentary terrains, from which hominin fossils come, indicated an age of 3.95±0.05 Ma (Ward et al., 2001).

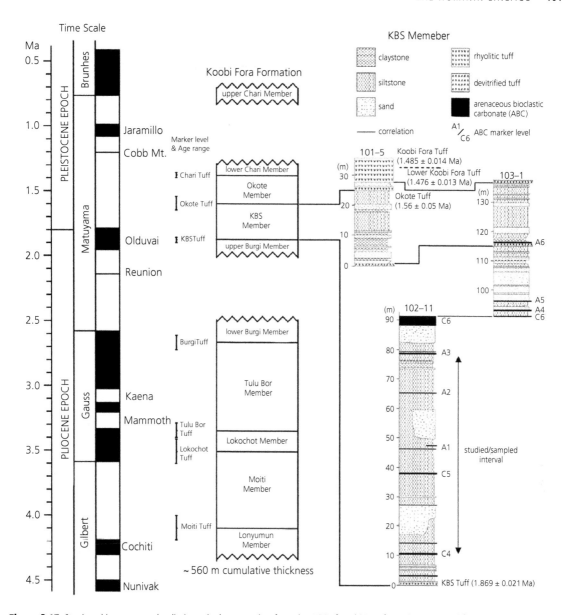

Figure 3.17 Stratigraphic context and radio-isotopic chronography of member KBS of Koobi Fora formation. Reprinted from Publication title, Lepre, C. J. & Kent, D. V. (2010). New magnetostratigraphy for the Olduvai Subchron in the Koobi Fora Formation, northwest Kenya, with implications for early Homo. *Earth and Planetary Science Letters*, 290, 290, 362–374, with permission from Elsevier.

3.3.11 Tugen Hills region

The Tugen Hills are one of the fossiliferous areas where very ancient hominin specimens have been found. They are in Baringo District (Kenya). Its stratigraphic reference corresponds, with respect to the fossils that interest us, to the Lukeino formation.

The geological conditions of the Lukeino formation have been described by Martin Pickford (1975), and Andrew Hill and collaborators (1985; Hill, 1999, 2002). Its lower limit is marked by the Kabarnet Trachyte Formation, with an age between 6.7±0.3 and 7.2±0.3 Ma by the $^{40}K/^{40}Ar$ method (Pickford, 1975). Andrew Hill

---

Box 3.8 KBS dating

In the fossil description, Richard Leakey (1973a) argued that the KBS tuff dating, under which the specimen was found, had been accurately fixed at 2.6 Ma. To the skull, found below KBS, Leakey assigned a probable age of 2.9 Ma. If the estimated age of KNM-ER 1470, of 2.9 Ma, is correct, a notable discrepancy of about half a million years occurred with the age of similar fauna, such as that of Hadar (Ethiopia) and Koobi Fora (Kenya; Maglio, 1972). Basil Cooke (1976), specialist on suids' fossils, warned, after a comparative examination of suids' fossils from both sites, about the need to revise the numbers given to KBS tuff, tentatively granting a maximum of 2.0 Ma.

Box 3.9 Trace elements

"Trace elements" or "minor elements" are defined in the geological context as every element except the most abundant in rocks: oxygen, silicon, aluminum, iron, calcium, sodium, potassium, and magnesium (Thrush, 1968). The concentration of trace elements is specific to each terrain, which allows performing comparisons.

(2002) attributed to the Lukeino formation an age between 6.2 and 5.6 Ma (Box. 3.10).

From the Lukeino formation on Tugen Hills come the Upper Miocene specimens of *Orrorin tugenensis* (see Chapter 4).

3.3.12 Olduvai Gorge

Olduvai (Tanzania), a site discovered by Wilhelm Kattwinkel, professor of Bayerische Palaeontologische Staatssammlung, Munich (Germany) in 1911 (Protsch, 1974), is remarkable in many ways. It was the first place, where radiometric techniques were applied for dating terrain. It was the site that launched the history of specimen discoveries of our tribe, Hominini, within the Rift Valley. There were found the first fossils equivalent to the "gracile" and "robust" hominins of South Africa. There the first specimen appeared to which the name *Homo habilis* was granted. From Olduvai came the evidence of the oldest identified lithic culture, called for that reason, olduwan. And, of course, Olduvai is a site forever linked to the names of Louis and Mary Leakey.

Kenyan by birth and son of a missionary couple, Louis Leakey published his first paper on Olduvai as a note in the journal *Nature*—in which, by the way, he gave to the site its German name, *Oldoway*. His first excavation campaign took place in 1931 and since then he worked at Olduvai—together with his wife Mary Nichols (Mary Leakey)—although continuously only since 1951. In the mid-1960s Mary Leakey took charge of the works (Figure 3.19).

The site of Olduvai ("the place where agave grows" in the local language) is located within the Serengeti plain, near the huge crater of the extinct volcano Ngorongoro, and just under 200 km from Arusha (Figures 3.20 and 3.21). It is a gorge with a depth of up to 100 m in some places and stretches roughly from east to west for 40 km. The canyon was carved by a disappeared river in sedimentary terrains with deposits of lacustrine, fluvial, and aeolian origin, in which are also found volcanic tuffs from the Ngorongoro.

Olduvai sediments were deposited on a basalt layer, the IB tuff (visible only in a place near VEK-FLK locations), with an age of 1.84±0.03 Ma, and are numbered sequentially from bottom to top as Beds I–IV, followed by the Masek, Ndutu, and Naisiusiu Beds (Figure 3.21 and Box 3.11).

The first dating of the volcanic tuff of Olduvai Bed I base by the potassium/argon method ($^{40}K/^{40}Ar$) was 1.7 Ma (Leakey et al., 1961); by comparing 10 tuffs, they obtained a valid average age of 1.75 Ma. A few years later, Jack Evernden and Garniss Curtiss (1965) were the architects of the first absolute time series of a hominin site, corresponding to Olduvai Bed I. As Bernard Wood (1997a) said in his obituary of Mary Leakey, Olduvai Gorge soon became the standard that allowed calibration of the ages of other sites.

Since then numerous dating techniques have been used at the site, i.e., sedimentation rate, geomagnetic polarity, amino acids racemization, and uranium fission tracks. Mary Leakey and Richard Hay (1982) documented the available information on the stratigraphic sequences and Olduvai's ages. Following the studies of these authors, Michael Day (1986) assigned the following dates for Olduvai beds: Bed I, 2.1–1.7 Ma; Bed II, 1.7–1.15 Ma; Bed III, 1.15–0.8 Ma; Bed IV, 0.8–0.6 Ma; Masek, 0.6–0.4 Ma; Ndutu, 400,000–32,000 years; Naisiusiu, 22,000–15,000 years.

The Olduvai stratigraphy spans two million years, but within the Plio-Pleistocene there are only two beds, I and II, with hominin findings. The separation between the two beds is arbitrary. They are distinguished by a layer of flagstones that Hans Reck and

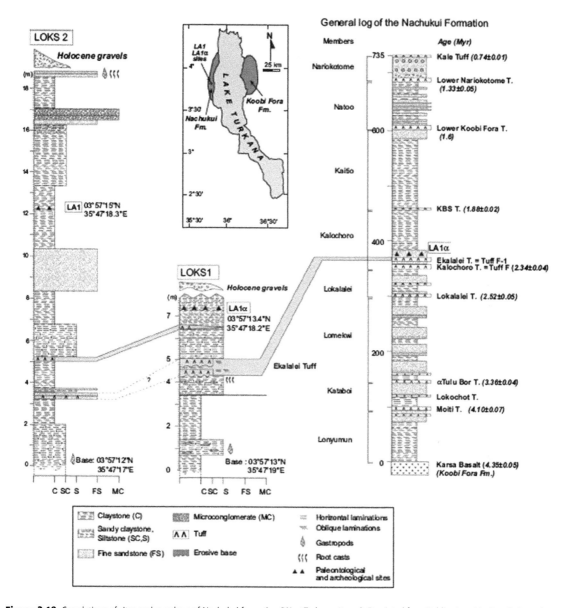

Figure 3.18 Correlations of sites and members of Nachukui formation (West Turkana, Kenya). Reprinted from Publication title, Prat, S., Brugal, J. P., Tiercelin, J. J., Barrat, J. A., Bohn, M., Delagnes, A., . . . Roche, H. (2005). First occurrence of early Homo in the Nachukui Formation (West Turkana, Kenya) at 2.3–2.4Myr. *Journal of Human Evolution*, 49, 230–240, with permission from Elsevier.

Louis Leakey identified in 1931 and established as the boundary, assuming that the geological and faunal changes coincided with the presence of this layer. Today we know that this is not so: from the point of view of a proper reconstruction, the lower part of Bed II is a unit with Bed I. However, for reasons of clarity, here we refer to "Bed I" and "Bed II" as if their separation corresponded to actual changes.

Bed I (from the base to 1.71 Ma) contains deposits corresponding to a salty lake. Bed II ranges from about 1.7 to just over 1.2 Ma, and includes the presence of IIA tuff, allowing calculation of its lower limit. Important geological and ecological changes occurred associated with the deposits of the second bed, such as salinization and reduction of the lake, and its subsequent replacement by open savanna.

Figure 3.19 Left: Louis Leakey and his wife Mary digging in Olduvai Gorge (photo by Robert F. Sisson © National Geographic Society). Right: Olduvai today (photograph by C. J. Cela-Conde).

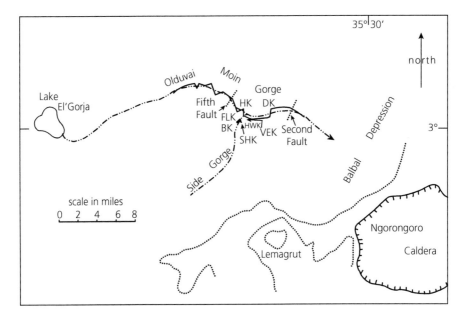

Figure 3.20 Location of Olduvai Gorge. Below, on the right, Ngorongoro Crater. Hay, R. L. (1963). Stratigraphy of Beds I through IV, Olduvai Gorge, Tanganyika. *Science*, 139, 829–833.

Box 3.10 Tugen Hills dating

The chronostratigraphy of Tugen Hills lands has been confirmed by Deino and collaborators (2002). The techniques used were ^{40}Ar/^{39}Ar, ^{40}K/^{40}Ar, and paleomagnetism. The Lukeino, Kaparaina Basalt, and Chemeron formations form a sequence spanning from 6.56 to 3.8 Ma. The Upper Lukeino formation in Kapcheberek was limited to the range of 5.88 to 5.72 Ma. The combination of ^{40}Ar/^{39}Ar and paleomagnetism techniques fixed the age of the Chemeron formation in Tabarin at 4.63 to 3.837 Ma.

The discovered fauna in Olduvai corresponds well with geological data. Numerous micro- and macro-mammals, with primates, bovines, equines, and carnivores among the latter, are present in Bed I, but there are also amphibians and fish that indicate a lacustrine environment (Leakey, 1967; Andrews, 1983). The differences that can be detected in the fauna of Bed I have been attributed to the onset of a climate change leading to the particularities of Bed II, but it has also been suggested that these differences could be due to the presence of different types of predators (Fernández Jalvo et al., 1998).

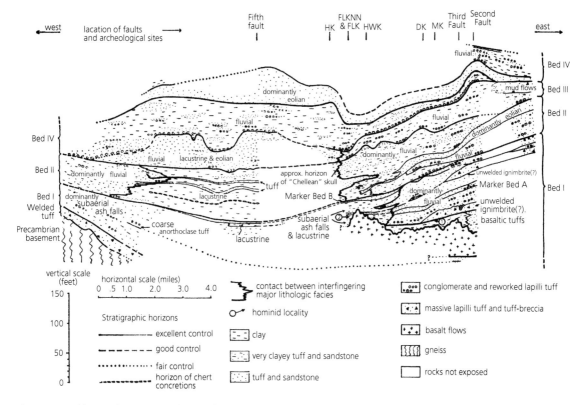

Figure 3.21 Olduvai Beds I–IV stratigraphy, according to Hay (1963). Hay, R. L. (1963). Stratigraphy of Beds I through IV, Olduvai Gorge, Tanganyika. *Science*, 139, 829–833.

Box 3.11 Olduvai locations

VEK means Vivien Evelyn (Fuchs) Korongo, and FLK, Frida Leakey Korongo. The names of Olduvai Gorge locations usually carry the letter K corresponding to "Korongo," which means in Swahili, "creek." The complete list of these names appears in the Mary Leakey report contained in L. S. B. Leakey (1967b).

Bed II contains a typical fauna Villafranchian: large herbivores and carnivores, with the presence of primates and hippopotamus (Leakey, 1967). That content indicates a diverse habitat in which the extension of open savannas was an important factor. It was important, of course, for the evolution of hominins for reasons that we will examine in later chapters.

The first human fossil specimen discovered in Olduvai was a skeleton, OH 1, which appeared on a slope of Bed II, about 3–4 ft from the top of the gorge when a trench was opened for Olduvai dating work (Protsch,

1974). The skeleton's appearance was without doubt very modern, with features of modern humans, including the bulbous cranium, absence of prognathism, and protruding chin.

But Hans Reck (1914), in charge of the research, placed the skeleton as being from Bed II, without taking into account possible taphonomic interpretations about its presence there. Louis Leakey (1928), in his note on the specimen, accepted the attribution made in the first studies: *Homo sapiens*. The various data did not fit because, if the presence of a modern human in Bed II was true, the age of our species would increase by between a half million to one million years. Five years later, the puzzle was solved: it was an intrusion. The skeleton came from a burial at the base of Bed V and, depending on the associated cultural elements, would have the same age as Gamble's Cave (Upper Paleolithic, Kenya), with a similar cultural level (Leakey et al., 1933).

OH 1 was the first in a vast array of specimens from the human lineage discovered in the Rift Valley. In Chapter 4 we will examine the most notable.

3.3.13 Laetoli region

Forty kilometers south of Olduvai, in Tanzania, is the Laetoli fossil area, which is located on the Eyasi Plateau, northwest of the lake of the same name. Eruptions of Sadiman, a volcano about 20 km from Laetoli, were deposited on Eyasi during the Pliocene in successive ash beds; these form the volcanic tuffs called Laetolil Layers, which are very extensive in both area (around 1,500 km²) and height (130 m at Laetoli) (Leakey & Hay, 1979), to which we have already referred (Figure 3.22).

The following sedimentary beds have been identified in Laetoli (Deino, 2011):

- Lower Laetoli Beds, 4.36/3.85 Ma.
- Upper Laetoli Beds, 3.85/3.63 Ma.
- Lower Ndolanya Beds, 3.58 Ma.
- Upper Ndolanya Beds, 2.66 Ma.
- Naibadad Beds, 2.155/2.057 Ma.
- Olpiro Beds (<2.057 Ma).

Paleoclimatology indicated for Upper Laetoli Beds a habitat with extensive presence of rivers, which only flowed during the rainy season. The river system maintained a complex vegetation mosaic, including gallery forests and prairies, with an abundance of lakes and ephemeral wells in the wet season (Ditchfield & Harrison, 2011). The absence of aquatic animals in the fauna indicates that Upper Laetoli Beds and Upper Ndolanya Beds did not have permanent water coverage. However, the presence of a *Crocodylus* specimen at the Lower Laetoli Beds—whose paleoclimatological characteristics are similar—makes us think that during this time range there were at least shallow lakes.

From Upper Laetoli Beds come *Australopithecus afarensis* specimens (see Chapter 4) and the fossil trails we discussed in Chapter 2. From Upper Ndolanya Beds is *Paranthropus aethiopicus* (see Chapter 5).

3.3.14 Malawi: Chiwondo Beds

The Hominid Corridor Research Project began to study the Malawi Rift in 1983, with the purpose of examining the area of southeastern Africa to better understand how fauna dispersal, including hominin, occurred. The first specimen of our lineage attributed to *Homo* was found in 1991 on the site of Uraha at the Chiwondo Beds, Karonga district (Malawi). In 1999, a second specimen appeared and was classified as *Paranthropus* (Sandrock, 1999) at the Malema site, a bit to the north of Uraha (Figure 3.23).

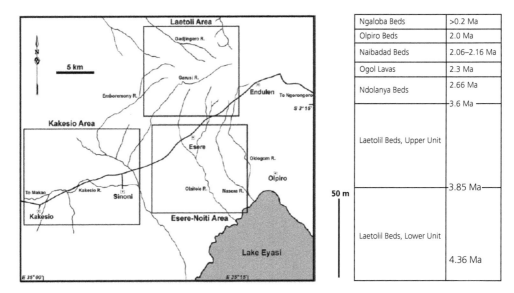

Figure 3.22 Left: Location of fossiliferous areas at Eyasi Plateau (Tanzania) (Harrison and Kweka, 2011). Right: schematic stratigraphy of Laetoli area sedimentary beds (Harrison, 2011a). Left: Harrison, T. and A. Kweka. 2011. Paleontological Localities on the Eyasi Plateau, Including Laetoli.In: T. Harrison (ed.), Paleontology and Geology of Laetoli: Human evolution in Context (pp. 17–45. Volume 1: Geology, Geochronology, Paleoecology and Paleoenvironment. Right: Harrison, T. (2011a). Hominins from the Upper Laetolil and Upper Ndolanya Beds, Laetoli. In T. Harrison (ed.), Paleontology and Geology of Laetoli: Human evolution in Context (pp. 141–188). Dordrecht: Springer Netherlands. With permission of Springer.

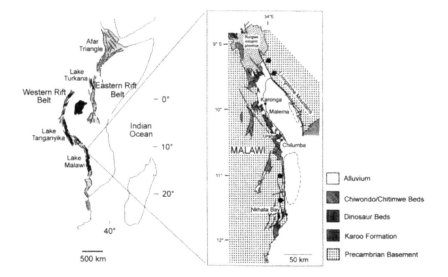

Figure 3.23 Chiwondo Beds (Karonga, Malawi). Left: Lake Malawi location with respect to Turkana and Afar. Right: Chiwondo Beds location, including the site of Malema (Uraha, further to the south) (illustration by Schrenk et al. (2002)).Schrenk, F., Kullmer, O., & Sandock, O. (2002). Early Hominid diversity, age and biogeography of the Malawi-Rift. *Human Evolution*, 17, 113–122.

The sites of the Chiwondo Beds were dated by Friedemann Schrenk and collaborators (1993) by comparing the local fauna associated with hominin remains—unit 3A of the Chiwondo Beds—to other biochronological units in Eastern Africa. Schrenck et al. (1993) suggest an ≈2.5 Ma age or, more conservatively, 2.3 Ma, giving an age of 2.5 to 2.3 Ma for strata from which hominins of Uraha and Malema come. Timothy Bromage and collaborators (1995) extended the Chiwondo faunal comparison, confirming correlations established by Friedemann Schrenk et al. (1993) with the Shungura formation (see Box 3.12). Christian Betzler and Uwe Ring (1995) described the paleoenvironment corresponding to Uraha hominins as that of swamp or alluvial channels.

3.4 South Africa sites

The discovery of the first australopith, the Taung Child (Dart, 1925), was in South Africa, paving the way to the genus *Australopithecus*. Other South African sites, Sterkfontein, Makapansgat, Swartkrans, and Kromdraai, subsequently provided specimens of this taxon and of the more robust *Paranthropus*. These are the best-known South African locations where Plio-Pleistocene remains have been found. You might call these sites "classic," others like Drimolen, and Malapa, which we will also discuss, were added later to this list.

3.4.1 Structure and dating of African sites

The locations of South Africa with Plio-Pleistocene specimens are limestone terrains (Figure 3.24).

Box 3.12 Chiwondo Beds dating

Alan Deino and Andrew Hill (2002) criticized the estimate made of the Chiwondo Beds age, arguing that Schrenk et al. (1993) and Bromage et al. (1995) had misinterpreted the data of Andrew Hill, Ward, Deino et al. (1992). But neither is the reference of Hill, Ward, Deino et al. (1992) given correctly in the paper of Deino and Hill (2002), nor did Schrenk et al. (1993) mention in any place the paper of Hill, Ward, Deino et al. (1992) or of any other author. Bromage et al. (1995) did make a reference to Hill, Ward, Deino et al. (1992), but even if they had confused the Chemeron fauna, Kenya (described by Hill, Ward, Deino et al., 1992), that fact has little effect on the confidence regarding the age of Chiwondo, as the main comparison is made with the suids of the Shungura formation (Omo).

They are originally dolomitic caves formed in the pre-Cambrian.

Dolomitic caves are excavated by water action, which dissolves the limestone terrain and forms the well-known stalactites and stalagmites inside the caves (Figure 3.25). It is common also that the cavern later filled with a cluster of rocky materials, waste, bones, and carbonate compounds, called "breccia," in which fossils are embedded. The original, very ancient structures usually appear damaged to varying degrees of destruction. Hence, there are difficulties in carrying out a correct stratigraphy that is able to reveal their age.

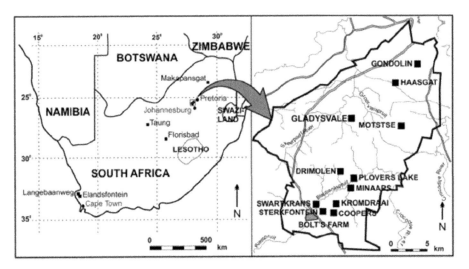

Figure 3.24 Several African sites with Pliocene and Pleistocene hominins (Pickering et al., 2007). Reprinted from, Pickering, R., Hancox, P. J., Lee-Thorp, J. A., R., G., Mortimer, G. E., McCulloch, M., & Berger, L. R. (2007). Stratigraphy, U-Th Chronology, and Paleoenvironments at Gladysvale Cave: Insights Into the Climatic Control of South African Hominin-Bearing Cave Deposits. *J Hum Evol*, 53, 602–619, with permission from Elsevier.

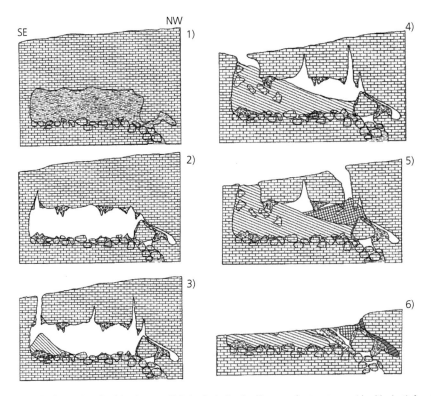

Figure 3.25 Phases in the destruction of a dolomite cave: 1) Dolomite is dissolved by groundwater at a considerable depth forming a well. 2) The water filters down and, in the resulting cave, large dolomite boulders begin to detach. 3) The cracks in the ceiling of the cave begin to open to the outside. 4) The outer opening widens. Speleothems occupy the cave interior. 5) New openings are created through which more speleothems infiltrate. 6) The erosion destroys part of the original dolomite and the infiltrated deposits allowing speleothems to surface. Drawing refers to Swartkrans Cave (South Africa). We have omitted some of the intermediate steps indicated by the author.Brain, C. K. (1993). Structure and stratigraphy of the Swartkrans cave in the light of the new excavations. In C. K. Brain (Ed.), Swartkrans: A Cave's Chronicle of Early Man (pp. 23–34). Pretoria: Transvaal Museum Monograph 8.

In some cases, such as that of Kromdraai, deposits were found on the surface of a hill when the cave collapsed entirely, and an excavation was necessary to identify the original geological structure. Swartkrans exhibits a completely eroded external cave, and another interior cave that still maintains the ceiling. Sterkfontein retains a cave opening to the outside surface and successively deeper chambers, containing different types of breccia. With regard to the Taung Child, its only discovery (in addition to a few much younger tools and bones), was made in the deepest and innermost part of a cavern inside a small hill. Mining works had completely destroyed the original cave, whose structure is therefore unknown. A brief but enlightening summary of the tormented geology of South African sites appears in Michael Day (1986).

Such a complex geological distribution casts shadows on the South African sites when assigning an age to the sediments in which specimens appear. The use of paleomagnetism has provided, in recent years, a new pathway for direct dating; however, the first paleomagnetic results of classic sites were, with the exception of Makapansgat, not consistent (Partridge, 1982). If one adds the absence of viable materials for using radiometric methods (tuffs), in most cases it was necessary to be guided by studies of associated fauna and its correlation with other, easier to date, deposits, as are those of East Africa. Fortunately, recent efforts have yielded more reliable and precise datings.

3.4.2 Sterkfontein

The Sterkfontein cave stratigraphy has been studied in more detail and serves both as a reference for calibrating other sites' dates and as an example of the difficulty of describing and dating a dolomitic cavern (Figure 3.26).

The Sterkfontein stratigraphic sequence has been widely described in the work of Timothy Partridge (Partridge, 1975, 1978, 1982, 2000; Partridge & Watt, 1991; Partridge et al., 1999, 2003; Clarke et al., 2003). Partridge (1978) distinguished six members, from M1 to M6, by order of antiquity, to which Kuman and Clarke (2000) added a final member, post-M6, much more modern (Figures 3.27 and 3.28): M1, M2, and M3 are the deepest; M4, M5, M6, and post-M6 are in the open area of the cave, or near it. But it is necessary to understand that the different members are distributed not only in the vertical plane, one above the other, but also in the horizontal, extending through the different chambers so that initially they were distinguished by

the color of their filling breccias. Cases of special interest for our purposes are those of the Silberberg Grotto, where the nearly complete skeleton, StW 573, was found, and Jacovec Cavern, from which come recently found specimens (see Chapter 5), as they are the oldest terrain with hominins in South Africa.

The first age estimates of the deepest Sterkfontein members with human remains (M2, M4) were performed by biochronology—comparative fauna—with somewhat diverse results, which sometimes only established the great antiquity of Sterkfontein in relation to other South African sites (Ewer, 1956). We have already mentioned how Elizabeth Vrba's studies on Sterkfontein's antelopes (1974, 1984) pointed to a detectable change from the environment of M4, forested, to the drier of M5. This change would correspond to the cooling process that led to the extension of savannas across Africa (Brown, McDougall et al., 1985; Behrensmeyer et al., 1997; Bobe & Behrensmeyer, 2004; deMenocal, 2004), which gives to the transition between M4 and M5 an age of around 2.5 Ma, or somewhat older. However, the first direct datings indicated more recent times. Henry Schwarcz, Rainer Grün, and Phillip Tobias (1994), using ESR applied to bovine tooth enamel, obtained an absolute age of 2.1±0.5 Ma for M4.

The paleomagnetic column of Sterkfontein obtained by Partridge and collaborators (1999) (Figure 3.28) gave to the member 2D—the youngest member inside M2—3.0/2.58 Ma, included (by its normal polarity) in the upper part of Gauss chron. The member 2A, the deepest of M2, would have 4.19/3.6 Ma, attributed to Gilbert chron (reverse polarity) below the Cochiti inversion. Given that M2 is a deeper member than M4, the discrepancy of dates given by Schwarcz, Grün, and Tobias (1994) and Partridge et al. (1999) is evident.

The age of member 2 proposed by Partridge et al. (1999) was rejected by Lee Berger, Rodrigo Lacruz, and Darryl de Ruiter (2002). The interpretation of the fauna, the archaeometric results, and the Sterkfontein magnetostratigraphy made it unlikely, in the words of Berger et al. (2002), that any of the members of that cave exceeds 3.04 Ma.

Partridge and collaborators (2003) recognized that, although paleomagnetic measurements in member 2, conducted by his team in 1999, were apparently accurate, a proper dating depends—as we already mentioned—on the correct identification by means of absolute dating of the magnetic inversions present in the sequence. The paleomagnetic age assigned is also subject to uncertainties both in fauna correlations, as well as sedimentation rates,

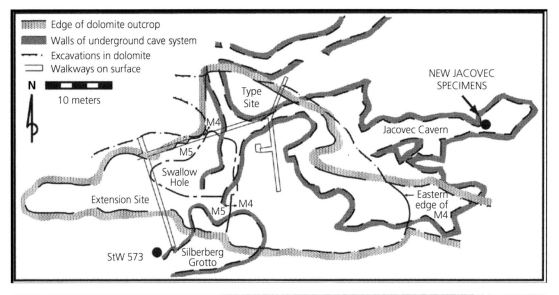

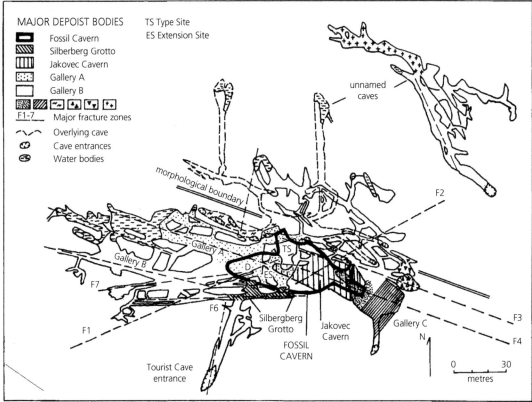

Figure 3.26 Above: Sterkfontein map, according to the studies by Timothy Partridge (Partridge et al., 2003), indicating the location of skeleton Stw573 and the Jacovec Cavern. Below: The cave system of the site according to Wilkinson (1983). (Bottom) Reprinted from the publication, Wilkinson, M. J. (1983). Geomorphic perspectives on the Sterkfontein australopithecine breccias. *Journal of Archaeological Science*, 10, 515–529, with permission from Elsevier. (Top) From Partridge, T. C., Granger, D. E., Caffee, M. W., & Clarke, R. J. (2003). Lower Pliocene Hominid Remains from Sterkfontein. *Science*, 300(5619), 607–612. Reprinted with permission from AAAS.

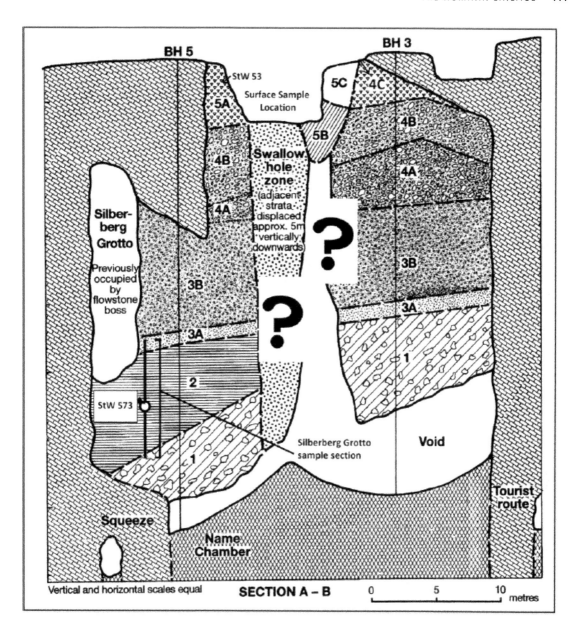

Figure 3.27 Section of Sterkfontein with an indication of its content and different members according to Partridge (2000) (illustration by Herries & Shaw, 2011). Reprinted from the publication, Herries, A. I. R. & Shaw, J. (2011). Palaeomagnetic analysis of the Sterkfontein palaeocave deposits: Implications for the age of the hominin fossils and stone tool industries. *Journal of Human Evolution*, 60(5), 523–539, with permission from Elsevier.

thus requiring independent verification. In order to carry out such verification, Partridge et al. (2003) conducted a radiometric survey by the cosmogenic nuclides' (^{26}Al/^{10}Be) method, obtaining for three samples from the Silberberg Grotto (M2) dates between 4.72 and 3.78 Ma.

But the results obtained using U/Pb radiometry by Joanne Walker, Robert Cliff, and Alfred Latham (2006) added new discrepancies. Speleothem samples from M2, located immediately above and below StW 573, gave ages of 2.17±0.17 Ma (outflow 2B) and 2.24±0.09/0.07 (outflow 2C; see Figure 3.29, left).

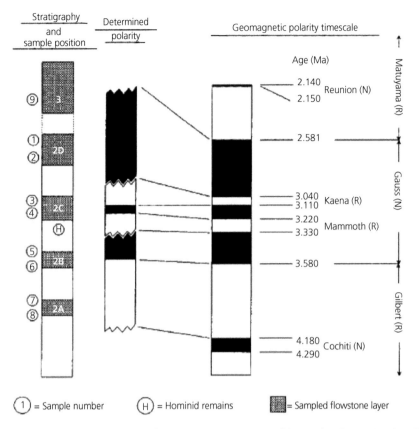

Figure 3.28 Sterkfontein paleomagnetic sequence. The left column indicates the position of the samples taken. To the right is the scale of geomagnetic polarity timescale (—), with the existing inversions in the long, temporal scope sequence of Gilbert–Gauss–Matuyama. The middle column shows the correlation of Sterkfontein paleomagnetism with the geomagnetic scale. Broken lines indicate magnetic reversals whose position cannot be specified due to being outside of the samples (Partridge et al., 1999).

Box 3.13 Sterkfontein fauna

The work of Travis Pickering, Ron Clarke, and Jason Heaton (2004), on the paleoenvironmental context of Sterkfontein member 2 indicated by the fauna, revealed the abundance of cercopithecoids (*Parapapio* and *Papio*) and felines (*Panthera pardus, P. leo, Felis caracal,* and *Felidae indet*). The authors pointed out, however, that M2 taphonomy indicates that these animals entered either voluntarily or accidentally into the cave and became trapped, so they are not a representative sample of the general fauna. With this caution in mind, they indicated as most likely a landscape of hills covered with forest and scrub and surrounded by open spaces. But that is a distribution that portrays well the Sterkfontein paleoecological conditions since c. 4.0 Ma, attributed by Pickering, Clarke, and Heaton (2004) to M2 until the ensuing aridification throughout Africa around 2.5 Ma ago, so it is of little value for estimating the dates of the site.

Box 3.14 Speleothems/breccia

The deposit of speleothem or breccia corresponds to the deposition of materials in karstic caves by water filtration. Naturally, a massive deposit of speleothems is indicative of having had a rainy climate, such as a tropical one.

The Jacovec Cavern has led to similar dissenting dating. Partridge and collaborators (2003) obtained for Jacovec Cavern similar results to those of Silberberg Grotto (Table 3.4).

Such disparity between the different dating studies is due largely to the distinctive Sterkfontein characteristics, with processes of erosion and filtration in the caves that led to the presence, within the same level, of different outflows covered by dolomitic masses. Trying to minimize these difficulties, Robyn Pickering

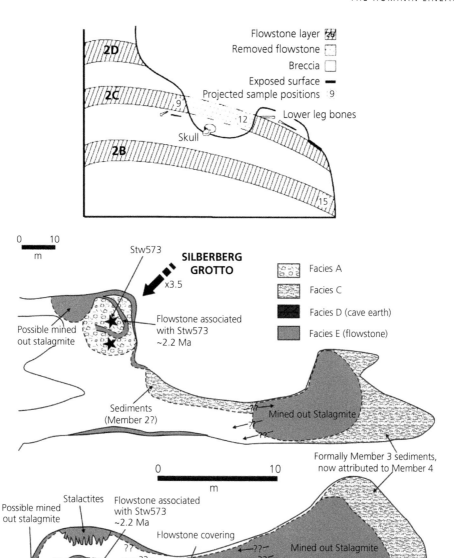

Figure 3.29 Top: Outflow surrounding the StW 573 specimen at the Silberberg Grotto (Sterkfontein, M2). Below: Recreation, in front view (above) and in profile (bottom) of the different materials of sedimentation and filtration in Silberberg Grotto. The original artwork has been cropped. (Left) From Walker, J., Cliff, R. A., & Latham, A. G. (2006). U-Pb Isotopic Age of the StW 573 Hominid from Sterkfontein, South Africa. *Science*, 314(5805), 1592–1594. Reprinted with permission from AAAS. (Below) Reprinted from publication, Pickering, R. & Kramers, J. D. (2010). Re-appraisal of the stratigraphy and determination of new U-Pb dates for the Sterkfontein hominin site, South Africa. *Journal of Human Evolution*, 59(1), 70–86, with permission from Elsevier.

Table 3.4 Dates of samples obtained in Silberberg Grotto and Jacovec Cavern indicating the depth (from the surface) at which each sample was obtained

Sample	Depth (m)	Silberberg Grotto (Ma)	Depth (m)	Jacovec Cavern (Ma)
Upper	24	4.72±1.02	32	4.02±0.27
Middle	25	4.17±0.14		
Lower	26	3.78±0.31	35	3.76±0.26

Data from Partridge et al. (2003).

and Jan Kramers (2010) conducted a detailed study of radiometry (U/Pb and U/Th methods) to narrow the absolute age of Sterkfontein members. In particular, for the dating of Grotto Silberberg (see Box 3.15), it was taken into account what might have been the distribution, and hence the origin, of the different sedimentary contributions, before the excavation removed from the interior a great stalagmitic pillar, scattering materials throughout the cave (Figure 3.29, right).

In Silberberg and Jacovec, Pickering and Kremers (2010) obtained more modern results than those found by Partridge et al. (2003; Table 3.4), in any case less than 3 Ma.

A new paleomagnetism analysis of Sterkfontein conducted by Andy Herries and John Shaw (2011),

Table 3.5 Ages of Sterkfontein members

Member	Ma (P&K, 2010)	Ma (H&S, 2011)	New interpretations by the authors
Post-6		0.5/0.3	Various infiltrations (H&S).
M5	<2.0	M5C 1.3/1.1 M5B 1.4/1.2	Various infiltrations (H&S). M5A, from which Stw 53 comes, is the oldest, although it is not old enough to be part of M4 (H&S).
M4	2.6/2.0	M4C 2.2/2.0 M4B 2.6/2.2 M4A 2.8/2.2	Main contents of the cave (P&K). Various infiltrations (H&S).
M3	2.6/2.0		Abandoned unit; sediments previously described as M3 are the best option to be considered as the distal equivalent of M4 (P&K).
M2	2.8/2.6	2.6/1.8	Brief sediment accumulation punctuated with humid episodes indicating several layers of outflows (P&K).
M1	>2.6		Dolomite by-product mixed with sand and organic material from external sources (cave floor) (P&K).

Ages of Sterkfontein members obtained by:
Pickering and Kremers (2010) (P&K) (U/Pb radiometry);
Herries and Shaw (2011) (H&S) (paleomagnetism).
Comments made by the authors are included.

Box 3.15 Silberberg Grotto dating

In the opinion of Herries and Shaw (2011), the ages attributed through paleomagnetism by Partridge et al. (1999) were based on the incorrect assumption that the deposits in Silberberg Grotto were more than 3 Ma, given their distance from the surface M4 and M5 members. The depth and faunal comparisons were instrumental in establishing the great antiquity of the Silberberg Grotto and that of all Sterkfontein members. The age obtained by radiometry depends, on the other hand, on the specific member in which samples are taken. As we have seen, in the Silberberg Grotto, various infiltrations occurred, which may lead to discrepancies between sediments analyzed by radiometry and those containing fossils.

together with the consideration by the authors of the results of ESR and U-Pb, provided quite similar figures—even lower in some cases—to those reported by Pickering and Kremers (2010): between 2.8 and 2.2 Ma for levels with hominins of member 4, and between 2.6 and 2.2 for those of M2 (Table 3.5).

The most likely age for Sterkfontein members containing older hominins (M2 and M4) seems to be, therefore, between 2.8 and 2.0 Ma. That would be the most accurate date—with the reservations already indicated—for the first presence of the human lineage in South Africa. Which species are involved is an issue that will be addressed in Chapter 5.

3.4.3 Makapansgat

We have focused on Sterkfontein as an example of the difficulties of dating South African caves. Similar uncertainties exist about different sites with hominin remains.

Makapansgat fossils come from the disposal site of the Limeworks' Mine—presently abandoned—with five sedimentary members numbered from M1 to M5. Most hominin remains are from the Gray Breccia (Partridge, 1979) corresponding to member 3, of around 100 m, which together with M2 and M3 occupies the *Classic Section* of the mine (Latham et al., 2002) (Figure 3.30).

The comparison of Makapansgat fauna led Basil Cooke (1964) to determine the age of deposits 3 and 4 at 3.0 to 2.5 million years. The paleomagnetic study of Phillip McFadden, Andrew Brock, and T. Partridge (1979) indicated a discontinuity in the stratigraphic section corresponding to member 3. Consequently, McFadden, Brock, and Partridge (1979) accepted two possible alternatives for M3: (i) it is previous to the Mamut inversion of 3.06 Ma; (ii) it is prior to the Kaena inversion of 2.90 Ma.

Timothy Partridge (1982), after a review of available studies, gave Makapansgat member 3 an age of 3 Ma, which is compatible with the results of paleomagnetism. Contrary to this idea was the comparison of a suids' (swine) series with their equivalent in East Africa made by Tim White, Donald Johanson, and William KImbel (1981). According to these authors, Makapansgat member 3 would only be 2.6 Ma. With regard to member 4 of this site, its fauna is similar to that of Sterkfontein member 4 (Partridge, 1982). Alf Latham and collaborators (2002) indicated that Makapansgat would have had a tropical or semitropical climate, judging by the abundance of speleothems. Both Makapansgat members 3 and 4 have provided specimens of *A. africanus* (see Chapter 5).

3.4.4 Swartkrans and Kromdraai

The Kromdraai and Swartkrans sites have been linked to the hominin species *Paranthropus robustus* (see Chapter 5) since the discovery in 1938 of specimen-type, TM 1517, in Kromdraai A, a brown-colored breccia outcropping at the surface. A second brown breccia (Kromdraai B) also provided hominins associated with an abundant wildlife.

The Swartkrans cave was explored by Robert Broom and John Robinson starting in 1948. Geological studies

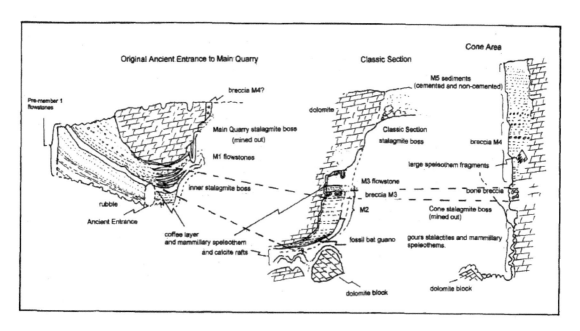

Figure 3.30 *Classic Section* of Makapansgat. Latham, A. G., Herries, A. I. R., Sinclair, A. G. M., & Kuykendall, K. (2002). Re-Examination of the Lower Stratigraphy in the Classic Section, Limeworks Site, Makapansgat, South Africa. *Human Evolution*, 17, 207–214. With permission of Springer.

of Charles Brain (1958) provided the first stratigraphic scheme indicating the presence of two clogging breccias (pink, in the Exterior Cave; brown, in the Inner Cave). The sedimentary members, M1 and M2, correspond to the pink and brown breccias, respectively, with a somewhat intricate final distribution due to the destruction of the cave.

Swartkrans and Kromdraai were initially dated by comparison of fauna with Sterkfontein. To Partridge (1982), member 1 of Swartkrans and member 3 of Kromdraai B belong to the period, between 2 and 1 Ma, with the specimens from Swartkrans being the oldest. Swartkrans member 2 would be about half a million years. C. K. Brain and Andrew Sillen (1988) estimated the age of Swartkrans at c. 1.5/1.0 Ma, while Eric Delson (1988) determined an age of 1.8/1.6 Ma. Based on a comparative study of the Cercopithecus fauna between East and South Africa, Delson (1988) assigned to Swartkrans (members 1–3) and Kromdraai, an age of about 1.8/1.6 million years.

Francis Thackeray, Joseph Kirschvink, and Thimothy Raub (2002) conducted a paleomagnetic study in Kromdraai B under the assumption, derived from the living fauna, that it belonged to the range 2.0/1.5 Ma. Based on the identified inversions in accordance with this assumption, Thackeray et al. (2002) assigned an age for the deposits in which the specimen type of *P. robustus*, TM 1517, was found, which corresponds to the beginning of the Olduvai subchron, that is, 1.9 Ma.

The most detailed study of Swartkrans up to now has been done by Morris Sutton and collaborators (2009) during work on the surface of member 4. In the course of this work the northern part of Talus Cone Deposit (TCD) was excavated to a depth of 10 m, a task that yielded the stratigraphy of member 4 (Figure 3.31), which includes surface excavation with tools from the Middle Paleolithic (*Middle Stone Age*, MSA; see Chapter 9) and the underlying levels SWK4—lava outflow that allowed accurate dating—and TCD, Talus Cone Deposit with hominin specimens.

Radiometric analysis, by the U/Th method, assigned to SW4 an average age of 110.330±1.980 Ma, which is therefore the maximum age for deposits with MSA tools. Three dental specimens appeared in the TCD level, which were attributed to *P. robustus* (see Chapter 5) (Sutton et al., 2009).

Cooper's D deposit, located partway between Sterkfontein and Kromdraai (Bloubank Valley, South Africa), made it possible to date, by U-Pb radiometry, the speleothems associated with fossils attributed also to *P. robustus*. The oldest age estimated for the analyzed speleothems is 1.526 (±0.088) Ma, which limits the age of the associated hominins to 1.5–1.4 Ma (de Ruiter et al., 2009).

3.4.5 Taung

The circumstances of the Taung 1 discovery (see Chapter 5) and the destruction of the cave by stonework led to inferring its age, initially, by paleoecological indirect considerations. The cave had a fauna that indicated two distinct phases: one old, of dry climate and open extensions, corresponding to savanna; and one more modern, of a forest setting. To which of the two would the hominin specimen belong? Since its description by Raymond Dart (1925), Taung 1 was associated with the arid phase, until its reinterpretation by Karl Butzer (1974), which placed it among the fauna of the wet phase; that meant denying the very old age of the specimen. Butzer (1974) found that it would be younger than *A. africanus* from Sterkfontein and Makapansgat, and a contemporary of *P. robustus* from Swartkrans and Kromdraai.

Timothy Partridge (1982) also used indirect evidence from the original fauna to establish the age of the deposit, attributing to Taung a similar age to member 3 of Kromdraai B, between 2 and 1 Ma.

However, contrary to the opinion of Butzer and Partridge, Eric Delson (1988) argued that the fauna corresponding to Taung 1 could be older than 2 Ma and less than 2.5 Ma, more likely an age of 2.3 Ma for the site. The work of Delson (1988) expanded and detailed the cercopithecoid specimen samples, correlating them

---

Box 3.16 Swartkrans' dating

The combined analysis of uranium series and spin resonance (ESR), conducted by Darren Curnoe and collaborators (2001) on two teeth fragments of *P. robustus* from Swartkrans member 1, provided an age of 1.63±0.16 Ma. However, the specimens were obtained from a collection in custody, not in situ, so their origin is uncertain. The results of these same techniques applied to bovine teeth of the same collection, theoretically from member 2, yield an age incompatible with that assumption (more modern).

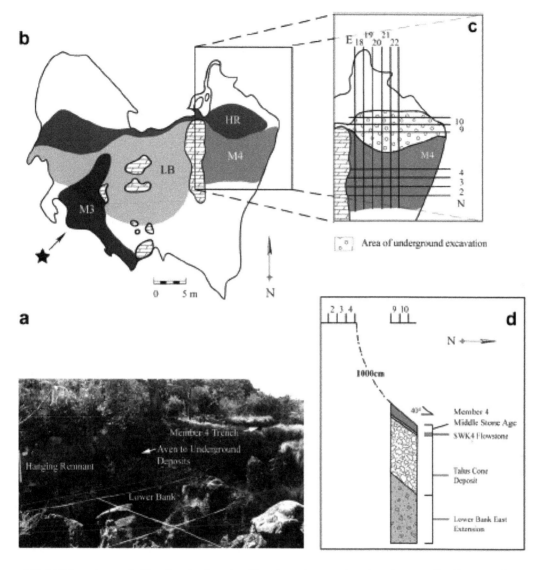

Figure 3.31 (a) Oblique photograph of Swartkrans' surface, viewed from the southwestern shore site—the arrow with a black star in the chart. (b) Floor plan indicating some of the deposits on the surface of Swartkrans Formation (HR = member 1 Hanging Remnant; LB = member 1 Lower Bank; M3 = member 3; M4 = member 4; member 2 disappeared almost entirely in previous excavations and is not on the chart; member 5, of 11,000 years of age, is not shown). (c) Details of M4, with the excavation grid of Sutton and collaborators overlaid. (d) Schematic profile of M4 area stratigraphy. (Illustrations from Sutton et al., 2009.) Reprinted from publication, Sutton, M. B., Pickering, T. R., Pickering, R., Brain, C. K., Clarke, R. J., Heaton, J. L., & Kuman, K. (2009). Newly discovered fossil- and artifact-bearing deposits, uranium-series ages, and Plio-Pleistocene hominids at Swartkrans Cave, South Africa. *Journal of Human Evolution*, 57(6), 688–696, with permission from Elsevier.

with the same species from East Africa. To this end, Delson took into account the chronological estimates of the Turkana depression formations (Koobi Fora, Nachukui, and Shungura) given by Brown, McDougall et al. (1985).

A statistical comparative study of tooth fossil microstriae from the fossil *Papio* species, compared with living *Papio ursinus*, led Frank Williams and James Paterson (2010) to determine the nutritional preferences of fossil monkeys and the Taung paleohabitat. Comparing the

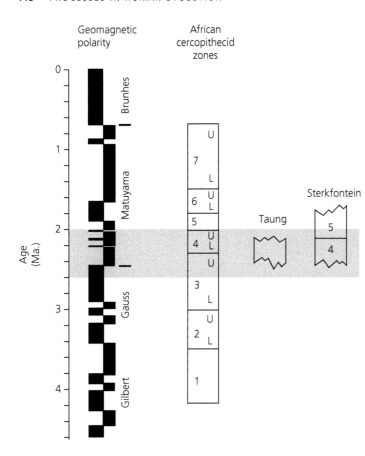

Figure 3.32 On the left, column indicating detected changes in magnetic polarity between 4.5 Ma and the present. On the middle, African cercopithecid zones established by Delson (1984) (where U indicates the upper and L indicates the lower levels of each zone). On the right, the temporal position of Taung in relation to Sterkfontein, deduced from the cercopithecoids' zones (middle column). Chart modified from Williams & Patterson (2010). Williams, F. L. & Patterson, J. W. (2010). Reconstructing the paleoecology of Taung, South Africa from low magnification of dental microwear features in fossil primates. *Palaios*, 25, 439–448.

Box 3.17 Gladysvale

Located 13 km from the group formed by Sterkfontein, Kromdraai, and Swartkrans, Gladysvale has the historical significance of being the first South African location in which new hominin specimens appeared, after the classic sites described earlier in the chapter. In 1993, excavations conducted by the Palaeo-Anthropological Research Unit of the University of Witwatersrand (Johannesburg, South Africa) led to the discovery of two teeth of doubtful identification (GVH-1 and GVH-2) with abundant associated fauna, characteristic of the Plio-Pleistocene (Berger et al., 1993). The authors did not assign these specimens to any species: they simply noticed the aspect of the mesial buccal side of the pieces as prominent, although without reaching the extreme of *Paranthropus robustus*. A year later a tooth, GVH-7, and a phalanx, GVH-8, appeared in an upper stratigraphic level, in the Pink Breccia. The site has three large chambers (Schmid, 2002), above which there is one very eroded chamber, GVED (*Gladysvale Internal Deposit*, Pickering et al., 2007), where the hominin specimens were found.

characteristics of the paleoecology, the authors point to member 4 of Sterkfontein as having had a climate more similar to Taung. Taking into account the areas defined by Delson (1984) for the temporal sequence in the evolution of the cecopithecoids, Williams and Paterson (2010) set a date between 2.5 and 2 Ma for Taung (see Figure 3.32)—although that was not the main objective of their work—a date that does not differ much from that assigned by Delson (1988).

3.4.6 Drimolen

Discovered by Andrew Keyser in July 1992, Drimolen is one of the new sites added to the classic list of Taung, Sterkfontein, Makapansgat, Swartkrans, and Kromdraai, with numerous discoveries, mainly of *P. robustus*. Particularly interesting is a skull with hinged jaw, which will be discussed in Chapter 5. The site is located 7 km north of Sterkfontein, in the district of Krugersdorp, Gauteng, South Africa.

This is, once again, a ruined dolomitic cave whose stratigraphic relationships are shown in Figure 3.33. It was subjected to limestone extraction work, so

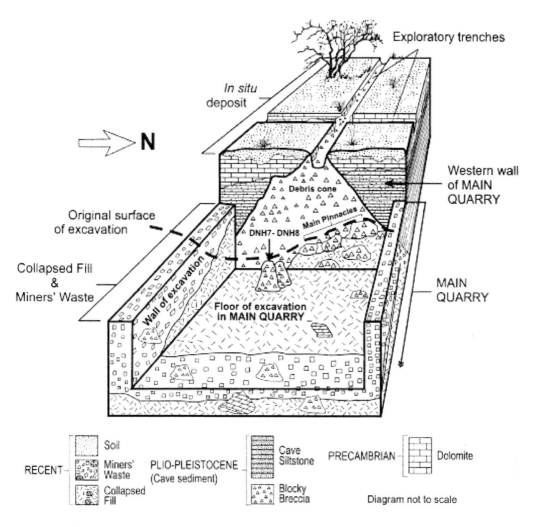

Figure 3.33 Diagram of Drimolen stratigraphic relationships. Keyser, A. W., Menter, C. G., Moggi-Cecchi, J., Pickering, T. R., & Berger, L. R. (2000). Drimolen: a new hominid-bearing site in Gauteng, South Africa. *South African Journal of Science*, 96, 193–197.

sediments became exposed in about 20 holes made by miners. The fossiliferous sediments are of two types: deposits in situ and those of the *Collapsed Fill* (CF); both are indicated in the diagram. To the CF sediments, among which are the main hominin specimens, the authors attributed 2.0/1.5 Ma based on the associated fauna (Keyser, 2000; Keyser et al., 2000). This suggests a mixed environment with the presence of open prairies.

From Drimolen come numerous specimens of *Paranthropus boisei* (see Chapter 5), associated with bones that could have been used as tools for foraging (Backwell & d'Errico, 2008).

3.4.7 Malapa

The site of Malapa, last of the destroyed dolomitic caves of South Africa to be described, is 15 km north-northeast of Sterkfontein, in the province of Gauteng. From Malapa come two partial skeletons of australopithecines, described in Chapter 5. The site consists of five sedimentary facies labeled from the base to the top: facies A, B, C, D, and E, separated by outflows. The layout of the Malapa facies appears in Figure 3.34.

The skeletons appeared in the facies D. Over 200 non-hominin fossils were also found at the site.

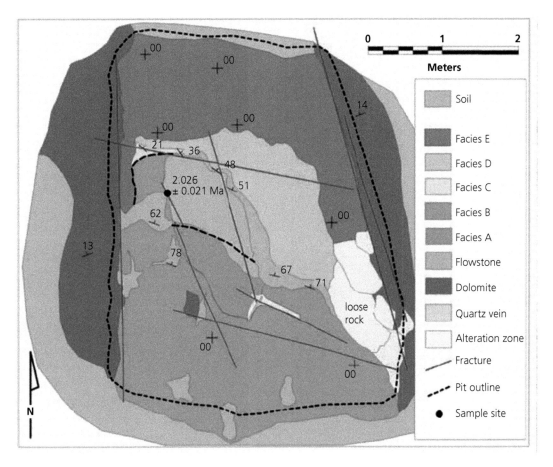

Figure 3.34 Malapa sedimentary facies A–E.From Dirks, P. H. G. M., Kibii, J. M., Kuhn, B. F., Steininger, C., Churchill, S. E., Kramers, J. D., . . . Berger, L. R. (2010). Geological Setting and Age of *Australopithecus sediba* from Southern Africa. *Science*, 328(5975), 205–208. Reprinted with permission from AAAS.

The outflow facies C and D were dated by the U/ Pb technique. Two independent laboratories obtained identical dates of 2.02±0.02 Ma. To more precisely date the fossils, a paleomagnetic column was established between the outflow base—of reverse polarity, except for a subchron interpreted as Huckleberry near its upper part—and the overlapping facies—of normal polarity, interpreted as the beginning of the Olduvai subchron (Dirks et al., 2010), dated as known, between 1.95 and 1.78 Ma.

3.5 Sites to the north of the Rift Valley

Two places located north of the Rift Valley are of great importance when studying the early evolution of hominins: Toros-Menalla (Tchad) and Dmanisi (Georgia). It could be said that they open and close the story of the first hominins, because the former contains the oldest fossils of our lineage and the latter documents what was the first dispersal of hominins out of Africa.

3.5.1 Toros-Menalla

The fossiliferous area of Toros-Menalla (Tchad), studied since the 1990s by the Mission *Paléoanthropologique Franco-Tchadienne*, is in the basin of the Holocene lake Mega-Tchad, the largest paleolake in the Sahara, from which the present Lake Tchad is residual (Figure 3.35). There are four main fossiliferous zones, from which two, Toros-Menalla and Koro-Toro, have provided hominins. A large fauna indicates that both the Miocene and Pliocene conditions were more humid,

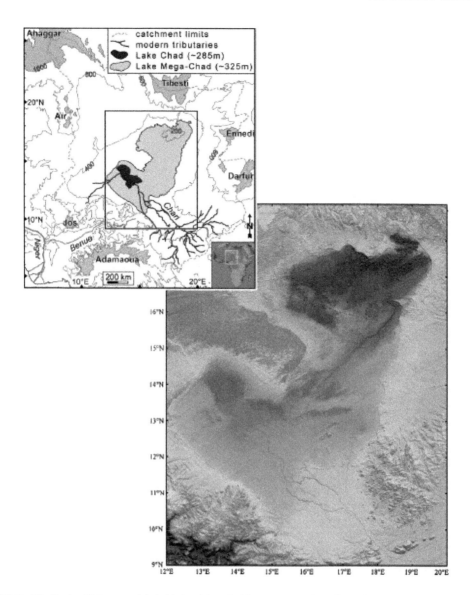

Figure 3.35 The Tchad basin with its current lake in black and the paleolake in gray. Reproduced from Schuster, M., Duringer, P., Ghienne, J.-F., Roquin, C., Sepulchre, P., Moussa, A., . . . Brunet, M. (2009). Chad Basin: Paleoenvironments of the Sahara since the Late Miocene. *Comptes Rendus Geoscience*, 341(8–9), 603–611, published by Elsevier Masson SAS.

although with recurrent desertic episodes in the north of the basin (Schuster et al., 2009), as is the case today. Especially valuable is the fossil insect series (termites of families Hodotermitidae and Macrotermitinae) (Schuster et al., 2009).

The pattern of lake–desert succession is a general framework in which very diverse local conditions occur.

The Miocene climatic conditions for the area 266 of Toros-Menalla correspond to a landscape mosaic similar to the current Okavango (Angola–Botswana), with lakes and riparian habitats, wetlands, forest areas, wooded islets, wooded savanna, and grassland within a general area of semiarid savanna conditions (Brunet, 2010). Upper Miocene fauna includes suids (swine) and loxodonta (elephants), similar to those of the Lothagam formation in Kenya, dated 7.4–6.5 Ma, as well as carnivores (like Hienidae), cats (*Dinofelis*), and cercopithecoids (Colobinae).

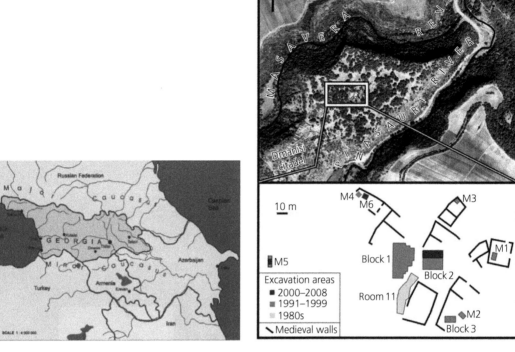

Figure 3.36 Left: Dmanisi location (http://donsmaps.com/dmanisi.html). Right: Aerial view and graphic of the medieval ruins.Left: Reprinted from Gabunia, L., Vekua, A., & Lordkipanidze, D. (2000). The environmental contexts of early human occupation of Georgia (Transcaucasia). *Journal of Human Evolution*, 38(6), 785–802, with permission from Elsevier. Right: Ferring, R., Oms, O., Agustí, J., Berna, F., Nioradze, M., Shelia, T., . . . Lordkipanidze, D. (2011). Earliest human occupations at Dmanisi (Georgian Caucasus) dated to 1.85–1.78 Ma. *Proceedings of the National Academy of Sciences*, 108, 10432–10436.

Box 3.18 TM266 vertebrates

A complete list of the vertebrate fauna of the TM 266 location in Chad appears in Le Fur et al. (2009). It is dominated by bovids (six species and 58% of specimens). Carnivores are fairly diversified (nine species and 11%).

Fauna comparison studies give an age of 6–7 Ma to Toros-Menalla (Vignaud et al., 2002). A radiometric dating by the $^{10}Be/^9Be$ method give to a total of 28 samples, corresponding to the sedimentary horizon in which hominin fossils were found, a range of 7.2–6.8 Ma (Lebatard et al., 2008). Various hominin specimens attributed to the genus *Sahelanthropus* come from the areas TM 266, TM 247, and TM 292 (see Chapter 8).

3.5.2 Koro-Toro

The Pliocene site 12 KT has provided a specimen of *Australopithecus* (see Chapter 4). The KT 13 site, similar to KT 12, contains at least 30 vertebrate fossil taxa, collected on the surface or in situ, of which 22 are carnivorous mammals (four species), proboscids (two species), Perissodactyla (*Hipparion* and *Ceratotherium*), artiodactyls (five species), and a majority of bovids, both in diversity and number of specimens (Beaulieu, 1998). Fish and reptiles—mainly crocodiles—abound. Faunal comparisons indicate an age of 3.5–3 Ma, based on the similarities with the Hadar fauna (Ethiopia) of that range. The landscape deduced from the fauna is, again, a mosaic within the general environment of lake or river banks.

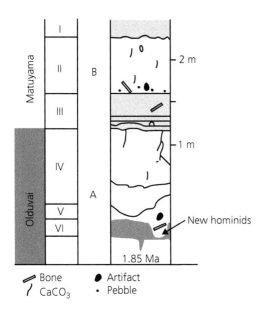

Figure 3.37 Paleomagnetic column with the situation of Dmanisi specimens and artifacts reported in 2002 (Vekua et al., 2002).

3.5.3 Dmanisi site

We have left for the end of the chapter the locality of Dmanisi for two reasons: first, because the fossils of hominins that appear there are at the boundary between the Pliocene and Pleistocene; and, second,

because it is the only location outside of Africa, in the so-called Levantine corridor, which is the same as saying at the gates of Asia.

The site of Dmanisi (Georgia) is located under the ruins of the medieval town of that name, an important urban center from the eighth to the twelfth centuries of our era. The city, 85 km southwest of Tbilisi, was built on alluvial deposits up to around 4 m over a basaltic base. The sediments contain mammal fossils of late Villafranchian. The excavations, in successive campaigns in the 1980s, from 1991 to 1999, and 2008, concentrated on a promontory located between the Masavera and Pinasauri rivers (Figure 3.36).

The site contains two beds, A and B, with different units indicated in the stratigraphy of Figure 3.37, located on the basalt deposit of Masavera dated at 1.85 Ma.

The basalt base was dated by the method of $^{40}K/^{39}Ar$ to 1.8±0.1 Ma (Gabunia & Vekua, 1995) and by $^{39}Ar/^{40}Ar$ (Schmincke & Van den Bogaard, 1995) at 2.0±0.1 Ma. By a paleomagnetic study it was determined Bed A belongs to the Olduvai subchron and Bed B to the Matuyama chron (Gabunia & Vekua, 1995).

Fossils of genus *Homo* (see Chapter 8) and stone carvings have appeared in the two beds, A and B. In late 1991 the first fossil was found, a jaw with teeth, D211. This was the first indication that Dmanisi is an important enclave in the migration of hominins out of Africa.

CHAPTER 4

Miocene and Lower Pliocene hominins

4.1 Miocene hominins

4.1.1 The time and place of the appearance of hominins

The emergence of the Hominini tribe (Table 4.1) probably occurred at the end of the Upper Miocene or the beginning of the Pliocene, as an outcome of a cladogenetic event in which the previous ancestral lineage of humans and chimpanzees diverged.

As we indicated in Chapter 1 (Table 1.4), the Upper Miocene extends between 11.6 and 5.3 Ma. The hominin–chimpanzee separation event, within constraints due to our limited knowledge of the Upper-Miocene hominoids, can be summarized as follows:

- About 7–8 million years ago in Africa, the evolutionary branches that led ultimately to the gorillas, chimpanzees, and modern humans separated.
- Molecular techniques indicate that the sister group of hominins is the chimpanzee.
- *Samburupithecus, Ouranopithecus, Nakalipithecus,* and *Chororapithecus* are genera close to the separation event of current hominoids, although it is very difficult to determine the particular lineage to which they belong.

The separation event of sister groups within our tribe, Hominini, and that of the chimpanzees, Panini, is illustrated in the graph of Figure 4.1.

The most reliable date obtained by molecular methods for the cladogenesis event that separated humans and chimpanzees is given by Galina Glazko and Masatoshi Nei (2003): 6 Ma, within a range from 5 to 7 Ma. We take the last date, 7 Ma, as the most plausible. Unfortunately, specimens immediately preceding that date are insufficiently informative. There is a fragment of a right jawbone with three molars, of which only one retains the crown (KNM-LT 329), obtained in 1967 in Lothagam Hill (Kenya) (Patterson et al., 1970), but their interpretation is difficult. The age obtained by the K/Ar method for the Lothagam fauna in Lothagam

Bed I places it between two basalt intrusions of, respectively, 8.5±0.2 and 3.8±0.4 Ma (Brown, McDougall et al., 1985). Biostratigraphic analyses of Andrew Hill, Stephen Ward, and Barbara Brown (1992) grant an age for the mandibular fragment older than 5.6 Ma. If its traits are human and the age is correct, we would be facing a very ancient hominin and not an ancestor of the clade *Homo + Pan*. In fact, Hill, Ward, and Brown (1992) classified the specimen as *Australopithecus* cf. *afarensis* (Box 4.1).

The KNM-SH 8531 maxilla of the Namurungule formation, found at Samburu Hills (Kenya) (Ishida et al., 1984), and thought to be 9 million years old, shares certain traits with current gorillas (large size, prognathism, and shape of the nasal aperture), but not others (thick enamel, low and rounded cusps). Some traits suggest a resemblance with gorillas, others with chimpanzees, others with humans, while the rest are primitive traits. It was classified by Hidemi Ishida et al. (1984) as *Samburupithecus kiptalami* (Figure 4.2).

Although the Samburu Hills' specimen shows no pongid-derived traits (the family including orangutans and their direct ancestors), some features are distinctive of the gorilla + chimpanzee + human set. But, the specimen lacks specific derived traits of each of these three branches. Thus, Colin Groves (1989) concluded that this individual lived before the separation of gorillas, chimpanzees, and hominins. In that case, the phylogenetic role of the Samburu Hills' specimen would be similar to the one of *Ouranopithecus*, as a common ancestor of African great apes and hominins.

There is an absolute void between the Samburu Hills' remains and current African great apes. We know of none of their remote or close ancestors, with the exception of the already mentioned Tugen Hills' exemplar, classified as a member of the chimpanzee lineage, which we will discuss later (McBrearty & Jablonski, 2005). Conversely, there are abundant fossil hominin specimens and, as years go by, their number increases. There are some authors, such as Leonard Greenfield (1983) and Russell Ciochon (1983) who,

Processes in Human Evolution. Francisco J. Ayala and Camilo J. Cela-Conde, Oxford University Press (2017).
© Francisco J. Ayala and Camilo J. Cela-Conde.
DOI 10.1093/acprof:oso/9780198739906.001.0001

Table 4.1 Taxa of the Hominini tribe and their time of emergence

Genera	Species	Age (Ma)	Emergence (the date given is the beginning of the period in Ka)
Sahelanthropus	S. tchadensis	7.2–6.8	Upper Miocene (11,606)
Orrorin	O. tugenensis	6	
Ardipithecus	Ar. kadabba	5.8	
	Ar. ramidus	4.4	Lower Pliocene (5,332)
Australopithecus	A. anamensis	4	
	A. afarensis	3.8	
	A. deriyemeda	3.5/3.3	
	A. platyops*	3.5	Middle Pliocene (3,600)
	A. bahrelghazali	c. 3.5	
	A. garhi	2.5	Upper Pliocene (2,588)
	A. africanus	2.2	
	A. sediba	2.0	
Paranthropus	P. aethiopicus	2.5	
	P. boisei	2.2–1.9	
	P. robustus	1.9	
Homo	H. rudolfensis	2.5–2.3	
	H. habilis	2.3	
	H. gautengensis	2.0	
	H. erectus	1.9	
	H. georgicus	1.8	Lower Pleistocene (1,806)
	H. erectus	c. 1.7	
	H. floresiensis	? >0.18	
	H. antecessor	0.8	
	H. heidelbergensis	c. 0.7–0.4	Middle Pleistocene (781)
	H. neanderthalensis	0.3	
	H. sapiens	0.2	

*Originally proposed as Kenyanthropus platyops.

agreeing with Darwin, maintain that we will never identify the fossils corresponding to organisms that lived immediately before the separation of African apes and hominins. To put it another way, we would not recognize them if we had them in front of us. Their traits would not have differentiated enough to allow their identification as members of the chimpanzee, gorilla, or human lineages.

A scene of emptiness in the fossil record prior to the appearance of our own tribe is the starting point for the presentation of the earliest hominin specimens.

They are African and have all been found in East Africa (the Rift Valley) and South Africa, with one important exception: Sahelanthropus tchadensis, from Central Africa. The temporal order in which the discoveries were made does not coincide with the age of the fossil finds. In fact, because research has been directed toward progressively older terrains, the situation has turned out to be rather the opposite. Thus, the findings may be presented in a historical sequence, starting with those that were found first or following the real chronological order. We have opted for this second possibility.

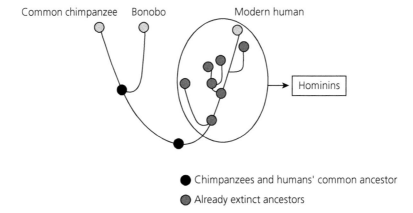

Common chimpanzee   Bonobo   Modern human

Hominins

● Chimpanzees and humans' common ancestor

◐ Already extinct ancestors

Figure 4.1 The term hominin applies to current humans and their extinct exclusive direct and collateral ancestors. Exclusive means that they are not also ancestors of our sister group, common chimpanzees and bonobos. The ancestors included in the figure are hypothetical; they are meant as an illustration. There are many known hominin ancestors but, besides the exception of two central upper incisors from the Kapthurin Formation of the Tugen Hills (Baringo, Kenya), dated to around 545,000 years ago (McBrearty & Jablonski, 2005), there is no known ancestor of current chimpanzees.

Box 4.1 The Lothagam specimen

The Lothagam specimen was initially classified by Patterson and colleagues (1970) as *Australopithecus* cf. *africanus*, but Eckhardt (1977) saw in it a possible pongid. Studies by Kramer (1986) and Hill, Ward, & Brown (1992) reveal traits resembling *A. afarensis*, such as molar width. McHenry and Corruccini (1980) agree that it shows hominin-derived traits, but due to the lack of some notorious *A. afarensis'* traits, they recommend its classification as Hominini indet. White (1986) classified the specimen within *Australopithecus afarensis*. Wood and Richmond (2000) have pointed out its affinities both with *A. afarensis* and *Ar. ramidus*. In this chapter we will see the age and characteristics of specimens from all these taxa.

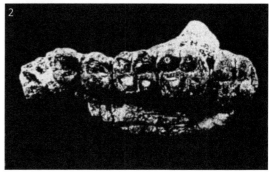

Figure 4.2 The KNM-SH 8531 maxilla, *Samburupithecus* (Ishida & Pickford, 1997).

4.1.2 Toros-Menalla

In 2002, the *Mission Paléoanthropologique Franco-Chadienne*, directed by Michel Brunet, reported the discovery of a fossil of an age very close to the time of the chimpanzee and hominin separation (Brunet et al., 2002). It came from the TM 266 locality of the fossiliferous area of Toros-Menalla, not far from Bahr el Ghazal (Chad).

As we saw in Chapter 3, biochronological studies dated the site of Toros-Menalla between 6 and 7 Ma (Vignaud et al., 2002), but a subsequent radiometric dating placed the specimen in the range of 7.2–6.8 Ma (Lebatard et al., 2008). This date has led us to accept 7 Ma for the separation of the chimpanzee and human lineages.

Initial findings of Toros-Menalla hominins included a nearly complete skull (TM 266–01-60–1) that lacks much of the occiput and six mandibular fragments (Figure 4.3). The skull received the colloquial name Toumai ("life hope" in Goran language).

TM 262–01-60–1 morphology shows a mixture of primitive and derived features, which always indicates a mosaic evolution within the overall process of change in a given lineage. The very small cranial capacity, the great supraorbital arch, and the highly developed mandibular ramus are traits that bring the specimen closer to the African apes. By contrast, small canines (which, according to the discoverers, belonged to a male), reduced subnasal prognathism, and molar enamel thickness, halfway between gorillas and chimpanzees and that are shown by later taxa (*Australopithecus*), bring the specimens closer to hominins. The authors who described the specimens of Toros-Menalla proposed, therefore, a new genus and species: *Sahelanthropus tchadensis* (Brunet et al., 2002) (see Box 4.2).

There are no postcranial elements of *S. tchadensis*. Soizic Le Fur and collaborators (2009) conducted an analysis of the fossils of Toros-Menalla TM 266 locality, whose results indicate—if locomotion and dietary patterns converge, as argued by the authors—the presence of a fauna of forest and open savanna with swamps and grasslands. Fifty-seven percent of the mammals from TM 266 have a compulsory terrestrial locomotion and the rest are semi-arboreal, with no finding of any strictly arboreal species. This conclusion favors giving *S. tchadensis* a terrestrial locomotion.

Toros-Mendalla localities TM 247, TM 266, and TM 292 later provided three new specimens of *S.*

Box 4.2 What is *Sahelanthropus*?

As noted by Bernard Wood (2002) in his commentary on the discovery, *Sahelanthropus tchadensis* is a strange morphological combination: the back of its skull resembles that of an ape while the face, with its large supraorbital arch and its moderate subnasal facial projection, brings it closer to hominins. But, it does not fit with the australopiths, as would be expected if it is one of their ancestors, but rather with *Homo erectus*, which came much later. For reasons having to do with the way in which traits spread along a lineage, *H. erectus* could have inherited the facial morphology of *S. tchadensis* directly, without such features appearing in the various *Australopithecus* and intermediate *Homo* species.

Milford Wolpoff and colleagues (2002) stated that *S. tchadensis* could actually be not a hominin, but an ancestor of African apes, perhaps of the gorilla. The more gracile features would be justified by the fact that it is a female. If its smaller teeth are due to sexual dimorphism, then *Sahelanthropus*' smaller canines lose importance as an evolutionary event—it can be expected that a female ancestor of gorillas also would have them. However, the Michel Brunet team attributed male gender to the Toros-Menalla specimens, which supports placing them within the hominin lineage.

---

tchadensis—two mandibular fragments (TM 292–02-01 and TM 247–01-02) and a right upper premolar (TM 266–01-462) (Brunet et al., 2005). Michel Brunet and collaborators (2005) described the presence in these

a   b   5 cm

Figure 4.3 TM 2662–01-60–1, Sahelanthropus tchadensis: frontal, lateral, and basal views (Brunet et al., 2002).

Table 4.2 Hypodigm of *Sahelanthropus*

Catalogue number	Specimen	Finding date	Discoverer
(A)			
TM 266–01-060–1 (holotype)	Cranium	2001	D.A.
TM 266–01-060–2	Fragmentary mandibular symphysis with incisive and canine alveoli	2001	Group
TM 266–01-447	Right M^3	2001	Group
TM 266–01-448	Right I^1	2001	Group
TM 266–02-154–1	Right mandible, (P3) P4-M3	2002	D.A.
TM 266–02-154–2	Right C	2002	D.A.
(B)			
TM 292–02-01	Mandibular fragment	2002	MPFT
TM 247–01-02	Fragment of the right mandible	2001	MPFT
TM 266–01-462	Right P^3	2001	MPFT

(A) Fossils included by Brunet et al. (2002); (B) fossils included by Brunet et al. (2005). D.A., Djimdoumalbaye Ahounta; MPFT, *Mission Paléoanthropologique Franco-Chadienne*.

materials of some derived traits, such as the C/P3 complex formed by the canine and the first premolar, and the radial enamel thickness, intermediate in both cases between those of chimpanzees and australopiths. These traits support the hominin character of the taxon (Table 4.2).

4.1.3 Tugen Hills

During the months of October and November of 2000, the Kenya Paleontology Expedition (KPE), organized by the Collège de France (Paris) and the Community Museums of Kenya (Nairobi) found up to 12 hominin mandibular, dental, and postcranial fragments at four sites belonging to the Lukeino formation (Cheboit, Kapsomin, Kapcheberek, and Aragai). An inferior molar (KNM LU 335), found by Martin Pickford at the Cheboit site in 1974, was included together with those new fragments (see Box 4.3).

Senut and colleagues (2001) noted that its thick dental enamel, its small dentition relative to body size, and the shape of the femur indicate that they are hominins. These hominins are different from *Ardipithecus* (with thin enamel) and from *Australopithecus*, with larger dentition and femora less *Homo*-like than *Orrorin*'s (see Box 4.4) Senut et al. (2001) named the new genus and species *Orrorin tugenensis*. The BAR

Figure 4.4 Upper strata of the Lukeino Formation at Tugen Hills (Kenya) (photograph by C. J. Cela-Conde).

Box 4.3 Millennium Man

Brigitte Senut and Martin Pickford referred to the Lukeino specimens as Millennium Man in the announcement of the discovery. There was a dispute regarding which research group possessed the authorized excavation permits to carry out research at Baringo (Butler, 2001). But, be that as it may, Millennium Man was a very important discovery, because of the specimen's age and because of the presence of dental and postcranial remains that allow to define our tribe's primitive traits.

Table 4.3 Hypodigm of *Orrorin* (Senut et al., 2001)

Catalogue number	Locality	Specimen	Discoverer	Discovery date
KNM LU 335	Cheboit	Lower molar	Martin Pickford	1974
BAR 349'00	Kapcheberek	Hand phalanx	Evalyne Kiptalam	10/13/ 2000
BAR 1000'00	Kapsomin	2 mandible fragments	Kiptalam Cheboi	10/25/2000
BAR 1002'00	Kapsomin	Left femur	Martin Pickford	11/04/2000
BAR 1004'00	Kapsomin	Right humeral shaft	Brigitte Senut	11/05/2000
BAR 1003'00	Kapsomin	Left proximal femur	Dominique Gommery	11/05/2000
BAR 1001'00	Kapsomin	Upper central incisive	Samuel Chetalam	11/10/2000
BAR 1215'00	Aragai	Right proximal femur	Martin Pickford	11/01/2000
BAR 1390'00	Kapsomin	P4	Samuel Chetalam	11/13/2000
BAR 1425'00	Kapsomin	Right upper incisive	Kiptalam Cheboi	11/16/2000
BAR 1426'00	Kapsomin	Left M^3	Evalyne Kiptalam	11/17/2000
BAR 1900'00	Kapsomin	Right M^3	Joseph Chebet	11/23/2000

1000'00 fragmentary mandible in two pieces constitutes the holotype and the remaining specimens, including the KNM LU 335 molar, found in 1974, are paratypes (Figure 4.5). *Orrorin* means "original man" in the Tugen language, while the species' name honors the toponym of the hills in which the fossils were found (Table 4.3).

Senut and colleagues (2001) emphasize that the appearance of the locomotor apparatus of *Orrorin* is more modern than that of australopithecines. The justification of the new genus *Orrorin* rests on locomotion and not just on dentition. However, Aiello and Collard (2001) question such an early divergence between two different kinds of locomotion. As we saw in Chapter 3, the bipedal locomotion of the first hominins is a

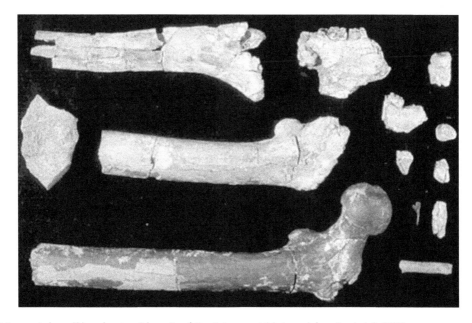

Figure 4.5 Fragmented mandible and postcranial remains of *Orrorin tugenensis* (photograph from Senut et al., 2001).

Box 4.4 What is *Orrorin*?

The hominin status of *Orrorin* was questioned by Yohannes Haile-Selassie (2001) because the specimens showed, in the upper canines, the primitive feature of low crowns. With regard to locomotion, Haile-Selassie argued in 2001 that its status was still uncertain at that time because the *Orrorin* description contained no information about traits used for the diagnosis of bipedalism. To Haile-Selassie, nothing prevented accepting that *Orrorin* represented "the last common ancestor, and thereby antedating the cladogenesis of hominids [. . .], a previously unknown African hominoid with no living descendants, or an exclusive precursor of chimpanzees, gorillas or humans" (Haile-Selassie, 2001, p. 180).

However, the question whether *Orrorin* is an ancestor of panids, gorillids, and hominins, or a member of Hominini, cannot be settled in terms of primitive features such as canines. Considering derived traits and, in our opinion, the proximal end of the femur, there is enough evidence to consider *Orrorin tugenensis* as a biped—consequently, a hominin. It was so understood later by Yohannes Haile-Selassie, Suwa, & White (2004), as we will see when reviewing *Ardipithecus ramidus*. *Orrorin* locomotion was discussed extensively in the French Academy of Sciences, session of September 2004 in Paris (Ann Gibbons, 2004). Later, Richmond and Jungers (2008) conducted a detailed analysis of *Orrorin*'s femur, concluding that it provides solid evidence of belonging to a biped.

particular form of adaptation that contrasts with current bipedalism, acquired by the genus *Homo*.

Based on tomography scans of the neck–shaft junction of BAR 1002 00, Galik et al. (2004, p.1450) conclude that "the cortex is markedly thinner superiorly than inferiorly, differing from the approximately equal cortical thicknesses observed in extant African apes." Accordingly, Brigitte Senut suggested, at a symposium on "Prehistoric Climates, Cultures, and Societies" (Paris, France, September 13–16, 2004), that "*Orrorin*'s gait was more humanlike than that of the 2- to 4-million-year-old australopithecines" (see Gibbons, 2004).

An interesting aspect brought to light by the KPE's research team is the size of *Orrorin*'s femur and humerus. It is 1.5 times larger than that of A.L. 288–1, an *A. afarensis* found at Hadar, which we will review later (Section 4.4.1). According to Senut et al. (2001) this fact contradicts the widespread idea that our first ancestors were small-sized. However, A.L. 288–1 corresponds to

a female, which raises the important question of sexual dimorphism.

4.1.4 Middle Awash

In 1994, Tim White, Gen Suwa, and Berhane Asfaw published the results of the research campaigns of the two previous years at the Aramis site, within the Middle Awash region (Ethiopia). The findings included 17 possible hominin specimens from sites 1, 6, and 7, which at the time were the oldest documented hominin remains (White, Suwa, & Asfaw, 1994). Sixteen of them, and a great number of fossils of other vertebrates (more than 600), were found in strata between two markers, the complex of vitreous tuffs known as Gàala Tuff Complex (GATC; *Gàala* means "dromedary" in the Afar language) and the basalt tuff Daam Aatu Basaltic Tuff (DABT; *Daam Aatu* means "monkey" in the same language), with an average of 4 m of sediments between them (WoldeGabriel et al., 1994) (Box 4.5).

The hominin specimens of Aramis include, among other remains, three fragmentary bones of a left arm (ARA-VP-7/2), found half a meter above DABT, associated dentition (ARA-VP-1/128; the holotype ARA-VP-6/1), a deciduous molar (ARA-VP-1/129) (Figure 4.6), a complete right humerus (ARA-VP-1/4), and cranial fragments (ARA-VP-1/125; -1/500) (White et al., 1994). White and colleagues initially classified all those specimens in *Australopithecus*, but in a different species, suggesting *Australopithecus ramidus* for the Aramis exemplars. However, one year later they decided to elevate the differences to the rank of genus, adding the taxon *Ardipithecus ramidus* (*ardi* means "ground" or "floor" in the Afar language) for the Aramis specimens (White et al., 1995) (Box 4.6).

Aramis hominins share a wide range of characters with *A. afarensis*, also found in Ethiopia. But they are

Box 4.5 The age of Aramis

By means of the ^{39}Ar/^{40}Ar laser fusion method, the GATC tuff was estimated to be 4.387±0.031 Ma, setting the maximum age for the hominin remains at the site (WoldeGabriel et al., 1994). The DABT tuff could not be dated due to the high contamination of the Miocene soils. However, WoldeGabriel and colleagues dated the strata containing hominins between 4.48 and 4.29 Ma by biochronology and paleomagnetism. The rounded estimate usually attributed to the Aramis specimens is 4.4 Ma.

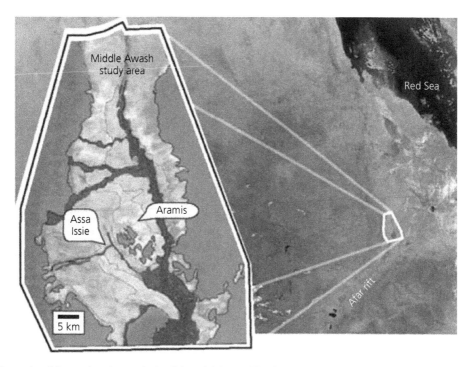

Figure 4.6 Location of the Aramis and Assa Issie sites (Ethiopia) (White, WoldeGabriel, et al. 2006).

Box 4.6 A new genus

The proposal of *Ardipithecus* as a genus to fit the Aramis specimens was done almost telegraphically. The new *Ardipithecus*' characteristic traits were defined on the base of a mandible discovered toward the end of 1994. Michael Day (1995) protested against the addition of a new genus in the tribe Hominin without offering a full argument. He also added, somewhat ironically, that the hurry to name the new genus before someone else did seemed directed to achieve priority rather than scientific clarity.

---

different in some features, mainly those relative to dentition. The first deciduous molar of ARA-VP-1/129 is closer to chimpanzees than to any hominins. Regarding adult dentition, some traits (the area of the ARA-VP-6/1 canine crown, for instance) are also similar to chimpanzees, but others are not. Incisors do not have the great width of current chimpanzees (the relationship between the incisors and molars and premolars

of Aramis specimens is typical of Miocene hominins and gorillas) and the morphology of canines is different from apes. The position of the foramen magnum, indicative of posture and, thus, of the possibility of bipedal locomotion, is close to that observed in the rest of hominins and distant from chimpanzees (White et al., 1994). Senut and colleagues (2001) expressed an opposite view, arguing that the bipedalism of *Ardipithecus* cannot be demonstrated on the basis of the described specimens.

New specimens, corresponding to nine hominins, were discovered in the Asa Duma site (Gona, Ethiopia) in 2004. They were found in soils dated between 4.5 and 4.3 Ma—estimation obtained by means of paleomagnetism and $^{40}Ar/^{39}Ar$. The exemplars include a partial right mandible (GWM3/P1), a left mandibular fragment (GWM5sw/P56), and other dental and postcranial fragments. The dentition allows these specimens to be included in *Ardipithecus ramidus*, according to Sileshi Semaw et al. (2005). The authors infer its bipedal character from the dorsal orientation in the transversely broad oval proximal facet of GWM-10/P1, a quite complete manual left proximal phalanx.

The Asa Duma specimens support the view that *Ardipithecus* are a group of peculiar hominins that combine bipedalism with ape masticatory traits (Box 4.7). Unfortunately, the site's Pliocene climatology corresponds to a mosaic of environments, which prevents consideration of those ardipithecines as organisms adapted to a precise habitat.

In 2001, Haile-Selassie reported the finding of 11 *Ardipithecus* specimens whose interpretation support their belonging to the hominin lineage. The new specimens were dated between 5.2 and 5.8 Ma and come from five localities in the Ethiopian part of Middle Awash (Saitune Dora, Alíala, Asa Koma, and Digiba Dora on the western margin of the Middle Awash; Amba East from the Kuserale member of the Sagantole Formation of the Central Awash Complex) (Haile-Selassie, 2001).

The 2001 Middle Awash specimens include postcranial fragments—like a manual phalanx (ALA-VP-2/11), a pedal phalanx (AME-VP-1/71), and arm bones—that lend support to the notion that *Ardipithecus* were bipedal (Figure 4.7). In particular, the dorsal orientation of the AME-VP-1/71 phalanx indicates,

according to Haile-Selassie, a similar pedal morphology to that of *A. afarensis*—in mosaic, with traits shared with apes—and indicative of a similar locomotion to that of *A. afarensis* and *Ar. ramidus*. Dental remains, composed by a right mandible with associated teeth (ALAVP-2/10) and other dental material, show a mixture of primitive traits, shared with apes, and hominin features—lower canines with developed distal tubercles and expressed mesial marginal ridges. Haile-Selassie (2001, p. 179) stated that "studies of enamel thickness are underway, but the available broken and littleworn teeth suggest that molar enamel thicknesses [. . .] were comparable to, or slightly greater than, those of the younger Aramis samples of *Ar. ramidus*." The primitive traits that separate them from the Aramis *Ar. ramidus* led Haile-Selassie (2001) to attribute the Middle Awash exemplars to a new subspecies: *Ardipithecus ramidus kadabba*.

Regarding the phylogenetic position of *Ar. ramidus*, Haile-Selassie (2001) defined the taxon as a hominin close to the divergence from chimpanzees. Because the ages attributed to Middle Awash *Ardipithecus* and *Orrorin tugenensis* places both of them very close to the

---

Box 4.7 What is *Ardipithecus*?

Henry Gee (1995) advanced a pair of prophecies regarding the *Ar. ramidus* specimens. First, by the year 2000, *Australopithecus ramidus*, as they were know at the time, would have been placed into another genus. It was not necessary to wait so long: a few months later White and colleagues (1995) introduced the taxon *Ardipithecus*. Indeed, ardipithecines are such special hominins that it is reasonable to classify them in a separate group. Peter Andrews described them as "ecological apes" (Andrews, 1995), meaning that their enamel must be related with an adaptation to the tropical forest, as is the case with chimpanzees and gorillas. But, were they bipedal? Should the genus *Ardipithecus* be included among other hominins or in the chimpanzee or gorilla lineages? Gee's (1995) second prophecy forecast that, again around the year 2000, *ramidus* would be considered a member of the "ramidopithecines," the common ancestors of chimpanzees and humans. Senut and colleagues (2001) considered them in such a way.

The most controversial trait of *Ardipithecus* is the thinness of its enamel. This trait is often used in the discussion regarding hominoid phylogenesis and classification as a criterion to identify hominins. *Ramapithecus* was previously considered as a hominin precisely because of its thick

molar enamel (Pilbeam, 1978). The appearance of such a very early being as *Ardipithecus ramidus* and its thin molar enamel raised doubts. Other hominins have thick enamel while the enamel of chimpanzees and gorillas is thin, and that of orangutans is of an intermediate thickness. Thus, *Ardipithecus* led to a new edition of the discussions concerning the value of enamel thickness for the determination of lineages. Peter Andrews, for instance, noted that thick dental enamel seems to be shared by hominins and by other 10 Ma fossil apes, which suggests that the chimpanzee thin enamel is a derived trait (cited by Fischman, 1994). Haile-Selassie (2001) thought that the controversy regarding the thin enamel of *Ar. ramidus* was not very important, given the great variability of the trait, even within a single species.

White and colleagues (1994) and Fischman (1994) have warned of the need for accounting for global features of teeth shape and other masticatory aspects when classifying specimens. The specific trait of enamel would, then, lose much of its significance. However, as Ramirez Rozzi (1998) has shown by means of the study of its microstructure, enamel retains relevance when specimens found in the same or close sites are compared.

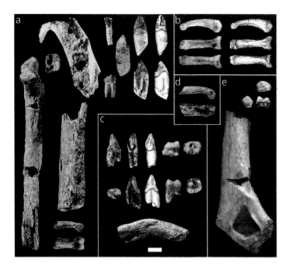

Figure 4.7 *Ardipithecus ramidus kadabba* from the Upper Miocene Middle Awash deposits (Haile-Selassie, 2001). (a) ALAVP-2/10, mandible and all associated teeth; ALA-VP-2/120, ulna and humerus shaft; ALA-VP-2/11, hand phalanx. (b) AME-VP-1/71, lateral, plantar, and dorsal views of foot phalanx. (c) STD-VP-2, teeth and partial clavicle. (d) DID-VP-1/80, hand phalanx. (e) ASKVP- 3/160, occlusal, mesial, and buccal views; ASK-VP-3/78, posterior view. All images are at the same scale: scale bar, 1 cm (Haile-Selassie, 2001).

cladogenesis that originated the tribe Hominini, it is striking that they are so different. However, it must be recalled that Haile-Selassie (2001) stated that it has not been demonstrated that *Orrorin* is a hominin.

The question of the relative positions of *Orrorin tugenensis* and *Ar. ramidus* changed notably after the finding in 2002 of six new *Ardipithecus* teeth in the Asa Koma Locality, with an age of between 5.6 and 5.8 Ma (Haile-Selassie, Suwa, & White, 2004). Specifically, the specimens' age ranges between the 5.77±0.08 Ma of basaltic tuff Ladina (LABT), close to the basal member, and 5.63±0.12/5.57±0.08 of two measurements made at the Witti Mixed Magmatic Tuff (WMMT) corresponding to Asa Koma (Wold-eGabriel et al., 2001).

In the article that christened the taxon *Ar. ramidus kadabba*, Haile-Selassie (2001, p. 178) pointed out the "possible absence of a fully functional honing canine/premolar complex in Ardipithecus." The Asa Koma sample also included a lower canine. After comparing available canines with those of chimpanzees and australopithecines, Haile-Selassie, Suwa, and White (2004, p. 1505) stated: "the projecting, interlocking upper and lower canines, and the asymmetric lower P3 with buccal wear facet imply that its last common

ancestor with chimpanzees and bonobos retained a functioning C/P3 complex. But wear on the upper and lower canines of *Sahelanthropus* and the lower canine of *A. kadabba* from Alayla suggest a lack of consistently expressed functional honing in these earliest hominins." Haile-Selassie, Suwa, and White (2004)) suggested that the scarce but meaningful dental-derived traits confirm the hominin condition of the Alayla and Asa Koma samples. However, definition of the new species *Ardipithecus kadabba* was based on primitive characters.

4.1.5 The Aramis skeleton

The information available on Miocene hominins multiplied dramatically with the publication of a set of detailed studies on the ARA-VP-6/500 exemplar, a nearly complete skeleton of *Ardipithecus ramidus* from the Aramis site (Ethiopia), whose first indications prompted the proposal of the genus (Lovejoy, Latimer et al., 2009; Lovejoy, Simpson et al., 2009; Lovejoy, Suwa, Simpson et al., 2009; Lovejoy, Suwa, Spurlock et al., 2009; Suwa, Asfaw et al., 2009; Suwa, Kono et al., 2009; White, Ambrose et al., 2009; White, Asfaw et al., 2009) (Figure 4.8 and Box 4.8).

ARA-VP-6/500 analysis occupied the entire issue of the journal *Science* of October 2, 2009. It gathers the details of the discovery of the successive materials that compose the exemplar, and very exhaustive studies on its anatomical features as well as the ecological environment and evolutionary meaning of *Ardipithecus*, studies carried out by comparing the exemplar with other specimens of different hominins from the Upper Miocene and Lower Pliocene.

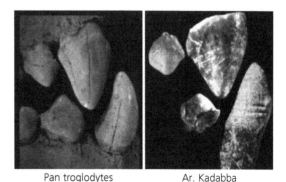

Pan troglodytes   Ar. Kadabba

Figure 4.8 Lateral views of a female common chimpanzee (left) and Ar. kadabba (right) upper and lower canines and premolars (upper canine ASK-VP-3/400, lower canine STD-VP-2/61, upper premolar ASK-VP-3/160 reversed, lower premolar ASK-VP-3/403 reversed). Photographs from Haile Selassie et al. (2004).

Box 4.8 Age and hypodigm of ARA-VP-6/500

The extensive study of ARA-VP-6/500 did not include new dating for the specimen. In the introduction to the issue of the journal *Science* of October 2, 2009, Brooks Hanson referred to the 4.4 Ma, repeated by Bruce Alberts in the editorial, by Ann Gibbons' commentary in the *News Focus* section, and in various articles of the collaborators in the monograph. White, Asfaw et al. (2009) indicated that the stratigraphic horizon of *Ar. ramidus* at Aramis was comprised between the volcanic tuffs Gàala ("Camel" in Afar language) and Daam Aatu ("baboon" in Afar language), dated 4.4 Ma, and was indistinguishable from them. WoldeGabriel et al. (2009, also in the same monograph of *Science*) adjusted the dates of these two tuffs referring to previous studies: 4.419±0.068 Ma for Gàala (WoldeGabriel et al., 1994) and 4.416±0.031 Ma for Daam Aatu (Renne et al., 1999), obtained in both cases by the $^{40}Ar/^{39}Ar$ method. The figure of 4.4 Ma for *Ar. ramidus* of Aramis is therefore justified.

It is impossible to provide a table of the hypodigm of *Ardipithecus*, since it consists of over 300 specimens. The whole list appears in the supporting online material of White, Asfaw, et al. (2009).

(http://www.sciencemag.org/content/326/5949/64/suppl/DC1).

The postcranial elements of ARA-VP-6/500 include a forelimb, femur, pelvis, and foot, and show a combination of primitive and derived features that, taken together, lead to two important conclusions deduced from two separate analyses. First, *Ardipithecus* locomotion appears to be a combination of bipedalism and tree-climbing. In the opinion of the team directed by Tim White (White, Asfaw et al., 2009), *Ardipithecus* performed both tree-climbing and a kind of terrestrial bipedalism more primitive than *Australopithecus* (Lovejoy, Latimer et al., 2009; Lovejoy, Simpson et al., 2009).

These results provided some important clues for explaining how the separation of human, chimpanzee, and gorilla lineages took place.

4.2 The role of locomotion in the divergence of hominoid lineages

In Chapter 3, we reviewed various cladistic proposals on hominoid evolution, leaving as pending new clues that fossils from the Upper Miocene might offer. As we have seen, neither the specimens of *Sahelanthropus* nor of *Orrorin* are sufficiently informative. By contrast, the hypodigm of *Ardipithecus* provides numerous elements of comparison that establish the polarity—primitive or derived trait—of characters involved in locomotion.

A trend going back to Arthur Keith (1903) saw brachiation as the primitive trait, which caused the development of postcranial features capable of leading, through several stages, to bipedal locomotion. But, what is the significance of a plesiomorphy when considering the development of the different forms of locomotion distinctive of apes and modern humans?

Since the work of George G. Simpson (1953), evolution has been seen as a directional vector in which certain traits (primitive) are retained atavisms and others (derived) characterize the new form of adaptation. Following this idea, Bruce Latimer (Latimer & Lovejoy, 1989; Latimer, 1991) considered that the functional value of the plesiomorphies related to hominoid locomotion is very low because no arboreal adaptation based on primitive traits remains intact in Upper-Miocene African apes (Box 4.9).

The position of Latimer (1991) is entirely correct but should be clarified. Musculoskeletal plesiomorphies of humans and African apes do not indicate the type of behavior that is undoubtedly bipedalism in the first case and knuckle-walking in the second. However, that does not mean that they are superfluous. Atavisms offer important clues for understanding how the phylogenesis took place in the return to the ground from tree life—behavior in which both African apes and humans coincide. Determining the answer raises serious difficulties.

Box 4.9 Traits in mosaic

The problem confronting any attempt to identify functional traits such as locomotion from anatomical features present in specimens with a mosaic evolution—as is the case with almost all Miocene and Pliocene hominins—was well explained in the work of Carol Ward (2002) regarding the posture and locomotion type of *A. afarensis*. Ward (2002) pointed out that the presence of traits in mosaic demands that the function of each trait be separately determined. Primitive characters might be traits that natural selection has not removed but whose functional significance turns out to be poor and of not much relevance.

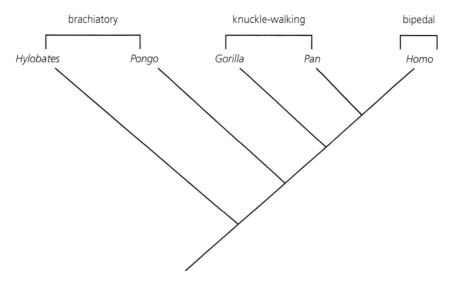

Figure 4.9 Cladogram of current hominoids with their respective modes of locomotion.

Let's see why. A glance at the cladogram establishing the proximity of the current hominoid lineages reveals the problem (Figure 4.9). Although it might seem that there is a coherent evolutionary order regarding the type of locomotion, with a brachiation–knuckle walking–bipedalism sequence, it is not the case at all.

In the cladogram of Figure 4.9, *Hylobates* is the sister group of all other hominoids. Actually, the whole clade of Asian higher apes, African apes, and humans has three different modes of locomotion: brachiation, knuckle-walking, and bipedalism.

What cladograms arise from considering one or the other trait, brachiation or bipedalism, as a plesiomorphy? Let us start with the first hypothesis already advanced about brachiation as a primitive trait, since it is the behavior of gibbons. This is the case of Figure 4.10(a).

In order to highlight the problem more clearly, in this figure we have placed sister groups *Pan* and *Homo* in a different way than they are usually represented. This does not imply any consequence either taxonomic or phylogenetic. Considering the meaning of what sister groups are on a cladogram, it is immaterial on which side to place humans or chimpanzees.

If brachiation is a plesiomorphy, knuckle-walking has to be a derived trait, and it would have been fixed independently twice, in parallel, in *Gorilla* and in *Pan*. We will later return to this question in light of the evidence provided by *Ardipithecus*.

There are alternatives, of course. One is to hold that the primitive trait is knuckle-walking, which was lost in the human lineage. If so, orangutans' brachiation would be a derived trait (Figure 4.10b). But, once again, this trait would have appeared twice separately: in

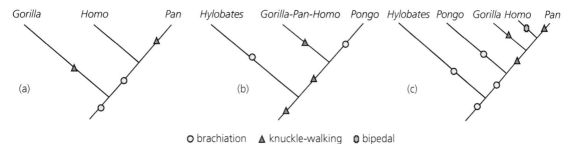

○ brachiation △ knuckle-walking ◓ bipedal

Figure 4.10 (a) Brachiation hypothesis as a primitive trait. Knuckle-walking of chimpanzees and gorillas, not shared by humans, must be a homoplasy. (b) Knuckle-walking hypothesis as a primitive trait. Brachiation of "gibbons + siamangs" and orangutans, which is not present in the clade of "African apes + humans," has to be a homoplasy in Asian apes. (c) Hypothesis of the emergence of knuckle-walking as a primitive trait of gorillas and humans + chimpanzees.

Hylobates and *Pongo*. Gibbons and orangutans, would now share brachiation, which would be a homoplasy. Could we settle the matter by arguing that brachiation is the primitive trait, that knuckle-walking developed in the ancestors of African apes, and, finally, that the tribe Hominini, with the human ancestors, is bipedal? This is the scheme shown in Figure 4.10(c).

The idea that knuckle-walking would have appeared as a primitive trait shared by gorillas and chimpanzees—and would be, therefore, the locomotion of pre-humans—was maintained by John Napier (1967) and Sherwood Washburn (1967), although without providing adequate anatomic evidence of that evolutionary fact. Brian Richmond and David Strait (2000), after examining australopith specimens, have provided evidence showing that ancient hominins retained certain morphology in the wrist indicative of previous knuckle-walking (Figure 4.11).

Richmond and Strait's proposal is compatible with the idea of knuckle-walking as an apomorphy common to African apes, replaced in hominins by bipedalism, but this is an hypothesis advanced for theoretical reasons, as well as for lack of sufficient evidence to settle the issue (Dainton, 2001; Lovejoy et al., 2001; Kivell & Begun, 2006; Patel & Carlson, 2007; Crompton et al., 2008).

If knuckle-walking is a primitive trait in gorillas, chimpanzees, and the first hominins, the cladogram proposed by molecular geneticists such as Morris Goodman is more than plausible (Figure 4.12). This was recognized by Mark Collard and Leslie Aiello (2000) in their commentary on the paper by Richmond and Strait (2000).

Nonetheless, Brigitte Senut (1991) warned about drawing any conclusions from the comparative analysis between fossil specimens and the great apes that exist today. The combination of features of fossil species and, hence, their form of locomotion would be very different from the features and specialized locomotion of current higher apes. This is the same argument used by Bruce Latimer (1991) that we saw recently regarding the plesiomorphies of the locomotor apparatus. It is, therefore, necessary to analyze the locomotion of early hominins for possible clues about which primitive trait—knuckle-walking or brachiation—was replaced by derived bipedalism.

4.2.1 Primitive and derived traits of locomotion

The interpretation of the postcranial elements of ARA-VP-6/500 done by the Tim White team led to a surprising result that contradicted the position most

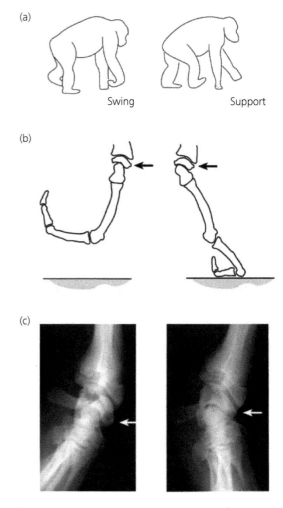

Figure 4.11 The wrist joint during the swing phase (left-hand panels) and support phase (right-hand panels) of knuckle-walking (Richmond & Strait, 2000).

commonly accepted about the evolution of locomotor apparatus in hominoids. This interpretation maintained that:

- As a general guideline, orthogrady and suspension are primitive traits fixed in hominoids in the Miocene.
- From these primitive traits, derived locomotor patterns developed both in hominins and higher apes.

For C. Owen Lovejoy and collaborators (Lovejoy, Latimer et al., 2009; Lovejoy, Simpson et al., 2009; Lovejoy, Suwa, Simpson et al., 2009), the primitive traits of *Ar. ramidus* indicate that the LCA (last common

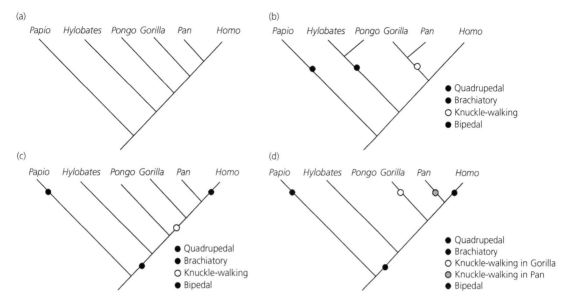

Figure 4.12 (a) Most-parsimonious cladogram according to hominoid genetic distances. (b) Most-parsimonious cladogram if knuckle-walking is a derived trait characteristic of chimpanzees and gorillas. (c) Most-parsimonious cladogram if knuckle-walking is a primitive trait of African apes, modified in hominins. (d) If the cladogram established by molecular distances has to be made compatible with knuckle-walking as a derived trait of chimpanzees and gorillas, then this form of locomotion must be a homoplasy fixed separately in *Gorilla* and *Pan*.

ancestor) of gorillas, chimpanzees, and humans lacked suspension adaptations common to all current apes. Lovejoy, Simpson et al. (2009) maintain that the examination of *Ar. ramidus* postcranial elements reveals a mix of retained plesiomorphies showing that LCA locomotion corresponds to an arboreal palmigrade locomotion (quadrupedy using the palms of the hands and the soles of the feet). Such traits of the ancestor of gorillas, chimpanzees, and humans would mean, therefore, an exaptation for both suspension and vertical climbing of gorillas and chimpanzees, as well as for human bipedalism (Lovejoy, Suwa, Simpson et al., 2009).

Is this evolutionary model acceptable? The scheme of the team led by Tim White, which analyzed the postcranial remains of *Ar. ramidus*, actually contains two different proposals:

(1) The postcranial elements of *Ardipithecus ramidus* do not have any primitive feature indicating in LCA the existence of knuckle-walking as the means of locomotion.
(2) The primitive features of the postcranial remains of *Ardipithecus ramidus* indicate that tree-climbing, and not suspension or brachiation, was the locomotion of LCA (Figure 4.13).

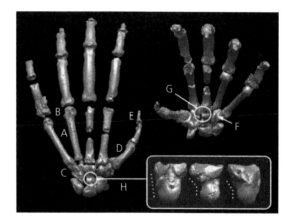

Figure 4.13 Digitalized views of the left hand of *Ar. ramidus* from elements coming from the specimens *ARA-VP-7/2* and *ARA-VO-6/500*. Primitive traits absent in apes: (A) short metacarpals; (B) lack of knuckle-walking grooves; (C) extended joint surface on fifth digit; (D) thumb more robust than in apes; (E) insertion gable for long flexor tendon (sometimes absent in apes); (F) hamate (carpal bone), which allows palm to flex; (G) simple wrist joints; (H) capitate head (largest carpal bone) that promotes a sharp palmar flexion. From Lovejoy, C. O., Simpson, S. W., White, T. D., Asfaw, B., & Suwa, G. (2009). Careful Climbing in the Miocene: The Forelimbs of Ardipithecus ramidus and Humans Are Primitive. *Science*, 326(5949), 70. Reprinted with permission from AAAS.

Box 4.10 Evidence of knuckle-walking

The adaptive key to the ancestral behavior of knuckle-walking would be shown particularly in the hands and especially in the medial metacarpal bones, excluding that of the thumb (i.e., of Mc2 to Mc5). These bones should present evidence of having previously carried out a locomotor function. Indications of the primitive locomotion of knuckle-walking could be the expansion of the medial metacarpal—to increase surface and supporting force—or the presence of prominent crests and/or grooves in the distal end of Mc2–Mc5—related with the insertion of developed muscles.

Although Lovejoy and collaborators linked both interpretations in their various studies published in 2009, the evidences provided are not the same, or carry the same weight. Let us examine one-by-one the two hypotheses.

Indicative plesiomorphies of an eventual knuckle-walking

In this regard, the most informative postcranial elements in *Ar. ramidus* are those of the anterior limbs. The foot has already suffered modifications toward bipedestation, which are apomorphic for that very ancient hominin; what we are looking for are primitive features (Box 4.10).

In relation to the anterior limbs, as part of the skeleton set ARA-VP-6/500 of *Ar. ramidus*, the most informative remains are those of ARA-VP-7/2 (anterior limbs), and ARA-VP-6/500 (in relation to the lower arm and both hands). Lovejoy and collaborators (Lovejoy, Simpson et al., 2009) have indicated that morphological characteristics of primitive knuckle-walking are completely absent in these specimens of *Ar. ramidus*. On the contrary, the metacarpals show marks of dorsal invaginations—a trait reflecting an extreme dorsiflexion—similar to that appearing in some cercopithecoids and ceboids (monkeys of the Old and New World), as well as Miocene hominoids such as *Proconsul*. From such a condition, Lovejoy, Simpson et al. (2009) deduced that knuckle-walking could not have been the locomotion of the human and chimpanzee LCA, or of humans, chimpanzees, and gorillas, because no trace of any plesiomorphy appears in the hands of *Ar. ramidus*.

That conclusion seems hard to deny. In fact, analysis of gait biomechanics on the whole human lineage, conducted by Crompton and collaborators (2008), supported the hypothesis of the appearance of terrestrial bipedalism from the tree orthograde position, aided by hands—such as in the case of the orangutan—discarding knuckle-walking as a primitive trait (Box 4.11).

Indicative plesiomorphies of arboreal palmigrade and/or exclusionary of vertical climbing, suspension, or brachiation

The second hypothesis of the Tim White team has as its best source of evidence the fossils of anterior limbs (ARA-VP-7/2 and ARA-VP-6/500), but, important with respect to this second hypothesis are those relative to the morphology of the humerus and ulna, the complex of the central-midcarpal joint and the radiocarpal joint (wrist–radius).

The phylogenetic meaning extracted by Lovejoy, Simpson et al. (2009) from these morphological differences suggests, as we have noted, that the primitive trait of the LCA of African apes and humans is the arboreal palmigrade; hence, the orthograde posture of all the great apes and humans would have developed as a homoplasy in each clade: *Pongo, Gorilla, Pan*, and *Ardipithecus*.

Lovejoy, Simpson et al. (2009) argued that the wrist–radius joint is the main evidence in favor of primitive palmigrady, which in *Ar. ramidus* shows plesiomorphies contrasting with derived traits from *Pierolapithecus* and great apes, including *Pongo* (the character of radius retraction indicated in Box 4.11). If so, the apomorphies of that joint present in *Pongo, Gorilla*, and *Pan*, as well as of the different apes from the Middle Miocene—not being primitive traits—would have fixed in parallel and independently in all these taxa.

The interpretation of Lovejoy, Simpson et al. (2009) of the orthogrady and suspension of current apes—and those of the Middle Miocene—requires a substantial number of homoplasies, which contradicts the parsimony criterion. Their conclusion would be inevitable if there was no alternative for interpreting the locomotion inferred from the postcranial traits of *Ardipithecus ramidus*.

But, in fact, there are several alternatives.

In a commentary about the various studies by Tim White and colleagues on *Ar. ramidus*, which appeared in the previously mentioned monographic issue of *Science* (October 2, 2009), Terry Harrison (2010) argued that *Ardipithecus* has an enigmatic phylogenetic position with respect to other hominins, which raises the question of whether it might be just one mere ape

Box 4.11 Locomotion of *Ar. ramidus*

The main indicators of a primitive locomotion attributed to *Ar. ramidus* by the team of White are the following:

- Humerus—The sample of *Ar. ramidus* includes a proximal humerus with its own axis (ARA-VP-7/2-A) and a humeral axis (ARA-VP-1/4). These exemplars maintain a rugose thick deltoid crest that, in contrast, is reduced in apes (in all of them, it should be emphasized, including in those with knuckle-walking). Therefore, the deltopectoral morphology may serve, according to Lovejoy, Simpson et al. (2009), as a key indicator of the locomotor behavior of fossil hominins.
- Ulna—Retroflexion of the trochlear notch (articular notch of the ulnar trochlea, also called greater sigmoid cavity or sigmoidal notch) was widely accepted as an early sign of suspensory locomotion. *Ar. ramidus* shows a different trait: a notch in the anterior side that, for Lovejoy, Simpson et al. (2009), is assumed to be a plesiomorphy associated with arboreal adaptation and bridging.
- Radius and CJC (central joint complex)—As Lovejoy, Simpson et al. (2009) indicate, all higher apes showing suspensory behavior have a thickening of the CJC ligaments. The expression of this bony thickening is reflected in the isolation of the facies of the dorsal joint and palmar intermetacarpals. *Dryopithecus* has the condition of the current great apes but this is absent in *Proconsul, Ar. ramidus, Australopithecus,* and *Homo,* which indicates that none of these taxa have robust carpometacarpal ligaments. The scaphoid of *Ar. ramidus* differs from that of cercopithecoids, *Proconsul,* and other Miocene apes such as *Pierolapithecus,* as indicated by Lovejoy, Simpson et al. (2009), for traits such as the fusion of *os centrale.* This fusion is also present in African apes, but not in *Sivapithecus* or in *Pongo.* However, the retraction of the radius is present in hominin, as well as in the great apes, including *Pongo.*

among the tangle of branches forming the basal part of the hominin lineage tree. In this case, *Ardipithecus* would be a terminal taxon that retains plesiomorphies of its very ancient origin, prior then to *Pierolapithecus,* a fact that tells us nothing about the apomorphies of human bipedalism. Such an assumption, contradicting

the evidence related to the features of the pelvis and posterior limbs of *Ardipithecus ramidus* (Lovejoy, Latimer et al., 2009; Lovejoy, Suwa, Spurlock et al., 2009), implies the removal of *Ardipithecus* from the human lineage. In this book we have rejected that taxonomic solution.

Another alternative is to argue that human hands are largely plesiomorphic. Thus, compared to African apes' apomorphies, hominins would have retained primitive traits characteristic of Middle Miocene hominoids. Sergio Almécija, Salvador Moyà-Solà, and Davis Alba (2010) referred that condition to *Orrorin,* not to *Ardipithecus*—a taxon about which they express agreement with Harrison (2010). But the trait examined by Almécija et al. (2010) was the distal phalanxes, not the wrist–radius joint. That the hands of hominins are in a large part primitive is precisely the argument used by Lovejoy, Simpson et al. (2009) in favor of the LCA palmigrady. Could we conceive that the hominin hands—including *Ardipithecus ramidus*—are primitive and indicate a palmigrady, and also that this is compatible with an orthograde posture and a suspensory capacity?

Such an interpretation is entirely consistent with the analysis made by Biren Patel and collaborators (2009) about terrestrial adaptations reflected in the hands of KNM-TH 28860, *Equatorius africanus.*

Considering knuckle-walking as a derived trait rather than primitive leads to an interesting question. As shown in Figure 4.14, the way lineages split indicates that shared traits of the chimpanzee and gorilla must also be found in *Ardipithecus.* To the extent that the clade "chimpanzee + *Ardipithecus*" has a common history as a sister group of *Gorilla,* derived traits that may exist in gorillas and chimpanzees but not *Ardipithecus* must be homoplasies, features that evolved independently retaining only a casual similarity. In other words, if knuckle-walking is not a primitive trait, then in chimpanzees and gorillas it must be an analogous trait.

The hypothesis of a convergent evolution that has led to homoplasies related to knuckle-walking in gorillas and chimpanzees has been maintained by, among others, Kivell and Schmitt (2009). However, the study of Scott Williams (2010) raised doubts about the morphological complex necessary to perform a locomotion compatible with the idea of two convergent and independent evolutions. Williams (2010) analyzed the morphological integration of the different traits of the locomotor apparatus that, taken together, allow functional knuckle-walking. He concluded that the null hypothesis of the knuckle-walking homology—that

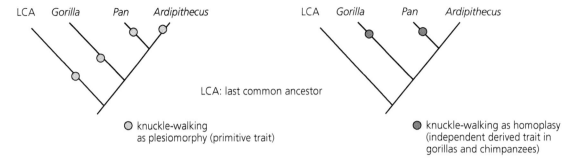

Figure 4.14 If knuckle-walking is a primitive trait (left), it must be present in *Ardipithecus*. If it is a derived trait in gorillas and chimpanzees (right), this must be an analogous feature fixed independently in each lineage.

is to say, its character as a primitive trait of the LCA (last common ancestor) of gorillas, chimpanzees, and humans—could not be discarded at the moment.

As we see, the role of plesiomorphies may be very important for understanding the evolutionary process. The role inferred from the presence of primitive traits related with arboreal palmigrady in *Ardipithecus* is not limited only to support the idea of knuckle-walking appearing independently in gorilla and chimpanzees. Lovejoy, Suwa, Simpson et al. (2009) argued that, although the inference of a common orthograde as the origin of bipedalism has been accepted as obvious, *Ar. ramidus* morphology indicates something else. Since arboreal climbing is a primitive trait of African apes and humans, the advanced orthograde had to evolve in parallel and independently in *Ardipithecus*, *Pan*, and *Gorilla*. Therefore, the hominoid cladogram becomes full of homoplasies: knuckle-walking would be so for gorillas and chimpanzees, and orthograde for African apes and humans.

4.3 Change in the Lower Pliocene: genus *Australopithecus*

Ardipithecus is the only Miocene genus properly identified. The Lower Pliocene ranges from 5.33 to 3.60 Ma. For hominins, the phylogenetic transit during this time period implies the emergence of a new taxon. *Australopithecus* derived through evolutionary changes from *Ardipithecus*, which in particular affected two traits: bipedalism, which evolved toward a more terrestrial locomotion, and the masticatory apparatus, showing the first tendencies of megadontia. The transition is well documented in the Rift Valley, as well as in the Chad. In the genus *Australopithecus* during the Middle Pliocene—3.60/2.58 Ma—hominins extended

their range to South Africa. Finally, in the Upper Pliocene—2.58/1.80 Ma—the hominin lineage saw the emergence of two new genera, *Paranthropus* and *Homo*, which we will explore in the ensuing chapters. Figure 4.15 shows the phylogenetic relationships among taxa that we accept, without including the reasons to support them.

Numerous species have been assigned to the taxon *Australopithecus*. We will accept here those appearing at the beginning of this chapter in Table 4.1, repeated here in Table 4.4 indicating the author(s) of each taxon and the year of the proposal.

In this chapter, we will address the *Australopithecus* species of the Lower Pliocene, i.e., *A. anamensis* and *A. afarensis*. The remaining species will be described in Chapter 5.

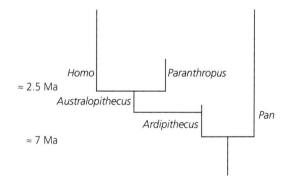

Figure 4.15 Phylogenetic tree of the genera *Ardipithecus*, *Australopithecus*, *Paranthropus*, and *Homo*. No timescale (except the references to the left) or range of each taxon's existence is indicated; only the idea that *Australopithecus* proceeds from phyletic evolution (without branches) of *Ardipithecus* is reflected, while *Paranthropus* and *Homo* are sister groups emerging from *Australopithecus* by cladogenesis. All these statements will gradually be documented.

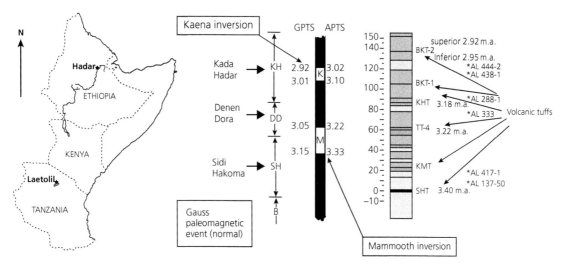

Figure 4.16 Left: Localization of Hadar (Ethiopia) and Laetoli (Tanzania) sites. Johanson, D. C., White, T., & Coppens, Y. (1978). A New Species of the Genus Australopithecus (Primates: Hominidae) from the Pliocene of Eastern Africa. *Kirtlandia*, 28, 1–14. Right: Kada Hadar (KH), Denen Dora (DD), and Sidi Hakoma (SH) members in the Hadar formation. APTS, astronomical polarity time scale; GPTS, geomagnetic polarity time scale. (Figure 3.10 is repeated here for reader convenience.).

Table 4.4 Species of the genus *Australopithecus*, with age, place, and year of taxon designation

Taxon	Age (Ma)	Place	Author and year of the proposal
A. anamensis	4.1	Rift	Leakey et al., 1995
A. afarensis	3.85–3.63	Rift	Johanson et al., 1978
A. bahrelghazali	c. 3.5	Chad	Brunet et al., 1996
A. platyops	3.5	Rift	Leakey et al., 2001
A. africanus	2.2	South Africa	Dart, 1925
A. garhi	2.5	Rift	Asfaw et al., 1999
A. sediba	1.98	South Africa	Berger et al., 2010

4.3.1 Apomorphies of *Australopithecus*

Although more modern than Miocene *Ardipithecus*, the historical circumstances of the discovery of the australopiths, described long ago, made them the "most ancient humans" for many decades. The first named australopith—*A. africanus*—is from 1925, while *Ardipithecus* was first described in 1994. For almost three-quarters of a century *Australopithecus* was synonymous with "the first human."

The species in Table 4.4 were named over a long period of time: 85 years elapsed since the proposal of *A. africanus* to that of *A. sediba*. During the decades between the two proposals, the requirements for dating, taphonomic analysis, and taxonomic rigor did not maintain the same criteria. Therefore, almost every new taxon proposal was discussed and, in some cases, rejected to the point of being forgotten. The taxa we maintain as *Australopithecus* species are generally accepted, although by no means free of doubts.

These doubts influence the definition of *Australopithecus'* apomorphies. As expressed by White, WoldeGabriel, et al. (2006) when reporting the discovery of the specimens of *A. anamensis* in Middle Awash: 20 years earlier, *Australopithecus afarensis* was the epitome of primitivism, in particular in relation to its locomotor apparatus, but since the proposal of *Ar. ramidus* and *A. anamensis*, the posterior limb traits of *A. afarensis* are considered as apomorphies. Derived characters needed to define *Australopithecus* are, therefore, not those mentioned in the proposals of the different species, but those described in later reinterpretations. If we add that any study of this kind needs some time after the original species' proposal, and that the last species of Table 4.4 was named in 2010, it is possible to think that there is no updated general study on the apomorphies of *Australopithecus*. Those available come from analysis done on fewer taxa than are generally accepted today. What we have the most are descriptions of each particular species. Nevertheless, there is enough agreement to consider two traits as distinctive of *Australopithecus* in comparison to the earlier *Ardipithecus* and the later *Homo*:

- Development of the locomotor apparatus toward bipedalism, maintaining to some extent climbing capabilities.
- Development of the masticatory apparatus toward megadontia, although less than that of *Paranthropus*.

These apomorphies do not present the same degree of development in each species of *Australopithecus* but indicate a common trend. The progression within this trend and the phylogenetic relationships drawn from such developments will be addressed when we review the phylogeny outline in the Pliocene.

4.4 Australopithecus afarensis

Australopithecus anamensis, as the most ancient species, is the first taxon of Table 4.4. However, due to historical reasons about the way the different australopiths were discovered in the Rift Valley, which gave important clues about the interpretation of the evolutionary process, we think it is convenient to change the descriptive order and start with *A. afarensis*. In Chapter 3 we described the sites in relation to the specimens of the different species of *Australopithecus*, where precise details can be consulted. Nevertheless, we have repeated here some of the maps and stratigraphic schemes.

4.4.1 Hadar specimens

If a map localizing the sites of hominins in the Rift Valley is made, we will see that Laetoli (Tanzania) and Hadar (Ethiopia) form the southern and northern extremes of a vast extension (Figure 4.16). But Laetoli, as well as Hadar, remain linked to one of the best known specimens placed among the Pliocene hominins from the morphological point of view.

The first discovery of fossil hominin remains at Hadar took place on October 30, 1973. It included four associated fragments of lower limb bones (left femur, A.L. 128–1, and right tibia, A.L. 129–1), which permitted reconstruction of the knee of an individual that, judging from this articulation's morphology, was bipedal (Johanson & Taieb, 1976; Johanson & Coppens, 1976). The specimens were found in the lower part of

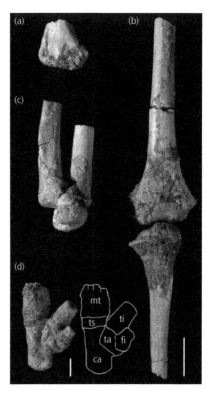

Figure 4.17 DIK-1-1 juvenile postcrania. a, Right distal humerus. b, Left distal femur and proximal tibia. c, Right distal femur and proximal tibia in a flexed position, connected by matrix, with the tibial and femoral diaphyses pointing upwards. d, Left foot and its outline including metatarsals (mt), distal tibia (ti), distal fibula (fi), talus (ta), calcaneus (ca) and tarsals (ts). Scale bars, 1 cm for the foot (d) and 2 cm for the limb bones (a–c). Reprinted by permission from Macmillan Publishers Ltd: Alemseged, Z., Spoor, F., Kimbel, W. H., Bobe, R., Geraads, D., Reed, D., & Wynn, J. G. (2006). A juvenile early hominin skeleton from Dikika, Ethiopia. Nature, 443(7109), 296–301.

Box 4.12 Lucy's discovery

The circumstances of Lucy's discovery are well known. On November 30, 1974, while exploring Hadar Locality 162, Donald Johanson and Tom Gray, came across numerous bone fragments that at first sight seemed to belong to a singe individual (Johanson & Edey, 1981). The finding was registered as A.L. 288–1 and named Lucy. That same night, during the celebration of the discovery, there were drinks, song, and dance at the Hadar campsite. The magnetophone played the Beatles' song *Lucy in the Sky with Diamonds* over and over. No one remembers when or who suggested it, but the skeleton was baptized with the popular name it has been referred to since (Johanson & Edey, 1981).

the Sidi Hakoma member, just above the SHT tuff, in soils dated around 3 Ma. The following year, 1974, ten additional specimens offered a much broader vision of the Hadar hominins. Among them was the famous, almost complete, A.L. 288–1 skeleton, Lucy (Box 4.12). The 1974 specimens include, in addition, a complete palate with all its teeth, A.L. 200-1a, and the right half of a maxilla, A.L. 199–1, together with other remains: mandibles and teeth (A.L. 666–1, A.L. 188–1, A.L. 277–1, A.L. 198–1, A.L. 198–18, and A.L. 198-17a) and femora (A.L. 211–1 and A.L. 228–1).

The retrieval of fossils at Hadar was initially done on the surface, taking advantage of the cleaning of the ground by sparse but torrential rains. The remains of A.L. 288–1 were collected later in three weeks of work that led to the retrieval of the well-conserved partial skeleton of a single individual, an adult female, from sediments in the inferior part of the Kada Hadar (KH) member, just above the KHT tuff. This is the most complete Pliocene hominin skeleton known to date: up to 80% if we include lateral symmetry. The preservation of fossils at Kada Hadar is excellent. Even fossilized tortoise and crocodile eggs have been found (Johanson and Taieb, 1976). A.L. 288–1 provided, thus, an exceptional opportunity to study the morphology of very early hominins. According to Walter and Aronson's (1982) estimate, Lucy would be 3.5 Ma. A later revision estimated the KHT tuff age to be 3.18 Ma, which makes A.L. 288–1 about 3.1 Ma.

In 1975, the discoveries at Hadar were complemented with up to 13 individuals of different ages and sexes, A.L. 333. They are fragmentary and incomplete specimens compared to Lucy, but conserved well enough to allow certain determinations relative to dimorphisms and to juvenile and adult forms (Johanson, 1976; Johanson & White, 1979). The series 333 was found under the KHT tuff and, thus, would be about 3.2 Ma.

4.4.2 A.L. 288–1: morphology and classification

The excellent conservation of the A.L. 288–1 specimen allows us to get a clear picture of the morphology of this 3-million-year-old ancestor. It was a small individual, between 1.10 and 1.30 m high. This height is confirmed by other Hadar remains, such as A.L. 128 and 129, while specimens belonging to the 333 series indicate a larger size. The relation between the length of its humerus and femur, which gives an idea of how long the arms are in comparison to the legs, is greater than in current humans. The hand bones of A.L. 288–1 and the 333 series, as well as the feet of the latter, are different from current morphology. The structure of Lucy's hip suggests a bipedal posture. A significant element of the morphology of A.L. 288–1 is its cranium.

One of the most notorious missing pieces of A.L. 288–1 is the face. The specimen includes only a few cranial fragments. The absence of crania is a common circumstance in all Hadar discoveries of the 1970s; this is why the reconstruction of the sample's cranium was done by grouping fragments from different specimens. Under such conditions the calculation of the cranial capacity is not very precise but, in any case, it is obvious that these individuals had a very small cranium. As Donald Johanson says of Lucy, "It was not more than three and a half feet tall, had a tiny brain, and yet walked erect" (Johanson & Edey, 1981, p.180).

How to classify Lucy? The book by Johanson and Edey (1981) on the Hadar discoveries devotes three chapters (13–15) to answering this question. Johanson recounts interesting details of his discussions with Tim White regarding the species in which the Hadar remains should be placed. The alternatives were clear: either Lucy belonged to one of the known species (and, in that case, which one?) or to a new one. The problems involved in placing the Hadar hominins in a known species, leaving aside the fact of their great age, were that their molars were as small as those of *Homo habilis* but their other features had little to do with the genus *Homo*. Thus, Johanson and White decided, in the first place, that it was an australopithecine. That decision, which may now seem obvious, required rejecting some of the taxonomic alternatives put forward for the Hadar remains. Johanson and Taieb (1976) did not suggest any formal classification for the specimens in their initial description; they only pointed out affinities and similarities: "On the basis of the present hominin collection from Hadar it is tentatively suggested that some specimens show affinities with *A. robustus*, some with *A. africanus* (sensu stricto), and others with fossils previously referred to *Homo*" (p. 297). This diagnosis covered all imaginable possibilities except a new species. It was a hasty statement that Johanson admits to having regretted very soon (Johanson & Edey, 1981). As he asserted, a more careful examination determined that all the specimens found at Hadar corresponded to a single taxon (Johanson et al., 1978). Johanson suggested the taxon *Australopithecus afarensis* for the specimens found at Hadar at a Nobel symposium held in May 1978, in Stockholm. Johanson et al. (1978) published a detailed description of *A. afarensis* that same year. However, the suggested type specimen was not any of the Hadar exemplars, but Laetoli specimen L.H.-4, described by Mary Leakey and colleagues (1976).

Box 4.13 What kind of bipedalism did
A. afarensis have?

The scope of A. afarensis' bipedalism is a much debated topic (see Chapter 5). In general, many authors noted the existence of mosaic traits in pointing to a particular kind of locomotion: bipedalism, but retaining some climbing capability. The finding in Hadar of 49 new postcranial remains during the campaigns of 1990–2007 led Carol Ward et al. (2012) to argue that the shape of the thorax of A. afarensis would be more "human" than previously accepted. The presence in the material of a fourth metatarsal—a foot element not present until then in the hypodigm of the species—with axial torsion, dorsoplantar expansion, and a domed head similar to the traits of modern humans, was evidence for Ward et al. (2012) of a heel movement similar to our current gait.

Johanson's emphasis on Lucy's bipedalism must be toned down, in favor of a different style from modern humans' locomotion. The Hadar discoveries proved the existence of bipedalism with chimpanzee-like cranial size. During the 1970s, when the main Hadar discoveries were made, no paleontological authority accepted the "Piltdown man" fraud. But if the influence of such a trick had lasted until then, Lucy would have provided the irrefutable proof that human evolution involved an early appearance of bipedal locomotion and a much later increase of cranial capacities (Box 4.13).

4.4.3 Derived traits of A. afarensis

The Hadar organisms are an excellent sample of mosaic evolution. Some traits, such as hip shape, are apomorphies indicating a notable proximity to later hominins. Others, such as the dental arcade (V-shaped, unlike our own, which has a parabolic form), are primitive characters, as are the length of the anterior limbs or the small and robust cranium (Johanson & White, 1979). It was, thus, "an ape-brained little creature with a pelvis and leg bones almost identical in function with those of modern humans" (Johanson & Edey, 1981, p. 181).

The type of hominin *Australopithecus afarensis* could be inferred, at least partly, in accordance with the paleoecological patterns of their place of origin. Kaye Reed (2008) conducted a reconstruction of the habitat of each member of the Hadar formation based on its fauna. The results obtained were as follows: the lower member, Sidi Hakoma, is a closed habitat. Denen Dora shows meadow expansion. Kada Hadar contains

indications of having had a more arid climate and, therefore, an open habitat. Although the discovered hominins corresponded to Sidi Hakoma, Reed maintained that A. afarensis would have lived in a variety of habitats throughout the Hadar formation, highlighting its condition as a non-specialized species (eurytopic). The paleoenvironmental study of James Aronson, Million Hailemichael, and Samuel Savin (2008), based on the analysis of isotopes ^{18}O and ^{13}C, concluded that the biomass for A. afarensis would have been primarily forest grasses, which is precisely what Reed (2008) attributed to Sidi Hakoma.

4.4.4 Extension of the temporal range of *Australopithecus afarensis*

A. afarensis' characteristics received new light after the discovery of the first cranium with the face conserved, the specimen A.L. 444–2, described in 1994 by William Kimbel, Donald Johanson, and Yoel Rak (1994). It was a specimen of lesser antiquity than the others, found in the intermediate part of the Kada Hadar member, approximately 3 Ma. Because it was an adult male, it soon received the popular name of "Lucy's child." The A.L. 444–2 cranium is very broad—the broadest in all the species' samples—and shows a weak sagittal crest. The mandibular body is wide, but less robust than the average. Additionally, the facial projection is considerable.

Overall, A.L. 444–2 retains typical morphological features of other Hadar A. afarensis specimens, which fact led Kimbel and colleagues (1994) to argue that A.L. 444–2 refuted the notion that the reconstructions carried out with different specimens of the site involved the superposition of two different kinds of contemporary hominins. Lucy's child certainly supports the hypothesis of a single species, although highly variable, at Hadar. Other findings speak in favor of large sexual dimorphisms. For instance, a maxilla associated with a partial mandible and cranial base fragments (A.L. 417-1d) from the middle of the Sidi Hakoma member (close to 3.25 Ma), which probably belonged to a female (Kimbel et al., 1994), shows smaller canines and less prognathism than A.L. 444-2.

Kimbel et al. (1994) have also described an ulna (A.L. 438–1) from the Kada Hadar member, a bit older than the A.L. 444–2 cranium (which is actually part of a skeleton that we mention in the next paragraph) and the A.L. 137–50 very robust humerus (from the lower part of the Sidi Hakoma member).

In 2005, Michelle Drapeau and collaborators reported the finding in Kada Hadar of a partial skeleton consisting of a mandibular piece, a fragment of frontal bone, a complete left ulna, two second metacarpals, a third metacarpal, and parts of the clavicle, humerus, radius, and right ulna, forming the exemplar A.L. 438–1. An age of 3 Ma placed it as one of the most modern *A. afarensis*. It is also one of the few specimens of Pliocene hominins known up to today with associated cranial and postcranial elements. As a novelty, A.L. 438–1 also included the first complete ulna of an *A. afarensis* adult, and the first metacarpal associated with the anterior limbs (Drapeau et al., 2005).

4.4.5 How many species of *A. afarensis* are in the Hadar sample?

The first findings of Hadar raised doubts about whether *Australopithecus* specimens belonged to a single species. According to the heterogeneity of the findings, it might be necessary to divide them into two or more taxa: A.L. 288–1 and A.L. 417-1d are very small exemplars, while the 333 series, A.L. 444–2 and A.L. 438–1, found later, are larger.

Box 4.14 Sexual dimorphism in *A. afarensis*

The sexual dimorphism indicated by the sample of *A. afarensis* (series 333 and other well-known specimens at that time) when compared with that of modern humans, has been subjected to scrupulous analysis with opposing results. The alternative is to attribute to ancient hominins wide differences in most of their features, with the exception of canines—very similar in both sexes—or a degree of dimorphism similar to the current one. As the results of studies depend on the estimation of body mass within a chosen sample, and hypotheses on similarities or differences in size are based on even more difficult issues to answer empirically (as is the probable monogamy of certain species of hominin), it is not surprising that interpretations are diametrically opposed. Michael Plavcan and collaborators (2005) argued that sexual dimorphism in *A. afarensis* is remarkable. Philip Reno et al. (2003, 2010) considered it similar to that of modern humans.

The analysis of Martin Häusler and Peter Schmid (1995) of the pelvis of A.L. 288–1 introduced another element in the controversy: the possibility that Lucy was a male and, therefore, the morphological differences in the Hadar sample could not be due only to sexual dimorphism. However, the hypothesis of Häusler and Schmid is based on a speculative argument—the attribution to the large specimens of Hadar of a head size at birth that would prevent passage through the birth canal of A.L. 288–1. Lovejoy and Johanson criticized these calculations (Shreeve, 1995).

The most obvious differences in size and weight could be attributed, of course, to an eventual dimorphism between males and females (Box 4.14). Considering the morphological similarity of the parts belonging to Hadar's large and small specimens, Johanson and White (1979) argued that the variability was due only to dimorphic content. Palates, mandibles, and distal end of femora, would then be replicas at a different scale, but identical from a morphological point of view (Johanson & White, 1979). Following this idea, Kimble and White (1988) carried out the reconstruction of a "whole" cranium of *A. afarensis*, combining skull and facial fragments from different individuals, which led to accusations that they mixed together remains of more than one species (see Shreeve, 1994).

The comparative study of Todd Olson (1985) on some aspects of cranium morphology (braincase and nasal region) of Hadar and South Africa specimens concluded that these localities would be influenced during the Pliocene by selective pressures, first by bipedalism and later by food specialization. Large and small Hadar specimens, according to Olson, would not be explained by sexual dimorphism but by those selective pressures. Some of the specimens, such as A.L. 333–45 and A.L. 333–105, would have developed specializations, which even allow classifying them as *Paranthropus* clade members, i.e., as robust australopiths. Thus, the idea that the two lineages *Paranthropus* and *Homo* separated 2.5 Ma ago will be discarded, for the simple reason that the robust specimens present in Hadar would have an age of at least half a million years older.

The postcranial elements of *A. afarensis* permit deducing its locomotion. According to Kimbel and collaborators (1994), humerus joint A.L.438–1 and ulna A.L. 137–50, as well as A.L. 288–1 Lucy, indicate that the anterior limbs of *A. afarensis* are closer to the arm length of chimpanzees than to humans. But, as stated by Leslie Aiello (1994)—who seems to agree with the idea of a very variable species in Hadar—ulna A.L. 438–1 has no characteristics of knuckle-walking. Its trait assembly is a mosaic, with simultaneous presence of primitive and derived traits that, together with the robust shape of the humerus, would be adequate for a creature to walk upright on the ground, but also to climb trees. As we see, it is a similar diagnosis to that seen in Chapter 3 on *Ar. ramidus*. Carole Ward (2002) pointed out that the most preferable hypothesis for *A. afarensis* would be habitual bipedalism retaining primitive traits of arboreal behavior, although recognizing that, at that time, there was not enough empirical evidence to support or reject the hypothesis.

The comparison of the large size specimen A.L. 438–1 of Kada Hadar and A.L. 288–1 revealed, according to

Michelle Drapeau and collaborators (2005), that the functional morphology of the forelimb of *A. afarensis* is similar in all cases, despite body size. This result supports the hypothesis that the Hadar sample includes only one hominin species. On the one hand, almost all apomorphic traits of forelimbs A.L. 438–1 are shared with humans. Only the ulna of A.L. 438–1 is more curved than that in modern humans—a trait diverging from A.L. 288–1—and the manipulative capabilities of its hands seem less than the current one, although this is a derived trait with respect to other primates. After a discriminant function analysis in which the metric data of current apes and humans was considered together with hominin fossils, Drapeau et al. (2005) concluded that natural selection imposed some features on the anterior limbs of *A. afarensis* that were less effective for tree-climbing than those of apes. The counterpart would be the manipulative capability of *A. afarensis*, which would have been a selective advantage. We have, then, as the most likely alternative, a single species present in the hypodigm of *A. afarensis*, and a functional interpretation pointing to a bipedal locomotion with climbing capabilities. The findings of Hadar 1990–2007 (Ward et al., 2012), with intermediate specimens between the largest and the smallest documented previously, support the presence of a single, highly variable species of *A. afarensis* at that site.

4.4.6 Laetoli specimens of *A. afarensis*

The taxon *Australopithecus afarensis* was proposed by Johanson at the Nobel symposium celebrated in May of 1978 in Stockholm. That same year, Donald Johanson, Tim White, and Yves Coppens (1978) published the formal description of *A. afarensis*. However, the type specimen indicated was not one from Hadar, but the specimen L.H.-4 of Laetoli (Tanzania) described by Mary Leakey and collaborators (1976) (Box 4.15).

The Laetoli site, in northern Tanzania, is located some 40 km south of Olduvai (see Chapter 2 for the fossil footprints). Since the first discovery made by Kohl-Larsen in 1938–39, Laetoli has yielded some very early hominin specimens. Dating by the $^{40}K/^{40}Ar$ method yields an age estimation of 3.5–3.8 Ma (Harris, 1985) for the sediments that contain hominin remains, the same age as the fossil footprints. The new dating carried out by the $^{40}Ar/^{39}Ar$ method of the lower and upper beds of Laetoli produced a result of 4.36–3.85 and 3.85–3.63 Ma, respectively (Deino, 2011). According to the calculated sedimentation rate, the ages of *A. afarensis* specimens are 3.85–3.63 Ma, and that of the fossil footprints 3.66 Ma.

Box 4.15 Type specimen of *A. afarensis*

The proposal of a hominin from Hadar as the type specimen of *A. afarensis* would surprise those who are not familiar with the ins and outs of human paleontology. Lewin's (1987) narration of the episode includes the reasons behind this decision: Mary Leakey's annoyance; the letters exchanged on the subject of the new species' name; the inclusion of Mary Leakey as co-author of the *Kirtlandia* article in which the new species *A. afarensis* was proposed and her demands for her name to be removed, even if the number was already printed. There occurred a frontal collision between Donald Johanson and Mary Leakey, motivated by various reasons, regarding the Laetoli hominin remains.

Up to 30 hominin specimens, including mandibles, maxillas, isolated teeth, and a partial juvenile skeleton, were discovered between 1938 and 1979 (Day, 1986). The research team led by Mary Leakey found 13 of them, consisting of teeth and mandibles, during the 1974 and 1975 campaigns. Among them is the L.H.-4 specimen, constituted by a relatively undistorted mandible without ramus and partial adult dentition (Leakey et al., 1976). Mary Leakey and colleagues pointed out the similarity between this and the other specimens, on one side, and gracile australopithecines, specifically early *Homo*, on the other; they also noted differences regarding robust australopithecines (Box 4.16). They did not assign the remains to any species or genus, but underlined their "strong resemblance" with the East African *Homo* specimens. However, in the description of the specimens found between 1976 and 1979 Tim White included a brief note suggesting that they should be ascribed to *Australopithecus afarensis* (White, 1980b).

We face difficulties related to the characterization of *A. afarensis*. By designating a specimen from Laetoli as the holotype, *A. afarensis* becomes bound to that exemplar, and, therefore, the Hadar specimens would be paratypes. To what extent can such geographically distant specimens be regarded as members of the same species? Phillip Tobias (1980) noted in his criticism of the *A. afarensis* proposal that between the Hadar and Laetoli specimens there is a distance of 1,600 km and a time gap of 800,000 years, in addition to their morphological differences, which strengthens the case for alternative proposals. After comparatively examining the morphology of the Hadar, Laetoli, and Transvaal (Sterkfontein and Makapansgat) samples, Tobias

> **Box 4.16 L.H.-4 as type specimen of**
> **A. afarensis**

It is prudent to distinguish two different questions related to the debate surrounding the description of *A. afarensis* with L.H.-4 as holotype. The first is whether paleontological data from other authors can be used without their permission. The second refers to the procedures to be followed when defining a new species. The problems involved in the description and characterization of *Australopithecus afarensis* did not end with the initial confrontation between Mary Leakey and Johanson. The controversy also reached the taxonomic level, given that the species *A. afarensis* was suggested in an irregular but valid way by a reporter without reference to a holotype. Thus, and in accordance with the rigorous rules of taxonomic procedures, L.H.-4 could not be used later as holotype (Day, 1986). Some authors like Strait and colleagues (1997) and Wood and Collard (1999) have suggested that *A. afarensis* should, in actual fact, be called *Praeanthropus africanus*, which Hans Weinert (1950) gave to certain exemplars later included in *A. afarensis*. To use the genus *Praeanthropus* to refer to australopithecines before the division of hominids into gracile and robust has several advantages. Particularly, it complies with the rules of the International Code of Zoological Nomenclature, which requires using the first name suggested for any taxon, except when there are well-founded reasons not to do so. However, the extended use of the name *Australopithecus afarensis* has deterred us from following the change suggested by David Strait et al. (1997), while defending its appropriateness.

(1980) maintained that all these specimens belonged to the species described as *Australopithecus africanus* by Dart (see Chapter 5), while remaining open to the possibility that they may constitute subspecies. Thus, Tobias suggested that the species *A. africanus* would include, in addition to *A. africanus transvaalensis*, the subspecies *A. africanus afarensis*, for the Laetoli specimens, given that Johanson et al. designated the paratype of *A. afarensis* on the basis of one of them, and *A. africanus aethiopicus* for the Hadar specimens (Tobias, 1980). However, most of the initial problems regarding the interpretation of *A. afarensis* have mitigated, and the criticisms of Tobias and the Leakeys directed at the new species have not had much success. *A. afarensis* is at present widely accepted as the species that includes the Hadar specimens.

4.4.7 Woranso-Mille specimens of *A. afarensis*

Of special interest to discussions about the locomotion grade of *A. afarensis* is the specimen KSD-VP-1/1 from the locality Korsi Dora of Woranso-Mille site (in the western region of the central part of Afar, Ethiopia). This is a partial skeleton without cranium (Figure 4.17), which provided the first association of anterior and posterior limbs found since A.L. 288–1, Lucy. The KSD-VP-1/1 is an old specimen, dated at 3.60–3.58 Ma by radiometric and paleomagnetic methods. It made it possible to determine the evolution of the proportions between the limbs and the shape of the trunk of *A. afarensis*, the species to which the specimen was attributed (Haile-Selassie, Latimer et al., 2010). Judging from the pelvis, it is a male, large enough to enter the range of modern humans.

Yohannes Haile-Selassie, Bruce Latimer, and collaborators (2010) noted in the specimen study that it had the same morphological pattern of Lucy, corresponding to an obligatory terrestrial biped with minimal climbing capabilities—if any at all.

The most plausible type of locomotion attributed to *A. afarensis* was evaluated in greater detail thanks to the finding of a left foot fourth metatarsal belonging to the series 333, exemplar A.L. 333–160. In their description, Carol Ward, William Kimbel, and Donald Johanson (2011) argued that the specimen clarifies the shape of the longitudinal and transverse arches of the foot to which it belonged, further removed from that of African apes and closer to modern humans. Thus, the finding supports the hypothesis of an obligatory bipedal gait of *A. afarensis*. Its possible scope has been clarified to a greater degree by comparing some traits of the adult specimen KDS-VP-1/1 of Woranso-Mille with those of DIK-1–1 of Dikika, a juvenile. This allows taking into account aspects of development, and provides clues about the locomotor patterns of the species (Box 4.17).

In the same fossiliferous region of Woranso-Mille, although in localities other than Korsi Dora, an important collection of dental and mandibular exemplars appeared (Haile-Selassie, Saylor et al., 2010). However, the authors related the exemplars, at least partially, to *A. anamensis*; we will explain the exemplars in Section 4.5.

4.4.8 Dikika specimens of *Australopithecus afarensis*

If specimen A.L. 444–2 allows the presence of *A. afarensis* to be extended to the upper member of Kada Hadar, a finding in the Dikika locality (Ethiopia) has provided

Box 4.17 Locomotion of *A. afarensis*

In a similar way to Ward et al. (2011), a study by Adam Sylvester, Mohamed Mahfouz, and Patricia Kramer (2011) argued about the mechanics of locomotion related to the quadricep muscle group in the exemplar of a distal femur A.L. 129-1a, the first one discovered in Hadar, which, together with A.L. 129-1b, formed the knee joint. The analysis revealed that the specimen would have possessed the effective mechanical advantage (EMA) of a hominin with the same body mass. The sample from which the average was obtained by comparison included femora of 71 women and 88 men of modern humans, *Homo sapiens*. The work of Jeremy DeSilva and Zack Throckmorton (2010) on the relationship between the ankle and the foot posterior arch in *A. afarensis* maintained that, although the modern type curvature was already present in the taxon, foot AL 288–1 would be more flat and not for pathological reasons. The conclusion of the authors was that the traits present variability similar to modern humans in the hypodigm of *A. afarensis*.

Figure 4.18 AL 288-1 skeleton cast from *Museum national d'histoire naturelle*, Paris (Image via Wikimedia Commons, http://schools-wikipedia.org/images/1476/147694.jpg.htm).

the oldest evidence of the taxon until now. It is a left mandible fragment with teeth, DIK-2–1, belonging to the basal member of the Hadar formation, below Sidi Hakoma, i.e., older than 3.4 Ma of age (Alemseged et al., 2005). Mandibular and dental traits of DIK-2–1 in general coincide with those of the taxon *A. afarensis* and deviate from those of *A. anamensis* (Alemseged, 2005). However, some details of the mandibular body are peculiar. Alemseged and collaborators (2005) assigned the specimen to *A. afarensis*, indicating that it increased the already considerable variation of the mandibular samples of the species.

One year later, Zeresenay Alemseged and collaborators (2006) reported the description of a partial juvenile from the same locality and belonging to the Sidi Hakoma member, DIK-1–1, with an age of 3.3 Ma (Figure 4.18).

DIK-1–1—which received the popular name of "*Selam*," "peace" in the local dialect—is a partial skeleton formed by a cranium and associated postcranial remains, corresponding to a female exemplar of around 3 years old at the time of death. Facial size and proportions are similar to other specimens of *A. afarensis*, such as A.L. 333–105, but also to the Taung Child, *Australopithecus africanus* (see Chapter 5). Nonetheless, the morphological details of the face deviate from Taung specimens. Nasal and maxillar traits of DIK-1–1 are part of what is expected of an *A. afarensis* (Alemseged et al., 2006) (Box 4.18).

Box 4.18 Reconstruction, age, and sex of DIK-1–1

Alemseged spent most of his summer days for five years describing the DK-1–2 fossil and cleaning it "under a microscope with dental instruments, because I decided not to use acid treatments that could destroy it." He asserts that fully exposing and isolating the many postcranial elements is a complex task that will take several more years to complete (Nature, 2006; Alemseged et al., 2006).

DIK-1–1 is so well preserved that it includes the hyoid bone, one of only three fossil hyoids found, the others being from Atapuerca (Spain) and Kebara (Israel). DIK-1–1 age and sex were determined with the use of a scanner—CT—by measuring the degree of development of the yet-to-emerge teeth, and assuming that the specimen belonged to *A. afarensis* (Alemseged et al., 2006). As an infant specimen, the problems for determining its morphology are always thorny. What are great anatomical differences in comparing adult exemplars of diverse hominin species, and even hominoids, can be still quite similar traits when dealing with immature individuals.

An important aspect was noted by Alemseged and collaborators (2006): the infancy of DIK-1–1 permits establishing, when compared to adult exemplars, some ontogenetic patterns of development in *A. afarensis*. The degree of brain development in comparison with the volume of an adult cranium, for example, points out that DIK-1–1, as well as A.L. 333–105, deviated from the chimpanzees' values, placing them in line with what we know from infantile exemplars of modern humans. This fact is particularly interesting if we consider that the adult cranial capacity of *A. afarensis* is not much different from that of a chimpanzee. This means a small increase in volume from a relatively large infant to a relatively small adult, suggesting that the process of brain development was slow in the australopiths.

The scapulas of *A. afarensis*—two juveniles from Didika and one adult from Woranso-Mille—provided a rare opportunity to establish the ontogenetic patterns of development of the shoulder area, of great significance to infer arboreal behavioral patterns. David Green and Zerenesay Alemseged (2012) carried out a comparison of juvenile and adult scapulas of modern humans, apes (*Pongo, Gorilla*, and *Pan*), and *A. afarensis* (specimens DIK-1–1, juvenile; KDS-VP-1–1, adult), including in the sample an adolescent *Homo erectus* (KNM-WT-15000).

The scapula has an infraspinous region—its lower part—which is large in modern humans, both in juveniles and adults. African apes have a relatively smaller infraspinous region, which grows during ontogenetic development. This type of development can be related to a change in behavior from an arboreal juvenile to a knuckle-walking adult—orangutans, which do not perform knuckle-walking, maintain an adult pattern nearer to the juvenile. *A. afarensis* has a juvenile form closer to that of African apes, but in its adult phase—indicated by the scapula of Woranso-Mille and also of A.L. 288–1 of Hadar—its development is not similar to that of gorillas and chimpanzees (Green & Alemseged, 2012). In the view of Green and Alemseged (2012), *A. afarensis* would supplement bipedalism with a good amount of arboreal behavior.

4.4.9 Maka specimens of *Australopithecus afarensis*

At the time, it was a surprise that the proposal of *A. afarensis* would include as type specimen a Laetoli fossil, given the considerable distance separating Ethiopia from Tanzania. However, later findings, besides expanding the hypodigm of *A. afarensis* and extending its temporal scope, highlighted the presence of specimens of this taxon in localities between Laetoli and Hadar. Let's review them.

Maka site, placed in the Middle Awash region—Ethiopia, to the South of Hadar—has provided specimens attributed to *A. afarensis*. Among them, there is an almost complete mandible (MAK-VP- 1/12, Figure 4.19), other mandibular and dental materials, a partial (proximal) femur, an ulna, and an almost complete humerus, with an age of 3.4 Ma (White et al., 1993). White et al.'s (2000) study of the mandibles dealt with the evolutionary significance of the C/P3 complex. As we saw, the C/P3 is an apomorphy in *Ar. ramidus*. Thus, it is one of the dental derived traits that first evolved from characters shared with the great apes. The presence of large and small specimens in the Maka site supports Johanson and White's (1979) idea of a single highly variable *A. afarensis* taxon.

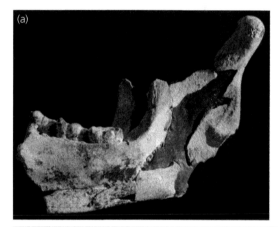

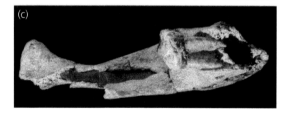

Figure 4.19 The MAK-VP-1/12 mandible; lateral (a), superior (b), and anterior views (c). White, T. D., Suwa, G., Simpson, S., & Asfaw, B. (2000). Jaws and teeth of Australopithecus afarensis from Maka, Middle Awash, Ethiopia. *Am J Phys Anthropol.*, 111, 45–68.

This feature of MAK-VP- 1/12 dentition reveals the evolution of the essential derived trait of *Australopithecus*: the megadontia. In Section 4.5 we will describe the characteristics of *A. anamensis*, a very ancient species from the genus (around 4 Ma), whose dentition is of a comparatively modest size. Thus, megadontia would have experienced an increment from *A. anamensis* to *A. afarensis*, indicating the incorporation of increasingly hard material in the diet.

4.4.10 West Turkana and South Turkwel specimens of *Australopithecus afarensis*

Kenya also has some localities that have provided specimens of *A. afarensis*. The West Turkana site, for example, is well known for supplying specimens such as KNM-WT 17000, *P. aethiopicus*, KNM-WT 40000, *K. platyops*, and KNM-WT 15000, *Homo erectus*. From the Lomewki member of West Turkana come exemplars attributed to *A. afarensis* KNM-WT 16006, a right mandibular body with some teeth, of 3.35 Ma, and KNM-WT 8556, 8557, and 1603, and more recent fragmentary mandibles (Brown et al., 2001) (Figure 4.20).

South Turkwel, Kenya, is among the possible places for the presence of *A. afarensis*, but with some

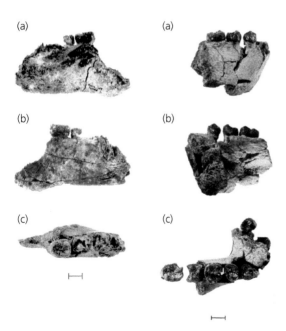

(a)

(b)

(c)

⊢⊣

(a)

(b)

(c)

⊢⊣

Figure 4.20 Mandibles KNM-WT 16006 (left), and KNM-WT 8556 (right), lateral (a), medial (b), and oclusal views (c). Reprinted from Brown, B., Brown, F. H., & Walker, A. (2001). New Hominids from the Lake Turkana Basin, Kenya. *Journal of Human Evolution*, 41(1), 29–44, with permission from Elsevier.

uncertainties. In the mentioned site, KNM-WT 22936 juvenile mandibular fragment and other associated fossils, including postcranial remains (KNM-WT 22944), were found. Its age, obtained by means of geological correlations and faunal analysis, could be close to 3.5 Ma. With regard to taxonomy, Carol Ward and colleagues (1999) only noted that the specimens' morphology is reminiscent of *A. afarensis* and *A. africanus*. One specimen, BEL-VP-1/1, found in 1981 in Belohdelie, on the Ethiopian side of the Middle Awash area, consisting of a partial frontal with a small fragment of the left parietal, had been estimated as being 3.9 Ma. But Asfaw's (1987) study of the specimen's morphology by specular imaging revealed that it was a generalized hominin, close to the divergence between hominins and African apes. He did not suggest it belonged to any particular species, though he did indicate that it shared some notorious traits with *A. afarensis*.

4.5 *Australopithecus anamensis*

Although *A. afarensis* may be the most representative taxon of the australopithecines, it is not the oldest. Allia Bay (on the eastern shore of Lake Turkana) and Kanapoi (on the western shore and somewhat south of the same lake; Figure 4.21) are two Kenyan sites that have provided the oldest *Australopithecus*' specimens known to date.

The first discovery of a specimen of hominin in Kanapoi was in 1967, the year in which Bryan Patterson and W. W. Howells (1967) described a fragment of humerus with an estimated age of 2.5 Ma, although without attribution to any species; the authors only indicated that it looked more like an *Australopithecus* than a *Paranthropus*. Older are the nine specimens from Kanapoi and the 12 from Allia Bay described by Mary Leakey and colleagues (1995), and morphologically and phylogenetically interpreted by Carole Ward, Meave Leakey, and Alan Walker (2001). With an age that ranges from 4.1 Ma for the former to 3.9 for the latter, they push back the existence of australopithecines by half a million years.

4.5.1 Hypodigm and age of *Australopithecus anamensis*

The KNM-KP 29281 mandible (Figure 4.21) and the KNM-KP 29283 maxilla come from the inferior stratigraphic level (between 4.17 and 4.12 Ma), while the NM-KP 29285 tibia, the KNM-KP 271 humerus, two

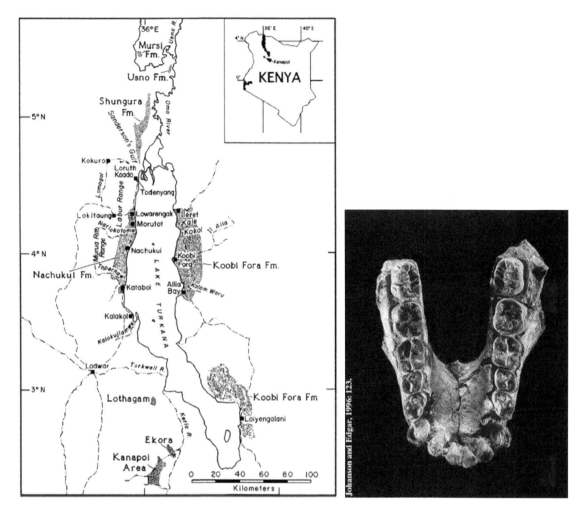

Figure 4.21 Left: Location of the fossiliferous zone of Kanapoi, to the south of Lake Turkana (Kenya) (map by Harris and Leakey, 2003). Next, sedimentary interval scheme of Kanapoi (Leakey et al., 1998). Right: The KNM-KP 29281 mandible, holotype of *Australopithecus anamensis* (photography of Johanson & Edgar, 1996). Harris, J. H. & Leakey, M. G. (2003). Introduction. In J. H. Harris & M. G. Leakey (Eds.), Geology and Vertebrate Paleontology of the Early Pliocene Site of Kanapoi, Northern Kenya (Vol. Contributions in Science Number 498, pp. 1–7). Los Angeles, CA: Natural History Museum of Los Angeles County.

mandibular fragments (KNM-KP 29281), and a large mandible (KNM-KP 29287), presumably male and found in the higher level, are between 4.1 and 3.5 Ma (Leakey et al., 1995; Ward et al., 2001). Allia Bay fossils—numerous dental and mandibular materials, and some postcranial elements such as the left radius KNM-ER 20419—are of an earlier age and come from the same locality 261-1, below the Moiti Tuff, dated 3.94 Ma (Leakey et al., 1995; Ward et al., 2001).

The Kanapoi and Allia Bay specimens fill the wide temporal range of 1 million years between *Ar. ramidus* and *A. afarensis*. Meave Leakey et al. (1995) described the morphology of the new specimens as a mixture of primitive and derived traits, confirming the view, already quite established, that evolution within the tribe Hominini took place in a mosaic fashion during the Pliocene. For instance, the study of the KNM-KP 29285 tibia indicated a bipedal posture. Also, the KNM-KP 271 humerus includes, according to Meave Leakey et al. (1995), many hominin-derived traits. However, the persistence of some plesiomorphies in the tibia, which are shared with African apes and *A. afarensis*, show that bipedal locomotion was at a different stage from that of *Homo*. All these features had already been detected in *A. afarensis*; but, together with the traits that place the Kanapoi and Allia Bay specimens close to *A.*

afarensis, there are other traits that move them apart, especially Hadar. These differences led Meave Leakey and colleagues to reject the idea of a long stasis in *A. afarensis*, the presence of the species for an extensive lapse of time with little variation. Rather, Meave Leakey et al. (1995) suggested a new species for the Kanapoi and Allia Bay findings: *Australopithecus anamensis* (*anam* means "lake" in the Turkana language). The type specimen (holotype) is the KNM-KP 29281 mandible, which retains all its teeth, found by Peter Nzube in 1994. The paratype is constituted by the remaining specimens of both sites.

The Kanapoi campaigns of 1995–97 supplied a juvenile specimen with associated teeth and cranial pieces (KNM-KP 34725), more pieces to complete the anterior mandible KNM-KP 29287, a proximal phalanx of a hand, and a maxillar. The discovery of new specimens attributed to *A. anamensis* in Middle Awash widened its geographic range to Ethiopia.

Aramis (Middle Awash, Ethiopia), the same site in which *Ar. ramidus* was identified for the first time, has also provided exemplars of *A. anamensis*. In locality 14

of the upper member, Adgantole, appeared a partial maxilla with dentition ARA-VP-14/1, attributed to *A. anamensis* by White, WoldeGabriel, et al. (2006). This meant the extension of the geographic range of the taxon more than 1,000 km to the northeast. The age of the specimens of Adgantole was fixed by the $^{40}A/^{39}A$ method and stratigraphy in 4.14 Ma.

In the same paper, White et al. indicated a few more exemplars from the localities ASI-VP-2 and ASI-VP-5 of Asa Issie, a site close to Aramis. Among them were the upper teeth ASI-VP-2/334 and left fragmentary femur ASI-VP-5/154 (Figure 4.22). The age of all of them is approximately 4.2–4.1 Ma by fauna comparison with Aramis (White, WoldeGabriel, et al. 2006).

Finally, in 2010, from the localities Am-Ado (AMA), Aralee Issie (ARI), Mesgid Dora (MSD), and Makah Mera (MKM) of Woranso-Mille, 26 hominin remains, dated between 3.8 and 3.6 Ma, were described as mainly consisting of dental and mandibular materials with some postcranial fragments (Haile-Selassie, Saylor et al., 2010).

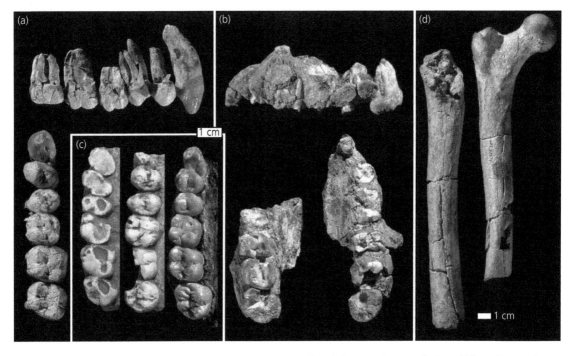

Figure 4.22 (a) ASI-VP-2/334 right maxillary dentition. (b) ARA-VP-14/1 maxilla with dentition. Alignment of right and left maxillary arcades is approximate. (c) *A. anamensis* (KNM-KP 29283 and KNM-ER 30745, left and middle, respectively; casts, reversed) and *A. afarensis* (A.L. 200–1, right) dentitions. (d) Comparison of the ASI-VP-5/154 right femoral shaft with the smaller but otherwise morphologically similar left proximal femur of A.L. 288–1 (Lucy; *A. afarensis*). Reprinted by permission from Macmillan Publishers Ltd: White, T. D., WoldeGabriel, G., Asfaw, B., Ambrose, S., Beyene, Y., Bernor, R. L.,. . . Suwa, G. (2006). Asa Issie, Aramis and the origin of Australopithecus. *Nature*, 440(7086), 883–889.

4.5.2 Derived traits of *Australopithecus anamensis*

According to Meave Leakey et al. (1995, 1998), *A. anamensis* can be distinguished from *A. afarensis* in certain dental traits, such as the longer canines and robust roots. But they are also different form *Ardipithecus ramidus* in that the enamel of *A. anamensis* is much thicker and similar to that of *A. afarensis* (Table 4.5).

Bernard Wood and Brian Richmond (2000) stated that some aspects of *A. anamensis* dentition are more primitive than in *A. afarensis*, such as the asymmetrical premolar crowns or the posterior less-inclined face of the canine roots, which could be expected from a taxon half a million years older than *A. afarensis*. However, other traits, according to Wood and Richmond (2000), would be more developed in *A. anamensis*, thus more similar to *Paranthropus*, which appeared 1.5 Ma later.

Yohannes Haile-Selassie, Beverly Saylor, and collaborators (2010) carried out a metric and morphological comparison of the specimens of Woranso-Mille with equivalents of *A. afarensis*—including the entire Hadar sample—*A. anamensis* and *Ar. ramidus* (Box 4.19). The comparative analysis indicated that the dental morphology of the Woranso-Mille species is intermediate between *A. anamensis* of Allia Bay/Kanapoi and *A. afarensis* from Laetoli. Thus, the authors considered that these specimens cannot be assigned to either taxa, indicating that they represent strong evidence in favor of reducing these two species to a single taxon, or of considering them as chronospecies subjected to anagenetic evolution (Haile-Selassie, Saylor et al., 2010). The finding of a postcranial fossil—a partial foot BRT-VP-2/73 (Haile Selassie et al., 2012)—from Burtele, another site of Woranso-Mille, Burtele, of c. 3.4 Ma, casts further doubts.

4.6 Miocene human genera

The specimens of Miocene hominins come from at least three localities, very remote from each other (Toros-Menalla, Tugen, and Middle Awash). This was not the reason for including them in three different genera, but rather, as we have seen, it is the consideration of morphological differences between the recovered remains. They are comparable only to a limited degree, as shown in Figure 4.23, because of the absolute disproportion between the number of known features of *Sahelanthropus*, *Orrorin*, and *Ardipithecus*. Based on the data available today, is it justifiable to consider the dispersal of Miocene hominins into three different genera?

After the Asa Koma discoveries, Haile-Selassie, Suwa, & White (2004) believed the early hominin phylogenetic sequence was as follows:

- *Sahelanthropus*—a taxon we will discuss further on—*Orrorin*, and *Ardipithecus kadabba* provide important outgroup comparisons to younger *Ar. ramidus* and *A. anamensis*.
- Metric and morphological variation within the available small samples of Upper Miocene teeth attributed to *A. kadabba*, *O. tugenensis*, and *S. tchadensis* is no greater in degree than that seen within extant ape genera.
- The interpretation that these taxa represent three separate genera, or even lineages, can be questioned.
- It is possible that all these remains represent specific or subspecific variation within a single genus.

If the taxonomic suggestion made by Haile-Selassie, Suwa, and White (2004) of a single taxon common to all Upper Miocene hominins becomes a formal proposal, nomenclatural convention requires grouping them in the first genus named: *Ardipithecus*. However,

Table 4.5 Distinctive traits of *A. anamensis* (Wood & Richmond, 2000)

Small external ear canal of elliptic form
Narrow mandibular arch with parallel mandibular bodies
Sloping mandibular symphysis
Long and robust canine roots
Upper molar crowns wider in mesial direction than in distal
Small marrow cavity in humerus

Box 4.19 The unity of *A. anamensis*

Andrews (1995) criticized the taxonomic solution of classifying the Kanapoi and Allia Bay hominins in a single species. The stratigraphic sequence of Kanapoi and Allia Bay separates early fluvial sediments from later ones originated in the ancestral Lake Lonyumun. Andrews pointed out that the more recent specimens from the lake sediments could belong to a different and more derived species, which would explain the relative robusticity of the KNMP-KP 29287 mandible. But, if this was the case, the locomotor characteristics of *A. anamensis*—inferred from the KNM-KP 29285 tibia, belonging to the same lithic interval as the cited mandible—would not be applicable to the earlier specimens.

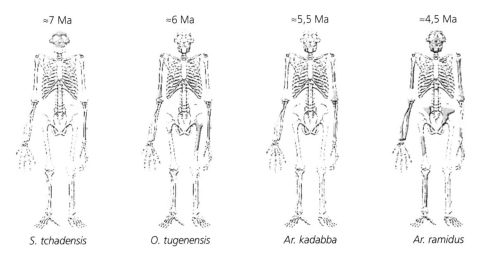

| ≈7 Ma | ≈6 Ma | ≈5,5 Ma | ≈4,5 Ma |

| *S. tchadensis* | *O. tugenensis* | *Ar. kadabba* | *Ar. ramidus* |

Figure 4.23 Known remains of different types of Miocene hominin.

to decide whether or not such a group makes sense, it would be necessary according to Mayr's concept of adaptive genus to indicate which was the adaptive strategy of each different Miocene taxa.

Ernst Mayr's criteria for genera identification has been modified by Bernard Wood and Mark Collard (1999b), introducing two conditions to be met by taxa included in the same genus: to be monophyletic (i.e., to belong to the same lineage or clade), and to share the same adaptive strategy indicated by six parameters: size and body shape, locomotion, mandibles with teeth, development, and size of the brain. The purpose of Wood and Collard (1999b) was to verify that the different species included in *Homo* satisfied those exigencies.

Wood and Collard (1999b) concluded that there is no evidence of an adaptive particularity in *Sahelanthropus* and/or *Orrorin* that is dissimilar in *Ardipithecus*. All of them have an equivalent locomotor and mandibular apparatus—the other traits, size and body shape, development and size of the brain, are not comparable. With respect to membership in the same lineage, there are no cladistic studies that would include these three taxa.

Consequently, the most-parsimonious decision seems to be to maintain all Miocene hominins in a single genus, unless it is proved they occupy different clades or maintain peculiar adaptive strategies. While no further evidence is provided on the locomotion and/or differential dentition of the specimens of Toros-Menalla and Tugen Hills, all Miocene hominins should be included in the genus *Ardipithecus*, with *Ardipithecus ramidus* as type species. Despite this, we maintain here the original names of *Sahelanthropus*,

Orrorin, and *Ardipithecus* only to avoid confusion, as it is habitual to consider them as different genera.

4.7 First phylogenetic changes in the tribe Hominini

The identification of the first hominins is related to the availability of specimens that, in contrast to the locomotion of African apes, initiated the typical synapomorphy of the human lineage: bipedalism. The synthesis of our knowledge of these first hominins corresponded to patterns already indicated:

- The hominin lineage emerged during the African Upper Miocene (between 11.60 and 5.33 Ma).
- Its sister group is the chimpanzee.
- The main derived trait that characterized hominins is bipedalism, different from that of modern humans, but with a higher degree of tree-climbing in the Miocene hominins.
- Available fossils show differences among them, but it is unlikely these justified their assignment to different genera.
- Adaptative strategies of *Ardipithecus* correspond to the resource exploitation of tropical forest ground.

Within this overall picture, the only sufficiently informative evidence of the traits that hominin and Panini might have shared after the separation of the two lineages comes from specimens of *Ardipithecus*, and, in particular, from the skeleton ARA-VP-6/500. Based on this evidence, Tim White and collaborators (2015) have provided a hypothetical scheme for the separation process. The starting point is to acknowledge that all primate species share a fairly similar repertoire of posture and locomotion, but each species has a distinctive dominant locomotor behavior, which is adaptively crucial for the species. That of *Ardipithecus* can be deduced by a comparative morphological analysis between its traits, those characteristic of modern chimpanzees—which are derived traits with respect to the common ancestor of panins and hominins—and those of Pliocene humans. Thus, *Ardipithecus'* hand is not intermediate but closer to that of modern humans. The foot is different from African apes and humans. Limb proportions—long arms, short legs in current simians, in contrast to current humans—are in *Ardipithecus* similar to that of Miocene apes, while the lower back and pelvis indicate bipedalism. In short, *Ardipithecus'* locomotion would be different than the bipedalism of *Australopithecus* and current humans, including occasional and deliberate arboreal suspension (White et al., 2015).

Regarding craniodental evolution, *Ardipithecus* shows a cranial structure of small volume (25% less than *Australopithecus* and similar to *Sahelanthropus*). Molar enamel is thin compared with most *Australopithecus*—though the enamel isotope signature is indistinguishable from the earliest *Australopithecus*. Incisors and molars are relatively unspecialized, such as in a range of Miocene apes and *Australopithecus*, pointing to opportunistic omnivory. Consequently, Tim White et al. (2015, p. 4882) state that "*Ar. ramidus* obviously postdates the CLCA [the common ancestor we shared with chimpanzees], it does not exhibit the derived features characteristic of either chimpanzees or gorillas. In contrast, *Ar. ramidus* shares four major, newly evolved, morphogenetically independent character complexes with later *Australopithecus* (as judged by comparison with fossil and modern outgroup apes): (i) a nonhoning C/P3 complex and feminized male canine (. . .), (ii) a short, broad cranial base (. . .), (iii) a broadened, shortened and 'twisted' ilium (. . .), and (iv) tarsal and metatarsal/phalangeal specializations related to upright walking. No other fossil or modern ape shares such evolutionary derivations. It is therefore likely that *Ardipithecus* lies within the hominid clade (. . .). At 4.4 Ma, it is obviously not an ancestral chimpanzee. Nor is it likely to be a relict ape that bears no special relationship to hominids."

One of the most important features of the discovery of *A. anamensis* specimens in Aramis is the evidence provided from the same stratigraphic succession about a possible evolutionary connection: *Ar. ramidus–A. anamensis*, with the hypothesis of a rapid evolution (about 0.2 Ma) between the two genera. As the specimens of *A. anamensis* from Middle Awash are intermediate

from a morphological point of view between *Ar. ramidus* and *A. afarensis*, a possible ancestral relationship is supported between the last two taxa (White, Woldegabriel et al., 2006). Thus, we have, with some empirical support, the evolutionary sequence: *Ar. ramidus–A. anamensis–A. afarensis*. An evolutionary relationship of this kind seems reasonable for the range of the various taxa (Figure 4.24). The proximity of sites in which specimens of these taxa are found, all in the Rift Valley, support the proposed lineage.

However, some morphological studies do not support such a lineage. The conclusion reached by Meave Leakey et al. (1995) is that *A. anamensis* represents a new 4 Ma ancestor leading to *Homo*. But this lineage,

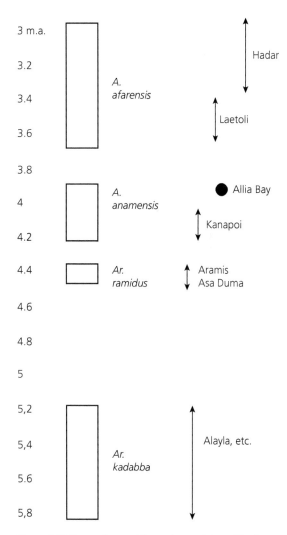

Figure 4.24 Temporal range of the specimens of *Ar. ramidus*, *A. anamensis*, and *A. afarensis*.

according to Meave Leakey et al. (1995), cannot consist of the descendants of *Ar. ramidus*. These authors suggest that *Ar. ramidus* formed a lateral branch, a sister species of the Kanapoi and Allia Bay specimens and all subsequent hominins. In their 1998 article, Meave Leakey et al. proposed the following alternative: either all hominins between 4.4 Ma and close to 3 Ma consist of an evolving single species, or there are three separate species (*Ar. ramidus*, *A. anamensis*, and *A. afarensis*) whose phylogenetic relations are imprecise. After a more detailed analysis of the morphology of the Kanapoi and Allia Bay exemplars, Carol Ward, Meave Leakey, and Alan Walker (2001) indicated that with the available material no autapomorphy of *A. anamensis* was detected, precluding consideration of the taxon as ancestor of *A. afarensis*—an indirect way to admit that the hypothesis of the evolutionary relationship between them was feasible.

According to White, Woldegabriel et al. (2006), two alternative hypotheses are possible concerning the three taxa:

- The first hypothesis derives *A. anamensis* phyletically from *Ar. ramidus* within a 200,000-year interval.

- The second involves cladogenesis of *A. anamensis* from an ancestor (presumably *Ardipithecus* or some close relative) even deeper in the Pliocene or Upper Miocene.

Under the latter hypothesis, *Ar. ramidus* would represent a relict species in an ecological refugium.

A more detailed morphological comparison between the species *Ar. ramidus*, *A. anamensis*, and *A. afarensis* (Figure 4.25) has been carried out with only dentomandibular materials; a necessary limitation given the scarcity of postcranial elements of *A. anamensis*. Teaford and Ungar (2000), Kimbel et al. (2006), White, Asfaw et al. (2009), Suwa, Kono et al. (2009), and Ward et al. (2010) agreed, in accordance with the morphological analysis of teeth and jaws, to placing *A. anamensis* as a taxon between the earlier ancestors and the more evolved *Australopithecus*.

The objective of the analysis by Mark Teaford and Peter Ungar (2000) was to determine the diet of the first hominins. The comparison group with the australopiths were the higher apes (the complete study of *Ar. ramidus* was published later). Teaford and Ungar (2000) stressed the intermediate character of *A. anamensis*. For these authors, the architecture of the mandibular body

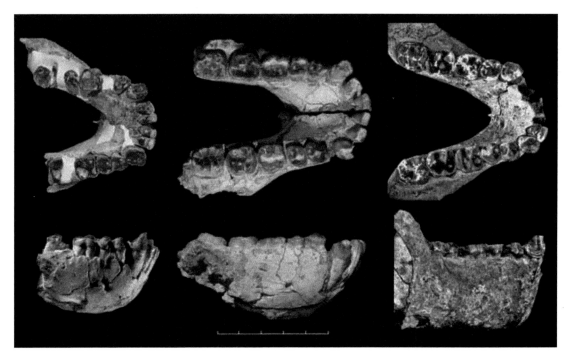

Figure 4.25 Oclusal and lateral view of the mandibles ARA-VP-1/401 (4.4 Ma) *Ar. ramidus*; KNM-KP 29281 (4.12 Ma) *A. anamensis*; and MAK-VP-1/12 (3.4 Ma) *A. afarensis*. From White, T. D., Asfaw, B., Beyene, Y., Haile-Selassie, Y., Lovejoy, C. O., Suwa, G., & WoldeGabriel, G. (2009). Ardipithecus ramidus and the Paleobiology of Early Hominids. *Science*, 326(5949), 64–86. Photos © and countesy of Tim White.

and the molar enamel thickness suggested that *A. anamensis* could have been the first hominin with a suitable masticatory apparatus for the consumption—habitual or occasional—of tougher materials (Box 4.20). *A. afarensis*, with a significant increase in mandibular robustness, would be a further step in that same direction.

If we accept that the transition from the Upper Miocene hominins (*Ardipithecus*) to those of the Lower Pliocene (*Australopithecus*) is an anagenetic process related to the development of australopith-derived traits—megadontia and bipedal changes—we find the following most probable trends:

- Partial loss in *Australopithecus* of the climbing capabilities of its anterior limbs compared with those of *Ardipithecus*, with the counterpart of an increase in manipulative capacity (evidence mainly derived from the sample of Hadar), although the specimen of Dikika indicates that bipedalism would be complete, while retaining certain arboreal behavior.
- The beginning of megadontia, of great variability but with a tendency to an increment in size of the masticatory apparatus from *A. anamensis* to *A. afarensis*, which lead to the inference of a diet of gradually tougher materials (evidence in particular from Maka sample).

4.8 Phylogenetic relationships of the Miocene and Lower Pliocene hominins

We have reviewed in this chapter the most significant specimens of the genera *Ardipithecus* and *Australopithecus*. Their fossils were distributed in two distant geographical areas, the Rift Valley and the Chad. Is it possible to reach a plausible model about their evolutionary relationship?

Box 4.20 Changes in dentition

Tooth changes distinguishing *A. anamensis* from *A. afarensis* would reflect, in the most common hypothesis, parallel changes in diet. However, Peter Ungar and collaborators (2010) maintained that a study of microstriations indicates that *A. anamensis*, as well as *A. afarensis*, had the same type of slightly soft diet, due to similarities in microstriations to the current apes, *Alouatta palliata* and *Trachypithecus cristatus*, both folivoros, and to *Theropithecus gelada*, which feeds on grass.

We can consider, as a starting point, the most commonly shared ideas:

(1) *A. anamensis* is the first taxon of *Australopithecus* and, thus, from which Miocene hominins came.
(2) *A. afarensis* is the direct successor of *A. anamensis*.

Let's examine the support for both hypotheses.

4.8.1 Evolutionary relationships between *Ar. ramidus*, *A. anamensis*, and *A. afarensis*

The only phylogenetic episode studied directly and in depth in the Pliocene is one corresponding to the hypothesis that *A. afarensis* is the result of anagenetic evolution of *A. anamensis*; that is to say, that they are chronospecies of a lineage that evolved without branches.

An evolutionary relationship of *Ar. ramidus*–*A. anamensis*–*A. afarensis* seems reasonable in agreement with their range. The proximity of sites in which specimens were found, all of them in the Rift Valley, also supports the idea of the proposed lineage. However, as we said in Section 4.7, the proponents of the taxon *A. anamensis* do not agree.

The evolutionary relationship between *A. anamensis* and *A. afarensis*, and its stratigraphic sequence, were supported by William Kimbel and collaborators (2006) after carrying out a phylogenetic analysis from the polarization (establishment of the primitive and derived condition) of 20 morphological dental and maxillar traits of four fossil samples (Kanapoi, Allia Bay, Laetoli, and Hadar).

Gen Suwa and collaborators (Suwa, Kono et al., 2009) conducted a study, also based on dental materials, on the phylogenetic placement of *Ar. ramidus* compared with *Ar. kadabba* as the most primitive group, and with *A. anamensis* and *A. afarensis* as derived groups. The authors considered paleobiological aspects deduced from the dentition, including the possible alimentary niche of each studied species. In particular, the canines indicated morphological changes in the series *Ar. kadabba*–*Ar. ramidus*–early *Australopithecus*, with a more primitive polymorphic condition in *A. anamensis*, which becomes derived in *A. afarensis* and is maintained in posterior hominins (Suwa, Kono et al., 2009). For their part, Carol Ward, J. Michael Plavcan, and Fredrik Manthi (2010) also accepted the hypothesis of a monophyletic lineage *A. anamensis*–*A. afarensis*, with changes in the transition from one to the other taxon, affecting the anterior dentition, the mandibular structure, and the molar shape (DeSilva, 2011).

Tim White, Berhane Asfaw, and collaborators (2009) focused directly on the hypothetical differences that can be established in the evolutionary transition between *Ar. ramidus*, *A. anamensis*, and *A. afarensis*, as shown in Figure 4.26.

White, Asfaw et al. (2009) recognized that, with such little information, it was impossible to detail the relationships among species. Hence, they proposed in their scheme a "sheaf" formed by "strings" of demes—populations exchanging alleles—with a certain continuity in time. The first hypothesis argues that only one species evolved through a phyletic process in which the presence of chronospecies corresponding to the taxa is noted: *Ar. ramidus*, *A. anamensis*, and *A. afarensis*. The second scenario shows a similar evolution from *Ardipithecus* to *Australopithecus*, pointing out that it took place around 4.5–4.2 Ma within a small regional group of populations—from Afar and/or Turkana. The third hypothesis suggests a process of a cladistic generation of *Australopithecus*, while maintaining the mother

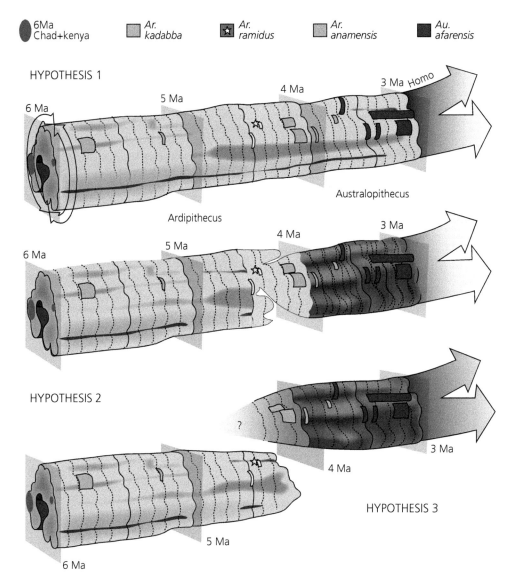

Figure 4.26 Alternative hypotheses about the phylogenetic relationship of *Ardipithecus* and *Australopithecus* examined by White, Asfaw et al. (2009). See explanation in the text. From White, T. D., Asfaw, B., Beyene, Y., Haile-Selassie, Y., Lovejoy, C. O., Suwa, G., & WoldeGabriel, G. (2009). Ardipithecus ramidus and the Paleobiology of Early Hominids. *Science*, 326(5949), 64–86. Reprinted with permission from AAAS.

clade. In this third hypothesis, *Australopithecus* would be the result of an allopatric speciation that emerged from an isolated peripheral population.

None of these hypotheses is more plausible than the others with the data available today. In fact, they need not be considered to be too radical alternatives. However, the third hypothesis is an excellent example of the reasons that promote the proposal of the reformed cladistics (see Chapter 1). Without this mechanism of interpretation, the third hypothesis would be reduced to either one of the two previous hypotheses.

4.8.2 The diversity of Pliocene hominins

The postcranial, dental, and mandibular exemplars of Woranso-Mille (see Chapter 8) (Haile-Selassie, Latimer et al., 2010; Haile-Selassie, Saylor et al., 2010) offer evidence of considerable diversity among hominins in the period from 4 to 3 Ma.

As may be recalled, Haile-Selassie, Latimer, and collaborators (2010) included the poscranial elements of Woranso-Mille, which are the specimen KSD-VP-1/1, in the species *A. afarensis*. But they preferred not to classify dental materials, such as the mandible MSD-VP-5/16, because they are intermediate between *A. anamensis* and *A. afarensis*, and cover the intermediate temporal range between both taxa. Consequently, Haile-Selassie, Saylor et al. (2010) maintained that the separation between *A. anamensis* and *A. afarensis* might not be real, or necessary. The entire sample could be included in a single taxon, although the authors acknowledged that the use of two different chronospecies could be useful to promote unambiguous communication among specialists.

The discovery of a partial foot—metatarsals and phalanxes (BRT-VP-2/73)—in the locality of Burtele (Woranso-Mille, Ethiopia) (Haile-Selassie et al., 2012), with an age around 3.4 Ma, has provided evidence of the complexity of distinguishing lineages from evolutionary processes—related to locomotion, at least—in East Africa. This exemplar is formed by eight very well preserved bones from the right foot (Figure 4.27).

Although a contemporary of *A. afarensis*, certain traits of this foot, such as the length of the four metatarsals compared with the first and second, and the lateral projection of the big toe, bring BRT-VP-2/73 closer

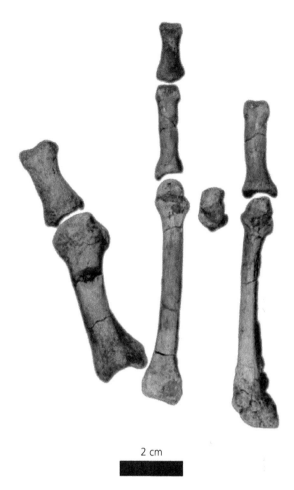

2 cm

Figure 4.27 The eight bones that form BRT-VP-2/73 (right foot). Reprinted by permission from Macmillan Publishers Ltd: Haile-Selassie, Y., Saylor, B. Z., Deino, A., Levin, N. E., Alene, M., & Latimer, B. M. (2012). A new hominin foot from Ethiopia shows multiple Pliocene bipedal adaptations. *Nature*, 483(7391), 565–569.

to *Ar. ramidus* and indicate that its climbing capabilities would have been much greater than those of *A. afarensis*. Despite the taxonomic difficulties presented by the exemplar BRT-VP-2/73, it reveals, according to Yohannes Haile-Selassie and collaborators (2012), the existence of a variety of adaptative lineages in the interval between 4 and 3 Ma.

CHAPTER 5

Middle and Upper Pliocene hominins

The Middle Pliocene is the geological epoch ranging from 3.60 to 2.58 Ma. Most significant, with respect to hominin evolution, is the dispersal of the genus *Australopithecus* and its diversification into several species. It is common practice to include in *Australopithecus* all human fossils corresponding to this temporal range. However, as we shall see, this is a taxonomic simplification difficult to justify.

At the beginning of the Upper Pliocene, one of the most relevant evolutionary events of the human lineage took place: a cladistic separation that resulted in the emergence of the genera *Paranthropus* and *Homo*. *Homo* will be the subject of Chapter 6, but we will point out in the present chapter some of its key features, relating them to the features of robust australopithecines or paranthrops.

5.1 What can be included in *Australopithecus*?

We have argued that *Australopithecus* is a genus corresponding to a particular type of hominin that developed certain locomotion patterns and had a derived dentition when compared to Miocene hominins. Several taxa, beyond *Australopithecus anamensis* and *A. afarensis*, fall within such a characterization.

David Strait, Frederick Grine, and Marc Moniz (1997) have pointed out that the taxon *Australopithecus*, conventionally defined as the set of all hominins prior to *Homo*—except *Ardipithecus*—constitutes a paraphyletic group. That is, *Australopithecus* was at that time a hodgepodge in which to place all early hominins that did not fit elsewhere. Strait and colleagues (1997) took into account the following taxa in their study: *Australopithecus afarensis*, *A. africanus*, *A. aethiopicus*, *A. robustus*, and *A. boisei*. Table 5.1 repeats, for easier reference, some data already advanced in Table 4.1. We include *A. anamensis* and *A. garhi*, species not yet known in 1997, which in fact increase the paraphilia of *Australopithecus* pointed out by Strait et al. (1997). The consequence of

defining the genus *Australopithecus* in a broad way is that specimens are included together that cannot be grouped in a single evolutionary lineage.

One way to resolve the problem, so that genera can be defined that constitute true lineages, is to separate "robust" australopithecines from the other taxa. The distinction between robust and gracile hominins was set up as a consequence of the discovery of fossils that exhibited very different features, but were found in close South African sites during the first half of the twentieth century. It is necessary to introduce a historical note here. The cranial traits of the Taung specimen, used by Raymond Dart to define, in 1925, the genus *Australopithecus* (Dart, 1925), and those found later at Sterkfontein and Makapansgat did not appear massive, lacking a sagittal crest—see Box 5.1. Robert Broom (1938) later discovered much more robust specimens at Kromdraai, similar to those found later at Swartkrans. Although Broom suggested the genus and species *Paranthropus robustus* for them, many authors just distinguished between gracile *Australopithecus africanus* and robust *Australopithecus robustus*. However, later findings at Olduvai (Tanzania) required the revision of this relative concept of robusticity and gracility.

If we group robust australopithecines in another genus, the sets are more coherent with the evolutionary lineages. This leads to the reduction of the genus *Australopithecus* to the following taxa (see Table 5.1): *A. anamensis*, *A. afarensis*, *A. platyops*, *A. bahrelgazhali*, *A. garhi*, *A. africanus*, and *A. sediba*, ordered by decreasing age. We have seen the first two species, from the Lower Pliocene in Chapter 4. In this chapter we address the characteristics and specimens of the rest of the australopithecines of the Middle and Upper Pliocene, which offer a panorama of the dispersal and diversification of humans in those periods. The dispersal is documented by the discovery of fossils in Chad and South Africa, adding to those of the Rift Valley. In all these localities there are other australopithecines that cannot be included with the species of the Lower Pliocene.

Processes in Human Evolution. Francisco J. Ayala and Camilo J. Cela-Conde, Oxford University Press (2017).
© Francisco J. Ayala and Camilo J. Cela-Conde.
DOI 10.1093/acprof:oso/9780198739906.001.0001

Table 5.1 Species of the genera *Australopithecus, Paranthropus,* and *Homo* from the Middle and Upper Pliocene, and their time of emergence

Genera	Species	Age (Ma)	Emergence (the date given is the beginning of the period Ka)
Australopithecus	A. anamensis	4	Lower Pliocene (5,332)
	A. afarensis	3.8	
	A. platyops*	3.5	Middle Pliocene (3,600)
	A. bahrelghazali	c. 3.5	
	A. garhi	2.5	Upper Pliocene (2,588)
	A. africanus	2.2	
	A. sediba	2.0	
Paranthropus	P. aethiopicus	2.5	
	P. boisei	2.2/1.9	
	P. robustus	1.9	
Homo	H. rudolfensis	2.5/2.3	
	H. habilis	2.3	
	H. gautengensis	2.0	
	H. erectus	1.9	

*Originally proposed as *Kenyanthropus platyops.*

Box 5.1 Sagittal crest

The sagittal crest is a kind of vertical keel that runs lengthwise along the midline at the top of the skull. The masseter muscles are anchored to the sagittal crest, reaching to the lower mandible, which they move. This is a dimorphic trait: a crest indicates that the exemplar of *Paranthropus* is male.

5.2 Australopithecines found outside the Rift: South Africa

South African hominin specimens of the Pliocene have been classified in a variety of different genera and species. The various forms have been reduced to three genera: (1) australopiths, with the species *Australopithecus africanus* and *A. sediba*; (2) robust australopiths, with the species *Paranthropus robustus*; and (3) homo, with the species *Homo habilis* and *H. erectus*. In this chapter we will not consider these last two species.

5.2.1 Australopithecus africanus

Remains of *A. africanus* have been found at Taung, Makapansgat (members 3 and 4), Sterkfontein (member 4), and, with less certainty, at Sterkfontein member 5.

No fossil has more right to be considered as *A. africanus* than the one appearing in the description of the species as type-exemplar: Taung 1, The Taung Child, described by Raymond Dart (1925) in the paper proposing the new taxon (Box 5.2). However, subsequent discoveries have cast doubts on the true significance of the Taung specimen. The great amount of *A. africanus* specimens from Sterkfontein member 4 provides a hypodigm whose traits differ somewhat from Taung's (Box 5.3). Thus, the true species to which the latter belongs is an open question.

The most complete *A. africanus* cranium is Stw 505 (Figure 5.1), found in Bed B of Sterkfontein member 4. It includes most of the face, the left endocranium, the cranial anterior fossa, and part of the right middle cranial fossa. A partial right temporal bone, initially believed to belong to another specimen (StW 504), was later reinterpreted as part of Stw 505, receiving the name of StW 505b. This specimen was studied in detail 10 years later (Lockwood & Tobias, 1999). Lockwood and Tobias concluded that, despite its large size and robust complexion, the StW 505 specimen's morphological features are characteristic of *A. africanus*. Nevertheless, this specimen shows affinities with *A. afarensis* interpreted as plesiomorphies (a weak sagittal crest (see Box 5.1), for instance, which is similar to that of A.L. 444–2) and with *H. habilis* (a similar brow ridge to the one observed in the KNM-ER 1470 specimen, for example).

Box 5.2 Taung child discovery

The circumstances of the Taung child discovery, its repercussions, the impact of its proposal as an intermediate species between apes and humans, and the criticisms received until the proposal of *A. africanus* was generally accepted, are reviewed in the book *Senderos de la Evolución Humana* [*Trails of Human Evolution*] (Cela Conde & Ayala, 2001).

Box 5.3 Sterkfontein diversity

StW 505 is not the only specimen from Sterkfontein member 4 exhibiting robust features. The diversity of this member's sample supports the notion that there are different species at the site. Hence, specimens Stw 183 and Stw 255 show derived traits of robust australopiths, such as *P. robustus* from South Africa, but also *P. boisei* and *P. aethiopicus* from East Africa, which we will review later in this chapter (Lockwood & Tobias, 2002). Even Stw 252 could be included in this robust set from member 4 on the grounds of its dental features (Clarke, 1988). The specimen showing the most robust features is Stw 183, but its comparison with other South African forms is difficult because it is a juvenile specimen. Lockwood and Tobias (2002) note some derived traits of *P. robustus* in Stw 183, such as the rounded lateral portion of the inferior orbital margin, found in no other specimen from Sterkfontein member 4. In any case, Lockwood and Tobias (2002) maintain that there was a single species at Sterkfontein member 4, *A. africanus*. Ron Clarke (1988) rather favors that the cranial sample includes a robust species (Stw 252, Sts 71) and a gracile species (Sts 5, Sts 17, and Sts 52).

The computer-generated three-dimensional reconstruction of StW 505 yields a cranial capacity of around 515 cm^3, the highest of all *A. africanus* (Conroy et al., 1998). After a comparison between StW 505 and other hominin specimens from South Africa (Sts 71) and East Africa (OH 24, KNM-ER 1813, and KNM-ER 732;

see Chapter 6), Conroy and colleagues concluded that the brain volumes of these specimens had been overestimated.

An interesting specimen from Bed B of Sterkfontein member 4 is the partial skeleton StW 431 described by Michel Toussaint and collaborators (2003). It lacks the cranium, but the specimen is formed by a total of 18 postcranial bones, including a good part of the backbone. Once more, the specimen includes a mosaic of primitive traits—the elbow, the small size of the lumbar vertebrae, and the sacrum—and derived traits—a pelvis generally similar to modern humans (Toussaint et al., 2003). The authors of the specimen description assigned the exemplar to *A. africanus*, although indicating that the morphological similarity of StW 431 with respect to *A. afarensis* supported the proximity of both taxa. Toussaint and collaborators (2003) attributed to StW 431 a non-obligatory bipedal locomotion; therefore, different from that of modern humans.

Between 1995 and 2001, new exemplars appeared in Jacovec Cave (or Jakovec) of Sterkfontein, with cranial, mandibular, and postcranial materials (Table 5.2). Timothy Partridge and collaborators (2003) indicated some particularities that appeared in the Jacovec sample, such as a clavicle (StW 606), with morphology closer to that of chimpanzees than to modern humans or *Australopithecus*. The longer neck of the femur (StW 598) is close to the *Paranthropus* of Swartkrans and to StW 99 from member 4 of Sterkfontein, contrasting with the shorter neck of StW 522—another specimen from member 4 of Sterkfontein. For the authors, that difference suggested that the femora of the site could belong to a different species of *Australopithecus*. Thus, Partridge et al. (2003, p. 611) stated that, "All of the Jacovec fossils can be identified only as belonging to *Australopithecus* species." A more concrete identification is not possible due to the great antiquity of the Jacovec Cave.

5.2.2 STW 573, Little Foot

We have already mentioned, in Chapter 2, the fossil footbones found at Sterkfontein member 2, known technically as StW 573 and colloquially as *Little Foot*. Certain details of its discovery suggested that additional remains belonging to the same individual could have been preserved and were still contained within the rock. The search for them is due to Ron Clarke's initiative (see Box 5.5). The result was the discovery of quite a complete skeleton embedded in the breccia. The work to retrieve the remains is still underway.

The age of the exemplar was established at 3.5–3 Ma by Clarke and Tobias in agreement with the personal

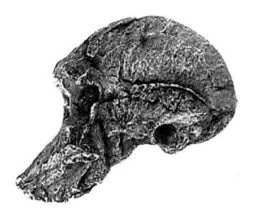

Figure 5.1 StW 505, *Australopithecus africanus* from Sterkfontein member 4 (Lockwood & Kimbel, 1999).

Table 5.2a Main hominins from Sterkfontein (Herries & Shaw, 2011)

Exemplar	Suggested age (Ma)	Member	Species
StW 585	0.5/0.3	Post-M6	*Homo* sp.
StW 80	1.3/1.1	M5C	*Homo* sp.
StW 584	1.4/1.2	M5B	*P. robustus*
StW 566	1.4/1.2	M5B	*P. robustus*
Stw 53	1.8/1.5	M5A	*Homo* sp.
Sts 5	2.2/2.0	M4C	*A. africanus*
StW 573	2.6/2.2	M2	*Australopithecus* sp.
Sts 14	2.6/2.2	M4B	*A. africanus*
Sts 431	2.6/2.2	M4B	*A. africanus*
StW 505	2.6/2.2	M4B	*A. africanus*
Sts 71	2.6/2.2	M4B	*A. africanus*

Table 5.2b Specimens from Jacovec Cave (Partridge et al., 2003)

Exemplar	Description	Date of collection
StW 578	Partial cranium	1995
StW 590	Right P^3 and P^4	1998
StW 598	Left proximal femur (fragment)	1998
StW 599	Deciduous upper canine	1998
StW 600	Lumbar vertebrae	1999
StW 601a	Half right M1 (buccal)	1999
StW 601b	Half right M1 (lingual)	2001
StW 602	Left distal humerus (fragment)	1999
StW 603	Left P4	2001
StW 604	Right M2	2001
StW 605	Hand Phalanx	2001
StW 606	Lateral half of left clavicle	2001

communication of Partridge, based on faunal and stratigraphic comparison (Clarke & Tobias, 1995). Nonetheless, Joanne Walker, Robert Cliff, and Alfred Latham (2006), applying radiometry based on the ^{238}U/^{206}Pb method, reduced that number to 2.2 Ma. The estimation of Andy Herries and John Shaw (2011) for a most probable age of StW 573, between 2.3 and 2.2 Ma, is within the same range.

In addition to the footbones, the StW 573 remains described to date include the cranium (Clarke, 1998) (Figure 5.2) and a complete left arm and hand (Clarke, 1999)

Box 5.4 Makapansgat

Unable to compete in fossil richness with Sterkfontein, Makapansgat has also yielded many exemplars of interest since Raymond Dart (1948, 1949a) published the first findings. Although Dart defined the species *Australopithecus prometeus* for these fossils, Makapansgat hominins are more generally accepted today as *A. africanus*. Tafline Crawford and collaborators (2004) provided a synthesis of previous works at the site, in addition to their own recent work. Making an exception to the general rule, which recognizes all specimens of Makapansgat as *A. africanus*, is Emiliano Aguirre, who argued for the possible presence of *Paranthropus* at the site, referring to the traits of the mandible MLD 2 (Aguirre, 1970).

Some of the specimens of *A. africanus* have been attributed to the species *Homo habilis*. We comment about these proposals in Chapter 6.

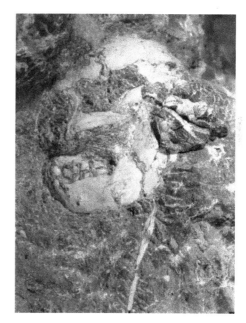

Figure 5.2 Left side of the StW 573 skull partially excavated from the breccia (Clarke, 1998).

(Figure 5.3). Given that these fossils are still embedded in the rock, Ron Clarke warned about the provisionality of the studies. With regard to the cranium, he noted that it corresponds to a mature adult with a zygomatic arch that is much more massive than that of *A. africanus*. Additionally, it has a small sagittal crest on the parietals, and the nuchal plane is very muscular with a pronounced, pointed inion. Clarke (1998) argued that in no way does it conform to the morphology of the *A. africanus*' specimens from Sterkfontein member 4.

Clarke's (1999) preliminary observations of the hand and the arm reveal a mosaic of traits. Some of them (such as the heads of both radius and ulna) more closely resemble those of apes than they do those of modern humans (Box 5.5). The size and length of radius, ulna, and humerus are within the range of average modern humans and of chimpanzees, and are not elongated like those of orangutans, although some features observed on the distal humerus and proximal ulna resemble those of orangutans rather than those of chimpanzees and humans. Metacarpal length is similar to that of *Australopithecus* and *Homo*. The proximal phalanges of the thumb and forefinger display a curvature like that of the phalanges of the *A. afarensis* specimens from Hadar (Ethiopia). The first metacarpal and its proximal phalanx indicate a thumb similar to that of modern humans. Conversely, the trapezium of the wrist is different from those of modern humans, chimpanzees, and orangutans, showing unique features. Clarke (1999) has noted that the significance of

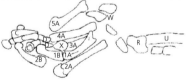

Figure 5.3 Handbones of StW 573. Legend for interpretive sketch: 1A, first metacarpal (thumb); 2A, second metacarpal (index finger); 3A, third metacarpal (middle finger); 4A, fourth metacarpal (ring finger); 5A, fifth metacarpal (little finger); 1B, proximal phalanx of thumb; 2B, proximal phalanx of index finger; C, middle phalanges; D, terminal phalanges; W, wrist bones; R, damaged radius; U, damaged ulna; X, unknown. The trapezium is at the lower right-hand edge of picture (Clarke, 1999).

Box 5.5 The mystery of the radius associated with StW 573 (narrative of Ron Clarke)

"There is one other bone present that, while appearing to be undoubtedly part of the *Australopithecus* skeleton, has three uncharacteristic features. It is a left radius that lies next to the left femur in a position that might be expected for a minimally disturbed skeleton. As there are no other animal fossils apart from an occasional small fragment within a wide vicinity of the skeleton, there was no reason to think that it was not part of the hominid. Furthermore, apart from slight crushing at the distal end, this slender bone is virtually intact and unbroken, despite being adjacent to and downslope of a large rock. Thus, it seems unlikely to have rolled down the slope from elsewhere (...). As the skeleton could not have had two left radii, it was necessary for me then to expose more of this previous left radius that was still largely encased in breccia next to the femurs. As soon as I had uncovered the proximal end of the shaft, it became clear that it was morphologically like a large cercopithecoid, although the distal end did not quite match any of our fossil or modern cercopithecoids. It appears an extraordinary coincidence that this is the only virtually complete animal bone in the vicinity of the hominid skeleton. It is also in the correct position relative to the femurs to have belonged to the hominid if the latter's arm had been resting by its side. Although it might simply be coincidence, one can at least consider the possibility that the *Australopithecus* may have been carrying that radius when it died" (Clarke, 1999, p. 479).

---

the unusual thumb joint on the StW 573 specimen "has still to be determined."

How can the taxonomy of such a specimen be resolved? Clarke's (1988, p. 462) answer: "I prefer to reserve judgment on the fossil's exact taxonomic affinities, although it does appear to be a form of *Australopithecus*." Preliminary descriptions suggest StW 573 could be regarded as a very early species, older than *A. africanus* and closer, in morphological terms, to contemporary Ethiopian australopiths. As Clarke notes, the hand–arm set makes possible the study of an *Australopithecus* forelimb, as a complete unit, for the first time, allowing interpretation of its function. The strong opposable thumb, the curved phalanges, and orangutan-like elbow joint are suggestive of arboreality, as the footbones that inspired the name *Little Foot* did earlier. Clarke (1999) agrees with Sabater Pi and colleagues (1997) that australopiths might have nested in trees. Clarke added that they might also have spent part of the day feeding in trees, as orangutans and chimpanzees do.

5.2.3 Australopithecus sediba

The exemplar MH1 comes from the South African site Malapa. It is formed by a juvenile cranium, UW88-50, a fragmented mandible, UW-88–8, and various parts of the skeleton that were assigned as the holotype of a new species, *Australopithecus sediba*, by Lee R. Berger and collaborators (2010). To this species was also attributed a second partial skeleton of an adult, MH2, thought to be a female (Figure 5.4).

The dating results of the specimens of Malapa were imprecise at first, as happens with all the South African sites. The associated fauna indicated 2.36–2.5 Ma. The sedimentary deposits in which the fossils were found—safeguarded from the possible activities of scavengers and predators—were estimated by paleomagnetic studies to be of an age corresponding to the Olduvai subchron, i.e., between 1.95 and 1.78 Ma (Dirks et al., 2010). New radiometric studies (U/Pb), combined with paleomagnetism and stratigraphic analysis of the sediments, provide a consistent dating of 1.977±0.002 Ma (Pickering et al., 2011).

MH1 traits deviate from the usual interpretations of the relationships of specimens found since 1925 in South Africa, and raise certain questions about the meaning of "robust" and "gracile" *Australopithecus*.

In the description by Berger and collaborators (2010), *A. sediba* is distinguished from *A. anamensis*, *A. afarensis*, and *A. garhi* by the lack of typical apomorphies of those australopiths, such as megadontia and facial projection (prognathism). *A. sediba* also lacks the apomorphies of robust australopiths: a large mandibular apparatus, extreme megadontia of premolars and molars, and pronounced marks of muscle insertions in the cranium. In the opinion of Berger et al. (2010), the specimens of Malapa resemble more *Australopithecus africanus*, in particular in the lower cranial capacity of *A. sediba*. But, the anatomical differences with respect to *A. africanus* are important, because the facial features and cranium of *A. sediba* bring it closer to the ancient specimens of genus *Homo*. The postcranial remains that maintain the general form of *Australopithecus* are also close to *Homo* in some pelvic traits, as argued by Berger et al. (2010). Despite these similarities to *Homo*, the authors assigned the specimens of Malapa to the genus *Australopithecus*, naming a new species: *A. sediba*.

The morphological characteristics of *A. sediba*, which include primitive traits typical of *Australopithecus*, as well as derived traits that bring the taxon closer to

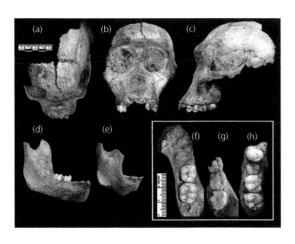

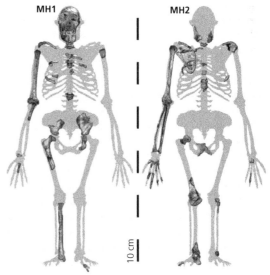

Figure 5.4 Left: *A. sediba* (a)–(c) cranium UW88-50; (d) and (f) juvenile mandible UW-88-8; (e) and (g) adult mandible UW88-54; (h) maxillar of UW88-50. Right: partial skeletons MH1 and MH2. From Berger, L. R., de Ruiter, D. J., Churchill, S. E., Schmid, P., Carlson, K. J., Dirks, P. H. G. M., & Kibii, J. M. (2010). *Australopithecus sediba*: A New Species of Homo-Like Australopith from South Africa. *Science*, 328(5975), 195–204. Reprinted with permission from AAAS.

Homo, were supported with new evidence a year after the first findings. Based on the reconstruction of the materials in 2010 and the discovery of new exemplars, the partial pelvises of *A. sediba* indicate the primitive character of traits such as the large ramus—but in a higher position—and the small sacrum. Derived traits closer to *Homo* include a more vertical and robust ilium and a shorter ischium (Kibii et al., 2011).

A right distal tibia, U.W. 88–97, a right talus U.W. 88–98, and a right calcaneus 88–99, belong to the specimen MH 2, which was found embedded in a calcareous matrix. The attempt to separate these three bones risked the destruction of the specimen (Zipfel et al., 2011). Berhard Zipfel and collaborators (2011) opted for scanning the whole specimen to examine the morphology of its components. Analysis of these bones, along with the rest of the exemplars forming the distal end of the leg and foot of MH 1 and 2, pointed in the direction of a mosaic of primitive traits associated with climbing—such as the gracile calcaneus—and derived traits pointing to bipedalism—such as the ankle joint.

The hand and wrist of *A. sediba* show an identical panorama: traits characteristic of *Australopithecus* that point toward an arboreal locomotion—robust flexor apparatus—together with traits found in *Homo*—long thumb and short fingers (Kivell et al., 2011). These last features suggest to Tracy Kivell and collaborators (2011) a precision grip, which could indicate the use of tools.

The virtual reconstruction of the endocranium of MH1 carried out by Kristian Carlson and collaborators (2011) indicated once more the pattern: circumvolutions distinctive of *Australopithecus*, particularly in the frontal area, with a tendency toward the orbitofrontal shape and organization, which will appear in *Homo*.

5.3 Australopithecines found outside the Rift: Chad

Every fossil of *Australopithecus* known at the beginning of the last decade of the twentieth century came mainly from the Rift Valley, and also from South Africa. The discovery of a very early australopithecine specimen, more than 3 million years old, in Chad, a considerable distance from all previously known deposits, represented a challenge. The authors of the article in which the new specimen was announced (Brunet et al., 1995; see Morell, 1995) argued that this finding completely changed the scene regarding humanity's birthplace. After the addition of *Sahelanthropus tchadensis*, which

we described in Chapter 4, the necessity to consider the findings from Chad became unavoidable.

In January 1995, Michel Brunet discovered a mandibular fragment of an adult hominin (KT 12/H1) at the locality KT 12 of the Bahr el Ghazal region ("river of gazelles" in classic Arab), in the Djourab desert, 45 km east of Koro Toro (Chad). The fragment includes an incisor, two canines, and two premolars from both sides of the mandible. The morphological examination carried out by Brunet et al. (1995) indicated that the specimen was similar to *A. afarensis*, even though some features differentiated it from that species and other australopithecines. The provisional classification of KT 12/H1 was *Australopithecus* aff. *afarensis* (Brunet et al., 1995). One year later, and after the comparison with the *A. afarensis* specimens kept at the National Museum of Addis Abeba (Ethiopia), that assignation was revised. Some derived traits of KT 12/H1, such as those related with the morphology of the mandibular symphysis, indicate a less prognate face than *A. afarensis*. The decision was made to name the taxon *Australopithecus bahrelghazali*, with KT 12/H1 as the holotype, and a premolar, discovered in 1996 (KT 12/H2), as the only paratype for the moment (Brunet et al., 1996) (Figure 5.5).

As a homage to a late colleague, Abel Brillanceau, Brunet and his team informally baptized KT 12/H1 as *Abel*. No radiometrically measurable soils were found, but the fauna associated with the specimen shows a close similarity with that of Hadar soils between 3.4 and 3.0 Ma (Brunet et al., 1995). Paleontological and sedimentological studies pointed toward a lakeshore habitat, with a mosaic vegetation: gallery forests and brushy savannas with open grasslands;

Figure 5.5 KT 12/H1, *Australopithecus bahrelghazali* (Brunet et al., 1996; image from http://www.scienceinafrica.co.za/pics/origin5a. gif). Reproduced from Brunet, M., Beauvilain, A., Coppens, Y., Heintz, E., Moutaye, A. H. E., & Pilbeam, D. (1996). Australopithecus bahrelghazali, une nouvelle espèce d'Hominidé ancien de la région de Koro Toro (Tchad). *C.R. Acad. Sci. Paris*, t. 322, série II a, 907–913, Elsevier Masson SAS. All rights reserved.

that is to say, a similar habitat to Hadar at the time. This coincides, certainly, with the ecological model usually related with the first hominins. But the notion that *A. bahrelghazali* was contemporary to the Hadar *A. afarensis* and, thus, represents one of the first hominins, stumbles with the diversification model of hominoids put forward by Yves Coppens, one of the authors of the articles that introduced the new species. The geological transformation of East Africa as a consequence of the separation of the African and Arabian tectonic plates, which produced the Rift Valley chain of faults, led to important changes in the ecosystem, which could have been a source of selective pressures for African late-Miocene hominoid populations, leading to the separation of the different lineages we know today.

Coppens (1994) and Jean Chaline et al. (1996) have explained such a separation by means of an allopatric model (see Chapter 3). The geographic separation between the different populations would be: pre-gorillas to the North of the barrier constituted by the River Zaire, pre-australopithecines in the Levant region of the Rift, and pre-chimpanzees in Central and Western Africa. A separate evolution in each geographical location would fix specific, derived traits (autapomorphies) in each of the branches. This allopatric speciation model would explain why there are no remains of chimpanzee ancestors in hominin sites.

However, the finding of KT 12/H1 suggests an alternative because it indicates that the presumed continuity of thick forests to the West of the Rift did not exist during the Pliocene. Brunet et al. (1995, 1996), believe that the forests and savannas, capable of housing the first hominins, extended from the Atlantic Ocean, through the Sahel, along East Africa, and up to the Cape of Good Hope. The reason why no specimens have been found in intermediate zones is the lack of Pliocene and Lower Pleistocene soils. The fast appearance and expansion of hominins would render speculations about their birthplace useless.

This is an attractive idea, and a recent discovery speaks in favor of it. Sally McBrearty and Nina Jablonski found three teeth, KNM-TH 45519, KNM-TH 45520, and KNM-TH 45521, which represent the first known chimpanzee fossil specimens (McBrearty & Jablonski, 2005) in the Kapthurin formation (Tugen, district of Lake Baringo, Kenya). The Kapthurin formation has been dated using the ^{40}Ar/^{39}Ar method (Deino & McBrearty, 2002), attributing to the chimpanzee fossils an age of 545±3 kyr.

5.4 The diversification of *Australopithecus* in the Rift Valley during Middle and Upper Pleistocene

5.4.1 Kenyanthropus (Australopithecus) platyops

The fieldwork carried out at the Lomekwi basin, belonging to the Nachukui formation, West Turkana (Kenya), during 1998 and 1999, yielded important discoveries. The research team, led by Meave Leakey, discovered a very complete but deformed cranium (KNM-WT 40000), a temporal bone, two fragmentary maxillas, and some isolated teeth that rendered a distinctive gracile hominin morphological picture (Leakey et al., 2001). However, its age was much greater than any other specimen belonging to the gracile lineage. KNM-WT 40000 appeared between two volcanic tuffs: the 3.57 Ma Lokochot tuff and the 3.40 Ma Tulu Bor tuff. The dating was profusely discussed and correlated by Leakey and colleagues (2001). The age assigned to the fossil was 3.5 Ma, a million years older than the first *Homo* specimens, regarded as the main specimens in the cladistic event separating *Homo* and *Paranthropus*. After an anatomical comparison with the rest of the Pliocene hominin species, Leakey et al. (2001) introduced a new genus and species, *Kenyanthropus platyops*, emphasizing the most notorious feature of KNM-WT 40000: its flat face (Figure 5.6). This specimen was considered as the species' holotype, and KNM-WT 38350—a partial left maxilla found by B. Onyango in 1998—as the paratype. The remaining

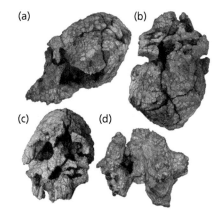

Figure 5.6 KNM-WT 40000 *Kenyanthropus platyops.* (a) Lateral view; (b) frontal view; (c) superior view; (d) occlusal view of palate (Leakey et al., 2001).

specimens found during the Lomekwi campaigns were left without being classified due to insufficient evidence. Nevertheless, Leakey and colleagues (2001) pointed out the affinities existing between *K. platyops* and previously found fossils, such as the KNM-WT 8556 partial mandible, attributed to *A. afarensis*.

K. platyops shows a mosaic of primitive and derived traits. The former, which draw it close to an early australopithecine, such as *A. anamensis*, include the small size of the external auditory pore; its small cranium is also typical of *Australopithecus*; its premolar and molar enamel thickness is similar to that of *A. afarensis*. A notorious derived trait of *K. platyops* is the very vertical plane below the nasal orifice, which gives it the flat-faced look its name refers to.

Is it necessary to add a new genus? The reasons for the proposal of *Kenyanthropus* lie in the morphological particularity of a moderate prognathism, which distance KNM-WT 40000 from the rest of the australopiths, with their great facial projection. None of the paratypes, KNM-WT 38350 and, possibly, KNM-WT 8556, are useful to confirm or deny this trait. However, the distinctive adaptive strategy deduced from the platycephaly—necessary to define a new genus—is not clear. Furthermore, the dental materials of the *Kenyanthropus* holotype bring it closer

---

Box 5.6 About the genus of Lomekwi specimens

The authors of this book have previously argued the consideration of *Australopithecus africanus* as the first paranthrop—*P. africanus*—defended by authors such as Skelton and McHenry (1992) when the date of c. 3.5 Ma was accepted for the most ancient exemplars of Sterkfontein. That supposition and the platycephaly of KNM-WT 40000 are an argument in favor of assigning this last specimen to the genus *Homo*, accepting the beginning of the Middle Pliocene as the most probable date for the cladistic separation of *Paranthropus/Homo*. However, the assignation of a much lesser antiquity to *A. africanus* makes that taxonomic proposal unworkable. The morphology of the hypodigm of *Kenyanthropus*, indicated here, supports the inclusion of those exemplars in *Australopithecus*.

---

to *Australopithecus*, to the point that KNM-WT 8556 could be attributed, at least tentatively, to *A. afarensis* (White, 2003). Nonetheless, we prefer to maintain the distinction of the taxon at the species' level of *Australopithecus: A. platyops*.

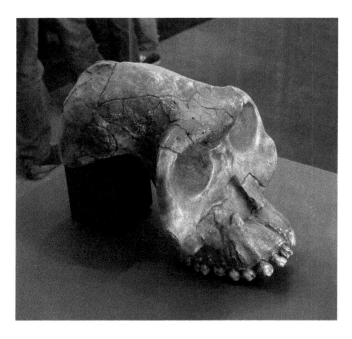

Figure 5.7 Addis Abeba, Musée National d'Ethiopie: reconstructed cranium of *Australopithecus garhi* from elements found in 1997 (picture from Wikimedia Foundation). Picture from Ji-Elle, Wikimedia Commons, https://pt.wikipedia.org/wiki/Australopithecus_garhi.

5.4.2 Australopithecus garhi

The Hata member of the Bouri formation in Middle Awash (Ethiopia) has yielded hominin remains since 1990. The most complete specimens were described in Asfaw et al. (1999), with the proposal of yet another australopithecine taxon: *Australopithecus garhi*. The holotype of the species is the specimen BOU-VP-12/130, a set of cranial fragments, including the frontal, parietals, and maxilla with dentition, found in 1997 by Haile-Selassie (Figure 5.7). The volume of an endocranial cast made by Ralph Holloway was 450 cc. The age obtained by means of the ^{40}Ar/^{39}Ar method, paleomagnetism, and associated fauna, is about 2.5 Ma (2.496±0.008, from radioisotopic studies; de Heinzelin et al., 1999). Other postcranial remains, including a left femur and right humerus, radius, and ulna, were found during the 1996–98 campaigns.

The small cranial capacity, together with wide premolars and molars, led Asfaw and colleagues (1999) to argue that the Bouri remains belonged to an australopithecine posterior to *A. afarensis*. The differences between the specimen and *A. africanus* or *P. aethiopicus* (*A. aethiopicus*, according to the authors) justify, in the view of Asfaw et al. (1999), the creation of a new species.

5.5 Adaptation; an Upper Pliocene difference

In Chapter 4, we echoed the hypothesis of Elizabeth Vrba and Phillip Tobias on the hominin evolution of the Upper Pliocene to be the result, in Africa, of the expansion of open savannas as the climate cooled. The Vrba and Tobias model of hominin evolution by adaptation to Pliocene climatic change makes it necessary to identify what were the adaptive strategies of *Paranthropus* and *Homo* that derive from the traits of *Australopithecus*, for confronting the new ecological horizons of the savanna. Several authors accept a simple interpretation, which attributes to *Paranthropus* a very robust development of the masticatory apparatus, allowing them to take advantage of the tougher vegetal materials of open prairies. Lacking that resource, *Homo* would have used the lithic culture to incorporate—thanks to scavenging, associated or not with hunting—increasing quantities of a carnivorous diet.

When trying to explain a phylogenetic change, the first evidence to consider is the traits of previous specimens. The characteristics present in the original lineage will remain, to some extent, as primitive traits in the new clades. These, in turn, will present their own derived traits, which serve to identify them. Therefore, in order to arrange the specimens, it is necessary to compare traits and establish directional vectors.

The traits characteristic of *Australopithecus*—the original genus—are, as we have seen in Chapter 4, an initial megadontia, small-size cranium similar to that of chimpanzees, very pronounced facial projection (prognathism), and locomotion that tends toward the development of obligatory bipedalism, although retaining some tree-climbing capabilities.

As primitive traits, *Paranthropus* retains, among other things, the prognathism and a lesser cranial capacity in relation to body size. *Paranthropus* apomorphies obviously affect a particular feature: the robustness of the masticatory apparatus, which requires the presence of large masseter muscles to move the jaws. *Paranthropus*' masseters are attached to the cranium, thus creating a very prominent sagittal crest, particularly in males.

The apomorphies of *Homo* are an increment of cranial capacity—which increases with the different species of the genus—a gradual loss of prognathism and a sharp decrease of the masticatory apparatus. The Pleistocene specimens of *Homo* show a bipedal locomotion similar to current humans, but those from the Pliocene still retain some primitive traits in the locomotor apparatus. Carving stone tools is another feature we associate with *Homo*.

5.6 Dental enamel and diet

For documenting the loss of *Australopithecus*' apomorphies and the incorporation of new ones in *Paranthropus* and *Homo*, it is necessary to detect those traits in the fossil record.

South African sites provide evidence of two ancient hominin types that, as we know, were informally called "gracile" and "robust" forms. Table 5.3 includes the main differences between these two forms, but for the first time the name is sufficiently explicit: *A. africanus* has a slight aspect in comparison to *P. robustus*. In Frederick Grine's words (1988), the election of terms by Robert Broom was prophetic in distinguishing these two types of hominins.

The teeth and facial structure of the hominin discovered in Kromdraai led Broom (1938) to name it "robust" in comparison with the specimens found a bit earlier in Sterkfontein. When the specimens of Swartkrans later emerged, their appearance, similar to those of Kromdraai, led them to be classified as *P. robustus* (Broom & Robinson, 1949). Actually, the teeth of the robust hominin of Swartkrans are even larger. Thus, tooth size became an important trait to measure the

Table 5.3 Differences between gracile and robust australopiths (After Wood and Richmond, 2000, modified)

	"Gracile"	"Robust"
Cranial		
1. Overall shape.	Narrow, with "unmistakable" forehead; higher value for supraorbital height index value for supraorbital height index (Le Gros Clark, 1950).	Broad across the ears; lacking a forehead; low supraorbital height index.
2. Sagittal crest.	Normally absent.	Normally present.
3. Face.	Weak supraorbital torus; variable degree of prognathism, sometimes as little as "robust" form.	Supraorbital torus well developed medially to form a flattened "platform" at glabella; face flat and broad, with little prognathism.
4. Floor of nasal cavity.	More marked transition from the facial surface of the maxilla into the floor of the pyriform aperture; sloping posterior border to the anterior nasal spine and lower insertion of the vomer.	Smooth transition from facial surface of maxilla into the floor of the pyriform aperture; small anterior nasal spine that articulates at its tip with the vomer.
5. Shape of the dental arcade and palate.	Rounded anteriorly and even in depth.	Straight line between canines, deeper posteriorly.
6. Pterygoid region.	Slender lateral pterygoid plate.	Robust lateral pterygoid plate.
Dental		
7. Relative size of teeth.	Anterior and posterior teeth in "proportion."	Anterior teeth proportionally small; posterior teeth proportionally large.
8. dm_1 (deciduous lower molars)	Small, with relatively larger mesial cusps. Lingually situated anterior fovea; large Protoconid with long, sloping buccal surface.	Large, molariform, with deeply incised buccal groove and relatively large distal cusps.
9. P-roots (premolars)	Single buccal root.	Double buccal root.
10. c̲ (deciduous upper canines)	Large, robust and symmetric crown with slender marginal ridges and parallel lingual grooves.	Small, Homo-like, with thick marginal ridges and lingual grooves converging on the gingival eminence.
11. c (deciduous lower canines)	Asymmetric crown with marked cusplet on the distal marginal ridge and marked central ridge on the lingual surface	More symmetric crown with parallel lingual grooves, weak lingual ridge and featureless distal enamel ridge.

Source: Wood and Richmond (2000).

degree of robustness (Robinson, 1954b, 1968). Larger and more robust teeth are the best indicator of *Paranthropus*' vegetarianism.

The dental enamel is another distinctive trait. Robust australopithecines have a very thick dental enamel in molars, in contrast to the thinner enamel—thick, but not extremely so—of *A. africanus*. This is a trait that, according to Frederick Grine and Lawrence Martin (1988), raises important phylogenetic and adaptive questions.

Talking about dental enamel is talking about dietary tendencies. Initial interpretations of the South African remains suggested that *Paranthropus* and *Australopithecus* specialized in different kinds of food (Box 5.7). Dart conceived of *A. africanus* as a hunter and, thus, a carnivore. He pictured *A. africanus* as using stone weapons to survive in the savanna in competition with other predators (Dart, 1925, 1949b, 1953, 1957). After a comparative study of the dentition of South African australopiths, Robinson (1954b) put forward his "dietary hypothesis" as an interpretation of the differences between gracile and robust hominins. According to Robinson, the diet of *Paranthropus* was vegetarian, whereas that of *A. africanus* was omnivorous and included an important amount of meat. Robinson believed the structure of the face and cranium reflected this dietary difference.

The studies of Frederick Grine on the microwear patterns left by food on teeth (Grine, 1981, 1987; Kay & Grine, 1985; Grine & Martin, 1988; Ungar & Grine, 1991) provided the first direct evidence about the dietary habits of robust and gracile australopiths. The microwear patterns vary with the hardness of the

Box 5.7 Australopiths' diet

Not everyone agreed with the dietary distinction between *Australopithecus* and *Paranthropus* grounded on dental traits. Milford Wolpoff (1971b), for instance, denied there was a great morphological disparity between the gracile and robust forms. All South African australopiths, according to Wolpoff, are much more similar than is often maintained. Furthermore, their dentition indicates that they shared a hard vegetarian diet. Wood and Strait (2004, p. 119) reached the same conclusion in their study of 11 traits related with diet, habitat preference, population density, and dispersion, among others. According to Wood and Strait, "*Paranthropus* and early *Homo* were both likely to have been ecological generalists."

materials that were chewed. Grine and colleagues examined the microwear patterns of deciduous and permanent molars under the microscope, and compared them with those of current primates. They concluded that *Paranthropus* and *Australopithecus* had different dietary habits. According to Grine's model, the diet of *A. africanus* was mainly folivorous. But the impossibility of comparing their microwear patterns with those of current frugivorous primates that do not also chew hard objects, led Grine to suggest that fruits could have also been part of the diet of *A. africanus*. Conversely, *P. robustus* would have chewed smaller and harder objects, such as seeds. The microwear studies on incisors revealed that the diet of *A. africanus* was more varied than that of *P. robustus*, which was more specialized (Ungar & Grine, 1991). Peter Ungar and colleagues (2006), which include Frederick Grine, have examined the correlation between microwear and diet. They examined 18 cheek teeth attributed to the genus *Homo* with preserved antemortem microwear from Ethiopia, Kenya, Tanzania, Malawi, and South Africa. Microwear features were measured for these specimens and compared with five extant primate species (*Cebus apella*, *Gorilla gorilla*, *Lophocebus albigena*, *Pan troglodytes*, and *Papio ursinus*), and two human foraging groups (Aleut and Arikara). Ungar et al. (2006, p. 78) concluded that "dental microwear reflects diet, such that hard-object specialists tend to have more large microwear pits, whereas tough food eaters usually have more striations and smaller microwear features."

The analysis of the strontium/calcium (Sr/Ca) ratios in tooth enamel is a second source of direct evidence of the effects food has on teeth. Andrew Sillen and colleagues (Sillen, 1992; Sillen et al., 1995; Sillen et al., 1998) have applied this technique to Swartkrans' fossil mammals. They found that Sr/Ca ratios correlated with feeding habits. Leaves and grass contain less strontium than fruits and seeds, so, to a first approximation, a high Sr/Ca ratio would suggest an herbivorous diet (but not only leaves) or carnivorous. Because the amount of strontium varies locally, migration patterns had to be taken into account. In any case, the studies showed that for *P. robustus*, Sr/Ca ratios were higher than for carnivores, but lower than for baboons, whose diet is omnivorous (Sillen, 1992). Gracile specimens from Swartkrans—*Homo habilis/Homo erectus*—showed a similar Sr/Ca ratio to baboons. Thus, it seems their diet was more varied than that of robust specimens.

The third direct source of dietary evidence is stable carbon isotope analysis. The ratio of carbon isotopes $^{13}C/^{12}C$ is lower in trees and shrubs than in grasses, reflecting the way they are incorporated into the trophic chain. Herbivores that feed on grass and their carnivore predators will have higher $^{13}C/^{12}C$ ratios (Sponheimer & Lee-Thorp, 1999). Julia Lee-Thorp and colleagues (Lee-Thorp & van der Merwe, 1993; Lee-Thorp et al., 1994) applied this analysis to Swartkrans *P. robustus* specimens. Their results suggested that the better part of their diet was food with a relatively low $^{13}C/^{12}C$ ratio, specifically about three-quarters of their total intake. *P. robustus* either fed mostly on fruits and leaves or on herbivores with this diet. Sponheimer and Lee-Thorp (1999) also applied stable carbon isotope analysis to *A. africanus*' specimens from Makapansgat, obtaining higher ratios than those expected for a frugivore. According to these authors, either *A. africanus* ate not only fruits and leaves, but also large quantities of foods such as grasses and sedges, or they fed on animals that ate these plants. In the first case, *A. africanus* from Makapansgat would have exploited the resources afforded by an open savanna, searching for food in woodlands or grasslands. The second scenario implies that a carnivorous diet had appeared before the genus *Homo* and Oldowan instruments (Sponheimer & Thorp, 1999).

For *Paranthropus*, Lee-Thorp and collaborators (2010) pointed out that in Eastern Africa (i.e., that of *P. boisei*) 80% of the diet consisted of the grasses and sedges of the open savanna—similar, then, to that of *A. africanus* of South Africa—making difficult the idea of an intake of tough materials. The author proposed for *Paranthropus* a change in diet to savanna grass materials prior even to *A. africanus*, although this is not

---

Box 5.8 Vegetation change in East Africa

Two carbon isotopes found in plants, C_3 and C_4, the proportions of which can be detected in marine sediments, make possible to analyze a vegetation change in East Africa. The vegetation has C_3 from 24 Ma until 10 Ma, at which time C_4 appears (Uno et al., 2016). Kevin T. Uno and collaborators assert that, "The response of mammalian herbivores to the appearance of C_4 grasses at 10 Ma is immediate, as evidenced from existing records of mammalian diets from isotopic analyses of tooth enamel. The expansion of C_4 vegetation in eastern Africa is broadly mirrored by increasing proportions of C_4-based foods in hominin diets, beginning at 3.8 Ma in *Australopithecus* and, slightly later, *Kenyanthropus*. This continues into the late Pleistocene in *Paranthropus*, whereas *Homo* maintains a flexible diet. The biomarker vegetation record suggests the increase in open, C_4 grassland ecosystems over the last 10 Ma may have operated as a selection pressure for traits and behaviors in Homo such as bipedalism, flexible diets, and complex social structure (Uno et al., 2016, p. 6355)."

---

supported by the isotopical studies of *Ardipithecus ramidus*. Consequently, the origin of this diet change would be between 3 and 4.4 Ma earlier.

Taken together, the $^{13}C/^{12}C$ ratio studies show that it cannot be assumed that the diet of *P. robustus* was specialized compared to the more varied one of *A. africanus*. Wood and Strait (2004) concluded, in their review of the different sources of evidence related with dentition and diet, that it is incorrect to associate stenophagy—reduced diet—with South African robust hominins (*P. robustus*) and euryophagy—broad diet—with gracile ones (*A. africanus* and *H. habilis/H. erectus*). Contrary to Robinson's "dietary hypothesis," the diets of South African early hominins must have been more intricate and complex.

The study of current conditions in the Serengeti, which are closely linked to climatic seasonal factors, makes scavenging only possible in times of abundant prey, so the need of scavenger hominins to supplement their diet by collecting plant materials seems certain (Jolly, 1970). The conclusion of a generalist behavior associated with *Paranthropus* was also later obtained by Darryl J. de Ruiter, Matt Sponheimer, and Julia A. Lee-Thorp (2008) after studying the habitat preferences of "robust" australopiths in Bloubank Valley, South

Africa. In addition, the study of dental wear marks of Olduvai hominins by Peter Ungar et al. (2011), led to the same conclusion: *P. boisei* does not present signs of wear of a specialist in a tough materials' diet. In other words, it appears that *Paranthropus* and *Homo* were opportunistic omnivores, but Thure Cerling and collaborators (2011) have indicated that, according to isotope analyses, it is likely that the masticatory apparatus of *Paranthropus*, and in particular of *P. boisei*, is due to the intake of large amounts of low-quality plants.

Interestingly, the work carried out by van der Merwe and colleagues (2003), on the carbon isotope ratios of ten *A. africanus* specimens from Sterkfontein member 4, concluded that their diet was highly diverse. It could have included grasses and sedges and/or the insects and vertebrates that eat these plants. But this variation existed among the sample's individuals, and it was "more pronounced than for any other early hominin or non-human primate species on record" (van der Merwe et al., 2003, p. 581). To put it another way, it might be necessary to study differences among individuals in order to advance our current knowledge of the diet of *A. africanus*.

Let us review what has been stated up to now. For reasons relating most probably to the Pliocene climate change in Africa, hominins faced the task of adapting to open grasslands. They did so through various strategies that led to the emergence of two different genera. So, if we intend to understand how the evolutionary step from *Australopithecus* to *Paranthropus* and *Homo* took place, it is necessary to find specimens with indications, around 2.5 Ma, of the following evolutionary trends:

- Evidence of evolution toward *Paranthropus*' sagittal crest extreme megadontia; in particular increase in robustness of the masticatory apparatus compared to *Australopithecus*.
- Evidence of evolution toward *Homo* loss of prognathism and other facial reduction changes of the masticatory apparatus with increment of brain capacity.

In this chapter we will address only the first of the two hominin lineages that faced the task of adaptation to open areas: those belonging to the genus *Paranthropus*.

5.7 The genus *Paranthropus*

The identification of *Paranthropus* exemplars was made for the first time in South Africa, when specimens notably different from *A. africanus* appeared and a new genus for them was proposed. During the 20 years that passed from the discovery of the Taung fossil to its final examination, Robert Broom, from the Transvaal Museum of Pretoria (South Africa), continued looking for fossils, fitting them into the evolution of our family. Many of the different genera and species used by Robert Broom and other authors to classify South African findings have eventually been abandoned. But it would be unfair to criticize the multiplication of genera and species of the past as if it were senseless. It would even be worse to believe we are the first ones to realize the problem. As different South African hominins were discovered, some researchers advocated considering them as belonging to a single genus, and even a single species. For example, Broom (1950) quoted S. H. Haughton and H. B. S. Cooke, who reintroduced an argument used against synapomorphies in many instances: if two very similar species were present at the same time and the same or similar places, then they must be regarded as a single species. But Broom (1950) gave sound arguments in favor of recognizing different South African genera. Taung corresponds to a juvenile specimen, and establishing it

as the holotype of *A. africanus* was, undoubtedly, an inconvenience. Hence, Broom's comparisons between the Taung specimen and other fossils available at the time had to be based on deciduous teeth.

Despite such problems, Broom (1950) reached an important taxonomic conclusion after examining South African hominin genera and species. He suggested there were two kinds of early hominins in South Africa. He placed them at the subfamily level: Australopithecinae, including the genera *Australopithecus* and *Plesianthropus*; and Paranthropinae, including the genus *Paranthropus*. Broom added a doubtful subfamily (Archanthropinae, encompassing *Australopithecus prometheus*).

5.7.1 Paranthropus robustus

The species *Paranthropus robustus* was proposed by Robert Broom in 1938, after examining the specimen TM 1517 from Kromdraai (Figure 5.8). It is a craniofacial fragment of five teeth and other associated skeletal fragments (Broom, 1938), which, in contrast with the specimens

Figure 5.8 Maxillar and mandible TM 1517 of Kromdraai (Modern Human Origins, http://www.modernhumanorigins.net/tm1517.html). Photograph by John Reader, modernhumanorigins.net.

identified by Broom in Sterkfontein, had a very massive appearance. Some traits (cranium, incisors, and canine sockets of small size) bring it closer to the previous "gracile" australopiths, while others (a large and projected face and wide molars) indicate that it was a different type of human fossil. The age attributed to TM 1517, according to paleomagnetic studies done at Kromdraai, is 1.9 Ma (Thackeray et al., 2002) (see Chapter 3). In the following years more specimens were found at the site, which were attributed to the same species (TM 1536, TM 1603).

In 1948, Broom found at Swartkrans an incomplete cranium with part of the face and the upper mandible (SK 46), which was classified as a new species: *Paranthropus crassidens* (Broom, 1949). Later, Robinson (1954a) reduced it to the level of a subspecies of *P. robustus*. Its robustness was reminiscent of the Kromdraai specimen. In 1950, another nearly complete cranium, with almost the whole face (SK 48), was found in the same site (Broom & Robinson, 1952). SK 48 revealed some previously unknown traits of *Paranthropus* (Box 5.9). Although the specimen broke during the excavation works, the reconstruction showed a pronounced brow ridge and glabella. The absence of a sagittal crest was attributed to sexual dimorphism: SK 48 was thought to correspond to a female. The discoveries at Swartkrans documented dental dimorphisms. A complete mandible (SK 23), which was very robust, with a high and solid ramus, wide molars and premolars, and small canines and incisors, was attributed to a female. Two mandibles, which were even more robust, were classified as male specimens (Day, 1986).

Since those early years, a great number of specimens of *P. robustus* appeared in Swartkrans and Kromdraai. Figure 5.9 illustrates those available in 1984.

Most of the materials assigned to *P. robustus* are from crania, mandibles, and dentition: 341 specimens versus 71 postcranial elements found in Swartkrans up to 2001, the year in which Randall Susman, Darryl de Ruiter, and C. K. Brain (2001) described new specimens: two femoral heads, a distal femur, four distal humerus, two distal radius, a talus, and six hand bones. But the great abundance of craniodental materials is something that can be said of almost any other taxon. Three new dental specimens were described in 2009 (Sutton et al., 2009).

The differences of Swartkrans specimens cannot be justified solely on the basis of sexual dimorphism. To the partial cranium, SKW 18, Ron Clarke (1977) associated the facial fragment, SK 52, to complete the same cranium (Figure 5.10). The morphological study by Darryl de Ruiter, Christine Steininger, and Lee Berger (2006) attributed it to a male gender. The authors noted that the features of the skull base show that the variation that SKW 18 introduced in *P. robustus* is extensive.

Box 5.9 Characteristic traits of *Paranthropus*

The characteristic robustness of *Paranthropus* is used as general evidence of its identification, but it cannot be considered enough to define this genus. It is necessary to specify details. Bernard Wood and Andrew Chamberlain (1987) indicated the following traits related with the cranium of *Paranthropus*: particularly its robust mandible, reinforced face with its own patterns, morphology of cranial sutures, and endocranial anatomy. With respect to dentition, the apomorphies would be "reinforced" premolars in the lower maxilla, and larger talonids with a higher incidence of distal accessory cusps on the lower molars. Moreover, all these features can be more detailed. For example, Wood and Strait (2004) related the condition "reinforced face" with the following traits: anterior pillars, concave and bulky zygomatic bones with more anteriorly positioned roots, supraorbital torus forming a slope, rounded inferolateral orbital margins, and a thick palate. Gen Suwa et al. (1997) utilized up to 40 derived traits to distinguish *P. robustus*, *P. boisei*, and *P. aethiopicus*—considering all of them as belonging to the genus *Australopithecus* by the authors—from *A. africanus* and *A. afarensis*. In order to avoid an even more cumbersome reading, these various features will be indicated here under the general term of "more robust," unless it becomes necessary to specify what we mean. Furthermore, because of sexual dimorphism, one must take extra care when speaking of the higher or lesser robustness of specimens. Within *Paranthropus*, females show a less robust appearance, both in the cranium and in the dentition and they are, besides, smaller in size.

But it was another South African site, different from those excavated by Broom, which would provide the specimen of *P. robustus* that is best preserved among all those known today. This is DN 7, a cranium with a mandible found in Drimolen in 1994, along with another mandible, DN 8. After Ron Clarke spent 5 years assembling the various parts of the cranium, André Keyser (2000) published the study of the specimen (Figure 5.11).

Showing in their major lines the massive morphology characteristic of *P. robustus*, DN 7 is a cranium with a prominent glabella, although it lacks a significant supraorbital torus. This feature, together with the absence of a sagittal crest, a dentomandibular apparatus smaller than DN 8, and its small size, made it attributable to a female (Keyser, 2000). But the modest

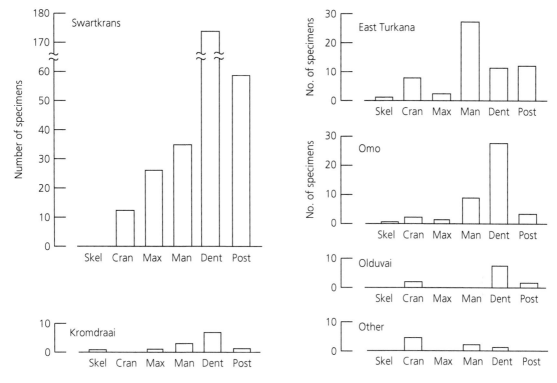

Figure 5.9 Numbers of "robust" australopithecine specimens, arranged according to skeletal part found at fossil sites in (left) southern and (right) eastern Africa. Categories are partial skeleton (Skel), intact or fragmentary crania, or endocranial casts (Cran), maxillae (Max), mandibles (Man), isolated or associated dental elements (Dent), postcrania (Post). Reprinted from Wood, B. & Chamberlain, A. T. (1987). The nature and affinities of the "robust" Australopithecines: a review. *J Hum Evol*, 16, 625–641, with permission from Elsevier.

size of the specimen cannot be due only to sexual dimorphism, because it is smaller than SK 48, also from a female. Keyser (Keyser, 2000; Keyser et al., 2000) argued that, based on the indications of DN 7 and other specimens from this site, the internal variability of *P. robustus* should be considered at least as high as that attributed to *P. boisei*.

A total of 80 specimens, besides DN 7 and DN 8, complete the sample coming from Drimolen (Keyser et al., 2000). The great majority of these specimens were attributed to *P. robustus*, although some of them—DNH 34 and DNH 35—were considered *Homo*, without assignation to any species, and with respect to others their ascription is dubious.

5.7.2 Paranthropus boisei

Besides the skeleton OH 1 mentioned in Chapter 3, and that belonging to *Homo sapiens*, the site of Olduvai yielded hominin remains as early as 1935, specifically belonging to *Homo erectus* (or *Homo ergaster*),

discovered by Mary Leakey. But, as Yves Coppens (1983) recalls, almost nobody took notice of those first specimens. Two 1.5 Ma teeth found by Louis Leakey in 1955 in Bed II (Leakey, 1958) did not have much repercussion either. The discovery made the following year suddenly placed Olduvai, and all of East Africa with it, at the center of attention of paleoanthropologists around the world.

Louis Leakey (1959) has recounted the finding in detail. On July17, Mary Leakey found a hominin cranium, partially exposed by the terrain's natural erosion, at the FLK site and about seven meters below the upper limit of Olduvai Bed I. The excavation works began the following day and, on August 6, an almost complete cranium was recovered. It was associated with animal bone fragments and tools belonging to a very primitive culture, which was named Oldowan. The specimen, one of the most famous pieces in the history of anthropology, is technically known as OH 5 (*Olduvai Hominid number 5*), and colloquially by the name of *Dear Boy* (Figure 5.12).

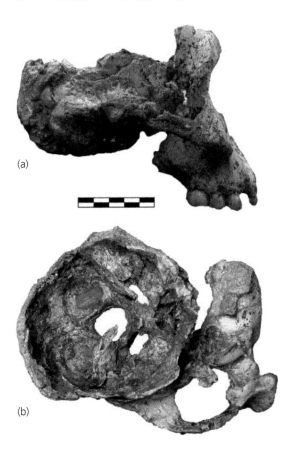

(a)

(b)

Figure 5.10 SKW 18 conjoined with SK 52 to form composite cranium. (a) Right lateral aspect; (b) superior aspect. Scale bar is 50 mm. de Ruiter, D. J., Steininger, C. M., & Berger, L. R. (2006). A Cranial Base of Australopithecus robustus from the Hanging Remnant of Swartkrans, South Africa. *Am J Phys Anthropol*, 130, 435–444.

The OH 5 cranium was found fragmented in small pieces, but the fragments had been preserved together, including some very fragile bits, like the nasal bones. The fossils of other animals associated with OH 5 were also fragmented, but they were shattered in a different way: they seemed to have been purposely broken to get to the bone marrow. Louis Leakey concluded that hominins such as *Dear Boy* lived there close to 2 million years ago, and had used stone tools to break open animal bones. Leakey rejected the idea that *Dear Boy* itself had also been a victim of predators, other hominins, or cannibalism (Leakey, 1959, p. 491): "Had we found only fragments of skull, or fragments of jaw, we should not have taken such a positive view of this. It therefore seems that we have, in this skull, an actual representative of the type of 'man' who made the Oldowan pre-Chelles-Acheul culture."

As soon as Tobias (1967a) reconstructed the cranium, it became apparent that *Dear Boy* was a robust exemplar; in fact, very robust. Leakey himself admitted from the start that it was an australopith resembling, in certain aspects, South African *Paranthropus*: a sagittal crest; reduction of canine and incisor size, placed in a straight line in front of the palate; and the general structure of the cranium. But, according to Leakey (1959), other traits observed in OH 5 were reminiscent of *Australopithecus africanus*: the size of the third upper molar, smaller than the second one, for instance. While admitting the inconvenience of multiplying taxa, Leakey chose a new genus of the subfamily Australopithecinae, giving OH 5 the name of *Zinjanthropus boisei*. "Zinj," for the ancient name for East Africa, and "boisei" in honor of Charles Boise, who contributed to financing the excavations at Olduvai. Owing to Phillip Tobias' (1967a) detailed study of the specimen, it was later included in the same genus as South African robust australopiths, but in a different species, namely, *Paranthropus boisei* (Box 5.10).

OH 5 was described as a specimen with certain *Paranthropus'* and *Australopithecus'* traits, but it is not an intermediate form between South African gracile and robust australopiths. Quite the opposite—it is even more robust than the Kromdraai and Swartkrans specimens, and this was the most important consideration in favor of the initial proposal of a new genus, in addition to the great distance between Olduvai and South Africa. The description of the cranium highlighted, among other distinctive traits, a continuous sagittal crest through the occipital bone in males, a very massive supraorbital torus, and a crest on the frontal bone with a very prominent anterior margin. This margin, which is keel-shaped, had not been documented even in the most muscular male specimens of South African *Paranthropus*. *Zinjanthropus* gave an overall impression of a much more muscular masticatory apparatus than robust australopiths known at the time when the discovery was made.

The last discoveries of *P. boisei* specimens in Olduvai correspond to OH 66, a lower premolar, and OH 74, an upper molar (Ungar et al., 2011).

Bed II of Olduvai has also provided, besides exemplars of *Homo habilis* (OH 13) and robust australopiths (OH 3, OH 38), certain specimens similar to Asian *Homo erectus*, associated with tools from a more advanced tradition than the Oldowan. The set indicates a transition very similar to that already noted in South Africa toward Middle Pleistocene *Homo erectus*. We will discuss it in upcoming chapters.

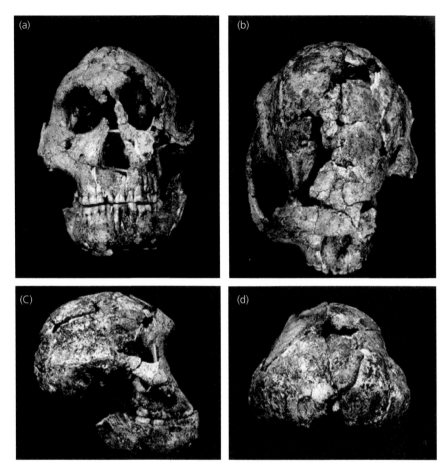

Figure 5.11 DN 7, *P. robustus* of Drimolen in (a) frontal, (b) upper, (c) lateral, and (d) occipital views. Keyser, A. W. (2000). The Drimolen skull: the most complete australopithecine cranium and mandible to date. *South African Journal of Science*, 96, 189–193.

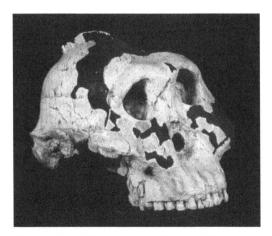

Figure 5.12 OH 5 Dear Boy, *P. boisei* (Leakey, 1959).

5.7.3 Hypodigm of *Paranthropus boisei*

Making use of the specimens from several sites in East Africa, whose correlation can be made thanks to studies of Omo, the hypodigm of *Paranthropus boisei* is very complete—in cranial and dental materials, at least—such that the taxon is properly characterized (Figure 5.13). However, some very old and very robust specimens have led some authors to distinguish another species, *P. aethiopicus* (see Section 5.7.4 and Box 5.14).

As we said when discussing in Section 5.7.2 the type specimen of *P. boisei*, the cranium OH 5 lacks the mandible. The first mandible, NMT-W64-160 (Peninj 1), was found by the Leakeys (husband and wife, L. S. B. Leakey and M. D. Leakey) in Lake Natron (Tanzania) in 1964, and was classified as *Zinjanthropus boisei*

Box 5.10 The rejection of genus *Paranthropus*

Phillip Tobias is among the authors who prefer to discard the genus *Paranthropus*, considering robust australopiths as a subgenus of *Australopithecus*. In this case, OH 5 would be classified as *Australopithecus boisei*.

(Leakey & Leakey, 1964). However, Koobi Fora is the site that has provided more informative exemplars.

The Koobi Fora site, on the eastern shore of Lake Turkana, in Kenya, has yielded a great number of hominins, comparable to the Tanzanian gracile and robust specimens. Up to 5,000 exemplars, among which there are some quite well-preserved crania, were discovered there during the 1970s. However, the *H. habilis'*

specimens found at Koobi Fora are older than those found at Olduvai.

Such paleontological treasure was linked from the beginning to Richard Leakey, son of Mary and Louis. When flying over the area during a trip to Nairobi that had been diverted because of a storm, Leakey noticed the presence of sandstones susceptible to containing fossils, in a place thought to have only volcanic soils. With *National Geographic* funds, and under the auspices of the *National Museum of Kenya*, Leakey began expeditions to the area in 1968. The *Koobi Fora Research Project* began its work in 1969. The most important finding during that first campaign was a robust australopithecine cranium, registered as FS-158 (later KNM-ER 406; the acronym means Kenya National Museum, East Rudolph). It was a nearly complete cranium, but without dentition, and very similar to OH 5, discovered by Mary Leakey in Olduvai. Richard Leakey (1970) classified the specimen as *Australopithecus boisei*, estimated to be 2.61±0.26 Ma, by means of the $^{40}K/^{40}Ar$ method.

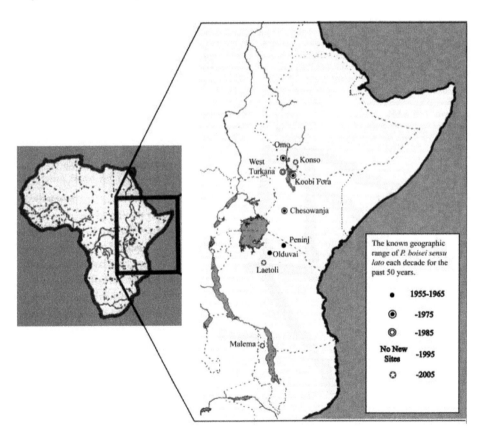

Figure 5.13 *P. boisei* sites, with dates of the discoveries. Wood, B. & Constantino, P. (2007). Paranthropus boisei: Fifty Years of Evidence and Analysis. *Yearbook of Physical Anthropology*, 134 (Supplement 45), 106–132.

Successive campaigns provided the skull KNM-ER 732, also from member KBS but smaller than 406. It was thought to be a female after a comparative analysis of 19 basicranial elements of *A. africanus*, *A. afarensis*, *P. robustus*, and *H. erectus* (Dean & Wood, 1988) and other materials from both Koobi Fora and West Turkana. It is common to consider that the differences in size between KNM-ER 406 and 732 indicate the highest degree of sexual dimorphism found in *P. boisei*. Despite being of very different size, they share traits such as a great mastoid process, an elongated mandible joint, and concave face (Aiello & Andrews, 2000).

Specimen KNM-ER 1500 is a partial and very fragmented skeleton of difficult interpretation attributed to *P. boisei* (Wood, 1991). It came from Upper Burgi and, with an age between 1.90 and 1.88 Ma, is older than the two crania 406 and 732.

Omo has yielded many early hominins, from the A to the L Shungura members, although subject to very different interpretations and classifications because the remains are mostly fragmentary. Coppens (1978b) provisionally attributed a great number of isolated teeth, mandibles, maxillas, and very modest (in his own words) cranial and postcranial fragments to *A. africanus*. Coppens (1978b) ascribed the most complete remains to *P. boisei*, including the Omo-323–1976-896 specimen, a partial hominin cranium, dated c. 2.1 Ma, from member G, Unit G-8 of the Shungura Formation (Alemseged et al., 2002). Omo-323 is made of fragments of the frontal, both temporals, occipital, parietals, and the right maxilla (Figure 5.14). A well-developed and completely fused sagittal crest, heavily worn teeth, and a relatively large canine suggest the specimen was a male *P. boisei*. It is one of the earliest specimens of that hypodigm. Alemseged et al. (2002) noted that Omo-323 shares certain traits with KNM-WT, *P. aethiopicus*. We will address this taxon in Section 5.7.4.

From the Chesowanja site, Chemoigut formation (Kenya), came KNM-CH1, a partial cranium with the right part of the face and the cranial base preserved (Carney et al., 1971), as well as other materials attributed to *P. boisei*.

A specimen from Konso (Ethiopia) is very interesting (Box 5.12). In 1997, Gen Suwa and collaborators reported the discovery of several robust specimens from the locality KGA10 of Konso. This locality is placed between two tuffs (KRT and TBT), which are, respectively, dated at 1.41±0.02 and 1.43±0.02 Ma by the ^{40}Ar/^{39}Ar method (Suwa et al., 1997). The age attributed to the fossils is, therefore, close to 1.42 Ma.

The specimens of KGA10 include the exemplar KGA10-525, which provided the first complete association of a cranium and mandible of *P. boisei* (Figure 5.15). A large part of the face was also present, with the exception of the frontal bone and the anterior cranial base. The specimen is an adult, large and presumably male, with a cranial capacity of 545 cm^3—somewhat larger than the previous robust specimens from South Africa and East Africa. The morphological interpretation by Suwa et al. (1997) suggests the possible integration of paranthrops in a single species.

If the mandible UR 501 had extended the geographic range of ancient gracile hominins to Malawi,

Figure 5.14 Right partial maxilla of Omo-323–1976–896, *P. boisei* (photographs from Alemseged et al., 2002).

Box 5.11 Other *Paranthropus* from Lake Turkana basin

In 2001, Barbara Brown, Frank Brown, and Alan Walker published the discovery of several mandibular and dental materials of *P. boisei* found in the basin of Lake Turkana. These were the fragmentary mandibles KNM-ER 25520 from Ileret, member KBS, of 1.85±0.05 Ma, KNM-WT-17396 from Kokiselei, member Natoo, of 1.8±0.1 Ma and KNM-WT 18600 from the same locality and age as the previous one (Brown et al., 2001).

Box 5.12 Konso site

Konso deposits were discovered by Berhane Asfaw and collaborators in 1991. Konso had already provided specimens of *Homo ergaster*—see Chapter 8—associated with abundant ancient instruments of the Acheulean culture (Asfaw et al., 1992). The stratigraphic horizon of those earlier findings was the same as that of the new discoveries.

a similar situation occurs with the robust specimens. Ottmar Kullmer and collaborators (1999) reported the discovery of the exemplar HCRP RC 911 in Malema, Chiwondo Beds (Malawi). This is a maxillar fragment with two molars, dated approximately 2.5 Ma by the associated fauna, whose characteristics correspond, according to Kullmer and collaborators (1999), to the size range of *P. boisei*. Taken together, the Konso and Malema fossils indicate the presence of *P. boisei* in a territory that stretches from Ethiopia to Malawi with a wide hypodigm.

5.7.4 Paranthropus aethiopicus

The controversy surrounding the number of species present at Lake Turkana is yet another episode of confrontation between paleontologists who accept only a single hominin evolutionary lineage (lumpers) and those who, on the contrary, believe there were several parallel lineages (splitters). The Koobi Fora discoveries provided solid arguments supporting

several simultaneous lineages, even some coexisting in the same geographical location. Thus, in the early 1980s, virtually all authors admitted at least the two lineages: gracile and robust. The latter was generally understood as a lateral branch, a late specialization in our family. Once a relation was established between robust mandibles, sagittal crest, great molars, and the intake of a hard vegetable diet, all the pieces seemed to fall into place. *A. africanus* would lead, through a specialization of its masticatory apparatus, to *Paranthropus robustus* (Rak, 1985) (Box 5.13). The more robust forms, such as the *Paranthropus boisei* exemplars from Olduvai, Turkana, and Omo, would represent a very specialized and late version of the robust clade, based to a great extent on the massive development of masticatory structures (Tobias, 1967b; Grine, 1985; Suwa, 1988). Some authors, such as Yoel Rak (1983), suggested that *P. boisei* could be a direct descendant of *P. robustus*, drawing, thus, the evolutionary lineage: *A. africanus*–*P. robustus*–*P. boisei*. Greater robusticity would tend to be associated with a later appearance in the fossil record.

Some findings on the Western shore of Lake Turkana in the mid-1980s rocked this simple scheme. In 1985, Alan Walker and his colleagues discovered a cranium, registered as KNM-WT 17000, in the Lomekwi basin. A mandible (KNM-WT 16005) was found a little to the south on the same western shore, at Kangatuku-seo. Both specimens were presumably attributed to the male sex. By correlation of two volcanic tuffs located above and below the place where the cranium was

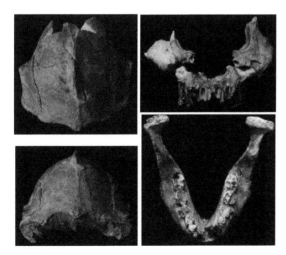

Figure 5.15 KGA10-525, *P. boisei*. Left: cranium; right: associated mandible (photographs from Suwa et al., 1997).

Box 5.13 Phylogenetic placement of *A. africanus*

Rak's notion that *A. africanus* is the first member of a robust clade that would later lead to *P. robustus* and *P. boisei* was not generally accepted. Phillip Tobias, for instance, was notoriously against it. Tobias considered *A. africanus* as the ancestor of both clades, the specialized (robust) and the gracile leading to *Homo*. This led him to reject *A. afarensis* as a distinct species and, additionally, ancestral to all the rest (Tobias, 1980). In his 1980 article, Tobias argued that the specimens attributed to *A. afarensis* should be classified as a subspecies of *A. africanus*, namely *A. africanus aethiopicus*. Tobias later included *A. afarensis* as a separate species in his writings, but placed *A. africanus* as the common ancestor of all hominins (e.g., Tobias, 1992).

located with the Lokalalei basin tuffs, which had been related with the CP submember of the Shungura formation, KNM-WT 17000 was estimated to be 2.50±0.05 Ma. The mandible would be slightly younger, around 2.45±0.05 Ma (Walker et al., 1986).

KNM-WT 17000 is a nearly complete cranium, which includes most of the anterior teeth, a molar, and part of the face (Figure 5.16). It was named "black skull"—the color due to the manganese of the soils. Walker and colleagues (1986) described it as a massive cranium, with a very broad face. The palate and the base of the cranium, also very broad, are similar to those of OH 5, found at Olduvai (*P. boisei*). The cranial capacity is low, close to 410 cm³, which represents the lowest cranial capacity for any adult fossil hominin, except maybe A.L. 162–28 from Hadar. The sagittal crest is huge, the largest among all hominins. The cranium corresponds, thus, to a hyper-robust specimen.

Some features of KNM-WT 17000 differ from other robust australopithecines. The specimen has a very marked prognathism (a forward facial projection, beyond the vertical plane that passes through the ocular orbits). Its dentition is also different from that of *P. boisei*. The conserved molar, the upper-right P3, is, in comparison with that of OH 5, larger in the longitudinal (mesodistal) direction and smaller in the transversal (bucolingual) direction. Furthermore, its robusticity is comparable to the most extreme instances of *P. boisei*.

What kind of hominin is KNM-WT 17000? When Walker and colleagues (1986) classified the West Turkana discovery, they emphasized its distinctive features, seeking to establish that it might be a species different from *P. boisei*.

It is common to classify KNM-WT 17000 as *P. aethiopicus*, though the identification of this specimen with the mandible found at Omo is not unambiguous (Box 5.14). The L7A-125 mandible is attributed to a female and, thus, is somewhat different from KNM-WT 17000 and KNM-WT 16005 (attributed to males). But it would seem unreasonable to assume that there were two very

---

Box 5.14 Taxonomy of *P. aethiopicus*

Several terminological clarifications are required here. Twenty years earlier, Arambourg and Coppens (1968) had classified a robust mandible, L7A-125, found at the Omo site of the Shungura formation, as *Paraustralopithecus aethiopicus* (Arambourg & Coppens, 1968). Although the initiative of Arambourg and Coppens to create a new genus was not sufficiently justified, Walker's team assigned the West Turkana specimens to that same species. The L7a-125 mandible from Omo initially received the identification number 18-1967-18. We have chosen to refer to this specimen with its current catalogue number, rather than by the one used in the mentioned papers. L7a-125 was assigned by Nicole Silverman, Brian Richmond, and Bernard Wood (2001) to *P. boisei* sensu stricto; as it is very old—2.2 Ma—it becomes the oldest specimen of the taxon. Meanwhile, Alan Walker and collaborators (1986) classified KNM WT-17000 as *Australopithecus aethiopicus* and not *Paranthropus aethiopicus*. However, because we have placed the robust australopithecines in the genus *Paranthropus*, we will retain the *P. aethiopicus* label.

---

similar species living at the same time at Omo and West Turkana, two very close sites.

In any case, Walker and colleagues attributed KNM-WT 17000 to *P. aethiopicus* tentatively, only if it could be convincingly shown that it was not a *P. boisei*. When Walker and colleagues compared the black cranium's morphological traits with the list of traits established for *P. boisei*, they noted that there were similarities between both taxa in most of the traits. Walker and Richard Leakey (1988) argued that the specimens from West Turkana (KNM-WT 17000, 16005, and 17400), Koobi Fora (like KNM-ER 406, 732, 1590, 3230, and 13750, among others), and the Omo mandible belonged to the same species as OH 5 from Olduvai. If so, *P. boisei* would be a highly variable taxon—extremely variable

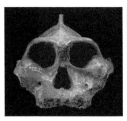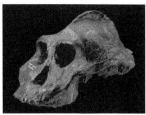

Figure 5.16 KNM-WT 17000, black skull, *Paranthropus aethiopicus* (photographs from http://www.mnh.si.edu/anthro/humanorigins/ha/WT17k.html).

in some traits, such as the sagittal and nuchal crests. This taxon's sexual dimorphisms would be comparable to those of current gorillas.

Walker and Leakey's (1988) main reason for not distinguishing two species was the danger of giving different names to a segment of short duration within one evolutionary lineage, which is a risk when the exemplars are so scarce. The set of robust australopithecine specimens from East Africa includes those found at Olduvai and Peninj (Tanzania), at Chesowanja, on both shores of Lake Turkana (Kenya), and Omo: 60 individuals, in addition to isolated dental remains. However, there are only 16 facial skeletons and the same number of crania that are informative enough to document a taxon ranging for more than a million years (Walker & Leakey, 1988). Therefore, it is difficult to decide whether the West Turkana and Omo specimens are within the variation range of *P. boisei* or whether they should be classified as *Paranthropus aethiopicus*. Faced with this state of affairs, it is more parsimonious, according to Walker and Leakey (1988), not to multiply species.

William Kimbel and colleagues (1988) favored a contrary point of view. These authors re-analyzed the features of KNM-WT 17000 highlighted in Walker et al.'s (1986) original article. They made a comparative study of:

- the apomorphies (derived traits) the specimen shares with:
  - *A. africanus, P. robustus,* and *P. boisei;*
  - *P. robustus* and *P. boisei;* and
  - only *P. boisei;*
- plesiomorphies (primitive traits) shared with *A. afarensis* (Kimbel et al., 1988).

Out of 32 traits, 12 were, according to Kimbel and colleagues, primitive characters shared with *A. afarensis;* six were derived and shared with *A. africanus, P. robustus,* and *P. boisei;* 12 were derived traits shared with *P. robustus* and *P. boisei;* they found only two derived characters exclusively shared with *P. boisei.* Kimbel et al. (1988) argued that their study supported the classification of KNM-WT 17000 as a separate species, *P. aethiopicus.*

So far, the only locality, apart from West Turkana, that has provided specimens of *P. aethiopicus* is Laetoli (Tanzania). In 1998, fieldwork was undertaken in the Upper Ndolanya Beds, with an age of 2.66 Ma, and in that same year a proximal tibia fragment was discovered, EP 1000/98, in the locality 22S of Nenguruk Hill. Most of the Laetoli specimens have been found at the surface, so to accurately estimate their age, a precise determination of the stratigraphic horizon from which they come is required. Terry Harrison (2011a) argued that in the explorations of the 1970s, the stratigraphic origin assigned to the fossils was simply the place where they were found, but considered that their real origin can be determined owing to the presence of numerous volcanic tuffs.

In 2011, a fragment of a right edentulous maxilla, EP 1500–1501, appeared in Silal Artum (Upper Ndolanya Beds) (Harrison, 2011a) (Figure 5.17). Following the discovery of the maxilla, assigned by Terry Harrison (2011a) to *Paranthropus aethiopicus,* the author maintained that it was very likely that the tibia EP 1000/98 also belonged to this taxon. A small fragment of zygomatic bone found by Mary Leakey in 1975 in the locality 7e of Upper Ndolanya Beds is of more doubtful identification.

EP 1500/01 is the oldest exemplar of *P. aethiopicus* available at present. Its age, and the distance between the sites of Turkana and Laetoli, indicated to Terry Harrison (2011a) that from the very early emergence of *P. aethiopicus* there was a significant dispersal of the species.

The ascription to just one species of the Olduvai fossils OH 5, the exemplars from West Turkana (KNM-WT 17000, 16005, and 17400), Koobi Fora (such as KNM-ER 406, 732, 1590, 3230 and 13750, among others), and the mandible of Omo, was defended by Alan Walker and Richard Leakey (1988), with the understanding that *P. boisei* would be a highly variable taxon—extremely so in some traits, such as the sagittal and nuchal crest. The sexual dimorphism of the taxon would be similar to that present today among gorillas.

Either as a single species, or two or even three, the distribution of robust australopithecines indicates a wide temporal and spatial presence. Between an age of 2.6 Ma of the exemplar L55-33 and the specimens Omo 18–18, Omo 18–31 7, and Omo 84–100 of 2.6 Ma, and those from Olduvai OH 3 and OH 38 with the age of 1.2 Ma, the period of existence of the robust branch is truly remarkable.

5.8 Consistency of the evolutionary scheme for the "gracile" and "robust" australopithecines in the Middle and Upper Pliocene

The genera *Australopithecus, Paranthropus,* and *Homo* are enough to accommodate the diversity of hominins in the Middle and Upper Pliocene, given that only three different adaptive strategies appeared at that time.

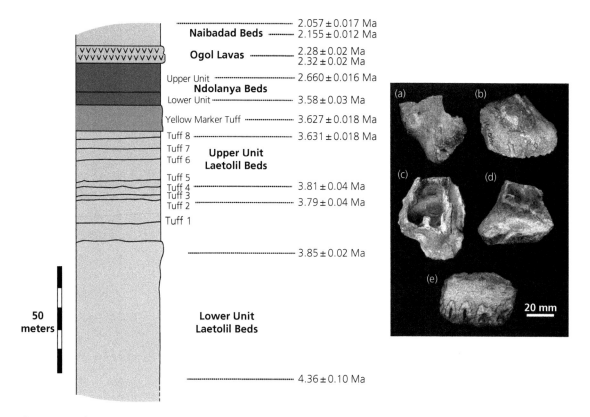

Figure 5.17 Left: Stratigraphic column and radiometric dating of the lower part of the sequence at Laetoli. Right: Edentulous right maxilla with the roots for I²-M². Harrison, T. (2011a). Hominins from the Upper Laetolil and Upper Ndolanya Beds, Laetoli. In T. Harrison (Ed.), Paleontology and Geology of Laetoli: Human Evolution in Context (pp. 141–188). Dordrecht: Springer Netherlands, with permission of Springer.

It remains for Chapter 6 to consider the genus *Homo*, but it should be anticipated that the evolutionary history of the hominins is simplified by considering only three genera for that span of time. The phylogenetic scheme can be as simple as the one we already provided in Chapter 2, but for the readers' convenience, it is repeated here as Figure 5.18. However, as we have noted repeatedly, taxonomy should follow as nearly as possible the process of lineage divergence, gathering species in genera that are monophyletic, i.e., which do not contain mixed species from separate lineages. Is this the case with the genera *Australopithecus* and *Paranthropus*? (The possible monophyly of *Homo* will be addressed in Chapter 6.)

5.8.1 Australopithecus monophyly

As indicated in Section 5.1, the study by Strait et al. (1997) revealed that the genus *Australopithecus*, as was understood in those years, turned out to be paraphyletic. In search of a solution, the authors analyzed different cladograms. The most appropriate, in line with the parsimony criterion, would include *A. africanus* as a sister group of the set *Paranthropus + Homo* (see Figure 5.19).

Although, as we know, reformed cladistics does not establish relations between ancestors and descendants (Chapter 1), it would be acceptable to consider *A. africanus* as the last common ancestor prior to the set formed by robust australopithecines (*A. robustus*, *A. boisei*, and *A. aethiopicus*, according to Strait et al., 1997) and *Homo*. In the years in which no other specimens of the genus *Australopithecus* were known, i.e., before the discovery of *A. anamensis*, *A. bahrelghazali*, *A. garhi*, and *A. sediba*, it was a reasonable solution. After the addition of the new taxa, we again run into the problem of an amalgam as a kind of catchall, which leads to a paraphyletic group.

The cladogram offered by Lee Berger and collaborators (2010) in the study accompanying the proposal

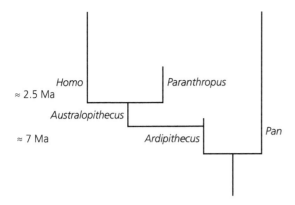

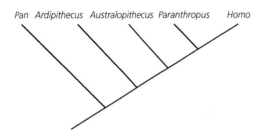

Figure 5.18 Above; phylogenetic tree of the Hominini tribe genera; below: cladogram of the Hominini tribe genera (Figure 2.4 is repeated here for reader convenience).

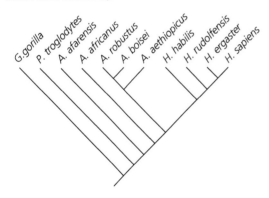

Figure 5.19 Most parsimonious cladogram favored by the Variable = Intermediate analysis. (The authors include robust australopithecines in the genus *Australopithecus*.) Reprinted from Strait, D. S., Grine, F. E., & Moniz, M. A. (1997). A reappraisal of early hominid phylogeny. *Journal of Human Evolution*, 32, 17–82, with permission from Elsevier.

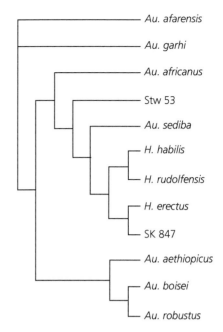

Figure 5.20 Most parsimonious cladogram using PAUP 4.0 (modified) (Berger et al., 2010; supporting online material). From Berger, L. R., de Ruiter, D. J., Churchill, S. E., Schmid, P., Carlson, K. J., Dirks, P. H. G. M., & Kibii, J. M. (2010). Australopithecus sediba: A New Species of Homo-Like Australopith from South Africa. *Science*, 328(5975), 195–204. Reprinted with permission from AAAS.

Box 5.15 A "bushy" evolution

Leslie Aiello and Peter Andrews (2000, p. 17) argued that: "At present it is impossible to resolve the phylogenetic relationships of the australopithecines with any degree of confidence. There is a growing realization of the 'bushy' nature of hominin evolution throughout the australopithecine period and also of the inevitability that additional early hominin species remain to be discovered." After more than a long decade, their words have been further confirmed.

---

of *A. sediba* appears in Figure 5.20. Apart from the doubts that one might raise about the choice of taxa and characters—neither the inclusion of *A. anamensis*, *A. bahrelghazali*, and *A. (Kenyanthropus) platyops*, nor their exclusion is justified—what is clear in this cladogram is that it is impossible to perform a monophyletic grouping of the included *Australopithecus*.

Adding the missing taxa would only worsen the situation.

Thus, the *Australopithecus* genus is paraphyletic. To avoid this it would first be necessary to name a new genus, grouping together *A. afarensis* and *A. garhi*, i.e., to join one of the oldest species with another of the most modern. In fact, to achieve a cladogram without paraphilias, more genera would be necessary. But, it seems obvious that this kind of taxa multiplication makes no sense and, in fact, no one has proposed it. For now, the paraphilia of *Australopithecus* cannot be solved—it is fair to admit this. The analysis proposed by Rolando González-José and collaborators (2008), which is the latest attempt to achieve cladograms appropriate for the actual phylogenetic processes—grouping traits into modules and considering continuous quantitative variations—only took into account the taxa *Australopithecus afarensis* and *Australopithecus africanus*. Despite this reduction, if these are grouped in the same genus, this becomes paraphyletic.

5.8.2 *Paranthropus* monophyly

Among all the genera of the Pliocene, *Paranthropus* is shown to be the most consistent, to the extent that its consideration as a monophyletic genus is quite generally accepted—although conflicting opinions can be found (e.g., Skelton and McHenry, 1992).

The most remarkable apomorphies of *Paranthropus* are related with the megadontia and, in general, with its craniofacial robustness. In his research on the possible monophyly of *Paranthropus*, Bernard Wood (1988) warned that the evolutionary convergence of traits related to complex functional structures—as are those of the masticatory apparatus—can often confuse, leading to conclusions about phylogenetic proximity. There are not enough reasons to reject the hypothesis of two separate clades among the "robust" South African specimens and those of the Rift, or to support the monophyly. Wood (1988) considered, therefore, that the most parsimonious solution is a monophyletic group that brings together all the *Paranthropus* species. The alternative would be, according to Wood, to maintain that *A. africanus* is the ancestor of the South African "robust" specimens and *A. afarensis* is the ancestor for those of East Africa.

Let us return to the cladistic approach of Strait et al. (1997), which appears in Figure 5.19. The grouping corresponding to the genus *Paranthropus* ("A. robustus + A. boisei + A. aethiopicus," in accordance with the taxonomy used by the authors) is monophyletic. But there is a problem that was highlighted by Walker et al. (1986). The cladogram of Figure 5.19 indicates, as noted above, that the taxon *A. africanus* is the sister group of the genera *Homo* and *Paranthropus*, both monophyletic. However, for Walker and collaborators (1986), KNM-WT 17000 (*P. aethiopicus*) is closer to *A. afarensis* than to *A. africanus*. Its antiquity would also make plausible the presence of a common clade for *P. boisei* and *P. aethiopicus*, akin to *A. afarensis* and separate from the evolutionary line that led to *P. robustus* (Box 5.16).

The difficulty we encounter in carrying out these groupings within the cladogram in Figure 5.19 is that clades *A. africanus* + *P. robustus* and *A. afarensis* + *P. boisei* are paraphyletic. Walker and collaborators (1986) argued that this difficulty derives from the fact that the common characteristics shared by the South African robust specimens and those from East Africa are actually homoplasies—analogous traits.

It would not be right, therefore, to speak only of the existence of a "robust clade," but of two: one clade with the succession *P. aethiopicus*–*P. boisei* and the second with *P. robustus*. If so, the different species would belong to two separate lineages and it would not be appropriate to group them into a single genus *Paranthropus*. The speculative cladogram reflecting this hypothesis appears in Figure 5.21.

The conclusion of the volume edited by Frederick Grine (1988) on the evolutionary history of robust australopithecines argued that the discovery of KNM-WT 17000 would have actually had the opposite effect of what was indicated by Walker et al. (1986). Grine argued that some of the traits shared by *P. robustus* and *P. boisei* must be considered synapomorphies

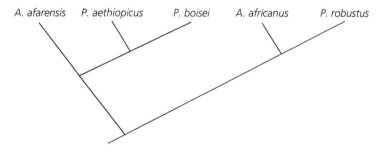

Figure 5.21 Tentative cladogram, not based on any analysis, offered as an illustration of the relations established between the taxa *A. afarensis, A. africanus, P. robustus, P. aethiopicus,* and *P. boisei,* according to the phylogenetic interpretation by Walker et al. (1986).

Box 5.16 Are *P. robustus* and *P. boisei* sister groups?

In the review by Bernard Wood and Terry Harrison (2011) of the evolutionary context of the first hominins, the authors noted that, although the vast majority of cladistic analysis suggests that *P. boisei* and *P. robustus* are sister groups, macrostructural (the morphology of the premolar roots, for example) as well as microstructural (daily ratios of enamel secretion) features speak against this consideration. But these same examples might actually indicate that, in very close taxa with similar adaptive strategies, overlapping combinations of traits produce homoplasies that confuse phylogenetic analysis. We call these overlapping traits parallelisms.

(homologous traits shared phylogenetically) and not homoplasies (convergences by analogy). In agreement with Wood (1988), for Grine (1988) the possibility of two clades existing among the robust specimens is greater than many authors acknowledge, but the stronger alternative still remains that of a single clade: a monophyletic group including all robust australopithecines, which justifies classifying them within the common genus *Paranthropus* (see Box 5.17).

The controversy over whether robust australopithecines form a monophyletic group took a turn due to a specimen from Konso described by Gen Suwa et al. (1997). The morphology of the specimen is, in the words of Suwa et al., undoubtedly that of a robust australopithecine, although with some traits (dentition, in particular) shared only by *P. boisei*. However, Suwa's team made an interesting interpretation of

its morphological meaning. KGA10-525 incorporates some traits, such as the beginning of the sagittal crest, uncharacteristic of *P. boisei* and *P. robustus* known hitherto, and only present in the specimen KNM-WT 17000 of *P. aethiopicus*. Other traits of KGA10-525 were also within, or even outside, the range of variation of the prior robust specimens. For Suwa and collaborators (1997), certain characters in the robust specimens were previously considered functional and adaptively significant, to the point of being considered essential in order to attribute the specimens to one or another species. But such traits might actually correspond to polymorphisms typical of large variation among different populations. The type specimen *P. boisei*, OH 5 of Olduvai, would thus be an "extreme" specimen (Delson, 1997) of the species, while the *P. boisei* from Turkana, as well as those found in Konso, would be specimens with intermediate morphology between the Olduvai and the South African *P. robustus* (Box 5.18).

As Eric Delson (1997) commented, Gen Suwa and collaborators (1997) could have perfectly attributed the exemplars of Konso to a new species. However,

Box 5.18 Variability of *P. boisei*

The question of the consistency of *P. boisei* as a taxon whose variability corresponds to what would be characteristic of a single species, has been analyzed by Nicole Silverman, Brian Richmond, and Bernard Wood (2001), comparing the mandibular hypodigm of *P. boisei* with that of current humans and apes. The authors concluded that the degree of variation in size of *P. boisei* does not appear in the samples of modern humans and chimpanzees, and is rare in gorillas, while common in orangutans. Variation in shape appears similarly in all cases, fossils and living.

Bernard Wood and Daniel Lieberman (2001) have argued that, according to conventional metric variables, the *Paranthropus* sample from Konso fits within the range of variability prior to the discovery of such specimens—contrary to the interpretation given by Suwa et al. (1997). For Wood and Lieberman (2001), the special status attributed to the specimens of Konso is due to those traits classified as "extreme" in that sample; these are cranial traits that tend to show high intraspecific variation. To assess the possible inclusion of all the robust australopithecines in a single species, it would be interesting to apply the same analysis of the craniodental variability of *P. boisei* made by Wood and Lieberman (2001) with Konso specimens to the full sample of *P. robustus* + *P. boisei* + *P. aethiopicus*.

Box 5.17 *Paranthropus'* molars

Frederick Grine and Lawrence Martin (1988) arrived at the same conclusion about the monophyly of *Paranthropus* after examining one of the most remarkable characteristics of robust australopithecines: their very thick molars. That thickness is achieved through similar developmental patterns in robust australopithecines from South Africa and East Africa. In the words of Grine and Martin (1988), this trait can be considered synapomorphic of the different robust species and separates them from *A. africanus*, as well as from any *Homo* taxa.

they opted for the opposite approach: to raise the possibility that all robust australopithecines (*P. boisei* and *P. robustus* at least) belong to the same species. The intermediate character of KGA10-525 can be related, furthermore, with the conclusions of the study of Bernard Wood, Christopher Wood, and Lyle Konigsberg (1994). Based on the review of numerous dental features, these authors argued that an abrupt change happened about 2.3/2.2 Ma, which led to the transformation of *P.* aff. *boisei* (WT-17000) into *P. boisei* sensu stricto (ER 403, 406, etc.). Wood et al. (1994) attributed to *P. boisei* a later stasis maintaining the same size of the molars over a considerable period of time (more than a million years). The cranial traits that distinguish *P. robustus* from *P. boisei* could be indicative, judging by KGA10-525, of accumulated variation in those traits during that long period of time, and not evidence of an evolutionary sequence between various species.

5.8.3 The hominin diversity of Middle and Upper Pliocene

The evolutionary relationships between different "gracile" australopithecines, which came after *Australopithecus afarensis*, have not been studied in the same detail as the evolution from *Ardipithecus* to *Australopithecus*.

With respect to *A. bahrelghazali* of Chad, phylogenetic analyses are very precarious. In a synthesis about the descriptions of the available specimens, Christophe Beaulieu (1998) indicated the morphological differences of *A. bahrelghazali* with respect to *Ar. ramidus*, *A. anamensis*, *A. afarensis*, *A. africanus*, and *P. robustus*. He gave as the only phylogenetic clue that *Australopithecus* from Chad seems to belong to an ancient and individualized branch prior to 4 Ma, but argued that certain features (premolar molarization and symphysis type) bring *A. bahrelghazali* closer to the characteristics of *Homo*. The only in-depth analysis that justified maintaining *A. bahrelghazali* as a separate species from *A. afarensis* was a comparative study of the shape of the mandibular symphysis (Guy et al., 2005).

Subsequent papers provided no new insights. Michel Brunet (2010) merely stated that the discoveries of *A. bahrelghazali* and *S. tchadensis* introduced a new phylogenetic paradigm without saying what it was. A lecture by the same author (Brunet, 2008) with a significant title, had earlier introduced paradigmatic alternatives, but reduced them to philosophical questions such as: "what are we, who is our ancestor and, where do we come from?" The only point made by Brunet (2008) with respect to *A. bahrelghazali* was to include

this taxon among the diverse australopiths that between 2 and 3 Ma precede *Homo*. Finally, the detailed studies of *Ar. ramidus* in the journal *Science* of October 2, 2009, mention only once the *Australopithecus* from Chad, placing it in an informal graphic as a younger specimen than *A. afarensis*, while adding a question mark (see Gibbons, 2009).

We have already indicated that the taxon *A. platyops* was proposed as a new species of the genus *Kenyanthropus* (Leakey et al., 2001). We have also given reasons, related to the morphology, as well as to possible inferred adaptive strategies, why it is advisable to include the specimen in *Australopithecus*.

Daniel Lieberman (2001), in a comment with regard to the finding of KNM-WT 40000, maintained that the specimen casts even more doubts about the phylogenetic interpretation of the Pliocene hominins. Certainly, the proposal of the new genus *Kenyanthropus* impacted other taxa. The close similarity of the facial anatomy of *K. platyops* and KNM-ER 1470 of Koobi Fora (Kenya), already made known by Leakey et al. (2001), caused Leslie Aiello and Mark Collard (2001) to ascribe KNM-ER 1470 and the other specimens classified up to then, such as *H. rudolfensis*, to the new genus proposed by Leakey et al. (2001): *Kenyanthropus rudolfensis*. But to place KNM-ER 1470 in *Kenyanthropus* runs into the problem of its similarity with other specimens of the Lower Pliocene, such as *H. habilis*. It is quite difficult to justify placing two taxa as similar as *H. habilis* and *K. rudolfensis* in two species, not to say placing them in two different genera.

Perhaps the issue is otherwise, so that it revolves around the warning made by Tim White (2003) about distortions suffered by the cranium of *K. platyops* in the process of fossilization. Being the only exemplar of the hypodigm, when the species was proposed it could not be compared to other fossils of the same taxon to assess the existing degree of deformity. White (2003) argued, as a possible interpretation of the Lomekwi specimen, that it might be an old variant of *A. afarensis*. To reach firmer conclusions it is necessary to find new specimens of *K. platyops*; while waiting for that to happen we have chosen to provisionally accept the most parsimonious solution indicated by White (2003): KNM-WT 40000 would be a member of the genus *Australopithecus*. However, is it possible to ascribe it to a new species or must it be included in *A. afarensis*? The arguments of Tim White against the excessive dispersion of the taxa are also applicable to the species, not only to the genus. But, if the genus is frequently subjected to taxonomic reorganization, it is different for proposals of species. Every new species is normally

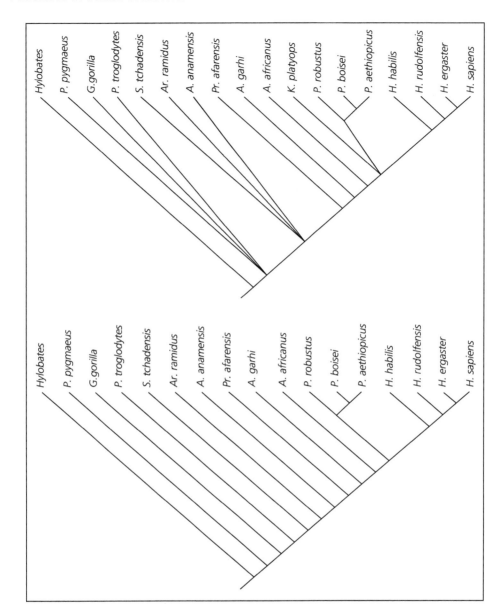

Figure 5.22 Cladograms of Strait and Grine (2004) with the inclusion and exclusion of *K. platyops*. Reprinted from Strait, D. S., Grine. F. E. (2004). Inferring Hominoid and early Hominid phylogeny using craniodental characters. *Journal of Human Evolution*, 47, 399–452, with permission from Elsevier.

accepted if there is no evidence of a lack of significance. Consequently, we have included KNM-WT 40000 in *Australopithecus platyops*. A different issue would be to establish the phylogenetic relationships that this taxon maintains with the rest of *Australopithecus*.

The cladistic analysis made by David Strait and Frederick Grine (2004) based on craneodental traits, besides rejecting a close relationship between *Homo*

rudolfensis and *K. platyops*, addressed a phylogenetic question. Strait and Grine (2004) considered alternative hypotheses, including and excluding *Kenyanthropus platyops*. When it was included, the most parsimonious cladograms placed this taxon close to the paranthrops (Figure 5.22).

In the analysis reached excluding *K. platyops*, Strait and Grine (2004) did not consider the specimens of its hypodigm as distinctive of the taxon *Australopithecus platyops*—the solution advocated here by us—so, the cladistic significance of our taxonomic proposal cannot be assessed according to that study.

5.8.4 The phylogenetic role of the South African "gracile" australopithecines

No other hominin taxon from the Pliocene has been subjected to so many morphological comparisons, so many systematic analyses, and evolutionary interpretations as *Australopithecus africanus*. It is normal that it should have happened this way. Raymond Dart's proposal was made in 1925 and, for more than 50

years the taxon *A. africanus* was the only one within its genus. *A. afarensis* was named in 1978 and, since then, the number of taxa has multiplied: *A. bahrelghazali* and *A. anamensis* in 1995; *A. garhi* in 1999; *A. platyops* (*K. platyops*) in 2001; *A. sediba* in 2010.

Half a century of discussion is no guarantee of a consensus and a phylogenetic explanation free of doubts—particularly if we consider the added circumstances in regard to the age of the South African sites.

The alternative between "gracile" and "robust" australopithecines was the starting point for interpretation of the phylogenetic role of *A. africanus*. The massive character of the craniodental features of *P. robustus* (which can be translated as more ape-like, i.e., more

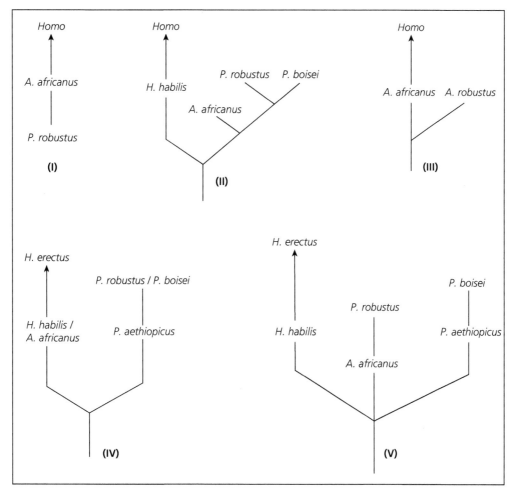

Figure 5.23 (I) Initial proposal for considering the phylogenetic relationships between *A. africanus*, *P. robustus*, and *Homo*. (II) After the Olduvai findings, with *A. africanus* as ancestor of *Paranthropus*. (III) After the Olduvai findings, with *A. africanus* as ancestor of *Homo*. (IV) After the findings of *A. afarensis* and *P. aethiopicus* with *A. africanus* as ancestor of *Homo*. (V) With *A. africanus* as ancestor of *P. robustus*.

primitive) led to an understanding that the evolutionary sequence would have *P. robustus* as the oldest taxon and, derived from it, *A. africanus* would emerge as the ancestor of *Homo* (Figure 5.23).

The most accurate estimates of the stratigraphy of the South African caves led to the understanding that *A. africanus* would be an older taxon than *P. robustus*, increasing the age of the former to 3.5 Ma. This dating completely contradicted the idea of *P. robustus* as the ancestor; in fact, it reversed the scheme by placing *A. africanus* in the ancestor role. Fitting those assessments with the model of an alternative occupation of the savanna by *Homo* and *Paranthropus*, and with the emergence of the first specimens in the Rift Valley, *A. africanus* came to be seen in a somewhat different way. Figure 5.23 shows the various possibilities.

The most recent dating of Sterkfontein places *A. africanus* in a temporal range that makes it difficult to maintain any of the alternatives of Figure 5.23. It is too young, with not enough derived traits with respect to *Paranthropus* or *Homo* to be considered a member of either of these two genera. With an age range of around 2.2 Ma, the least forced interpretation of *A. africanus* sees this taxon as a very advanced member of *Australopithecus*, which in its older epoch maintained traits characteristic of the genus but with homoplasies bringing it closer to *Homo*. Nevertheless, this phylogenetic solution must be compatible with the diagnosis made regarding the last taxon discovered in South Africa: *A. sediba*. We will address this question in Chapter 6.

5.8.5 Phylogenetic relationships between species of *Paranthropus*

To a large extent, the history of the possible evolutionary relationships existing between the robust australopithecines follows the lines already stated about *A. africanus*, based on the identification made in the first third of the twentieth century of the early "robust" hominins from South Africa. Changes in the phylogenetic interpretations based on the findings of specimens from East Africa followed a parallel path. But the coherence of the robust branch of the australopithecines, considered monophyletic by many authors, is an advantage when it comes to interpretation. The only doubt—beyond the geographical issue, i.e., the relationships between the taxa from South Africa and the Rift Valley—affects the evolutionary position of *P. aethiopicus* regarding *P. boisei* and *P. robustus*.

Table 5.4 Affinities of KNM-WT 17000 with different taxa

Primitive features shared with *A. afarensis*
Strong upper facial prognathism*
Flat cranial base*
Posterior–anterior temporalis large
Temporomandibular joint flat, open anteriorly
Postglenoid process anterior to tympanic plate*
Extensive temporal squama pneumatization
Strongly flared parietal mastoid angle (asterionic notch?)
Large horizontal distance between molars and temporomandibular joint
Absolutely large anterior tooth row*
Maxillary dental arch convergent posteriorly
Flat, shallow palate
Nasion coincident with high glabella
Derived features shared with all post-*A. afarensis* species
Short cranial base*
Vertically inclined tympanic plate inferosuperiorly concave*
Reduced medial inflection of mastoid process
Nasoalveolar contour protects weakly anterior to bicanine line
Derived features shared with *A. africanus*, *A. robustus*, and *boisei*
Maxillary lateral incisor roots medial to nasal aperture margins
Zygomaticoalveolar crest weakly arched in facial view
Derived features shared with *A. robustus* and *A. boisei*
"Dished" midface*
Zygomatic process forward relative to palate length
Guttered nasoalveolar clivus grades into nasal cavity floor
Anterior vomer insertion coincident with anterior nasal spine
Nasals widest superiorly
Supraorbitals in form of "costa supraorbitalis"
Receding frontal squama with "trigonum frontale"
Relatively enlarged postcanine toothrow
Incisors in bicanine line Petrous inclined coronally*[y]
Tympanic vertically deep, with strong vaginal processy
Mastoid bulbous, inflated beyond supramastoid crest
Derived features shared exclusively with *A. boisei*
Heart-shaped foramen magnum
Temporoparietal overlap at asterion[z]

Source: Kimbel et al. (1988).
*Variable in *A. africanus*.
[y]Also in Homo.
[z]Rak, personal communication.

William Kimbel, Tim White, and Donald Johanson (1988) carried out a re-analysis of the traits of KNM-WT 17000 indicated in the original paper of Alan

Box 5.19 To which species is KNM-WT 17000 closer?

The question about the species in which the black cranium should be classified (together with KNM-WT 16005, KNM-WT 17400, and the mandible L7a-125 from Omo) remains unresolved. Some of its features are close to those of *A. afarensis*. But, as Rak (1988) pointed out, it is risky to take anatomical characters and use them as taxonomic elements separated from their functional value. Rak indicated that the great prognathism shared by KNM-WT 17000 and *A. afarensis* is not sufficient to place them closer together, taxonomically speaking, than KNM-WT 17000 would be to orthognathous *P. robustus*.

Walker and collaborators about this specimen (1986). They carried out a comparative study of:

- shared apomorphies between KNM-WT 17000 and (1) *A. africanus*, *P. robustus*, and *P. boisei*; (2) *P. robustus* and *P. boisei*; and (3) only *P. boisei*;
- shared apomorphies with *A. afarensis* (Kimbel et al., 1988).

Of the 32 traits analyzed, 12 were primitive characters shared by KNM-WT 17000 and *A. afarensis*,

according to Kimbel and collaborators (1988). Six traits were apomorphies common between *A. africanus*, *P. robustus*, and *P. boisei*. Twelve traits were apomorphies shared by KNM-WT 17000 and *P. robustus* and *P. boisei*. And, finally, the authors found two derived traits shared exclusively with *P. boisei* (see Table 5.4). Kimbel and collaborators (1988) concluded that their study supports the classification of KNM-WT 17000 as a separate species, *P. aethiopicus* (see Box 5.19).

The notion maintained in this book of attributing to *P. robustus* and *P. boisei* an increase in robustness with the passage of time, allows these two taxa to be included in the same clade or lineage. But the discovery of a hyper-robust specimen with an age of almost two and a half million years, contradicts the idea of robustness as a late feature. The sagittal and nuchal crests of KNM-WT 17000 are much more developed traits than in *P. boisei* and *P. robustus*. The discovery of the black cranium, therefore, created the need to reinterpret the evolutionary history of the robust hominins and thus of all our tribe. Figure 5.23 shows two alternative ways of doing so, depending on the consideration of the taxon *P. aethiopicus* as an ancestor of *P. boisei* and *P. robustus*, or only of the latter. These are the two proposals made shortly after the appearance of the specimen KNM-WT 17000. The first was defended by Eric Delson (1986); the second, by Roger Lewin (1986).

CHAPTER 6

The emergence of the genus *Homo*

Homo is the second genus we will consider among those that emerged after the cladistic divergence of the hominins during the Pliocene. It is the sister group of *Paranthropus*—the taxon described in Chapter 5. *Pithecanthropus erectus*, proposed by Dubois in 1894 (see Chapter 8), after being renamed *Homo erectus*, was considered for more than half a century as the oldest member of the genus to which our species belongs. This view was held until the findings at the Olduvai site compelled Louis Leakey, Phillip Tobias, and John Napier (1964) to propose a new taxon for the *Homo* emerging in the Pliocene.

6.1 Homo habilis

Between 1960 and 1964, Louis and Mary Leakey's team found a series of specimens in Olduvai Beds I and II. Their interpretation was immediately very controversial. One of them, OH 7—Jonny's Child—(Leakey, 1961a), included a mandible, a parietal, and hand bones of a juvenile individual (Figure 6.1) from the FLKNN I site at Bed I, slightly older than the sediments in which the first important specimen from Olduvai, *Zinjanthropus boisei* (Leakey, 1959), had appeared a couple of years earlier. OH 8 (Leakey, 1961b) (Figure 6.1) was found at the same FLKNN I site. This specimen included two phalanxes, a molar fragment, and a set of adult foot bones. The age attributed to the new specimen was c. 1.7 Ma (Hay, 1971).

Other postcranial elements of Bed I found in the same years were: OH 10—a distal phalanx of the big toe; OH 35—tibia and fibula; OH 43—two metatarsals; OH 48—a clavicle; and OH 49—a fragment of radius (Wood, 1974). Dental and cranial materials were added, such as: OH 4—third lower molar inserted in a mandibular fragment; OH 6—teeth and cranial fragments; OH 13—mandible and cranial fragments; and OH 16—cranial fragments (Tobias, 1991c). All of them contributed evidence, in Olduvai Bed I, of a type of hominin different from *Zinjanthropus*.

Considering the gracile appearance of the cranium, mandible, and teeth of OH 7, especially by comparison to OH 5, Leakey et al. (1964) suggested including all those findings in the genus *Homo*, defining the new species *Homo habilis*. OH 7 (an immature specimen, unfortunately) constituted the type specimen; OH 13 (Figure 6.2), OH 16, OH 6, OH 8, and OH 4 were paratypes. The description of *Homo habilis* included Leakey's (1961a) analysis of OH 7, together with Tobias' (1964) calculations of cranial capacity and Napier's (1962b) study of hand bones. Overall, *H. habilis* exhibits certain features that represent changes in the cranium and dentition when compared with *Australopithecus*. Its face is less prognathic and its cranial capacity is larger. Its masticatory apparatus is smaller, especially molars and premolars, and dental enamel is slightly thinner. The shape of its dental arcade is parabolic, like later *Homo* specimens.

In order to classify specimens, taxonomy usually takes into account their morphology, above any other consideration. This is why morphological descriptions of the type specimen and paratypes were used by Leakey et al. (1964) to propose *H. habilis*. However, the genus *Homo* has become associated with features other than morphological traits, namely, the production of tools used for scavenging and hunting. This behavior requires a big enough brain to carry out the complex cognitive operations involved in those tasks. The proponents of the new taxon suggested in their 1964 article that *H. habilis* was the true author of the Oldowan culture, the lithic industry at Olduvai, while *Zinjanthropus*—the earlier candidate—was a mere intruder. Thus, *Homo* would be the genus that introduced the adaptive strategy of stone tool-making, and *H. habilis* its first representative. Following Dart's suggestion, the new species was christened *H. habilis* mainly for this reason: *"habilis"* means "able, handy, mentally skillful, vigorous" as noted by the authors in 1964.

The proposal of the taxon *H. habilis* was provocative at the time. Tobias (1992) explained later on that the accepted doctrine during the mid-twentieth century

Processes in Human Evolution. Francisco J. Ayala and Camilo J. Cela-Conde, Oxford University Press (2017).
DOI 10.1093/acprof:oso/9780198739906.001.0001

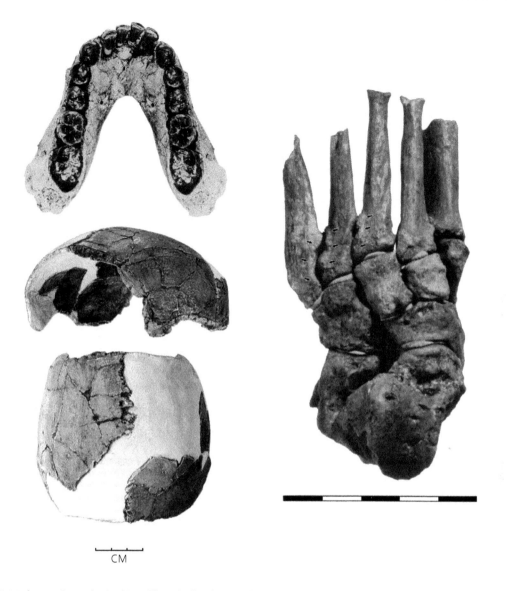

CM

Figure 6.1 Left: Lower jaw and parietal OH 7 (illustration by Tobias, 2003). Right: Foot OH 8 (illustration by Susman, 2008). Right: Susman, R. L. (2008). Evidence bearing on the status of Homo habilis at Olduvai Gorge. Am J Phys Anthropol, 137, 356–361. Left: From Tobias, P. V. (2003). Encore Olduvai. Science, 299, 1193–1194. Reprinted with permission from AAAS.

maintained that the "morphological distance" between *A. africanus* and the typical Middle-Pleistocene hominin, *H. erectus*, was not enough to accommodate any other gracile taxon, because they are very similar. Moreover, the genus *Homo* was considered characteristic of the Middle Pleistocene and thought to have a larger endocranial capacity than that of the Olduvai specimens. These two kinds of reasons against the *H. habilis* proposal were, in actual fact, incompatible: on

the one hand, it was too similar to *Homo erectus* and, on the other hand, it was not close enough to known *Homo* specimens. All of the morphological features of *Homo habilis* were cited with severity by some illustrious contemporary paleontologists during the discussion following its proposal, including Le Gros Clark, Clark Howell, B. Campbell, D. Pilbeam, E. Simons, and Robinson, among others. The letter sent by Le Gros Clark (1964a) to the editor of *Discovery*

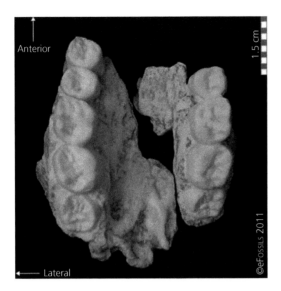

Anterior

1.5 cm

©eFossils 2011

←—— Lateral

Figure 6.2 Illustration of OH 13 (from http://www.efossils.org/page/boneviewer/Homo%20habilis/OH%2013).

Box 6.1 Age at death indicated by Olduvai fossils

Surprised by the great amount of remains corresponding to immature *H. habilis* individuals found at Olduvai, Tobias (1991a) studied the relationship between the age at death, demographic patterns, and environmental conditions, and concluded that only 59% of *Homo habilis* at Olduvai lived to become adults. This is similar to the figure calculated for *A. robustus* (56–57%), but it is much less than for *A. africanus* (75–81%). Given that child survival rate depends on environmental conditions, Tobias argued that those conditions were harder in the times of *H. habilis* and robust australopiths, when Africa had undergone a cooling process, than during the more temperate period of *A. africanus*.

---

shortly after the proposal of *H. habilis* summarizes these doubts. He believed the specimens discovered by Leakey at Olduvai were too similar to *Australopithecus* and too different from *Homo* so that without doubt it belonged to the former genus. An examination of the available evidence, including cladistic studies, has led Bernard Wood and Mark Collard to suggest the same idea 35 years later (Wood & Collard, 1999a, 1999b, 1999c).

According to Tobias (1992), the most devastating attack on *H. habilis* was written by Loring Brace and colleagues (1973). After criticizing the dental measures of the type specimen and the paratypes, Brace et al. concluded that, since the taxon *H. habilis* lacked a type specimen and paratypes, it constituted an inadequately proposed empty taxon and, therefore, deserved to be formally eliminated. Tobias (1992) has noted that in the 15 months following the description of the new species, the *H. habilis* specimens were reclassified by different authors as *Australopithecus africanus habilis*, *Australopithecus habilis*, *Homo erectus habilis*, or *Homo erectus* (unspecified subspecies); that is to say, in any of the possible ways that avoided admitting the new taxon as defined by Leakey et al.

6.1.1 Apomorphies of *Homo habilis*

The anatomical traits that characterized *Homo habilis* were indicated by the authors (Leakey et al., 1964) as follows:
- Cranial and mandibular traits:
  - Mandible and maxillar of smaller size than in *Australopithecus*, although equivalent to those of *H. erectus* and *H. sapiens*.
  - Larger brain size than *Australopithecus* but smaller than in *H. erectus*.
  - Muscle markings neither faint nor strong.
  - Parietal curvature in the sagittal plane ranging from mild (i.e., distinctive of *Homo*) to moderate (i.e., *Australopithecus*).
  - Curvature in a relatively open angle from the external sagittal to the occipital.
  - Receding chin, weak or no trigonum mentale.

- Dental traits:
  - Larger incisors compared to *Australopithecus* and *Homo erectus*.
  - Molar size overlapping the ranges of *Australopithecus* and *H. erectus*.
  - Larger canines in comparison with premolars.
  - Narrower premolars than *Australopithecus* and in the range of *H. erectus*.
  - Narrow teeth buccolingually and elongated in mesiodistal direction, in particular in premolars and molars.

- Postcranial elements:
  - Clavicles similar to those of *H. sapiens*.
  - Hand bones with broad terminal phalanges, metacarpophalangeal and capitate joints similar to those of *H. sapiens*, except the scaphoid and trapezium, the insertions of superficial flexor tendons, and the curvature of the phalanges.

- Foot bones similar to those of *H. sapiens* in strength and support of the big toe, but different in the shape of the talar trochlear surface and the third relatively robust metatarsal.

These features can be summarized in a more gracile cranium, mandible, and postcranial elements compared to *Australopithecus*, together with the trend of increasing cranial size. However, the locomotor apparatus retains certain plesiomorphies.

Let us consider the most remarkable examples that are part of the hypodigm of *Homo habilis*.

6.1.2 Olduvai specimens of *Homo habilis*

After naming the taxon as *Homo habilis*, other Olduvai fossils were attributed—not without question—to the same species. Standing out among them, for different reasons, are the specimens OH 24, OH 62, and OH 65.

In October 1968, P. Nzube discovered a partial, fractured, and squashed cranium at Locality DKE (Douglas Korongo East), corresponding to the lower part of Bed I of Olduvai. The specimen received the technical designation of OH 24 (Leakey, 1969) (Figure 6.3 and Box 6.2).

The cranial capacity of OH 24 is small, around 600 cc, which raised doubts as to whether it belonged to the taxon *H. habilis*. However, as Tobias (1991c) noted in his meticulous study of the specimen, OH 24 is one of the most remarkable *H. habilis* exemplars. Its profile is similar to that of *A. africanus*, such as Sts 5, StW 13, and MLD 6, with a marked prognathism. But

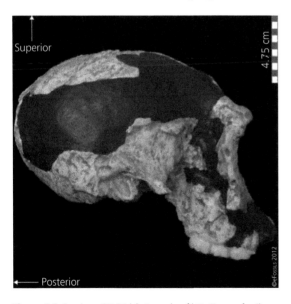

Figure 6.3 Specimen OH 24 (photography of http://www.efossils.org/page/boneviewer/Homo%20habilis/OH%2024).

the palate, which forms a parabolic arcade, is different from both *A. africanus* and *P. robustus*, and though the size of its molars (measured as the sum of their surfaces) is greater than the *H. habilis'* average, it is below that of *A. africanus*, *P. robustus*, as well as *P. boisei* (Tobias, 1991c). OH 24 seemed decisively to confirm the taxon *H. habilis*. However, its problems had just started.

6.1.3 OH 62 and the question of variability within *Homo habilis*

A specimen discovered by Donald Johanson et al. (1987) raised new doubts concerning the taxon *H. habilis*. The specimen was found at Dik Dik Hill, Olduvai Bed I, close to the FLK site of *Zinjanthropus*. It included up to 300 fragmented remains of an individual's face, palate, cranium, and jaw, together with a complete humerus and fragments of the radius, tibia, fibula, and femur. The assemblage was designated OH 62 (Figure 6.4). Johanson and colleagues believed OH 62 belonged to the species *H. habilis*, but this ascription is not unproblematic. For a start, it is a very short adult individual, one of the smallest among all hominins. Given that previous *H. habilis* specimens hardly allowed associating cranial and postcranial remains, OH 62 afforded a great opportunity for integrating those disperse traits in a single individual. But the result yielded more shadows than lights. Due to its modest size, OH 62 fell within the body size of the much earlier Ethiopian australopiths, *A. afarensis*. In addition, the OH 62 humerus, measured by Johanson et al. (1987), was 264 mm long. Hence, it was longer than the humeri of *A. afarensis*, such as AL 288-1, Lucy. We'll return to this issue later.

Olduvai continued providing interesting specimens of *Homo habilis*. In 1995, a maxilla, complete with its

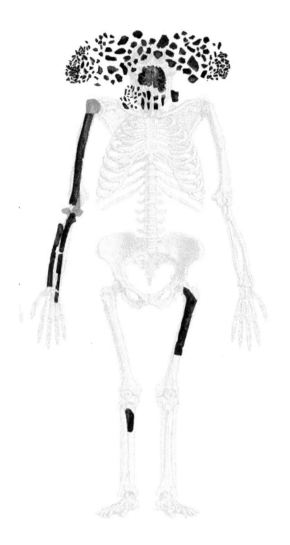

Figure 6.4 Drawing of the fossil fragments that make up exemplar OH 62, placed over the outline of AL 288-1 (drawing by Haeusler & McHenry, 2004; modified). Reprinted from Haeusler, M. & McHenry, H. M. (2007). Evolutionary reversals of limb proportions in early hominids? Evidence from KNM-ER 3735 (Homo habilis). *J Hum Evol*, 53, 383–405, with permission from Elsevier.

dentition and part of the face, was found. Designated as OH 65, it has been described by Robert Blumenschine and colleagues in 2003. The specimen came from Trench 57 excavated by the authors. These sediments were initially attributed to the lower part of Bed II. However, stratigraphic analyses and later dating with the ^{39}Ar/^{40}Ar method placed OH 65 in Bed I, with an age of 1.85±0.002 Ma. The authors of the description note that the orthognathic profile is reminiscent of *Paranthropus*. But the lower nasal region and,

specifically, the naso-alveolar clivus, is clearly different from that of robust australopiths. Blumenschine et al. (2003) believed that OH 65 was closer to the Koobi Fora (Kenya) specimens resembling *Homo habilis*.

6.1.4 Turkana specimens of *Homo habilis*

The proposal of *H. habilis* did not become generally accepted until similar exemplars appeared in other East African sites (Tobias, 1992). Koobi Fora, on the eastern shore of Lake Turkana (Kenya), in particular, yielded the largest number of hominins comparable to the graciles from Olduvai. The specimens found in Koobi Fora provided a better understanding of the features of *Homo habilis* and, being older than those of Tanzania, extended the temporal range of this taxon.

The team led by Richard Leakey in 1968 found a very fragmentary, cranium (FS-210), with only the parietals and the baso-occipital region. Leakey (1970) indicated that these specimens belonged to a gracile hominin, though he did not identify the species. In 1973, also from Koobi Fora, new gracile specimens were described, among them a cranium, KNM-ER 1470, a right femur, KNM-ER 1472, and a proximal fragment of a second right femur, KNM-ER 1475 (Leakey, 1973a, 1973b, 1974; Leakey & Wood, 1973). These specimens were found in the Upper Burgi member sediments, slightly below the KBS tuff, with a dating, as we have already explained (see Chapter 3, Section 3.3.8), that caused a notable controversy.

The 1470 cranium is the most popular of all the mentioned specimens. It was discovered in the area 131 by Bernard Ngeneo, Richard Leakey's assistant, who noticed a great number of fragments on the slope of an eroding gorge: up to 150 fragments from an area of 20 × 20 m^2 (Leakey, 1973a). After a laborious reconstruction, a cranium was obtained that lacked most of the base, the dental crowns, and part of the face; but the remains were sufficient to conclude that it had features that contradicted what would be expected for a hominin of its age (Figure 6.5).

Its cranial capacity was large, between 770 and 775 cc, its superciliary arches were not very protruding, and the face was long and flat, with hardly any subnasal prognathism (Walker, 1981; Day, 1986). All these traits were considered advanced in relation to gracile australopithecines and even to Olduvai *H. habilis*. Though the general appearance of the cranium showed none of the derived robust structures found in paranthropines, the alveoli were suggestive of very wide incisors and canines, larger than those of *Australopithecus* but also larger in size than *H. habilis*, thus,

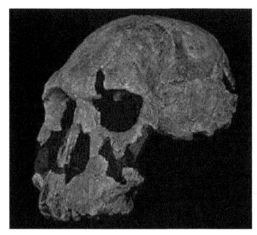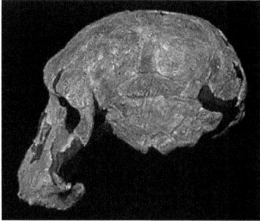

Figure 6.5 KNM-ER 1470, *Homo* sp. indet. (Leakey, 1973b). Left: Frontal view; right: lateral view (pictures from http://www.mnh.si.edu/anthro/humanorigins/ha/ER1470.html).

relatively nearer to *Paranthropus*. Nevertheless, the cranium's general aspect did not show any of the derived robust structures found in robust australopithecines. In other words, the specimen 1470 turned out to be a very old hominin with peculiar apomorphies.

Richard Leakey (1973a) immediately discarded the possibility that KNM-ER 1470 was an australopithecine. Morphologically, it did not fit *H. erectus* either, which included specimens that were much younger than the age attributed to 1470. Leakey also disregarded Olduvai *H. habilis* as the possible species in which to include his specimen, because of their younger age and lower cranial capacity. Hence, he decided to classify KNM-ER 1470 as *Homo* sp. indet.; that is to say, as belonging to the genus *Homo*, without further precisions (Leakey, 1973a) (Box 6.3).

The differences in the sample of gracile hominins of Koobi Fora Pliocene with respect to Olduvai specimens produced two opposite reactions. On the one hand, the Turkana findings supported the appropriateness of the taxon *Homo habilis*; but, they also contributed to the complication of the species panorama, confirming some of the doubts that arose when the fossils first appeared in Tanzania. Let us see the reasons for these complications.

Found in 1973 by Kamoya Kimeu in the Upper Burgi member of Koobi Fora (Leakey, 1974), KNM-ER 1813 has the same cranial shape as KNM-ER 1470 and it is slightly younger—between 1.8 and 1.9 Ma—but its cranium is small despite corresponding to an adult with a developed third molar (Leakey, 1974). This is the reason why it has sometimes been considered as *A. africanus* (e.g., Walker & Leakey, 1978) (Figure 6.6).

***

Box 6.3 Taxonomic status of KNM-ER 1470

Leakey was accused of not daring to take the logical step of assigning 1470 to the species *H. habilis* or, alternatively, defining a new species. It could be thought that the cause was the ongoing controversy regarding *Homo habilis*. However, Richard Leakey has said in an interview (Lewin, 1987) that he had not paid attention to the discussions about the meaning of *H. habilis* and that, for this reason, the controversy could not have affected his decision very much. It seems that it was lack of precision in the description of the *Homo habilis* holotype and paratypes that led Richard Leakey to suggest the classification of KNM-ER 1470 as *Homo* sp. indet. The specimen ended up with the default assignation of *H. habilis*.

Probably the difficulty of obtaining an accurate date for the KBS tuff contributed to distancing this fossil from those of Olduvai. As we will see, there are authors who prefer to place KNM-ER 1470 in another species, *Homo rudolfensis*.

***

The arched and rounded supraorbital torus and evidence of a transversal torus bring KNM-ER 1813 closer to the later *Homo erectus*, a proximity maintained also by the exemplar OH 13, according to Philip Tobias (Tobias & von Koegniswald, 1964); although it is worth recalling that Tobias, along with Leakey and Napier, included OH 13 as paratype of *Homo habilis* (Leakey et al., 1964).

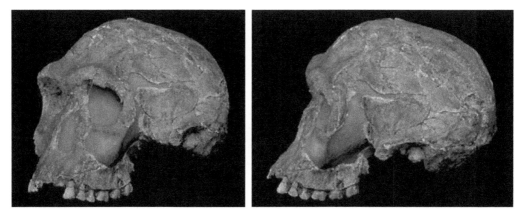

Figure 6.6 KNM-ER 1813, *Homo habilis* (Leakey, 1974). Left: Frontal view; right: lateral view (ilustrations by http://www.mnh.si.edu/anthro/humanorigins/ha/ER1813.html).

Other features, such as facial morphology, the shape and size of the parietal, the anterior mandible, and the small dentition, relate KNM-ER 1813 to the gracile australopithecines (Blumenschine et al., 2003). With almost 510 cc of cranial capacity (or even less, if the interpretation of Glenn Conroy and collaborators, 1998, is correct), this specimen is below the range of all known *Homo habilis* from Olduvai, and only a little over that of *Australopithecus afarensis*.

The contrast with KNM-ER 1470 is evident, but what explains these differences? Could both specimens belong in the same species?

KNM-ER 1805 (cranium and mandible) from Koobi Fora raises similar questions (Figure 6.7). It was discovered by Paul Abell, also in 1973, in the Upper Burgi member and, therefore, has a similar age to KNM-ER 1813. It is an adult exemplar, with a developed third molar. The cranium has an indication of a nuchal crest that brings it close to *Paranthropus*, but its dentition is very small for it to be considered a robust australopithecine.

Its cranial capacity is close to 600 cc. KNM-ER 1805 has been assigned to different species. It was included as paratype of *Homo ergaster* (see Chapter 8) but its cranial capacity and prognathism challenge this classification. F. Clark Howell (1978) and Milford Wolpoff (1984) considered it as *Homo erectus*; William Kimbel, Tim White, and Donald Johanson (1984), as *H. habilis*. Andrew Chamberlain and Bernard Wood (1987) preferred leaving it as *Homo* sp. In Wood's (1991) study of the cranial sample of Koobi Fora, it was assigned to *H. habilis*.

The exemplar KNM-ER 3735, a partial skeleton from the upper part of the member Upper Burgi, with the age 1.91–1.88 Ma, was found by Richard Leakey and Alan Walker (1985) and described in-depth by Richard Leakey et al. (1989). The authors placed the specimens in *Homo habilis*. However, their features, in particular its very small size, created problems for interpreting the homogeneity of the taxon—as happened with OH 62.

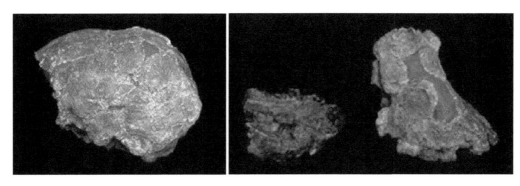

Figure 6.7 KNM-ER 1805, *H. habilis?* Left: partial cranium; right: partial mandible (pictures from http://www.mnh.si.edu/anthro/humanorigins/ha/ER1805.html).

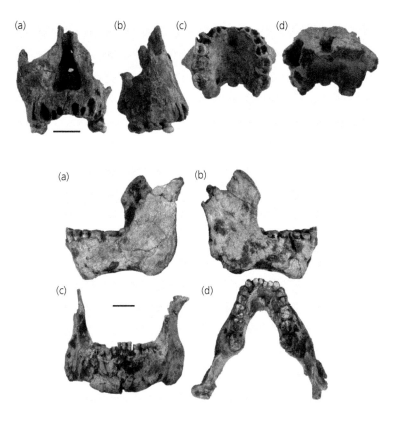

Figure 6.8 Top: Face KNM-ER 62000 in views (a) anterior, (b) right lateral, (c) inferior, (d) superior; bottom: mandible KNM-ER 60000 in views (a) left lateral, (b) right lateral, (c) frontal, (d) upper (illustrations by Leakey et al., 2012; modified; scale bar: 3 cm). Reprinted by permission from Macmillan Publishers Ltd: Leakey, M. G., Spoor, F., Dean, M. C., Feibel, C. S., Anton, S. C., Kiarie, C., & Leakey, L. N. (2012). New fossils from Koobi Fora in northern Kenya confirm taxonomic diversity in early Homo. *Nature*, 488, 201–204.

The last discoveries of hominins made in Koobi Fora up to now correspond to the fossils described by Meave Leakey and collaborators (2012). The face, KNM-ER 62000, is quite well preserved with a maxillar, some teeth, the palatal bones, and the right zygomatic arch, belonging to a juvenile specimen, almost an adult—the third upper molar, present in the fossil, had not erupted yet—with a geological age of 1.95–1.91 Ma, came from the area 131 (Figure 6.8). The mandible, KNM-ER 60000 from area 105, and a mandibular fragment, KNM-ER 62003 from area 130, are more modern but all the specimens were found below KBS tuff.

The size of KNM-ER 62000 is similar to the specimen 1813 of the same site. The teeth are smaller and the facial morphology resembles KNM-ER 1470 with its lack of subnasal prognathism.

Other East African sites have also contributed to broaden the hypodigm of *H. habilis* (or *Homo rudolfensis*)— but not without doubts. This is the case with the L.894-1 fragmentary cranium from the Shungura Formation in

Omo (Ethiopia), from a locality situated between tuffs estimated to be with the K/Ar method, between 1.93±0.10 and 1.84±0.09 Ma. The specimen was classified by Noel Boaz and F. Clark Howell (1977) as *H. habilis* or *H. modjokertensis*, because of its similarities with the OH 24, OH 13, and Sangiran 4 specimens. The partial mandible HCRP UR 501 discovered in the Uraha site, locality 3A (Karonga district, Malawi) (Schrenk et al., 1993) extended the presence of early hominins toward the Chiwondo corridor, where Oldowan tools had been retrieved since 1963 (Clark, 1995). With an age between 2.5 and 2.3 Ma, Timothy Bromage et al. (1995) classified the specimen as *H. rudolfensis*, asserting that the antiquity and morphology of UR 501 speak in favor of a cladistic event around 2.5 Ma ago that led to *Homo*. A new exemplar from Chiwondo, HCRP RC 911, and a mandibular fragment from the Malema site, very close to Uraha, were classified as *Paranthropus boisei* (Sandrock, 1999).

The age of the Chiwondo Beds and, therefore, of HCRP UR 501 was questioned by Alan Deino and

Andrew Hill (2002) (see Chapter 3, Section 3.3.14). However, the most detailed critique does not refer to the age of the different stratigraphic horizons of Chiwondo, but to the attribution of the fossil to the sedimentary unit. Robyn Pickering et al. (2011) indicated that the specimen, found at the surface within a limestone matrix containing a ferric cementation agent over the right-half of the fossil, was related to unit 3A, corresponding to a ferruginous calcimorphic paleosol. But, Pickering et al. (2011) argued that the same kind of paleosol is present in unit 3B, dated at 2.0–1.5 Ma, so the mandible UR 501 could be of that age. If so, the exemplar would continue to indicate the presence of *Homo habilis* in such a southern enclave, but it could no longer be considered one of the oldest specimens of the taxon. Pickering et al. (2011) also criticized the fauna comparison by which unit 3A was dated.

Ethiopia has also yielded exemplars with traits closer to those of *Homo habilis*, with the particularity of having been dated and found associated with lithic artifacts. Ali Yesuf and Maumin Allahendu, members of the research team led by William Kimbel, discovered in Hadar an almost complete upper maxilla (A.L. 666-1) on November 2, 1994, while examining an unexplored area of the upper portion of the Kada Hadar member. The specimen was fragmented in two pieces along the intermaxillary suture, and lacked some teeth, but conserved the subnasal zone. Additionally, 20 Oldowan flaked stone tools appeared in the same horizon. A.L. 666-1 was located in an outcrop immediately below the BKT-3 ash layer, estimated by the ^{40}Ar/^{39}Ar method to be 2.33±0.07 Ma (Kimbel et al., 1996). This is the age attributed by Kimbel's team to the specimen and the associated tools.

The morphology of A.L. 666-1 is different, according to Kimbel et al. (1996), from early Hadar *A. afarensis* and from any other *Australopithecus* (Figure 6.9). Rather, it seems closer to the genus *Homo*, on the grounds of the moderate subnasal prognathism, a relatively wide palate, the square profile of the anterior maxilla, the narrow dental crown of the first molar, and the second molar's rhomboid shape. The issue of the specific *Homo* species in which to allocate A.L. 666-1 is difficult. The traits that move the specimen close to *Homo* are derived, but at the same time they are shared by different species of the genus (they are synapomorphic traits). Kimbel et al. (1996) noted that the specimen's morphology was reminiscent of Olduvai *H. habilis*, so the adequate description would be *Homo* aff. *Homo habilis*. In the systematic study of the maxillar published later by William Kimbel et al. (1997), the same taxonomic ascription was maintained.

Robyn Pickering and collaborators (2011) disqualified the interpretation of A.L. 666-1 as *Homo habilis*, expressing doubts about the character of the traits considered by Kimbel et al. (1997). For Pickering and

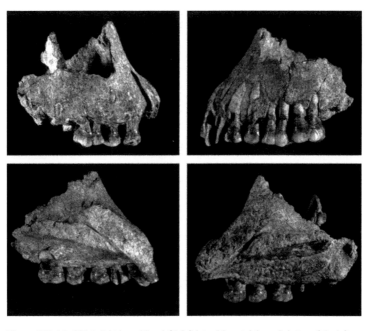

Figure 6.9 A.L. 666-1: right lateral (top left); left lateral (top right); medial view of the left side (lower left); medial view of right side (lower right). Photographs from Kimbel et al., 1997.

collaborators, some of these traits would also be present in *Australopithecus*, while others would not be found in all exemplars of *Homo habilis*. However, in our opinion, the critique is excessive. As we have said, the hypodigm of *Homo habilis* shows a great variability; thus it is possible to expect some features not to be always present in all exemplars. Moreover, the fact that among the ten apomorphic traits pointed out by Kimbel et al. (1997) some could be considered as plesiomorphies, does not disqualify the others. A single apomorphy is enough to indicate the presence of a new taxon. The fact that A.L. 666-1 is not *Homo habilis* but a very old *Homo*, whose accurate ascription is doubtful because, as we said, its derived traits are synapomorphics, still supports the idea of the presence around 2.33 Ma of the genus *Homo*. This is, while keeping in mind the tools associated with the exemplar, an issue to be discussed in Chapter 7.

For their age and the associated presence of tools, both A.L. 666-1 from Hadar and UR 501 from Malawi are evidence in favor of the emergence of the genus related to a cladogenesis 2.5 Ma ago. Pointing toward this same event is the exemplar KNM-WT 42718, a juvenile molar found in 2002 in the Lokalelei 1 site (Nachukui formation, West Turkana) (Prat et al., 2005) and attributed to *Homo*. The estimation of 2.34 Ma for Lokalelei 1 was obtained by correlation with the tuff F-1 of Shungura formation. Once more, Pickering et al. (2011) doubted the evidence placing KNM-WT 42718 as a very old *Homo habilis*, criticizing on this occasion the discriminant analysis done by Sandrine Prat and collaborators (2005). The fact of being an isolated tooth favored the critique; for that reason we will not address other similar fossils from the Shungura formation (Omo, Ethiopia) (Howell et al., 1987; Suwa et al., 1996) and the Nachukui formation (West Turkana, Kenya) (Prat et al., 2005), all of them attributed to *Homo habilis*.

Deserving of special mention is the temporal bone KNM-BC 1, from the BPR#2 site, the Chemeron formation of the Tugen Hills, in the district of Lake Baringo (Kenya) (Martyn, 1967). Initially classified as a hominin of indeterminate genus and species (Tobias, 1967c), due to the ambiguity of its traits, it was later assigned to *Homo* by Andrew Hill and collaborators (1992). A decade later, by a detailed anatomical analysis, Richard Sherwood et al. (2002) confirmed its attribution to the genus *Homo*. The temporal from Chemeron was dated at 2.43±0.02 Ma by a radiometric study (Ar^{39}/Ar^{40}) by Hill, Ward, and Brown (1992). A more detailed study with the same technique placed the fossil between 2.456±0.006 and 2.393±0.013 Ma (Deino & Hill, 2002), which made it the oldest specimen of *Homo* dated with acceptable reliability until the discovery of

the Ledi-Geraru mandible. It confirms the age of about 2.5 Ma for the emergence of the genus. Pickering et al. (2011) considered, however, that the origin of the specimen is not properly documented and, therefore, its age is imprecise.

A partial mandible with teeth, LD 350-1, coming from the site of Ledi-Geraru in the Afar region (Ethiopia), combines primitive traits present in the australopiths with derived traits characteristic of *Homo*. Its age is 2.8–2.75 Ma, obtained by radiometry owing to its association with the Gurumaha and Lee Adoyta tuffs (Villmoare et al., 2015). Brian Villmoare and collaborators (2015) classified LD 350-1 as an indeterminate *Homo* species, maintaining that it differs from *H. habilis* and *H. rudolfensis*, and would be a taxon "representing a likely phyletic predecessor to early Pleistocene *Homo*" (Villmoare et al., 2015, p. 3). This means that the Ledi-Geraru mandible becomes the oldest specimen of the *Homo* genus, pushing the dating of the *Paranthropus/Homo* cladogenesis back to c. 2.8 Ma.

6.1.5 Was *Homo habilis* in South Africa?

Several South African specimens have been included in *Homo habilis*, not without disagreement about the good judgment of that taxonomic operation—especially the cranium Stw 53 of Sterkfontein member 5A (although there are other examples, among which SK 847 and Sts 19 are the most remarkable).

Coming from the member M5a of Sterkfontein, with an age, therefore, a bit less than 2 Ma and found in 1976, Stw 53 was classified—tentatively—as *Homo habilis* by Alun Hughes and Phillip Tobias (1977). After extensive reconstruction of the specimen and a detailed analysis, Darren Curnoe and Phillip Tobias (2006) decided to keep it as part of that species (Figure 6.10).

Curnoe and Tobias (2006) conducted a comparison of the cranial, facial, mandibular, and dental traits of the specimen in order to assign it to a genus and species. With regard to the genus, the authors argued that it was certainly *Homo*. Although the size of the molar crowns matched that of *A. africanus*, the number of tooth roots was not the same in this species and in Stw 53. With respect to species, Curnoe and Tobias (2006) indicated that Stw 53 and *H. habilis* shared many traits of the neurocranium, viscerocranium, and dentition. Four characteristics brought Stw 53 closer to *Homo sapiens*: long wall of the mastoid, presence of molar tubercle, expansion of the lower canine crown, and the crown of the second lower molar elongated mesiodistally. But the high number of features different from those of *H. sapiens* led Curnoe and Tobias (2006) to opt for keeping Stw 53 as *H. habilis*, although recognizing that

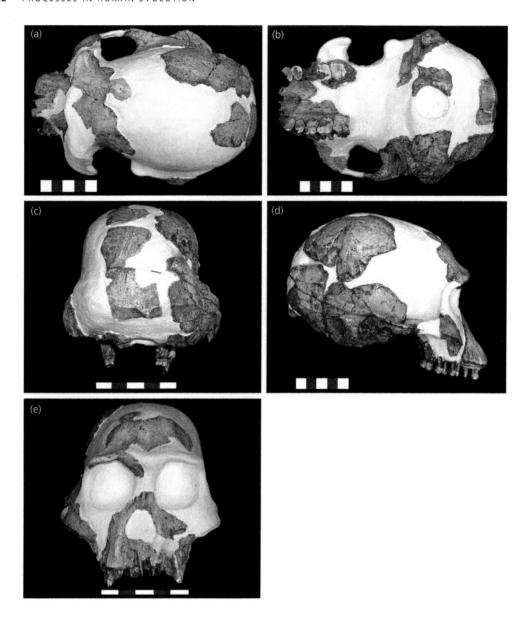

Figure 6.10 Cranium Stw 53, after the reconstruction by Curnoe and Tobias (2006); upper: lateral, frontal, and posterior views (scale bar: 5 cm; photography by Curnoe & Tobias, 2006). Reprinted from Curnoe, D. & Tobias, P. V. (2006). Description, new reconstruction, comparative anatomy, and classification of the Sterkfontein Stw 53 cranium, with discussions about the taxonomy of other southern African early Homo remains. *Journal of Human Evolution*, 50(1), 36–77, with permission from Elsevier.

this was not the last word. Regarding the specimen of SK 847—from member 1 of Swartkrans—attributed usually to *H. erectus* (Clarke et al., 1970; among others) but also to *H. habilis* (Howell, 1978), Curnoe and Tobias (2006) decided that it was a *Homo sapiens* sensu lato.

Both specimens, Stw 53 and SK 847, were subjected to statistical analysis—Euclidean distance (linear measurements) between crania pairs of the compared samples—by Frederick Grine, W. L. Jungers, and J. Schultz (1996) against other crania attributed to *Homo habilis* (OH 24, KNM-ER 1470, and KNM-ER 1813), *Homo erectus* (KNM-ER 3733 and KNM-WT 15000), three *Paranthropus* and abundant samples of *Homo sapiens, Pan troglodytes*, and *Gorilla gorilla*. The study concluded that Stw 53 and SK 847 might represent a different species of *Homo* than those from East Africa included in the sample, which we will discuss in Section 6.3 when we refer to *H. gautengensis*. The cladistic analysis by Heather Smith and Frederick Grine (2008) on the hominins of Sterkfontein and Swartkrans was contrary to the inclusion of SK 847 and Stw 53 in *Homo habilis*. We have already warned, however, that cladograms depend entirely on operations such as trait selection, assignment of specimens to taxa, and the election of species, which introduce difficult to avoid aprioristic bias—although the statistical analyses are also not free of problems. Kathleen Kuman and Ron J. Clarke (2000) showed support for the attribution of the cranium Stw 53 to *Australopithecus* for reasons having to do with the morphology of the specimen—cranial shape and size, dentition, and nasal skeleton.

Sts 19, a basicranium from member 4 of Sterkfontein, was assigned to *Homo* in his doctoral dissertation by Ron Clarke (1977), and was classified as *H. habilis* by William Kimbel and Yoel Rak (1993). However, the comparative morphological analysis by James Ahern (1998) considered that the specimen enters into the intraspecific variation range of *A. africanus*, a taxon that does not increase its variability by including Sts 19.

A series of teeth and mandibular fragments were rebuilt as a juvenile jaw to form the specimen Stw 151 of Sterkfontein (Moggi-Cecchi et al., 1998), coming from an isolated deposit to which Herries and Shaw (2011) granted an age of less than 1.8 Ma. Jacopo Moggi-Cecchi et al. (1998) argued that its features distanced the specimen from Sterkfontein *A. africanus*. Robyn Pickering et al. (2011) disagreed with that assessment, considering *A. africanus* to be the taxon that better accommodates Stw 151.

6.2 The taxon *Homo rudolfensis*

The extensive monographic study by Tobias (1991c) of the remains of Olduvai *H. habilis* maintained that there was no reason to amend the proposal made in 1964 of a single species for *H. habilis*. But as more specimens from other fossiliferous areas joined the hypodigm of *Homo habilis*, the problem of the sample's diversity increased, leading to conflicting opinions about how the morphological differences should be interpreted. As Bernard Wood (1992b) has said, while some authors seemed to easily accept the range of variation attributed to the species, others considered it excessive.

When including the samples of *Homo habilis* from Olduvai and Koobi Fora, at least two different types of

---

Box 6.4 Is it meaningful to include *A. africanus* in the genus *Homo*?

The cladistic study of Bernard Wood (1985) considered, among its main objectives, whether *A. africanus* as a whole must be considered as a member of *Homo*. The conclusions indicated that *A. africanus* shares a large number of characters both with *Homo* and with the clade of "robust" australopithecines; changing the genus of the taxon *A. africanus* does not make much difference from the point of view of the cladograms that are obtained. The work of Wood (1985) is prior to the discovery of the most recently described *Australopithecus*—*A. anamensis*, *A. bahrelghazali*, *A. garhi*, and *A. sediba*—but his comparative study between the characters of *A. africanus* and *Homo* is entirely valid.

---

morphology appear. On one hand, we find specimens with a relatively large cranial capacity—clearly larger than that of any *Australopithecus*—such as in the case of KNM-ER 1470 and numerous other specimens. In contrast, we find specimens with relatively small cranial capacity, such as OH 13 and KNM-ER 1813, a feature that—together with others, such as smaller teeth—brings them toward the gracile australopithecines (see Table 6.1). To distinguish the two groups, we will call the specimens of modest cranial size *Homo habilis* sensu stricto—as the taxon was defined according to them—reserving the name *Homo habilis* sensu lato to accommodate fossils of greater cranial volume. Although it is not just a matter of cranial size, *Homo habilis* sensu stricto has smaller dentition than *H. habilis* sensu lato (Box 6.5).

Box 6.5 The content of *Homo habilis* sensu lato

It should be clarified that, in formal terms, *Homo habilis* sensu lato should include the entire sample, i.e., also the specimens considered as *H. habilis* sensu stricto. However, *H. habilis* sensu lato is commonly used to refer to specimens that fall outside *Homo habilis* sensu stricto. As in Table 6.1, we will also use this term from now on, except when referring to the set of all specimens—in which case, to avoid confusion, we will rather use the expression, "whole sample." In respect to *Homo erectus*, we will use the same term when we distinguish between *H. erectus* sensu stricto and sensu lato.

The taxonomic solution to informally distinguish between *Homo habilis* sensu stricto and sensu lato is not the only way to attempt to resolve the problem of the variability of this taxon. With a somewhat incomplete description, the Russian anthropologist Valerii Alexeev (1986) proposed for the specimen KNMER-1470 a new species: *Homo rudolfensis*. The taxon *Homo rudolfensis* progressively gained recognition (e.g., Groves, 1986; Wood, 1991; Bromage et al., 1995; Collard & Wood, 1999; Wood & Collard, 1999b), so that today it is commonly—but not unanimously—used (Box 6.6).

The combination of primitive and derived features of KNM-ER 1470 led Bernard Wood (Wood, 1992c; Wood & Richmond, 2000) to establish certain criteria regarding mandibular facial areas that every specimen

should meet in order to be attributed to *H. rudolfensis*. These criteria have been widely used and their application has led to the inclusion in that taxon of the cranial remains KNM-ER 1590, KNM-ER 3732, KNM-ER 1801, and KNM-ER 1802, all of them from Koobi Fora. From outside this site, the mandible UR 501 from Malawi and the maxilla and cranial fragments OMO 75–14 from Omo (Ethiopia) have also been attributed to *H. rudolfensis*—although Bernard Wood and Daniel Lieberman (2001) interpreted this last specimen as typical of *P. boisei*.

Wood (1992c) attributed to *H. rudolfensis* and *Homo habilis* sensu stricto the specimens that appear in Table 6.1.

The proposal of two different taxa for the first specimens of *Homo* is consistent with respect to the high variability of the sample of these gracile specimens. However, it faces an obstacle: that of the so-called single species hypothesis (Mayr, 1950; Loring Brace, 1965; Wolpoff, 1971a), which rejects the presence of two contemporaneous species occupying the same locality—one of them would expel the other (see Box 6.8). To

---

Box 6.6 Is the taxon *H. rudolfensis* valid?

Valerii Alexeev attributed a male gender to KNM-ER 1470, arguing that KNM-ER 1813 would be a female of the same species *H. rudolfensis*. However, KNM-ER 1813 is often attributed to *Homo habilis*. Presently we will discuss the issue of the possible relationships between the taxa *Australopithecus*, *H. habilis*, and *H. rudolfensis*. But first a technical issue should be addressed. Is the taxon *Homo rudolfensis* well defined? The somewhat informal manner of Alexeev's proposal has led to the argument that it is an invalid taxon (Kennedy, 1999). However, Bernard Wood and Brian Richmond (2000, p. 41) came to the following conclusion in their taxonomic study of the human lineage: "In a presentation of the fossil evidence for human evolution, published in English in 1986, the Russian anthropologist Valery Alexeev (1986) suggested that the differences between the cranium KNM-ER 1470 and the fossils from Olduvai Gorge allocated to *Homo habilis* justified referring the former to a new species, *Pithecanthropus rudolfensis*, within a genus others had long ago sunk into *Homo* (…). Some workers have claimed that Alexeev either violated, or ignored, the rules laid down within The International Code of Zoological Nomenclature (…). However, there are no grounds for concluding that Alexeev's proposal did not comply with the rules of the Code, even if he did not follow all of its recommendations (…). Thus, if *Homo habilis sensu lato* does subsume more variability than is consistent with it being a single species, and if KNM-ER 1470 is judged to belong to a different species group than the type specimen of *Homo habilis sensu stricto*, then *Homo rudolfensis* (Alexeev, 1986) would be available as the name of a second early *Homo* taxon."

Box 6.7 What happened with the postcranial elements of *H. rudolfensis*?

The absence of postcranial elements associated with the holotype KNM-ER 1470 is a problem in properly characterizing the taxon *H. rudolfensis*. It has been proposed to include in the hypodigm of the species the femora KNM-ER 1472 and KNM-ER 1481 (Wood, 1992c). Rightmire (1993) also raised the possibility that the femora KNM-ER 1428 and the hip KNM-ER 3228 could be associated with KNM-ER 1470, but the relationship of the postcranial materials to the available specimens of *H. rudolfensis* is questionable.

Table 6.1 Specimens of *Homo habilis* sensu stricto and *Homo rudolfensis* from Olduvai and Koobi Fora (according to Wood, 1992c)

Locality	Taxon	Specimens
Olduvai	*H. habilis* sensu stricto	OH 4, 6, 8, 10, 13–16, 21, 24, 27, 35, 37, 39–45, 48–50, 52, 62
Koobi Fora	*H. habilis* sensu stricto	KNM-ER 1478, 1501, 1502, 1805, 1813, 3735
	H. rudolfensis	KNM-ER 813, 819, 1470, 1472, 1481–1483, 1590, 1801, 1802, 3732

Box 6.8 Hypodigm of *Homo habilis* and *Homo rudolfensis*

To the specimens of Table 6.1 should be added others that appeared after the study by Wood (1992c), such as UR 501 from Chiwondo (Malawi), attributed to *H. rudolfensis*, in contrast to A. L. 666-1 and A.L. 984-1 from Hadar (Ethiopia), Omo 75-14 (Ethiopia), SK 847 and Stw 53 (Swartkrans, South Africa), and other fossils pending attribution, e.g., KNM-ER 60000 and 62000. As we know, the attribution of specimens to a particular species, a necessary step to complete the hypodigm, is often questionable.

justify the simultaneous presence of two species, it is necessary to understand that they do not compete for the same adaptive resources, but if that is so, it would be impossible to understand why they have such a close morphology (Box 6.9). However, could it not be that KNM-ER 1470 and KNM-ER 1813 were in fact not contemporaneous but members of different chrono-species within a phyletic lineage?

Box 6.9 Single species and competitive exclusion

The hypothesis of the single species is based on a theoretical consideration. The Principle of Competitive Exclusion states that two closely related species cannot occupy the same territory. Although this idea could already be found in the work of Darwin (den Boer, 1986), the usual formulation of the principle is credited to Georgyi Frantsevitch Gause (1934, p. 19): "It is admitted that as a result of competition two similar species scarcely ever occupy similar niches, but displace each other in such a manner that each takes possession of certain peculiar kinds of food and modes of life in which it has an advantage over its competitor." In other words, two sympatric species cannot coexist competing for the same resources because one displaces the other. If hominins are beings generally cut from the same cloth, only one species could have existed in each location. Milford Wolpoff (1971a) used the "single species hypothesis" to group in one taxon the robust and gracile australopithecines. However, there is experimental evidence for the invalidation of the Principle of Competitive Exclusion (Ayala, 1970).

The most significant aspect indicate the relative age of KNM-ER 1470 compared to other specimens is the stratigraphic horizon in which the fossils appeared at Koobi Fora. Patrick Gatogo and Francis Brown (2006) noted that hominins of the locality belonging to the area 123 of the *Paleontological Collection*—which includes KNM-ER 1813—were attributed in 1970 to the strata above the KBS tuff, but later they were assigned to strata below it. Gatogo and Brown argued that their placement should be, categorically, in member KBS, which is located above the tuff of the same name, ranging from about 1.88 Ma to just over 1.6 Ma. In this case, the age difference between KNM-ER 1470 and KNM-ER 1813—from member Okote, more than 1.6 Ma—becomes small.

A much greater age of KNM-ER 1470 would facilitate considering, as did Gen Suwa and collaborators (2007), that the hypodigm of *Homo rudolfensis*/*Homo habilis* represents "a single early *Homo* lineage evolving through time." But the main features that separate KNM-ER 1470 from KNM-ER 1813—cranial volume and teeth size—led to the thought that evolution would have followed an opposite path leading from an australopithecine of gracile skull and small teeth to *Homo erectus* with a bigger brain and larger teeth.

6.2.1 Hypotheses about the variability of *Homo habilis* sensu lato

Once the possibility that we are dealing with the case of two chronospecies that follow each other is weakened, the great morphological variation of *Homo habilis* needs to be explained. Three different interpretations can be made concerning the reasons for such extensive variation. These would be in the order in which they were presented:

(1) We are facing another case of sexual dimorphism, so *H. habilis* sensu stricto would be female—with smaller teeth and crania—while *H. habilis* sensu lato would correspond to a male.
(2) There are two different, relatively contemporary taxa, i.e., two different species, the second being *H. rudolfensis*.
(3) The taxon *Homo habilis* is very variable, and all specimens can be placed within it.

Let's see what evidence can be provided for each of these hypotheses.

The variability of *Homo habilis* is due to sexual dimorphism

Both the volume of the cranium and the size of the teeth—the two main traits taken into account to separate *H. habilis* sensu stricto and sensu lato—usually vary between the sexes in primates, so to see the reflection of a sexual dimorphism in the two different samples of Table 6.1 is a widespread interpretation, particularly in the first fossil studies. The identification of these traits in the Olduvai specimens is often complicated by the fragmentation of the remains and the limited information that most of them provide about the shape of the cranium or facial structure (Rightmire, 1993). By contrast, two fossils from Koobi Fora, the already mentioned KNM-ER 1470 and 1813, allow a better assessment of their potential gender. As Bernard Wood (1985) stated, if they belong to the same species, it is reasonable to consider KNM-ER 1813 as a female and 1470 as a male.

How can it be decided which of the two explanations is more appropriate? The best starting point is to identify the sex of the specimens and to check to what extent it correlates with the size of the cranium and teeth. Phillip Tobias (1991c) indicated that it is likely that OH 7 and OH 16 were males and OH 13 and OH 24 females. Meanwhile, it is very common to consider KNM-ER 1470 as a male and KNM-ER 1813 as a female. But the fact that the cranial size is a dimorphic trait can lead us to a circular reasoning. If, from the start, we consider those specimens with smaller cranial volume to be female, we cannot claim that it is demonstrated that those specimens grouped under *Homo habilis* sensu stricto are female. It would be necessary to relate the gender with another feature and then test whether the males and females identified in that way have crania and teeth, respectively, larger and smaller.

In search of other indicative gender features, Daniel Lieberman, David Pilbeam, and Bernard Wood (1988) conducted a statistical analysis of the morphology of KNM-ER 1470 and 1813, comparing them with 20 males and 20 females of *Gorilla gorilla*, as these apes present the greatest sexual dimorphism in craniofacial features. The results indicate that in most of the considered traits, the probability that the differences between KNM-ER 1470 and 1813 correspond to sexual dimorphism typical of gorillas, or, in other words, that the two fossil specimens belong to the same species, is less than 5% (Table 6.2 and Box 6.10).

Lieberman et al. (1988) indicated that the main features that contradict the hypothesis of sexual dimorphism as the origin of differences between KNM-ER 1470 and 1813 are the cranial capacity and the prognathism of the lower facial part. With regard to the former, the difference between the two fossils (32%) far exceeds the degree of dimorphism of current gorillas (13%). Facial shape, in turn, raises a paradox in quantitative

Table 6.2 Features of KNM-ER 1470 and 1813 that make the probability of belonging to the same species <5%, when considering a degree of sexual dimorphism equivalent to that of current gorillas

Measure	Male gorilla	Female gorilla	KNM-ER 1470	KNM-ER 1813
Cranial capacity*	568.1	494.2	751	510
Glabella–prosthion	12.85	11.14	10.2	7.3
Nasion–rhinion	5.03	4.48	2.1	1.4
Supraorbital torus–prosthion	14.25	12.19	11.4	7.4
External width M2	7.27	6.59	8.6	6.5
Maxillar root of the height of the zygomatic process	1.85	1.57	1.1	1.5
Postorbital constriction	7.23	7.10	8.1	7.0
Bimastoid width	15.04	13.02	13.8	11.3

Table by Lieberman, Wood, and Pilbeam (1996), simplified (only the averages of male and female gorillas was considered).

*Calculated as the lowest estimation for KNM-ER 1470 and the highest for KNM-ER 1813 within the range indicated by Holloway (1978).

Box 6.10 Probability that KNM-ER 1470 and 1813 belong to the same species

The probability that these two fossils from Koobi Fora belong to the same species is ≤5% for 45% of the traits considered by Lieberman et al. (1988), and ≤15% for 67% of the traits.

---

terms. Both current apes and fossil hominins show sexual dimorphism, with males tending to be more prognathic than females. As a result, "the maxillary root of the zygomatic process is situated relatively more posteriorly in males than in females. In adult gorillas and 'robust' australopithecines, for example, the maxillary root of the zygomatic emerges above M^1 in females and between M^2 and M^1 in males. We see the opposite in KNM-ER 1470 and KNM-ER 1813. The zygomatic process emerges above P^4 for KNM-ER 1470 and above M^1 for KNM-ER 1813—the reverse of what one would expect if KNM-ER 1470 were a male and KNM-ER 1813 a female" (Lieberman et al., 1988, p. 509).

The most significant conclusion to be drawn from the statistical analysis by Lieberman et al. (1988) reveals that the two groups of *Homo habilis* sensu stricto and sensu lato do not correspond to females and males, respectively. Therefore, the differences would indicate, as the authors pointed out, that they are in fact two different species. Andrew Kramer and collaborators (1995) arrived at the same conclusion.

Philip Rightmire (1993) concurred in the idea that facial projection is the best feature to indicate the sex within the entire sample of *Homo habilis*. In current apes, the degree of prognathism is greater in males than in females. If the same criterion is applied to the large sample of *Homo habilis*, some of the specimens of smaller cranial volume—KNM-ER 1813, OH 24—have a considerable prognathism, while KNM-ER 1470, of a larger cranium size, shows a much smaller facial projection (Rightmire, 1993). Accordingly, it is in fact the case that in the sample with high cranial volume, as well as in that with the lowest volume, both sexes exist. Thus, the differences between *H. habilis* sensu stricto and sensu lato cannot be attributed to sexual dimorphism.

The large sample of *Homo habilis* (sensu stricto + sensu lato) contains two species

When opting to accept that the large sample of *Homo habilis* contains two species, the feature that best indicates the assignment to one or the other is the one that we have considered in Table 6.3: different cranial capacity (Box 6.11).

Considering the differences in cranial size using the coefficient of variation (CV) of the volume, Chris Stringer (1986) calculated a value of 12.4 for the entire sample of *Homo habilis*, a figure significantly higher than the 10.0 the author obtained for current apes, modern humans, and the most common species of fossil

Table 6.3 Comparison of cranial size and the coefficient of variation of samples of *Homo habilis*, *Homo sapiens*, and *Pan troglodytes*, considering one or two species within *H. habilis* (Booth, 2011)

Sample	Size	Average cranial capacity (cc)	Standard deviation	Coefficient of variation (CV) of the cranial capacity
H. habilis sensu lato	9	666.4	98.2	14.7
H. habilis	6	611.8	63.0	10.3
H. rudolfensis	3	775.7	42.7	5.5
Homo sapiens				
Naqada	30	1304.8	84.8	6.5
Teita	30	1273.8	116.2	9.1
Congo	30	1265.6	142.8	11.3
Borneo	30	1347.5	116.7	8.7
Tasmania	28	1245.0	112.9	9.1
Moorfields	30	1419	144.5	10.2
Pan troglodytes				
Frankfurt collection	18	367.7	19.8	5.4
Collections from the Powell-Cotton and Roth-child Museums	111	390	41.1	10.5

hominins in most cases. The conclusions of Stringer (1986) indicated, therefore, the need to consider two distinct species. Laura Booth (2011) also analyzed the cranial volume CV of a sample of modern humans, chimpanzees, and *Homo habilis* fossils, and compared the results to consider whether the fossils were from one species or two. The results are shown in Table 6.3.

From the work by Booth (2011) it is deduced that when the whole sample of *Homo habilis* is grouped together in a single taxon, an excessive CV is obtained for the cranial capacity feature, larger than any of the other samples included in the analysis. If, on the contrary, we distinguish between *H. rudolfensis* and *H. habilis* (sensu stricto, as the author added), the limited CV of the last taxon is still larger but in the range of the samples of *Pan troglodytes* and *Homo sapiens*.

A single (very variable) taxon is enough to encompass the whole sample of *Homo habilis*

The statistical analyses of the entire craniofacial sample of *Homo habilis* we have seen support the idea of a division into two species. But we have actually limited

---

Box 6.11 The cranial size criterion for *H. habilis*

The criterion for naming two species depending on cranial size, or keeping *Homo habilis* for specimens of small cranial size and leaving the other species unidentified, has been followed by various authors (Walker & Leakey, 1978; Wood, 1985; Stringer, 1986; Groves 1989; Wood, 1991; Wood, 1992c; to name just a few), prior to the work of Rightmire (1993), who also chose to split the full sample of *H. habilis* into two distinct species.

ourselves to a few studies; there are many other procedures that we have not yet mentioned. Thus, the work of Joseph Miller (2000), who reviewed the statistical analyses of the morphology of *Homo habilis*, indicated the following: endocranial coefficient of variation (Wood, 1985; Stringer, 1986); estimates of sexual dimorphism (Lieberman et al., 1988); multivariate analysis of the facial region (Bilsborough & Wood, 1988); analysis of craniofacial measurements (Kramer et al., 1995); multivariate pattern of sexual dimorphism (Wood, 1993); profiles of CV variability (Kramer et al., 1995); distance measurements by exact randomization (Grine et al., 1996); and quantitative arrangement (Grine et al., 1996).

The main conclusions reached by Miller (2000) after his statistical analyses are that the degree of variation of the whole *Homo habilis* sample has a close correspondence with that of current apes. Moreover, the different patterns shown by the specimens of *Homo habilis* sensu lato do not convey any particularity that requires splitting the sample into two taxa. In other words, no complete specimen of *Homo habilis* in the sample, either KNM-ER 1470 or any other, has, according to Miller (2000), characteristics that necessitate including it in another species (Box 6.12).

Why a conclusion so radically different from that obtained by the authors whose work was reviewed by Miller (2000)? Miller detailed the errors and omissions of each of the reviewed analyses. In addition, it

is possible to give a general idea of the inherent problems faced by any statistical study based on the fossil record. In the discussion of results obtained by her own analysis, Laura Booth (2011) indicated possible sources of error: from volume dispersion obtained by calculating cranial capacities, to biases introduced by the assignment of specimens to taxa, as well as the inherent difficulties of the small size sample. It is, indeed, risky to base taxonomic diagnosis on statistical procedures (Box 6.13).

Let us examine another calculation—posterior to Miller's (2000)—in which changing the variable under consideration supports the hypothesis of a single taxon for *Homo habilis*.

Juan Manuel Jiménez-Arenas, Paul Palmqvist, and Juan A. Pérez-Claros (2011) argued that, when comparing the variability of fossil samples, a different indicator of cranial capacity should be used. These authors applied the cranial index (Crindex; see Box 6.14), arguing that this variable related to the cranium is a better discriminator for carrying out comparisons among current apes. If so, Jimenez-Arenas et al. maintained, it is preferable to use the cranial index CV also in studies of fossil hominins.

The results obtained by Jiménez-Arenas et al. (2011) appear in Table 6.4.

As the authors indicate, the grouping in a single species of a specimen with a large cranial size along with others of small size (including a fossil from Dmanisi, to which we'll refer later) exceeds slightly the variation range that the Crindex obtained for modern humans and chimpanzees, but falls below the Crindex of gorilla. This result allows, according to Jimenez-Arenas

---

Box 6.12 Falsifying hypotheses

As stated clearly and concisely by Miller (2000, p. 122), "Failure to reject a null hypothesis is not evidence in its favor. Obviously, hypotheses can never be 'proved,' only supported, modified or rejected." The author added: "However, at present, the null hypothesis of intraspecific variation [in *Homo habilis*] does explain the data satisfactorily by means of the present tests. This does not preclude further testing and the possibility of rejecting the null hypothesis in the future" (Miller, 2000, p. 122).

Actually, the most trivial hypotheses may prove themselves. A hypothesis asserting that "there are white swans" is proved by finding one. But arguing that "there are no black swans" cannot be falsified by the repeated identification of many white swans; this may weaken the hypothesis, yes, but it remains provisional no matter for how long a black swan is not found. So, what is important for our purposes is that the hypothesis that the whole variation of *Homo habilis* is intraspecific has not been eliminated up to now, according to Miller (2000).

Box 6.13 Difficulties in the cranial volume estimation

As an example of the difficulties in the calculation of cranial capacity, that of OH 7 was estimated to be 700–750 cc by Ralph Holloway (1980), 590 cc by Milford Wolpoff (1981), and 674 cc by Phillip Tobias (1987), although in the latter case it was an estimation of the volume reached in maturity (this specimen, as you may recall, is a juvenile). One possible source of discrepancy is fossil cranium reconstruction. The available fragmentary remains oblige the use of other comparison techniques with endocrania—such as multiple regression, used by Holloway (1980)—whose validity depends on the chosen sample for comparison, as well as on the accuracy of the extrapolations (see Wolpoff, 1981).

Box 6.14 Cranial index

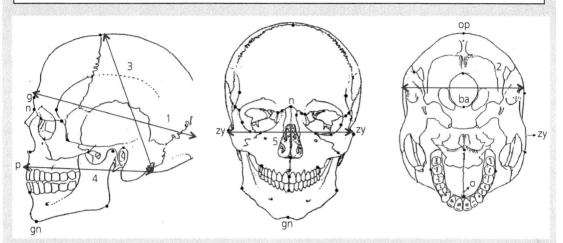

The cranial index is defined as the ratio between the geometric measurements of three neurocranial variables (GOL, XCB, and BBH) and three viscerocranials (BPL, ZYB, and NPH):

GOL: distance glabella–opisthocranium (1).
XCB: maximum width of the neurocranium (2).
BBH: distance basion–bregma (3).
BPL: distance basion–prosthion (4).
ZYB: maximum width between the lateral ends of right and left zygomatic arches (5).

NPH: distance nasion–prosthion (6).
Thus, the algorithm of the cranial index is:

Crindex = $(GOL \times XCB \times BBH)^{1/3}/(BPL \times ZYB \times NPH)^{1/3}$

Jiménez-Arenas et al. (2011) carried out analyses on different hominin specimens and taxa, from *H. habilis* to "archaic" humans, obtaining different groupings. Here we only consider those related to the taxa concerning the large sample of *H. habilis*.

et al. (2011), keeping within a single taxon the whole sample of *Homo habilis* sensu lato. If the specimen from South Africa Stw 53 is added, the variation coefficient of Crindex increases, but less than when specimens such as KNM-ER 3733 are added, which is normally placed within *Homo erectus*.

Table 6.4 Crindex obtained by different fossil groupings—the exemplar D 2282 came from the Dmanisi site (Georgia)

Grouping	Crindex Variation
KNM-ER 1470, KNM-ER 1813, OH 24, D 2282	4.772
KNM-ER 1470, KNM-ER 1813, OH 24, D 2282, Stw 53	5.890
KNM-ER 1470, KNM-ER 1813, OH 24, D 2282, KNM-ER 3733, Stw 53	5.333
Gorilla gorilla (N = 30)	6.10
Pan troglodytes (N = 30)	4.30
Homo sapiens (N = 30)	3.77

Data reconstructed from Jiménez-Arenas, Palmqvist, and Pérez-Claros (2011).

It seems, therefore, that statistical analyses cannot reach conclusive results. Jimenez-Arenas et al. cite 18 studies that support the acceptance of *Homo rudolfensis*, including several that we have not mentioned. They countered this view with ten other studies that maintain the viability of a single taxon: *Homo habilis*. Clearly, to choose one or the other formula is far from obvious.

The last specimens discovered in Olduvai and Koobi Fora added certain clues. OH 65, as we recall, comes from the same stratigraphic horizon, Bed I, as do the rest of the *H. habilis* specimens of the site—although from a different paleobasin, since OH 65 was found in the western part of Olduvai Gorge (Blumenschine et al., 2003). The study of the similarities between the parietal bone of KNM-ER 1470 and the type specimen of *Homo habilis* OH 7, along with the shared architecture of the maxilla in OH 7, OH 65, and KNM-ER 1470, led Robert Blumenschine and collaborators (2003) to conclude that all of them can be considered as members of a very variable species, difficult to consider in phylogenetic terms. Regarding KNM-ER 62000 and 60000

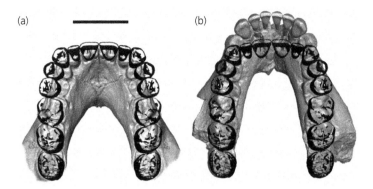

(a)   (b)

Figure 6.11 Comparison of the mandibles (a) KNM-ER 60000 and 62000; (b) KNM-ER 60000 and 1802. In (a) they are almost completely the same. In (b) the differences are very noticeable. (Illustration by Leakey et al., 2012.) Reprinted by permission from Macmillan Publishers Ltd: Leakey, M. G., Spoor, F., Dean, M. C., Feibel, C. S., Anton, S. C., Kiarie, C., & Leakey, L. N. (2012). New fossils from Koobi Fora in northern Kenya confirm taxonomic diversity in early Homo. *Nature*, 488, 201–204.

(Figure 6.11), Meave Leakey and collaborators (2012) argued that the new findings increased significantly the diversity of the gracile sample from Koobi Fora. However, the only taxonomic indication they gave is that KNM-ER 60000 "seems to be associated with the species to which KNM-ER 1470 and KNM-ER 62000 belong" (Leakey et al., 2012, p. 204), arguing that to decide whether this is *Homo habilis* or *H. rudolfensis* should be left for more relevant future studies of the association of OH 7 with other fossils.

As we have seen, the question of whether the sample of *Homo habilis* sensu lato can be reduced to a single species, or whether it is necessary to accept two, does not have a definite answer. In the current situation, and even more after the discovery of the last specimens from Olduvai and Koobi Fora that we have just considered, it is as justifiable to separate the whole sample into two species, *Homo habilis* and *Homo rudolfensis*, as to maintain that all the specimens fit into only one *Homo habilis* sensu lato. As Phillip Tobias (2003) argued in his commentary on the discovery of OH 65, the issue of the species related to *Homo habilis* and *Homo rudolfensis* is by no means settled. Therefore, and in accordance with the principle of parsimony, we have chosen to include *Homo rudolfensis* within the species *Homo habilis*, although referring to both taxa whenever necessary for clarity.

6.3 Homo gautengensis

Frederick Grine, William Jungers, and J. Schultz (1996) sought to clarify by quantitative analysis, based on the randomization and statistical distance method, the phenetic relationships existing between different specimens of *Homo* from East and South Africa. The authors took as external references 44 current humans, 50 *Pan troglodytes*, and 50 *Gorilla gorilla*. The main conclusion reached by Grine et al. (1996), apart from confirming the clear difference between *Homo habilis* sensu lato and *Homo erectus/ergaster*, was to determine the existence of three different groups within the broad sample of *Homo habilis* that, in the study of Grine et al. (1996), included Stw 53 and SK 847 from South Africa, and OH 24, KNM-ER 1470, and KNM-ER 1813 from the Rift Valley. The larger size of KNM-ER 1470 called for its own group, but among the smaller specimens the authors considered it convenient to separate, at the species' level, the specimens from South Africa OH 24 and KNM-ER 1813. The limitations of the analyzed sample make it problematic to accept the necessity of creating a single species for KNM-ER 1470, but we mention here this work because it includes a second suggestion: to create a new species within *Homo*, even if tentative, for placing Stw 53 and SK 847.

That proposal has been made in a more broad, detailed, and accurate manner by Darren Curnoe (2010), naming the new species *Homo gautengensis*, with Stw 53 as type specimen and a long list of South African specimens within the paratypes: SE 255, SE 1508, Stw 19b/33, Stw 75–79, Stw 80, Stw 84, and Stw 151 from Sterkfontein; SK 15, SK 27, SK 45, SK 847, SKX 257/258, SKX 267/268, SKX 339, SKX 610, and SKW 3114 from Swartkrans; and DNH 70 from Drimolen (Figure 6.12).

Such a large hypodigm implies extending the geographic range of the new species, but what increases considerably is particularly the temporal range, which, according to Curnoe (2010), from the 2.0 Ma of Swartkrans member 1 to 1.26/0.82 Ma of Sterkfontein member 5c. It is, therefore, the taxon of longest survival range

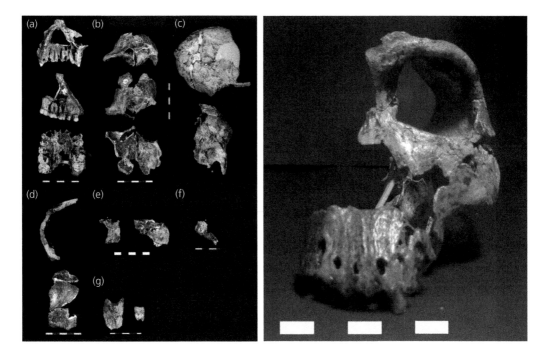

Figure 6.12 Left: Composition of the main fragments forming Stw 53, type specimen of *Homo gautengensis*: (A) maxillar; (B) frontal; (C) right parietal–temporal–occipital; (D) left parietal; (E) left margin of the *foramen magnum* and a left petromastoid fragment; (F) right zygomatic bone; and (G) fragment of mandibular ramus with the third upper molar. Right: Sk 847 from Swartkrans, paratype of *H. gautengensis*. Reprinted from Curnoe, D. (2010). A review of early Homo in southern Africa focusing on cranial, mandibular and dental remains, with the description of a new species (Homo gautengensis sp. nov.). *HOMO – Journal of Comparative Human Biology*, 61(3), 151–177, with permission from Elsevier.

among the ancient *Homo*, and probably the oldest of the whole genus according to Curnoe, since the author does not accept the formal attribution of the specimens KNM-CH 1 from Chemeron and A.L. Hadar 666-1 to *H. habilis* and thinks the age of *H. rudolfensis* is very doubtful.

Does the new species described by Curnoe make sense? In view of the phenetic distances indicated by comparing Stw 53 and SK 847 with *H. habilis* and *H. erectus*, there seems to be room for a new early *Homo* in South Africa, something that various authors previously maintained, in particular by considering dental studies (Grine et al., 2009), although without formal proposal (Table 6.5). The same reasoning would lead to the acceptance of *H. rudolfensis*. But in the case of *H. gautengensis* there is another factor to be taken into account: the geographic separation. It is necessary to understand that if *P. robustus* and *P. boisei* are acknowledged as two different species, it would add weight to the arguments for accepting a new South African early *Homo*. The alternative would be to place their specimens within *A. africanus.*

The biggest problem linked to *Homo gautengensis* is phylogenetic. If it is a very ancient *Homo*, what is its evolutionary relationship with other early *Homo*? And, particularly, what role does it play in the transition to *Homo erectus*? At the end of this chapter we shall return to these questions.

6.4 Homo naledi

Between November 2013 and March 2014, in the Dinaledi Chamber of the Rising Star cave system, 1,550 hominin exemplars were found, which belong to at least 15 individuals and include most of the bones in the skeleton (Berger et al., 2015). The Rising Star cave system lies in the Bloubank River valley, 2.2 km west of the Sterkfontein cave. The location of the Dinaledi Chamber is 30 m below the surface and 80 m, in a straight line, away from the current nearest entrance to the cave (Dirks et al., 2015).

Despite the presence of associated fauna (rodents), the alteration of the stratigraphic deposits does not allow biochronological dating of the Dinaledi Chamber. Radiometric dating has been attempted by the U/Pb method, but the limited amount of suitable clean flowstone has not yielded results. Consequently, the

Table 6.5 Cranial features of Stw 53 and SK 847 compared with Plio-Pleistocene *Homo* species (Curnoe, 2010)

	Stw 53	SK 847	H. habilis	H. erectus
Cranial vault				
Flattening of parieto-occipital region	Not present	-	Present	Present
Occipital bun	Not present			Present
Occipital profile	Rounded	-	Almost sub-angular	Angled
Lateral protrusion of mastoid-supramastoid-auriculare complex relative to temporal squama	Strong	-	Slight	Slight/moderate
Frontal mid-line ridge	Present	-	Absent	Present
Position of temporal lines/crests on parietal	Near mid -line	-	Widely spaced	Widely spaced
Bare area of the occiput	Large	-	Extensive	Extensive
Cranial bone thickness	Thin	Thin	Thin	Moderate-thick
Minimum frontal breadth (absolute)	Narrow	Narrow	Moderate/ broad	Moderate/broad
Mastoid length	Long	Short	Short	Long
Mastoid process/crest laterally swollen	Present	-	Absent/ present	Absent
Occipital torus	Small	-	Small	Large
Inion in relation to opisthocranion	Below	-	Below	Coincident
Difference between inion and endinion	<10 mm	-	<10 mm	<10–25 mm
Cranial base				
Anteroposterior length of mandibular fossa	Short	Moderate	Moderate	Long
Mediolateral breadth of mandibular fossa	Short	Moderate	Moderate	Moderate/broad
Mandibular fossa length relative to breadth (length-breadth index)	Short	Short	Short/moderate	Moderate/long
Tympanomedian angle	Low	Moderate	High	High
Mandibular fossa angle	Moderate	Low	Moderate	Moderate/high
Articular tubercle	Large	Large	Large	Small
Postglenoid tubercle size	Reduced	Reduced	Reduced	Enlarged
Medial recess of mandibular fossa	Absent	Present	Absent	Present
Tympanic merges with postglenoid tubercle	Present	Present	Present	Absent
Eustachian process (supratubal process)	Strong	-	Strong	Small/prominent
Petrous crest that merges with anterior mastoid	Present	Present	Present	Absent
Occipital condyle area	Small	-	Small	Large
Occipital condyle shape index	Low	-	High	Moderate
Facial skeleton				
Temporal lines encroach on lateral one-half of supraorbital rim	Present	Present	Absent	Absent
Salience of supraorbital torus	Marked	Marked	Marked	Moderate
Transverse extent of supraorbital torus	Low	Low	Moderate	High
Supraorbital torus vertical thickness	Thin	Moderate	Moderate	Moderate/thick
Location of thickest part supraorbital torus	Medial	Medial	Variable	Variable
Supratoral sulcus	Slight	Present	Absent/slight	Slight/present
Glabella	Moderate	Moderate	Small/prominent	Small

(continued)

Table 6.5 (*Continued*)

	Stw 53	SK 847	H. habilis	H. erectus
Supraorbital torus divided into two parts	Present	Present	Present	Mostly absent
Superior facial breadth (absolute)	Narrow	Moderate	Moderate	Broad
Midfacial breadth (absolute)	Narrow	Narrow	Narrow	Broad
Superior facial breadth exceeds midfacial breadth	Absent	Absent	Typically exceeds	Absent
Superior and inferior orbital margins lie in approximately same plane	Present	Absent	Present	Absent
Alveolar height	Low	Moderate	Low	Moderate/ high
Nasal projection	Low	Moderate	Low	Moderate/ high
Nasal bridge	Vertical	Flattened	Vertical	Flattened
Nasal cavity entrance	Smooth	Stepped	Smooth/stepped	Stepped
Palate				
Maxillo-alveolar length	Moderate	Moderate	Short	Moderate
Palate breadth (absolute): anterior	Narrow	Narrow	Modera te	Broad
Palate breadth (absolute): mid-palatal	Narrow	Narrow	Broad	Broad/v. broad
Palate breadth (absolute): posterior	Narrow	Moderate	Broad	Narrow/moderate
Palate breadth (relative): posterior vs. middle	Broad	Narrow	Narrow	Broad
Palatal height	Low	Low	Moderate/high	Moderate/high

age of the terrain containing the hominin fossils could not be determined so far (Dirks et al., 2015).

The interpretation of the Dinaledi specimens led Lee Berger and collaborators (2015) to propose a new species, *Homo naledi*, whose holotype is the exemplar DHI, consisting of a partial calvaria, partial maxilla, and nearly complete mandible (Figure 6.13). After comparison of this exemplar with the rest of the sample, the authors concluded that it is a male. The paratypes are the incomplete crania and mandible DH2, 3, 4, and 5, the mandibular fragment U.W. 101–377, the hand H1, jointed to various exemplars of the upper limb, and other materials that encompass almost the entire skeleton. The justification of a new species' proposal is based on differential cranial, mandibular, and dental traits, according to Berger and collaborators (2015). The cranial capacity is very small, between 560 and 465 cc, within the australopith range; with respect to body mass, *H. naledi* is among the small-bodied modern human populations, similar in stature to the largest australopiths. However, according to the authors, the lower limbs and, particularly, the hand show characteristics that are not present in any other hominin.

H1, a complete right hand, shows "a long, robust thumb and derived wrist morphology that is shared

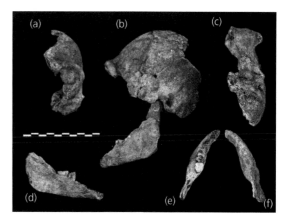

Figure 6.13 Holotype specimen of *Homo naledi*, Dinaledi Hominin 1 (DH1). U.W. 101–1473 cranium in (A) posterior and (B) frontal views (frontal view minus the frontal fragment to show calvaria interior). U.W. 101–1277 maxilla in (C) medial, (D) frontal, (E) superior, and (F) occlusal views. (G) U.W. 101–1473 cranium in anatomical alignment with occluded U.W. 101–1277 maxilla and U.W. 101–1261 mandible in left lateral view. U.W. 101–1277 mandible in (H) occlusal, (I) basal, (J) right lateral, and (K) anterior views. (Scale bar: 10 cm.) Berger, L. R., Hawks, J. de Ruiter, D. J., Churchill, S. E., Schmid, P., Delezene, L. K.,... Zipfel, B. (2015). Homo naledi, a new species of the genus Homo from the Dinaledi Chamber, South Africa. *eLife*, 4, http:// elifesciences.org/content/4/e09560v1.

with Neanderthals and modern humans, and considered adaptive for intensified manual manipulation. However, the finger bones are longer and more curved than in most australopiths, indicating frequent use of the hand during life for strong grasping during locomotor climbing and suspension" (Kivell et al., 2015, p. 1). This mix of primitive traits—the curved phalanges—shared with the australopiths and the first *Homo*, contrasts markedly with the derived traits of wrist and palm, which Kivell and collaborators (2015) relate to the Upper Pleistocene hominins. These latter features indicate that "much of the hand anatomy in *H. naledi* may be the result of selection for precision handling and better distribution of compressive loads during forceful manipulative behaviours such as tool making and tool use (although tools have not been recovered in the Dinaledi Chamber itself" (Kivell et al., 2015, p. 7). Other taphonomic evidence, which we will examine next, also points to very advanced behavior in this species, according to the authors who analyzed the *H. naledi* exemplars.

6.4.1 Taphonomy of the *Homo naledi* remains

Paul Dirks et al. (2015) carried out the site stratigraphy and its taphonomic interpretation; we will follow literally their descriptions. The Dinaledi Chamber is exclusively filled by flowstone and fine-grained sediment involving two depositional facies distributed across three stratigraphic units: units 1, 2, and 3. Unit 3 is the youngest stratigraphic unit in the Dinaledi Chamber. It is composed of largely unconsolidated sediment of facies 2 dominated by reworked orange mud clasts embedded in a brown muddy matrix.

Most hominin bones have been collected from Unit 3, but remains also occur in Unit 2. In Unit 3 the hominin bones were distributed "across the surface in almost every area of the chamber, including narrow side passages and offshoots, with the highest concentration of bone material encountered near the southwest end of the chamber, about 10–12 m downslope from the entry point, where the floor levels out (…). Thus, the fossils of *H. naledi* were probably deposited over an extended period of time during deposition and reworking of Units 2 and 3, and before deposition of Flowstone 2 (…). The distribution of Unit 2 below the current entry shaft into the Dinaledi Chamber, and the orientation of Flowstones 1a–e that cover Unit 2 and slope into the chamber, together strongly suggest that the current entry shaft has always been the main entry point for sediment into the chamber. This also indicates that the fossils of *H. naledi* entered via this

route" (Dirks et al, 2015, p. 26). The authors maintain that "there is no indication of a direct vertical passageway to surface into the Dinaledi Chamber." However, they also recognize that the "flowstone formation continues today (Flowstone 3), changing the morphology of cave passages. This makes it possible that a more direct access-way or easier passage may have existed when hominins entered."

In addition to the morphology of the limbs, taphonomic interpretations, in particular, allow the authors interpreting the meaning of *Homo naledi* to propose hypotheses that attribute a very advanced behavior to the taxon. According to Dirks et al. (2015, p. 28), "There are no signals of occupation debris or evidence of occupation within the Dinaledi Chamber, or anywhere else in the Rising Star cave. Based on our current assessment, occupation would have required accessing the chamber in the dark through an entrance similar to the current one, through an area inaccessible to other medium- to large-sized mammals. Thus, if hominins were traveling to the chamber, it is assumed that they would almost certainly have required artificial light."

If occupation of the cave is discarded, the main question is how so many hominin exemplars reached such an inaccessible place, a question the authors relate to the taphonomic interpretations of Sima de los Huesos in Atapuerca (Spain) (Berger et al., 2015; Dirks et al., 2015). Paul Dirks et al. (2015) examined the following hypotheses:

- Water transport.
- Predator accumulation.
- Mass fatality or death trap.
- Deliberate disposal.

The authors reject the first two hypotheses, arguing that "the Dinaledi Chamber has been a closed depositional system for a long time, and did not allow the sudden ingress of water and sediment; only fine-grained muddy sediment accumulated, which accessed the chamber through narrow cracks that filtered out coarser material. Thus, the accumulation of the hominin remains in the chamber does not fit the pattern of a flood or fluvial event" (Dirks et al., 2015, p. 28). "We have, thus far, found no trace of carnivore damage on the Dinaledi remains. Nor have we found any trace of carnivore remains or the remains of other likely prey animals. Thus, the predator would have had to select a single prey species—*H. naledi*—carrying into the chamber all age and size categories (Berger et al., 2015) without leaving a trace of its own presence. We consider this very unlikely." In summary, "Considering the geological and taphonomic context of the Dinaledi

Chamber, the occupation, predator accumulation and water transport hypotheses cannot adequately explain the fossil assemblage" (Dirks et al., 2015, p. 30).

The third hypothesis (mass fatality or death trap) has pros and cons. According to the authors, this hypothesis cannot be interpreted as a result of a single calamity, because "the sheer number of remains encountered in the Dinaledi Chamber, is hard to explain as a single event hypothesis." Similarly, "the large number of immature individuals (8 out of 13) does allow us to reject hypotheses that would strongly over-represent adults, such as repeated cave exploration by socially isolated adult males." Nevertheless, the authors assert: "We found no significant result when comparing the currently available distribution to either catastrophic or attritional mortality profiles, and therefore a mass death scenario involving some sort of calamity or death trap cannot be completely excluded to explain the Dinaledi assemblage (…). Further work in this regard will be required" (Dirks et al., 2015, p. 29).

The deliberate body-disposal hypothesis implies that the bodies found in the cave could have been thrown from the outside, as well as transported to the place in which they were found. Regarding the first possibility, the authors state that "none of the bone elements studied shows evidence of green fracture (…) indicating lack of trauma. Therefore, if bodies were dropped down the entrance, it is unlikely that they would have fallen rapidly, or landed with any force; perhaps because the entry is too irregular and narrow to allow a body to freefall and gain speed, or perhaps because a pile of soft muddy sediment had accumulated below the entry way, breaking the momentum of any falling object" (Dirks et al., 2015). As an alternative, "the hominins could have entered the chamber directly, carrying the bodies or dying there, which would explain, not only the absence of green fractures, but the presence of delicate, articulated remains in the excavation pit, deep in the chamber, well away from the entrance point, on the other side of floor drains" (Dirks et al., 2015).

For the members of the research team who analyzed the exemplars of *H. naledi*, both the hypotheses of the death-trap scenario and the deliberate disposal are plausible. However, they opt for the latter hypothesis: "our preferred explanation for the accumulation of *H. naledi* fossils in the Dinaledi Chamber is deliberate body disposal, in which bodies of the individuals found in the cave would either have entered the chamber, or were dropped through an entrance similar to, if not the same as, the one presently used to enter the Dinaledi Chamber" (Dirks et al., 2015). However, both

types of activity imply different hypotheses. Transportation of the bodies to the bottom of the cave could be considered an instance of deliberate burial, somewhat comparable to those that will be addressed in Chapter 11. This would imply an even higher cognitive level than the hypothesis of deliberate body disposal in Sima de los Huesos. The adaptive significance of *H. naledi*, even if we accept the most optimistic hypotheses about its advanced behavior, is difficult to clarify in the absence of a dating, even approximate, of the remains.

6.5 Homo georgicus

In the Upper Pliocene, at a date we will try to determine later, hominins left Africa and carried out the first colonization of Asia and later of Europe.

The natural exit of Africa is the Levantine corridor—the Near and Middle East—a path widely thought to be the one used by hominins for their various dispersals from the African continent. Located between the Black Sea and the Caspian Sea, Georgia is part of the transit zone between Africa, Asia, and Europe. A Georgia site, Dmanisi, has provided the best evidence to determine the exit of the first hominins from their continent of origin (Figure 6.14). The early hominin dispersals cannot be reduced to a single episode; we will seek to determine a bigger picture in the following chapters. Now, let us examine here the evidence provided by Dmanisi.

The site of Dmanisi has been described in Chapter 3, where we stated that the first hominin fossil found at the site was the mandible D211, with a nearly complete body, and a broken ramus (Gabunia & Vekua, 1995).

The morphological study of D211 by Léo Gabunia and Abesalom Vekua (1995) stated that this was a very old specimen of *Homo*, without attribution to any species, which had small teeth in a large mandibular body; its gender was uncertain. According to the authors of the analysis, the most distinctive feature of the mandible was the decrease of the molar area from M1 to M2, and from M2 to M3. In the words of Gabunia and Vekua (1995), D211 more closely resembled African than Asian *Homo erectus*. The comparative study by Antonio Rosas and José M. Bermúdez de Castro (1998) also showed the similarity of D211 with the latter.

The affinities attributed to the mandible of Dmanisi did not end there. The analysis by Günther Bräuer and Michael Schultz (1996) pointed out that D211 shared common derived traits with advanced *Homo erectus* of less than 1 Ma. Gabunia and Vekua (1995) came to the same conclusion: D211 was attributable to a *Homo*

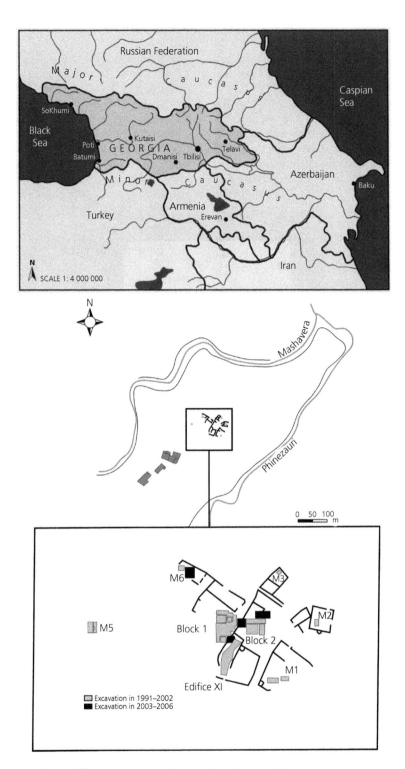

Figure 6.14 Top: Location of Dmanisi (illustration by Gabunia, Vekua, and Lordkipanidze 2000a); bottom: excavation areas of Dmanisi. Bottom: Reprinted in Mgeladze, A., Lordkipanidze, D., Moncel, M.-H., Despriee, J., Chagelishvili, R., Nioradze, M., & Nioradze, G. (2011). Hominin occupations at the Dmanisi site, Georgia, Southern Caucasus: Raw materials and technical behaviours of Europe's first hominins. *Journal of Human Evolution*, 60(5), 571–596, with permission from Elsevier. Top: Reprinted from Gabunia, L., Vekua, A., & Lordkipanidze, D. (2000a). The environmental contexts of early human occupation of Georgia (Transcaucasia). *Journal of Human Evolution*, 38(6), 785–802, with permission from Elsevier.

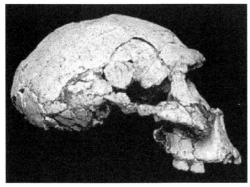
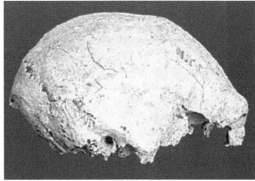

Figure 6.15 Specimens from Dmanisi in Georgia; left: D2282; right: D2280 (Gabunia, Vekua, Lordkipanidze, Swisher III et al. (2000)) (photographs from http://www.athenapub.com/13intro-he.htm; they are not at the same scale).

erectus who possibly prefigured the archaic *H. sapiens*. The authors also reported that it was the oldest evidence of the presence of hominins out of Africa.

David Dean and Eric Delson (1995) applauded the decision to include the specimen in *Homo erectus*, so that the Dmanisi's jaw would allow outlining a simple phylogenetic model. Between *Homo erectus* and "archaic sapiens" would be an ancestor–descendant relationship, in which the latter will replace the former from west to east. The Dmanisi specimen would be one example of this replacement.

The controversy about which species to attribute Dmanisi's mandible took a new direction, however, with the discovery, in the summer of 1999, of a nearly complete cranium (D2282) and a calotte (D2280) (Gabunia, Vekua, Lordkipanidze, Swisher III et al., 2000) in the same site and stratigraphic horizon as D211 (Figure 6.15). The morphology of D2282 and D2280, according to Léo Gabunia et al. (2000b), shows essential similarities to *Homo erectus* specimens from Koobi Fora—*Homo ergaster*—rather than to Asian *H. erectus*.

The Dmanisi hominins all come from Bed B, according to Reid Ferring and collaborators (2011), although Abesalom Vekua et al. (2002) and David Lordkipanidze et al. (2007) placed the finding of their specimens in Bed A. In any case, what we want is to determine their age. The detailed study of associated fauna, as well as the archaeological analysis and the paleomagnetic and geologic interpretation, fixed the date of the Dmanisi specimens to c. 1.7 Ma (Gabunia et al., 2000a). However, a new review of the archaeological and geological evidence pushed the age of these specimens back to 1.85–1.78 Ma (Ferring et al., 2011).

Reid Ferring and collaborators (2011) maintained that the stratigraphic study of the Dmanisi stone tool

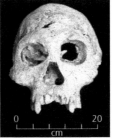
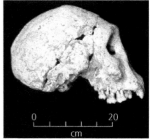

Figure 6.16 Dmanisi specimen D2700; left: frontal; right: lateral views (photographs from Vekua et al., 2002).

sets indicates that the site was repeatedly occupied during the last stretch of the Olduvai subchron, i.e., between the range of 1.85 and 1.78 Ma. According to these authors, this antiquity implies that the Georgia specimens precede the appearance of the African *Homo erectus / Homo ergaster*. But which species could it be?

6.5.1 A new taxon for Dmanisi

The volumes of the crania D2280 and D2282, 775 and 650 cc, are small—smaller than expected for *Homo ergaster*. The cranial capacity of D2280 and D2282 is very close to that of the Turkana and Olduvai *Homo habilis*. Three years later, in 2002, the existence of a cranium of the same age from Dmanisi was reported, D2700 (Vekua et al., 2002), which had an even smaller volume, 600 cc (Figure 6.19 and Box 6.15).

For Abesalom Vekua and collaborators (2002), the new findings meant that it was necessary to explain the different sizes of fossils D2280 and D2282 vs D27000 (see Box 6.16). The authors rejected the idea

---

Box 6.15 The cranial volume of D2700

D2700 appeared side by side with the associated mandible D2735. The set was attributed to a juvenile, perhaps a female, but the broad crowns and massive roots of the upper canines prevented an accurate sex determination (Vekua et al., 2002). To estimate the significance of the cranial volume of D2700, it should be noted that this is an immature specimen. In maturity, it could have had 618 cc or even 645 cc, depending on the degree of maturity attributed to the specimen (Sang-Hee, 2005).

---

of two different taxa, attributing the differences in size to sexual dimorphism. The statistical study made by Lee Sang-Hee (2005) of the variation of the Dmanisi cranial sample compared to modern humans, chimpanzees, and gorillas obtained a result that supports the hypothesis of a single species. Sang-Hee retained the sex ascription made by Vekua et al. (2002): D2700 and D2282, females; D2280, male. However, a similar comparison made by Matthew Skinner, Adam Gordon, and Nicole Collard (2006) on the mandibular body, the symphysis, and the first molar in the sample of Dmanisi provided an opposite result, indicating that the sex assignments were not justified and the idea of a single species should be reviewed.

If we accept that only one species appears in Dmanisi, specifying which one is no easy task.

D2700 shares affinities with specimens of *Homo habilis*, such as Koobi Fora KNM-ER 1813. However, Vekua et al. (2002) argued that, although very small, this specimen from Dmanisi has traits of *H. ergaster*:

- The keel along the sagittal midline.
- The low overall appearance of parietal areas and the shape of the temporal wall.
- The transverse expansion of the base relative to the low dome.

The systematic diagnosis of Abesalom Vekua et al. (2002) acknowledged the difficulties of classification, although suggesting that the Dmanisi hominins would be among the most primitive individuals attributed to *Homo erectus* or to any other species, but without question they belong to *Homo*. They also argued that the Dmanisi population was closely related to *Homo habilis* sensu stricto from Olduvai, Koobi Fora, and possibly Hadar (Vekua et al., 2002).

Over seven years, Dmanisi specimens passed from being placed in an undetermined species of *Homo*

(Gabunia & Vekua, 1995), later in *Homo ergaster* (Gabunia et al., 2000a), and finally with some uncertainty in *Homo habilis* sensu stricto (i.e., *Homo habilis* except for *H. rudolfensis* specimens). Such indetermination expresses well the difficulties in interpreting which ones were the hominins that left Africa for the first time.

A new finding led Dmanisi scientists to make a taxonomic step forward. It was the discovery in September 2000 of the mandible D2600, of rather robust appearance. After comparison with other *Homo* taxa of similar age, Léo Gabunia and collaborators (2002) proposed the new species *Homo georgicus*, with D2600 as its type specimen. The rest of the Dmanisi sample would form the taxon holotype.[1]

Two more specimens from Dmanisi, the cranium D3444 and associated mandible D3900, were discovered in the campaigns of 2002–2004 (Lordkipanidze et al., 2005, 2006). They resemble prior specimens of Dmanisi, with the particularity that D3900 is edentulous (Figure 6.18 and Box 6.17).

D3900's total absence of teeth represents an unusual trait from which can be inferred that such an individual must have been kept alive with the help of its own relatives. This hypothesis implies the attribution to very ancient *Homo* of a social structure, as well as very advanced adaptive strategies. David Lordkipanidze et al. (2005) argued that the consumption of soft tissues, such as bone marrow or brain, have increased the chances of survival for individuals with masticatory failure. This would have happened with the help of others and gone beyond what non-human primates could provide.

New postcranial exemplars from Dmanisi belonging to an adolescent—related to the cranium 2700/2735 for having been found in a nearby stratum—and three adults, one large and two smaller, were described in 2007 (Lordkipanidze et al., 2007). Although the authors did not assign the remains to a particular species, they indicated that this Dmanisi set lacks derived features characteristic of *Homo erectus*.

In the opinion of the authors of the discovery, the new materials expanded the available information on the evolution of body size, because the age of the Dmanisi specimens—1.77 Ma if we accept the most modern range or 1.85–1.78 Ma according to Ferring et al. (2011)—is midway between that of *Australopithecus afarensis* (Lucy), of around 3 Ma, and Nariokotome specimen KNM-W15000, *Homo erectus* of 1.45 Ma.

---

1 In the paper of 2002, the surname of Léo Gabunia was written in French style, Gabounia. To avoid confusion, we have preserved the common form—Gabunia—used in all other papers.

Box 6.16 The meaning of the cranial variability in Dmanisi

The variability found in the size of Dmanisi hominin crania has been interpreted in two ways: on the one hand, attributing the differences to sexual dimorphism (Wolpoff, 2002); on the other, suggesting that actually there are two different species (Schwartz, 2000). It is a very common problem, which we have encountered many times throughout this book. Lee Sang-Hee conducted a statistical analysis on the Dmanisi cranial variation, comparing these specimens with samples from modern humans, chimpanzees, and gorillas (Sang-Hee, 2005). Lee's results support the hypothesis of a single species.

By contrast, Jeffrey Schwartz (2000) pointed out that Dmanisi crania permit arguing for the presence around 1.7 Ma of a variability of hominin species even greater than that posed by the contrast between Asian and African *Homo erectus*. In his response, Léo Gabunia et al. (2000b) agreed about the great evolutionary diversity that Dmanisi reflects. The idea of Schwartz (2000) is consistent with the mechanism he offers as responsible for the speciation: a change in

one or a few developmental genes (Schwartz, 1999). Opposed to giving such importance to Dmanisi crania was Lorenzo Rook (2000), for whom these specimens are no more than another indication of the variability within the taxon *Homo erectus* sensu stricto present both in Africa and Asia.

Rook took the opportunity to complain about the methodology applied to taxonomic proposals of human fossils. The scarcity of the fossil record makes, according to Rook (2000), intraspecific differences—the typical variability of a species—unknown. Consequently, the presence of distinctive characters is interpreted as intertaxa differences, allowing definition of new species. This is certainly not just an inherent problem of the Dmanisi specimens, but a general issue in paleontology, although much more pronounced in human paleontology. Rook called for the formulation of new taxonomic criteria, something that many authors, including Schwartz, agree with, but up to now, as we shall see in Chapter 11, these new and necessary tools have not been formulated.

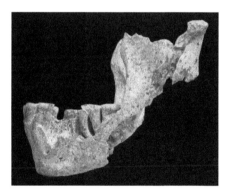

Figure 6.17 Dmanisi specimen D2600, lateral view (photograph from Vekua et al., 2002).

Box 6.17 Without teeth

"Edentulous" means without teeth. As David Lordkipanidze et al. (2005) have stated, the teeth of the maxillary D3900 were lost before its death, as evidenced by the complete reabsorption of the dental alveoli and an extensive remodeling of the alveolar process. In the mandible, all connectors, except for those of the canines, have been reabsorbed and only the left canine persisted until death. The body of the mandible is reabsorbed at the level of the foramen of the chin, and the projection of the symphysis region is likely the result of the remodeling after the loss of the incisors. Applying comparative clinical terms, the advanced atrophy of alveolar bone indicates the loss of substantial teeth for several years before death, as a result of aging, disease, or both (see Figure 6.18).

The postcranial specimens of adults such as D4167 and D3901—femur and right tibia associated with the humerus D4507—make it possible to calculate the height, body mass, and proportions of the Dmanisi hominins. The results obtained by Lordkipanidze et al. (2007) for the various available specimens indicated an average height of 145–166 cm and a weight of 40–50 kg. These data significantly depart from the Nariokotome specimen that, as discussed in Chapter 7, with its 150–169 cm height and 45–71 kg weight, is in the range of modern humans or even more than that of our species. Lordkipanidze et al. (2007) provided two

interpretations—not mutually exclusive—about the reason for the differences. Perhaps the body size of the Dmanisi specimen was a primitive character shared with *H. habilis* that became derivative in KNM-WT 15000. Or maybe the different stature that separates the West Turkana specimen from that of Dmanisi reflects the adaptation to different paleoecological contexts.

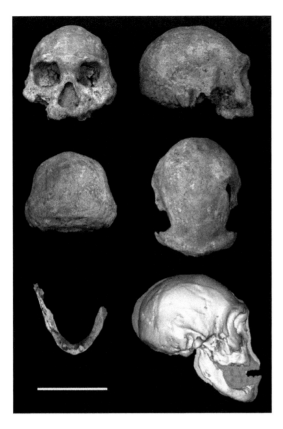

Figure 6.18 Dmanini specimens; cranium D3444 and associated mandible D3900, adult specimen (Lordkipanidze et al., 2007). Scale bar: 19 cm. Reprinted by permission from Macmillan Publishers Ltd: Lordkipanidze, D., Vekua, A., Ferring, R., Rightmire, G. P., Agusti, J., Kiladze, G.,... Zollikofer, C. P. E. (2005). The earliest toothless hominin skull. *Nature*, 434, 717–718.

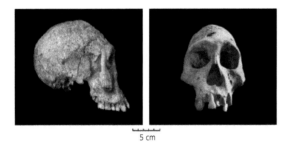

5 cm

Figure 6.19 Cranium D2700, a sub-adult specimen, in lateral and frontal views. Reprinted by permission from Macmillan Publishers Ltd: Lordkipanidze, D., Jashashvili, T., Vekua, A., de Leon, M. S. P., Zollikofer, C. P. E., Rightmire, G. P.,... Rook, L. (2007). Postcranial evidence from early Homo from Dmanisi, Georgia. *Nature*, 449(7160), 305–310.

6.6 The transition to *Homo*

The study of postcranial and craniodental material of the Bouri australopiths, *A. garhi*, led Berhane Asfaw et al. (1999) to argue that the phylogenetic relationship between australopiths and *Homo*, based for decades on the advanced morphology of *A. africanus*, should be reconsidered under the new taxon. *A. garhi* would be equivalent in the Rift Valley to the South African gracile australopith, with the particularity that the cranial anatomy of *A. garhi* brings him even closer to *Homo*. For reasons having to do with the logic of the phylogenetic process, there are different ways of relating *A. garhi* to *Paranthropus* and *Homo*. The first is to claim that *A. garhi* is part of the sister group of the set formed by *Paranthropus* and *Homo*. This is a hypothesis that Asfaw and collaborators (1999) represented by a polychotomous tree—with various branches from the node instead of the two corresponding to the sister groups of the classic cladograms—shown in Figure 6.20.

To choose between the various alternatives indicated by the detailed cladograms that solve the polychotomy is not feasible due to the insufficient data of the fossil record. Asfaw et al. (1999) argued that it is not possible to empirically support any of the alternative hypotheses. However, it is possible to translate these into phylogenetic trees (B, C, and D, shown in Figure 6.19), which correspond to the following evolutionary scenarios:

- B: *A. garhi* is an ancestor placed in the *Paranthropus* clade.
- C: *A. garhi* is an ancestor placed in *Homo*.
- D: *A. garhi* is a species of *Australopithecus*, a contemporary alternative to the first *Paranthropus* and *Homo*.

The suggested use of tools by *A. garhi* would be an argument in favor of B. But we know that this kind of circumstantial evidence must appear repeatedly in different sites to be able to make the association between fossil specimens and tools, which is not the case.

The age of *A. garhi*, around 2.5 Ma, coincides with that attributed to the cladogenesis of *Paranthropus*/*Homo*, a circumstance which, together with the morphology of *A. garhi*—closer to *Homo* but with a considerable megadontia—supports option D. However, it is possible to hypothesize that the megadontia is a plesiomorphy maintained in the most primitive forms of *Homo*. Therefore, the available evidence does not allow the exclusion of *A. garhi* from the *Homo* lineage. To consider the examplars of Bouri as belonging to *Paranthropus* seems the weakest of all the alternatives.

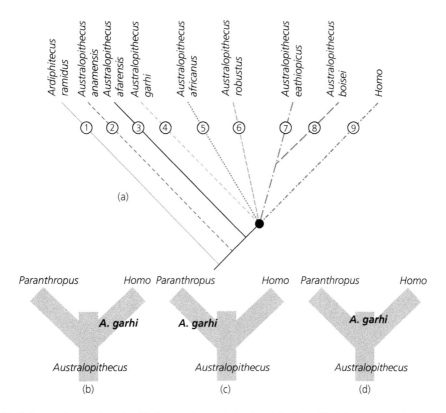

Figure 6.20 (A) Cladogram of ancient hominins; (B) phylogenetic tree placing *A. garhi* in *Homo*; (C) phylogenetic tree placing *A. garhi* in *Paranthropus*; (D) phylogenetic tree maintaining *A. garhi* within *Australopithecus* (the diagrams follow the proposal of Asfaw et al., 1999, with some modifications).

6.6.1 A South African transition?

The difference in the ages of *A. africanus* and *A. sediba* permits the interpretation that the latter taxon is a chronospecies in which the features of *A. africanus* are emphasized, bringing *A. sediba* closer to those of *Homo*. If *A. sediba* is maintained as an australopithecine, and *Homo habilis* as a legitimate member of *Homo*, those features should be considered as homoplasies in *A. sediba* and *H. habilis*, because *A. sediba* would have appeared *after* the emergence of the genus *Homo*.

As we have seen, both Lee Berger et al. (2010) and Robyn Pickering et al. (2011) rejected the idea that ancient specimens attributed to *Homo* (classified either as *H. habilis* or *H. rudolfensis*) could belong to this taxon. Either the specimens do not have apomorphies characteristic of *Homo*, or they are not as old as has been proposed. Bernard Wood has often supported, as we have repeatedly stated, that *H. habilis* and *H. rudolfensis* should be removed from *Homo*.

If *A. sediba* has *Homo* traits as apomorphies and not as homoplasies, and if *H. erectus* is the first authentic

Homo, then *A. sediba* has the same status as *H. habilis*— a taxon that includes *H. rudolfensis*—that is to say, they are all populations evolving toward *Homo*. This is the phylogenetic interpretation maintained by Berger et al. (2010) and Pickering et al. (2011) (Figure 6.21).

The hypothesis of a set of populations evolving toward *Homo* and ending in *H. erectus* is not to be readily dismissed; but it is not necessarily preferable, for two reasons. First, in order to reject the existence of a *Homo* species prior to *Homo erectus*, it is also necessary

---

Box 6.18 Grouping of *H. habilis* with *H. rudolfensis*

Asfaw et al. (1999) warned that doubts about whether to accept or reject a separation between *H. habilis* and *H. rudolfensis* affect the outcome of the cladograms. So, as a simpler solution for graphic representation, they reduced these two forms of ancient *Homo* to a single taxon.

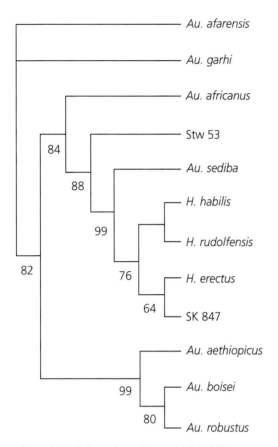

Figure 6.21 Cladogram by Lee Berger et al. (2010) (the genus *Australopithecus* appears abbreviated as *Au*). The clades figures indicate the number of *bootstraps* supporting that solution out of 10,000 alternatives. From Berger, L. R., de Ruiter, D. J., Churchill, S. E., Schmid, P., Carlson, K. J., Dirks, P. H. G. M., & Kibii, J. M. (2010). Australopithecus sediba: A New Species of Homo-Like Australopith from South Africa. *Science*, 328(5975), 195–204. Reprinted with permission from AAAS.

to object to the age and/or morphology of *H. georgicus*. Second, with respect to *A. sediba*, there are geographical considerations. It is highly unlikely that the ancestors of the oldest African, *H. erectus*, came from South Africa. At the end of this chapter we will return to the ancient hominin dispersals within the African continent.

6.7 Monophyly of the first *Homo*

In an analysis to determine the meaning of the genus *Homo*, Bernard Wood and Mark Collard (1999b) examined six cladograms proposed by Chamberlain and Wood (1987) (A, in Figure 6.21), Chamberlain (1987) (B), Wood (1991) (C), Wood (1992c) (D), Lieberman et al. (1996) (E), and Strait et al. (1997) (F). Considering

Homo habilis and *H. rudolfensis* as two species belonging to this genus, three cladograms yield a monophyletic result for *Homo* and three others result in a paraphyletic one, which, according to the authors, are outcomes that do not justify the unequivocal ascription of *H. habilis* and *H. rudolfensis* to *Homo*. However, the main reason for Wood and Collard (1999b) to suggest the removal of the early *Homo* taxa from this genus is the lack of evolutionary coherence, considered by the authors to be necessary. Considering body size and shape, skeleton traits related to locomotion, relative brain size, the rate of ontological development patterns, and the relative size of the masticatory apparatus—all conditions proposed by Wood and Collard (1999b)—we find that in most of these features, if not in all, *H. habilis* and *H. rudolfensis* are closer to australopithecines than to *Homo sapiens* (Figure 6.22). Wood (2011) has tempered the rejection of *Homo habilis* as a member of *Homo*, admitting as a hypothetical alternative, its assignation to the genus.

Let us examine in more detail one of these traits, locomotion, to assess the magnitude of the problem.

6.7.1 The locomotion of the first *Homo*

The phylogeny of bipedalism is understood as a process in which elongated anterior limbs, a big toe with a certain degree of thumb-like character (opposable to the other fingers), and a high mobility of joints, are some of the primitive traits. Taken together, these features, as we have seen, are largely characteristic of *Ardipithecus*, remaining partially as plesiomorphies in *Australopithecus*, which indicate a bipedalism compatible with tree-climbing. Such plesiomorphies disappear in the bipedalism of *Homo erectus*, a locomotion similar to that of modern humans.

What happened with the *Homo* species living before *H. erectus*? Did they practice a similar bipedalism or did they retain tree-climbing skills, similar to *Australopithecus*?

Let's review the case of *Homo habilis*. When proposing the species (Leakey et al., 1964), it was described on the basis of morphological features of the few available postcranial remains. It is not surprising, therefore, that the argument against criticisms focused on dental and cranial features (see Tobias & Clarke, 1996, in their discussion with Robinson, for example). But the significance of extending the genus *Homo* to the Pliocene included the distinct idea that, by that time, new beings, in adaptive terms, would have emerged. They colonized the savanna using tools that they themselves made and, in the process of adapting to the open lands,

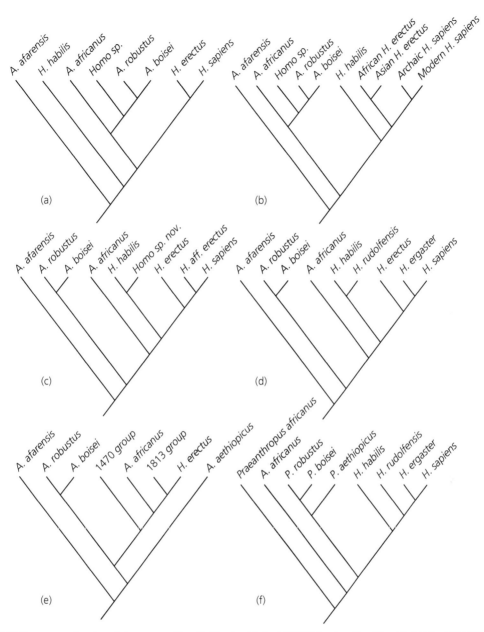

Figure 6.22 The most parsimonious cladograms considered by Wood and Collard (1999b). Reprinted from Wood, B. & Collard, M. (1999b). The Human Genus. *Science*, 284, 65–71. With permission from AAAS.

they experienced musculoskeletal transformations whose evidence was largely hypothetical due to the lack of postcranial remains.

The problem we face is that of specimens such as OH 62 and KNM-ER 3735, discovered after the proposal of *Homo habilis* but included in the taxon, which show proportions that do not fit with the image of the genus

Homo as it is indicated by *H. erectus*. Thus, the comparison of OH 62 with the morphology of current great apes led to the argument that it is a hominin with "ape-like style" (Hartwig-Scherer & Martin, 1991; Aiello, 1992).

KNM-ER 3735 added evidence in the same direction. Although its fragmentary condition does not allow the taking of general measurements, the available parts of

KNM-ER 3735 were compared by Richard Leakey et al. (1989) with the *Australopithecus* A.L. 288-1, using as references an exemplar of *Pan troglodytes* and one of *Homo sapiens*. The result showed greater length for the anterior limbs of KNM-ER 3735 than of A.L. 288-1, which indicates a greater degree of arboreal behavior in *Homo habilis* than in *A. afarensis* (Leakey et al., 1989). There is no easy way to explain an evolutionary sequence of this kind in the transition toward *H. erectus* bipedalism.

But the study of Richard Leakey et al. (1989) had at least two drawbacks. First is the sex assigned to the specimens: A.L. 288-1 is a female and KNM-ER 3735 is presumably a male (Figure 6.23). As there are no skeletons of *H. habilis* and *A. afarensis* of the same sex, this problem cannot be resolved. The other difficulty of the work by Leakey et al. (1989) lays in having carried out

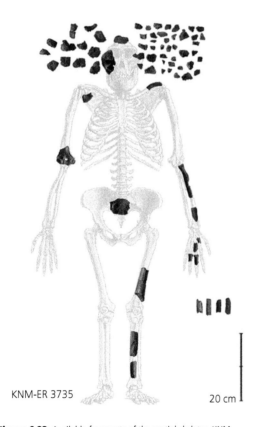

KNM-ER 3735

20 cm

Figure 6.23 Available fragments of the partial skeleton KNM-ER 3735, *H. habilis* (in black), laid over a skeleton drawing of *Australopithecus* (illustration by Haeusler & McHenry, 2007). Reprinted from Haeusler, M. & McHenry, H. M. (2007). Evolutionary reversals of limb proportions in early hominids? Evidence from KNM-ER 3735 (Homo habilis). *J Hum Evol*, 53, 383–405, with permission from Elsevier.

comparisons with only one modern chimp and one modern human.

Different analyses (Hartwig-Scherer & Martin, 1991; Haeusler & McHenry, 2007; Ruff, 2009; Pontzer et al., 2010) have followed the methodological patterns of the research by Leakey et al. (1989). Thus, Sigrid Hartwig-Scherer and Robert Martin (1991) concluded, agreeing with Leakey and collaborators (1989) on the existing relationships among the various measurements corresponding to the anterior limbs with respect to the posterior, that OH 62 is closer to current apes than to modern humans, while A.L. 288–1 shows more human-like proportions.

Martine Haeusler and Henry McHenry (2007) proposed amending the deficiencies of the original study by Richard Leakey and collaborators (1989) by repeating his analyses with a larger sample of chimpanzees and modern humans. Opposing the conclusions by Leakey et al. (1989), the results indicated that the two fossil specimens do not differ statistically. The relative size of the distal humerus, the radius head, and their joints fall within the range of variation of modern humans, both in the case of KNM-ER 3735 and A.L. 288-1; and the hand phalanges of the latter are similar in having a gracile base and a robust middle joint. However, Haeusler and McHenry (2007) did find differences in the robustness of the shoulder, particularly the scapular spine, and the relative size of the radius neck. In conclusion, they argued that A.L. 288-1 would have had an arm more similar to humans, while KNM-ER 3735 would have a more powerful shoulder—which conforms to an arboreal behavior. In other words, the main conclusions of Leakey et al. (1989) were supported.

Christopher Ruff's work (2009) analyzed the traits of OH 62, two African *Homo erectus* (KNM-WT 15000 and KNM-ER 1808), chimpanzees, and modern humans. The examination was focused on the strength of two points of the femur and two of the humerus, deduced from computational analysis of the shape and thickness of the bone sample in the transversal section. For each and every combination among the studied sections, the proportion between the strength of the OH 62 femur and that of its humerus falls into the chimpanzee range, while that same proportion for KNM-WT 15000 and KNM-ER 1808 is in the human range (Ruff, 2009) (Box 6.19).

The most logical conclusion taken from these studies suggests repeatedly that *H. habilis* should be described, according to the postcranial traits of KNM-ER 3735 and OH 62, as a partially arboreal hominin.

To define *H. habilis* as partially arboreal in the function of KNM-ER 3735 and OH 62 is a diagnosis in

---

Box 6.19 The femur of OH 62

Berhane Asfaw and collaborators (1999) have criticized the reaching of conclusions from the comparative morphology studies between OH 62 and A.L. 288-1 on the relationship between the length of the humerus and that of the femur. For Asfaw et al. (1999), the OH 62 femur size cannot be estimated accurately and, therefore, is not useful for comparison.

---

contradiction with the traits indicated in the proposal of the taxon. OH 8 was adopted as paratype of *H. habilis* by Leakey et al. (1964) and, in fact, the new species was described then as having a locomotion similar to that of a current human; thus, *H. habilis* was not taken as a partially arboreal creature. The specimen OH 6 was interpreted, therefore, as evidence that the tibia–fibula joint was similar in *Homo habilis* to that of modern humans (Davis, 1964; Lovejoy, 1975).

The character of the limbs of *Homo habilis* then stumble upon two opposite interpretations, unless the traits of the first postcranial remains discovered in Olduvai are also reinterpreted as plesiomorphies.

As we have seen in Chapter 2, Campbell Rolian, Daniel Lieberman, and Benedikt Hallgrimsson (2010) provided sound evidence that, in hominins, hands and feet changed during the process of coevolution. We would expect that the degree of bipedalism of *Homo habilis* will be revealed in the anterior as well as in the posterior limbs.

Randall Susman and Jack Stern (1979) and Françoise Jouffroy (1991) have argued that the OH 7 hand shows evidence of climbing (curved phalanges). Oxnard and Lisowski (1980) have also seen in OH 8 traits associated with a climbing function. Other authors, on the contrary, continued to classify OH 8 as a modern foot (Deloison, 1991; Susman & Stern, 1991; White & Suwa, 1987) (Box 6.20).

The problem of the plesiomorphy or apomorphy of the locomotion traits of *Homo habilis* could be extended to other old taxa of the genus. About *H. rudolfensis* nothing can be said, as there are not reliable postcranial elements. But, *Homo georgicus* does have these elements. What do they say to us? The examination by Herman Pontzer and collaborators (2010) of Dmanisi postcranial remains checked the hypothesis that the carnivorous diet led to increased bipedalism, and to a more efficient endurance running. The results indicated that the posterior limbs of the

Box 6.20 Clarifying the meaning of OH 8

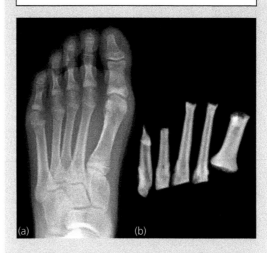

Radiography of the foot of a current 12-year-old female with the correspondent bones of OH 8 (Susman, 2008).

The study of Randall Susman (2008) aimed to compare the development of OH 7 and OH 8. It is common to consider OH 7 as an immature exemplar and OH 8 as an adult individual. After radiography of OH 8 and the foot of a 12-year-old current female to compare the similarities in development, the conclusion of Susman (2008) was not only that OH 8 is immature, but that the OH 7 (hand) and OH 8 (foot) belong to the same individual.

Dmanisi specimens were similar—with respect to the longitudinal plantar arch, and the larger size and ankle morphology—to those of modern humans from a functional point of view. But the retention of primitive traits—tibia torsion and metatarsal shape—reveals that at the time of the first exit from Africa (see Chapter 8), the evolution of the locomotor apparatus characteristic of *Homo erectus* was incomplete (Pontzer et al., 2010).

6.8 The geographical issue—dispersal of ancient hominins in Africa

The Pliocene represents for hominin evolution the morphological trends of change and functional patterns that we have been discussing. During this period a phenomenon was also present whose interpretation cannot be further delayed: the hominin dispersal over territory that covers almost all of Africa. Table 6.6

Table 6.6 Pliocene hominin taxa indicating the oldest specimen and the locality for each species. No distinction is made between *H. habilis* and *H. rudolfensis*. *H. gautengensis* and *H. naledi* are not considered

Taxon	Oldest specimen	Age (Ma)	Locality
A. anamensis	ARA-VP-14/1 (White, WoldeGabriel, Asfaw et al., 2006)	4.2–4.1	Kenya (Allia Bay and Kanapoi)
A. afarensis	Laetoli (Deino, 2011)	3.85–3.63	Ethiopia (Hadar, Woranso-Mille, Dikika, Maka)/Kenya (West Turkana, South Turkwell)/Tanzania (Laetoli)
A. deriyemeda	BRT-VP-3/1 (Haile-Selassie et al., 2015)	3.5/3.3	Ethiopia (Woranso-Mille)
A. bahrelghazali	Bahr El Ghazal (Brunet et al., 1996)	c. 3.5	Chad (Bahr El Ghazal)
A. platyops	KNM-WT 4000 (M.G. Leakey et al., 2001)	c. 3.5	Kenya (Lomekwi)
A. garhi	BOU-VP-12/130 (Asfaw et al., 1999)	2.5	Ethiopia (Bouri)
A. africanus	StW 573 (Clarke & Tobias, 1995)	2.2	South Africa (Taung, Sterkfontein, Makapansgat)
A. sediba	MH 11/12 (Berger et al., 2010)	1.98	South Africa (Malapa)
P. aethiopicus	KNM-WT 17000 (Walker et al., 1986)	c. 2.5	Kenya (West Turkana)/Tanzania (Laetoli)/Ethiopia? (Omo)
P. robustus	TM-1517 (Broom, 1938)	1.9	South Africa (Kromdraai, Swartkrans)
P. boisei	L7a-125 (Arambourg & Copppens, 1968) KNM-ER 3729 (Leakey & Walker, 1985)	2.2–1.9	Ethiopia (Konso, Omo)/Kenya (Koobi Fora, West Turkana, Chesowanja, Chemeron)/Tanzania (Olduvai, Peninj)/Malawi (Malema)
H. habilis	UR 501 (Schrenk et al., 1993)	2.5–2.33	Ethiopia (Omo, Hadar)/Kenya (Koobi Fora, West Turkana)/Tanzania (Olduvai)/Malawi (Uraha)
H. georgicus	D2600 (Gabunia et al., 2002)	1.85–1.78	Georgia (Dmanisi)

summarizes the different species of Pliocene hominins, the age of the oldest specimen of each of them, and the localities where they were found.

A quick glance at Table 6.6 shows that when interpreting the evolutionary process of Pliocene hominins it is not only necessary to consider the named genera and species, and the age of available specimens, but also their geographic location. The scheme in Figure 6.24 will be used to contextualize the possible hypotheses about the origin of the various populations.

Box 6.21   OH 86

The specimen OH 86, found by Manuel Domínguez-Rodrigo and collaborators (2015) in the new locality of Philip Tobias Korongo (PTK) and discovered in 2012 by The Olduvai Paleoanthropology and Paleoecology Project, increases the difficulties for interpreting the evolution of the postcranial elements of early *Homo*. PTK consists of three archeological levels. Levels 1 and 2 correspond to "Zinj," the level where OH 5 was found in the FLK 22 locality. The Level 3, where OH 86 was found, "underlies the Zinj clay, within the tuffaceous layer known as the 'Chapati Tuff,' and corresponds stratigraphically to the top of the Olduvai Bed I archeological level designated as FLK NN 2" (Domínguez-Rodrigo et al., 2015, p. 3). OH 86 is a manual proximal phalanx, which in the author's opinion "represents the earliest known hominin hand bone (>1.84 Ma) with MHL [modern human-like] appearance" (Domínguez-Rodrigo et al., 2015, p. 2). In the analysis of the fossil, Domínguez-Rodrigo et al. (2015, p. 5)

argue that "OH 86 represents a hominin species different from the taxon represented by OH 7, and whose closest form affinities are to modern *H. sapiens*. The functional morphology of OH 86 would seem to indicate that the paleoecosystem of Bed I (B 2.0–1.8 Ma) at Olduvai was characterized by the sympatry of a minimum of three distinct hominin species, *P. boisei*, *H. habilis* (s.l.) and the OH 86 morph—only the latter of which clearly exhibits phalangeal features indicative of more relaxed postural and locomotive selective pressures on the hand." Obviously, the hypothesis of a third hominin species at Olduvai around 1.8 Ma is hard to defend based on the evidence of only one phalanx. The authors recognize this and conclude that "the confirmation of lack of arboreal features in the hominin species to which the OH 86 phalanx belonged should await further discovery of more remains from other regions of its hand (and other anatomical regions)" (Domínguez-Rodrigo et al., 2015, p. 6).

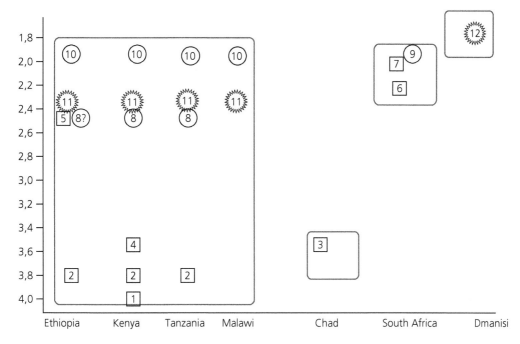

Figure 6.24 Geographic and temporal distribution of Pliocene hominins. Temporal scale in millions of years (attributing the oldest age to the species exemplars). With a square: *Australopithecus*; with a circle: *Paranthropus*; with a star: *Homo*. (1) *A. anamensis*, (2) *A. afarensis*, (3) *A. bahrelghazali*, (4) *A. platyops*, (5) *A. garhi*, (6) *A. africanus*, (7) *A. sediba*, (8) *P. aethiopicus*, (9) *P. robustus*, (10) *P. boisei*, (11) *H. habilis*, (12) *H. georgicus*. *H. gautengensis* and *H. naledi* are not included.

When crossing temporal, geographic, and taxonomic ranges we face the need to interpret some dispersions in order to account for the presence of some hominin species:

- *Sahelanthropus* in Chad, versus *Orrorin* and *Ardipithecus* (*Ar. ramidus* and *Ar. kadabba)* in the Rift Valley (an episode during the Miocene).
- *Australopithecus* in the Rift Valley (Ethiopia, Kenya, and Tanzania)/in Chad/and in South Africa.
- *Paranthropus* in the Rift Valley (including Malawi)/ and in South Africa.
- *Homo* in the Rift Valley (including Malawi)/in South Africa (Lower Pleistocene)/and in Georgia.

6.8.1 *Ardipithecus–Australopithecus* dispersal

The existence of ardipithecines in the Rift Valley (*Ardipithecus* and *Orrorin*), as well as in Chad (*Sahelanthropus*), plus that of *Australopithecus* in both regions, obliges us to consider two dispersals during the transition from the Miocene to the Pliocene. The oldest date for the ardipithecines is that of Chad; an initial dispersal D1 would have occurred from Chad toward the Rift Valley around 6 Ma. Also, we have accepted

the evolution from *Ar. ramidus* to *A. anamensis* as the most probable; therefore, it is plausible to maintain that *Australopithecus* emerged in the Rift. This proposal requires understanding that once in the Pliocene, two more dispersals took place: D2 from the Rift to Chad, prior to, or coinciding with, a date of 3.5 Ma, and another, D3, from the Rift to South Africa around 2.2 Ma.

After justifying the presence of hominins in Chad, the Rift Valley, and South Africa, various hypotheses should be proposed about the further evolution in these last two areas.

6.8.2 *Australopithecus–Paranthropus* dispersal

The phylogenetic transition *Australopithecus–Paranthropus* occurred (I) locally, so that in South Africa *A. africanus* gave way to *P. robustus* around 2 Ma ago, and the australopithecines from the Rift did so around 3–2.5 Ma with respect to *P. aethiopicus*/*P. boisei*, or (II) evolution took place in one of the two areas, with a later dispersal toward the other zone. The (I) hypothesis impedes consideration of the *Paranthropus* set as a monophyletic clade because two "robust" lineages would then have evolved separately. The existing homoplasies between forms as similar as *P. robustus* and *P. boisei* would be

Box 6.22 What if the hominins came from Saharan Africa?

If the origin of hominins and *Australopithecus* was Chad, as could be deduced from the interpretations of Brunet el. (1995, 2002), and if the evolution *Ar. tchadensis–A. bahrelghazali* is postulated, the migration scheme would not change much. A dispersal D2 Chad–Rift would still be necessary (although in the reverse direction to the one indicated in the text) and a D3 Rift–South Africa dispersal as well. A direct dispersal from Chad to South Africa has not been proposed, as far as we know, by any author.

many; a reason why it is most parsimonious to accept hypothesis (II). This implies accepting a new dispersal, which would have happened in either of the two directions; however, the age of *P. aethiopicus* favors the migratory direction Rift–South Africa. This means that *A. africanus* cannot be the ancestor of *P. robustus*.

One way to accept that *A. africanus* is the ancestor of *P. robustus* would be to introduce an alternative hypothesis about the dispersals around 2.2–2 Ma, and to assume that the flow between South Africa and the Rift Valley populations was continuous in both directions. If this were so, it would be possible to accept the presence of *A. africanus* in the sites of East Africa, and *A. sediba* would also be understood as a species that was present during that population transition.

CHAPTER 7

Lithic traditions: tool-making

The adaptive strategies of all taxa belonging to the genus *Homo* included the use of stone tools, although the characteristics of the lithic carvings changed over time. The earliest and most primitive culture, Oldowan or Mode 1, appears in East African sediments around 2.4 Ma in the Lower Paleolithic. Around 1.6 Ma, appears a more advanced tradition, Acheulean or Mode 2. The Mousterian culture or Mode 3 is the tool tradition that evolved from the Acheulean culture during the Middle Paleolithic. Finally—for the limited purposes of this chapter—the Aurignacian culture, or Mode 4, appeared in the Upper Paleolithic (see Box 7.1). The original proposal of cultural modes by Grahame Clark (1969) included a Mode 5 by differentiating some technical details, allocating to Mode 4 the punch-struck blades from prismatic cores of the Upper Paleolithic, while the Mode 5 was reserved for the microliths and compound tools of the Late Upper Paleolithic. We believe that this distinction is not necessary for the present chapter, whose aim is to relate cultural development to human evolution. A first approach attributes each cultural stage to a particular human taxon. Thus, the beginning of tool-making, i.e., Mode 1, is linked to *Homo habilis*, and "technology"—understood as the making of tools that requires a modern mind necessary for Mode 4—to *Homo sapiens*. Although we will also examine the technical advances assigned to Mode 5 by Clark (1969), these are part of an evolution that does not involve a change of species. In fact, the "technology" may be adding new modes due to the multiple technological advances that cultural evolution of *Homo sapiens* has achieved, starting with agriculture. However, in this book we will not address the cultural development that occurred after the acquisition of the modern mind.

We should already express a methodological caveat before proceeding: the scheme cultural mode = species, is too general and incorrect. The assumption that a certain kind of hominin is the author of a specific set of tools is grounded on two complementary arguments: (1) the hominin specimens and lithic instruments were found at the same level of the same site; and (2) morphological interpretations attribute to those particular hominins the ability to manufacture the stone tools. The first kind of evidence is, obviously, circumstantial. Sites yield not only hominin remains, but those of a diverse fauna. The belief that our ancestors rather than other primates are responsible for the stone tools comes from the second type of argument, the capacity to manufacture. This consideration is perfectly characterized by the episode involving the discovery and proposal of the species *Homo habilis*. As Louis Leakey, Phillip Tobias, and John Napier (1964, p. 9) have said, "When the skull of *Australopithecus (Zinjanthropus) boisei* was found [in Olduvai, Bed 1] no remains of any other type of hominid were known from the early part of the Olduvai sequence. It seemed reasonable, therefore, to assume that this skull represented the makers of the Oldowan culture. The subsequent discovery of *Homo habilis* in association with the Oldowan culture at three other sites has considerably altered the position. While it is possible that *Zinjanthropus* and *Homo habilis* both made stone tools, it is probable that the latter was the more advanced tool-maker and that the *Zinjanthropus* skull represents an intruder (or a victim) on a *Homo habilis* living site."

Here we have a clear example of the argumentative sequence. First, a *P. boisei* cranium and associated lithic instruments were discovered at the *F.L.K.* I site, Olduvai. Later, hominins with a notably greater cranial capacity, included in the new species *H. habilis*, were discovered at the same place. Eventually, stone tools were attributed to *H. habilis*, which was morphologically more advanced in its planning capacities. Leakey et al.'s (1964) paper included a cautionary note. Even though it is less probable, it is conceivable that *Zinjanthropus* also made lithic tools.

However, the attribution of capacities that identify *Homo habilis* as the author of Olduvai lithic carvings has some reservations. John Napier (1962a) had published an article on the evolution of the hand two years before, relating stone tools to the discovery of

Processes in Human Evolution. Francisco J. Ayala and Camilo J. Cela-Conde, Oxford University Press (2017).
© Francisco J. Ayala and Camilo J. Cela-Conde.
DOI 10.1093/acprof:oso/9780198739906.001.0001

Box 7.1 The names of cultural traditions

Cultural traditions are often named after the site in which they were first discovered. Before the development of techniques that characterize modern humans, the following cultures appeared, in chronological order: Oldowan (Olduvai, Tanzania), Acheulean (Saint Acheul, France), and Mousterian (Le Moustier, France). Following the proposal of Grahame Clark (1970), the numbering of modes is also used, ranging from the simplest (and oldest) to the more complex (and modern). Mode 0 is a backward extrapolation to name spontaneous tools of a pre-cultural stage. Modes 1–3 retain a considerable spatial and geographical uniformity—stasis. Despite certain internal advancements, each mode has its own identity separate from the next culture, something usually interpreted as a reflection of the different cognitive level of the tool-makers (Longa, 2009). Consequently, the replacement of a mode by another is not a gradual process, although there are certain exceptions to this rule. For example, the Levallois technique appears in Africa as part of Mode 2, but in a way somewhat sporadic and primitive; it became fully developed and widespread in the European Mousterian. Authors such as Fernando Díez Martín (2003) warned about the danger of using such a simple scheme as Clark's modes, given that cultural traditions are often multiple and with difficult-to-explain relationships. In particular, Mode 2 and its evolution in Europe toward Mode 3 present a complex scenario.

15 hominin hand bones by Louis and Mary Leakey at the site where *Zinjanthropus* had been found. According to Napier, "Prior to the discovery of *Zinjanthropus*, the South-African man-apes (Australopithecines) had been associated at least indirectly with fabricated tools. Observers were reluctant to credit man-apes with being tool-makers, however, on the ground that they lacked an adequate cranial capacity. Now that hands as well as skulls have been found at the same site with undoubted tools, one can begin to correlate the evolution of the hand with the stage of culture and the size of the brain" (Napier, 1962a, p. 57).

Napier (1962a; 1962b) and Leakey et al.'s (1964) interpretations of the Olduvai findings exemplify the risks involved in the correlation of specimens and tools. Both the skull of *Zinjanthropus* (OH 5) and the OH 8 collection of hand and feet bones (with a clavicle), all found by the Leakey team in the same stratigraphic horizon, could be related to lithic making. Sites yielding

tools and fossil samples of australopiths and *H. habilis* require a decision about which of those taxa made the tools. The widespread attribution of Mode 1 to *Homo habilis* is based on a set of indicators among which are hand morphology and size, as well as brain lateralization—an expression of the control capabilities of either hand (Ambrose, 2001; Panger et al., 2002).

7.1 Pre-cultural uses of tools

Regarding the use of stones or other materials for obtaining food, one must distinguish between two different operations. It is one thing to make use of pebbles, sticks, bones, or any available object to, for example, break nutshells and access the fruit; it is a different matter to manufacture very deliberately tools with a specific shape to carry out a precise function. Although we are speaking in speculative terms, it is conceivable that the spontaneous use of objects as tools preceded stone carving.

By means of the comparative study of the behavior of African apes, ethology has provided some interesting interpretations about how chimpanzees use, and sometimes modify, stones and sticks to get food (Figure 7.1). Since the first evidence of such behaviors collected by Jane Goodall and Jordi Sabater Pi (Goodall, 1964; Sabater Pi, 1984), many cases of chimpanzee tool use that can be considered cultural have been brought to light (see Box 7.2). Very diverse cultural traditions have been documented, including up to 39 different behavioral patterns related with tool use by chimpanzees (Boesch & Tomasello, 1998; Vogel, 1999; Whiten et al., 1999). Some of these patterns include the use of different tools in sequence, as it is done by the Loango chimpanzees (Gabon) for obtaining honey (Boesch et al., 2009). It is, of course, true that the use of tools includes different patterns in the case of humans, who carry out operational planning tasks and, in particular, technical improvement processes (Davidson & McGrew, 2005). But it is also true that chimpanzees are able to consider future uses of tools, which involves some planning (Mulcahy & Call, 2006). It has also been observed experimentally that chimpanzees may conform to "cultural" behaviors expressed by dominant individuals in the group, a behavior incipiently similar to human behavior (Whiten et al., 2005).

One of the most interesting aspects of chimpanzee behavior, in order to understand the evolution of the lithic traditions, is the production, at the beginning unintentional, of flakes that resemble those produced by the first human cultures. This "spontaneous" production appears when chimpanzees accidentally shatter a

Figure 7.1 Chimpanzee cracking nuts at Bassa Islands (Liberia), using stone hammer and anvil. Davidson, I. & McGrew, W. C. (2005). Stone tools and the uniqueness of human culture. *J. Roy. anthrop. Inst.*, 11, 793–817.

stone while trying to crack nuts; the result can lead to sets of cores and flakes that are reminiscent of those in the oldest hominin sites containing tools (Mercader et al., 2007; Mercader, Panger, & Boesch, 2002) (see Figure 7.2). It is reasonable to think that the hominins themselves would use, at least as much as chimpanzees, the spontaneous tools available (Panger et al., 2002); and that they would do it for a considerable time before starting to produce tools explicitly. This idea was expressed by John Robinson (1962) when he said that the australopiths did not produce the complex

carved stone found in Sterkfontein; but, for this author, this does not mean they lacked culture. When seeking food they could have used rocks, sticks, bones, and any other tools that would be useful for their purposes. Eudald Carbonell, Marina Mosquera, and Xosé Pedro Rodriguez (2007) have referred to these usages prior to tool production as the "biofunctional stage" or "Mode 0." Shannon McPherron and collaborators (2010) have identified, at the site of Dikika (Ethiopia), stone tool-inflicted marks on bones whose age is more than 3.39 Ma. Even though McPherron and

Box 7.2 Culture in other apes

Although the use of tools is often associated exclusively with chimpanzees, gorillas too, on rare occasions, make use of utensils such as wooden sticks. Elizabeth Lonsdorf and collaborators (2009) studied experimentally the gorilla and chimpanzee responses to stimuli (rewards in the form of food) to reinforce the use of tools to access simulated termite mounds, with results that initiated greater social involvement than in chimpanzees.

The extent of the similarities and differences in the capacity for culture between chimpanzees and humans was proven experimentally by Esther Herrmann and collaborators by cognitive test batteries in a large sample of subjects—chimpanzees, gorillas, and human children under two and a half years old without education. The results indicated that, although chimpanzees have skills similar to children when dealing with the physical world, they are well below children when it comes to solving issues of social involvement (Hermann et al., 2007).

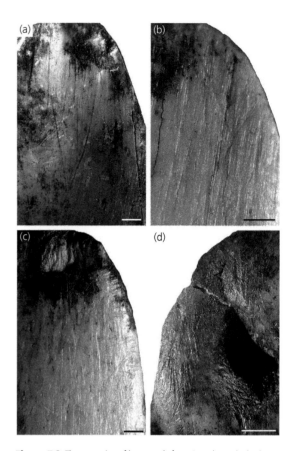

Figure 7.2 The worn tips of bone tools from Swartkrans (a, b, c) and Drimolen (d) showing wear patterns consisting of striations radiating from the tip, and oriented longitudinal to the main axis of the bone; scale: 1 mm. Reprinted from d'Errico, F. & Backwell, L. (2009). Assessing the function of early hominin bone tools. *Journal of Archaeological Science*, 36(8), 1764–1773, with permission from Elsevier.

collaborators (2010) found no tools in Dikika, Sonia Harmand presented, at the meeting of the Paleoanthropology Society in San Francisco on April 14, 2015, the finding at the site of Lomekwi (Lake Turkana, Kenya) of tools coming from sediments with an age of around 3.3 Ma (Callaway, 2015). There are, moreover, very heavy artifacts, some of them up to 33 lb (15 kg). Although at the time of writing this chapter the research on these tools has not been published, clues about the ancient use of stone tools are increasing.

It should be noted that suspicions about the existence of a distinct cultural level for australopiths were for a long time tied to evidence coming from Taung and Sterkfontein. The fractured bases of baboon skulls of Taung and other places, for example, indicated to Raymond Dart that they were cracked in order to consume their insides. Dart (1957) argued that the bones themselves had been used by australopiths as tools to strike, crush, and cut, giving rise to a tradition of using tools of "natural" origin, the osteodontokeratic culture, prior to the use of stone tools.

Although the osteodontokeratic culture was eventually considered as a misinterpretation, and taphonomic studies would tend to argue that the identified bones were not actually tools, studies such as that of Francesco d'Errico and Lucinda Backwell (2003) on the uses of bones from Sterkfontein—members 1–3, between 1.8 and 1.0 Ma—have shown indications, in the form of wear marks, of their being used in milling tasks. In a later work, d'Errico and Backwell (2009) studied the different uses of bones. Once again, for the functions assigned to bones as tools, the use of sticks by chimpanzees in tasks such as digging, to extract termites or to separate the bark of trees, can serve as a model. Optical interferometer analysis of terminal areas of bones used as tools has revealed different wear patterns on specimens from Swartkrans and Drimolen. d'Errico and Backwell (2009) concluded that the differences found indicate diverse activities, as well as contacts with abrasive particles of various sizes, that would point to tasks similar to those that have been observed in chimpanzees.

The use of bones as tools extends to the African Middle Paleolithic, and even to the Upper Paleolithic, although with very different purposes to those inferred for tools of Swartkrans' lower levels, as is evidenced, for example, by the small ivory points from the Upper Semliki Valley, Zaire (Brooks et al., 1995; Yellen et al., 1995). The markings found on the bones have been used as evidence of butchering activities (but see Box 7.3). If appropriate taphonomic considerations are taken into account, the markings observed on carcasses are an irrefutable proof of the use of cutting tools on them. However, according to Sherwood Washburn (1957) the accumulation of remains in the breccias of South African caves is unrelated to hominin butchering tasks. There is a predominance of mandibular and cranial remains because they are the bones most difficult to break, so that they tend to accumulate in the lairs of predators and scavengers. Ancestral hyenas are likely responsible for the accumulation of mandibular and cranial fossil remains from australopiths. It has been suggested that the Taung child itself was the victim of a predator, probably an eagle (Berger & Clarke, 1996), though this hypothesis has been criticized (Hedenström, 1995).

Manipulated stones cannot be attributed to predators. Many lithic instruments have been found at the Sterkfontein *Extension Site*—handaxes, cores, flakes, and even a spheroid—which are unequivocal signs of the manipulation of raw materials to obtain tools designed to cut and crush (Robinson & Manson, 1957). However, there are doubts regarding the association between stone tools and their authors. The sites that have provided *A. africanus*, Sterkfontein, Makapansgat, and Taung, are not the only ones that have provided samples of an early lithic culture. There is also a stone industry at Swartkrans (Brain, 1970; Clarke, 1970), though it was found a long time after Dart elaborated his idea of hominization. The interpretation of the possible stone artifacts found at Kromdraai is not easy (Brain, 1958). But even in Sterkfontein, the *Extension Site* belongs to member 5, whereas

member 4, older than 5, has provided a great number of *A. africanus* specimens but has yielded no lithic tool whatsoever.

If the accumulation of bones at Sterkfontein member 4 was due to scavengers, and if australopiths were the hunted and not the hunters, the question concerning the first tool-makers remains unanswered. The answer will depend on preconceptions regarding cognitive capacities and hominin adaptive strategies. New kinds of evidence have a bearing on this issue: the paleoclimate to which different genera and species were adapted; the morphology of certain key elements required for the intentional manipulation of objects, such as hands and the brain; the diet; and the taphonomic study of the relation at the sites of bones and tools.

7.2 Taphonomic indications of culture

Paleoclimatological conclusions regarding early hominin taxa suggest they were adapted to tropical forests. This is the case of *Australopithecus anamensis* (Leakey et al., 1995), *Ardipithecus ramidus* (WoldeGabriel et al., 1994), *Australopithecus afarensis* (Kingston et al., 1994), and *Australopithecus africanus* (Rayner et al., 1993). This argues against Raymond Dart's original hypothesis that related bipedalism, the expansion of open savannas, and the appearance of the first hominins. The first hominins would have emerged long before the expansion of the savanna in Africa and before any evidence of lithic tool use (Box 7.4).

But in Dart's time no Miocene hominins were known, so it was logical that he spoke of "the first humans," referring to those who colonized the savanna during the Pliocene–Pleistocene. Which of them first began to use stone tools? Again, we are facing the necessity to associate fossil remains to the lithic tools found at the sites.

We have said previously that the attribution of a particular hominin taxon to the making of a specific culture type is based on finding the hominin specimens and lithic instruments at the same level of the same site. However, we must avoid falling into circularity. Most especially, every precaution should be taken when attributing manipulation of ancient tools to hominins of different sympatric species. If we find two taxa, T1 and T2, present at the same site and stratigraphic horizon—as happens with *H. habilis* and *P. boisei* at Olduvai—and if we claim that the authors of the carvings belong to one of them, let's say to T1, because we assume that they have the ability to make tools, we will be falling into a circular argument. When finding the tools, we assume that those who carved them are precisely those individuals to whom we had previously attributed the possession of a cognitive level and manual capacity for manufacturing them.

Circularity can be broken if in sites where the alleged authors of the carvings are found, such as T1, tools are also found in a fairly widespread way, while such is not the case for T2 or any other taxon, which only sporadically are found associated with tools. In that case, it is reasonable is to accept T1 as the tool-maker. The systematic coincidence between a specimen of a particular morphology and lithic carvings of a specific cultural tradition is what has led us to consider the first species of the genus *Homo* as responsible for the oldest culture. However, when addressing the tools from Gona and Lomewki we will see that other possible interpretations of the oldest utensils exist.

With regard to South Africa, the issue is uncertain. Sterkfontein member 5 has yielded the StW 53 cranium, which, as we saw, is considered as either *H. aff. habilis* or an *Australopithecus* of an unspecified species; and it was considered as the specimen-type of *Homo gautengensis* by Darren Curnoe (2010). Swartkrans has also provided some exemplars attributed to *H. habilis* and, regarding Taung, the most widespread opinion argues that the stone tools are much more recent and that they were made by more evolved hominins.

The words "more evolved" obscure the circularity trap about which we spoke earlier. It makes us think that, as the carving of lithic tools imply high cognitive abilities, the presence of tools of that type leads us to conclude that their creators had reached a higher cognitive development. To accept that as a truth it is necessary to relate that "cognitive leap" with some other evidence aside from stone tools.

Taphonomic studies, which reconstruct the process of accumulation of available fossil evidence at a site, have enabled further progress in understanding the behavior of ancient hominins. Different sites in East Africa (Olduvai, Koobi Fora, Olorgesailie, and Peninj) provide evidence of hominin habitation with direct association of hominin fossil remains and manipulated stones. As a result of such studies, Raymond Dart's idea of hominins as hunters in the open savanna was followed by the hypothesis that the first stone tool-makers were scavengers who cooperated to a greater or lesser degree to obtain food (Isaac, 1978; Binford, 1981; Bunn, 1981; Blumenschine, 1987). The role of cooperation and the type of activity aimed to obtain meat are still controversial, and some authors advocate the idea that hominins associated to sites with ancient cultural presence were hunters rather than scavengers (e.g., Domínguez-Rodrigo et al., 2007). In fact, the controversy is not decisive. Once again, it is reasonable to think that the first *Homo* were opportunists and, in their carnivorous behavior, benefited from the available opportunities to both scavenge and hunt.

7.3 Mode 1: Oldowan culture

Taphonomic studies, aimed at reconstructing the process of accumulation of available fossil evidence at a site, have increased our understanding of the behavior of early hominins. Different East African sites (Olduvai, Koobi Fora, Olorgesailie, and Peninj) provide samples of hominin living sites with a direct association of hominin fossil remains and manipulated stones.

Although Olduvai Gorge was not the first place in which early stone tools were found, it gave its name to the earliest known lithic industry: Mode 1, also known as Oldowan culture. The excellent conditions of the Olduvai sites provided paleontologists and archaeologists with the chance to carry out taphonomic interpretations for reconstructing hominin habitats. Any lithic culture can be described as a set of diverse stones manipulated by hominins to obtain tools to cut, scrape, or hit. They are diverse tools obtained by hitting pebbles of different hard materials: silex, quartz, flint, granite, and basalt are some of the materials used for tool-making. In the Oldowan culture, the size of the round-shaped cores is variable, but they usually fit comfortably in the hand; they are tennis ball-sized stones. Many tools belonging to different traditions fit within these generic characteristics. What specifically identifies Oldowan culture is that its tools are obtained with very few knocks, sometimes only one—the resultant tools are misleadingly crude. It is not easy to hit the stones with enough precision to obtain cutting edges and efficient flakes.

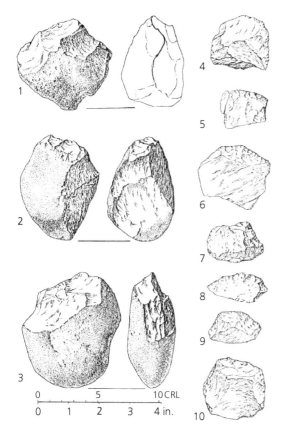

Figure 7.3 Oldowan tools: 1–3, lava choppers; 4–10, quarcite flakes (drawing by Leakey, 1971).

The Oldowan tools are usually classified by their shapes, with the understanding that differences in appearance imply different uses (Figure 7.3). Large tools include:

- Cobbles without a cutting edge, but with obvious signs of being used to strike other stones, with the very appropriate name of hammerstones.
- Cobbles in which a cutting edge was obtained by striking, which served to break hard surfaces such as long bones (to reach the marrow, for example)—they are called choppers.
- Flakes resulting from a blow to a core. Their edges are very sharp—as much as that of a metal tool—and their function is to cut skin, flesh, and the tendons of animals that need to be butchered; they can be retouched or not.
- Scrapers—retouched flakes with an edge, which recall, in some ways, a serrated knife, and whose function would have been to scrape the skin into rawhide.

- Polyhedrons, spheroids, and discoids—cores manipulated in various ways, as if flakes had been removed from their outer perimeter. Their function is uncertain; they may be nothing more than waste without particular utility.

It is not easy to arrive at definitive conclusions regarding the use of Oldowan tools. The idea we have of their function depends on the way we interpret the adaptation of hominins that used them, based on arguments that are often circular. Tool-makers can be seen, as Lewis Binford (1981) did, as a last stage in scavenging, when only large bones are available. If this were the case, the most important tools would be the handaxes that allow hitting a cranium or femur hard enough to break them. If, on the contrary, we understand that early hominins butchered almost whole animals, then flakes would be the essential tools. A functional explanation can be established between handaxes, manipulated with power grips, and flakes, which require handling them with the fingertips using a precision grip. It is not easy to go beyond this, but some authors, like Nicholas Toth (1985a, 1985b), have carried out much more precise functional studies. Toth argued that flakes were enormously important for butchery tasks, even when they were unmodified, while he doubted the functional value of some polyhedrons and spheroids (Box 7.5).

Several kinds of evidence have been used to resolve the question of how early hominins obtained animal proteins. One is the detailed analysis of the tools and their possible functionality. The microscopic examination of the edge of a lithic instrument allows inferring what it was used for—whether it served as a scraper

---

Box 7.5 Using non-manufactured tools

Nicholas Toth (1985a, 1985b) did not share the view that the hominins who made tools depended entirely upon them. By studying Koobi Fora sites and their Oldowan tool distribution, Toth argued that it was quite possible that these hominins often used non-manufactured tools such as broken shells, horns, wooden sticks, and even, quoting Brain (1982), bones. In other words, the idea expressed 20 years earlier by Robinson (1962) has had a remarkable effect. However, hominin opportunism does not only affect the obtaining of food; it is common sense to argue that any object with potential use as a tool will be used when necessary.

to tan skin, or as a knife to cut meat, or as a handaxe to cut wood. This affords an explanation of behavior that goes beyond the possibilities of deducing a tool's function from its shape. In certain instances lithic tools might have been used as woodworking tools. Indications of the use of wood instruments are not rare in the Upper Pleistocene. In the Middle Pleistocene, the finding of plant microremains (e.g., phytoliths, fibers) on the edges of Peninj (Tanzania) Acheulean bifaces is the earliest proof of processing of wood with artifacts (Dominguez-Rodrigo et al., 2001).

The examination of the marks that tools leave on fossil bones provides direct evidence of their function. Taphonomic interpretations of cutmarks suggest hominins defleshed and broke the bones to obtain food. This butchery function related to meat intake portrays early hominins as scavengers capable of taking advantage of the carcasses of the prey of savanna predators (Blumenschine, 1987). But in some instances the evidence suggests other hypotheses. Travis Pickering and collaborators (2000) analyzed the cutmarks inflicted by a stone tool on a right maxilla from locality Stw 53 at Sterkfontein member 5. The species to which the specimen belongs is unclear, but it is certainly a hominin. They noted that "The location of the marks on the lateral aspect of the zygomatic process of the maxilla is consistent with that expected from slicing through the masseter muscle, presumably to remove the mandible from the cranium" (Pickering et al., 2000, p. 579). In other words, a hominin from Sterkfontein member 5 dismembered the remains of another.

Are these marks indicative of cannibalistic practices, or are they signs of something like a ritual? The available evidence does not provide an answer to this question. It is not even possible to determine whether the hominin that disarticulated the Stw 53 mandible and its owner belonged to the same species. But cannibalistic behaviors have been inferred from Middle Pleistocene cutmarks; this is how the cutmarks on the Atapuerca (Spain) ATD6-96 mandible have been interpreted (Carbonell et al., 2005). It has also been suggested regarding the Zhoukoudian sample (Rolland, 2004). Cannibalism seems to have been common among Neanderthals and the first anatomically modern humans.

The Oldowan culture is not restricted to Olduvai. Stone tools have also been found at older Kenyan and Ethiopian sites, though in some occasions their style was slightly different. These findings have extended back the estimated time for the appearance of lithic industries (Table 7.1). For a list of Mode 1 main sites, see Plummer (2004).

Table 7.1 The oldest cultures

Name	Localities	Age (Ma)
Lokalalei	West Turkana	2.34
Shungura	Omo	2.2–2.0
Hadar	Hadar	2.33
Gona	Middle Awash	2.5–2.6

Close to 3,000 artifacts were found in 1997 at the Lokalalei 2C site (West Turkana, Kenya), with an estimated age of 2.34 Ma (Figure 7.4). They were concentrated in a small area, about 10 m², and included a large number of small elements (measuring less than a centimeter) (Roche et al., 1999). The tools were found in association with some faunal remains, but these show no signs of having been manipulated. Nearby sites, LA2A, LA2B, and LA2D, and the more distant LA1 have also provided stone tools; LA1 and LA2C, with an age of 2.34 Ma, are the oldest sites with utensils in Kenya (Tiercelin et al., 2010).

The importance of the Lokalalei tools lies primarily in the presence of abundant debris, which allows the establishment of the sequence of tool-making in situ. Helene Roche and colleagues (1999) have argued that the technique used by the makers of these tools required very careful preparation and use of the materials—previously unimaginable for such early hominins. This suggested that the cognitive capacities of those toolmakers were more developed than is usually believed. One of the cores was hit up to 20 times to extract flakes, and the careful choice of the materials (mostly volcanic lavas like basalt) indicates that those who manipulated them knew their mechanical properties well.

Lokalalei findings indicate that hand control and, therefore, brain development must have been already quite advanced nearly two and a half million years ago. The question of what species would have been responsible for manipulating these artifacts is a different matter, as we have repeatedly pointed out. James Steele (1999) raised the issue of the cognitive capacities and knowledge of the authors of the Lokalalei 2C tools. Steele admitted that the available evidence does not allow going beyond hypotheses similar to the one that attributed the Olduvai tools to *H. habilis* because of its larger cranial capacity compared with *Paranthropus boisei*. The Lokalalei findings indicate that almost two and a half million years ago the motor control of the hands, and thus, the development of the brain, must have been considerable. The identity of the species

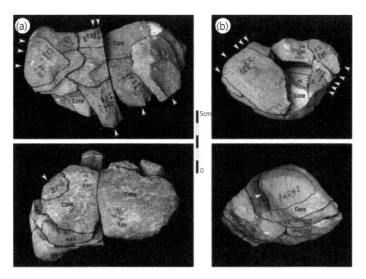

Figure 7.4 Lokalalei sets of complementary matching stone artifacts: (a) R35 refit (2 cores and 11 flakes); (b) R9 refit (1 core and 14 flakes). The main reduction sequence consists of unidirectional or multidirectional removals flaked on a single debitage surface, from natural or prepared platforms. Arrows indicate the direction of the removals. (Pictures from Roche et al., 1999.)

responsible for manipulating those artifacts is a different issue, difficult to answer. In his commentary about Roche and colleagues' discovery, Steele (1999) refused to give a definitive answer. He simply argued that we still have similar doubts to those of the authors who, in 1964, associated the tools found at Olduvai with the species *H. habilis*.

The Middle Awash region includes many sites that have yielded Oldowan and Acheulean—a culture that replaced the Oldowan over time—tools, described for the first time by Maurice Taieb (1974) in his doctoral thesis. A *Homo* maxilla (A.L. 666-1) was found in association with Oldowan tools at Hadar (Ethiopia), to the north of Middle Awash. The sediments from the upper part of the Kada Hadar member were estimated to 2.33 Ma; this was the earliest association between lithic industry and hominin remains (Kimbel et al., 1996). The 34 instruments found in the 1974 campaign (indicative of a low density of lithic remains) are typical of Oldowan culture: choppers and flakes. Additionally, three primitive bifaces, known as end-choppers, appeared on the surface, but it is difficult to associate these tools with the excavated ones.

The earliest known instruments have been found at the Gona site (Ethiopia), within the Middle Awash area, in sediments dated to 2.6–2.5 Ma by correlation of the archaeological localities with sediments dated with the $^{40}Ar/^{39}Ar$ method and paleomagnetism (Semaw

et al., 1997). Thus, they are about 200,000 years older than the Lokalalei tools.

Gona has provided numerous tools, up to 2,970, including cores, flakes, and debris (Figure 7.5). Many of the tools were constructed in situ. No modified flakes have been found, but the industry appears very similar to the early samples from Olduvai. Sileshi Semaw and colleagues (1997) attributed the differences, such as the greater size of the Gona cores, to the distance between the site and the places where the raw materials (trachyte) were obtained: these are closer in Gona than in other instances. As hominins have not been found at the site, it is difficult to attribute the tools to any particular taxon. Semaw et al. (1997) believed it was unnecessary to suggest a "pre-Oldowan" industry. Rather, the Oldowan industry would have remained in stasis (presence without notable changes) for at least a million years. The precision of the Gona instruments led Sileshi Semaw's team to assume that their authors were not novices, so even earlier lithic industries might be discovered in the future (Box 7.6).

Bernard Wood (1997b) wondered about the authors of the tools found at the site. The great stasis of the Oldowan culture suggested by the tools raises a problem for the usual assignation of the Oldowan tradition to *H. habilis*. Given that the latest Oldowan tools (Figure 7.6) are about 1.5 Ma old, this tradition spans close to a million years. This is why Wood (1997b) noted that

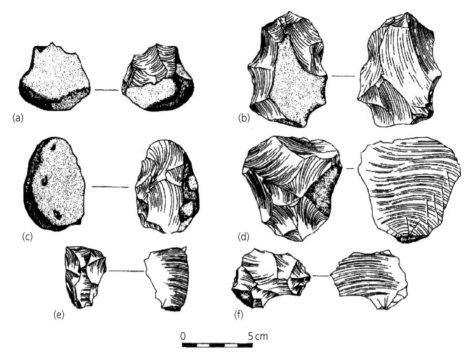

Figure 7.5 Tool assemblages from the Gona (Ethiopia) EG12 and EG10 localities. Flaked pieces: (a) unifacial side chopper, EG12; (b) discoid EG10; (c) unifacial side chopper, EG 10. Detached pieces: (d–f) whole flakes, EG 10. (Picture from Semaw et al., 1997.)

Box 7.6 Lomekwi tools

The future to which Semaw et al. (1997) referred seems to have arrived. The journal *Nature* (Callaway, 2015) has announced the discovery of stone cores and flakes, probably the result of carving, in the Lomekwi site located on the western shore of Lake Turkana (Kenya). The age of the terrain is 3.3 Ma, i.e., much older than *Homo habilis*. Thus, an even older hominin must have been responsible for making the tools that were just found by Harmand Sonia, a researcher at the University of Stony Brook (New York, United States), who presented them at a meeting of the Paleoanthropological Society in San Francisco on April 14, 2015. While waiting to publish their research, Harmand and collaborators named this industry: Lomekwian. The team noted at the meeting of the Paleoanthropological Society that the cores are huge, of approximately 15 kg (Callaway, 2015); somewhat surprising given the small size of the australopithecines. How would they have handled such large stones?

if Oldowan tools had to be attributed to a particular hominin, then the only species that was present during the whole interval was *Paranthropus boisei*. This is circumstantial evidence in favor of the notion that robust australopiths manufactured tools. But, as we have mentioned several times along this book, there is no need for making a close identification between hominin species and lithic traditions, because cultural sharing must have been quite common. In any case, de Heinzelin and colleagues (1999) attributed the Gona utensils to the species *Australopithecus garhi*, whose specimens were found at Bouri, 96 km south of where the tools come from.

The comparison between instruments from different sites has its limitations. As Glynn Isaac (1969) has noted, it is not uncommon to find that the differences between the Oldowan techniques found at different locations of the same age are as large as those used to differentiate successive Oldowan stages, or even larger. This problem illustrates that the complexity of a lithic instrument is a function of its age, but also of the needs of the tool-maker.

7.4 The transition Mode 1 (Oldowan) to Mode 2 (Acheulean)

Mode 2, or Acheulean culture, corresponds to a new carving procedure whose most characteristic element is the biface, "teardrop shaped in outline, biconvex in cross-section, and commonly manufactured on large (more than 10 cm) unifacially or bifacially flaked cobbles, flakes, and slabs" (Noll & Petraglia, 2003, p. 35). These tools, made with great care, were identified for the first time at the St. Acheul site (France), and are known as "Acheulean industrial complex" or "Acheulean culture" (Mode 2). Acheulean culture appeared in East Africa over 1.7 Ma, and extended to the rest of the Old World, reaching Europe, where the oldest Acheulean tools received the local name of "Abbevillian industry." The life of the Acheulean continued in Europe until about 50,000 years ago, although since 0.3 Ma, more advanced utensils could be found in Europe from other cultural traditions, the Mousterian or Mode 3.

Mary Leakey (1975) described the transition observable at Olduvai from perfected Oldowan tools to a different and more advanced industry. The oldest instruments of Olduvai, which come from Bed I, were in a level dated by the $^{40}K/^{40}Ar$ method at 1.7–1.76 Ma (Evernden & Curtiss, 1965). The first Acheulean tools are from Bed II. Between both beds there are tuffs, but the section that corresponds precisely to the time of the transition between the two cultures cannot be precisely dated (Isaac, 1969). If, in addition, we point out that both cultures overlapped in Olduvai for a considerable time, with the concurrent presence of utensils from both Mode 1 and 2, the difficulties of determining the precise moment of the cultural transition increase. Nevertheless, a gradual transition from Oldowan culture to Acheulean culture was justified by the sequence established by Mary Leakey for the Olduvai beds (Table 7.2 and Box 7.7).

Louis Leakey (1951) had previously considered the coexistence of cultures and the evolution of Oldowan instruments as evidence of gradual change. However, subsequent studies painted another scenario. Glynn Isaac (1969) argued that the improvement of the necessary techniques to go from the Oldowan to the Acheulean traditions could not have taken place gradually. A completely new type of manipulation would have appeared with Acheulean culture, a true change in the way of carrying out the operations involved in tool-making. A similar argument has been made by Sileshi Semaw, Michael Rogers, and Dietrich Stout (2009) when interpreting the sequence of cultural change. Depending on the archaeological record of Gona (Ethiopia) and other African locations, Semaw et al. (2009) concluded that the Mode 2 would have arisen rather abruptly by a rapid transition from the Oldowan technique.

If so, it would be important to determine exactly when that jump forward occurred and to establish the temporal distribution of the different cultural traditions. Such detailed knowledge is not easy to achieve. The Olduvai site does not reveal precisely when the cultural change took place. The earliest instruments, from

---

Box 7.7 Particularities of the advanced Oldowan

To Semaw et al. (2009), the artifacts assigned to the developed Oldowan tradition do not differ much from those of Mode 1. The differences are probably due to the shape of the raw stone used for carving. Marcel Otte (2003) argued that the natural limitations of physical type, as are those of the raw materials, forced the creation of similar shapes.

Table 7.2 Cultural sequence at Olduvai established by Mary Leakey (1975; modified)

Beds	Age (Ma)	Number of pieces	Industries
Masek	0.2	187	Acheulean
IV	0.7–0.2	686	Acheulean
		979	Developed Oldowan C
Middle part of III	1.5–0.7	99	Acheulean
		–	Developed Oldowan C
Middle part of II	1.7–1.5	683	Developed Oldowan A
I and lower part of II	1.9–1.7	537	Oldowan

Bed I, are found in a level dated to 1.7–1.76 Ma by the potassium/argon method (Evernden & Curtiss, 1965). The later Acheulean utensils appeared at the Kalambo Falls locality at Olduvai, in association with wood and coal materials. The age of these materials was estimated by the ^{14}C method at 60,000 years (Vogel & Waterbolk, 1967). There are other volcanic tuffs between both points, but the 1.6 Ma interval between the most recent level and Kalambo Falls limits the precision of the chronometry. This period corresponds precisely to the time of the transition between both cultures (Isaac, 1969). If we take into account the evolution within Mode 1, with developed Oldowan tools that overlap in time with Acheulean ones, the difficulties involved in the description of the cultural change increase.

The technical evolution from Mode 1 to Mode 2 can also be studied at other places, such as the Humbu formation from the Peninj site, to the west of Lake Natron (Tanzania). After the discovery made by the Leakys and Isaac in 1967, authors such as Amini Mturi (1987) and Kathy Schick and Nicholas Toth (1993) carried out research at the Natron area. Several Natron sites show a transition from Oldowan to Acheulean cultures close to one and a half million years ago (Schick &

Toth, 1993). The correlation of the Peninj and Olduvai sediments allows the identification of the Oldowan/Acheulean transition with the upper strata of Bed II from Olduvai. But neither Olduvai nor the western area of Lake Natron allow a more precise estimate of the time of the change.

Another site excavated after the works at Olduvai and Peninj, Olorgesailie (Kenya), provided precise dating (by means of the ^{40}K/^{40}Ar method) for the Acheulean tools from members 5 through 8 of that formation, but they are recent sediments, estimated to between 0.70 and 0.75 Ma (Bye et al., 1987). The precise time of the substitution of Oldowan by Acheulean tools cannot be specified. Any group of hominins capable of using Acheulean techniques could have very well employed, on occasions, simple tools to carry out tasks that did not require complex instruments.

An illustrative example is the large number of Acheulean artifacts found at locality 8 of the Gadeb site (Ethiopia) during the 1975 and 1977 campaigns. Altogether, 1,849 elements, including 251 handaxes and knives, were found at the 8A area, a very small excavation; whereas 20,267 artifacts appeared at 8E (Clark & Kurashina, 1979). The age estimates for the different Gadeb localities with lithic remains are imprecise: they range from 0.7 to 1.5 Ma. These localities contain, in addition to Acheulean tools, developed Oldowan utensils, which led J. Desmond Clark and Hiro Kurashina (1979) to conclude that two groups of hominins would have alternated at Gadeb, each with its own cultural tradition. But it is curious that the examination of the bones from Gadeb showed that the butchery activities had been carried out mostly with the more primitive handaxes, those belonging to developed Oldowan. This fact raises an alternative interpretation, namely, that tools obtained by advanced techniques are not necessary for defleshing tasks.

Konso-Gardula (Ethiopia), south of the River Awash and east of River Omo, has allowed the most precise dating of the beginning of the Acheulean culture. Additionally, it has provided the oldest known tools belonging to that culture. Since its discovery in 1991, Konso-Gardula has provided a great number of tools, which include rudimentary bifaces, trihedral-shaped burins, cores, and flakes, together with two hominin specimens, a molar and an almost complete left mandible (Asfaw et al., 1992). The sediments were dated by the ^{40}Ar/^{39}Ar method to 1.34–1.38 Ma (Asfaw et al., 1992). Berhane Asfaw and colleagues (1992) associate the Konso-Gardula hominin specimens with the *H. ergaster* specimens from Koobi Fora, especially with KNM-ER 992.

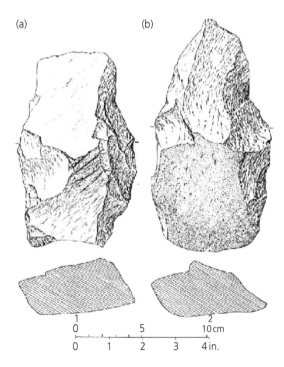

(a)   (b)

1   2
0   5   10 cm
0   1   2   3   4 in.

Figure 7.6 Developed Oldowan tradition from the upper part of Middle Bed II, Olduvai Gorge: (a) cleaver; (b) handaxe (picture from Leakey, 1971).

7.4.1 Mode 2: Acheulean culture

To what extent can the Acheulean tradition be considered a continuation or a rupture regarding Oldowan? Was developed-Oldowan a transition phase toward subsequent cultures? Mary Leakey (1966) believed that developed-Oldowan was associated with the presence of primitive handaxes, i.e., protobifaces that anticipated Acheulean bifaces. However, protobifaces cannot be strictly considered as a transitional form between Oldowan and Acheulean techniques. Marcel Otte (2003) argued that natural constraints (e.g., mechanical laws of the raw materials) forces the manufacture of similar forms, which thus may be considered successive stages of a single or very close elaboration sequence, although this may not always be the case.

The successive manipulation of a core, passing through several steps until the desired tool is obtained, is a task that Leroi-Gourhan (1964) named *châine opératoire* ("working sequence"). While a chopper and a protobiface respond to the same *châine opératoire*, the manufacture of Acheulean bifaces is the result of a completely different way of designing and producing stone tools (see Figure 7.7). The main objective of the Oldowan technique was to produce an edge, with little concern for its shape. However, Acheulean bifaces have a very precise outline, which evinces the presence of design from the very beginning. The existence of design has favored speculation about the intentions of the tool-makers.

In the tradition of Leroi-Gourhan, Nathan Schlanger (1994) suggested that the sequence of operations in the making of tools reflects an intention and a mental level of some complexity. One might, accordingly, distinguish between two types of "knowledge":

- Practical knowledge, necessary for any carving operation; it is what psychologists call "procedural knowledge," such as is needed to ride a bicycle without falling.
- Abstract knowledge, or posing problems and their solutions; this is closer to "declarative knowledge," such as designing a route for cycling around town from one place to another with the least risk.

The most common view holds that while a chopper and a protobiface belong to the same *châine opératoire*, obtaining Acheulean bifaces is the result of an entirely different approach when designing and producing a stone tool (Box 7.8) (see in Figure 7.8 an early, but true, biface). The most conspicuous novelty is the diversity of Acheulean instruments. Sometimes it is difficult to assign a function to a stone tool. We have already

> **Box 7.8 What was the Acheulean *châine opératoire*?**
>
> An easy way to distinguish between declarative and procedural knowledge is to understand that declarative knowledge can be transmitted through a spoken or written description, while the procedural knowledge cannot. However, as we will discuss, it is doubtful that the Acheulean culture involves accurate mental models of the tools that will be obtained, which brings into question the very *châine opératoire* of the Acheulean. In the opinion of Sheila Mishra et al. (2010), the manufacture of accessories for transporting objects such as stones would be the real innovation of Mode 2.

seen that Oldowan *châine opératoire* and flakes have been interpreted both as simple debris and as valuable tools. However, Acheulean tools include knives, hammers, axes, and scrapers, whose function seems clear. The materials used to manufacture lithic instruments are also more varied within the new tradition. But the most notorious difference associated with the Acheulean culture is the tool we mentioned before: the handaxe.

The work of Glynn Isaac (1969, 1975, 1978, 1984) in Olorgesailie and Peninj (Tanzania) showed the main role of handaxes in the form of large flakes (Large Flake Acheulean, LFA) of more than 10 cm, in African Lower Paleolithic tool production. The study by Ignacio de la Torre and collaborators (2008) on the amount of raw material used for manufacturing various tools within two lithic sets found in Peninj, RHS-Mugulud, and MHS-Bayasi, showed convincingly that the essential goal of hominins was to obtain large cutting tools, among which are the cleavers (without retouched edge), handaxes, and flakes, to use as knives. The carving technique used followed a characteristic pattern with a succession of steps not exclusive to Peninj carvings; it is the key to the Acheulean tradition:

(1) The transformation of raw materials to be converted into cutting instruments involved, first, selecting suitable large stones to carve. The availability or lack of stone quarries with such raw materials may lead to significant differences between the cultural content of different sites.
(2) Once the rock is selected, it is reduced by chipping off large flakes until obtaining still sizeable blocks with a suitable form to begin careful carving.

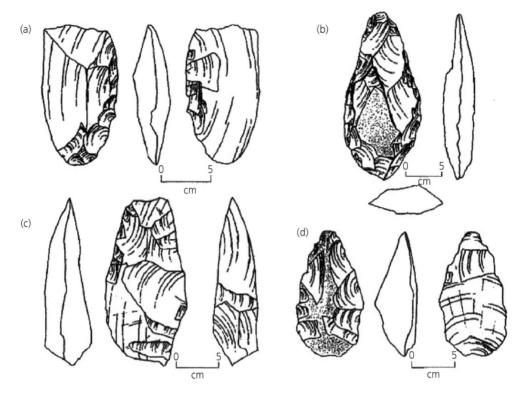

Figure 7.7 Bifaces for different uses: (a) cleaver; (b) handaxe; (c) knife; (d) pick (picture from Noll & Petraglia, 2003).

(3) The blocks are worked in *châine opératoire*, obtaining three different sizes of flake: small, medium, and large. The large flake, still of considerable size, is a handaxe in its basic shape, which still needs a sharpened edge.

(4) Larger flakes, which contrast with the intermediate ones by size, shape, and weight, are subjected to precision carving, with a number of successive blows to achieve its edge and final form: thus a LFA appears. The number and accuracy of the blows contrast with the less systematic and manipulated of the protobifaces.

Peninj handaxes weighed, once finished, about 1 kg, so the waste materials from the large initial stones are abundant. In the intermediate stages of the Acheulean *châine opératoire*, flakes of different sizes are obtained, which can be, in turn, simultaneously used as tools for further carving. Smaller chips come from preparing the blocks or from shaping the handaxe. LFA production is complex and in most handaxes there are notches of 2–3 cm long, which show that fragments, or chips, were knocked off—similar to those obtained

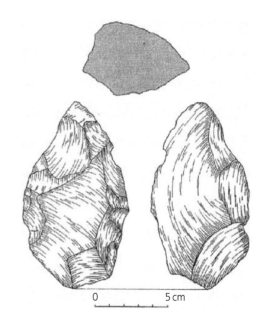

Figure 7.8 Early biface from Peninj (Natron Lake, Tanzania), c. 1.3–1.5 Ma (picture from Wynn, 1989).

intentionally in the Oldowan tradition—but they are actually the result of percussion while retouching. Medium-sized flakes from LFA carving were found both in Olduvai and Peninj, although they are larger in the latter location (de la Torre et al., 2008)—they are large but very thin flakes, so that their volume and weight are modest. They could have served for carving tasks or used as blades, just in the same condition as we have found them.

The most advanced Acheulean technique, with symmetrical bifaces and carefully carved edges, required a *soft-hammer* technique. This method consists in striking the stone core obtained in step 3 with a hammer of lesser hardness, such as of wood or bone. The blows delivered with such a tool allow a more precise control but requires, of course, much more labor. A detailed description of the process was provided by Nicholas Toth and Kathy Schick (1993).

Schick and Toth (1993) noted that bifaces can also be obtained, in the absence of sufficiently large raw material, from smaller cores similar to those that served as a starting point for the manufacture of Oldowan choppers. But the manipulation of large blocks of material (mostly lava and quartzite) to produce long flakes seems to have been the turning point for the development of the Acheulean culture. It would also have involved risk for those who had to manipulate stones of large size (Schick & Toth, 1993).

The oldest Acheulean tool presence documented corresponds to the Kokiselei 4 site of the Nachukui formation, West Turkana (Kenya). By radiometry of nearby volcanic tuffs ($^{40}Ar/^{39}Ar$), stratigraphic equivalence with Koobi Fora, and paleomagnetism, Christopher Lepre and collaborators (2011) assigned to the terrain of Kokiselei 4 an age of 1.76 Ma. The site has the added advantage of also containing Oldowan utensils, which supports the idea that Mode 1 and Mode 2 technologies were not mutually exclusive. Lepre et al. (2011) argued as alternative hypotheses for the presence of Acheulean tools at Kokiselei 4, that they were either brought there from another location—unidentified yet—or carved by the same hominins of the site that produced Oldowan tools.

7.4.2 Mode 2 and cultural dispersion

The occasional presence of Oldowan tools is not proof indicating that a certain group had a primitive cultural condition. It is possible to find simple carvings in epochs and places that correspond to a more advanced industry. There is no reason to manufacture a biface by a long and complex process if what is needed at a given time is a simple flake. But the argument does not work in reverse. The presence of Mode 2 clearly indicates a technological development.

As we have seen, Oldowan culture is generally attributed to *Homo habilis*. However, the identification of the Acheulean culture with the African *Homo erectus* is also very common. The strength of the bond of Acheulean/*H. erectus* led Louis Leakey to consider the emergence of Acheulean tools at Olduvai as the result of an invasion by *Homo erectus* from other localities (Isaac, 1969). Fossils of the taxon *Homo habilis* are African, but numerous exemplars that can be attributed to *Homo erectus* have appeared out of Africa (the *Homo erectus* taxon will be addressed in Chapter 8).

Obviously, the occupation of Asia began with one or more African hominin dispersals (see Box 7.10). The natural way out of Africa is the Levantine corridor—Middle and Near East—a path that is widely understood as the one used by hominins during their various departures from the African continent. Located between the Black Sea and the Caspian Sea, Georgia is part of the transit area between Africa, Asia, and Europe. As we saw in Chapter 6, a site in Georgia, Dmanisi, has provided the best existing evidence to characterize the first hominin exit from their continent of origin. Aside from the fossils found there, from Dmanisi also come stone artifacts and animal bones with cutting and percussion marks (Figure 7.9).

Box 7.9 Core size and Acheulean technique

The manipulation of large stones (mostly basalt and quartzite) for making handaxes seems to have been the turning point for the development of the Acheulean culture (but see Box 7.6). Incidentally, it would also create a significant risk to those who had to manipulate stones of large size (Schick & Toth, 1993).

Box 7.10 Migration and dispersal

It is possible to distinguish between "migration" as a term that refers to cyclical, often seasonal, movement of individuals, and "dispersal," which would be the permanent occupation of a new territory by a species (van der Made & Mateos, 2010). However, we will use in this book "migration" and "dispersal" as equivalent terms.

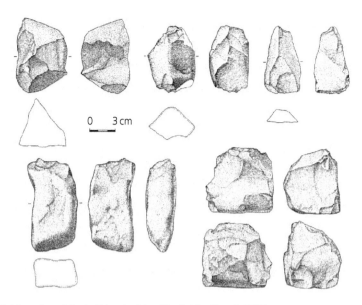

Figure 7.9 Dmanisi tools (picture from Gabunia, Vekua, Lordkipanidze, Swisher III et al., 2000).

More than 8,000—some choppers and scrapers, and abundant flakes—have been found in the two stratigraphic units, A and B, of the site. Reid Ferring and collaborators (2011) maintained that the stratigraphic study of the Dmanisi set of lithic utensils indicates that this place was repeatedly occupied during the last segment of the Olduvai subchron, i.e., between 1.85 and 1.78 Ma. In the authors' opinion, such an antiquity implies that the Georgia specimens precede the African *Homo erectus* or *Homo ergaster* emergence (see Chapter 8) (see Box 7.11).

But Dmanisi is not the only site providing us with evidence of the migration out of the African continent. In the Yiron site, to the North of Israel near the valley of the Jordan River, instruments were found in 1981 consisting of flakes of a very primitive appearance, adding to other more modern tools previously found. The primitive artifacts were found in the stratigraphic horizon below the basaltic volcanic intrusion dated by radiometry at 2.4 Ma, thus the age of Yiron tools was considered comparable to the oldest culture of Mode 1 from the Rift (Ronen, 2006).

At least four other sites in Israel have provided old lithic utensils. Chronologically, Yiron is followed by the Erk-el-Ahmar formation, a few kilometers south of Ubeidiya, also in the Jordan Valley, where cores and silex flakes were found (Tchernov, 1999). After a few failures to date it by paleomagnetism, the magneto-stratigraphy of the Erk-el-Ahmar formation made by Hagai Ron and Shaul Levi (2001) correlated the normal events of the area with the Olduvai subchron, attributing it thus an age of 1.96–1.78 Ma. However, the tools appeared at 1.5 km from the collected samples. Ron and Levi (2001) accepted an age for the silex utensils of 1.7–2.0 Ma, a date supported by faunal studies (Tchernov, 1987), which, by the way, is coincident with that of Dmanisi. Ubeidiya (Israel) is a locality between Yiron and Erk-el-Ahmar. Between 1959 and 1999, numerous lithic instruments were found, similar in age and appearance to those at the Oldowan/Acheulean transition of Olduvai Bed II, along with cranial fragments, a molar, and an incisor attributed *Homo* sp. indet. (Tobias, 1966), or to *Homo* cf. *erectus* (Tchernov, 1986). An additional incisor was described in 2002 (Belmaker et al., 2002). The horizon with hominin remains was dated by faunal comparisons and stratigraphic study at c. 1.4 Ma, on the basis of the deposits' age and tooth similarities with KNM-ER 15000 and the Dmanisi specimens. Miriam Belmaker and collaborators (2002) maintained that the last incisor found in Ubeidiya, UB 335, could

---

Box 7.11 Cultural changes in Dmanisi

Javier Baena et al. (2010) studied how the culture present in Dmanisi evolved as time passed. The authors show how progress occurred with the use of better raw materials that allowed expanding the range of usable materials.

be tentatively identified as *Homo ergaster*. The utensils from Gesher Benot Ya'aqov are 0.8 Ma of age, dated by paleomagnetism. Finally, utensils found in Bizat Ruhama are younger than Gesher Benot Ya'aqov, which belongs to the lower part of the Matuyama chron (in both cases the information comes from: Ronen, 2006).

All the tools found in these sites, which indicate the first dispersals out of Africa, are of Mode 1. Around 1.7–1.6 Ma, hominins undertook various successful dispersals throughout Asia, reaching the Southeast—Java (Indonesia)—and the Far East—China. This means that vast zones of the Asian continent were occupied without Acheulean utensils. However, later migrations brought Mode 2 out of Africa as well.

Acheulean instruments have a geographical eastern limit in the Indian subcontinent. With a detailed compilation of all the available evidence at that time, Hallam Movius (1948) established two areas: the first in Africa,

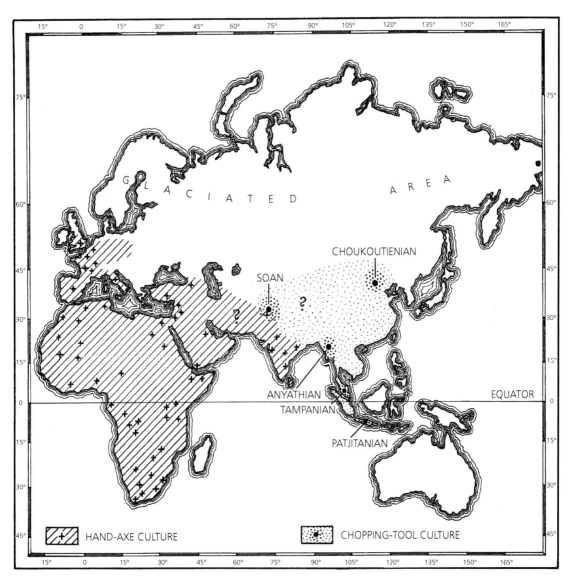

Figure 7.10 "Movius line." Distribution of Lower Palaeolithic handaxe and chopping-tool cultures in the Old World during Upper/Middle Pleistocene times. Movius, H. L. (1948). The Lower Paleolithic cultures of southern and eastern Asia. *Transactions of the American Philosophical Society*, 38, 330–420.

Box 7.12 Cultural traditions in India

With regard to the cultural traditions in India, it is common to distinguish between the Acheulean, with bifaces, and the Soanian, without them. The Soanian-type traditions are, paradoxically, in most cases, younger. Soanian tools in the Potwar region (Pakistan) and Acheulean tools in Hoshiarpur, Nangal (Punjab, India), and Jandori (Himachal Pradesh, India), with large flakes similar to the African cleavers (Soni & Soni, 2010) have been simultaneously found in the Holocene.

The abundance of utensils in India contrasts with the shortage of hominin remains. In the 1980s, at the Hathnora site in the Narmada River basin (Central India), a skullcap, two clavicles, and a rib fragment of an older age than 0.23 Ma were found. They have been attributed both to *H. erectus* and to an "archaic" *H. sapiens* (see Chapter 9), with the most prudent ascription being to *Homo* sp. indet. (Patnaik et al., 2009). Narmada utensils include ancient Acheulean pieces, including not very elaborated large cleavers.

West Asia, and West Europe; the second ranging from the Far East to Southeast Asia (see Box 7.12). During the Middle Pleistocene, both had lithic industries, corresponding to different technical levels: choppers, i.e., Mode 1, in the East, and bifaces, i.e., Mode 2, in the West. This is known as the "Movius line," the virtual limit that separates these two vast areas (Figure 7.10).

7.4.3 Cultural modes in Java and China

As we shall see in Chapter 8, the occupation of Java goes back to c. 1.6 Ma. The scarcity of tools found on the island presented a difficult situation to interpret. Surely, colonization would have been accomplished with the help of the utensils characteristic of such old hominins? In 1934, Gustav Heinrich von Koenigswald found many flakes in Ngebung (Sangiran), which he attributed to the Middle Pleistocene; however, that age was very much debated, and the idea of an absence of stone utensils on the island up to the Upper Pleistocene was generally accepted.

The clarification of the Javanese culture took a step forward thanks to the fieldwork in the Ngebung hills by François Sémah et al. (1992). Stones suitable for use as tools were transported there, but present little evidence of manipulation, while some large, clearly worked andesite pieces were also found (Figure 7.11, right). The stratigraphic study by Semah et al. (1992) reported a tentative age of 0.75–0.25 Ma. Subsequently, radiometric analyses have adjusted the range to 0.88–0.86 Ma (Falguères & Yokoyama, 2001).

No tool of less than 1 Ma has been found in Java sites (Box 7.13). The isolated flakes from Sangiran (Grenzbank) and Miri (Kedung Cumpleng) are approximately of that age. Truman Simanjuntak, Françoise Sémah, and Claire Gaillard (2010) claimed that around 0.8 Ma a noticeable change occurred in Java, when tools emerged that have been classified as Acheulean by these authors. Ngebung cleavers are the oldest indication, followed by the three human occupations of the Song Terus Cave (Punung, East Java), among which the "Terus period," of 0.3 to c. 0.1 Ma, is the oldest (Simanjuntak et al., 2010). Other Southeast Asian locations that also contain ancient lithic utensils, as examples of a Mode 2 late dispersal are the island of Flores (Mata Menge site), with an age of 0.88–0.8 Ma obtained by fission track (Morwood et al., 1998), Bukit Bunuh (Malaysia), Ogan (Sumatra), Sembiran (Bali), Nulbaki (West Timor), Wallanae (Sulawesi), and Arubo (Luzon, Philippines) (Simanjuntak et al., 2010). The overview of the various described industries of Southeast Asia led Sheila Mishra and collaborators (Mishra et al., 2010) to draw several somewhat controversial conclusions with regard to cultural dispersion. First, that Mode 2 reached Java and other Asian areas. Second, that an indicative sequence of an initial period with an absence of large handaxes does not actually exist in India; all occupations of Southeast Asia would have had the set *Homo erectus*/LFA as a protagonist. Third, a more bold assumption, that India might have been both the origin of Mode 2 and of *Homo erectus*, as well as the source of what later would become their African counterparts (*Homo ergaster* and the Acheulean technique) by a reverse dispersal from Asia to Africa.

Regarding China, the oldest tools come from the basins of Yuanmou and Nihewan. Majuangou, the eastern border of the Nihewan basin has four stratigraphic

---

Box 7.13 The carving of tools in Java

The interpretation of the tools found in Javanese sites is affected by the lack of quarries close to places of hominin habitation. The utensils found have been transported, not carved in situ, which impedes assemblage of cores and flakes to reveal the manufacturing process.

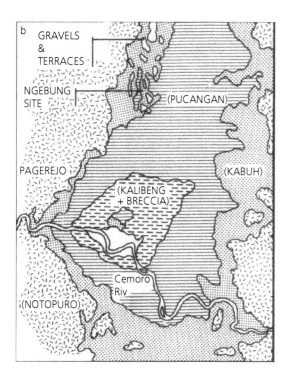 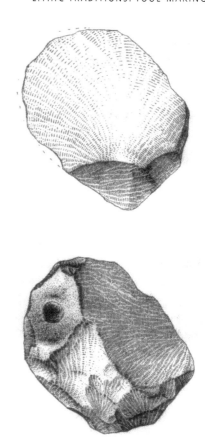

Figure 7.11 Left: Simplified geological map of the Sangiran dome, location of the Ngebung site. Right: Large flake tool (fine-grained andesite). Reprinted from Sémah, F., Sémah, A. M., Djubiantono, T., & Simanjuntak, H. T. (1992). Did they also make stone tools? *Journal of Human Evolution*, 23(5), 439–446, with permission from Elsevier.

horizons in which Mode 1 utensils have appeared, which are, from top to bottom, Banshan, MJG-I, MJG-II, and MJG-III (Table 7.3) (Figure 7.12).

In accordance with paleomagnetic studies, the four beds with tools of Majuangou are distributed over 340,000 years, between the Olduvai and Cobb Mountain subchrons. Fossils of mollusk shells and aquatic plants indicate a lacustrine environment. The lower bed is of 1.66 Ma (Zhu et al., 2004).

In the Yangtze riverbed, Sichuan province, is the Longgupo site, with Mode 1 instruments of an uncertain age. They could be 1.9–1.7 Ma, according to paleomagnetism (Wanpo et al., 1995), but electro spin resonance analysis on the cave specimens' dental enamel have indicated a much later date.

Cultural indications of very ancient human presence exist at Yuanmou, with four members, which are, from the oldest to the youngest, M1 (lacustrine and

Table 7.3 Stratigraphic horizons with lithic tools of Majuangou (Zhu et al., 2004)

Bed	Location (m)	Area (m²) × depth (in cm) of the excavation	Year	Number of tools
Banshan	44.3–45	2 × 70	1990	95
MJG-I	65.0–65.5	20 × 50	1993	111
MJG-II	73.2–73.56	40 × 36	2001–02	226
MJG-III	75.0–75.5	85 × 50	2001–02	443

fluvio-lacustrine deposits), M2 (fluvial), M3 (fluvial), and M4 (fluvial and alluvial) (Zhu et al., 2008). The member M4 in Niujiangbao has provided hominin remains and four stone tools, which were found in 1973: a scraper, a small biface core, and two flakes of Mode 1 with evidence of laborious production (Yuan et al., 1984).

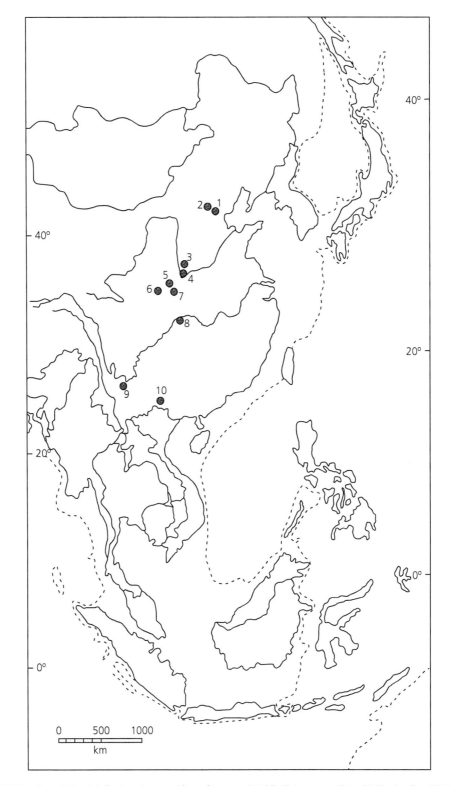

Figure 7.12 Major sites with hominin fossils and stone artifacts of Lower and Middle Pleistocene in China: (1) Zhoukoudian; (2) Nihewan; (3) Dingcun; (4) Xihoudu; (5) Lantian; (6) Hanzhong; (7) Quyuanhekou; (8) Longgupo; (9) Yuanmou; (10) Bose. Huang, W. & Hou, Y. (1997). *Archaeological evidence for the first human colonisation of East Asia. Indo-Pacific Prehistory Association Bulletin*, 16, 3–12.

All the utensils of the described Chinese sites belong to Mode 1 (Oldowan). However, it has been claimed that tools from more modern locations correspond to Mode 2. Thus, Hou, Yamei, and collaborators (Hou et al., 2000) pointed out the presence at various sites in Bose basin, Guanxi province in Southern China, of more advanced tools. Although two-thirds of the basin contain only monoface tools, from the western area of Bose—in which the adequate raw material exists—come LCT bifaces of 803,000±3,000 years of age, described by Hou and collaborators (2000). According to that presence, Hou et al. (2000, p. 1622) affirm that "Acheulean-like tools in the mid-Pleistocene of South China imply that Mode 2 technical advances were manifested in East Asia contemporaneously with handaxe technology in Africa and western Eurasia."

In the same Bose basin, but in its northern zone—Fengshudao site—were found an industry set with an abundance of handaxes, although smaller in size, with an age of 0.8 Ma, obtained by the ^{40}Ar/^{39}Ar method (Zhang et al., 2010). Pu Zhang, Weiwen Huang, and Wei Wang (2010) attributed to the Fengshudao industry these small handaxes and attributed the absence of cleavers to the lack of adequate raw material (big blocks). In their view, the tools correspond to the variability of the Acheulean, with its own particularities, such as unidirectional carving.

7.4.4 An ancient Mode 2 in Asia?

The cultural dispersal hypotheses, such as the one by Sheila Mishra and collaborators (2010) previously mentioned, or any other argument in favor of an ancient Mode 2 in the Far East (Java and China), such as that of Hou, Yamei, and collaborators (2000), stumble upon the idea of Asian colonization, which is widely accepted as the most probable and is based on the Movius model. In fact, the controversy between the idea of an ancient presence of LFA in Java and the Movius line is more substantial. Mishra and collaborators, in a paper of 2010 as well as in other previous works, denied the presence of different *chaînes opératoires*' characteristics of Mode 1 and Mode 2, which is tantamount to denying the distinction between these two techniques. If it is the same industry with a higher or lesser development encompassing the entire ancient world, then the Movius line lacks meaning. However, in spite of the limitations of a simple geographic scheme, the common view accepts the Movius line, although its meaning has been much debated and there

are still details to be explained, such as the absence of bifaces in Eastern Europe.

A cultural dispersal synthesis of the Middle Paleolithic highlights the following points, indicated by Ofer Bar-Yosef and Miriam Belmaker (2011):

- Absence of Acheulean culture in Southeast Asia.
- Presence in numerous locations of Mode 2 in Western Asia—Near East—decreasing abundance of bifaces as we approach the East—Caucasus and Anatolia.
- Discontinuity between the two areas with evident presence of Mode 2 in the Levant and India.
- Absence of Mode 2 in China, with the exception of Bose basin.

A satisfactory explanation for the observed patterns of cultural dispersal requires that migrations from and to the west were discontinuous, in subsequent waves. But the evidence in relation to Java, indicated by Sheila Mishra and collaborators (2010), cannot be thus justified. Critiques, like that of Parth Chauhan (2010), point to an incorrect age estimation due to inherent dating problems of the ^{230}Th/^{234}U technique. Although it is possible that the Acheulean culture arrived in Southeast Asia as early as the Brunhes–Matuyama boundary, i.e., 0.78 Ma, additional evidence is required.

Naturally, the problem of cultural dispersal doesn't end with the absence or presence of Mode 2 in the Far East. Indeed, local particularities lead to the need to make more precise distinctions to account for what was an evolution subject to large population movements. As Marcel Otte (2010) has said, the Movius line exists, rather, as a frontier; it is like a veil that moves as time passes by the hand of ethnic traditions. Neither should these be confused with carving techniques, nor is it possible to identify a "biface" or "handaxe" with an "Acheulean utensil." In the strictest sense, Mode 2 refers to a specific *chaîne opératoire*, which never is the first to appear in a site, nor is it required to be exclusive, because it can coexist with simpler carvings. Otte (2010) recognized that, exceptionally, bifaces are present in the Middle Paleolithic of China, but a close look at Bose handaxes led him to argue that they cannot be considered Mode 2 at all. They would be the result of a *discovery*: from cores of adequate origin, bifaces could be obtained with not much manipulation. The procedure is the opposite of the Olorgesailie or Peninj technique, in which a huge block is flaked to obtain LFA. In an unfortunate expression, Marcel Otte (2010) qualified the "Chinese Acheulean" as a "research artifact." In the best of cases, it could be considered as cultural parallelism.

7.5 Beyond tools: the use of fire

The very ancient colonization of continental China indicated by fossil specimens and cultural evidence shows that hominins had to face the consequences of the ice ages. Although the territories were occupied during warm periods, and the northern areas above the 40th parallel with hominins present were never covered by ice, the cold and aridity during glaciations indicate environmental conditions far different from the tropical climate of Java (Ciochon & Bettis III, 2009). The most common explanation for the adaptive success of hominins in this extreme environment is cultural: the use of fire.

A similar interpretation scheme explained the subsequent presence of hominins in European high latitudes. In the mid-twentieth century it was still assumed that the control of fire, documented at around 500,000 years ago, would have been a complement to the cultural progress related to lithic carving (Oakley, 1956). That point of view was firmly maintained for several years and tallied well with the idea of a late hominin entry to Europe.

But extreme climates did not occur only in Asia and Europe. Colonization during the Lower Pleistocene of elevations in East Africa, such as the Gadeb plateau (Ethiopia), located about 2,300 m high, must have presented similar adaptive problems to the hominins. Thus, the proposal of an experimental use of fire as an adaptive solution in Gadeb about 1.5–0.7 Ma (Clark & Kurashina, 1979). It is, however, a hypothesis based on indirect evidence.

The detection of evidence on the use of fire is not easy, because a sporadic campfire made with just a few combustible materials leaves almost no trace and, moreover, meteorological agents can eliminate it very quickly. The burnt stones and ash of the habitual use of fireplaces are strong indications when enough remains are preserved, and constitute the majority of the evidence that has been provided by taphonomic interpretations. An alternative means to detect small and sporadic fires is the identification of magnetic anomalies. The Earth's magnetic field undergoes a slight distortion in the environment near a fire, as the heat produces changes in the minerals containing iron oxide. Some of the altered particles that induce that minimal alteration after cooling retain only a weak remnant magnetism, but the effects on the Earth's magnetic field of the stone can be detected by means of a portable magnetometer (Barbetti. 1986; Box 7.14).

The magnetic activity of sediment or stone could be due to natural causes, as is the case whenever chrons are located in a paleomagnetic column. By accurate paleomagnetic analysis it is possible, however, to detect signs of anthropic fire (Barbetti, 1986). Magnetic particles, composed of iron oxides or hydroxides, are dispersed in small amounts in almost all rocks and sediments. The sum of all magnetic moments of the particles contained in a specimen (a rock or a sediment sample) is called natural remanent magnetism (NRM), which is used to estimate the dating in a paleomagnetic column.

In the normal state, the magnetic direction of the particles is "frozen" and quite stable, with a distribution that produces a weak NRM. In Figure 7.13(a), pairs 1 and 2 have opposite magnetic moments and cancel each other, contributing little to the NRM. But, if a heating episode is followed by cooling, the magnetic moments are released and reoriented with respect to the external magnetic field, resulting in a thermo-remanent magnetism (TRM) that is higher than the original NRM. In Figure 7.13(b), pair 2 has been realigned to contribute more to the new TRM; pairs 3 and 4 are the result of particles of dehydrated hydroxide that, when forming oxides, acquire new moments.

Intense heat, above 700°C, completely eliminates the magnetic moments of NRM and when cooled, the magnetic particles tend to align with respect to the Earth's magnetic field. A gentle heating of around 400°C, leads to an episode of partial TRM, which can be examined in taphonomic terms. In Figure 7.13(c), stones A, B, C, and D do not follow the direction shared by the rest and to explain that discrepancy it can be said that A and B would have been recently subjected to heat from fire, C has moved slightly, and D probably rolled away from the fireplace during the erosion process (Barbetti, 1986). By a laboratory analysis of demagnetization and remagnetization it is possible to study the possible TRM character of a specimen's NRM, which we suspect has been subjected to fire.

Paleomagnetic studies have allowed the evaluation, in certain cases, of the heating that led to TRMs. Michael Barbetti (1986) highlighted the following:

• Koobi Fora (Kenya). Site FxJj20 East (1.5 Ma), with Mode 1 tools and the mandible KNM-ER attributed to *P. boisei*. Discolored sediments from areas 130 and 131 were analyzed. The 130 area supplied scant TRM evidence; the area 131 provided more but it was inconclusive (Clark & Harris, 1985). In addition to the case referred to by Barbetti, there are signs of thermal alterations in Mode 1 tools from the site FxJ150

Box 7.14 Magnetic activity of wood ash

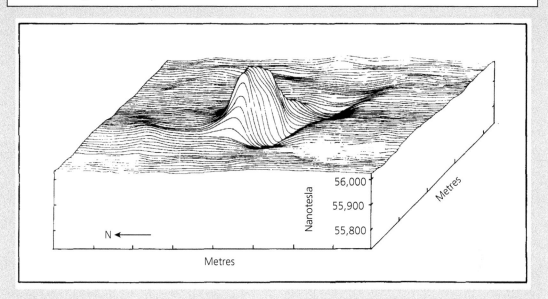

Magnetic anomaly recorded over an experimental fireplace, on sediment derived from weathered cherts and mudstone with some contribution from nearby basalts. A fire was lit in a pit 0.7 × 0.7 m and 0.2 m deep and allowed to burn out overnight; the pit was then re-filled with the original topsoil (figure from Barbetti, 1986, reprinted with permission from Elsevier).

Richard McClean and W.F. Kean (1993) have shown experimentally that the ash of wood used as firewood also has a certain magnetism that may be the cause of some of the anomalies. The reason for this magnetic activity of ash could be due to the phytoferritin contained in plants.

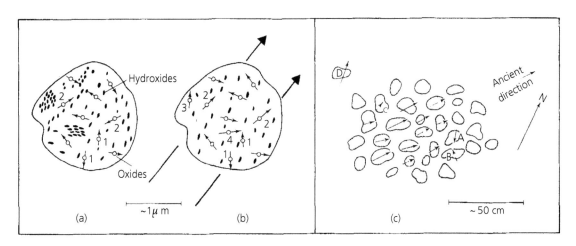

Figure 7.13 NRM anomalies due to ancient heating (see text) (figure of Barbetti, 1986). Reprinted from Barbetti, M. (1986). Traces of fire in the archaeological record, before one million years ago? *Journal of Human Evolution*, 15(8), 771–781, with permission from Elsevier.

(1.6–1.5 Ma) (Isaac, 1984), but no archeomagnetism analysis has been carried out.

- Gadeb (Ethiopia). Acheulean site Gadeb 8E. Of the ten analyzed specimens (stones with a gray and reddish discoloration as signs of having been subjected to fire), six indicated magnetic anomalies that were compatible in the most obvious cases with ancient temperatures of about 500°C (Barbetti et al., 1980). The stones' NRM could have had a volcanic origin but, in that case, one would expect that the direction of magnetic fields would change from one stone to another in a pile of rocks. The constancy in the direction of magnetization in Gadeb favors the idea of a cultural fire, although the weight of that evidence is not very great due to the difficulties in identifying the original position of the stones in the site (Barbetti, 1986). According to Barbetti et al. (1980) and James et al. (1989), the Gadeb paleomagnetic analyses are inconclusive because the recent volcanic activity prevents conducting studies of demagnetization and remagnetization.
- Chesowanja (Kenya). Oldowan site situated below a basaltic tuff dated 1.42±0.07 Ma. Fifty-one red clay clasts were found there with signs of having been subjected to fire, associated with artifacts of Mode 1 and with animal bones; the whole set was interpreted as remains of an ancestral fireplace (Clark & Harris, 1985). Paleomagnetic analysis has identified abnormalities in a clay sample, estimating a temperature of 400°C for the ancient fire, its most likely source is anthropic, but only hypothetically due to the absence of solid evidence (James et al., 1989).
- Middle Awash (Ethiopia). Specimens from an Oldowan site (Bodo-A4) and from an Acheulean site (Har-A3) were analyzed in the laboratory. The demagnetization and remagnetization analysis indicated that in both samples the TRM was produced by a temperature of 700°C. However, in the absence of an identifiable fireplace structure, taphonomic interpretation of materials pointed to a possible wood fire, so human action is questionable (Clark & Harris, 1985).

Steven James and collaborators (1989) conducted a review of ancient fire evidence and its taphonomic interpretations. Their argument is that either the indications can be attributed to natural fire, as is the case in member 7 Olorgesailie (Kenya), or there is a lack of solid evidence of the association between the use of fire, carved stones, and burnt bones that could point to the existence of an ancestral fireplace. The presence of burnt clay near stone tools is not sufficient proof,

---

Box 7.15 Keeping fire alive

J. Desmond Clark and John Harris (1985) argued that burning tree stumps could have been a means to preserve latent fire, in a way similar to that still in use today in India. Insofar as these tree stumps were located near termite nests, Clark and Harris think that the burnt clay would be the result of a fortuitous association.

---

according to James et al. (1989), as there are numerous places in Middle Awash where baked clay occurs as a result of natural fire in the absence of choppers and flakes; their simultaneous presence may, therefore, be fortuitous.

The case of South Africa is different. There is tangible evidence of the use of fire in Swartkrans, Kalambo Falls, Cave of Hearts, and Montagu Cave (James et al, 1989). The most solid proof comes from the first of these sites.

Abundant fossil bones that look as if they have been burned appeared in Swartkrans member 3 (South Africa). After analyzing them, Charles Brain and Andrew Sillen (1988) drew the conclusion that the temperature to which they were subjected had to be similar to that of a campfire. As we saw in Chapter 3, the dating of the members below Swartkrans member 4 is very imprecise, but member 3 could have a date of c. 1.6 Ma, implying a very ancient use of fire. Members 1 and 2, of similar age, contain numerous remains of robust australopithecines and some *Homo*, while member 3 supplied only nine specimens attributed to *P. robustus*. Brain and Sillen (1988) argued that the introduction of fire should have taken place in the short interval that separates member 2 and 3, when *P. robustus* still existed. However, the authors considered the available evidence on the control of fire insufficient to be able to assign this skill to one particular species.

7.5.1 Cultural dispersal of fire

The oldest evidence of controlled fire out of Africa is the Acheulean site of Gesher Benot Ya'aqov (Israel), where burnt wood, seeds, and flints appeared (Goren-Inbar et al., 2004). The locality is above the Brunhes–Matuyama boundary; Naama Goren-Inbar and collaborators (2004) attributed to it an age of approximately 700,000 years.

The presence of fire was indicated at six localities in China (Zhoukoudian, Xihoudu, Lantian, Yuanmou,

Jinniushan, and Huanglong) and one in Java (Trinil). The reservations about the most ancient sites in Africa also apply to the evidence from these Asian sites: more precise analyses to confirm their cultural origin is necessary. However, many archaeologists have interpreted the presence together of burnt bones, ash, charcoal, and calcinated stone artifacts, as well as a structure that could be a fireplace, to be evidence of the use of fire in situ at Zhoukoudian (James et al., 1989). Thus, Davidson Black (1931) stated that, based on chemical analyses in locality 1 of Zhoukoudian, the use of fire by *Sinanthropus* was beyond reasonable doubt. Later, the same idea was often maintained (Zhang, 1985) due to evidence of the use of fire from levels 4 and 10, in particular, of locality 1. Nevertheless, doubts have been expressed about the interpretation of such samples, such as in the case of the "burnt earth cave," collected by Breuil in 1931. Steven James and collaborators (1989), after reviewing the available evidence on the use of fire in the Lower and Middle Pleistocene, rejected the idea that the evidence from the "burnt earth cave" was ash. According to James et al. (1989), it is sediment, bone, and other kinds of organic material, including traces of carbonized wood. The evaluation of the evidence by Steve Weiner and collaborators (1998) concluded that, given the absence of ash or of ash remains (siliceous aggregates), and taking into account the characteristics of the supposed fireplace, there is no direct evidence of the use of fire in situ at Zhoukoudian (Box 7.16).

The basic question regarding a possible use of fire refers to the association between hominin specimens, lithic tools, and evidence of the effects of fire. If due to human activity, it would be expected to appear in the same locations as the tools. But this association is not easy to make in locality 1 due to the complexity of Zhoukoudian taphonomic interpretations. A good example of the difficulties of this task is when one must

---

Box 7.16 Was anthropic fire at Zhoukoudian?

The conclusion by Weiner et al. (1998) of a lack of evidence on the use of fire at Zhoukoudian was discussed by Wu Xinzhi (1999) based on several historical references that originated with Black (1931). All these references interpreted the evidence of burnt material at locality 1 as the result of anthropic activity.

---

consider the fact that scavengers scattered the hominin remains.

Noel Boaz and collaborators (2004) have provided, by a three-dimensional reconstruction of locality 1, the best evidence for understanding the anthropic activities that Zhoukoudian sedimentary sequences show (Figure 7.14).

Locus G of level 7, identified in 1931, was the first to offer an association of hominin specimens and tools (Black et al., 1933). In the same Locus G, or nearby, evidence of fire of anthropic origin was also detected (Pei & Zhang, 1985). Nonetheless, Boaz and collaborators (2004, p. 519) stated the following: "Carbon on all the *Homo erectus* fossils from Locus G, a circumscribed area of 1-meter diameter, earlier taken to indicate burning, cooking, and cannibalism, is here interpreted as detrital carbon deposited under water." According to Boaz et al. (2004), the marks on the bones are possibly due to the activity of hyenas. However, other authors concluded that, despite the fact that geochemical and sedimentary studies have modified previous interpretations on the use of fire in locality 1 of Zhoukoudian, the distribution of stone tools, as well as burnt bones, points to the transitory presence of hominins who used fire inside the cave.

Perhaps the most prudent conclusion is that of Liu Wu et al. (2009): whether Zhoukoudian's evidence indicates the ability of humans to control fire or is only the result of a naturally caused fire, remains an unresolved issue.

7.6 The transition Mode 2 (Acheulean) to Mode 3 (Mousterian)

The last Acheulean utensils of East Africa, i.e., the most recent, are from the Kalambo Falls location (Tanzania), associated with coal and wood materials. The age of these materials was fixed by ^{14}C method at 60,000 years (Vogel & Waterbolk, 1967). With regard to South Africa, tools attributed to the Late Acheulan appear in various sites—Cape Hangklip, Canteen Kopje (stratum 2A), Montagu Cave, Wonderwerk Cave, and Rooidam, Duinefontain 2, for example—with an age of c. 0.2 Ma (Kuman, 2007).

Mode 3, or Mousterian culture, is the lithic tool tradition that evolved in Europe from the Acheulean culture during the Middle Paleolithic. The name comes from the Le Moustier site (Dordogne, France), and was given by the prehistorian Gabriel de Mortillet in the nineteenth century, when he divided the Stone Age, known at the time in different periods, according

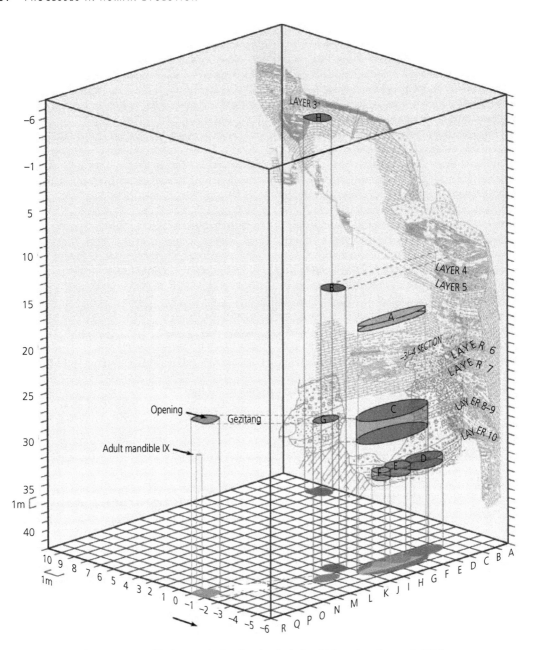

Figure 7.14 A three-dimensional map of locality 1 at Zhoukoudian, showing loci A–G (picture from Boaz et al., 2004).

to the technologies he had identified (Mortillet, 1897). Mortillet introduced the terms Mousterian, Aurignacian, and Magdalenian, in order of increasing complexity, to designate the tools from the French sites of Le Moustier, Aurignac, and La Magdalene. However, almost all the sites belonging to the Würm glacial period contain Mousterian tools. In many instances, their lower archaeological levels also show the transition of Acheulean to Mousterian tools, and even from the latter to Aurignacian ones. If speaking of a simple transit from Mode 1 to Mode 2 is difficult, the complexity increases when addressing the change to Mode 3, even if we restrict the analysis to the European case.

Table 7.4 Principal European sites with lithic industry, after MIS 15 and before MIS 7

Country	Site	Age (MIS or Ma)	Cultural tradition	Reference
France	Caune de l'Arago	MIS 14, 13, and 12	Mode 2	Barsky & de Lumley (2010)
Hungary	Vértesszöllös	c. 0.35 MIS 11/9	Mode 2	Despriée et al. (2011)
Germany	Bilzingsleben	MIS 7 or MIS 11	Mode 2	Haidle & Pawlik (2010)
Spain	Cuesta de la Bajada	MIS 11	Mode 2	Santonja & Pérez-González (2010)
France	Arago G	0.50–0.45	Mode 2	Falguères et al. (2010)
Italy	Visogliano	0.48–0.44	Mode 2	Falguères et al. (2010)
France	La Celle	MIS 11	Mode 2	Limondin-Lozouet et al. (2010)
Italy	Fontana	0.46	Mode 2	Abbate & Sagri (2011)
Italy	Ceprano	0.45	Mode 2	Giovanni Muttoni et al. (2010)
Italy	Castel di Guido	c. 0.43	Mode 2	Francesco Mallegni et al. (1983)
Germany	Bilzingsleben	0.4	Mode 2	Abbate & Sagri (2011)
France (Loir Valley)	(Grouais-de-Chicheray)	0.4	Mode 2	Despriée (2011)
France	Arago	c. 0.4	Mode 2	Abbate & Sagri (2011)
France	St. Acheul	0.4	Mode 2	Abbate & Sagri (2011)
England	Swanscombe	0.4	Mode 2	Abbate & Sagri (2011)
Italy	Terra Amata	0.4	Mode 2	Abbate & Sagri (2011)
Spain	Torralba	0.4	Mode 2	Falguères et al. (2006)
Spain	Ambrona	0.4–0.3	Mode 2	Falguères et al. (2010)
France	La Micoque	0.4–0.3	Mode 2	Falguères et al. (2010)
France (Cher Valley)	La Morandiére	0.37	Levallois	Despriée (2011)
France	Orgnac	0.35	Mode 2	Falguères et al. (2010)
England	Hoxne	0.3	Mode 2	Abbate & Sagri (2011)
Italy	Torre in Pietra	c. 0.3	Mode 2	Despriée (2011)
Italy	Valdarno	0.3	Mode 2	Abbate & Sagri (2011)
France (Creuse Valley)	Champs-de-Chaume	0.3/0.2	Levallois	Despriée (2011)
France (Loir Valley)	Villeprovert	0.24	Levallois	Despriée (2011)

The permanent colonization of Europe began with a series of migratory waves, which we will describe in upcoming chapters. In several sites that provide evidence of human presence, Mode 2 artifacts appeared in contrast with the Mode 1 utensils associated with evidence of the oldest occupation of the continent. Many European sites contain, starting from Marine Isotope Stage (MIS) 14 (see Box 7.17) and, in particular, from MIS 11, more advanced tools than the choppers, flakes, and simple cores of the Oldowan culture. Table 7.4 shows the most important cases.

The Levallois technique appeared during the Acheulean period, and was used subsequently (Figure 7.15). The oldest Levallois carvings are probably c. 400 Ka and come from the Lake Baringo region (Kenya) (Tryon, 2006). Its pinnacle was reached during the Mousterian culture. The purpose of this technique is to produce flakes or foils with a very precise shape from stone cores that serve as raw material. The cores must first be carefully prepared by trimming their edges and removing small flakes until the core has the correct shape. Thereafter, with the last blow, the desired

Box 7.17 Marine Isotope Stages (MIS)

Oxygen isotopes present in the calcite of marine animal shells contained in sediments provide information about climate cycles. During cold periods, large amounts of ^{16}O—lighter and more abundant in the snow produced by the evaporation of ocean water—accumulates on glaciers, while in the oceans (i.e., in the shells) there remains a higher content of ^{18}O, which is heavier. In warm periods, the content of ^{18}O in the shells decreases.

The identification of a MIS starts with the interglacial we are living in now, MIS 1, numbering in reverse chronological order the succession of cold and warm periods, so that MIS 2 corresponds to the last "classic" glaciation (Würm). The others are identified as MIS 6 (Riss), MIS 12 (Mindel), and MIS 16 (Günz), with 19 major climatic alternations since 1.78 Ma, at the beginning of the Middle Pleistocene (see Chapter 8).

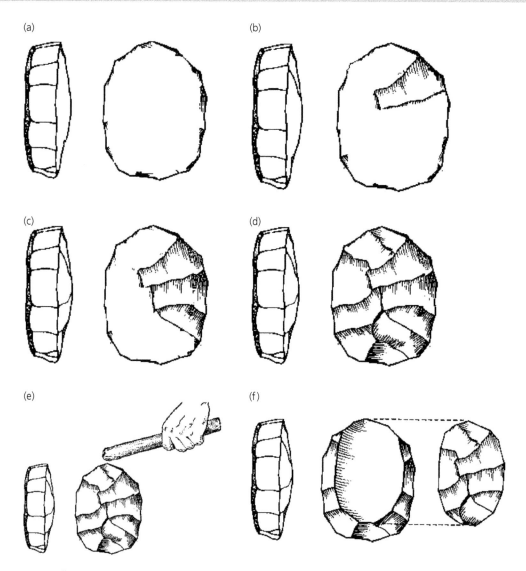

(a) (b) (c) (d) (e) (f)

Figure 7.15 Levallois technique. Phases in the construction of a tool by flake removal: (a) preparation of an adequately shaped core; (b and c) removal of flakes; (d) the prepared platform is obtained; (e) a last blow with a soft hammer separates the tool from the core (f) (drawings from http://www.hf.uio.no/iakh/forskning/sarc/iakh/lithic/LEV/Lev.htm).

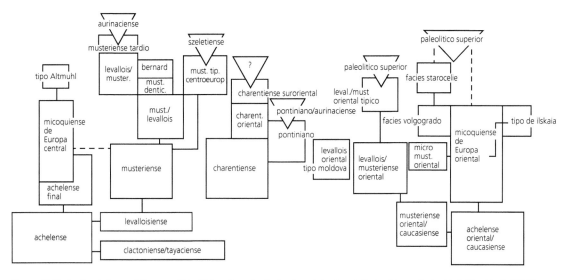

Figure 7.16 Regional phases of European Paleolithic cultures beginning from Acheulean industry. Scheme by José Luis Díez Martín (2003) based on Otte (1996). The names of the industries are in Spanish (*musteriense* = Mousterian; *achelense* =Acheulean). Díez Martín, F. (2003). La aplicación de los "modos tecnológicos" en el análisis de las industrias paleolíticas. Reflexiones desde la perspectiva europea. SPAL. Revista de prehistoria y arqueología de la Universidad de Sevilla, 12, 35–51.

Box 7.18 Relationship between the cultural developments of Neanderthals and modern humans

David Brose and Milford Wolpoff (1971) described as "transitional industries" those which, according to the authors, indicate a gradual evolution of technological level from the Middle Paleolithic to the Upper Paleolithic. However, we are considering here a previous transition that appeared in Europe up to the Mousterian industry, and which constitutes the starting point of the "transitional industries" of Brose and Wolpoff (1971). Included in these are the Châtelperronian tradition, an evolution of the Mousterian culture, and the Aurignacian, which is one of the characteristic techniques of Upper Paleolithic modern humans. Their possible relationships will be discussed in Chapter 11.

---

flake, a Levallois point, for instance, is obtained. The final results of the process, which include points and scrapers, among many other instruments, are subsequently modified to sharpen their edges. The amazing care with which the material was worked constitutes, according to Bordes (1953), evidence that these tools were intended to last for a long time in a permanent living location. From cultural Mode 2, European

hominins developed more advanced techniques that can be outlined by the sequence Acheulean–Levallois–Mousterian (see Box 7.18). However, the process of change is very complex, as could be imagined. There are several techniques of lithic carving that in different European locations indicate either an evolution to Mode 2 or an alternative to the more characteristic Acheulean tools.

Figure 7.16 shows a diagram of the cultural diversity in Europe during the Middle Paleolithic. The scheme combines both a temporal perspective of evolution (indicated from the bottom up, by the technological development), and a geographical diversity (with the process from Eastern Europe located on the right side of the figure).

However, even a diagram such as that in Figure 7.16 could be considered too simple. François Bordes (1953) indicated up to five different cultural facies in the Mousterian industry: typical Mousterian; of Acheulean tradition—types A and B; Quina-Ferrassie; denticulate; and Vasconian. The main technical system in each industry should also be considered as a variable. Taking this into consideration, Emmanuelle Vieillevigne and collaborators (2008) analyzed the temporal change sequence of the lithic industries in southwestern France between stages MIS 10 and MIS 3, crossing them with the cultural facies defined by Bordes (1953), and adding three more: ancient Middle Paleolithic; Châtelperronian; and ancient Aurignacian, with the

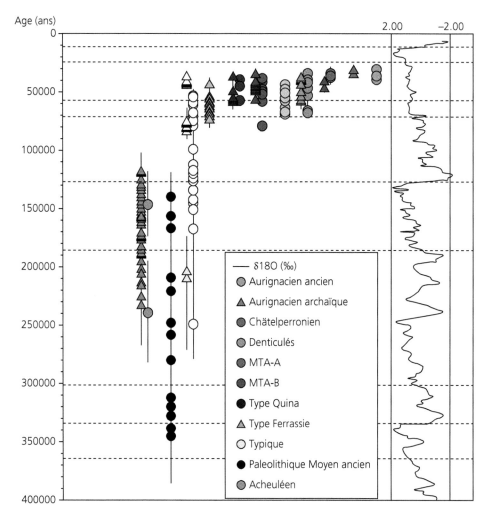

Figure 7.17 Temporal change sequence of lithic industries in southwestern France between MIS 10 and MIS 3 (Vieillevigne et al., 2008; the industry names are in French): Acheulean; ancient Middle Paleolithic; typical (typical Mousterian); Ferrassie type (Mousterian type Ferrassie); Type Quina (Mousterian type Quina); MTA (Mousterian of Acheulean tradition, types A and B); denticulate (denticulate Mousterian); Chatelperronian; archaic and old Aurignacian. Vieillevigne, E., Bourguignon, L., Ortega, I., & Guibert, P. (2008). Analyse croisée des données chronologiques et des industries lithiques dans le grand sud-ouest de la France (OIS 10 à 3) Paléo [En ligne], 20.

technical systems Levallois, discoid, Quina, laminar, and *façonnage*. The results are shown in Figure 7.17.

The graph by Vieillevigne et al. (2008), corresponding to southwest France, shows clearly the existence of a process of remarkable technological diversity. Needless to say, if we consider the entire continent, the list of described industries is almost endless. The reason we focus on the various cultural modes is not archaeological but anthropological. It is to understand to what extent it is possible to trace within the cultural diversity a shared evolution of carving techniques that highlights the possible replacement of populations.

The evolutionary explanation of a process of overlapping, diversity, and cultural change over a long period, as is the one to which the European transition studies point to, is somewhat precarious. Along with many other authors who have interpreted the meaning of culture for human evolution, we understand lithic industries as a medium—socially shared, utilized, and transmitted—first to obtain resources for survival, but which cannot be separated from other functional aspects arising from the organization of each group. As such, cultural diversity depends not only on general patterns spread over wide areas and over considerable periods of time, but also on a multitude of factors strongly influenced by ecological pressures and having local temporal and geographical components. The climatic alternatives, availability of raw materials, the required use in a precise location and time, the regional carving traditions, the bonds of social norms, and the historical contingencies—which include the relationships between each group and others nearby—are all variables to consider when assessing the effect of the succession of cultural modes and the role, within this general meaning, of the possible deviations present in each locality.

It is possible to add that the functional use of lithic tools could lead to a confluence in sizes and shapes as needs dictate. The example given by Marcel Otte (2010) regarding the bifaces is very illustrative. If, following common practice, we identify "biface" with the typical tools of the Acheulean tradition, then—argued Otte—this name cannot be used to refer to the Micoquian utensils abundant in Central Europe: asymmetric axes carved by alternating retouches in each of the two faces, which have little to do either with Acheulean *chaîne opératoire* or with the shape of true bifaces. Similarly, Neanderthal craftsmen, versed in the Mousterian technique, were able to achieve very refined blades, highly symmetrical, and with very sharp edges, following a *chaîne opératoire* that in its early stages started with already shaped cores. If such an artifact was left abandoned without continuing the carving process, it could now be mistaken for a "biface" (Otte, 2010).

It is advisable to consider the technological evolutionary process outlined in Figure 7.16, leaving aside what may be local particularities and focusing on the most general traditions. Within this general segment, the Clactonian industry is an alternative technique to the European Acheulean technology transition from Mode 1, and the most noticeable difference between the two is the lack of bifaces in the Clactonian industry. As Fernando Díez Martín (2003) reiterated, the Clactonian industry was interpreted for years as a preparatory stage to the Acheulean, whereas now it is considered as a particular response, simultaneous to that of the Acheulean tools, to the lack of suitable raw materials. As such, the Clactonian industry must be accepted as belonging to Mode 2 and, therefore, characteristic of the populations that contributed to permanent settlement in Europe.

7.6.1 Mode 3: Mousterian culture

The archaeological richness and sedimentary breadth of some of European sites like La Ferrassie, La Quina, and Combe-Grenal, grants them a special interest for studying the interaction between cultural utensils and adaptive responses. Most European sites belonging to the Würm glacial period contain Mousterian tools. Similar utensils have appeared in the Near East, at Tabun, Skuhl, and Qafzeh.

Mousterian culture is, strictly speaking, a lithic tool tradition that evolved from the Acheulean culture during the Middle Paleolithic. The splendor of the Mousterian culture occurred in Europe and the Near East during the Würm glaciation, which was the last one. The term "Mousteroid" is occasionally used to highlight differences between the Middle Paleolithic tools found in Africa and the Far East, where there are no Neanderthals, and the European Mousterian tradition (Bever, 2001).

Mousterian techniques changed over time. Geoffrey Clark's (1997a) study of the Middle and Upper Paleolithic cultural stages convincingly demonstrated how wrong it is to speak about "Mousterian" as a closed tradition, with precise limits; as a unit with precise temporal boundaries. Even so, we will talk about a Mousterian style, as Clark himself did, which becomes apparent when compared with the Upper Paleolithic technical and artistic innovations that constitute Mode 4. However, to understand the magnitude of Mode 3, we must extend the consideration of "Mousterian culture" from lithic tools to other products and techniques that appear at Mousterian sites. In a broad sense, Mode 3 culture includes controversial features, such as objects created with a decorative intention and indications of funerary practices. We will return to this question in Chapter 11. Let us begin with the Mousterian tool-making techniques. They were used to produce tools that were much more specialized than Acheulean ones. The most typical Mousterian tools found in Europe and the Near East are flakes produced by means of the Levallois technique, which were subsequently modified to produce diverse and sharper edges. Objects made from bone are less frequent, but up to 60 types of flakes and stone foils can be identified, which served different functions (Bordes, 1979).

Tools obtained by means of the Levallois technique are typical of European and Near East Mousterian sites. Bifaces, on the contrary—so abundant in Acheulean sites—are scarce. The difference has to do mostly with the manipulation of the tools; scrapers were already produced using Acheulean, and even Oldowan, techniques. The novelty lies in the abundance and the careful tool retouching.

Both in spatial and temporal terms, the Mousterian culture coincides with Neanderthals (to be addressed

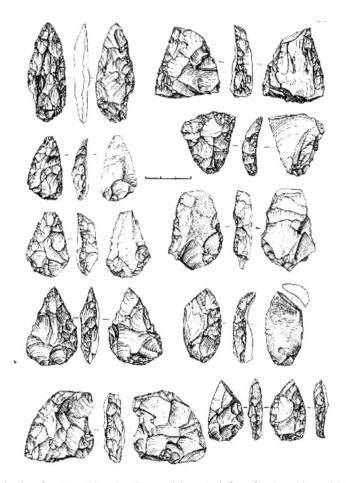

Figure 7.18 Mousterian handaxes from Mezmaiskaya Cave (Caucasus), layers 3–2b (figure from Doronichev and Golovanova, 2003).

in Chapter 10). This identification between the Mousterian culture and *H. neanderthalensis* has been considered so consistent that, repeatedly, European sites yielding no human specimens, or with scarce and fragmented remains, were attributed to Neanderthals on the sole basis of the presence of Mousterian utensils. Despite the difficulties inherent in associating a given species with a cultural tradition, it was beyond a doubt that Mousterian culture was part of the Neanderthal identity—exclusively?

This perception changed with the reinterpretation of the Near East sites (Bar-Yosef & Vandermeersch, 1993). Scrapers and Levallois points, which were very similar to the typical European ones, turned up there. Neanderthals also existed there, of course, but in contrast with European sites, a distinction could not be drawn between localities that had housed Neanderthals and anatomically modern humans solely on the grounds of

the cultural traditions. The more or less systematic distinction between Neanderthal–Mousterian and Cro-Magnon–Aurignacian helped to clarify the situation in Europe. But it could not be transferred to the Near East, where sites occupied by Neanderthals and those inhabited by anatomically modern humans, proto-Cro-Magnons, yielded the same Mousterian-tradition utensils.

This coincidence implies several things. First, that cultural sharing was common during the Middle Paleolithic, at least in Levant sites. Second, that during the initial stages of their occupation of the eastern shore of the Mediterranean, anatomically modern humans made use of the same utensils as Neanderthals. Hence, it seems that at the time Skuhl and Qafzeh were inhabited, there was no technical superiority of modern humans over Neanderthals. The third and most important implication has to do with the inferences that can be made because Neanderthals and *Homo sapiens* shared identical tool-making techniques. As we have already seen, the interpretation of the mental processes involved in the production of tools suggests that complex mental capabilities are required to produce stone tools. We are now presented with solid proof that Neanderthals and modern humans shared techniques. Does this mean that Neanderthal cognitive abilities to produce tools were as complex as those currently characteristic of our own species? We will address this question in Chapter 11.

7.6.2 The transition Mode 3 (Mousterian) to Mode 4 (Aurignacian): the "modern mind"

The Mode 3, i.e., the Mousterian culture characteristic of the Neanderthals in Europe and Near East Asia during the Middle Paleolithic, ranged from about 100 to 40 Ka. Around these dates, more developed technocomplexes appeared in Europe. Industries called "transitional," to contrast them with the "real" Mode 4, a set of cultural traditions of the Upper Paleolithic, coincided with the entry of the first modern humans, the Cro-Magnon, into Europe between 40 and 28 Ka. Traditions of the Upper Paleolithic include not only tools that are more precise and sophisticated than those from the earlier Mousterian culture. There also are abundant representations of real objects in the form of engravings, paintings, and sculptures; realistic representations that display significant differences in favor of the development of Mode 4. A good example is the large mammal paintings of the Chauvet cave, dated by radiocarbon calibration (^{14}C) at c. 36 Ka (von Petzinger & Nowell, 2014). Such a realistic intensity of the polychromes of the Upper Paleolithic have led to the argument that modern humans achieved an artistic revolution explainable only by a corresponding cognitive revolution attaining what we call the "modern mind." Does this cognitive revolution appear suddenly and exclusively in our species?

As McBrearty and Brooks (2000) pointed out, the proposal of a cognitive revolution repeats a scenario introduced in the nineteenth century with the Age of the Reindeer (Lartet & Christy, 1865–75). Around the 1920s, the Upper Paleolithic was generally characterized by the presence of sculptures, paintings, and bone utensils. But according to McBrearty and Brooks (2000), the evidence used to determine the changes between the Lower, Middle, and Upper Paleolithic was always taken from the Western Europe archaeological record. During the last glacial period, the human occupation of that area was irregular, as F. Clark Howell (1952) pointed out, with populations periodically reduced, or even extinguished. McBrearty and Brooks (2000) argued that the "revolutionary" nature of the European Upper Paleolithic is mainly due to the discontinuity in the archaeological record, rather than to cultural, cognitive, and biological transformations, as suggested by advocates of the "human revolution." Instead, McBrearty and Brooks (2000) hold that there was a long process that gradually led to the European Aurignacian richness.

Was the transition to the European Aurignacian gradual or sudden? To analyze the process of change from Mode 4 to Mode 5, which is the same as specifying the time and mode of the emergence of the modern mind, we need to clarify a number of interrelated processes:

- The appearance of *Homo sapiens.*
- Its dispersal from the place of origin.
- Cultural development leading to the industries of the European Upper Paleolithic.

In early manifestations all these processes are related with certain evolutionary events that took place in the southern part of Africa. In this chapter we only address the cultural evolution in Africa that is not related to the European Acheulean–Levallois–Mousterian transit. Africa has its own sequence of changes.

7.7 The African Middle Stone Age

As we have said repeatedly, the Mousterian culture is often identified with Neanderthals, European hominins that did not occupy Africa. Neither did the Mousterian industry arrive in the African continent. Even so, the traditional consideration of the African archaeological record was influenced by the scheme used to create the European sequence of Lower, Middle, and Upper Paleolithic stages. The cultural phases of Africa were consonantly grouped into Early, Middle, and Late Stone Age (ESA, MSA, and LSA, respectively). ESA encompasses not only Mode 1 but also Mode 2, so that the difference between the cultural level of ESA and MSA comes from innovations that go beyond the tools of the Acheulean culture. If ESA is linked especially with large bifaces (LFA handaxes), the MSA has been traditionally characterized by the absence of large bifaces, an emphasis on Levallois technology, and the presence of points (Goodwin & Van Riet Lowe, 1929).

Both East Africa and South Africa contain evidence of an ancient presence of MSA technocomplexes.

In addition to the findings on the surface, whose age is imprecise, about 60 sites in East Africa susceptible to dating and which contain MSA utensils (Basell, 2008) have been described. According to the review by Laura Basell (2008), their ages range from <200 Ka to about 40 Ka. The beginning of MSA could be even much older. Jayne Wilkins (2013) indicated that there are MSA utensils at Kathu Pan 1 site (Northern Cape, South Africa) of an age of up to c. 500 Ka.

In principle, ESA and MSA could be distinguished simply by the presence of handaxes or points. But, as clarified by Sally McBrearty and Christian Tryon (2006), the sites normally lack tools capable of leading to a formal classification. The problem is that "formal tools are vastly outnumbered at nearly all sites by flakes, cores, and expedient tools, and the basic flake and core artifact inventories [of the Acheulean and MSA], are in many cases indistinguishable" (McBrearty & Tryon, 2006, page 261). In the absence of reliable dating, we face the fact that the method of direct percussion is not an accurate chronological marker. However, MSA can also be associated with "blade and microlithic technology, bone tools, increased geographic range, specialized hunting, the use of aquatic resources, long distance trade, systematic processing and use of pigment, and art and decoration" (McBrearty & Brooks, 2000, p. 453). And among the innovations related to hunting are procedures and resources that were attributed initially to Mode 4 of European Cro-Magnon, as is the use of hafting adhesives, identified in the South African MSA (Charrié-Duhaut et al., 2013).

Considering the novelty of compound-adhesive manufacture, the age of the emergence of the oldest MSA is fixed at c. 300 Ka (Wadley, 2010; Henshilwood & Dubreuil, 2012). Dates of that range are similar to those of the lower horizon of the Gademotta formation (Ethiopia), 276±4 Ka (Morgan & Renne, 2008) and the bedded tuff member of the Kapthurin formation (Kenya), 284±24 Ka (Deino & McBrearty 2002), both obtained by the ^{40}Ar/^{39}Ar method. However, as argued by Robert Foley, José Manuel Maillo-Fernández, and Marta Mirazón Lahr (2013, p. 3): "The majority of MSA sites postdate 130 Ka, and it is from the beginning of MIS5 (c. 130–74 Ka) and during MIS4 (74–60 Ka) that the classic African MSA becomes widespread and abundant."

7.7.1 The transition MSA to LSA

South African sites have allowed the study in greater detail of the development of the MSA, especially in its final stages, and the transition to LSA (Figure 7.18).

According to Richard Klein and collaborators (2004), the Mediterranean artifacts from Western Europe and North Africa that are characteristic of the range from 250–200 to 40–35 Ka can be considered "Mousterian," while the utensils that appear in the same age range in Southern Africa—generally in the sub-Saharan part of the continent—are assigned to the African MSA. Klein (2008) warned about what can bring confusion to a technological classification such as this one, since the internal variability of both Mousterian artifacts, like those of the MSA, is greater than the differences between the two types of technology. However, Klein (2008) provided an interesting criterion, shared by the Mousterian mode and MSA, that distinguishes these traditions from the culture of the modern humans entering Europe. Thus, the modes previous to the African LSA lack:

- unequivocally artistic objects;
- artifacts of bone, ivory, tortoiseshell, or horn carved with care;
- tombs with irrefutable evidence of funeral ceremony or ritual.

All these "advanced" elements refer to the development of the "modern mind," which we will address in Chapter 11. There we shall return to the question of the presence or absence of these signs of symbolism in the southern Africa MSA.

There are several concurrent processes at the temporal range of 80–60 Ka: the expansion of modern human populations, their exit from Africa, and the emergence of technological and symbolic innovations. Naturally, the possibility of specifying dates, in particular for new tools, becomes the key to relate all those events. The final stages of cultural evolution within the MSA correspond in South Africa to the traditions of Still Bay (SB) and Howieson's Poort (HP), widespread in the southern cone of Africa (Figure 7.19). The study of Zenobia Jacobs and colleagues (2008, p. 733) characterizes SB as "flake-based technology includes finely shaped, bifacially worked, lanceolate points that were probably parts of spearheads." On the other hand, HP is described as "blade-rich (. . .) associated with backed (blunted) tools that most likely served as composite weapons, made of multiple stone artifacts." However, both traditions share "associated bone points and tools, engraved ochres and ostrich eggshells, and shell beads." Four South African sites are particularly useful to detail the scope of SB and HP phases: Diepkloof, Sibudu, Blombos, and Klasies River (see Figure 7.20); the first two because they have tools of both traditions; Blombos, because of the abundance of engravings,

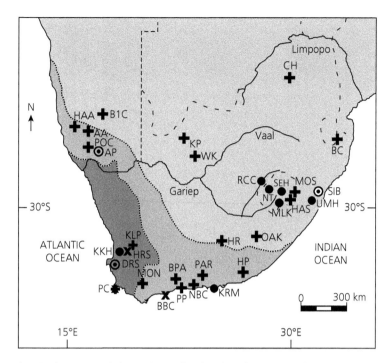

Figure 7.19 Locations of South Africa sites at which SB and HP artifacts have been found. Solid circles indicate those sites where HP deposits have been dated in the study by Jacobs et al. (2008), whereas open circles with a central dot denote study sites that contain both dated SB and HB industries. The symbols × and + indicate other known (or claimed) occurrences of SB and HP, respectively; these sites may have associated independent age estimates. Also shown are the modern rainfall zones: winter (dark gray); all year (medium gray); and summer (light gray) (Jacobs et al., 2008). Reprinted in Tribolo, C., Mercier, N., Valladas, H., Joron, J. L., Guibert, P., Lefrais, Y., Selo, M., Texier, P.-J., Rigaud, J.-Ph., Porraz, G., Poggenpoel, C., Parkington, J., Texier, J.-P., Lenoble, A. (2009). Thermoluminescence dating of a Stillbay–Howiesons Poort sequence at Diepkloof Rock Shelter (Western Cape, South Africa). *Journal of Archaeological Science*, 36(3), 730–739, with permission of Elsevier.

pigments, and perforated beads; and Klasies River, because the fossil remains led to clarification of which hominins were responsible for the transition from MSA to LSA. This is something of special interest because both SB and HP already show different innovations, which previously were associated only with the most advanced culture of the Upper Paleolithic.

Sibudu cave, located in KwaZulu-Natal North coasts, near Durban (South Africa), include a remarkable sequence of MSA occupations extending over a short timeframe. The first MSA tools appeared in MIS (Marine Isotope Stage) 4, i.e., they are older than c. 61 Ka. From that point, phases follow one after another: pre-Still Bay, Still Bay, Howieson's Poort, post-Howieson's Poort phase, and late and final MSA phases directly overlain by Iron Age occupation (Backwell et al., 2008). The final phases, post-Howieson's Poort, of Sibudu have been studied by Manuel Will, Gregor Bader, and Nicholas Conard (2014), who attributed MIS 3 to around 58 Ka. This proliferation of different traditions over a short period may have been related

to both climate change and the tendencies of hunter-gatherers in regard to their use of local resources.

Diepkloof Rock Shelter shows a similar succession. It is located on the west coast of South Africa (Western Cape Province), 14 km from the Atlantic at the Table Mountain Group, very close to various sites with MSA industry. The excavation completed up to 2013 reveals the following sequence: MSA (type "Mike")–Pre-SB–SB–Early HP–MSA (type "Jack")–Interm. HP–Late HP–Post-HP (Porraz, Texier et al., 2013). The change from SB to HP in Diepkloof is abrupt and that rapid shift has been interpreted in three ways: a population replacement, a discontinuity in the archaeological record, and/or a fast innovation, with a new way of hafting (and using) tools (Igreja & Porraz, 2013). However, Porraz, Texier, and collaborators (2013) disagree with the hypothesis connecting the appearance of the SB and the HP to the arrival of new populations. Instead, the authors support a scenario based on local evolution with distinct technological traditions that coexisted in Southern African during MIS 5. As argued by Porraz,

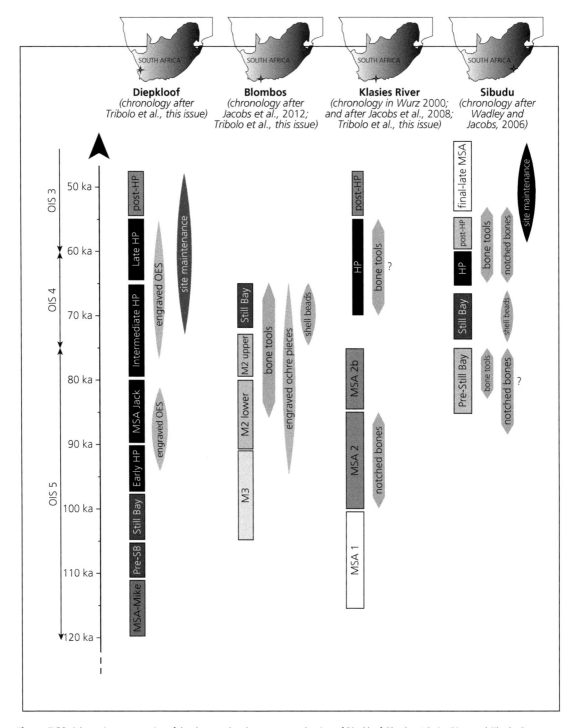

Figure 7.20 Schematic representation of the chrono-cultural sequences at the sites of Diepkloof, Blombos, Klasies River, and Sibudu. Bone technology, abstract manifestations, and main phases of site maintenance activities are figured. Absolute chronology combines mean ages, standard deviations, and stratigraphic orders. This representation illustrates the diversity and the arrhythmic tempo of appearance of cultural innovations recorded during the southern African MSA (Porraz, Texier et al., 2013). Reprinted in Porraz, G., Parkington, J. E., Rigaud, J.-P., Miller, C. E., Poggenpoel, C., Tribolo, C., . . . Texier, P.-J. (2013). The MSA sequence of Diepkloof and the history of southern African Late Pleistocene populations. *Journal of Archaeological Science*, 40(9), 3542–3552, with permission from Elsevier.

Parkington et al. (2013, p. 3542), during MIS 5 there was "the coexistence of multiple, distinct technological traditions. We argue that the formation of regional identities in southern Africa would have favored and increased cultural interactions between groups at a local scale, providing a favorable context for the development and diffusion of innovations (. . .). The southern African data suggest that the history of modern humans has been characterized by multiple and independent evolutionary trajectories and that different paths and scenarios existed towards the adoption of 'modern' hunter-gatherer lifestyles." Within that independent evolution, SB and HP traditions appear and disappear at the sites in very short periods. But on the whole these technocomplexes "are neither of short duration in time, nor homogeneous across space" (Porraz, Texier et al., 2013, p. 3398). Consequently, for Porraz, Texier et al. (2013), the traditions SB and HP cannot be considered as horizon markers.

Who were the authors of the African MSA and its development in the SB and HP traditions? What is the relationship between these industries and the decorative objects associated with the cultural manifestations of the European Upper Paleolithic? To get into these issues we must describe the fossils associated with these industries and establish the relationships between diverse populations and cultural traditions. We will address these points in Chapters 10 and 11.

CHAPTER 8

Middle and Lower Pleistocene: the *Homo* radiation

The evolution of our lineage during the Pleistocene reaches heights of great consequence. After their exit from Africa, hominins soon occupied Asia and half a million years later reached Europe (Table 8.1) (Figure 8.1).

The consequences resulting from dispersals of such magnitude are varied; thus, it is not easy to determine how the different Pleistocene events are related. In this chapter we will refer only to the early events, those involving the beginning of the hominin radiation out of Africa in the Lower Pleistocene and its diversification up to the Middle Pleistocene. To analyze this process we have various pieces of information regarding the age, morphology, location, and lithic techniques of the respective specimens. The sum of all these form a mosaic whose pieces we must fit into place (Table 8.2).

Although the reconstruction of any evolutionary process always proves difficult, the human lineage during the Middle Pleistocene appeared to be an exception during the mid-twentieth century. At the time it was common to argue that, given the diversity in the Pliocene of the tribe Hominini—understood back then as the family Hominidae—humans were reduced in the Lower Pleistocene to a single species, *Homo erectus*, present not only in Africa, but also in Asia and, with some exceptions, in Europe. The other protagonist of the Pliocene cladogenesis, *Paranthropus*, had already become extinct around 1 Ma.

But the problems raised by such a simple interpretation appeared as soon as the meaning of the typical taxon of the Middle Pleistocene was contemplated, starting with the question of whether it is actually a well-defined species according to the criteria we are working with.

8.1 Is *Homo erectus* a well-defined species?

At the end of the nineteenth century, the Dutch doctor Eugène Dubois began fieldwork on the Island of Java—a Dutch colony at the time—hoping to find the human ancestors hypothetically described by Darwin. The work began at the River Solo, near the Kending Hills in the area named Trinil by Dubois, containing a similar fauna to Siwaliks in Pakistan. The work soon paid off. In 1891 a skullcap (technically, this is called a "calotte"), known today as Trinil II, was found in Trinil, and a femur the year after (Trinil I; Figure 8.2 and Box 8.1). Dubois attributed both specimens to the same individual, though this identification has often been questioned: *Pithecanthropus erectus* was born. Ernst Mayr (1950) included the taxon *Pithecanthropus erectus* in *Homo erectus*.

Only Neanderthals had been identified as possible ancestors of current humans at the time of the Trinil discoveries. Not surprisingly, the features of *Pithecanthropus* suggested this was an intermediate being between chimpanzees and us. The Trinil 1 femur was almost identical to ours, indicating from the first moment that the fossil belonged to a biped. The cranial capacity—slightly over 800 cc—was double that of chimpanzees, but much smaller than ours, which led to the specimen being named "ape-man".

When demarcating the taxon hypodigm of *Homo erectus*, G. Philip Rightmire (1990) restated that the formal diagnostic of a species must be based on apomorphic traits—or derived—and not on plesiomorphic or primitive traits, inherited from previous taxa. Thus, to characterize *Homo erectus*, the trait alluded to by the name of the first discovered specimen is of no use. The condition "erectus" granted by Dubois (1894) to *Pithecanthropus* corresponds in fact to a primitive trait.

Then, which are the apomorphies defining *Homo erectus*?

When Dubois named the taxon for the first time, the current methods required to formally describe any taxonomic proposal were not in force. So, actually, *H. erectus* was never appropriately defined, i.e., indicating

Processes in Human Evolution. Francisco J. Ayala and Camilo J. Cela-Conde, Oxford University Press (2017).
© Francisco J. Ayala and Camilo J. Cela-Conde.
DOI 10.1093/acprof:oso/9780198739906.001.0001

Table 8.1 Main evolutionary events of the Pleistocene

Early Pleistocene	1.806–0.781 Ma	*Homo* left Africa Extinction of *Paranthropus*
Middle Pleistocene	0.781–0.126 Ma	Presence in all continents (Africa and Eurasia)
Late Pleistocene	0.126–0.0118 Ma	*Homo sapiens* arrived in Asia and Europe The remaining *Homo* species disappear

their apomorphies. The huge number of subsequent studies pointed out several typical traits of the species. Gail Kennedy (1991) conducted a polarity analysis of 12 traits considered broadly as autapomorphies of this taxon, i.e., *H. erectus'* traits not present in any other previous or subsequent species. Kennedy compared the distribution of those characters within the

genus *Homo*, taking as *outgroup* the African apes *Pan troglodytes*—common chimpanzee—and *Gorilla gorilla*. As mentioned by the author, the method was appropriate because the monophyly of the three genera *Homo*, *Pan*, and *Gorilla* is not in question.

The results obtained by Kennedy (1991) showed that none of the 12 traits considered were autapomorphies of *Homo erectus*. Eight of the nine non-metric characters (qualitative) were symplesiomorphies, inherited characters that appeared in the most recent common ancestor. The ninth trait, the presence of an occipital torus, is also found in hominins other than *Homo erectus*, such as the Neanderthals. The remaining three traits studied by Kennedy (1991) are: endocranial volume, cranial wall thickness, and tympanic plate thickness—high in all cases. These three characters are metric—a number can express them. Therefore, according to the strict interpretation of the original cladistic rules, which is

Table 8.2 *Homo* species from the Plio-Pleistocene, with age, location, and cultural mode (European specimens have not been included)

Species	Oldest specimen	Age (Ma)	Location	Cultural mode
H. habilis	UR 501 (Schrenk et al., 1993)	2.5–2.33	Ethiopia / Kenya / Tanzania / Malawi	1
H. georgicus	D2600 (Gabunia et al., 2002)	1.85–1.78	Georgia	1
H. ergaster	sample from Koobi Fora	1.9	Africa	2
H. erectus (sensu stricto; see Box 8.4)	Mojokerto (von Koenigswald, 1938)	c. 1.8	Java / China	1

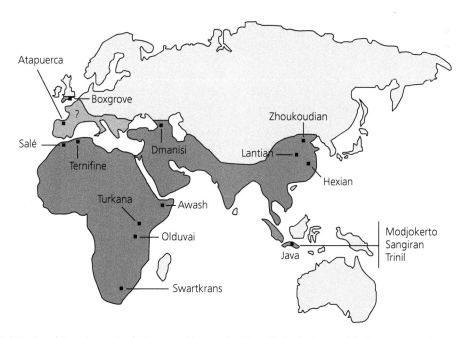

Figure 8.1 Main sites of the entire sample of *H. erectus*, with a questionable attribution in the case of the European exemplars.

Figure 8.2 Trinil I (left), and Trinil II (right), *Pithecanthropus erectus* (Dubois, 1894; drawings by the author).

based on qualitative traits, the more noticeable feature when comparing between *H. erectus* and earlier hominins, that of cranial volume, cannot be considered and is disqualified as a possible apomorphy.

Box 8.1 Trinil femur

The discovery of the Trinil remains attributed by Eugène Dubois to *Pithecanthropus erectus* led to the consideration of *Homo erectus* as a creature with a very primitive skull, although, from the nape of the neck down its anatomy was similar to ours. This interpretation was questioned, however, after the meticulous study of its femur by Franz Weidenreich (1941). Weidenreich indicated that *Homo erectus* has a combination of anatomical features clearly distinguishable from those of *Homo sapiens*, and attributed the Trinil femur to an exemplar of our own species. Aside from these considerations, it cannot be maintained that bipedalism, to which the term "erectus" refers, has the status of a derived trait in a Pleistocene hominin. As we know, bipedal locomotion is a feature that appeared during the Miocene and is a synapomorphy—a shared apomorphy—for all the taxa of the human lineage. Today this fossil is not considered in this way.

The first name given to the Trinil specimen was *Anthropopithecus erectus*, because of its similarity with a hominoid found in 1878 in Siwaliks (Pakistan). That name clearly reflected the idea of an intermediate between ape and man. Dubois changed the genus for *Pithecanthropus* three years later, recovering Ernst Haeckel's (1868) proposal of a hypothetical *Pithecanthropus alalus*. Thus, the Trinil fossil was initially referred to as *Pithecanthropus erectus* (Dubois, 1894). However, the Java specimens and other *H. erectus* known at the time—*Sinanthropus*, *Meganthropus*, and *Telanthropus*—were placed in *Homo erectus* by Ernst Mayr (1944, 1950). The lumping tendency continued with the inclusion of OH 9 and other similar African specimens in *Homo erectus* by Le Gros Clark (1964b).

Box 8.2 Cranial volumetric continuity

The values indicating the cranial volume overlap between the various taxa considered in Kennedy's analysis (1991)—*Pan, Gorilla*, and some *Homo*—make it impossible to define a limit to indicate when a "low" volume becomes "high." It was precisely because of the continuity shown by metric traits that the original cladistics of Hennig (1966) rejected them as useful for indicating apomorphies, although they were later accepted by the reformed cladistics (Platnick, 1979; see Chapter 1).

Regarding the second trait, besides having the same inconvenience of its metric expression, Gail Kennedy (1991) indicated that it is due to two processes: one primitive—lateral expansion of the lower part of the cranial vault—and another derived—thickness of the upper part of the cranial vault. *Homo erectus* and Neanderthals share this last process. To conclude, the third trait—the tympanic plate thickness—is also primitive and once again shared with the Neanderthals.

Kennedy (1991) concluded that the lack of autapomorphies (exclusive and characteristic apomorphies) made *Homo erectus* a poorly defined species and, therefore, invalid; a judgment in which she agreed with the analysis by Chris Stringer (1984) and Jean-Jacques Hublin (1986).

Affirming that *Homo erectus* is an invalid species implies that it does not have a particular and distinctive entity, so it should be incorporated into another preexistent taxon. But, which one? Kennedy (1991) admitted the derived character of *Homo erectus* with respect to *Australopithecus*, so it cannot be reduced to an australopithecine: *Homo erectus* apomorphies with respect to *Australopithecus* indicate that a new type of hominin has appeared. As we recall, Bernard Wood and Mark Collard (1999b) argued that *Homo erectus* is the first acceptable taxon of the genus. Then, how could it be a non-valid species?

Homo erectus would be a non-valid species if, according to taxonomic rules, it can be accommodated in another previously defined taxon. The species named at the time when Dubois proposed *Pithecanthropus* were *Homo neanderthalensis* and, obviously, *Homo sapiens*. Jean-Jacques Hublin (1986) argued that *Homo erectus* specimens lacked their own distinct traits that would separate them from modern humans, so they should be included in *Homo sapiens*. In his study, Kennedy (1991, p. 407) accepted that the *Homo erectus* taxon was badly

Box 8.3 The biological reality of *Homo erectus*

The underlying problem of the absence of autapomorphies was well indicated in the analysis of Kennedy (1991); it is not so much related with the biological reality of *Homo erectus* as with its taxonomic consideration. Kennedy (1991) admitted the possibility that phylogenetic systematics, based on cladistic methods—something the author does not say, but implies—lack sufficient discrimination to be useful at lower categories than genus. If so, in the case of *Homo erectus* we will have to settle for a phenetic analysis or, at most, with a description of the entire morphological pattern. Nevertheless, the taxon may face other problems such as an excessive variability or the separation of the African and Asian populations.

Box 8.4 Taxonomy of *Homo erectus*

A reflection of the extent of the differences between African and Asian populations in the Lower and Middle Pleistocene is their taxonomic interpretation. Some authors divide them into two distinct species, *Homo erectus* and *Homo ergaster*. Others prefer not to use the taxon *Homo ergaster* for the African specimens, preferring to use for them the name of *Homo erectus* sensu lato versus the Asian *Homo erectus* sensu stricto. Here we accept the taxon *Homo ergaster*, although the reasons for it will not be exposed until later.

defined, maintaining, however, that "just as the constituency of *H. erectus*, as presently understood, cannot be defended, dissolution of the species may also be indefensible. It may well be that clear autapomorphs will be identified in some grouping of middle Pleistocene hominids which will justify the recognition of specific integrity." Hublin's proposal (1986) is in the minority. Neither Kennedy nor Stringer, to name only a few of the mentioned authors, proposed removing *H. erectus* from the list of *Homo* species and including it in *H. sapiens*. Today it is common to consider the taxon *Homo erectus* as a full member of the human lineage (Box 8.3).

8.2 The first exit out of Africa

In Chapter 6 we finished explaining human evolution in the Pliocene, mentioning the hominin dispersals throughout the African continent and the options that placed hominins at the exit doors of Africa, just at the point to begin the colonization of Asia. Now we will try to understand the meaning of the evidence indicating the dispersal process in the Middle Pliocene. To do so, we first need to place it in the context of its evolutionary pattern.

After the departure of the African hominins from the continent, two separate populations appeared that, because of the distance between Africa and Asia, could maintain little contact. This fact led each one to develop its own particular traits (Box 8.4).

In accordance with the interpretation we made when talking about its remains, the Dmanisi site plays an important role in regard to the various migratory

flows that took place between 1.85 and 1.78 Ma (Ferring et al., 2011) (Box 8.5). Moreover, there is evidence indicating that this dispersal was not the first exit from Africa. As we noted in Chapter 7, when explaining the cultural traditions in the Yiron site (Jordan Valley, Israel), many Mode 1 lithic artifacts were found of an age close to 2.4 Ma (Ronen, 2006), which indicate a previous dispersal.

The tools appearing in the strip of land near the Jordan River show that between 2.4 and 0.8 Ma there could have been at least five exits from Africa, of which the first one is significantly previous to the one detected in Dmanisi. This fact raises questions about which hominin group could have originated the populations of *Homo erectus* in Asia, since in the Israeli corridor site no sufficiently informative fossil has been found. Emiliano Aguirre (2000) rightly argued that it is somewhat difficult to maintain that *H. georgicus*,

Box 8.5 Migration corridors out of Africa

The Dmanisi and the Jordan Valley enclaves are not the only corridors permitting the hominin migration out of Africa. Two sites of Anatolia (Turkey), Dursunlu and Kaletepe Deresi 3—the first with an age of 1.1–0.9 Ma obtained by paleomagnetism and faunal comparison, and the second with an insufficiently accurate dating, pointing in any case to less than 1.1 Ma—indicate another path of dispersal (Kuhn, 2010). The proximity of Dmanisi and Ubeidiya to Anatolia implies, however, that it is not an alternative route. As we stated in Chapter 7 (Box 7.10), in this book we use "migration" and "dispersal" as equivalent terms.

with *H. ergaster'* traits that Asian specimens lack, can be considered as the ancestor. It would be possible that the Asian *H. erectus* came from a previous migration. A similar argument has been made by different authors in reference to *Homo floresiensis*, a taxon discussed in Chapter 9 and whose morphology seems to derive from an ancient member of the genus *Homo* (Morwood et al., 2009; Wood, 2011).

8.2.1 Population dispersal variables

Whatever the date of the first exit from Africa, it is necessary to interpret the causes of population movements of such magnitude. Later in this and subsequent chapters we will examine the possible migratory routes to Asia and Europe, as well as the paleontological and archeological evidence to support them; let us begin now with those traits that give hominins such a wide-ranging migratory capacity.

Susan Antón, William Leonard, and Marcia Robertson (2002) offered a model of the first hominin exit from Africa using a two-parameter calculation: size of the occupied territory (HR, home range size) and the dispersion coefficient of the population (DQ, dispersion quotient). Unable to obtain the data directly from the fossil record, the authors estimated them by comparison with other fossil species of non-human primate protagonists of known dispersals. With respect to HR, Antón et al. (2002) arrived at the results that appear in Table 8.3.

Despite the sensitivity of the estimate of body weight with respect to estimates of age, development, and size of the specimens—we will see the example of KNM-WT 15000 later on—Table 8.3 indicates a tendency to increase occupied territory, and shows a big step forward precisely with *Homo erectus*, the species to which

cultural Mode 2 can be attributed. As the dispersion coefficient is a function of HR, this can be assumed in this case.

The result obtained by Antón et al. (2002) implies a certain paradox: the extension of open savannas—the environment in which the first hominins with cultural traditions appeared—led, according to Antón et al. (2002), to the decrease of primary productivity due to growing aridity, so there would be less vegetation resources for herbivores. But *Homo habilis*, *Homo ergaster*, and *Homo erectus* show an increment in body mass, as well as in HR and DQ parameters. The solution to the paradox is a change of diet, with a greater intake of meat, and a tendency to extend the home range to be able to obtain it. To follow the herbivores' migratory movement would have become much more necessary for those hominins than in previous epochs, while also leading to an increase of the dispersal tendency. The graphic representation of the model by Antón et al. (2002) appears in Figure 8.3.

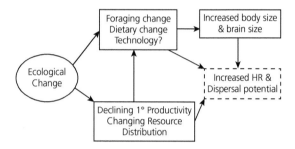

Figure 8.3 Ecomorphologic model by Antón et al. (2002) relating the factors involved in the population dispersal out of Africa. Reprinted from Antón, S. C., Leonard, W. R., & Robertson, M. L. (2002). An ecomorphological model of the initial hominid dispersal from Africa. *Journal of Human Evolution*, 43, 773–785, with permission from Elsevier.

Table 8.3 Estimated body weight and Home Range (HR)

Species	Male weight (kg)	Female weight (kg)	Average (kg)	HR/Apes*	HR/Human**
A. afarensis	44.6	29.1	37	40	247
A. africanus	40.8	30.2	35.5	38	234
P. robustus	40.2	31.9	36.1	39	239
P. boisei	48.6	34.0	44.3	51	316
H. habilis	51.6	31.5	41.6	47	290
H. erectus	63.0	52.3	57.7	73	452
H. sapiens	65	54	59.5	76	471

*Assuming a quality of diet similar to that of current apes.
**Assuming a quality of diet similar to that of current humans (Antón et al., 2002).

It is possible to relate, as Antón et al. (2002) did, the migrations of *Homo* with meat consumption. But many authors point to exogenous patterns—environmental changes, coincidence with other mammal migrations—as the leading cause of human population movements. Jan van der Made (2011) has provided the most comprehensive model of the origin and impact of migratory flows between Africa and Asia, beyond climate changes and the accessibility of corridors with open landscapes. For their study, van der Made (2011) compared the sequences of dispersal of various mammals by the inspection of the fossil record in different localities, considering the following variables:

- the fauna origins, its transit areas, as well as final destinations;
- the presence of new species by immigration or local evolution;
- the climatic and environmental changes.

According to the guidelines we have considered so far, the hominin exit from Africa took place around 1.8 Ma ago, or even earlier. During this temporal range, the exchange of fauna between Africa and Eurasia was hindered for a considerable time, apparently due to a documented increase in aridity in North Africa and the Arabian Peninsula (van der Made, 2011). *Homo* populations were, thus, established in East Africa around 2.5 Ma ago and the question to address is what circumstances would force hominins to migrate to the gates of Eurasia when the rest of the fauna found it difficult to complete this transit. The answer given by the model of van der Made (2011) is: biological evolution, or social and cultural advancement. There is evidence of these phenomena: the appearance and rapid dispersal of *Homo erectus* (Antón, 2002, 2003), and the signs of cooperative behavior in the care for disabled members of the group, indicated by the Dmanisi edentulous skull (Lordkipanidze et al., 2005).

Box 8.6 A further limitation due to the paucity of the fossil record

Van der Made (2011) warned about the limitations of any study attempting to be globally significant, given how fragmentary and discontinuous the fossil record is, with some areas better studied than others. This caution must always be considered when addressing any interpretative model of evolutionary processes.

The model of Jan van der Made (2011) argues that the arid barrier preventing the exchange of fauna between Africa and Eurasia disappeared about 1.2 Ma, when open spaces began to expand cyclically in Central Europe. The hominin arrival in Europe, as we shall see, occurs at that time, and seems to be related to the migration of the large herbivorous fauna to open habitats. The expansion of the Acheulean culture, which appeared in Africa around 1.6 Ma, must have been much later in Eurasia, as its presence can be estimated at c. 0.9 Ma for Southwest Asia and at c. 0.6 Ma for Europe (Figure 8.4).

8.2.2 Climatic cycles

The chronology of climatic oscillations in Africa follows the patterns indicated in Chapter 3, with three intervals in which changes toward more open and varied habitats occurred: 2.9–2.4 Ma, 1.8–1.6 Ma, and 1.2–0.8 Ma (deMenocal, 2004). In Figure 8.5 we repeat the graphic included in Chapter 3 (Figure 3.8), relating climatic cycles, cultural traditions, hominin speciation processes, transformation of forest into open savanna, and glacial and interglacial periods.

The diagram in Figure 8.5 provides a context for the process of rhythmic evolution, but does not show to what extent human migrations, on their exit from Africa, adjusted to climate changes. According to Ofer Bar-Yosef and A. Belfer-Cohen (2001), the available archaeological evidence indicates that in the interval between 1.8 and 0.7 Ma there were at least three migration waves. The first, with instruments of Mode 1 around 1.7–1.6 Ma, a second, with Acheulean tools—Mode 2—about 1.4 Ma, and the last, also of Mode 2 and an abundance of leaf-shaped flakes, about 0.8 Ma. Thus, that succession of migrations does not seem to bear any relation to climatic cycles. Table 8.4 contains the temporal periods of the climatic cycles, the dates of detected dispersions to Asia, and the stone tool culture of the migrant population.

From Table 8.4 no pattern can be deduced: neither a cultural level nor a climatic condition seems to be a determinant factor for the hominin migration to Asia. Which other factors can be pointed out as determining population movements?

Various authors have lately agreed in pointing to intrinsic hominin conditions as the driver for dispersal toward Asia (Carbonell et al., 2010; Agustí & Lordkipanidze, 2011; Bar-Yosef & Belmaker, 2011; van der Made, 2011). The hominin's exceptional characteristics would set them apart from patterns seen in other fauna, providing them with a certain behavioral

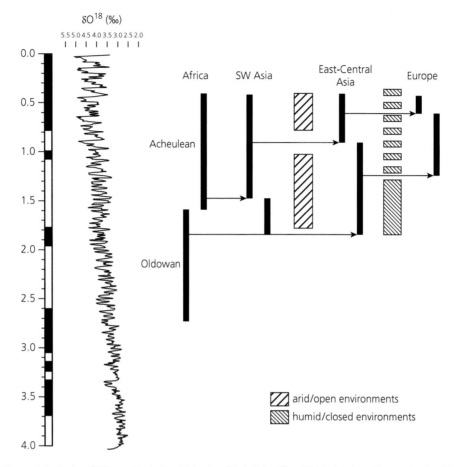

Figure 8.4 Temporal distribution of Oldowan (Mode 1) and Acheulean (Mode 2) in Africa, SW Asia, E Asia, and Europe. Reprinted from van der Made, J. (2011). Biogeography and climatic change as a context to human dispersal out of Africa and within Eurasia. *Quaternary Science Reviews*, 30(11–12), 1353–1367, with permission from Elsevier.

Box 8.7 Isolation of populations

A clarification should be made regarding the scheme by Jan van der Made in Figure 8.4. Archaeological evidence, the absence of Acheulean tools to the East of the Movius line (see Chapter 7), indicates that the separation of Asian and African populations must have occurred before the appearance of African Mode 2 and maintained for a long period of time. If that technological change is related to the most advanced tools of Ubeidiya, then the date of c. 1.4 Ma marks the age at which the isolation of Asian populations from the African ones took place. The ages of the various specimens of *H. erectus* in the Far East bring the realization that the colonization of Java and China had to be achieved by the pre-Ubeidiya dispersals.

freedom and, of course, a greater adaptability through culture to changing conditions. Within this general idea, many different "causes" to facilitate migration have been indicated:

- Mechanical—related to an efficient gait and long-distance endurance (Bramble & Lieberman, 2004).
- Biological—a volumetric brain as well as cognitive increment (Aiello & Wheeler, 1995).
- Cultural—technological efficiency (Carbonell et al., 2010).
- Groupal—social behavior (Kroll, 1994) with evidence even of altruistic care (Lordkipanidze et al., 2005).

The evolutionary path of *Homo* can be linked closely, more or less, to all these factors, but the final result is at the very least difficult to interpret. There are hominin

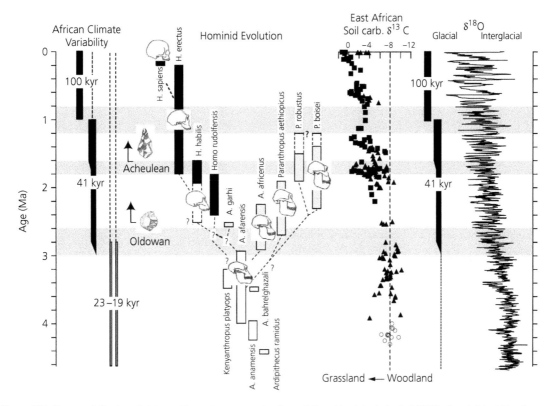

Figure 8.5 Diagram of climate and phylogenetic events accompanying the evolution of hominins in the last 4.5 Ma. From left to right, columns indicate temporal periods of climatic alternation of dry and wet conditions, cultural sequences, hominin taxa, transformation of dense forests into open grasslands, and a succession of glacial/interglacial conditions. Reprinted from deMenocal, P. B. (2004). African climate change and faunal evolution during the Pliocene-Pleistocene. *Earth and Planetary Science Letters (Frontiers)*, 220, 3–24, with permission from Elsevier.

Table 8.4 Existing relationships between climatic cycles and hominin dispersal to Asia

Cycles (Ma)	Condition	Migrations	Culture
1.2–0.8	arid-cold	0.8	Mode 2
1.4–1.2	humid-cold	1.4	Mode 2
1.8–1.6	arid-cold	1.7	Mode 1

fossils from the Middle Pleistocene that are very different from those of the various species of the Pliocene. But there are some particular differences between the populations of *H. ergaster* in Africa and *H. erectus* in Asia. We will talk about them later (leaving the European specimens for the end of the chapter). Before beginning a description of the various African and Asian specimens, we will give a general idea about the morphology of *Homo erectus* as a whole, which includes *H. ergaster*.

8.3 *Homo erectus* characterization

The list of apomorphies attributed to a generic *Homo erectus* (Figure 8.6) would be very long, as so many authors took up the task of describing the taxon, particularly in the 1980s, proposing numerous distinctive traits. Gail Kennedy (1991) made a first attempt at pruning the list to obtain the most "rigorous" cranial apomorphies: those selected following methods consistent with the proper characterization of a clade.

According to Kennedy (1991), it is possible to indicate the following apomorphies of *Homo erectus* (we have eliminated those more difficult to understand):

- Cranial capacity greater than that of australopithecines, between 700 and 1,250 cc.
- Elongated cranial shape (platycephaly).
- Thickening (pneumatization) of cranial walls.
- Glabellar and occipital torus.

- Thick tympanic plate.
- Ectocranial keel.
- Mastoid fissure.

The ectocranial keel trait reveals the problem of a qualitative characterization suitable to define *Homo erectus*. As such, it is present in the control group (African apes) of Kennedy's analysis (1991), as well as in all *Homo* members, although with different dimensions. Even if the overlapping sample makes a metric analysis of the "global aspect" of *Homo erectus'* cranium unreliable, it still appears to be very characteristic. If an excessive formalism compels us to indicate objective differences and not only "global aspects", we could use Kennedy's (1991) observation of a particularity related to the ontogenesis of the ectocranial keel: only juveniles and females of *H. erectus* have this trait among all the taxa analyzed by the author.

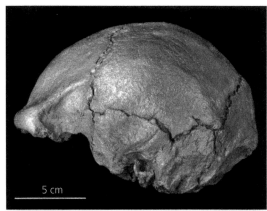

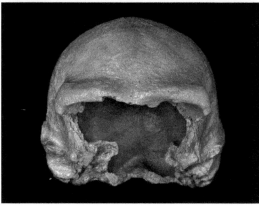

Figure 8.6 Left lateral (above) and frontal views of Sm 3 (cast), which could be considered as representative of *H. erectus*. Delson, E., Harvati, K., Reddy, D., Marcus, L. F., Mowbray, K., Sawyer, G. J.,. . . Márquez, S. (2001). The Sambungmacan 3 Homo erectus Calvaria: A Comparative Morphometric and Morphological Analysis. *The Anatomical Record*, 262, 380–397.

Homo erectus also retains some primitive traits (plesiomorphies), such as facial prognathism, somewhat projected canines, and relatively large molars, although smaller than in the case of the australopithecines—to the point that Chris Stringer (1987) took dental reduction as a derived character.

8.3.1 Size and body mass of *Homo erectus*

The cranial and mandibular morphology of *Homo erectus* was clarified in detail by comparing the numerous Asian and African fossils. Postcranial traits were more difficult to establish due to the scarcity of remains. In a paper devoted to studying the postcranial elements of *Homo erectus*, Michael Day (1984) mentioned the femur from Trinil (Indonesia) described by Dubois, the femoral fragments from Zhoukoudian (China), the OH 28 femur from Olduvai (Tanzania), the KNM-ER 737 femur, and the KNM-ER 3228 pelvic fragments from Koobi Fora (Kenya), and a European specimen: Arago XLIV (France). The list is remarkable for being so limited; the doubts about the presence of *Homo erectus* in Europe will further reduce it.

In the same year of 1984, an ulna, three vertebrae, a rib, and part of the innominate bone, and the hand and foot of a *Homo erectus* were found in Yingkou (China) (Wu Rukang, 1988), increasing the available postcranial samples. But, until the discovery of the specimen of Nariokotome—KNM-WT 15000 (Brown, Harris et al., 1985), a fairly complete skeleton to which we will soon refer—not enough information was available. The measurements of KNM-WT 15000 were surprising. By the degree of dental development it was estimated that the individual's age at death would have been about 12 years. Despite this early age, the length of the long bones of the skeleton was similar to that of a North American adult of white ethnicity; that means that the height of KNM-WT 15000—about 1.60 m—would have reached about 1.85 m at the end of his development, exceeding by far the average of modern humans (Brown, Harris et al., 1985; Brown, McDougall et al., 1985). The subsequent study of Alan Walker and Richard Leakey (1993) confirmed these first impressions.

James Ohman and collaborators (2002) disagreed with the estimation of the height. Considering the axial/appendicular disproportion and an age—according to the dentition—older than indicated at first, they estimated the height of KNM-WT 15000 at 1.47 m, with the possibility of being even shorter. The authors did not indicate the possible adult height, but if the extrapolation made by Brown, Harris et al. (1985) is correct, the final height would be around 1.70 m.

However, Ronda Graves and collaborators (2010) estimated that the developmental patterns of KNM-WT 15000 indicate that he would have reached a mature age at 12.3 years old. The size corresponding to that age in the calculation made by Graves et al. (2010), is 1.63 m, not 1.85 m.

The uproar caused by the discovery of KNM-WT 15000 resulted in scrutiny through a new prism of the existing sample of Middle Pleistocene postcranial remains—very limited, as already mentioned. Thus, Alan Walker (1993), in a study of six African *Homo erectus* specimens on which it was possible to calculate the necessary measurements, drew the conclusion that the average height of the species would be 1.70 m (between 158 and 185 cm) and the average weight 58 kg (between 51 and 68 kg). *Homo erectus*, therefore, would fall within the 17% of larger size and weight modern humans.

The estimation of Walker (1993) did not make us think, however, that all the members of such a broad and diverse set as *Homo erectus* shared the same height range. Bernard Vandermeersch and María Dolores Garralda (1994) argued that the sexual, individual, and population variability we see today was fixed in *H. erectus*. Consequently, if we observe the conditions of current humans, the height of *Homo erectus* would be very variable. Bernard Vandermeersch and María Dolores Garralda (1994) estimated it at between 1.50 and 1.80 m, adding that, in relative terms, *Homo habilis* as well as *Australopithecus afarensis* would have even a greater variability.

In support of the hypothesis of a great variability in *Homo erectus*, more even than that calculated by Vandermeersch and Garralda (1994), was the discovery of a pelvis in Gona (Ethiopia).

The specimen BSNN49/P27 is the almost complete pelvis of an adult female, which includes the last lumbar vertebrae (Figure 8.7). The fossil was found in the site of Gona, belonging to the Busidima formation of the Afar zone in Ethiopia. The locality BSN49 in Gona is placed between the Silbo tuff, with 0751±0022 Ma, and the lower part of the Matuyama chron, 1.778 Ma (Simpson et al., 2008).

The reconstruction of the pelvis from Gona revealed two interesting features. The first, that it is a specimen of very modest size: between 1.26 and 1.46 m, according to Scott Simpson et al. (2008). This contrasts significantly with the data from KNM-WT 15000, even if one takes into account the reduced estimation by Ohman et al. (2002). As we know, several African *H. erectus* specimens are of small size. In a subsequent study, Christopher Ruff (2010) calculated for the Gona female

a body mass of 33.2 kg. This figure is well below the average for *Homo* specimens (Ruff obtained 70.5 kg on average, with a range between 52 and 82 kg), and falls into the range of *Australopithecus* (34.1 kg and 24–51.5 kg, respectively). Consequently, Ruff (2010) argued that the Gona fossil increased the range of sexual dimorphism in *Homo erectus* to the point of being larger than that of any other living primate species.

The second feature to note in BSNN49/P27 is shape. The bi-iliac width of 288 mm is greater than the current human pelvis. This width plus the height of the Gona specimen produces a very wide and short trunk: only the pelvises from Sima de los Huesos in Atapuerca (Spain), Jinniushan (China), and Kebara (Israel) are wider than that from Gona (Simpson et al., 2008).

After analyzing the specimen, Scott Simpson and collaborators (2008) maintained that during the Middle Pleistocene the human pelvis suffered anatomical changes that are not related with a bipedal locomotion useful for enduring long distances—already fixed in previous *Homo*, according to the authors. With what are those changes related? The size and aspect of the trunk and pelvis are traits with great ecogeographic variability in current humans. Populations adapted to cold weather normally exhibit a smaller size, with a wider trunk and pelvis; ethnicities of tropical zones show the opposite pattern: a long and thinner trunk. However, with the exception of KNM-WT 15000—whose reconstruction seems uncertain according to Simpson et al. (2008)—all *Homo erectus*' pelvises are, in lateral view, similar to that of Gona. It is a condition maintained in all the previous specimens until the Near East modern humans of c. 100,000 years ago, such as those from Skhul (Israel). Despite the diversity in habitat, *Homo erectus* seems to have fixed a somewhat uniform body structure that remained in stasis over a long period of time (Simpson et al., 2008).

If so, what would the adaptive reasons be for such a pronounced and important change? The study by Scott Simpson and collaborators (2008) on the Gona pelvis pointed to an obstetric explanation. The authors calculated the neonate brain size of a Lower Pleistocene *Homo erectus* at 34–36% of the adult size; a value in between the 40% of chimpanzees and the 28% of modern humans. This percentage corresponds, if we start from an adult *Homo erectus* brain volume of 600–1,067 cc—880 cc in average—to a newborn brain of 299–316 cc. Thus, the Gona pelvis would have allowed giving birth to offspring of up to 315 cc of cranial volume (Simpson et al., 2008).

Simpson et al. (2008) indicated in their study that the two theoretical anatomical solutions for giving birth to

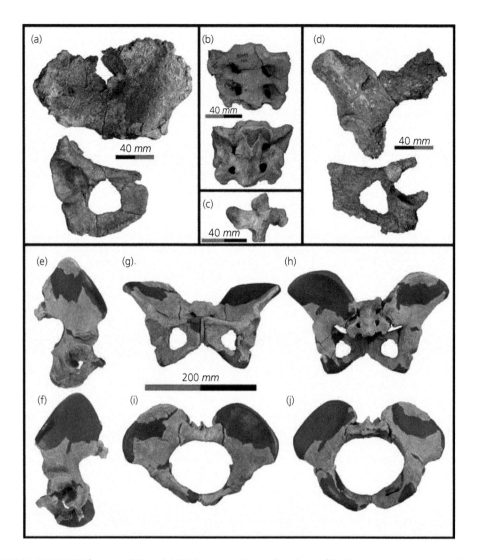

Figure 8.7 Pelvis BSNN49/P27 from Gona (Ethiopia). (a) Right coccyx with anterolateral view of the ilium and post-lateral view of the ischiopubic fragment. (b) Anterior and posterior view of sacrum. (c) Right lateral view of lumbar vertebrae. (d) Posterolateral views of ilium and anterolateral view of ischiopubic fragment. (e–j) Views of the rebuilt pelvis. In dark gray, reconstructed or restored areas. Scales: (a–d) 40 mm; (e–j) 200 mm. From Simpson, S. W., Quade, J., Levin, N. E., Butler, R., Dupont-Nivet, G., Everett, M., & Semaw, S. (2008). A Female Homo erectus Pelvis from Gona, Ethiopia. *Science*, 322(5904), 1089–1092. Reprinted with permission from AAAS.

offspring with bigger brains are: a larger sized female body permitting an isometrically larger pelvis, or the development of a wider pelvis in females, which implies a significant sexual dimorphism. The Gona pelvis is evidence of the second solution, but the first solution could be considered as a real alternative. The same isometric increment that would increase the pelvis size linked to an increment in height, will lead to an isometrically bigger cranium—without considering the

extra-allometric increase of brain size in *Homo erectus*. It seems that the only anatomical solutions are those indicated by the Gona pelvis, or newborns with less mature brains, as occurs in modern humans.

8.3.2 The footprints of Koobi Fora

Further evidence of the size and body mass of *Homo erectus* comes from fossil trails of footprints: those of area 103 of Koobi Fora, studied by Anna Behrensmeyer and Léo Laporte (1981), and those of the FwJj14E site, belonging to the Okote member in Ileret (Kenya), studied by Matthew Bennett and collaborators (2009). We have referred to both footprint series in Chapter 2.

The Koobi Fora footprints correspond to a foot around 26 cm long and 10 cm wide, which would make the footprint-makers close in height to an average modern human in North America. The height would be approximately 1.685 m, if we take American males as a reference group, both of European and African origin, and it would increase up to 1.80 m if the foot were compared with the San Bushmen. The length of the step is, however, smaller than that of modern humans, but that circumstance was interpreted by Behrensmeyer and Laporte (1981) as the consequence of faltering steps on a slippery slope.

The average height of the makers of Ileret footprints was estimated at 1.75±0.26 m for those individuals who left their footprint on the upper surface and at 1.76±0.26 m for those who walked on the lower surface, excluding from the calculation of this latter series an individual, thought to be a sub-adult, whose footprints provided an estimate of 0.92±0.13 m (Bennett et al., 2009) (Figure 8.8). Based on these weight and height data, Matthew Bennett and collaborators (2009) attributed the tracks to *Homo erectus/Homo ergaster*.

If we accept the estimates for Koobi Fora footprints and, in particular, those for Ileret, the hominin height and weight should have reduced in the last stages on the path to become modern humans. That is the conclusion of Christopher Ruff and collaborators (1997) after studying the dimensions of 163 skeletons belonging to an age range from 10,000 years to 1.95 Ma. In this study, Pleistocene specimens were on average up to 7.4 kg heavier than modern humans (65.6 kg vs. 58.2 kg), representing a 12.7% difference in body mass. So the human body, at least in the genus *Homo*, was smaller in the Lower Pleistocene, reaching a peak at the beginning of the Upper Pleistocene and decreasing afterward. The brain size also experienced a decrease, according to Ruff et al. (1997), from the beginning of the Upper Paleolithic and continuing the decline during the Neolithic, at least in Europe. The trend was reversed in recent centuries in general terms but, today, a large body mass equivalent to that of the Paleolithic ancestors is only reached in some populations from the high latitudes of the Northern Hemisphere.

Mark Cohen and George Armelagos (1984) linked the loss of body mass with diet, i.e., with reduced dependence on meat due to the emergence of agriculture. Vincenzo Formicola and Monica Giannecchini (1999) also indicated nutritional factors to explain changes in height, without ruling out that variations of gene flow would have played a role.

The decrease in body mass was interpreted by John Kappelman (1997) using a different explanation than usual for human adaptive success, which is often linked to an extra-allometric brain growth, with higher increments in relation to that of the body. But the selection of smaller bodies, favored perhaps by cooperative systems and a more effective communication, might

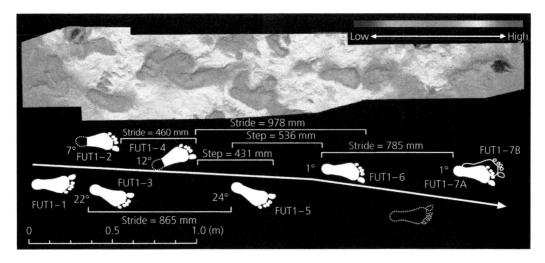

Figure 8.8 Fossil footprints in Ileret (Kenya) (scheme by Matthew Bennett et al., 2009). From Bennett, M. R., Harris, J. W. K., Richmond, B. G., Braun, D. R., Mbua, E., Kiura, P.,. . . Gonzalez, S. (2009). Early Hominin Foot Morphology Based on 1.5-Million-Year-Old Footprints from Ileret, Kenya. *Science*, 323(5918), 1197–1201. Reprinted with permission from AAAS.

have led, according to Kappelman (1997), to an increase in relative brain size by the simple fact that the body mass was lower.

8.4 African specimens (*Homo ergaster*)

If we agree that *Homo habilis* led to more advanced hominins, *Homo ergaster* and *Homo erectus*, and possibly the chronospecies *H. georgicus*, the best available evidence in favor of that idea should be in Africa, since the first of these two taxa is exclusively African.

Specimens of *Homo ergaster* have appeared in the Rift Valley, in South Africa, and in the Maghreb, but in historical terms these were not the first-discovered fossils of this taxon; this honor belongs to *Pithecanthropus erectus* from Java. Nonetheless, due to the evolutionary meaning of the Lower Pleistocene, as well as the criteria followed up to now in this book—which is to address the oldest specimens first—we will start with the African fossils.

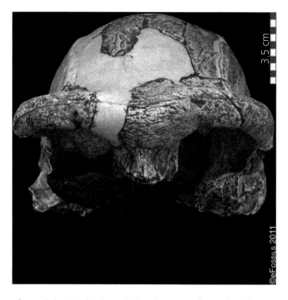

Figure 8.9 OH 9 (Leakey, 1971) replica. Image from Wikimedia Commons.

8.4.1 Olduvai

As in the case of *P. boisei* and *H. habilis*, the findings in Olduvai provided the first solid data about the morphology of the African *Homo erectus* and its possible phylogenetic relationships. The specimens from Olduvai Bed I and from the lower part of Bed II, classified as *Homo habilis*, were followed in stratigraphic terms by more modern specimens, such as OH 9. Placed in the upper part of Bed II, OH 9 has an estimated age of 1.25 Ma (Leakey, 1971). The OH 9 specimen was studied in detail by Rightmire (1979a), who later revised his initial conclusions (Rightmire, 1990). OH 9 is a partial specimen (Figure 8.9), which includes the supraorbital structures and the base of the cranium, but provides no facial information below the nasal bones. Its cranial capacity was estimated to be 1,067 cc (Holloway, 1973). OH 12, on the other hand, was found by Margaret Cropper in 1962 on the surface of terrains attributed to Bed IV (Leakey, Clarke, & Leakey, 1971). It is the posterior part of a small cranium, close to 700–800 cc (Holloway, 1973), to which certain facial fragments studied by Susan Antón (2004) were later added. Based on these fragments, Antón (2004) considered that OH 12 shares traits in part with OH 9 and with KNM-ER 3733 from Koobi Fora (Kenya).

OH 22 is a fairly well-preserved right mandible, with premolars and the first two molars. It was found in sediments belonging either to Bed III or Bed IV, and was estimated to be no less than 0.62 Ma (Rightmire, 1980). There are also other mandibular fragments

(OH 23, Bed IV; OH 51, Beds III–IV), and some post-cranial remains from Bed IV, such as a left femur and an unidentified bone, which form the OH 28 specimen (Day, 1971). Day classified OH 28 as *H. erectus*, owing to its similarity with the Peking specimens. An almost complete ulna, OH 36, is older, found in the upper part of Bed II, and its morphology is more robust (Day, 1986).

As it is often the case, the differences among those Olduvai remains, which span an interval of 750,000 years, were interpreted either as indicative of sexual dimorphism (Rightmire, 1990) or as evidence of different species (Holloway, 1973). The relation with Asian specimens has also been a constant source of controversy.

8.4.2 Koobi Fora

The sites on both shores of Lake Turkana offer a large sample of *Homo* specimens, all those we have seen—including the exemplar on which is based the proposal of *H. rudolfensis*—as well as other more modern specimens of *H. erectus* sensu lato. The latter include the best-preserved crania and the most complete skeleton.

The 3733 specimen (Figure 8.10) was found in situ in the upper member of Koobi Fora. It is a complete skullcap with the better part of the face—including zygomatic and nasal bones—together with the alveoli of anterior teeth and some molars and

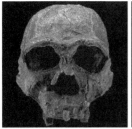

Figure 8.10 KNM-ER 3733, frontal (left) and lateral view (right), *Homo erectus* (Leakey & Walker 1976) (photography by http://www. msu.edu/~heslipst/contents/ANP440/ergaster.htm).

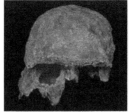

Figure 8.11 KNM-ER 3883, frontal (left) and lateral view (right), *Homo erectus* (Leakey & Walker 1976; photography by http://www. msu.edu/~heslipst/contents/ANP440/ergaster.htm).

premolars (Leakey, 1976; Leakey & Walker, 1976). Richard Leakey and Alan Walker (1976) concluded that the cranial capacity of KNM-ER 3733 was close to 800–900 cc and that all the cranial traits resemble those of Peking *H. erectus*. Not surprisingly, Leakey and Walker (1976) classified 3733 as *H. erectus*. This incidentally, was a very solid argument against the "single species hypothesis" (see Box 6.9). Compared with Koobi Fora crania such as KNM-ER 406—which belongs to a paranthropine—3733 shows that more than one hominin species had lived at the same place and time. In fact, it was the purpose of opposing the single species' perspective that fueled Leakey and Walker's (1976) article in which they defended that 3733 is *H. erectus.*

Many authors have supported the relation between 3733 and the Peking *H. erectus* noted by Leakey and Walker. We have cited textbooks of human paleontology that agree with that point of view. However, Colin Groves (1989) reached a different conclusion after studying the best cladograms that expressed the relations between the different Turkana specimens attributed to *Homo*. Groves disagreed with the idea that morphometric comparisons between 3733 and the Peking specimens indicated they were similar, and thus, that they should be classified in the same species. Groves concluded that 3733 does not show the autapomorphic traits of *H. erectus* and that, therefore, the specimen cannot be classified as such. This had been noted by Wood (1984), who suggested two alternatives: either to change the definition of *H. erectus* to accommodate the Koobi Fora 3733 and 3883 crania, or to exclude these from *H. erectus* sensu stricto.

KNM-ER 3883 is a slightly older and more robust cranium, with certain differences that Wolpoff (1980) attributed to sexual dimorphism (Figure 8.11). Thus, 3733 would belong to a female and 3883 would correspond to a male. In any case, and as Groves (1989)

noted, both their cranial capacities are within the variation range of the crania of Chinese *H. erectus*. The volume of 3883, supposedly a male, is lower than the 3733 specimen, which has a more gracile appearance.

8.4.3 West Turkana

The other shore of Lake Turkana, the western one, has also provided very interesting hominin remains belonging to *Homo erectus*. From the Kaitio member of the Nachukui formation, with an age of 1.9–1.65 Ma, come numerous utensils of Mode 1 and five teeth, of which four were attributed to *P. boisei* and one, tentatively, to *Homo* sp. aff. *ergaster* (Prat et al., 2003). The most informative specimen from West Turkana and, in fact, of the whole hypodigm of Middle Pleistocene African hominins, is the immature individual from Nariokotome.

As Francis Brown, Harris et al., 1985 have reported, Kamoya Kimeu found a small fragment of a hominin frontal bone on the surface of the Nariokotome III site on the west shore of Lake Turkana during the 1984 campaign. Many other facial, cranial, mandibular, and postcranial remains appeared after cleaning the terrain and excavating a 5 × 6 m² area. The remains presumably belonged to the same individual, but they had been dispersed before their fossilization and showed signs of having been transported by water and having been trampled by large mammals, though there are no indications of the action of scavengers. After its reconstruction, the specimen was catalogued as KNM-WT 15000 and was assigned to *H. erectus* (Brown, Harris et al., 1985). It turned out to be one of the most complete early hominin specimens available (Figure 8.12). The 15000 specimen appeared just above a volcanic tuff, part of the Okote tuff from Koobi Fora, which is 1.65 Ma, so the specimen's age was estimated to 1.6 Ma (Brown, Harris et al., 1985).

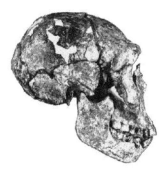

Figure 8.12 Detail of KNM-WT 15000 cranium (Brown, Harris et al., 1985). Image from http://www.talkorigins.org/faqs/homs/15000.html.

The large size and weight estimated for KNM-WT 15000 led to rethinking previous ideas about the size of *Homo erectus*.

8.4.4 Other specimens from the Rift Valley

The African sample of *Homo erectus* is not limited to the specimens from Olduvai and Turkana. There are many fossils that have been attributed to the taxon; some of the latest discoveries, such as from the sites of Daka and Olorgesailie, and the skull KNM-ER 42700 in Koobi Fora, will be addressed in a later epigraph, since they increase the uncertainty about the meaning of the taxon. Of the rest, we will mention only the most informative.

William Howells (1980) included remains from the late Middle Pleistocene in *Homo erectus*: a mandible from the Kapthurin beds in Baringo (Kenya) and a cranium from Ndutu (Tanzania). He also included other, somewhat older specimens of about 700,000 years age, such as the parietal from Melka Kounturé, Gomboré

site, Awash (Ethiopia), and a very old humerus fragment of approximately 1.5 Ma: the Gomboré IB fossil.

From the Kayle and Karat members of the Konso formation (Ethiopia), near the Omo region, come the specimens listed in Table 8.5 (Suwa et al., 2007). The ages shown in Table 8.5 are from the geological formation survey by Shigegirho Katoh and collaborators (2000) (see Chapter 3).

Among the Konso specimens, the mandible KGA10-1, of c. 1.4 Ma, stands out. Its great robustness and size place the specimen among the more extreme hominins of that time, something that Suwa et al. (2007) attributed to the preservation of plesiomorphies. It also presents apomorphies, such as a weak development of the lateral prominence and a greater reduction of the postcanines, the reason why Suwa et al. (2007) ascribed the specimen to the African *H. erectus*. The other dental materials show compatibility, according to the authors, with this consideration.

From the Danakil formation, in the Afar region close to Buia village (Eritrea), comes a very complete cranium with portions of the face and the roots of some molars and premolars (UA 31) (Figure 8.13), as well as two incisors (UA 222 and 369), and two pelvic fragments (UA 173) (Abbate et al., 1998).

Despite the presence of volcanic material at the site, the attempt to date Danakil the remains by the ^{40}Ar/^{39}Ar method failed due to contamination of the ashes. However, by faunal comparisons and paleomagnetic methods, the horizon in which the hominin remains appeared, near the top of the Jaramillo subchron, was dated to about 1 Ma (Abbate et al., 1998). A fission track study in an intercalated tuff of the sedimentary terrain (Bigazzi et al., 2004) and faunal comparisons (Martínez-Navarro et al, 2004) confirmed this dating.

Table 8.5 *Homo erectus* specimens from the Konso formation. TBT, BOT, BWT, LHT, and HGT are tephra layers of the formation

Specimen	Element	Stratigraphic placement	Estimated age	Discovery date	Discoverer
KGA4-14	Right upper M3	Just below TBT	1.45 Ma	October 4, 1991	Yohannes Haile-Selassie
KGA7-395	Right occipital fragement	Just below BOT	1.3 Ma	August 17, 1997	Kampiro Kairante
KGA8-150	Left lower P4 fragment	Just above BWT	1.3–1.4 Ma	July 31, 1994	Alemu Admassu
KGA10-1	Left mandible with P4-M3	Just above LHT	1.42 Ma	October 8, 1991	Gen Suwa
KGA10-620	Right parietal fragment	Circum LHT	1.42 Ma	August 3, 1994	Alemu Admassu
KGA10-656	Right parietal and frontal fragment	Just above LHT	1.42 Ma	August 5, 1994	Yonas Beyene
KGA11-350	Left upper M1	Just above LHT	1.42 Ma	August 22, 2000	Kebede Geleta
KGA12-970	Right upper dm2	Just below HGT	1.25 Ma	August 25, 1996	Hideo Nakaya

Suwa et al. (2007).

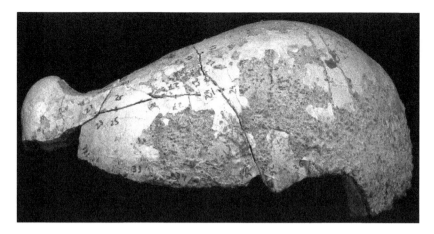

Figure 8.13 Calotte of UA 31, Buia (Eritrea) (photography by Macchiarelli et al., 2004).

The UA 31 cranium exhibits features that are characteristic of *H. erectus*, such as elongated shape, thick supraorbital torus, and cranial capacity close to 750–800 cc, together with other derived traits, characteristic of *H. sapiens*, such as the thinning at the base and the verticality of the parietal region. Abbate and colleagues (1998) noted that the specimen shows signs of transition toward anatomically modern humans. In a detailed study, Macchiarelli et al. (2004) simply assigned the specimen to an *erectus*-like type.

A similar transitional role was attributed to the Bodo parietal after Asfaw's (1983) initial proposal. But because the Buia specimen is 300,000 years older, the process of the transformation, *H. erectus* to *H. sapiens*, was extended further back in time. Wolpoff noted that the Danakil cranium confirms the difficulties involved in establishing a definitive frontier between fossils belonging to the *erectus* grade and *H. sapiens* (see Gibbons, 1998). We will review in Chapter 9 the alternative proposals concerning the transition from *Homo erectus* to modern humans.

8.4.5 Northern Sahara

Exemplars belonging to *Homo ergaster* have also been found north of the Sahara at Ternifine, or Tighenif, (Algeria) and Sidi Abderraman, Rabat, and Salé (Morocco). The Ternifine specimens include three mandibles, one parietal, and some isolated teeth, which, owing to the fauna present at the site, were attributed to the Middle Pleistocene (Arambourg, 1955). The mandibles are rather robust, they lack a chin, and their similarities with Peking *erectus* specimens were noticed (Arambourg, 1954), but Arambourg preferred to define a new species, *Atlanthropus mauritanicus*, which

gained little acceptance. As Day (1986) mentioned, the Ternifine specimens are generally considered to belong to *H. erectus*.

The Salé specimen consists of a cranium, lacking the face, and a left maxilla, with some teeth. It was found in 1971 in an open quarry close to El Hamra, north of Salé. In the initial description, Jaeger (1975) noted the specimen's clear affinities with *H. erectus*, though the occipital area is much more prominent and rounded than what could be expected, conferring it a modern appearance. Some have explained this anomaly as a pathology, but Hublin (1985) argued that it should be considered a primitive *H. sapiens* that had retained some *H. erectus*' traits. According to Hublin, the North African Middle Pleistocene remains show a mosaic of *H. erectus*' and *H. sapiens*' traits, which underlines the need for interpreting the evolutionary relationships between those species. The controversy regarding the evolutionary meaning of the Salé specimens, as Hublin (1985) sharply noted, is the same one that could be applied to the Ngandong and Zhokoudian specimens, or even all *H. erectus*. These considerations led Hublin to reject that the *Homo erectus* species was well described.

8.4.6 South Africa

The first identification of *Homo erectus* fossils in Africa, using fragments of mandibles, maxillars, and other bones found in the Swartkrans site (South Africa), occurred prior to naming the taxon *Homo ergaster*, which at first Robert Broom and John Robinson (1950) ascribed to a new genus, *Telanthropus*.

These hominins are now considered *H. ergaster*, together with other specimens such as the SK 847 partial cranium, reconstructed by putting together different fragments, and assigned by Clarke and colleagues (Clarke et al., 1970) to *Homo* sp. It seems likely that this specimen belongs to *H. ergaster* rather than *H. habilis*, but, in any case, it is anatomically different from australopiths. Something similar can be said about the SK 45 mandible. These are early specimens; doubts regarding their classification in one or another species are understandable.

8.5 Asian specimens of *Homo erectus*

The fossils from Java (Indonesia) were, in historical terms, as we have seen, the protagonists of the *Homo erectus* proposal. Asian specimens similar to *Homo erectus* appeared in other places, such as a partial cranium from Narmada, Madhya Pradesh, India (de Lumley & Sonakia, 1985). However, further studies attributed this specimen to an "archaic" *Homo sapiens* (Kennedy et al., 1991; we will explain the meaning of "archaic" in Chapter 9). Other evidence has also appeared, such as abundant Acheulean lithic utensils in India, extending the scope of the hominin occupation in Asia, including *Homo floresiensis*. Nevertheless, when addressing Asian *Homo erectus* we will refer exclusively, first, to Java specimens and, second, to those from China.

8.5.1 The island of Java (Indonesia)

There are several difficulties in understanding the evolutionary meaning of the Java specimens from the Lower and Middle Pleistocene. The initial findings were usually very old and the descriptions given in those days lacked the precision necessary to locate the exact place where the remains were collected. The complex process of the volcanic material sedimentation does not help to clarify the dating either.

O. Frank Huffman and collaborators (2005) gave an example of how uncertain the available interpretations are when they described the discovery of the "Mojokerto child" (Box 8.8). On February 13, 1936, Andoyo, a collaborator of the geologist Johan Duyfjes—both working for the Java Mapping Program of the Bureau of Mining of the Netherlands Indies—found a skullcap, which a month and a half later was described by Gustav Henry Ralph von Koenigswald (1936), who attributed it to *Pithecanthropus erectus*. However, there is incongruence with respect to the finding place: either on the surface, according to the initial description by von Koenigswald, or 1 m underground, as later interpretations of the same author and Duyfjes stated (see Huffman et al., 2005). The difference is not trivial, because if it was found on the surface, there is no way to relate the fossil to the terrain stratigraphy and, as we will see next, that fact affects the possible age of the fossil.

8.5.2 Java colonization

The selection of Trinil for field work was due to the similarity of its fossils to those from Siwaliks. However, Trinil fossiliferous deposits include two different sedimentary beds; thus, its fauna does not form a unity (Day, 1986), providing few clues to their dating. But the problem of assigning age to Java fossils is not exclusive to Trinil. Four sedimentary formations have traditionally been recognized in the island, each one with its respective fauna: Kalibeng, Pucangang (now called Sangiran, but we maintain the old denomination so the Sangiran formation is not confused with the site of the same name), Bapang (prior Kabuh), and Notopuro (Pope, 1988; Wolpoff et al., 1988). Most of the hominin fossils found at different Javanese sites belong to the Bapang formation (Middle Pleistocene) but some of them—like *Pithecanthropus* VII—could belong to Pucangan, which reaches an age of 2 Ma. In sites like Sangiran, at the foot of the Lawu volcano, this formation is separated by lava and ash intrusions, which allow dating of the sediments. In other distant sites, such as Mojokerto (Perning), dating accuracy depends on being able to estimate the temporal relationship between tuffs and sedimentary materials.

Such difficulties affect the task of estimating the age of the colonization of Java. Did it occur in the Lower or in the Middle Pleistocene?

Box 8.8 "Mojokerto child" discovery

We mentioned before that the main obstacle to establishing the age of fossils is in determining the specific location from which they come. It is believed that the oldest Javanese specimen is the "Mojokerto child" (Mojokerto I or Perning I) (Figure 8.14). The specimen was discovered in 1936 in the Perning Valley, west of Trinil, by Tjokrohandojo, a native who worked for Johan Duyfjes. The story of the discovery has been narrated by O. Frank Huffman et al. (2005). The Mojokerto child was initially described by von Koenigswald (1938). Swisher et al. (1994) identified the place of the discovery based on Duyfjes' descriptions and the presence on the specimen's skull of volcanic materials similar to the pumice stone used to estimate the site's age. No other Mojokerto stratum contains volcanic material.

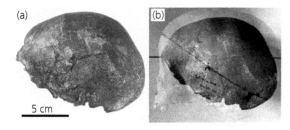

Figure 8.14 The Mojokerto child skull: (a) left-lateral view; (b) left lateral view compared to a one-and-a-half-year-old modern-child skull of about the same length (photographs by von Koenigswald, 1936 and 1938; from Huffman et al., 2005).

The age of Mojokerto by the ^{40}Ar/^{39}Ar method yields an average of 1.80±0.07 Ma (Swisher et al., 1994; see Box 8.10). The paleomagnetic study attributed the sediments containing the specimen to a positive event. Swisher and colleagues (1994) chose to assign it to the Olduvai subchron, slightly under 2.0 Ma. If this were the case, the Mojokerto child would be at least as old as the KNM-ER 3733 cranium from Koobi Fora (Kenya). If this conclusion is accepted, *Homo erectus* would have arrived in Southeast Asia before 1.8 Ma. This would mean that the departure out of Africa would have taken place before the emergence of the Acheulean culture (see Chapter 7), given that its earliest manifestations appeared 400,000 years later. Swisher and colleagues (1994) argued that the evolution of *H. erectus* took place out of Africa, but they add that the assignation of the Pucangan formation specimens to a specific species is not easy.

The attribution to Mojokerto of such older dates has been questioned. Vos and Sondaar (1994) noted that there is a contradiction between the age assigned by Swisher et al. (1994) and that obtained by careful magnetostratigraphical dating studies at Mojokerto by Hyodo et al. (1993) that correlate well with the ages obtained using fission-track dating. The study of Hyodo et al. (1993) placed the fossil terrains in the Jaramillo subchron, with an age of 0.97 Ma, very close in age, then, to the Trinil cranium (*Pithecanthropus* I). A review by O. Frank Huffman (2001), however, suggests that Swisher and colleagues' (1994) estimate was correct, and concludes that "recent fieldwork and archival research strongly support the conclusion that the Perning *H. erectus* was found *in situ* in the upper Pucangan Formation, as defined by Duyfjes (1936). Although the excavation spot has not been relocated and detrital materials were used by Swisher et al. (1994) to radiometrically date the site, lithologic and paleogeographic evidence from the Pucangan indicates that

the *H. erectus* is likely to be 1.81±0.04 Ma" (Huffmann, 2001, p. 358). Although they accepted the location of the fossil to be correct, Morwood and collaborators (2003) estimated a younger age for the fossil's terrain, 1.49–1.43 Ma, by a different radiometric procedure, a fission-track analysis of embedded zircons in pumice.

A later study by Huffman and collaborators (2006) sought to determine accurately the discovery location of the Mojokerto cranium. They used photographs of the time of discovery and all kinds of complementary testimony, as well as a reconsideration of the origin of the large pebbles that fill the specimen. These authors relied on the preservation of the specimen to establish as a proven fact that the fossil was found underground, as none of the several hundred vertebrate fossils of similar size and fragility found on the surface at Perning show a similar state of preservation. The relocated discovery bed would be, they say, about 20 m stratigraphically above the 1.8 Ma horizon. But, Huffman et al. (2006) recognized that, due to the discrepancies between different studies and the difficulties of establishing the origin of the various volcanic intrusions, it is difficult to determine accurately the age.

8.5.3 Sangiran: a reference for Asia colonization

Sangiran is the locality that allows the best reconstruction of the stratigraphic history of *Homo erectus* in Java. The first reason for this is because volcanic intrusions are very abundant, although this fact does not solve a basic problem in Javanese sites: the difficulty of establishing the time between eruption and the deposit of sedimentary remains (Huffman et al., 2006). The advantage of Sangiran in this regard is that it has a much more precise stratigraphy and, in particular, it is possible to identify the relationship between the stratigraphic horizon of each specimen, at least if it belongs to the Pucangan or Bapang formations. Because of the abundance of fossils, around 80 *Homo erectus* specimens, of which 50 have a precise localization (Larick et al., 2001), Sangiran is the best reference to understand not only the process of *Homo erectus* dispersals to Java, but also the possible relationships between local populations and those of another Asian region with abundant Middle Pleistocene hominins: Zhoukoudian, China.

The sedimentary cycles of the Bapang formation (Kabuh), typical of terrains from which Sangiran hominins come, were established by Roy Larick and collaborators (2001), using the scheme shown in Figure 8.15.

The Bapang formation has in the fossil localities of Sangiran, five deposition cycles (indicated by C1 to C5

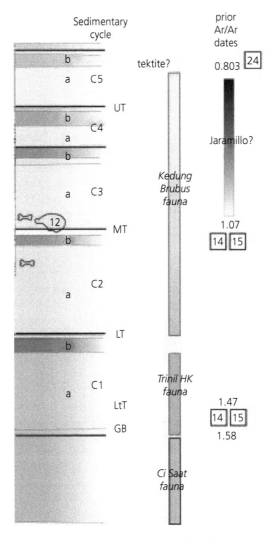

Sedimentary cycle

prior Ar/Ar dates

Figure 8.15 The middle column indicates the fauna found in several cycles (C1–C5), followed by a column indicating previously established ages. Boxes to the right display reference icons, including those corresponding to hominin remains. Larick, R., Ciochon, R. L., Zaim, Y., Sudijono, Suminto, Rizal, Y.,. . . Heizler, M. (2001). Early Pleistocene 40Ar/39Ar ages for Bapang Formation hominins, Central Jawa, Indonesia. *Proceedings of the National Academy of Sciences*, 98(9), 4866–4871. Copyright (2001) National Academy of Sciences, U.S.A.

in the scheme), each of which consists of two facies (a and b, with a as the oldest in each case). Bapang C1a cycle, corresponding to the bed traditionally called "Grenzbank" (GB in the scheme), is located immediately above the Pucangan formation (Sangiran, in the scheme). The oldest hominins indicated in the diagram correspond to C2a, although as we shall see, there is

also a C1a specimen (Grenzbank). The most modern specimens were found on terrains that Larick et al. (2001) assigned to C5a. Plagioclase and hornblende from the pumice intrusions present in the facies a permitted accurate dating by the $^{40}Ar/^{39}Ar$ method; in some cases the volcanic materials were found in close association with hominin remains. Although the age of the oldest remains from C1a cannot be assessed with great accuracy, Sangiran indicates a date of c. 1.5 Ma for the colonization of Java.

After reducing the age of the colonization of the island from 1.8 Ma, as determined by Swisher et al. (1994), to 1.5 Ma by Roy Larick et al. (2001), this later date has been challenged as still too old by Masayuki Hyodo and collaborators (2002). With paleomagnetism analysis and faunal considerations, Hyodo et al. (2002) attributed Sangiran's oldest hominin fossils to the Jaramillo subchron, giving them an age of 1.1 Ma. The older dates obtained previously, according to Hyodo et al. (2002), were due to mistakes made in the identification of the dated volcanic tuffs, as well as in the correlation made between intrusive materials and the sediments in which *Homo erectus* specimens were found. However, it seems unlikely that the numerous samples of pumice used by Larick et al. (2001) were all subject to these errors. Thus, we accept the estimation of a presence of *Homo erectus* on the island dating back to c. 1.5 Ma. This is the reference used by Yahdi Zaim et al. (2011) in their description of the maxillar Bpg 2001.04 from Sangiran, which is one of the most recent analyses.

Working with the Dutch Geological Survey, G. H. Ralph von Koegniswald retrieved more specimens at the Sangiran site. We will focus on several of them. First, a skullcap, *Pithecanthropus* II (von Koenigswald, 1938), which lacks the cranial base. It is very similar to *Pithecanthropus* I morphologically. The *Pithecanthropus* IV specimen (von Koenigswald & Weidenreich, 1939) confirms the posterior part of a cranium, including the inferior portion of both maxillae. It was retrieved at the

Box 8.9 Fossil associations and Java tools

In the locality of Sendangbusik, inside horizon C2a of the Sangiran site, a pumice (intrusion Sbk-1) of 1.24±0.04 Ma was identified associated with a calotte, one incisor, and two occipital-parietal fragments. In Pucung, another pumice of 1.02±0.13 Ma was found associated with the hominin fossil concentration of C3a (Larick et al., 2001).

Pucangan formation and its features relate it to *erectus* specimens from Java and East Africa, which favored its consideration as the same species.

Three skullcaps, Bukuran, Hanoman 1, and Sangiran 38, come from the most eastern part of the Sangiran zone, close to Sendangbusik town. All of them belong to the lower part of the Bukand formation (C1a), but could even come from the upper part of Pucangan (Indriati & Antón, 2010).

Pithecanthropus VIII cranium (Sangiran 17) (Sartono, 1971) (Figure 8.16), discovered by Sastrohamidjojo Sartono in 1969, is the most complete and well preserved of those found at Java, lacking only the left zygomatic region. It exhibits typical *Homo erectus*' traits, with a distinct supraorbital torus and occipital crest. Four additional remarkable specimens appeared later: the Tjg-1993.05 cranium, the Gwn-1993.09 skullcap, Sbk-1996.02 skullcap, and the Brn-1996.04 occipital bone (Larick et al., 2001).

The reconstruction of the cranium Sangiran IX (Tjg-1993.05), carried out by Yousuke Kaifu and collaborators (2011), had the goal of eliminating the morphological distortion of a very fragmentary exemplar, allowing new comparisons of sexual dimorphism and of the facial variability of *Homo erectus*. In most of its traits the reconstructed specimen belongs in the sample of the Bapang formation, although in the lower extreme with respect to size. This size difference, very obvious if comparing S 17 and IX, could be explained if these were exemplars of a different sex. The flatter frontal bone of S 17—male—and the vaulted cranium of IX—female—would correspond to current dimorphism. Nonetheless, the diversity of the Java sample,

with old and modern specimens as well as local variations, handicaps accepting a dimorphic norm.

The last discovery made in Sangiran was the maxilla Bpg 2001.04 (Figure 8.17), from the base of the Bapang formation (facies C1a-Zone, traditionally) and described by Yahdi Zaim and collaborators (2011). The stratigraphic horizon of the finding corresponds to C1a (Figure 8.15) and was found 2 m above a hornblende dated by the $^{40}Ar/^{39}Ar$ method at 1.51±0.08 Ma.

As Yahdi Zaim et al. (2011) pointed out, Sangiran is the only Southeastern Asian site with a complete population of *Homo erectus* specimens able to indicate evolutionary tendencies in the temporal range of 1.51–0.9 Ma. Comparison of the sample with the specimens from Zhoukoudian (China), Dmanisi (Georgia), and East Africa, led Zaim et al. (2011) to argue that the largest morphological differences are exhibited in the oldest Sangiran fossils—closer in their dental morphology to those from Africa—and in the Zhoukoudian specimens. These latter show more derived patterns in the relation of molars/premolars (upper M1 and M2 of smaller size). As we will soon see, the specimens from Zhoukoudian are approximately 0.8 Ma younger than the oldest from Sangiran. However, since the younger Sangiran fossils of *Homo erectus* coming from Bapang C5a have a similar morphology to those from Zhoukoudian (Kaifu, 2006), it is difficult to determine the evolutionary relationships between the populations of *Homo erectus* in Java and China.

Zaim and collaborators (2011) attributed the origin of Sangiran and Zhoukoudian *Homo erectus* to different hominin exits out of Africa, although admitting the possible alternative that the specimens found in China came from the youngest hominins of Sangiran. This possibility is somewhat remote, according to the respective ages of the first specimens from Java and China.

8.5.4 *Homo erectus*' variability in Java

Probably the great variability of *Homo erectus* in Java is due to different aspects with local—geoecological—as well as temporal patterns, but may also be due to sexual dimorphism and even individual characteristics.

The morphological differences between specimens of older and younger age from Sangiran point, however, toward a process of phyletic evolution. Here the two questions to settle are: first, whether the uniformity of the species would be affected by this evolution so that it would indicate the presence of two distinct chronospecies; and, second, whether the late Java specimens might be considered members of the "archaic"

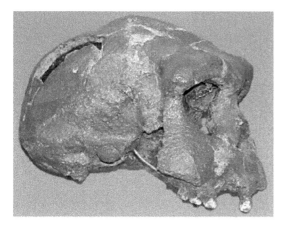

Figure 8.16 Sangiran 17, "*Pithecanthropus* VIII" (Sartono, 1971) (photography: *Athena Review* http://www.athenapub.com/13intro-he.htm).

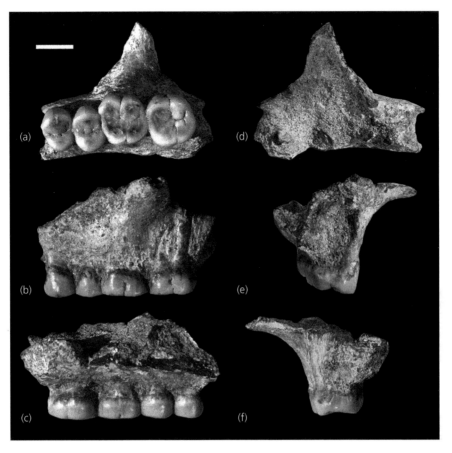

Figure 8.17 Bpg 2001.04 exemplar from Sangiran. Scale: 1 cm. Reprinted from Zaim, Y., Ciochon, R. L., Polanski, J. M., Grine, F. E., Bettis Iii, E. A., Rizal, Y.,. . . Marsh, H. E. (2011). New 1.5 million-year-old Homo erectus maxilla from Sangiran (Central Java, Indonesia). *Journal of Human Evolution*, 61(4), 363–376, with permission from Elsevier.

sapiens' grade that marks the transition of *Homo erectus* to *Homo sapiens* (see Chapter 9).

The proposal of a different chronospecies for the more modern specimens from Java was made based on the exemplars from the Ngandong site, which are some of the most recent in Java. In Ngandong up to 12 skulls classified by W. F. F. Oppennoorth (1932) as *Homo soloensis* were found. This sample includes massive-looking specimens with large cranial capacity (average 1,210 cc). However, specimens from other sites in Java of an older age and with slightly smaller cranial volume support the idea of a single evolving taxon. That is the case with the exemplar Sm 4 from Sambungmacan, a sizable calotte of 1,006 cc, probably of a male, discovered in 2001, whose general appearance is typical of the Javanese *H. erectus* (Baba et al., 2003). Very well preserved, the fossil allows us to observe that the flexion of the skull base is similar to that of modern

humans, a condition already mentioned regarding Olduvai OH 9, although with some uncertainty due to the poor state of the specimen. Hisao Baba and collaborators (2003) concluded, after studying Sm 4, that its condition is intermediate between the sample from Trinil/Sangiran and that from Ngandong. Contrary to this view, Valery Zeitoun and collaborators (2010) argued that the evolutionary changes of *Homo erectus* in Java would have led to the emergence of a new species on the island.

8.5.5 China

On the whole, the arguments about the difficulty to adequately characterize the fossil specimens in Java could be repeated when discussing China. A considerable part of the Chinese sample of *Homo erectus* (Figure 8.18) was discovered at a time when neither the

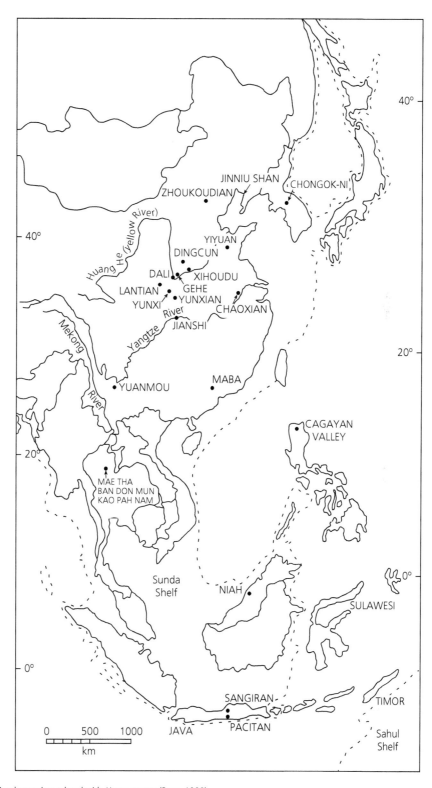

Figure 8.18 Southeast sites related with *Homo erectus* (Pope, 1988).

precise location of the finding nor the antiquity of the original terrain was subject to great effort.

As in the case of Java, however, important progress has been made to specify the age of the deposits and, therefore, the process of colonization of the Far East. With the particularity that, unlike Java, the presence of stone tools is abundant, allowing the completion of the available information.

8.5.6 The colonization of China

Zhoukoudian is, by far, the best-known Chinese site, but its age, as we soon see, is less than a million years. Other enclaves in mainland China have provided the oldest evidence of hominin presence. Two areas of comparable age or even older than the Javanese occupation, correspond to the Nihewan depressions, in the basin of the Yellow River, and the Yuanmou depression in that of the Yangtze River. Table 8.6 offers a synthesis of the sites and its attributed ages.

The first specimens that pointed to an older occupation of China (Figure 8.19) were found in Chenchiawo or Chen-Chia (or Jia) Wo, as transcribed by Poirier (1987), and Kungwangling or Gongwangling, belonging to the Lantian (Lan-T'ien) district, province of Shaanxi. A well-preserved mandible (Lantian 1) was discovered in 1963 in Chenchiawo by Wu Rukang (1964). A tooth, a cranium, and part of the facial skeleton (Lantian 2) appeared the following year at Kungwangling hill (Wu Rukang, 1966). Both specimens were attributed to a new species, *Sinanthropus lantianensis*, described by Wu Rukang as morphologically more primitive than the Zhoukoudian specimens, and even than the Trinil (Java) *erectus*. The estimated cranial capacity for the Lantian 2 cranium is around 780 cc (Day, 1986). However, Wu Rukang himself later reclassified these specimens as *Homo erectus* (see Wu Rukang, 1980).

Box 8.10 What is the Longgupo fossil?

One of the signers of the paper by Huang Wanpo et al. (1995) on the Longgupo exemplar, Russell Ciochon (2009, p. 910), published a reinterpretation 14 years later in the following terms: "Now, in light of new evidence from across southeast Asia and after a decade of my own field research in Java, I have changed my mind. Not everyone may agree; such classifications are always open to interpretation. But I am now convinced that the Longgupo fossil and others like it do not represent a pre-*erectus* human, but rather one or more mystery apes indigenous to southeast Asia's Pleistocene primal forest. In contrast, *H. erectus* arrived in Asia about 1.6 million years ago, but steered clear of the forest in pursuit of grassland game. There was no pre-*erectus* species in southeast Asia after all."

The morphological analysis of the hominin remains and the faunal comparative study indicated for Lantian an age of c. 0.7 Ma, established by biostratigraphy (Aigner & Laughlin, 1973), which is quite consistent with later paleomagnetism studies. Research using paleomagnetism and lithostratigraphy by An Zhisheng and Ho Chuan Kun (1989) moved back the age of the Lantian specimens to 1.15 Ma.

A greater age was attributed to Longgupo exemplars. Longgupo Cave, also known as Wushan site, is located 20 km south of the Yangtze River in the east end of Sichuan province. From the middle of the cave (levels 7–8 of the excavation) came a mandible fragment, an upper incisor, and two stone tools of the Oldowan tradition (Huang Wanpo et al., 1995).

Huang Wanpo and colleagues' (1995) study of the Longgupo dental fragments identified certain affinities

Table 8.6 Sites with the oldest human presence in China

Basin	Site	Location	Remains	Ma	Reference
Yellow River	Donggutuo	150 km East of Beijing	Mode 1 utensils	1.2/1.1	Zhu et al., 2004
	Kungwangling	Lantian district	cranium	1.15	An & Ho, 1989
	Xiaochangling	150 km East of Beijing	Mode 1 utensils	1.26	Li, Yang et al., 2008
	Majuangou	150 km East of Beijing	Mode 1 utensils	1.66	Zhu et al., 2004
Yangtze River	Niujiangbao	Shangnabang, Yunnan	V1519 incisors Mode 1 utensils	c. 1.7	Zhu et al., 2008
	Longgupo	Chongqing, Wushan	fragmentary mandible, incisor Mode 1 utensils	1.9/1.8	Huang et al., 1995

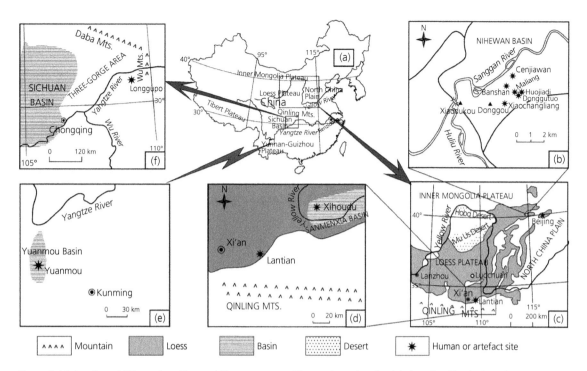

ʌʌʌʌ Mountain		Loess		Basin		Desert	✳ Human or artefact site

Figure 8.19 Locations of Chinese sites with very old human presence. Nihewan depressions (b–e), in the Yellow River basin and Yuanmou depressions (f), in the Yangtze River basin (illustration Rixiang Zhu et al., 2003). Reprinted from Zhu, R., An, Z., Potts, R., & Hoffman, K. A. (2003). Magnetostratigraphic dating of early humans in China. *Earth-Science Reviews*, 61, 341–359, with permission from Elsevier.

with African early specimens of *Homo*. Consequently, these authors thought that the specimens should not be classified as *H. erectus*. According to Huang Wanpo et al., they belong to an almost 1.9 Ma pre-*erectus*. This would imply two things: that the genus *Homo* left Africa almost at the time as its diversification, close to 2 Ma, and that the evolution toward *H. erectus* occurred in situ within the Asian continent. Jeff Schwartz and Ian Tattersall (1996b) doubt that one of the dental fragments belongs to a hominin (it could be close to the orangutan). In regard to the other fragment, Schwartz and Tattersall doubt that it can be compared with any specific hominin species. Dennis Etler, Tracy Crummett, and Milford Wolpoff (2001) argued that the mandible fragment from Longgupo is indistinguishable from those belonging to Asian apes, particularly to *Lufengpithecus*, while the incisor is identical to Asian modern humans and likely is due to an intrusion into the deposit. However, Wood and Turner (1995) accept the hypothesis that the Longgupo hominins preceded *H. erectus* sensu stricto and, thus, that the departure out of Africa was initiated by early members of the *erectus* grade. This hypothesis had already been suggested by Bernard Wood (1992a).

The discrepancies in Longgupo dates are remarkable. The paleomagnetic column obtained by Huang Wanpo et al. (1995) attributed to the middle levels containing hominin remains an age corresponding to the Olduvai subchron (1.96–1.78 Ma). However, electron spin resonance applied to the dental enamel yields a much more recent age (1.0±0.1 Ma) (Zhu Rixiang et al., 2003). Bernard Wood and Alan Turner (1995), in their commentary on the Longgupo discovery, argued that the fauna of the site seems older than that of the indicated date.

The Niujiangbao site, in the Yuanmou depression, has provided two hominin incisors—V1519, jointly—found in 1965 and attributed to *Homo erectus yuanmouensis* (Hu, 1973). The stratigraphic horizon of the incisors and associated tools—member M4— was dated, according to paleomagnetism studies and the calculated sedimentation rate, to c. 1.7 Ma (Zhu et al., 2008).

According to Rixiang Zhu and collaborators (2008), the incisors V1519 have numerous traits identical to those of Zhoukoudian *Homo erectus* and are quite similar to African specimens such as KNM-WT 15000, *H. erectus* sensu lato and KNM-ER 1590B, *H. habilis*. These

Box 8.11 The names of Chinese authors

A cautionary note about Chinese names—it has become common, particularly in English-speaking countries, to follow the Chinese Government recommendation to transcribe Chinese names into Western languages in a different way than was traditionally done. Thus, "Pekin"—Latin form, commonly used in Spanish—or "Peking"—commonly used in English—is now transcribed as Beijing. In this book we have followed the traditional English spelling of "Peking." With regard to the first name and surname of the authors, the order in China is the reverse: first the surname and then the first name; for example, Wu Rukang. This formula may cause confusion because some scientific journals observe the practice, while others follow the Western use: Chinese surname, first name—with a comma between (Wu, Rukang)—or the first name followed by the

surname without comma (Rukang Wu). As an example, the journal *Nature* has published papers by a reputed Chinese paleontologist as "Wu Rukang"—the usual Chinese order, with the surname first—while *Science* published a commentary by Wu's colleague using the common Western style of first name (Xinzhi) followed by surname (Wu). A clue often available about which is the first name and which is the surname is to remember that, usually, Chinese family names—which we call surnames—have one syllable and what we call first names usually have two syllables. In this book we have followed the criteria of maintaining the order used by the authors in their publications, which typically list the surname followed by the first name, with the first name's two syllables sometimes separated by a hyphen.

authors maintain that the Yuanmou specimens can be attributed to a very old *Homo* sp., and that it is possible to include them in the large hypodigm of *Homo erectus* until more informative remains are found.

If the remains found in the depressions of Yuanmou and Nihewan are considered together—with the most northerly site of Majuangou in the latter—it is possible to infer that the hominin dispersion through mainland China occurred in a very short time, probably due to a warm climate phase, which would also explain the rapidity with which these hominins reached the Far East (see Figure 8.20).

8.5.7 Zhoukoudian specimens

Among the Chinese sites with Middle Pleistocene hominins, of particular note is Zhoukoudian, a site near the town of the same name, 40 km southeast of Beijing.

The excavation at Zhoukoudian cave, about 25 miles southwest of Beijing, began in 1921 and again in 1923. It was there that Otto Zdansky found a much worn human molar. This is how the dragon teeth received the desired paleontological context.

In October 1927, Birger Bohlin, one of Davidson Black's colleagues, found a large human molar at Zhoukoudian. It looked like no other specimen known at the time. Black described a new hominin species based on it, *Sinanthropus pekinensis*, also known as "Peking man" (Black, 1927). In 1940, Weidenreich included these specimens in the same species as the Javanese Middle Pleistocene exemplars, in the subspecies *Homo erectus pekinensis* (Weidenreich, 1940).

All the Zhoukoudian human fossils have been found at the so-called Locality I, a demolished cave

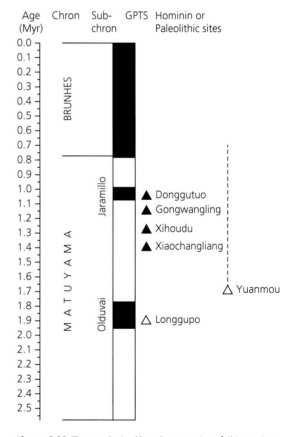

Figure 8.20 The age obtained by paleomagnetism of Chinese sites with the oldest hominin evidence; black triangles indicate precise dates; white triangles, uncertain estimates. Reprinted from Zhu, R., An, Z., Potts, R., & Hoffman, K. A. (2003). Magnetostratigraphic dating of early humans in China. *Earth-Science Reviews*, 61, 341–359, with permission from Elsevier.

Box 8.12 Dragon teeth

The initial identification of the Chinese specimens is an excellent example of the adventures of early twentieth-century paleoanthropologists. The story was told in a lively and elegant literary style by von Koenigswald (1981). What follows is taken from that work. During the early twentieth century, Chinese chemists sold fossils—advertised as "dragon teeth" (*lung tse*) and "dragon bones" (*lung ku*)—that were said to possess efficient aphrodisiac effects. Given that teeth were believed to be more effective than bones, fossil mandibles were destroyed following an understandable commercial objective. The bones were smashed and ground, because they were used in small doses, and their price was very high. K. A. Haberer, a German naturalist who traveled to north China between 1899 and 1901, was able to acquire a considerable collection of such dragon bones and teeth, described in 1903 by Max Schlosser. The specimens gathered by Haberer belonged to many different species. One of them, an upper molar purchased from a Beijing chemist, belonged to a hominin. Schlosser tentatively attributed it to a Pliocene "grad. et sp. indet." being (see Antón, 2002).

Based on the hints from Haberer, Davidson Black tried to track down the dragon bones, following clues of ancient Chinese pharmacopoeia, dating back to the Wei period (seventh century BC). Also, the Swedish Academy sent researchers such as Johan Gunnar Andersson to China with the intention of collecting fossils. Andersson and the Austrian paleontologist Otto Zdansky identified "Dragon Bone Hill" as a possible place of origin of the fossils (Shapiro, 1976). The Zhoukoudian site is located on the northern side of this hill.

---

of considerable size (Figure 8.21) The study by Huang Wanpo (1960) of the locality deposits described six main layers, assigning the first one (layer I) to the Pleistocene, layers II and III, for now, to the Middle Pleistocene, and layers IV–VI to the late Middle Pleistocene. Liu Zechun (1985), however, distinguished up to 17 layers.

Following the stratigraphic map and the unpublished notes of the excavations by Jia Lanpo, director of the excavations in the 1980s, Noel Boaz and collaborators (2004) conducted a three-dimensional distribution of the different loci (from Locus A to Locus O) with hominin presence (Figure 8.22).

During the 1920s, the cave of Zhoukoudian yielded a considerable number of remains, some of which were very well preserved. Close to 40 different individuals were identified, including 14 crania, 11 mandibles, more than 100 teeth, and a few postcranial remains.

The first cranium—ZKD III—appeared in layer 11 of Locus E, in 1929 (Pei Wenzhong, 1929), Davidson Black reconstructed its endocranial cast. In 1930, various fragments of Locus D (layers 8–9) were put together to form cranium ZKD II. Three more cranial exemplars were recovered in 1936 from the same layers (8–9), but at Locus L: ZKD X, XI, and XII (Grün et al., 1997).

The Zhoukoudian collection of human fossils was described by Davidson Black (a very detailed study of a juvenile cranium was inserted in Black, 1930) and, in 1936, by Weidenreich (1936) who also made some casts (Figure 8.23). It was fortunate that Black and Weidenreich carried out this work because most of the early Zhoukoudian remains were lost during World War II.

Five additional teeth and two limb bone fragments were found in 1949 and 1951 (Woo Jukang, 1980), and a mandible, together with other dental and postcranial remains, were found later, in 1959 (Woo Jukang & Chao Tzekuei, 1959). Finally, in 1966, members of the Institute of Vertebrate Paleontology and Paleoanthropology (Academia Sinica, Beijing) found a frontal and an occipital.

Zhoukoudian Locality 1 has been excavated systematically and regularly since 1979. The report offered by Wu Rukang (1985) included references to over 5,000 lithic artifacts, but no additional hominin findings. One of the latest works is the reconstruction of the skull Zhoukoudian V (Figure 8.24), carried out by Xiujie Wu, Lynne Schepartz, and Wu Liu (2010), from materials of 1934 and 1966, obtaining a new endocranial cast of the *Homo erectus* brain. It adds to the endocranial specimens from Zhoukoudian II, III, X, XI, and XII, previously available. Another specimen was found in 1993 in Hulu Cave (Nanjing, Jiangsu Province, Southern China) by the Tangshan Archaeological Excavation Team, Nanjing 1, with the endocranium, with an age of between 0.62 and 0.58 Ma, obtained by ESR and uranium series (Xiujie Wu et al., 2011).

The synthesis of the Zhoukoudian findings makes this site an exceptional enclave: 50 hominins and 17,000 lithic utensils had been recovered up to 2009 (Ciochon & Bettis III, 2009).

The chronology of Locality 1 has been the subject of a considerable debate that the latest dating techniques have failed to entirely resolve. The possibility of specifying the dates of Locality 1 depended on the principle of land correlation with other sites and, according to this calculation, Björn Kurtén (1959) suggested that the second glaciation (Mindel, c. 450,000 years ago) would be the most appropriate time to place the deposits with the oldest hominin specimens. Liu Zechun (1985), meanwhile, applying the oxygen isotopes' method, obtained for the 17 layers of the site ages ranging from around 590,000–128,000 years.

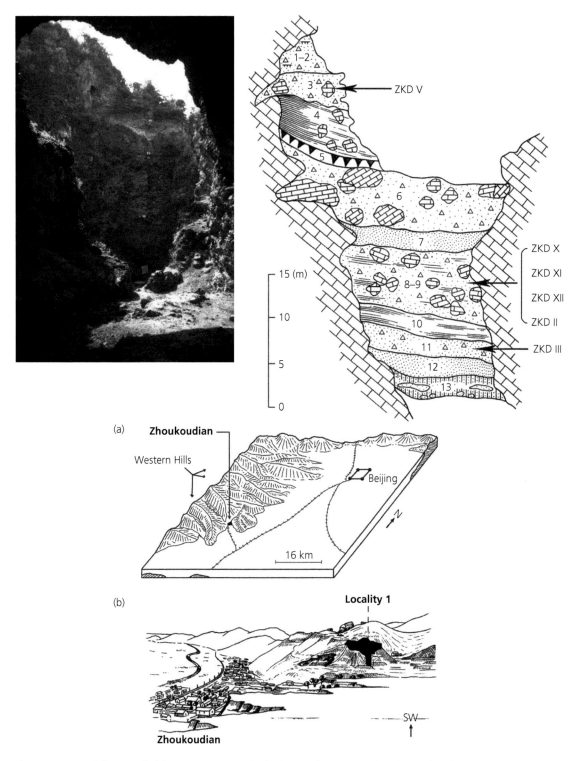

Figure 8.21 Upper left: West wall of the main excavation at Zhoukoudian, Locality 1; upper right: side view of Locality 1 with the stratigraphic position of the specimens from which endocrania have been obtained; bottom: Locality 1 on the map. Upper left:Reprinted by permission from Macmillan Publishers Ltd: Ciochon, R. L. & Bettis III, E. A. (2009). Asian Homo erectus converges in time. *Nature*, 458(7235), 153–154. Upper right: Reprinted from Wu, X., Schepartz, L. A., & Norton, C. J. (2010). Morphological and morphometric analysis of variation in the Zhoukoudian Homo erectus brain endocasts. *Quaternary International*, 211, 4–13, with permission from Elsevier. Bottom:Reprinted from Boaz, N. T., Ciochon, R. L., Xu, Q., & Liu, J. (2004). Mapping and taphonomic analysis of the Homo erectus loci at Locality 1 Zhoukoudian, China. *Journal of Human Evolution*, 46(5), 519–549, with permission from Elsevier.

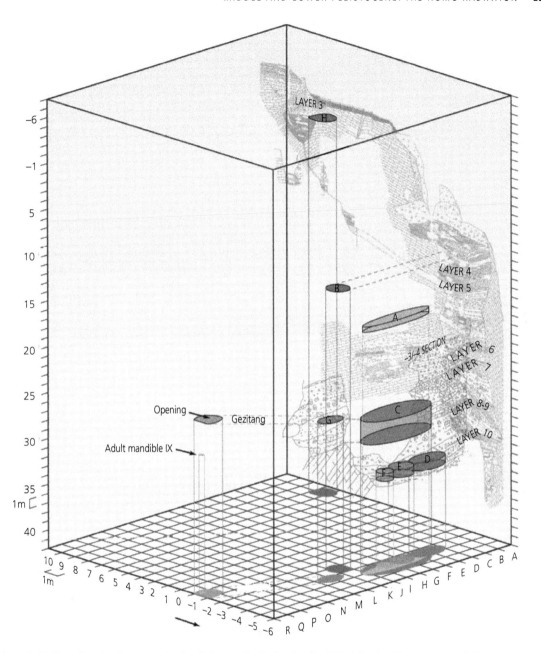

Figure 8.22 Three-dimensional representation from loci at Locality 1 in Zhoukoudian (China). Reprinted from Boaz, N. T., Ciochon, R. L., Xu, Q., & Liu, J. (2004). Mapping and taphonomic analysis of the Homo erectus loci at Locality 1 Zhoukoudian, China. *Journal of Human Evolution*, 46(5), 519–549, with permission from Elsevier.

From the late 1970s, new dating methods were applied to Zhoukoudian, yielding a series of absolute dates compiled by Rukang Wu (1985), indicated in Table 8.7. Chen Tiemei and Yuan Sixun (1988), using uranium series, later granted to the Zhoukoudian Locality I a much more recent age: 0.29–0.22 Ma.

The paleomagnetic measurements indicate that the Brunhes–Matuyama boundary is located between the layers 13 and 14, which means that the human remains are younger than 0.78 Ma. Using electron spin resonance (ESR) applied to deer teeth, Huang Pei-Hua and collaborators (1993) established a range of 0.28–0.58

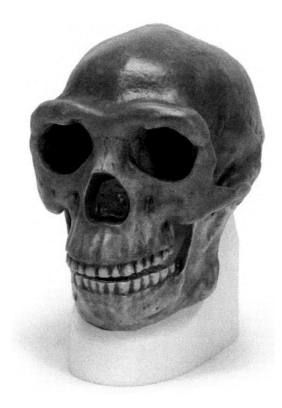

Figure 8.23 *Sinanthropus pekinensis* (Black, 1927)—*Homo erectus* (cast from a replica from the collection at the Institute of Anthropology and Human Genetics for Biologists, Wolfgang Goethe University, Frankfurt, Germany; https://www.3bscientific.es/manualde-producto,pm.html (cast HYPERLINK https://www.3bscientific.es/product-manual/VP750-1.pdf"VP750/1 [1001293])

Ma for the samples collected between the layers 3–12, with a span of human occupation in Zhoukoudian of about 300,000 years; dates confirmed using the same technique by Rainer Grün et al. (1997). Xiujie Wu, Lynne Schepartz, and Christopher Norton (2010), in a study designed to evaluate the morphological variety of the cranium from Locality 1, summarized the dating studies, and attributed the specimens tentatively to an age somewhat older: 0.5–0.4 Ma for the most modern skull of Zhoukoudian (ZKD V) and up to 0.8 Ma, or even more, for the oldest cranium (ZKD III). That means that human presence in the cave could have extended to nearly half a million years.

The last available analysis on the age of Locality 1 is that by Shen Guanjun and collaborators (2009) using the ^{26}Al/^{10}Be method. Lower levels with lithic utensils (7–10) obtained an average in six measurements of 0.77±0.08 Ma; that is to say, a much older age than obtained by uranium series (Chian Tiemei &Yuan Sixun,

1988) and inside the previous range estimated by Shen Guanjun et al. (2001).

As Russell Ciochon and Arthur Bettis (2009) indicated in a comment on the last dating of Locality 1, if Zhoukoudian's nearby sites, close to latitude 40° N, are taken into consideration, an important view arises of hominin presence in the extreme climate of the continent, very far from tropical Java. The fossils from the Nihewan and Yuanmou formations prolong considerably the temporal and geographic range of *Homo erectus*. However, the Zhoukoudian specimens are very similar to the Javanese. The similarity is so striking that many authors, such as Ernst Mayr (1950), Chris Stringer (1984), Bernard Wood (1984), and G. Philippe Rightmire (1990), have felt the need to group all those specimens in *Homo erectus*. There are, however, dissenting views (Aguirre & de Lumley, 1977; Aguirre et al., 1980; Rosas & Bermúdez de Castro, 1998; Aguirre, 2000). The greater cranial capacity of Zhoukoudian Locality 1 specimens, and certain differences in dental traits and other cranial features between the Chinese and Javanese specimens, was interpreted by "lumpers" as evolution expected by the greater age of the latter. In any case, the population set comprising the specimens from Zhoukoudian to the oldest exemplars from Nihewan, Yuanmou, and Java, show a morphological proximity that contrasts with the later specimens similar to *Homo sapiens*.

8.5.8 Environment and adaptation of *Homo erectus* in Asia

The type of environment prevalent in the Lower and Middle Pleistocene in Java is strongly reminiscent of the place of hominin origin: a mosaic of open savanna, prairies, and tropical forest. It is not surprising that hominins had a certain facility to adapt to the island when they arrived after spreading through Asia. However, the case of China is different. With a continental climate, its localities—particularly the northernmost—were much more affected by a harsh climate with a succession of glaciations. During the Pleistocene there were many glacial and interglacial cycles, integrated into four "classic" glaciations (Gunz, Mindel, Riss, and Würm, listed in Table 8.12; see also Box 8.24). It is likely, however, that the hominin arrival in China initially, and in Java a bit later, took place during an interglacial period. The presence of later climatic alternatives lead us to believe that the island could have served as a "refuge" for the population of *Homo erectus* in cooler times, a population that would return to China as soon as conditions improved. It is the "source

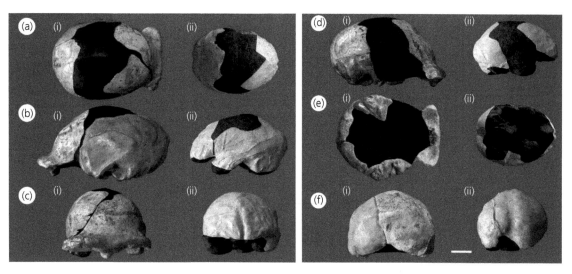

Figure 8.24 Upper: Zhoukoudian V skull—an endocast reconstruction in different views: (a) superior; (b) left lateral; (c) anterior; (d) right lateral; (e) basal; and (f) posterior; scale: 4 cm (illustration by Xiujie Wu, Schepartz, & Wu Liu 2010). Bottom: Fragments of Nanjing 1; (a) anterior part of the cranial vault; (b) left parietal with lower squama occipitalis (illustration by Xiujie Wu et al., 2011). Top: Wu, X., Schepartz, L. A., & Liu, W. (2010). A new Homo erectus (Zhoukoudian V) brain endocast from China. *Proceedings of the Royal Society of London – B*, 277(1679), DOI: 10.1098/rspb.2009.0149, by permission of the Royal Society. Bottom: Wu, X., Holloway, R. L., Schepartz, L. A., & Xing, S. (2011). A New Brain Endocast of Homo erectus From Hulu Cave, Nanjing, China. *American Journal of Physical Anthropology*, 145, 452–460.

and sink" model proposed by Dennell, Martinón-Torres, and Bermúdez de Castro (2011) for the European Middle Pleistocene (see Section 8.6.7) and used by Julien Louys and Alan Turner (2011) for Asia. Java would act as "source" and China as "sink" through a fairly steady flow of people. Population movements between Central Asia and Northern China would have been easier in the Pleistocene, when cold periods were shorter and less severe (Martinón-Torres et al., 2011). In any case, it is important to note that localities such as Shanxi (Central China), with a continuous occupation over ten glacial/interglacial cycles, indicate that the

Table 8.7 Zhoukoudian absolute ages

Bed		Dating method
1–3	230,000 256,000	Uranium series
4	290,000	Thermoluminescence
6–7	350,000	Uranium series
7	370,000–400,000	Paleomagnetism
8–9	420,000 462,000	Uranium series Fission track
10	520,000–620,000	Thermoluminescence
12	>500,000	Uranium series
13–17	>730,000	Paleomagnetism

According to Wu, R. (1985); modified.

hominin presence was not occasional, occurring only during mild climatic seasons, but a permanent occupation (Ciochon & Bettis III, 2009).

8.5.9 Alternatives in the hominin taxonomy during the Lower and Middle Pleistocene

The morphological diversity of hominin specimens from the Lower and Middle Pleistocene in Asia and Africa—we will leave the European case for later in the book—follows a general pattern. There are different interpretations about how these hominins should be classified. We will reduce the alternatives to the two most common, which are mutually exclusive. The first alternative states that all specimens belong to the same species, *Homo erectus*, although Asian populations would be considered informally *Homo erectus* sensu stricto, while the African populations would be *H. erectus* sensu lato. The second taxonomic proposal distributes the specimens into two different species: *Homo ergaster* for African specimens and *Homo erectus* for Asians.

8.5.10 *Homo ergaster*

The species *Homo ergaster* was proposed by Groves and Mazák (1975) in a study on Villafranchian gracile hominin taxonomy based almost exclusively on dental traits. The examination of the specimens from the upper levels—between 1.4 and 1.8 Ma—of East Rudolf (Lake Turkana, Kenya) led to the definition of the new species, whose holotype is KNMER 992: two associated hemi-mandibles, with a complete dentition except for the first incisors. They were described by

Leakey and Wood (1973) and classified as *Homo habilis* by Leakey (1974). The hypodigm is constituted by specimens such as KNM-ER 730, 731, 803, 806, 807, 808, 809, 820, and 1480, all dental or mandibular fragments, the KNM-ER 734 parietal, and the 1805 cranium. According to Groves and Mazák, the premolars and, most of all, the smaller molars, the thick and massive mandible, and the greater cranial capacity, represent differences regarding *Homo habilis* and justify the proposal of *Homo ergaster*.

What is the point of placing the African specimens of *H. erectus* in a species different from the Asians? As we have noted, certain features of the Asian specimens, such as a pronounced occipital torus, are absent in some African fossils and present in others. To what extent that is a significant difference could only be established through precise analyses.

The comparative study by Gen Suwa and collaborators (2007) of the specimens attributed to *Homo* in Omo, Konso, and Turkana led to some interesting taxonomic considerations. First, the authors rejected the division into two species, *Homo habilis*/*Homo rudolfensis*. They also considered that there was no temporal overlap between the specimens of *Homo habilis*, even when considered broadly, and those of *Homo erectus* in the area. Finally, they asserted the advantages of grouping into a single species all specimens of *Homo erectus* sensu lato from Asia and Africa.

Parsimony rules would favor such an interpretation. However, as we shall see, in some cases the grouping is somewhat excessive.

Some specimens of *H. ergaster* recently discovered—those from Daka and Olorgesailie—belong to the range 0.5–1 Ma, with few African exemplars known

---

Box 8.13 Cranial differences between *H. ergaster* and *H. erectus*

Figure 8.25 (from Manzi, 2004) outlines the morphological differences between the crania of *H. ergaster* and *H. erectus*. The taxon has gained credit in recent years, but there are distinguished authors, such as Tobias, Wolpoff, Rightmire, Conroy, and Stringer, who deny the existence of *Homo ergaster* as separate from *Homo erectus*. Groves (1989)—who proposed the new taxon—did not attribute to the species *H. ergaster* either the crania 3733 and 3883, or the South African exemplar from Swartkrans, classified by Broom and Robinson as *Telanthropus*. He only referred to a *Homo* sp. (unnamed) to accommodate them.

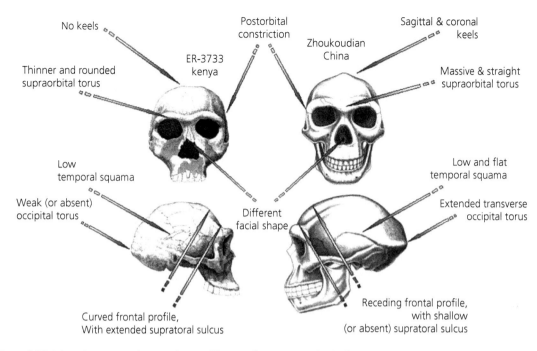

No keels

Thinner and rounded
supraorbital torus

ER-3733
kenya

Postorbital
constriction

Zhoukoudian
China

Sagittal & coronal
keels

Massive & straight
supraorbital torus

Low
temporal squama

Weak (or absent)
occipital torus

Different
facial shape

Low and flat
temporal squama

Extended transverse
occipital torus

Curved frontal profile,
With extended supratoral sulcus

Receding frontal profile,
with shallow
(or absent) supratoral sulcus

Figure 8.25 Schematic comparison between the cranial features of *H. ergaster* (or "early African *H. erectus*") on the left, represented by KNM-ER 3733 from Koobi Fora, Kenya, and *H. erectus* s.s. from Eastern Asia on the right, represented by Weidenreich's reconstruction of a female type-specimen from Zhoukoudian (China) (illustration from Manzi, 2004). Manzi, G. (2004). Human evolution at the Matuyama-Brunhes boundary. *Evolutionary Anthropology*, 13(1), 11–24.

up to now. Daka specimens, a calotte, three isolated femurs, and a proximal tibia, come from the member Dakanihylo—"Daka"—of the Bouri formation in Middle Awash (Ethiopia). The dating by the ^{40}Ar/^{39}Ar method of a pumice unit at the base of the member yielded an age of 1.042±0.009 Ma (De Heinzelin et al., 2000). A paleomagnetic study assigned a reverse polarity to Dakanihylo, which is likely to have occurred earlier than 0.8 Ma.

The state of preservation of BOU-VP-2/66 is good, although the specimen was a little distorted (Figure 8.26 and Box 8.14). The vault and the supraorbitals show damage from perimortem scraping. The frontal and parietal bones are marked by groups of subparallel striations, each with internal striations, attributed by the discoverers to having been gnawed by animals. The endocranial capacity is 995 cc (Asfaw et al., 2002).

The systematic interpretation of BOU-VP-2/66 by Berhane Asfaw et al. (2002) included a cladistic analysis of the 22 traits most commonly used in previous, similar studies on *Homo erectus*, dividing the sample into operational taxonomic units obtained by paleodemes. According to the results, the authors argued that the skull measurements overlap with the ranges

of both African and Asian samples and are not useful to consistently distinguish the fossil of Daka. Asfaw et al. (2002) argued that, both in metric and morphological terms, the Daka cranium confirmed previous suggestions that the geographical division of the first *Homo erectus* into separate species is misleading from a biological point of view, and artificially multiplies the species' diversity of the Lower Pleistocene. The resemblance between the Daka specimen and its Asian counterparts indicates that the hominin fossils in Africa and Eurasia represent populations within an extended paleospecies.

Despite such strong opinion, the morphological meaning of the Daka skull is unclear. Asfaw et al. (2002) considered the specimens from Daka and Buia (Ethiopia) as very similar, but as Susan Antón (2003) pointed out, these two fossils differ in some important morphological points. Ernesto Abbate et al. (1998) argued something similar with respect to the description of the Buia specimen, indicating that it exhibits a mosaic of characters from *Homo erectus* and *Homo sapiens*. Returning to the hominin from Daka, Antón (2003) mentioned the lack of some important apomorphies of the Asian *Homo erectus*. The most conspicuous

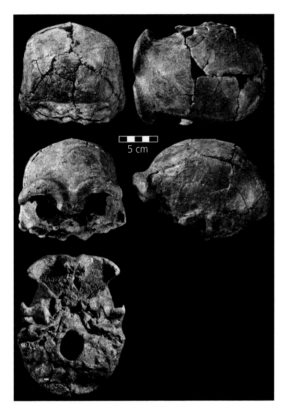

Figure 8.26 BOU-VP-2/66 cranium: posterior, superior, frontal, and lateral views (Asfaw et al., 2002).

is, to our knowledge, the nonexistent occipital torus in BOU-VP-2/66. As Asfaw et al. (2002) indicated, the occipital wall rises vertically and bends forward. Viewed from behind, the undistorted parietal walls would have been vertical. The morphological study of Giorgio Manzi, Emiliano Bruner, and Pietro Passarello (2003) brought BOU-VP-2/66 closer to KNM-ER 3733 and 3883, asserting that they all support their belonging to *Homo ergaster*, while OH 9, the Dmanisi specimens and the Ngandong *Homo erectus* (Java) would be intermediate exemplars in the evolution to *H. sapiens*.

8.5.11 The variability of the *Homo erectus* samples

The first problem facing the proposal of a single species for the hominins of the Middle Pleistocene is to assess the range of variability observed in such a broad hypodigm. The literature in this regard is immense and one can only refer to some of the studies.

The most objective method to estimate the extent of the morphological distance between the various specimens is to perform a statistical analysis comparing them. James Kidder and Arthur Durband (2004) conducted a multivariate analysis—canonical correlation—to compare 20 of the most complete skulls of the *Homo erectus* hypodigm, including African specimens, in order to determine the degree of particularity possessed by the Zhoukoudian specimens (China) compared to the rest (Box 8.15). Comparisons were made on eight cranial measurements identified as essential in two studies by W. W. Howells (1973, 1989). The results confirmed the particularity of Zhoukoudian, but grouped together in the same set specimens from Africa, Java, and the Hexian fossil (China). The most logical interpretation for the obtained metric relationships, attributes to the Zhokoudian crania its own particularity; but as Kidder and Durband (2004) argued, the sample variation is better explained by local differences due to environmental adaptations or genetic drift than to a multi-species hypothesis. Consequently, Kidder and Durband (2004) argued as the best option a species unity for *Homo erectus*.

Brian Villmoare (2005) undertook a statistical study addressing first the analysis of the sample of Asian and

African *Homo erectus* based on 23 metric characters—most of them according to Howells (1989)—compared with a sample of 1,000 modern humans. In the second part, the analysis established the Euclidean average between Asian and African specimens compared to the same average of 1,000 samples of current Asian and Africa humans, randomly generated. Finally, Villmoore (2005) used just ten non-metric characters, which covered the totality of the morphological differences found in the fossil sample and a sample of 221 crania of modern humans (Box 8.16).

The hypothesis of a single species of *Homo erectus* was not falsified in any of the first two analyses (metrics) by Villmoore (2005), although the differences found were greater than those exhibited by modern humans in Asia and Africa. By contrast, the analysis of non-metric characters indicated a notable difference between the African and Asian fossil samples, with the supramastoid crest, the angular torus, and the metopic prominence as the characters of greater distance between them.

The discrepancy between the metric and non-metric analyses is interesting because, as Villmoore (2005) indicated, most authors who have opted for the hypothesis of two different species, *Homo erectus* and *Homo ergaster*, used non-metric characters in their morphological studies. To the extent that non-metric traits are assigned by their status (present/absent), a very large interval is introduced, which is much more diminished in metric studies.

Brian Villmoore (2005) indicated that it would be interesting to transform non-metric characters into metric using three-dimensional landmarks able to define curves. Such an analysis by quantifying non-metric characters of the *Homo erectus* cranium has been done by Claire Terhune, William Kimbel, and Charles Lockwood (2007), and by Karen Baab (2008).

Terhune and collaborators (2007) studied the morphology of the temporal bone using 83 three-dimensional landmarks in a total of 520 crania of modern and fossil species (Table 8.8).

Existing distances in current humans were calculated using a Generalized Procrustes Analysis for obtaining the interspecies and inter-individual variations within the species and its various groups (Box 8.17). Afterward, Terhune et al. (2007) obtained the Procrustes distances between the various fossil specimens and between the groups defined a priori (Africa, Eurasia, Indonesia, and mainland Asia).

The distance comparison by Terhune et al. (2007) showed that the variation of forms in the full sample of *Homo erectus* is usually greater than in current species, which supports the differentiation as species between *Homo erectus* and *Homo ergaster*. However, the authors encountered the same obstacle that appears in qualitative analysis: OH 9 and D2280 differed from the Koobi Fora *Homo ergaster*, but were closer to *Homo erectus* sensu stricto. The conclusion of Terhune and collaborators (2007) was somewhat convoluted: *Homo erectus* is composed of multiple species of complex differentiation so that the available samples cannot be grouped either geographically or temporally, which is not so different from saying that it is not possible to propose an appropriate taxonomy for *Homo erectus*. This observation has led to the use of a different concept than species for Middle Pleistocene hominins, which is to include all of them in an "*erectus* grade."

Julian Huxley suggested the concept of "grade" to express the notion that different taxa might have reached a certain evolutionary stage (Huxley, 1958). Bernard Wood and Mark Collard (Collard & Wood, 1999; Wood & Collard, 1999a, 1999b) rescued Huxley's grade in order to face the difficulties involved in dealing with *Homo* (Box 8.18).

Box 8.16 Specimens included in the study by Villmoore (2005)

The specimens of the fossil sample of Villmoore (2005) were: KNM-ER 3733; KNM-ER 3883; OH 9; OH 12; KNM-WT 15 000; ZKD 3; ZKD 10; ZKD 11; ZKD 12; Sambungmacan; Sangiran 2; Sangiran 3; Sangiran 4; Sangiran 17; and Trinil 2. The non-metric characters examined were: sagittal keel, bregmatic eminence, tympanic-mastoid fissure, supernumerary bones at lambda, occipital torus, metopic prominence, angular torus, and supramastoid crest.

Box 8.17 Generalized Procrustes Analysis

The Generalized Procrustes Analysis (GPA) takes its name from the mythical story of the Bed of Procrustes. It is a geometrical transformation that allows the statistical distribution of different forms to be studied. By identifying points on the surface of objects, three-dimensional forms are obtained for comparison, a reference form is arbitrarily generated, and one by one the various forms to be compared are superimposed, calculating the Procrustes or Procrustean distance—the distance between each form and the reference (Zelditch et al., 2004).

Table 8.8 Sample used in the study by Terhune et al. (2007): above, comparative sample of current crania; next page, fossil specimens

Species	Population/subspecies	Source*	Female	Male	Total
H. sapiens	Nubian (Egypt)	ASU	21	22	43
	Native American (Utah)	AMNH	10	10	20
	East African (Tanzania)	AMNH	11	8	19
	Southeast Asian (Singapore)	AMNH	11	10	21
	Medieval European (Hungary)	AMNH	10	11	21
	Australian Aborigine	AMNH	11	10	21
	Native Alaskan	AMNH	10	10	20
Pan troglodytes	*P. t. troglodytes*	CMNH, PCM	39	38	77
	P. t. schweinfurthii	RMCA	20	20	40
	P. t. verus	PM	24	24	48
P. paniscus	Zaire				
		RMCA	21	19	40
Gorilla gorilla	*G. g. beringei*	NMNH RMCA	6	11	17
	G. g. gorilla	CMNH, PCM	36	36	72
Pongo pygmaeus	*P. p. pygmaeus*	NMNH	21	15	36
	p. p. abelli	NMNH	5	5	10
Total			256	249	505

The same technique of GPA was used by Karen Baab (2008) to quantify the overall variation in cranium shape, going beyond the study of a particular region. The sample analyzed included 15 specimens of *Homo erectus* from Asia and Africa in which the appropriate features were identifiable (Table 8.9), 392 current humans of 11 regions from Africa, East Asia/Oceania/North America, and Europe/West Asia, with the addition of fossil skulls of modern humans: Skhul 6 and Qafzeh 6 in the Near East, Abri Pataud in France, and Fish Hoek in South Africa. Also added were samples of non-human primates *Pan, Gorilla, Pongo, Papio, Macaca,* and fossil exemplars of *Theropithecus*.

Baab's study (2008) indicated that the variation in *Homo erectus* is comparable to that in individual species of cercopithecoids (baboons), as well as of the genus *Pan*, including both species of chimpanzee. Rather than

species' identification, Baab (2008) intended to establish the intraspecific variability patterns due to ecological factors—geographic and temporal. According to those patterns, he argued that his work favors the consideration of a single species for *Homo erectus*. However, Baab considered other possibilities worth mentioning.

8.5.12 What species' concept are we talking about?

The controversy about the presence of one or two species of *Homo erectus* in Asia and Africa seems insoluble if, as Brian Villmoare (2005) pointed out, the studies of comparative morphometry yield different results depending on whether the analyses are linear or of discreet characters. However, the differences that appear between the use of metric or non-metric

Speciman no./name	Abbreviation	Locality	Age (Ma) +	Source*	Original or cast
Africa					
KNM-ER 3733	3733	Koobi Fora, Kenya	1.78	NMK	Original
KNM-ER 3883	3883	Koobi Fora, Kenya	1.5–1.65	NMK	Original
KNM-WT 15000	15000	West Turkana, Kenya	1.51–1.56	NMK	Original
OH 9	OH9	Olduvai Gorge, Tanzania	1.47	IHO	Cast
Eurasia					
Dmanisi 2280	D2280	Dmanisi, Georgia	1.7	AMNH	Cast
Indonesia					
Sangiran 4	San4	Sangiran, Java	>1.6	AMNH	Cast
Sangiran 17	San17	Sangiran, Java	1.3	AMNH	Cast
Sambungmacan 1	SM1	Sambungmacan, Java	0.1–0.05	AMNH	Cast
Sambungmacan 3	SM13	Sambungmacan, Java	0.1–0.05	AMNH	Cast
Ngandong 6**	Ng6	Ngandong, Java	0.1–0.05	IHO	Cast
Ngandong 7**	Ng7	Ngandong, Java	0.1–0.05	AMNH	Cast
Ngandong 12**	Ng12	Ngandong, Java	0.1–0.05	AMNH	Cast
Continental Asia					
Zhoukoodian III	SinIII	Zhoukoudian, China	0.58	AMNH	Cast
Zhoukoodian XI	SinXI	Zhoukoudian, China	0.42	AMNH	Cast
Zhoukoodian XII	SinXII	Zhoukoudian, China	0.42	AMNH	Cast

*AMNH: American Museum of Natural History, New York, NY; ASU: Arizona State University, Tempe, AZ; CMNH: Cleveland Museum of Natural History, Cleveland, OH; PCM: Powell-Cotton Museum, Birchington, UK; RMCA: Royal Museum for Central Africa, Tervuren, Belgium; PM: Peabody Museum, Harvard University; IHO: Institute of Human Origins, Tempe, AZ; NMK: National Museum of Kenya, Nairobi, Kenya; NMNH: National Museum of Natural History, Washington, DC.
**The numbering system of the Ngandong specimens used here follows that outlined by Oakley et al. (1975) where Ngandong 6 1/4 Solo V, Ngandong 7 1/4 Solo VI, and Ngandong 12 1/4 Solo XI.

Box 8.18 Grades of hominization

In a comparative study of the Asian and African samples of *Homo erectus*, Tobias and von Koenigswald (1964) defined a number of "grades of hominization." Australopithecines were "Grade 1" and *H. habilis* "Grade 2." Specimens from Sangiran (Java), along with some exemplars (such as OH 13) from Olduvai (Tanzania), corresponded to "Grade 3." "Grade 4" was represented by Olduvai OH 9 and Ternifini specimens (Morocco), Kabuh (Java), and the samples from Zhoukoudian (China). Such a series of hominization grades leads to paradoxes. For example, OH 13 had been considered by Leakey and collaborators (1964) as paratype of the species *H. habilis*. But, as we have said, the concept of "grade" does not attempt to maintain the coherence of a species.

Seeking a greater correlation between "grades" and "clades," Bernard Wood and Nicholas Lonergan (2008) defined six grades of humanization: (1) possible and probable hominins, *S. tchadensis, O. tugenensis, Ar. ramidus* s.s., *Ar. kadabba*; (2) archaic hominins, *Au. anamensis, Au. afarensis* s.s., *K. platyops, Au. bahrelghazali, Au. africanus, Au. garhi*; (3) archaic hominins with megadontia, *P. aethiopicus, P. boisei* s.s., *P. robustus*; (4) transitional hominins, *H. habilis, H. rudolfensis*; (5) *Homo* pre-modern, *H. ergaster* s.s., *H. erectus, H. floresiensis, H. antecessor, H. heidelbergensis, H. neanderthalensis*; and (6) *Homo* anatomically modern. Attributing to the whole *Homo erectus* sample a "pre-modern" grade avoids the problems we have indicated, but grouping it with the other four species prior to *Homo sapiens* makes the concept of "grade" as a tool useless. In a previous paper (Cela-Conde & Ayala, 2007), the authors of this book used the category "*erectus* grade," defined by Collard and Wood (Collard & Wood, 1999; Wood & Collard, 1999b), to place Asian and African *Homo erectus*.

Table 8.9 Specimens used in the study of cranial shape variation in *Homo erectus*

Speciments by site	Age (Ma)	Original/cast
East Turkana		
KNM-ER 3733 and 3883	1.78–1.58	Original
Olduvai Gorge		
OH 9	1.30	Original
Bouri		
Daka	1.00	Original
Dmanisi		
D2280	1.77	Cast
Sangiran		
S 17	1.30	Cast
Ngandong		
Ng 6, 10, 11, 12	0.10–0.03	Original
Sambungmacan		
Sm 3	1.00–0.03	Original
Zhoukoudian		
Zkd 3, 5, 11, 12	0.80–0.45	Cast

Baab (2008).

characters (including in the latter the quantitative study by Terhune et al., 2007) can be interpreted in two ways. One would assert that there are different species that retain a cranial plane similar in many aspects, while varying in the presence/absence of certain characters—so that the presence or absence of certain traits would be due to reproductive isolation. This would indeed be the case if they were sympatric species, but if we are talking about remote populations in space or time, there is a second possible explanation: that of a species that displays local adaptations because the ecological environment is different. That is the conclusion reached by Karen Baab (2007) in her doctoral thesis on cranial variations of *Homo erectus*.

To derive a taxonomy from two possible explanations for the distribution of traits in the *Homo erectus* set, it is necessary to choose which species' concept we are going to use. Part of the controversy about the number of species necessary to explain the fossil evidence of *Homo erectus* really depends on the concept being used. After concluding that the statistical analysis favored unity, Karen Baab (2008) agreed that under an evolutionary species' concept (ESC) it

would be possible to distinguish between two taxa within *Homo erectus* sensu lato, with strict geographical separation.

Let's address the second of the major problems posed by the Middle Pleistocene fossil record. The distinction between *Homo ergaster* and *Homo erectus* would account for the different cultural traditions of the Middle Pleistocene in Asia and Africa, but it forces us to confront once again the appearance in African sites of specimens that closely resemble *Homo erectus*; this is the case of OH 9 and the Daka skull.

The study by Karen Baab (2008) showed the existence of a greater morphological affinity between OH 9 and the African sample, in contrast with what many previous morphological analyses had pointed out (Wood et al., 1991; Clarke, 2000; Terhune et al., 2007; carried out by GPA). The reason that led to the previous grouping of OH 9 with the Asian sample, according to Baab (2008), is that the weight of geographic variation is lesser in the full sample of *Homo erectus* than in the current populations; thus, individual differences are highlighted. It may be, therefore, that OH 9 shares a general cranial shape with other African specimens (as indicated by Baab, 2007), but at the same time some isolated features bring it closer to the Asian sample.

The meaning of a proposal like Karen Baab's of two species according to the ESC is clearer if the taxonomic perspective is abandoned in an attempt to comprehend

---

Box 8.19 Allotaxa

Based on the distribution of Old World apes, such as *Papio* and *Macaca*, Clifford Jolly (2001) defined the term "allotaxa" to characterize morphologically distinct groups that are not yet reproductively isolated and form a geographic mosaic without overlap. From the point of view of the ESC, allotaxa may be considered different "species" that, however, belong to only one species in terms of the BSC (Jolly, 2001). Neanderthals and other "pre-modern" populations could be considered allotaxa according to Jolly (2001). Susan Antón (2003) has argued that, for the general set of *Homo erectus*, the concept of "allotaxa" is preferable to "species" if an endless debate is to be avoided. The "geographic-temporal varieties" of Karen Baab (2007, 2008) for *Homo erectus* correspond to the very similar idea of "allotaxa" and, indeed, Baab used Jolly's work on baboons as a reference.

the phylogenetic process. After the exit of hominins from Africa, both African and Asian populations remained separate to a very high degree. It is true that several different dispersals to Asia occurred during the Plio-Pleistocene, but they can still be considered essentially sporadic (Bar-Yosef & Belfer-Cohen, 2001). In any case, the differences between African and Asian lithic utensils in the Middle Pleistocene permit acceptance of the scheme of separation of Asian and African populations of *Homo erectus* since the beginning of the Acheulean culture. Such a spatial and temporal range of separation is enough to produce geographical variations, different adaptations and ways of living, with no correlation from one to the other continent. In accordance with that idea, the need to identify the "evolutionary tendencies and the historical destiny" (Wiley, 1978) of each group makes it convenient to use the ESC.

It makes sense, then, to distinguish between the species *Homo erectus* (Asia) and *Homo ergaster* (Africa), if we stick to an ESC and not to a biological one. According to morphometric analyses, Middle Pleistocene hominins may very well belong to the same biological species, but the support or falsification of this hypothesis will not happen until sequencing the genomes of the involved populations, something which, as discussed in Chapter 9, is out of reach of current techniques. In any case, when carrying out the reconstruction of the hominin phylogeny in the Lower and Middle Pleistocene it does not make any difference whether we consider two different species or two populations that remained completely separate.

8.5.13 Specimens of smaller size

Although we have opted for the separation of the Middle Pleistocene hominins in *Homo ergaster* and *Homo erectus*, the taxonomic necessity to accommodate such a diverse sample may force the multiplication of taxa. For example, the exemplar from Olorgesailie (Kenya) adds even more uncertainty about the evolutionary meaning of the African hominins from the early Lower and Middle Pleistocene. The specimen consists of the frontal and left temporal bones and nine fragments of the cranial vault, KNM-OL 45500 for the whole set (Figure 8.27). The frontal bone was found in situ by the research team directed by Richard Potts in June 2003, within a sedimentary block excavated and removed from the site in August 1999. Other parts of the cranium were found near where the frontal bone appeared between July and August 2003. The same

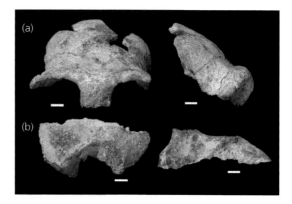

Figure 8.27 KNM-OL 45500, frontal bone (a) and left temporal bone (b) from Olorgesailie (Kenya). From Potts, R., Behrensmeyer, A. K., Deino, A., Ditchfield, P., & Clark, J. (2004). Small Mid-Pleistocene Hominin Associated with East African Acheulean Technology. *Science*, 305(5680), 75–78. Reprinted with permission from AAAS.

stratigraphic level had previously supplied a large accumulation of Acheulean artifacts and mammal fossils (Potts et al., 2004).

The age of KNM-OL 45500 corresponds to the limit between members 6/7 and 5 of the Olorgesailie formation. The pumice of the lower part of member 5 was dated by the ^{40}Ar/^{39}Ar, method yielding an age of 974±7 Ka (Deino & Potts, 1990). The ash of the upper part of member 8 produced an age of 747±6 Ka. By paleomagnetism, the transition Matuyama–Brunhes was registered near the base of member 8 (Tauxe et al., 1992). Potts et al. (2004) estimated the age of the limit of members 5/7 and the specimen KNM-OL 45500 between 970,000 and 900,000 years.

KNM-OL 45500 is a small but adult or nearly adult individual. Various cranial traits correspond to that expected from an *erectus* adult, or even more developed (Potts et al., 2004). Nevertheless, KNM-OL 45500 shows the lower measurements of the frontal bone width, of the thickness and width of the supraorbital torus, and the temporal size from all of the Lower and Middle Pleistocene crania. In addition, some cranial features diverge from previously known *Homo erectus*. Thus, the supraorbital torus is thinner in its vertical dimension than that of an adult *Homo erectus*. The mandibular fossa is the smallest of all the hominin adults of the Middle Pleistocene despite its position under the relatively large root of the zygomatic bone. The mastoid process is thin and is not associated with a supramastoid crest, typical of the large adult specimens of *Homo* in the Pleistocene (Potts et al., 2004).

> **Box 8.20 Fossils included in the comparative study of KNM-OC 45500**
>
> The specimens of *Homo ergaster* and *Homo erectus* compared to KNM-OC 45500 in the study by Potts et al. (2004) were: KNM-ER3733, ER3883, WT15000 from Turkana, Kenya; OH9, OH 12 from Olduvai, Tanzania; Daka, Ethiopia; Buia, Eritrea; Bodo, Ethiopia; Ndutu, Tanzania; D2280, D2282 from Dmanisi, Georgia; Ceprano, Italy; Atapuerca, Spain; Zhokoudian, China; Sangiran, Java and Ngandong, Java; besides those from Kabwe, Zambia, and Saldanha, South Africa.

This set of characteristics produces a condition of extreme gracility.

Moreover, despite its very small volume, the Olorgesailie cranium has features noted by Potts et al. (2004) that appear in the largest specimens of *Homo erectus*; for example, the torus supraorbital double arch, more developed in *Homo ergaster* such as KNM-ER 3733, Bodo, Daka, or Kabwe, than in the Asian *Homo erectus*. KNM-OC 45500 also seems to lack the typical deep mandibular fossa of the Asian specimens (Box 8.20).

The KNM-ER 42700 cranium from Koobi Fora is an even smaller cranium than the specimen from Olorgesailie—691 cc estimated by tomography TC—whose age is older than the specimens from Daka and Olorgesailie: 1.55 Ma (Spoor et al., 2007) (Figure 8.28). Although its cranial volume corresponds to the pattern of *Homo habilis*, the frontal-parietal keel, the narrowness of the temporal-mandibular joint, and the mid-sagittal profile led Fred Spoor and collaborators (2007) to include it in *Homo erectus* (the authors did not consider the *H. ergaster* taxon).

A multivariate analysis on the cranial dimensions confirmed the relationship of the skull KNM-ER 42700 with *Homo erectus* (Spoor et al., 2007). However, the cranial vault, the thick supraorbital torus, and the sharp angulation of the occipital bone characteristic of *Homo erectus* are missing in KNM-ER 42700. An explanation for this absence could be that these are allometric traits dependent on cranial size and, therefore, less expressed in smaller specimens.

The specimen further increases the variability of *Homo erectus*, which appears to present a case in favor of the consideration by Schwartz (2004) of the taxon *Homo erectus* as an "historical accident," which

doesn't correspond to any biological reality. However, KNM-ER 42700 is not placed at an extreme that would facilitate the multiplication of potential species in the *Homo erectus* hypodigm but, on the contrary, serves as a nexus between different populations. Thus, its characteristics impose a certain continuity in geographical and temporal terms. Its size and allometric traits with regard to the cranial volume bring the specimen closer to the Dmanisi sample. The cranial keel and mastoid process are shared characters with Asian *Homo erectus,* and the glabellar arch divided into two resembles that of Sambungmacan 3, of a much younger age.

8.6 The colonization of Europe

Europe holds its own paradox related to the culture of the first hominins that arrived in the continent. The central element of Acheulean culture, the handaxe, is absent in many early European sites with signs of human presence (Italy, France, Germany, Czech Republic, and Spain). It was not until a second colonizing wave, which took place about half a million years ago, that Mode 2 handaxes were introduced. Sites corresponding to this time interval include Torralba and

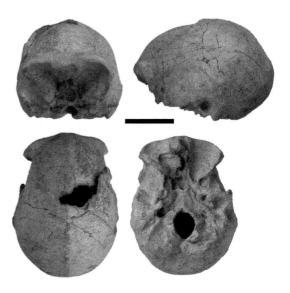

Figure 8.28 KNM-ER 42700: anterior, lateral, upper, and lower views; scale: 2 cm. Reprinted by permission from Macmillan Publishers Ltd: Spoor, F., Leakey, M. G., Gathogo, P. N., Brown, F. H., Anton, S. C., McDougall, I.,. . . Leakey, L. N. (2007). Implications of new early Homo fossils from Ileret, east of Lake Turkana, Kenya. *Nature,* 448(7154), 688–691.

Ambrona (Spain), St. Acheul and Abbeville (France), Swanscombe, Boxgrove, and Hoxne (England), Torre di Pietra and Venosa-Notarchirico (Italy). What could have caused such a cultural sequence?

8.6.1 "Short" chronology vs. ancient occupation

With such an early date as c. 1.8 Ma for the exit of the hominins from Africa, the question of whether the almost immediate dispersal to Asia could have its counterpart in the parallel colonization of Europe was a constant debate in the last decade of the twentieth century.

Based on an examination of the available archaeological and chronological evidence, Wil Roebroeks and Thijs van Kolfschoten (1994) proposed the "short chronology" as the best hypothesis about the hominin occupation of Europe. The short chronology places at around 0.5 Ma an unquestioned and continued presence of members of our lineage in the European continent; the possible previous incursions had been sporadic, without continuity. In support of this claim, the authors provided a scheme—reproduced here (Table 8.10)—summarizing the difference between the previous evidence and those after 500,000 years.

That was the situation in 1994. Only 2 years later, the proposal of the short chronology had to be clarified because of the evidence found in Atapuerca—including human remains—of an occupation prior to half a million years (Dennell & Roebroeks, 1996). Since then, the panorama is now more complete and reveals a remarkable ancient presence in several European enclaves (Table 8.11).

Most of the evidence of hominins in Europe listed in Table 8.11 is cultural, and some is subject to a variety of questions. But even discarding the more doubtful,

there is sufficient evidence of the presence in Europe of *Homo* before what the hypothesis of the "short chronology" indicates (Carbonell et al., 1996).

But, of what type would the oldest dispersals to Europe be? To what degree can an occupation be labeled as "permanent" or "sporadic"? Only a continuous presence in a continental enclave would cause the "short chronology" proposed by Roebroeks and van Kolfschoten (1994) to be discarded.

Whether the actual colonization of the European Continent took place c. 0.5 Ma or even up to c. 1 Ma, hominins had to face climate changes—as happened during the dispersals to the Asian Continent—which brought extreme cold during ice ages. Therefore, two problems regarding the arrival of our lineage to Europe are: with which succession of glacial/interglacial episodes are the early presence of hominins in Europe related, and, which migration routes were used in their entry into the continent under each condition (cold or warm climate)? The examination of alternatives for human dispersion related to these variables will serve as a theoretical model for the possible character of early European occupations.

8.6.2 Glacial cycles

The climatic alternatives we know under the term "glaciations" correspond to cycles in which the planet's temperature lowers to an extreme, which causes the accumulation of large masses of ice on the continents, especially those in the northern hemisphere (Box 8.22).

The pioneering studies by Albrecht Penck and Eduard Bruckner (1909) in the Alps identified four major advances of ice, which they named: Gunz, Mindel, Riss, and Würm. These names corresponded to valleys of Danube River tributaries, on whose fluvial terraces were found evidence of glaciers at different times of the Middle Pleistocene, with the last glaciation coinciding with the beginning of the Upper Pleistocene (Figure 8.29).

Further studies in different continents showed the planetary scale of glacial cycles, with each detected episode given its own name, creating a need to clarify to what extent it was a local or global glaciation. The capacity to accurately identify the climatic alternatives came when the analysis of the content of oxygen isotopes in fossil corals began in the 1950s (Box 8.23). Thanks to this method, Wallace Broecker and Jan van Donk (1970) uncovered in detail some of the Pleistocene climate fluctuations.

Figure 8.30 shows the "International Stratigraphic Chart"—only the last 2.5 Ma—with indications of established correlations between geological ages, marine

Table 8.10 Scheme of the differences in the Paleolithic record in Europe before and after c. 500,000 years

Before 500,000 years old	After 500,000 years old
Small series of isolated pieces selected from a natural pebble background	Large collections from excavated knapping floors with conjoinable material
Disturbed context (coarse matrix)	Primary context sites (fine-grained matrix)
Contested "primitive" assemblages	Uncontested Acheulean and non-Acheulean industries
No human remains	Human remains common

Roebroeks & Van Kolfschoten (1994).

Table 8.11 Evidence of human occupations in Europe prior to 0.5 Ma

Country	Locality	Age (Ma)	Evidence	Reference
France	St. Eble	c. 2.2	Mode 1	Boivin et al. (2010)
France	Chilhac	c 2.1	Mode 1	Boivin et al. (2010)
Italy	Valdarno	1.95	Mode 1	Abbate & Sagri (2011)
Bulgaria	Kozarnika	1.6/1.4	Mode 1	Sirakov et al. (2010)
France	Lézignan-le-Cèbe	1.57	mode 1	Crochet et al. (2009)
Italy	Pirro Nord	c. 1.5	Mode 1	Arzarello & Peretto (2010)
Italy	Monferrato	1.0/0.9	Mode 1	Siori & Sala (2007)
France	Le Vallonet	1.2	Mode 1	Abbate & Sagri (2011)
Spain	Sima Elefante (Atapuerca)	1.2/1.1	Fossil / Mode1	Carbonell et al. (2008)
France	La Terre-des-Sablons	1.1/0.9	Mode 1	Despriée et al. (2011)
Germany	Untermassfeld	1.07	Mode 1	Landeck (2010)
Spain	Barranco León	c. 1	Mode 1	Oms et al. (2000)
Spain	Fuente Nueva	c. 1	Mode 1	Oms et al. (2000)
France	Crozant	c. 1	Mode 1	Despriée (2011)
France	Le Pont-de-Lavaud	c. 1	Mode 1	Bahain et al. (2007)
Germany	Kärlich	MIS* 23	Mode 1	Haidle & Pawlik (2010)
France	Soleihac	1–0.9	Mode 1	Abatte & Sagri (2011)
England	Happisburgh	0.99–0.78	Mode 1	Parfitt et al. (2010)
France	Moncé	0.93	Mode 1 (handaxes)	Despriée (2011)
Spain	Solana	0.9	Mode 1	Abatte & Sagri (2011)
Italy	Monte Poggiolo	MIS* 21	Mode 1	Muttoni et al. (2011)
Spain	Vallparadís	0.8	Mode 1	Duval (2011)
Spain	Atapuerca (Gran Dolina)	>0.78	Cranium / Mode 1	Bermúdez de Castro, Arsuaga et al. (1997)
Ukraine	Korolevo	>0.78	Mode 1	Koulakovska, Usik, & Haesaerts (2010)
Germany	Dorn-Dürkheim	>0.78	Mode 1	Haidle & Pawlik (2010)
Italy	Ceprano	c. 0.8	Cranium	Ascenzi et al. (1996)
Germany	Sarstedt	c. 0.7	Fossils	Czarnetzki, Gaudzinski, & Pusch (2001)
France	La Genetiere	0.7–0.6	Mode 1 (handaxes)	Despriée (2011)
France	La Morandiére	0.7–0.6	Mode 1 (handaxes)	Despriée (2011)
France	La Noira	0.7–0.6	Mode 1 (handaxes)	Despriée (2011)
France	Le Pont-de-la-Hulauderie	0.7–0.6	Mode1	Despriée (2011)
England	Pakefield	c. 0.65	Mode 1	Parfitt et al. (2005)
Italy	Isernia La Pineta	MIS* 15	Mode1	Coltorti et al. (2005)
Greece	Polemistra	Lower Paleolithic (?)	Mode 1	Harvati et al. (2008)
Greece	Karpero	Lower Paleolithic (?)	Mode 1	Harvati et al. (2008)

*Marine Isotope Stage.

Box 8.21 Doubts about an early occupation of Europe

The doubts that led us to dismiss some evidence of an early occupation of Europe refer, first, to its possible natural character due to volcanic activity and, therefore, of non-anthropic manipulation of the supposed utensils in St. Eble and Chilhac (Muttoni et al., 2010), and Dorn-Durkheim 3 (Haidle & Pawlik, 2010). There are also uncertainties about the age that Ernesto Abbate and Mario Sagri (2011) gave to the human presence in Valdarno, which came from an impossible to verify personal communication. Paul Peter Anthony Mazza and collaborators (2006) placed the Valdarno utensils—very modern, advanced Acheulean—over the Matuyama–Brunhes boundary. Finally, Giovanni Muttoni, Giancarlo Scandia, and Dennis Kent (2010) pointed out potential analytical errors in the radiometric and paleomagnetic measurements of Le Vallon and Pirro Nord. Although we consider this latter evidence to be valid, we will not take into consideration the ancient ages indicated in Table 8.11.

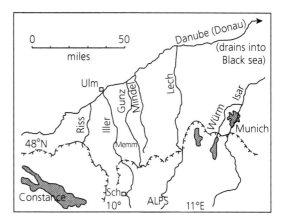

Figure 8.29 Locations of Danube River valleys after which glaciations studied by Penck and Bruckner (1909) were named.

Box 8.22 The discovery of glaciation

In historic terms, it was the Swiss biologist Louis Agassiz—Cuvier's disciple—who identified periods of intense cold in remote times. In 1837, Agassiz presented his idea to the Swiss Society of Natural History in Neuchâtel, and 3 years later (Agassiz, 1840) gave the term "glaciation" to these cold episodes. Adhemar and James Croll (see Elias, 2007) provided the first explanation on the origin of the climatic changes. However, the Serbian mathematician Milutin Milankovitch (1941) explained them by developing a model of glaciation based on a cyclic variation of Earth orbit eccentricity, precession, and obliquity.

Box 8.23 Measure of glaciations by oxygen isotopes

The study of ancient climatic events is done by measuring the ratio of two oxygen isotopes—$^{18}O/^{16}O$—present in calcareous microfossils in tropical seas. When the weather cools, much of the ^{16}O, which is lighter and therefore abundant in evaporation clouds of ocean water, precipitates as snow on glaciers and, therefore, the content of ^{18}O becomes greater in the oceans. Thus, an increase in the relative amount of ^{18}O in oceanic sediments implies the presence of a cold period, possibly an ice age. The situation described by Michael Prentice and George Denton (1988) based on this method, indicates an abrupt climatic change 14 Ma, in the Middle Miocene, an age in which the previous high temperatures tended toward a gradual cooling—acute in average—although with some fluctuations (glacial and interglacial periods). The process accelerated in the Middle Pleistocene leading to the so-called "ice age."

isotope stages (MIS), paleomagnetic events, loess-soil depositional sequences in China, and the various successions of glacial/interglacial episodes.

As we can see, the column of the MIS in Figure 8.31 shows that the climatic alternatives were many more than those established in the "classical" consideration of four glaciations during the Middle and Upper Pleistocene, indicating precisely the latest climatic oscillations and their correspondent paleomagnetic chrons (Box 8.24).

Whenever a glacial episode occurred, some of the water of the oceans is deposited as ice, as a result of the cooling of the planet, in inland areas of Europe, Asia, and America. Up to 5.5% of the existing water on the planet accumulated on the mainland during the glaciations, compared with the current 1.7% (Williams et al., 1993). Glaciers covered one-third of the land in the period of their greatest extension, which coincided with the second major glaciation (Mindel, MIS 12) (Figure 8.32).

Global chronostratigraphical correlation table for the last 2.7 million years
v. 2010

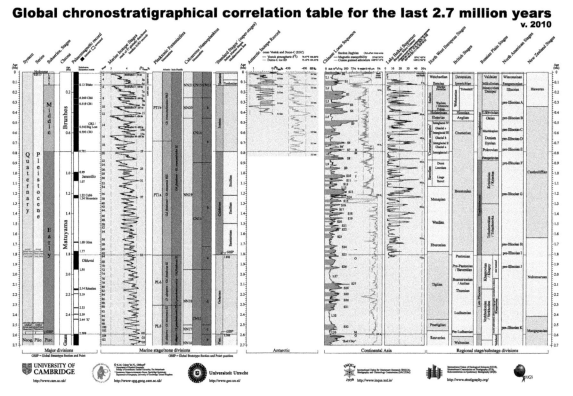

Figure 8.30 International Stratigraphic Chart, with the existing correlation between the various indicators (see text) Quaternary Paleoenvironment Group, Godwin Institute for Quaternary Research, University of Cambridge (United Kingdom); picture from http://www.stratigraphy.org/index.php/ics-chart-timescale). Graph by Gibbard & Cohen, 2008; Cohen & Gibbard, 2010; modified www.stratigraphy.org.

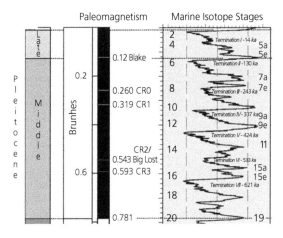

Figure 8.31 Details of MIS from the Middle and Upper Pleistocene, with the stages corresponding to the times of maximum severity of the four "classic" glaciations (derived from the International Stratigraphic Chart).

Box 8.24 "Classic" glaciations

Despite the succession of numerous glacial and inter-glacial periods that affected Europe, we will maintain as much as possible the terminology of the four glaciations, Günz, Mindel, Riss, and Würm, as a general framework to review the hominin dispersal process in the continent. For the convenience of the reader, the dates of the beginning and end of each of the "classic" glaciations are indicated in Table 8.12. However, this periodization is totally inadequate as a reference for human dispersals in Europe. For example, the Mindel–Riss interglacial, which is given in Table 8.12 as 150,000 years, actually consists of several cycles that reduce the warm period. When necessary, we will indicate the specific MIS episode to which we are referring.

Table 8.12 "Classic" glaciations in Europe. If the cold intermediate stages are not taken into consideration, the interglacials seem to cover a broad temporal range, but, in fact, that is not so. They were interrupted by glacial episodes MIS 14 in the Günz–Mindel interglacial (known as Cromerian), MIS 10 and 8 in the interglacial Mindel–Riss (Hoxnian) and MIS 4 in the Riss–Würm (Ipswichian). The interglacial we are experiencing now is called Flandrian. Although glacial episodes encompass the planet, they have received local names because of their identification in different places

Episode	Age (Ka)	Glacial/interglacial
End	11	Würm (MIS 2)
Beginning	120	
End	120	Riss–Würm (MIS 5 and 3)
Beginning	140	
End	140	Riss (MIS 6)
Beginning	300	
End	300	Mindel–Riss (MIS 11, 9 and 7)
Beginning	400	
End	400	Mindel (MIS 12)
Beginning	450	
End	450	Günz–Mindel (MIS 15 and 13)
Beginning	750	
End	750	Günz (MIS 16)
Beginning	1100	

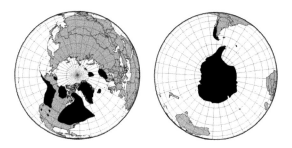

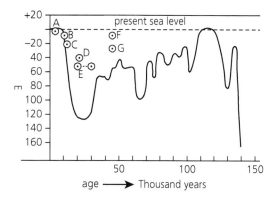

Figure 8.32 Upper: Maximum extent of ice on terrestrial hemispheres. Because of the unequal distribution of continents, in the southern hemisphere glaciers are concentrated in Antarctica. Bottom: Variation of isobaths corresponding to sea level, signaling the lowering of sea levels corresponding to the different glacial periods (illustration by Flemming et al., 2003; modified). Top: Reprinted from Ehlers, J. & Gibbard, P. L. (2007). The extent and chronology of Cenozoic Global Glaciation. *Quaternary International*, 164–165, 6–20, with permission from Elsevier. Bottom: From Lambeck, K. & Chappell, J. (2001). Sea Level Change Through the Last Glacial Cycle. *Science*, 292(5517), 679–686. Reprinted with permission from AAAS.

The environmental consequences of the advancing ice encompass an interlaced chain of events that reach their peak during the ice age:

- Desiccation of shallower seas (the ocean level is lower by up to 150 m).
- Opening of contact zones between continents.
- Facility for migration of terrestrial animals, including hominins.
- Continentalization of the climate and a decrease of average rainfall with increasing aridity in temperate zones.
- Profound changes in fauna: large extensions of coniferous forests, open tundra, and associated fauna of large herbivorous mammals (moose, mammoths, and woolly rhinos) and their predators.

The average annual temperature in northern Europe was around –2°C during the glaciations, with only 30–40 days without frost throughout the year in areas not covered by glaciers. The conditions of the Mediterranean Sea during the last glaciation were dry and arid, with no water supply other than snowmelt in the summer; however, this is a general panorama during which some humid interglacial episodes occurred (Rosselló & Verger, 1970).

It is easy to understand what adaptive problems hominins would have faced in the Europe of the Middle and Upper Pleistocene, because, as primates emerging from tropical forests, they were adapted to hot and humid weather. Obviously, climate change affected the entire flora and fauna, not only the hominins. The study by Marta Mirazón Lahr and Robert Foley (1994) indicated that the slow accumulation of continental ice sheets at high latitudes, followed by a rapid thaw leading to short periods of interglacial conditions, would have modified the existing natural barriers for migration. Both in Africa and Europe the fauna of large mammals during interglacial periods had ample room for expansion. However, during glacial episodes, only the Eurasian fauna could have modified its range of occupation; the African fauna would have been subject to an endemic isolation because of the arid belt that the Sahara represents (Figure 8.33). During the glaciations, European animals would be forced to move to the South because of the permafrost present in the periglacial areas; nevertheless, the Sahara would have meant an impassable barrier. Hence, the migratory movement in the West–East direction until arriving at the no-exit alleyway of the

Middle East. The distribution of the current fauna reflects, in fact, the conditions imposed by glaciers (Mirazón Lahr & Foley, 1994).

An empirical confirmation of the work of Mirazón Lahr and Foley (1994) was provided by Giovanni Muttoni, Giancarlo Scardia, and Dennis Kent (2010), indicating possible migration routes of large herbivores (and hominins) around the first major glaciation of the Pleistocene MIS 22 (0.87 Ma) (Figure 8.34). According to Muttoni et al. (2010), *Mammuthus meridionalis* is a

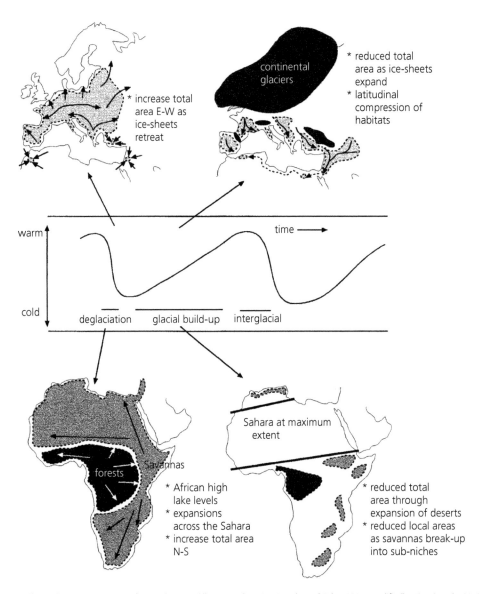

Figure 8.33 Effects of the glacial cycle in Africa and Europe (illustration by Mirazón Lahr and Foley, 1994; modified). Mirazón Lahr, M. & Foley, R. A. (1998). Towards a Theory of Modern Human Origins: Geography, Demography, and Diversity in Recent Human Evolution. *Yearbook of Physical Anthropology*, 107(S27), 137–176.

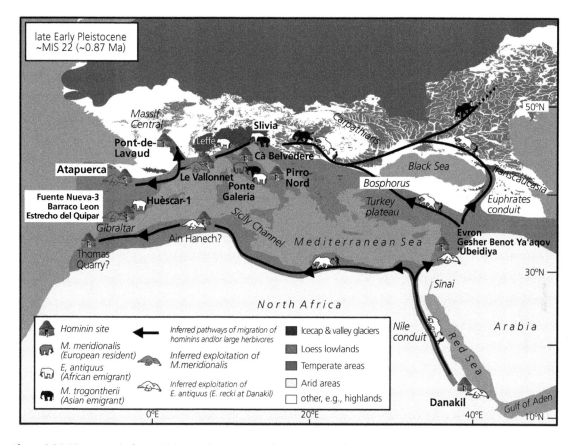

Figure 8.34 Migratory path of Lower Pleistocene (M22, c. 0.87 Ma). During MIS 22, the coast was in the isobath of −120 m. Reprinted from Muttoni, G., Scardia, G., & Kent, D. V. (2010). Human migration into Europe during the late Early Pleistocene climate transition. *Palaeogeography, Palaeoclimatology, Palaeoecology*, 296(1–2), 79–93, with permission from Elsevier.

European species that has been found in the TD6 level of Gran Dolina, Fuente Nueva 3, and Barranco León (Spain), and Le Vallonnet (France). *M. trogontherii*, immigrant from Asia, appears in Cúllar Baza 1 (Spain) and Soleilhac (France). *E. antiquus*, from Africa, has been found in Soleilhac (France), Ceprano, Fontana Ranuccio, and Notarchirico (Italy), Gesher Benot Ya'aqov, and possibly Evron (Israel). In all cases these are sites with indications of human presence.

Both the model of Mirazón Lahr and Foley (1994) and the empirical study of Muttoni et al. (2010) indicate that the dispersions between the African and Eurasian continents were limited during the Pleistocene to migratory patterns in the W–E direction. Still, the transit between Africa and Europe offers some alternatives that could have been exploited by hominins. This is the second question posed at the beginning of this chapter: what were the different migration routes for entering the Eurasian continent?

8.6.3 Obstacles to European migration: the Saharan barrier

The study of Ernesto Abbate and Mario Sagri (2011) on the geological, climatic, and environmental determinants that could have influenced the dispersal of the *Homo* genus from Africa to Eurasia during the Lower Pleistocene make a case for the strategic character of the Danakil depression (Afar Rift, Ethiopia) as a starting point. From this enclave, the alternatives to reach the Mediterranean shores are, from West to East, some of the following routes:

• Western Sahara: beds of the Niger and Senegal rivers, from the mega-lake Chad.
• Central Sahara: large Saharan lakes and/or fluvial belts, from the same origin: mega-lake Chad.
• North route along the Nile Valley.
• Crossing the Strait of Bab el Mandeb using an eastern route into the Arabian Peninsula.

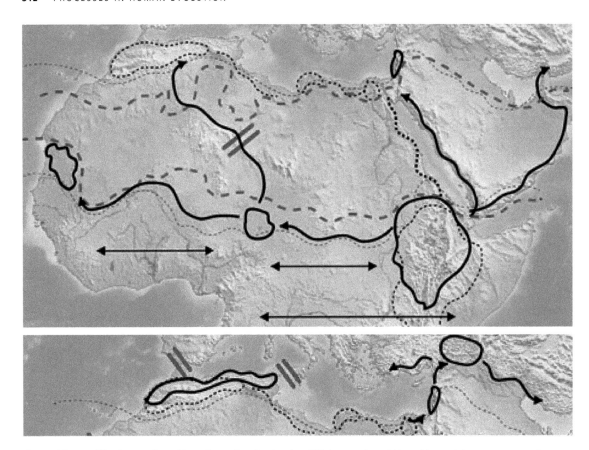

Figure 8.35 Possible exit routes from Africa to Europe. Top: Crossing the arid Saharan region, which would extend the westernmost option northward along the Atlantic coast. Thick lines indicate the hypothetical narrowing of the Sahara during a humid phase of the Upper Pliocene; compare the maximum expansion of the desert indicated by fine lines. Bottom: Crossing of the Mediterranean at Gibraltar, through the Sicilian Channel, or by traveling through the Levantine Corridor, which would then also venture into Asia. Mirazón Lahr, M. (2010). Saharan corridors and their role in the evolutionary geography of "out of Africa I". In J. G. Fleagle, J. J. Shea, F. E. Grine, A. L. Baden, & R. E. Leakey (Eds.), Out of Africa I. The First Hominin Colonization of Eurasia, Vertebrate Paleobiology and Paleoanthropology (pp. 27–46). Berlin: Springer.

The ease of travel on these four routes out of Africa is not equivalent. As Marta Mirazón Lahr (2010) pointed out, the hominin European colonization can be seen as a byproduct of a prior dispersal, and a more difficult one than the abandonment of East Africa and overcoming the geographical barriers within the African Continent: in particular, the Sahara (Figure 8.35). With regard to the temporal range of the Upper Pliocene and Lower Pleistocene, Mirazón Lahr (2010) argued that a hypothetical "humid episode," in which the Sahara desert would have latitudinally reduced, would have created corridors and potential shelters that facilitated the hominin dispersal.

When identifying such corridors, an important empirical support is the available evidence of human occupation in North Africa. Figure 8.36 shows different

sites with lithic carvings indicating that presence. In northern Algeria, Ain Hanech is the oldest town with Mode 1 tools of a questionable age ranging between 1.95–1.78 and 1.2 Ma. (Mirazón Lahr, 2010); as we shall see, this is a similar temporal range to the first occupations of southern Europe. But, how would these hominins have reached the African shore of the Mediterranean?

The basin of the mega-lake Chad is a strategic point to disperse from East Africa. Tools of Modes 1 and 2 found in the area do not have an adequate dating but, as one might recall, from this basin come the specimens of *Sahelanthropus tchadensis* and *Australopithecus bahrelghazali*, with ages separated by millions of years, indicating the feasibility—at least intermittent—of Chad as a place of refuge. From the mega-lake, two routes

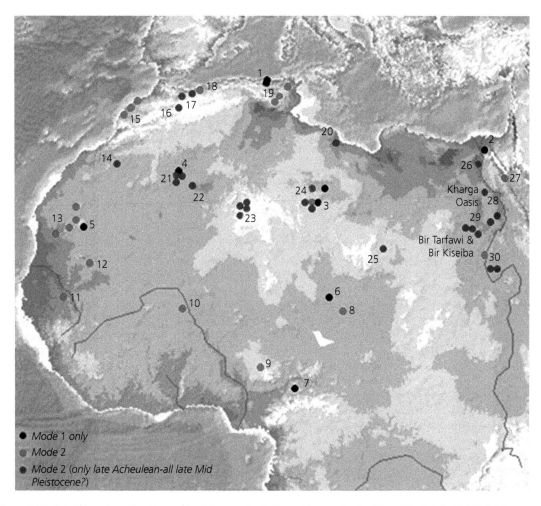

Figure 8.36 Main African sites with evidence of hominin occupation in the corridors crossing the Sahara. Mirazón Lahr, M. (2010). Saharan corridors and their role in the evolutionary geography of "out of Africa I". In J. G. Fleagle, J. J. Shea, F. E. Grine, A. L. Baden, & R. E. Leakey (Eds.), Out of Africa I. The First Hominin Colonization of Eurasia, Vertebrate Paleobiology and Paleoanthropology (pp. 27–46). Berlin: Springer.

would permit reaching the Mediterranean during wet, interglacial episodes (see Figure 8.36). The first, that of Central Sahara through mega-lakes and associated paleo-rivers (Chad, Tibesti, Hoggar-Tassili, and South Atlas) would have been available only for very short periods of time during the glacial cycle changes of the Middle Pleistocene, but no localities present lithic carvings supporting this claim. By contrast, the second route, the Western Sahara—through Senegal and Mauritania to reach the Atlantic coast and thence along it toward the North—includes shelter facilities in the beds of the Niger and Senegal rivers, and shows deposits with tools, particularly of Mode 1 (Mirazón Lahr, 2010).

Box 8.25 Paleoenvironmental reconstruction

The model of Marta Mirazón Lahr is based on the reconstruction of Saharan paleoenvironments prevalent during the second phase of the last glacial cycle, but in hypothetical terms it can be applied to any humid interglacial episode, so that the scant presence in North Africa of Mode 1 tools shown by Figure 8.35, which would correspond to the first occupation of Europe, is not a big obstacle.

Geomorphological and paleontological evidence indicates that the Nile River in the Middle Pleistocene (c. 0.7 Ma) joined the water sources from East Africa to the Mediterranean Sea, but in the Plio-Pleistocene this did not happen due to the arid climate (Mirazón Lahr, 2010). Therefore, the hominin populations of Dmanisi, if they reached the Levantine Corridor across the African Continent, would have used the routes of the Central or Western Sahara.

A different way to reach the eastern shore of the Mediterranean has been proposed by Ernesto Abbate and Mario Sagri (2011): a direct path from the Afar depression, crossing the strait of Bab el Mandeb, with the possibility of crossing the Arabian Peninsula either by the eastern or western zone.

The Red Sea was formed as a result of the separation between the tectonic Danakil microplate and the Arabian plate. The current separation is 29 km, with the island of Perm (a Miocene volcano) facilitating direct crossing. If a rate of separation of the plates of 3–5 mm/year is accepted, the width of the straits would have been narrower in the Pleistocene—although never less than 4 km—assisting the passage. Further North, the Hanish and Zukur islands also help to cross the waters. The abundance of sites with Acheulean tools and, to a lesser extent, Oldowan utensils (those numbered 2, 9, and 11 in Figure 8.37, bottom) demonstrates the ancient human occupation of the Peninsula.

Geoffrey Bailey and Geoffrey King (2011) have studied in detail the geological history of the Red Sea basin as a route for the hominin dispersal out of Africa. The map in Figure 8.38 shows the distribution of the steepest terrain and of the more accessible plains around the Ethiopian plateau—with the Afar depression and the coast of Eritrea—and the eastern and western flanks of the Arabian Peninsula and the Sinai.

The map in Figure 8.38 shows terrains with large flows of basaltic lava, which therefore could supply raw material for carving stone tools. Moreover, these lava extensions present indications of fluvial sediment suggesting a greater availability of water in the Pleistocene than today—with a very arid climate—providing, in addition, favorable corridors for human occupation.

Considering all these variables, the two most favorable passages for the hominin eastern exit out of Africa are the direct connection from the Afar area to the flanks of the Arabian mountains and the North African route by land up to the Levantine corridor. The distribution of basaltic terrains led Bailey and King (2011) to consider the first, that of the access to the Arabian Peninsula from the Afar depression, as the most viable

route. In the eastern part of the Arabian ridge there is a pass that reaches the European threshold through Syria and Jordan—with the Jordan Valley as an important reference—and then bends to the East along the southern foothills of the Taurus Mountains in Turkey and to Zagros, in Iraq–Iran (Box 8.26).

8.6.4 Migration routes to Europe: crossing the Mediterranean

Considering the possible trajectories for migration to Europe after reaching the Afro-Asian shore of the Mediterranean, there are the following pathways:

- The Levantine Corridor.
- Channel of Sicily.
- The Strait of Gibraltar.

The Levantine route is the easiest way and is often accepted as the likely point of hominin entry into Europe from Africa. The Levant is also the natural access from Asia, if the colonizers originated there. Even so, the Strait of Gibraltar and the Channel of Sicily, which separate Spain and Italy from the North African coast, have also been proposed as hypothetical pathways for crossing the Mediterranean (e.g., Blanc, 1940; Alimen, 1975; Flemming et al., 2003; Derricourt, 2005), in particular because of the similarity of cultural remains on each side of the straits. The work of Ernesto Abbate and Mario Sagri (2011) has specified the geological conditions of those possible pathways.

The Sicilian Channel connects the eastern and western basins of the Mediterranean Sea, establishing a break in the Apennine–Maghreb orogenic belt. Currently, it is 140 km wide, but more than half of it is part of the continental shelf of Sicily that, in this place, constitutes an area of shallow waters (less than 100 m depth); this is the Adventure Bank (A.B. in Figure 8.39). The African coast of Tunisia also has a shelf of shallow water, so the deeper area of the channel (over 200 m depth) is only 40 km wide. In addition, a string of islands and reefs are spread along the western area of the strait.

During the Pliocene and early Pleistocene, geological changes could have further reduced the Sicilian Channel. When the sea level fell, a cluster of nearby islets and emerging banks would have permitted the passage from island to island in the central part of the Channel, where Tunisia and Sicily are closer, in particular during glacial periods (Abbate & Sagri, 2011). With the rise of the isobaths, possible traces of this and other pathways crossing the Mediterranean are

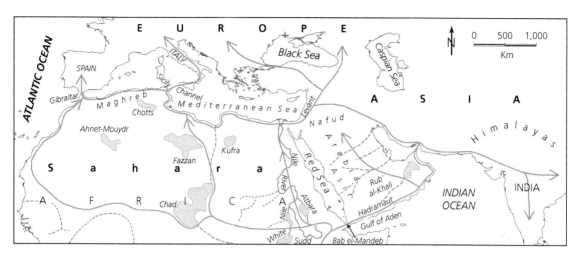

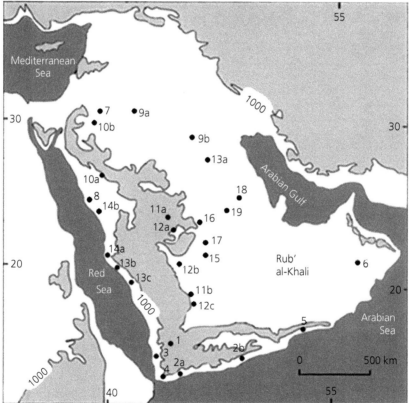

Figure 8.37 Upper: Out of Africa through a direct crossing to the Arabian Peninsula. Bottom: Lower Paleolithic localities on the Arabian Peninsula; in gray, the Hejaz–Asir–Hadramaut mountain chain. Top: Reprinted from Abbate, E. & Sagri, M. (2011). Early to Middle Pleistocene Homo dispersals from Africa to Eurasia: Geological, climatic and environmental constraints. *Quaternary International*, 267, 3–19, with permission from Elsevier. Bottom: Petraglia, M.D. (2003). The Lower Paleolithic of the Arabian Peninsula: Occupations, Adaptations, and Dispersals. *Journal of World Prehistory*, 17(2), 141–179, with permission of Springer.

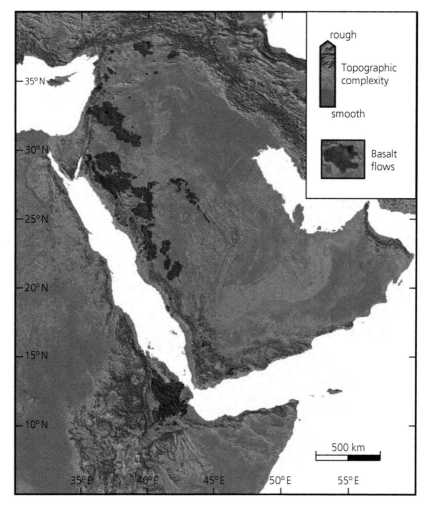

Figure 8.38 Map of steep areas and the easier access transit zones in the passage from Africa to Asia, with indication of basaltic terrains (Bailey & King, 2011).

Box 8.26 Transit passage for the Arabian Peninsula

As Bailey and King (2011) pointed out, there is a close association between the basaltic terrains of the Arabian Peninsula and the localities with Pleistocene lithic tools, although, unfortunately, these are mainly surface sites, which have not been accurately dated. The Arabian Peninsula has local enclaves that are not visible in maps with little detail. The indicated pathways are the principal ones, but this does not exclude other sporadic passages, which could occasionally have some relevance.

now under water, but some submerged sites have been identified that give evidence of this (Figure 8.40) (Flemming et al., 2003) (Box 8.27).

The Gibraltar Strait that separates southwestern Europe from the north coast of Africa is only 14 km wide but with a considerable depth of about 900 m. As this Strait opened in the Lower Pliocene, 5.33 Ma ago, after the Messinian Salinity Crisis, it became a significant barrier to hominin crossing. But there is no shortage of hypotheses about the possibilities of crossing, as is indicated by a similar fauna on both sides (e.g., Arribas & Palmqvist, 1999). Ernesto Abbate and Mario Sagri (2011) mentioned a possible crossing route between Tangier and Punta Camarinal, to the west of Algeciras.

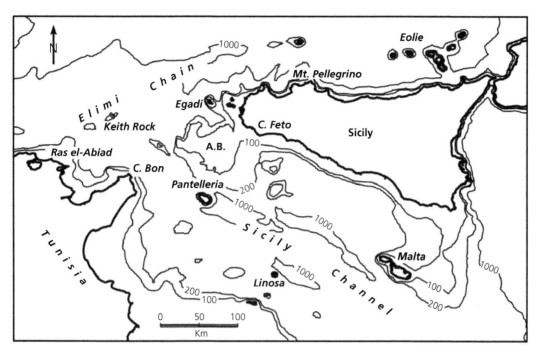

Figure 8.39 Simplified bathymetry of the Sicilian Channel. Reprinted from Abbate, E. & Sagri, M. (2011). Early to Middle Pleistocene Homo dispersals from Africa to Eurasia: Geological, climatic and environmental constraints. *Quaternary International*, 267, 3–19, with permission from Elsevier.

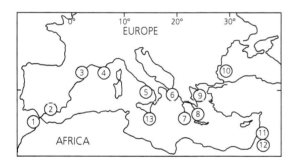

Figure 8.40 Mediterranean sites under water; unless otherwise indicated, all belong to the Paleolithic. 1, Hercules Cave (Tangier). 2, Gorham's Cave and the Punta de Europa (Gibraltar). 3, Leucate (Neolithic, France). 4, Cosquer (France). 5, Palinuro (Italy). 6, Aghios Georghios (Corfu, Greece). 7, Metoni (Bronze Age, Greece). 8, Pavlo Petri (Bronze Age, Greece). 9, Aghios Petros, Sporadhes and Franchti Cave (Neolithic, Greece). 10, Bulgarian Coast (Bronze Age sites). 11, Atlit (Neolithic, Israel). 12, Hof Dado and Tel Harez (Neolithic, Israel). 13, Malta's caves. Flemming, N., Bailey, G., Courtillot, V., King, G., Lambeck, K., Ryerson, F., & Vita-Finzi, C. (2003). Coastal and marine palaeo-environments and human dispersal points across the Africa-Eurasia boundary. In C. A. Brebbia & T. Gambin (Eds.), The maritime and underwater heritage (pp. 61–74). Southampton: Wessex Institute of Technology.

Box 8.27 A bridge to Sicily

The possible existence of a completely emerged bridge to allow passage on foot from North Africa to Sicily during the glaciations has been controversial. Abbate and Sagri (2011) indicated the presence of a safe passage between Tunisia and Sicily in the Pleistocene, which has been rejected by Villa (2001) and Mirazón Lahr (2010). But, as Marcel Otte (2010) stated, it makes no sense to discuss whether the strait between Sicily and Tunisia closed completely or not. It may have been enough that the other side was visible from a hill to encourage hominins to cross the water. One might add that a column of smoke rising from a forest fire can be an indication of the existence of land across an arm of the sea, visible even from afar. The strong currents of Gibraltar would make the western passage difficult (Straus, 2001), but the occupation of Flores Island by *Homo erectus* demonstrates the ability of early humans to colonize new lands, overcoming inlets with strong currents.

In this area the continental shelves are large and are surrounded by a number of banks and reefs that do not exceed 150 m of depth—the bathymetric threshold of Camarinal. These seabeds originated through landslides of gigantic masses from the continental shelves; flows that partially obstructed the inlet of the Strait in the Lower Pliocene.

8.6.5 The first European hominins

The studies on possible pathways from Africa to Asia only point out hypothetical migration routes. It has yet to be clarified when the dispersals that led hominins into Europe occurred and, in particular, what species would have been the protagonist of the colonization.

Only some evidence, listed in Table 8.12, would allow us to appreciate the morphology of the first human settlers in Europe. These are specimens from the Gran Dolina and Sima de los Huesos in Atapuerca (Spain), a cranium from Ceprano (Italy), and some cranial fragments from Sarstedt (Germany). Let us examine their characteristics.

Atapuerca sites

The sedimentary area of Atapuerca—approximately 14 km east of Burgos (Spain)—runs between the towns of Atapuerca and Ibeas de Juarros along some low hills. In Atapuerca, up to 25 sites have been catalogued, nine of which are caves more than 20 m long, and among the latter we highlight the Cueva Mayor and the Cueva del Silo (Aguirre, 1995) (Figure 8.41). Atapuerca research was led by Emiliano Aguirre until 1991 and thereafter by Juan Luis Arsuaga, José María Bermúdez de Castro, and Eudald Carbonell.

In the late-nineteenth century, The Sierra Company, Ltd., mining coal and iron ore in the area, opened a pass in Atapuerca with the intention to build a railroad. Although the train project was abandoned, the excavation known as Trinchera del Ferrocarril [the Railroad Trench] allowed easy access to the sites. Various deposits have been excavated: Gran Dolina, Galería, Sima del Elefante, and Penal in Atapuerca, and outside Trinchera del Ferrocarril but with great importance for European hominin history, the Sima de los Huesos. The first human fossils were found in 1976 in this latter site, and were attributed to the Middle Pleistocene. Two decades later, the discovery of the oldest hominins associated with stone tools of very primitive manufacture, would take place in another Atapuerca locality: Gran Dolina. Finally, in 2008 the oldest specimens identified so far in Europe appeared in the Sima del Elefante.

Sima del Elefante

The Sima (or Trinchera) del Elefante has a sedimentary depth of 25 m, extending under the ground of the railway trench (Figure 8.42). The lithic levels, numbered from the bottom, reach to E21; the most interesting are those located below E16.

The paleomagnetic study conducted by Josep M. Parés et al. (2006) revealed the existence in the Sima del Elefante of a change in magnetic polarity between the stratigraphic levels E16 (reversed polarity) and E17 (normal polarity). A faunal comparison identified the reversed episode as belonging to the Matuyama chron; but the normal episode could correspond either to the Brunhes chron or to the Jaramillo subchron (Figure 8.42b). The first option would assign to the tools of lower sedimentary levels an age >0.78 Ma; the second, an age >1.07 Ma. In both cases, the maximum age from the Matuyama chron to the Olduvai subchron is 1.77 Ma.

The application of the cosmogenic nuclides method (Chapter 1.14) on the content of ^{26}Al and ^{10}Be in the quartz samples allowed the age of level E9 to be refined to 1.22±0.16 Ma (Carbonell et al., 2008). It is an important detail because, in addition to stone tools, that level has provided human fossils.

From the sedimentary levels E9–E14 come stone tools of Mode 1; Mode 2 objects were recovered from level E18, and Mode 3 from E19 (Parés et al., 2006). Eudald Carbonell et al. (2008) mentioned, for lower levels (Mode 1), a lithic set of 32 objects: four simple flakes, five waste flakes (debris), and 23 pieces of indeterminate flint (working material available within 2 km from the site). The authors indicated the suggested evidence of in situ carving, deduced from two flint flakes associated with debris, which seem to belong to the same core; this is a carving technique of Mode 1.

From the same level (E9) comes the specimen ATE9-1, a mandibular fragment with the roots of seven teeth and the fragmentary crowns of three of them, plus the associated lower premolar LP4 (Carbonell et al., 2008); it is the oldest hominin specimen found so far in Europe (Figure 8.43, right). Later (Section 8.7.6) we will give its taxonomic affiliation and possible phylogenetic affinities.

Gran Dolina

Gran Dolina—a site close to Sima del Elefante and excavated prior to it (see Figure 8.42)—is a karst deposit of 18 m of sediments, with 11 lithic levels numbered from bottom to top. The levels TD3–4, 5, 6, 7, 10, and 11 contain abundant stone tools (Carbonell et al., 1995).

The age of the Gran Dolina levels were obtained by paleomagnetism by Josep M. Parés and Alfredo

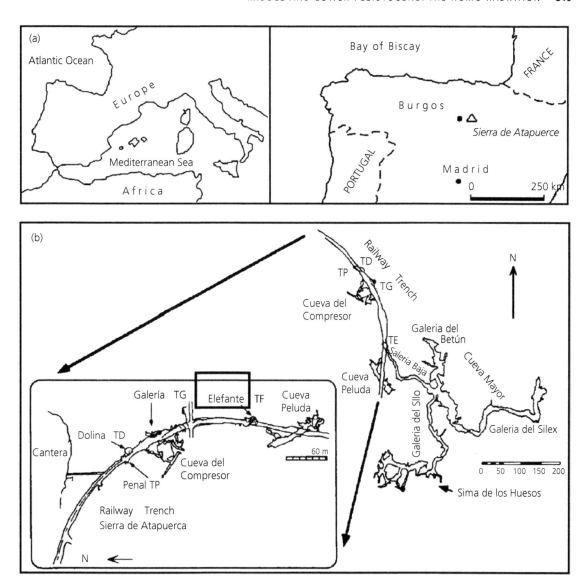

Figure 8.41 (a) Location of the Atapuerca sites; (b) plan of Trinchera del Ferrocarril, including the Sima del Elefante site. Reprinted from Parés, J. M., Pérez-González, A., Rosas, A., Benito, A., Bermúdez de Castro, J. M., Carbonell, E., & Huguet, R. (2006). Matuyama-age lithic tools from the Sima del Elefante site, Atapuerca (northern Spain). *Journal of Human Evolution*, 50(2), 163–169, with permission from Elsevier.

Pérez-González (1995) (Figure 8.44). The TD6 level, to which we will refer next, was located by those authors inside the reversed Matuyama chron, a meter below its boundary with that of normal polarity, Brunhes, so it must be older than 780,000 years. For the lower levels of TD (3, 4, 5, and 6) an older age of 730,000 years was proposed by Carmen Sesé and Enrique Gil (1987) when studying the local microfauna.

In 1994, 100 stone objects of limestone, sandstone, quartzite, and two varieties of silex were recovered within the TD6 level in Aurora stratum (Carbonell et al., 1995). Ten percent of the set consisted of cores and hammers (Figure 8.45 and Box 8.29).

More than 30 specimens of cranial, mandibular, and dental hominin fossils, belonging to at least four individuals, were also found in the Aurora stratum in 1994. The most complete of all these is a large piece of the frontal bone, ATD6-15, with parts of the glabella and the right supraorbital torus, perhaps belonging to a teenager (Carbonell et al., 1995; Figure 8.46). By

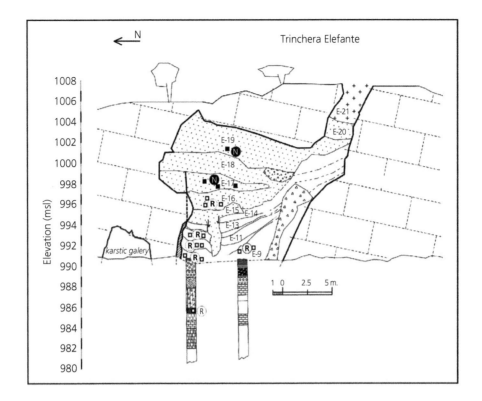

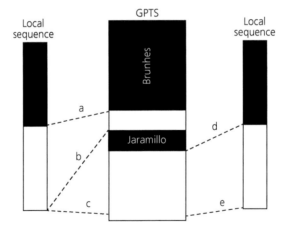

Figure 8.42 (a) Sedimentary record of Sima del Elefante; circles with letters R and N indicate, respectively, reversed or normal magnetic polarity (illustration by Parés et al., 2006). (b) Possible interpretation of the detected sequence of reversed and normal polarity (illustration by Parés et al., 2006); the local sequence is compared with that of the Global Polarity Time Scale (GPTS), with the alternatives that assign a "normal" polarity either to Brunhes or to Jaramillo. (c) Lithostratigraphy and chronology; symbols correspond to: (1) Mesozoic limestone, (2) speleothems, (3) lutite, clay, (4) bat guano, (5) clay and laminated sandy silt, (6) marlstone, (7) gravel and pebbles, (8) crossed lamination, (9) main stratigraphic discontinuity. The latitude of the virtual geomagnetic pole (VGP) and ages obtained by cosmogenic nuclides are indicated; the arrow indicates the level E9 with fossils and tools (illustration by Carbonell et al., 2008). 8.43a & 8.43b: Reprinted from Parés, J. M., Pérez-González, A., Rosas, A., Benito, A., Bermúdez de Castro, J. M., Carbonell, E., & Huguet, R. (2006). Matuyama-age lithic tools from the Sima del Elefante site, Atapuerca (northern Spain). *Journal of Human Evolution*, 50(2), 163–169, with permission from Elsevier. 8.43c: Reprinted by permission from Macmillan Publishers Ltd: Carbonell, E., Bermudez de Castro, J. M., Pares, J. M., Perez-Gonzalez, A., Cuenca-Bescos, G., Olle, A.,. . . Arsuaga, J. L. (2008). The first hominin of Europe. *Nature*, 452(7186), 465–469.

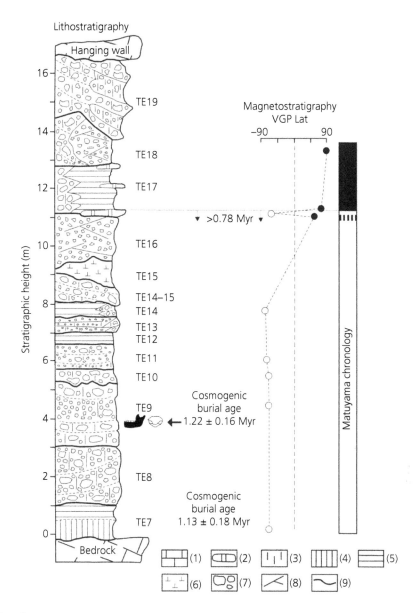

Figure 8.42 *continued*

Box 8.28 Discrepancy of E7 and E9 between fauna and age

Samples corresponding to the E7 level of the Sima del Elefante, with a microfauna similar to that of E9, yield an age by the cosmogenic nuclides method of 1.13±018 Ma (Carbonell et al., 2008)—a number that differs from the one obtained for the E9 level.

1996 there were already up to 80 specimens from six individuals.

The morphological study conducted by Eudald Carbonell et al. (1995) indicated that the ATD6-15 fragment corresponded to a frontal bone of a size much larger than those from *Homo erectus*, both African—ER-3733 and 3883 from Turkana—and Asian—Sangiran 2 and Trinil from Java—all of them with a cranial capacity over 1,000 cc. Thus, the estimated cranial volume for ATD6-15 would be similar to Sangiran 17, to the

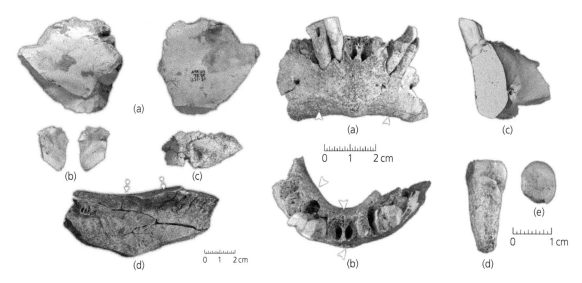

Figure 8.43 Left: (a)–(d) Stone tools of Mode 1 from the E9 level of Sima del Elefante; arrows indicate percussion marks (Carbonell et al., 2008). Right: Specimen ATE9-1. (a) Front view: the arrows indicate the position of the anterior marginal tubercle (left) and incision submental. (b) Top view: arrows indicate the mental protuberance, the subvertical alveolar plane and the weak alveolar prominence. (c) Three-dimensional reconstruction made by computer tomography of the symphysis. (d) Distal view of the lower premolar LP4 of ATEP-1. (e) Occlusal view of LP4. Reprinted by permission from Macmillan Publishers Ltd: Carbonell, E., Bermudez de Castro, J. M., Pares, J. M., Perez-Gonzalez, A., Cuenca-Bescos, G., Olle, A.,. . . Arsuaga, J. L. (2008). The first hominin of Europe. *Nature*, 452(7186), 465–469.

specimens from Sambungmacan (Java), and to the smaller exemplar from Ngandong (Java). Moreover, the supraorbital torus with double arch was clearly distinguishable from that typical of *Homo erectus*, both Asian and African (OH 9). The teeth also separated the TD6 specimens from *Homo habilis* sensu stricto and indicated a continuity with the later European remains of the Middle Pleistocene. The lithic industry was pre-Acheulean, with no handaxes. The first conclusion drawn by Carbonell and collaborators (1995) did not go further than to establish the presence of *Homo* in Europe during the Pleistocene.

New specimens found in 1995 allowed the facial morphology of the hominins of level TD6 (Bermúdez de Castro, Arsuaga et al., 1997) to be established. Thus, a partial face of a juvenile specimen, with part of the cranium and some teeth, ATD6-69, was described as completely modern in its midfacial topography (Figure 8.46). According to Bermúdez de Castro, Arsuaga et al. (1997), no other exemplar prior to the Upper Pleistocene specimens from Djebel, Irhoud 1 Skhul, or Qafzeh (Israel), which are all modern humans, has similar characteristics, as we shall see in the next chapter.

Another specimen, ATD6-58, this time attributed to an adult, retained, according to the authors, juvenile characteristics that are not present in *Homo ergaster*

(WT 15000) or in the later Neanderthals, which lack those facial characteristics even in juvenile exemplars (Bermúdez de Castro, Arsuaga et al., 1997). Thus, the TD6 specimens seem to have derived traits that separate them from *H. erectus* (including *H. ergaster*) and Neanderthals (Table 8.13 lists the apomorphies). Bermúdez de Castro, Arsuaga et al. (1997) proposed, therefore, a new species, *Homo antecessor*, designating as holotype the specimen ATD6-5 (fragment of the right mandibular body with three molars found in July 1994) and a set of associated teeth belonging to the same individual. In the paratype of the new species were included various specimens, up to 38, all from level TD6. The phylogenetic meaning of the *Homo antecessor* will be addressed (Section 8.7.5).

A specimen found in 2003 in the lower part of the ASS—"Aurora Stratigraphic Set" (see Figure 8.47, left), which correlates with the Aurora stratum of the southern zone of Gran Dolina (Bermúdez de Castro et al., 2008)—enlarged the hypodigm of *Homo antecessor*, albeit at the cost of casting doubt on the evolutionary relationships. This is ATD6-96, the left-half of a very gracile mandible attributed first to a female (Carbonell et al., 2005).

Other relevant specimens coming from the TD6 level are a mandibular fragment from a child, ATD6-112,

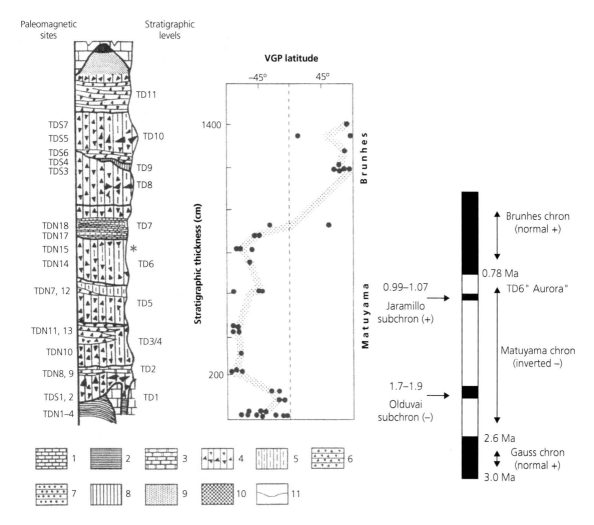

Figure 8.44 Left: Stratigraphy and the corresponding paleomagnetic poles in the Gran Dolina. The asterisk indicates the position of the Aurora stratum—named in honor of a senior member of the Atapuerca research team. The latitude VGP of each paleomagnetic sample is indicated: (1) limestone and dolomite; (2) finely laminated clay loam; (3) speleothems; (4) breccias of angular clasts of sandy mud; (5) siltstones; (6) breccias of fine-grained gravel; (7) oblique layers of calcarenites; (8) bat guano; (9) silty clay; (10) terra rossa; (11) major stratigraphic discontinuities. Right: Outline of the paleomagnetic events of the last 2.6 Ma, showing the location of the Aurora stratum. The poles identified in the Gran Dolina correspond to the Brunhes chron (levels TD11–TD8) and the Matuyama chron (levels TD7–TD1) prior to the Jaramillo subchron (see Box 8.29). From Parés, J. M. & Pérez-González, A. (1995). Paleomagnetic Age for Hominid Fossils at Atapuerca Archaeological Site, Spain. *Science*, 269, 830–832. Reprinted with persion from AAAS.

found in 2006 in the upper part of this level, and a left mandibular fragment from an adult with the molars M2 and M3, ATD6-113 (Figure 8.47, right), which was found 30 cm lower, in the Aurora stratum. The stratigraphic cut showing the location of the different specimens ATD6-96, 112 and 113 appears in Figure 8.47 (left).

The presence of the TD6 mandibles, ATD6-5 (child), ATD6-96 (female), and ATD6-113 (male) allowed comparisons that proved that the size of the specimen attributed to a child, upon reaching adulthood, would be similar to the male specimen and larger than the female exemplar. The metric variability of the sample is not too large. The similarity between ATD6-113 and ATD6-96 led the authors to argue that these are individuals of the same biological population (Bermudez de Castro et al., 2008); thus, all mandibular specimens from level TD6 are included in the hypodigm of the *Homo antecessor*.

Box 8.29 The age of TD6

A study of the paleomagnetism of TD6 conducted later by Parés and Pérez-González (1999) confirmed the observations made in 1995. The only significant contributions of the new study for the content of this chapter are: (1) a detailed examination of the reasons that led to attributing a change in polarity at the top of TD7 to the Matuyama–Brunhes boundary, and (2) the identification of a positive paleopole in TD1. The authors attributed the paleopole of TD1 to the Jaramillo subchron or even to an earlier and shorter subchron—which we have not included in the graph in Figure 8.44—the subchron Kamikatsura (not without considering that this paleopole might actually be an artifact, i.e., the result of an instrumental error). In the first case—the paleopole of TD1 corresponds to Jaramillo—the Aurora stratum would necessarily be between 0.78 Ma (Matuyama–Brunhes transition) and 0.98 Ma (early Jaramillo). In the second case—the paleopole that corresponds to Kamikatsura—the maximum age of Aurora would be 0.85 Ma.

The Kamikatsura subchron has been identified in various parts of the world. As Robert Coe and collaborators (2004) indicated, Kamikatsura is a well-identified paleomagnetic marker of high resolution, allowing precise specification of the changes of events in the last segment of the Matuyama chron. Its age was established at 900.4±4.6 Ka through radiometric dating by the ^{39}Ar/^{40}Ar method in the Haleakala Maui caldera (Hawaii, USA) (Coe et al., 2004).

The last specimens from Gran Dolina are nine ribs from at least three individuals attributed to *Homo antecessor*, found in level TD6 (Gómez-Olivencia et al., 2010).

Ceprano site

The Anagni and Ceprano depressions are in Latina Valley, 90 km southeast of Rome (Italy), in the Sacco River basin (Figure 8.48). Sites Fontana Ranuccio 1 and 2 belong to Anagni, both with Acheulean tools, 458±5.7 Ka; Ceprano sites 1 and 2 are in the second depression of the same name. The hominin specimen, which came to represent the first populations occupying Europe, was found in Ceprano 1 (Manzi et al., 2001).

The age of the specimen from Ceprano is difficult to determine. The first fragments appeared by chance in March 1994, outcropped and also damaged by a bulldozer (Ascenzi et al., 1996); Italo Bidditu identified these fragments as a human fossil. After collecting those fragments that could be recovered, the reconstruction of the specimen was made with the usable parts that could be fitted together—those that were more damaged, smaller, and attributable to the facial area, could not be placed acceptably—so a fragmentary skullcap was obtained, initially classified as *Homo erectus* and given an approximate age of over 0.7 Ma, suggesting as the most probable 0.8 Ma (Ascenzi et al., 1996) (Figure 8.49). When addressing the hominin dispersal model to Europe and the phylogenetic affinities of early settlers, we will return to the dates that most likely correspond to the Ceprano fossil.

The interpretation of the specimen by Antonio Ascenzi and collaborators (1996) indicated that it could have suffered some pathological deformation (oblique groove in the right supraorbital torus); however, features such as a low cranial vault, bone thickness, lateral wall shape, and a massive and prominent superciliary arch, which extends to form a single glabellar torus, are coincident with the features of Asian *Homo erectus*. The estimated cranial capacity was large at 1,185 cc—larger, even, than the average *H. erectus*—and the fossil had modern features, such as the absence of a sagittal keel, the development of frontal sinuses, and a relative reduction of cranial bone mass in the cranial vault with respect to its base. The authors attributed the skullcap to a male, given its massive appearance, and, according to the cranial sutures, gave him an age between 20 and 40 years.

A reconstruction of the Ceprano cranium by Ron Clarke (2000) (Figure 8.50) eliminated the supposedly pathological distortion, as well as some errors—from Clarke's point of view—in the initial restoration work.

After his reconstruction, Ron Clarke (2000) confirmed the presence in the Ceprano calotte of apomorphies that were typical of Asian *Homo erectus*:

- Prominent occipital crest.
- Well-developed and extended supraorbital torus.
- Greater wall thickness in the base of the cranium.

Further considerations regarding the reconstruction of the calotte made by Marie-Antoinette de Lumley et al., led the Ascenzi team (2000) to argue for expansion of the hypodigm of *Homo erectus* sensu stricto if the Ceprano calotte should be placed in that species. The cranium of *H. erectus* sensu stricto is more elongated than that of Ceprano, which shows a very broad frontal part over the supraorbital torus, broader than that of OH 9. The mediolateral expansion of the Ceprano specimen is also outside the *H. erectus* sensu stricto measurements. However, Ascenzi et al. (2000)

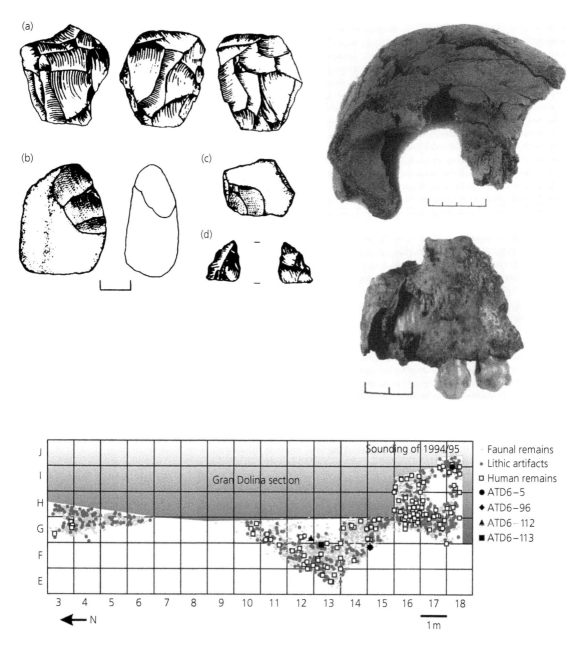

Figure 8.45 Upper left: Lithic industry of TD6. (a) Core (FGNB) made on neogene flint, with multiple rotation on preferential axes. (b) Unifacial heavy-duty tool (FGNB) made on limestone. (c) Flake (PB) made on quartzite. (d) Denticulate (SGNB) made on cretaceous flint. Scale bar: 3 cm. Upper right: Frontal bone fragment ATD6-15 (scale bar: 2 cm) and left maxillar fragment ATD6-14 (Scale bar: 1 cm). Bottom: Schematic plan of the TD6 level, indicating the various fossils and lithic objects found up to 2008. The test pit (south area) was excavated in 1994–95. Bottom: Reprinted from Bermúdez de Castro, J. M., Pérez-González, A., Martinón-Torres, M., Gómez-Robles, A., Rosell, J., Prado, L.,. . . Carbonell, E. (2008). A new early Pleistocene hominin mandible from Atapuerca-TD6, Spain. *Journal of Human Evolution*, 55(4), 729–735, with permission from Elsevier. Upper left and upper right: From Carbonell, E., Bermúdez de Castro, J. M., Arsuaga, J. L., Díez, J. C., Rosas, A., Cuenca-Bescós, G.,. . . Rodríguez, X. P. (1995). Lower: Pleistocene Hominids and Artifacts from Gran Dolina site.

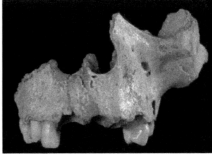

Figure 8.46 ATD6-96. Holotype of *Homo antecessor* (on the left, buried image). (Bermúdez de Castro, Arsuaga et al., 1997).

Table 8.13 Apomorphies of *Homo antecessor*

Cranial traits

1. The topography of the midfacial third shows an entirely modern pattern: the infraorbital surface is oriented with respect to the crowns downward and backward (true canine fossa), with a horizontal lower edge and deeply rooted.
2. Supraorbital torus in double arch, frontal view.
3. Convexity of the superior border of the temporal squama (arched).
4. Presence of a styloid apophysis.
5. Cranial capacity greater than 1,000 cc.

Mandibular traits

6. The mylohyoid line extends almost horizontally forward and reaches the mandibular body up to the levels M2/M3.
7. The thickness of the mandibular body is clearly less than that of *H. ergaster, H. habilis* sensu stricto, and the specimens from Baringo and Java, and OH 22.
8. Absence of alveolar prominence at level M1.
9. Narrow extramolar sulcus.
10. Lateral smooth prominence, limited to level M2.
11. The design of the internal face of the corpus is defined by moderately hollowed subalveolar fossa, but well developed and a clear oblique internal line, similar to those of the European Middle Pleistocene fossils.

Dental traits

12. Lower incisors expand in buccolingual direction with respect to *H. habilis* s.s., Zhoukoudian, and specimens such as KNM ER 992 and those of Dmanisi, although to a lesser degree than in *H. heidelbergensis* and *H. neanderthalensis*.
13. Postcanines are smaller than in *H. habilis* s.s., and are in the range of *H. ergaster, H. erectus,* and *H. heidelbergensis.*
14. Upper incisors have a shovel shape.
15. The lower canine is short in mesiodistal direction.
16. The vestibular faces of lower premolars show marginal crest and mesial and distal sulcus connecting with a cingulum in a shell shape.
17. The shape of P3 crown is strongly asymmetrical.
18. P3 presents a well-developed talonid.
19. Size sequence of the crown area P3-P4, in the upper and lower premolar.
20. The upper and lower premolars are broader in buccolingual direction.
21. The molar M1 is larger in buccolingual direction than in *H. ergaster.*
22. Size sequence of the crown area of the upper and lower molar in the series M1-M2.
23. M3 is obviously reduced with respect to M1.
24. M1 and M2 show a Y-pattern in the buccal and lingual sulcus which separate the five principal cusps.
25. Upper premolars show roots, buccal and lingual, well separated.
26. P3 and P4 present a complex root system formed by one flat root MB with two pulp canals and a root DL with a single canal.
27. Roots of upper and lower molars are well separated and are divergent, presenting a moderate taurodontism.
28. The teeth root system is small with respect to the dimensions of the crown.
29. The enamel of the postcanine occlusal surface is moderately to very undulated.

Bermúdez de Castro, Arsuaga et al. (1997).

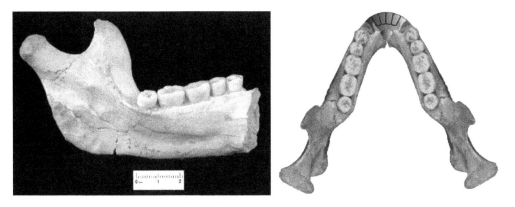

Figure 8.47 ATD6-96 specimen. Left: medial view; right: drawing based on a photographic restoration; the illustrations are not at the same scale (Carbonell et al., 2005).

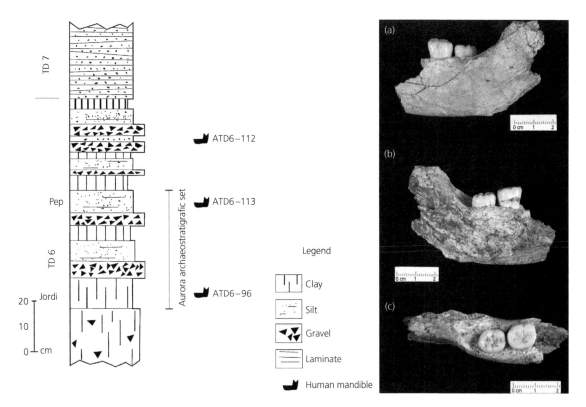

Figure 8.48 Left: Stratigraphy of level TD6 showing the location of the specimens ATD6-96, 112 and 113. Right: Mandible ATD6-113 (Bermúdez de Castro et al., 2008). Reprinted from Bermúdez de Castro, J. M., Pérez-González, A., Martinón-Torres, M., Gómez-Robles, A., Rosell, J., Prado, L., . . . Carbonell, E. (2008). A new early Pleistocene hominin mandible from Atapuerca-TD6, Spain. *Journal of Human Evolution*, 55(4), 729–735, with permission from Elsevier.

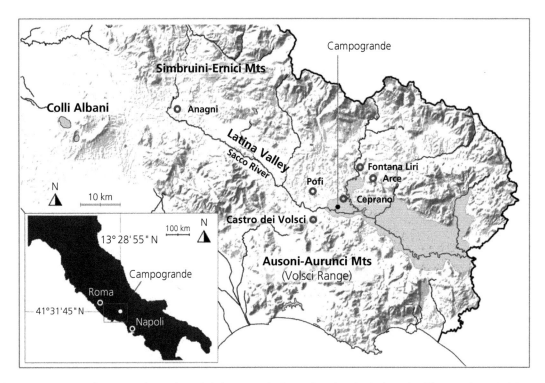

Figure 8.49 Location of Campogrande area, from which the Ceprano fossil came (Manzi et al., 2010). Reprinted from Manzi, G., Magri, D., Milli, S., Palombo, M. R., Margari, V., Celiberti, V., . . . Biddittu, I. (2010). The new chronology of the Ceprano calvarium (Italy). *Journal of Human Evolution*, 59(5), 580–585, with permission from Elsevier.

maintained that the specimen from Ceprano belongs to *Homo erectus*. As we shall see, this is not the latest taxonomic proposal for the fossil.

Sarstedt site

Alfred Czarnetzki, Sabine Gaudzinski, and Carsten Pusch (2001) reported the finding of three cranial fragments of old hominins in the basin of the Leine River, near Sarstedt (Hildesheilm, Germany). They are a fairly complete temporal bone (Sst I) attributed to a female, an occipital fragment (Sst II), and a fragment of a left parietal (Sst III). The first and last fragments were recovered in 1999 by suction dredge in a quarry belonging to the construction company Kies und Bau Union, while the second fragment comes from a different gravel pit belonging to the company Steinhäuser, and was found in 1997 (Czarnetzki et al., 2001). Nine lithic utensils were found in the same location by amateur archaeologists (Otrud and Karl-Werner Fragenberg, from whose collection the fossils come). After a morphological study of the specimens, the authors attributed them to *Homo neanderthalensis*, noting the stone industry as typical of the Middle Pleistocene.

Two new temporal bone fragments (Sst IV and V) were obtained in the same manner by Otrud and Karl-Werner Fragenberg in 2002 and 2004, respectively. Their morphological study allowed Czarnetzki and collaborators (2007) to relate them with the crania from Trinil I (Java) and Dmanisi 3444 (Georgia). By morphological comparison and general considerations of Sarstedt geology, Czarnetzki et al. (2007) gave the specimens an approximate age of 1.5–1.3 Ma.

8.6.6 Colonization and extinction: climatic evidence

The fossil specimens from Atapuerca, Ceprano, and Sarstedt point, according to the age attributed in the first descriptions of the fossils, to a much earlier occupation of Europe than that indicated by the "short chronology." But they are clearly insufficient to support a model of how the colonization of the continent took place. The evidence in this respect is to be found in the archaeological record.

In the analysis of the Gran Dolina level TD6, which closed a double issue of the *Journal of Human Evolution*

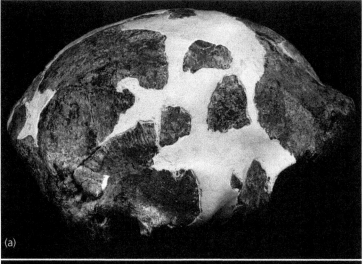

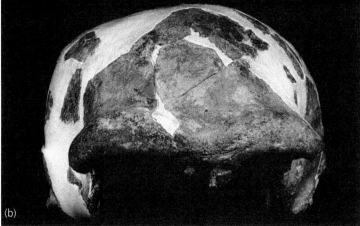

Figure 8.50 Ceprano calotte after the initial reconstruction (Ascenzi et al., 1996). Left: (a) Lateral view; (b) frontal view. Right: (a) Occipital view; (b) sagittal view. Reprinted from Ascenzi, A., Biddittu, I., Cassoli, P. F., Segre, A. G. & Segre-Naldini, E. (1996). A calvarium of late *Homo erectus* from Ceprano, Italy. *Journal of Human Evolution*, 31, 409–423, with permission from Elsevier.

in 1999, José María Bermúdez de Castro, Eudald Carbonell, and collaborators (1999) examined the question of the arrival in Europe of *H. antecessor* under the following premises:

- One million years ago, or even earlier, a human population from the Rift Valley migrated to Europe, reaching the Iberian Peninsula approximately 0.8 Ma.
- This migrant population carried with them tools of Mode 1 (Oldowan), similar to those found at TD6 level.
- However, utensils of Mode 2 (Acheulean) have existed in the Rift since 1.4 Ma.

The explanation about why the entry to Europe could be achieved with a more primitive cultural mode than existed then in Africa is not so simple. The possibility raised by Hallam Movius, which could be applied to Europe, is that populations leaving Africa had a cultural level of Mode 2, which later was lost or degenerated into a carving technique reminiscent of Mode 1. But this is not a very plausible speculation since we lack evidence of any reversion process. The argument that the Acheulean technique is complex and difficult and, therefore, vulnerable to discontinuities in space and time (Schick, 1994), could explain a detectable technical reversal in the archaeological record. But, in its absence, the hypothesis that hominins left Africa with the same tools found in the oldest European sites and, for that matter, in Dmanisi, is much firmer.

Bermúdez de Castro, Carbonell et al. (1999) rejected a cultural "return to the past," and promoted another explanation for the primitive tools of Atapuerca, starting from the fact that the Acheulean and Oldowan techniques coexisted in Africa, originating different systems of survival. Competing populations with different cultures would lead, in the long run, to the emigration of those with Mode 1.

The hypothesis of "unfair competition" forcing the weaker to emigrate is far more solid than the "return to the past." It is very common that migratory pressures heighten in peripheral populations, particularly if they are in an inferior economic position. But consideration of this second hypothesis weakens the strength of technological and cultural transmission, which, as discussed in coming chapters, goes beyond the level of species.

The most convincing theory is also the simplest: Asian populations with an Oldowan culture carried out the occupation of Europe. If this was the way in which the arrival of hominins took place, the explanation about why they possessed a Mode 1 culture leaves no loose ends.

8.6.7 Episodes of occupation

Utensils of Mode 2 are abundant in Europe from around 0.5 Ma. Where would they have originated? The interpretation of Bermúdez de Castro, Carbonell et al. (1999) states that the European Acheulean—which was already present c. 0.5 Ma, or even earlier—did not evolve in situ from Mode 1, but came in a new migration wave from Africa. As such, we have as the less risky explanation that of two successive occupations of Europe: the first would be reflected in level TD6 (Gran Dolina, Atapuerca, Spain), with a probable Asian origin, and having utensils of Mode 1; the second would be indicated by the numerous European sites of the Middle Pleistocene, probably of African origin, with utensils of Mode 2.

But, is it possible to limit to two the migrations of the human population to Europe? The study by Mark White and Danielle Schreve (2001) on the paleogeography and occupation of England in the Lower Paleolithic established a succession of phases that include in each glacial/interglacial episode various occupational possibilities:

I. Glacial episode, meaning the disappearance of geographical barriers—England became a peninsula of Europe, but with an extreme climate that did not allow occupation.
II. End of the glacial period and the beginning of the next interglacial, bringing conditions that made human presence viable.
III. Interglacial: isolating the existing populations, which are now subjected to the recovered island conditions.

Due to the cyclical nature of these phases, human presence would be intermittent. A first colonization during the warmer period would be followed by the extinction of the population group with the return of the cold period, and later by a recolonization in the next warm phase. Successive phases of colonization–extinction–recolonization would have occurred, therefore, since the first occupation of southern Great Britain during the Cromerian Stage (Günz–Mindel interglacial). But how many times would this process have occurred? Robin Dennell, María Martinón-Torres, and José María Bermúdez de Castro (2011) have outlined the colonization, extinction, and recolonization of the northern European latitudes (England, in particular) through five episodes, shown in Figure 8.51. By "extinction" Dennell et al. (2011) mean both the disappearance in Great Britain of the hominin groups—the most likely case for the authors—and a possible migration to warmer latitudes.

The successive phases shown in the graph by Dennell, Martinón-Torres, and Bermúdez de Castro (2011) (Figure 8.51) can be related to the work by Mark White and Danielle Schreve (2001) and White, Scott, and Ashton (2006), who analyze the available evidence about the occupation of England, summarized in Table 8.14.

The intermittent occupation is accentuated in the period from MIS 11 to MIS 7, corresponding to a long interglacial of about 100,000 years, separated into several segments (Hoxnian, Purfleet, and Avery interglacials) through two stages of intense cold, one of short duration (MIS 10) and one more extensive (MIS 8). A time of special importance is that which goes from MIS 9 to MIS 7. It is, in the words of White, Scott, and Ashton (2006), a critical segment for the occupation of Europe because it coincides with technological change from the Acheulean handaxes to the Levallois *châine opératoire* of reduced core.

8.6.8 Colonization and extinction: cultural evidence

In a paper analyzing the conditions of human occupation during the Hoxnian interglacial (MIS 11), Ashton and collaborators (2006) warned that the established model for South England is not of mandatory application to the general European context. One might add that it is particularly not so when referring to the southernmost localities in Italy and Spain, which could have served as a refuge during periods of extreme cold.

The first colonizing sequences can be established more precisely with the presence of Mode 1 utensils

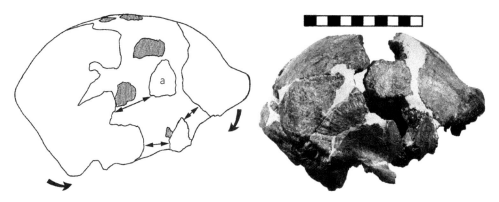

Figure 8.51 Left: Scheme of the original reconstruction, lateral view, of the Ceprano cranium pointing out the modifications made by Ron Clarke. The shaded parts were removed. The piece (a) is a parietal fragment in contact with the frontal bone. Arrows indicate Clarke's assembled bones; curved arrows indicate rotations introduced in the skull (Clarke, 2000). Right: Final appearance of the new reconstruction (Clarke, 2000). The images are not at the same scale. Reprinted from Clarke, R. J. (2000). A corrected reconstruction and interpretation of the *Homo erectus* calvaria from Ceprano, Italy. *Journal of Human Evolution*, 39, 433–442, with permission from Elsevier.

Table 8.14 Glacial and interglacial periods with evidence of human presence in England. The dates are given in thousands of years (Ka). From left to right: MIS period; approximate age of onset; episodes of colonization (C), extinction (E), and recolonization (R) corresponding to those shown in Figure 8.55; geographic status: island or peninsular character for each temporal segment; and human occupation evidence

MIS	Age (Ka)	C	Character	Geographic status	Occupation evidence
MIS 6	140–200		Glacial (Riss)		Abandonment
MIS 7	c. 210	R5	Aveley interglacial (Mindel–Riss)	Island, except for a short time in the middle of the period (fauna)	Levallois
MIS 8	c. 300	E5	Glacial	Peninsular (inferred)	Later Levallois Abandonment at the high point of the glacial Beginning Acheulean
MIS 9	c. 340	R4	Purfleet interglaciar (Mindel–Riss)	Peninsular, with an intermediate marine presence	Later Acheulean Set without handaxes at the beginning
MIS 10	c. 350	E4	Glacial	Peninsular (inferred)	Set without handaxes at the end Abandonment at the high point of the glacial Beginning Acheulean
MIS 11	c. 400	R3	Upper Hoxnian interglacial (Mindel–Riss)	Island with marine invasions that start in the North, in Hox-IIc (uncertain). Island in IIb	Only Acheulean
		R2	Middle Hoxnian interglacial (Mindel–Riss)	Peninsular from Hox-IIb to lower III (pollen)	Only Clactonian until Hox-IIb. Later Acheulean since IIc (uncertain)
		R1	Lower Hoxnian interglacial (Mindel–Riss)	Imprecise status, with an initial peninsular situation	Only Clactonian
MIS 12	400–450	E1	Anglian Glaciation (Mindel)	Peninsular	Set without handaxes at the end Abandonment at the high point of the glacial
MIS 13–15	450–620	C	Cromerian Interglaciar (Günz–Mindel)	Island/peninsula successively	Acheulean

Compiled from White and Schreve (2001) and White, Scott, and Ashton et al. (2006).

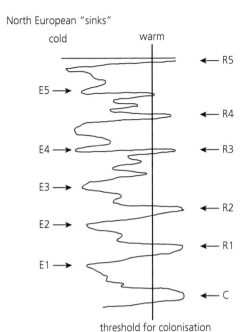

North European "sinks"

Figure 8.52 Graphic of the successive colonizations of North Europe (Dennell et al., 2011). Reprinted from Dennell, R., Martinón-Torres, M. & Bermudéz de Castro, J. M. (2010). Hominin variability, climatic instability and population demography in Middle Pleistocene Europe. *Quat. Sci. Rev*, with permission from Elsevier.

in Southern and Central Europe. The sites we will consider are those in Figure 8.52.

Kozarnika Cave is a place that offers interesting evidence about the hominin entry into Europe (Figure 8.53). It is located in the northwestern part of the Balkans, near the Danube Valley, about 3 km from the city of Oreshets, Belogradchik district (Bulgaria). Although known since 1931, the excavations carried out since 1994 revealed a human occupation from the oldest age levels, 13–11, to the modern, 4–3a (Sirakov et al., 2010).

Kozarnika paleomagnetism studies did not produce useful results, but its biochronology has permitted establishment of a continuous range of human occupation between 1.6–1.4 Ma and 0.5 Ma (Sirakov et al., 2010).The lower levels of the cave (13–11) have provided nearly 10,000 stone objects, composed mostly of small debris chips. They are not bifaces or more primitive axes ("pebble-tools"); an absence that Nikolay Sirakov and collaborators (2010) did not attribute to the difficulties in obtaining raw materials

but to the lack of a tradition of carving. Levels 10a–9a of Kozarnika indicate Mousterian industry. Level 5b, with volcanic ash dated at 31,237±389 and 26,490±270 years, present a modern industry of stone-baked sheets.

Kozarnika reveals that there was a continuous and long-term permanent occupation in Eastern Europe. The stone tools indicate a very slow development of an industry based on obtaining flakes from cores, but lacking choppers and Acheulean handaxes (Box 8.30). In any case, the Balkans manifest the presence of a stable population, which does not fit with the idea of various and successive waves of migration separated by periods of extinction.

It is, however, a unique and particular case. Kozarnika could be regarded as a refuge from which the dispersals into Europe originated. Giovanni Muttoni and collaborators (2010, 2011) have provided an interpretation of how this transition could have occurred. In the period after 1 Ma (0.98–0.87 Ma), the MIS 24 and MIS 22 glaciations caused a prolonged cooling that led to the extension of aridity in North Africa and Eastern Europe. MIS 22 glaciation was the first one of great magnitude in the Pleistocene, leading to a "climate revolution" (Berger et al., 1993), affecting, in particular, the European continent and its environment. These circumstances would have been what drove the fauna to migrate toward refuges in Southern Europe.

Searching for southern shelters, the Po Valley played a strategic role as a corridor allowing passage between the Alps and the Mediterranean Sea (Figure 8.55). The Po Valley would have been immersed during most of the Lower Pleistocene, then emerged due to the MIS 22 episode, easing transit for both large mammals and humans (Muttoni et al., 2011). The faunal transformation of the European Lower Pleistocene (Azanza et al., 2000) would be the result of that combination of migration pressures and opportunities to travel E–W.

The site in Monte Poggiolo, located precisely in the Alps–Mediterranean corridor, provides evidence of human presence in the MIS 21 interglacial, following the great migration pressure of MIS 22. According to Giovanni Muttoni et al. (2011), many of the most ancient sites of Southern Europe would follow the same pattern. That is the case with Vallparadís, Gran Dolina (Atapuerca), and the French localities of Soleihac or the Loire Valley. Tentatively, with the necessary caution when dating sites for which there is no possibility of using radiometry, we would assert that all of these are somewhat below the Brunhes–Matuyama boundary (0.78 Ma).

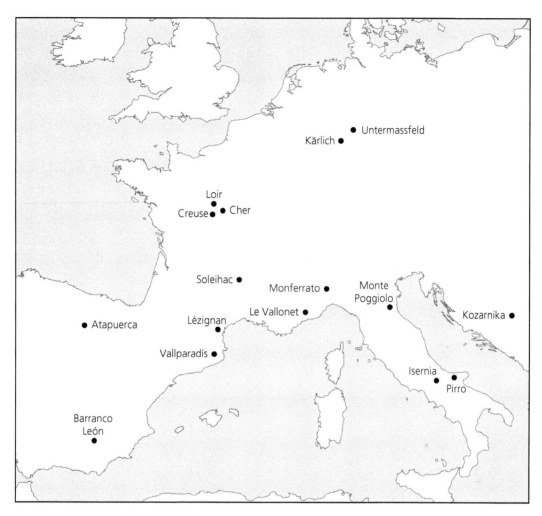

Figure 8.53 European sites with Mode 1 utensils dated prior to c. 0.5 Ma. We have not included those from England, nor from Eastern Europe. Sites such as those of Ceprano, without stone objects, Sarstedt, of dubious age, and Chilhac, Val d'Arno, and St. Eble, whose evidence of human occupation has been doubted, have not been taken into consideration (see Muttoni et al., 2010). Regarding Southern Spain, only the site of Barranco León is shown.

Box 8.30 Choppers and flakes

Sirakov et al. (2010) showed that the absence of choppers in the lower levels of Kozarnika requires the consideration that the core-chopper and core-flake industry are not part of the same tradition, contrary to what is often argued, and that the distinction between cultural modes (1, 2, etc.) is inadequate to specify the type of "*châine opératoire.*" The old Kozarnika industry would correspond to Mode 1, although they wondered whether it is necessary to use such terminology.

Nevertheless, as we know, there are European sites with signs of human presence before the big MIS 22 climate change. In both Italy (Pirro Nord and Monferrato), France (Lézignan-le-Cébe, Soleihac, Le Vallonet, Creuse, and Cher valleys), and Spain (Sima Elefante in Atapuerca; Barranco León and Fuentenueva in Orce), and even in places as far north as Germany (Untermaßfeld and Kärlich), it is possible to find evidence of an earlier occupation. Population incursions would have happened prior to the large migration flow proposed by Muttioni et al. (2010, 2011). According to the indications of Lézignan (Crochet et al., 2009), previous dispersions may be related to ancient faunas such as those corresponding to MNQ 18–19 (Box 8.31).

Figure 8.54 Left: Location of Kozarnika Cave; right: lower levels of Sector II (both illustrations from Sirakov et al., 2010). Reprinted from Sirakov, N., Guadelli, J. L., Ivanova, S., Sirakova, S., Boudadi-Maligne, M., Dimitrova, I., . . . Tsanova, T. (2010). An ancient continuous human presence in the Balkans and the beginnings of human settlement in western Eurasia: A Lower Pleistocene example of the Lower Palaeolithic levels in Kozarnika cave (North-western Bulgaria). *Quaternary International*, 223–224(0), 94–106, with permission from Elsevier.

Naturally, the occupation of the northern areas (such as the sites of Untermaßfeld and Kärlich) could only have taken place following a pattern of colonization–extinction–recolonization, such as the one in England proposed by Dennell et al. (2011). The MIS sequences attributed to the evidence of human occupation of these locations are MIS 31 for Untermaßfeld (Landeck, 2008) and MIS 23 to Kärlich (Haidle & Pawlik, 2010). The fauna is, of course, typical of temperate or warm environments; since the interglacial episodes are widely separated in time, the most reasonable hypothesis is that of extinction (in the sense of Dennell et al., 2011) of the human populations in Germany during glacial periods, with subsequent recolonization.

Meanwhile, the sites of Lézignan (Crochet et al., 2009), Pirro Nord (Arzarello et al., 2007), and Monferrato (Siori & Sala, 2007) indicate an arid climate with areas of higher humidity during times of human occupation—conditions that correspond to the beginning and end of an ice age. Taken together these various circumstances point toward a dispersal model that corresponds to:

- An area of origin, either in Eastern Europe or North Africa, with permanent human occupation from dates coinciding with those of the Dmanisi population.
- Various sporadic migrations to the West and even Northern Europe.
- A human presence distributed throughout Southern Europe (Italy, France, and Spain) during episodes of intense cold, reaching Central Europe occasionally during warmer intervals.

Box 8.31 Mammal zones

The biochronological identification usually refers in Europe to the division of "mammal zones" proposed by Pierre Mein (1975). Those corresponding to the colonization under consideration are: MNQ 23, 24, and 25, above the Matuyama–Brunhes boundary; MNQ 22 coincides with the limits between the chrons; MNQ 21 above and MNQ 20 below the Jaramillo subchron.

The model of intermittent colonization requires us to understand that one cannot refer to the arrival of hominins in Europe as a migration process such that one could speak of a European human deme, as we speak of the African and Asian demes. Although, the phylogenetic trees may give such an impression, the fact is that the "European" hominins did not constitute a population made up by an early arrival to the continent, evolving in situ, and giving way to successive species of the Middle and Upper Pleistocene. The evidence from both the detailed studies on the colonization of England and the less informative ones about Southern Europe, manifests that various migratory waves occurred. Although the most commonly accepted origin is that of Eastern Europe, the available data do not allow the rejection of the routes by the Sicilian lands or the Gibraltar Strait as corridors used by hominins.

8.7 An evolutionary model for the hominins of the Lower and Middle Pleistocene

As we have seen, the Lower and Middle Pleistocene mean the hominin exit from Africa and the occupation of the Asian and European continents. The way in which the dispersals occurred was different in the two colonized continents. The Far East colonization is very old and it is well known how it happened, as are the relationships between the fossil specimens that document the colonization of Java and China. Europe is a different case, with more uncertainties about the successive migrations and their places of origin. Therefore, one cannot advance a single evolutionary model for the first human migrations out of Africa.

In cladistic terms, the hypothesis of a cladogenesis that separates African and Asian *H. ergaster* and *H. erectus*, was rejected by the analysis of Berhane Asfaw et al. (2002). This rejection would persist even if, as the authors argued, the last Asian demes—the most recent ones—were eliminated from the analysis.

An opposite view to that adopted by Asfaw et al. (2002) is that of Jeffrey Schwartz (2004), after analyzing the specimen KNM-OL 45500 from Olorgesailie, Kenya. This is, as we shall see, an exemplar with *Homo erectus'* traits, or even more modern, but of a very small size. Richard Potts and collaborators (2004), after a detailed comparison of KNM-OL 45500 with other skulls of *Homo erectus/ergaster*, merely argued that KNM-OL 45500 had a different set of traits, indicating the hominins' wide variation between 1.7 and 0.5 Ma. But, in the commentary accompanying the initial description of the specimen, Schwartz (2004) argued that the task

of understanding the relationships between the various specimens from Africa, Asia, and Europe would be easier by recognizing that *Homo erectus* may be more an historical accident than a biological reality. In the opinion of Schwartz (2004), the morphology of the different specimens attributed to *H. erectus* exceeds the limits of individual variations, which are well documented in the sample from Trinil/Sangiran (Figure 8.55). Another example of a very small specimen lacking some of the typical traits of *H. erectus*, but classified as such, is the KNM-OL 45500.

With regard to South African exemplars, their consideration is not straightforward. Heather Smith and Frederick Grine (2008) conducted a cladistic analysis of 99 morphological traits detectable in the SK 847 specimen from Swartkransand Stw 53 from Sterkfontein—type specimen of *Homo gautengensis*—considering them either as a single or two separate OTU (Operational Taxonomic Units), in contrast to *H. habilis*, *H. rudolfensis*, and *H erectus* (Figure 8.56). In both cases, the most-parsimonious analyses point to an identity among the South African forms (either as a single or two separate OTUs), positioned between *H. rudolfensis* and the set *H. habilis* + *H. erectus* + *H. sapiens*. Although cautioning about the weakness of an analysis done exclusively on two South African specimens, Smith and Grine (2008) interpreted the results in two ways:

- The cladistic distribution indicates the presence of a South African lineage separate from the other *H. erectus* sensu lato.
- Lack of shared apomorphies among South African specimens and both *Homo habilis* and *Homo erectus* indicates a more general morphology in the South African specimens that would be the ancestral condition for all other members of *Homo*.

Which is the better option for understanding the phylogeny of the hominins that appeared in the Middle Pleistocene? The possible alternatives are listed in Figure 8.57. Alternative (a) would maintain the idea of a single very variable species in the Middle Pleistocene, which would encompass both African and Asian specimens, with the intermediate chronospecies *H. georgicus*. Aside from the question of accepting that such a great degree of variation is possible, this first hypothesis encounters the problem of cultural evidence. We have already commented that Mode 2, with its characteristic bifaces, so abundant in the Middle Pleistocene of Africa, does not appear in Asia to the East of the Movius line. Given the ease of cultural diffusion, an absence of Mode 2 in the Asian *Homo erectus* sites is difficult to understand if the population flow was strong enough to maintain the unity of the species.

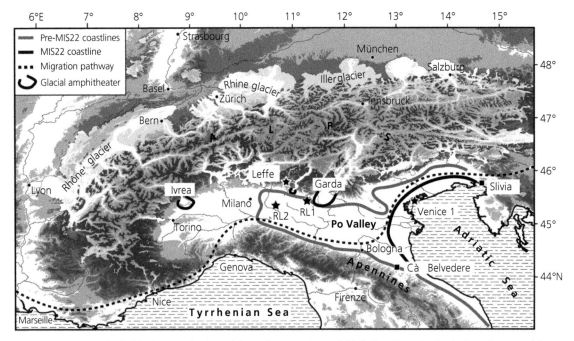

Figure 8.55 Corridor of Valle del Po between the Alps and the Mediterranean Sea available for East-West migration in the environment of the MIS 22 glaciation. RL1 and RL2 refer to the localities of Pianengo and Ghedi (Muttoni et al., 2010). Reprinted from Muttoni, G., Scardia, G. & Kent, D. V. (2010). Human migration into Europe during the late Early Pleistocene climate transition. *Palaeogeography, Palaeoclimatology, Palaeoecology*, 296(1–2), 79–93, with permission from Elsevier.

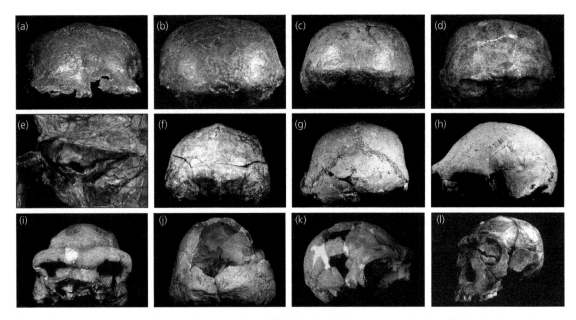

Figure 8.56 Several specimens attributed to *Homo erectus*. (A and B) Trinil cranium in frontal and posterior views; (C) Sangiran 2, posterior view; (D) Sangiran 12, posterior view; (E) Sangiran 4, internal view of the right petrous bone; (F) Sangiran 4, posterior view (according to Schwartz, 2004, these specimens establish the range of individual variation in *Homo erectus*); (G) Dmanisi D2282, posterior view; (H) D2280; (I and J) OH 9 in frontal and posterior views; (K) Ceprano in 3/4 view; (L) KNM-WT 15000 in 3/4 view (images are not to scale; illustration from Schwartz, 2004). Schwartz, J. H. (2004). Getting to Know *Homo erectus*. *Science*, 305(5680), 53–54.

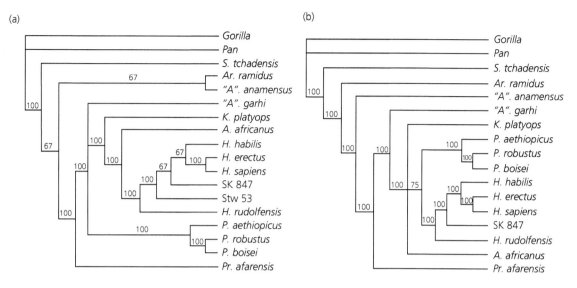

Figure 8.57 Cladograms obtained by Smith and Grine (2008) using the criterion of the "50% majority rule consensus," taking: (a) SK 847 and Stw 53 as two different OTUs; (b) SK 847 and Stw 53 as a single OTU (with the name of the first one). Reprinted from Smith, H. F. & Grine, F. E. (2008). Cladistic analysis of early Homo crania from Swartkrans and Sterkfontein, South Africa. *Journal of Human Evolution*, 54(5), 684–704, with permission from Elsevier.

Box 8.32 Cladograms below the level of genus

W. Henry Gilbert, Tim White, and Berhane Asfaw (2003) criticized, as we have also done throughout this book, conducting cladograms below the level of genus, because the geographic, temporal, and individual variations can be taken for morphological variations that would indicate evolutionary divergence. According to Gilbert et al. (2003), the case of *Homo* in the Middle Pleistocene would be a relevant example. In order to avoid circular reasoning, Asfaw et al. (2002) chose to do the analysis based on paleodemes. However, the paleodemes selected by Asfaw et al. (2002) were criticized by Manzi and collaborators (2003) for grouping together the specimens OH 9 and "Daka." The paper by Gilbert et al. (2003) aims to defend the superior utility of paleodemes in contrast to the use of individual specimens in cladograms.

Alternatives (b) and (c) are based on the separation of species in the African and Asian specimens. Option (b) is based on the evidence from Dmanisi. Option (c) takes into account the difficulty of explaining the derived traits of *Homo georgicus* absent in *Homo erectus*.

Species' separation under any of the last two options would require genetic isolation mechanisms, which could explain the presence of diverse morphological features, eliminating the difficulties of an excessive variability in the taxon. Cultural differences would also be easy to justify in this way. However, we run into a geographical conundrum: some specimens from Africa show a very high morphological resemblance to *Homo erectus*, having features that the defenders of *H. ergaster* attribute to the Asian taxon. This applies, in particular, to OH 9 (Groves, 1989) and the cranium from Bouri, "Daka" (Asfaw et al., 2002). If these individuals are assigned to *Homo erectus*, then we must accept that this species, in principle Asian, was also present in Africa.

8.7.1 Toward a reasonable systematic of *Homo erectus*

In this chapter we have seen how the taxon *Homo erectus* can be seen in two contrasting ways. The first is to consider it as an example of extreme variability whose evolutionary significance can be understood by grouping the different specimens in a single species. The second option suggests that the possible taxon does not correspond to any coherent form of existing hominin with a strong biological significance and, therefore, the attempt to bring together the various fossils in the same taxon is to be considered a "historical accident" (Schwartz, 2004). In the first case, we are talking about *Homo erectus*. In the second, the most we could acknowledge is an "*erectus* grade."

However, the evolutionary perspective of two populations with local history of adaptation to a particular environment that can be very remote both in time and geography—defended by Karen Baab (2007)—can sustain a scheme whose scope can solve many of the taxonomic and systematic questions raised in this book.

Many of the discrepancies that exist when considering the characters used in both statistical and cladistic analyses come from methodological difficulties, such as the use of inappropriate reconstructions, a biased ascription of a specimen to a taxa, or an inappropriate selection of traits. The work of Rolando González-José and collaborators (2008) on the modular integration into "mega-traits"—the expression is ours—with quantitative variations in a multivariate space, can solve some of these difficulties. However, another source of problems is the presence of specimens having remarkable differences within a narrow geographical and temporal range. The most obvious example is the fossils from Olorgesailie and Daka. In this case, the value of local differences between populations of the same species is somewhat diminished.

However, if we still have the need to consider local populations that, for whatever reason, remain integrated with little dispersion, we find that in a modest geological timeframe they could accumulate perceptible morphological differences. This may be the origin of some of the most relevant distinctive traits that appear in certain fossils encompassed in *Homo erectus* sensu lato, which led to huge extremes of variation.

Box 8.33 Trait variation among species

Rolando González-José and collaborators (2008) have shown that the evolutionary novelties reflected in the skeleton—those registered in the fossil record—are largely the result of changes in developmental processes that are integrated into different hierarchical levels. Examples of this phenomenon would be the transformations of the masticatory apparatus when discussing the thickness of the molar enamel. Consequently, observable traits in different species show more quantitative than qualitative variations in a multivariate space of integrated modular units. The authors proposed taking into account these types of traits, reducing, for example, the multitude of craniofacial features used in a cladistic analysis to four modular characters (cranial base flexion; facial retraction; neurocranial globularity; shape and relative position of the masticatory apparatus) and distinguishing by morphometric methods the status of those characters in the different taxa.

Once the local populations separated, could we consider them as a biological species (BSC), even though, frequently, we only have one specimen with which to judge? This is impossible to ascertain. Is it worthwhile then to assign them to different evolutionary species (ESC)? We have argued that, in the case of *Homo erectus* vs. *H. ergaster*, the answer is in the affirmative. But it might not be so when we refer to the fossils from Olorgesailie vs. Daka, and we have not considered it necessary, of course, in the case of *H. rudolfensis* vs. *H. habilis*.

It is likely that some of these local populations could have evolved toward a new ESC, and possibly even quite a few of them did so. The case of the South African specimens is relevant in this regard. *Homo gautengensis* could then be an acceptable ESC without forcing the general phylogenetic interpretation. Suppose that SK 847 and SK 45 are the result of the evolution of *H. gautengensis*. If we want to also include the two specimens from Swartkrans in *Homo erectus*, one must accept that the whole taxon comes from *H. gautengensis* and its members moved to other African locations. But this hypothesis does not really agree with the evidence of Dmanisi, so the alternative would be to consider that SK 847 and SK 45 are specimens of *H. erectus* that do not have *H. gautengensis* among their ancestors, or to reduce the latter taxon to *H. habilis*.

The example of *H. gautengensis*, showing how local populations may occasionally reach the rank of ESC, provides a very coherent solution. The large number of South African hominin remains of the Upper Pliocene and the Lower Pleistocene points to this process. But, in the absence of specimens that indicate how the local phylogenetic processes might have occurred, it is preferable not to multiply the number of taxa in use, and to consider such specimens as "extreme" or "intermediate," as variations within a broad taxon.

Australopithecus is another case in which the population approach may be useful but, with regard to the taxon, it would also be appropriate to make cladistic analyses under the formula of the "mega-traits" of González-José and collaborators (2008). To finish the evolutionary history of the hominins of the Middle Pleistocene, we would suggest the following guiding principles:

- Morphological diversity is remarkable.
- It is accentuated by specific adaptations of local populations.
- Dispersals were important but probably limited to short, isolated episodes, as a result of sporadic population pressures.

- Within *Homo* the cultural level attained is a strong indication of population isolation.
- The proposal of ESCs should be limited to cases where this taxonomic resource is needed to clarify the phylogenetic process.
- Using cladograms for species' level may not be possible for the same reason, unless performing an important grouping of traits in modular units.

8.7.2 The European case

We have seen that the comparison between African and Asian hominins helps to clarify what it means to use the species' concept in phylogenetic studies. However, the European specimens are a different matter. To describe how the European colonization took place is very difficult and subject to many uncertainties, and to specify which species of hominin would have been the protagonist of the entry into Europe is even more difficult. The hope to explain the meaning in phylogenetic terms of a hominin entry into Europe much earlier than the one established by the "short chronology" is somewhat diminished after reviewing the available evidence. It is an analysis that has to adjust to the fossil evidence and, with regard to the oldest specimens, is limited to those from Sarstedt (Germany), Ceprano (Italy), and Atapuerca (Spain), with questions about the age of the specimens in the first two cases.

8.7.3 Sarstedt

The chance discovery of Sarstedt fossils during the extraction of gravel made it very difficult to place them in a particular stratigraphic level and therefore to estimate their age. Czarnetzki et al. (2001) indicated that the terrain could come from an interglacial previous to the Saale glaciation (MIS 6).

None of the aforementioned circumstances would suggest that the Sarstedt specimens could be considered older than 0.5 Ma, therefore, prior to the "short chronology." However, the morphological study of the new fragments of the temporal bone found later (SST IV and V) led Czarnetzki and collaborators (2007) to relate them to Trinil I (Java) and Dmanisi 3444 (Georgia), classifying Sst IV and V as *Pithecanthropus erectus europaeus*. According to the morphological comparison and to Sarstedt general geological considerations, Czarnetzki et al. (2007) gave those specimens a possible age of up to 1.5–1.3 Ma. Miriam Noël Haidle and Alfred F. Pawlik (2010), following Czarnetzki et al. (2007), stated that the age of the specimens had been estimated at 700,000 years or even older.

Such assumptions seem exaggerated. The lithic industry of Sarstedt was described as typical of the Keilmesser complex (Czarnetzki et al., 2002). It is a variation of the Micoquien tradition, a carving technique corresponding to MIS 3. The study of Martin Street, Thomas Terberger, and Jörg Orschiedt (2006) provided a better estimation of the age of the Sarstedt specimens in a range between 117,000 and 25,000 years. Although the diagnosis was made before the discovery of Sst IV and V, the attribution of an older age in specimens of such vague localization requires evidence that at the moment does not exist.

As we may recall, after corresponding morphological study of specimens Sst I, II, and III, their discoverers related them to *Homo neanderthalensis*; if this is so, they could not be previous to the "short chronology". In the review of the fossil record of the German Paleolithic hominins, Street et al. (2006) merely stated that the circumstances of the discovery of these specimens provide insufficient information on the geological or archaeological context necessary to determine their age.

8.7.4 Ceprano

After reconstruction of the specimen (Clarke, 2000), Giorgio Manzi, Francesco Mallegni, and Antonio Ascenzi (2001) stated that the Ceprano cranium was a morphological bridge between *Homo ergaster/erectus* and the Middle Pleistocene specimens (*H. heidelbergensis/rhodesiensis*). The authors rejected its ascription to *Homo antecessor* because of the lack of verifiable material in the specimens of Gran Dolina, and claimed for the Ceprano specimen the condition of LCA of Neanderthals and modern humans. Two years later, after a morphometric comparison and cladistic analysis, Francesco Mallegni and collaborators (2003), named a new species: *Homo cepranensis*.

Ceprano's credibility as common ancestor of Neanderthals and modern humans depends not only on its morphology but also on its age and, in this regard, there is some doubt about such an old age estimation as that granted by Ascenzi et al. (1996).

The paleomagnetic analysis conducted by Giovanni Muttoni and collaborators (2009) showed that the various stratigraphic units of Ceprano unit 1 belong to a normal polarity interval, detecting the earliest possible, but dubious, reversal at 45 m below, already in unit 2 (Box 8.34).

The authors assumed that unit 2 corresponds to a lacustrine episode, whereas the lithic nature of unit 1 indicates a more arid and dry interval. As the lacustrine sedimentary sequence of the Latina Valle yielded, by the K/Ar method, an age of either 0.354±0.007 or 0.583±0.0010 Ma (Carrara et al., 1995), Muttoni et al. (2009) attributed to the Ceprano cranium c. 0.45 Ma, an estimation matching the archaeological level of Rannucio Fontana.

Giorgio Manzi and collaborators (2010) evaluated the paleomagnetic data by Muttoni et al. (2009) taking into account various geological and paleobiological evidence from subsequent excavations of the site. The paleovegetation features led the authors to argue that the specimen of Ceprano corresponds to the interglacial MIS 11, with an age between 430 and 380 Ka.

The latest estimation of the fossil dates further lowered that figure. Sébastien Nomade and collaborators (2011) obtained a direct dating of the terrains of Ceprano unit 1 by the $^{40}Ar/^{39}Ar$ method, from the feldspar K that it contained. The age obtained was 353±4 Ka, and the volcanic materials' origin was attributed to an eruption of the Pofi, with significant volcanic activity between 0.41 and 0.345 Ma. This date situates the Ceprano fossil in MIS 10. Thus, their phylogenetic meaning seems to correspond with the process that we will examine in Chapter 9.

8.7.5 Atapuerca: the transition to the Neanderthals

The species *Homo antecessor* was proposed in 1997, as we saw when speaking about the Gran Dolina site, from specimens discovered at level TD6 (Bermúdez de Castro, Arsuaga et al., 1997). Four years later, José María Bermúdez de Castro and Juan Luis Arsuaga identified in the sample up to six individuals by comparative study of their dental material (José María Bermúdez de Castro & Arsuaga, 2000–01; Table 8.15).

Table 8.15 Identified individuals of *Homo antecessor* up to 2001 by dental materials (Bermúdez de Castro & Arsuaga, 2000–01). Later on other specimens appeared such as the mandible ATD6-96 (Carbonell et al., 2005); this was attributed to an adolescent 15–16 years old

Specimen	Remains	Age at death
H1	Right mandibular fragment with M1-M3 found in situ Maxillar fragment (C-P3 *in situ*) Left mandibular teeth: C, I2 and right: C, P3, P4 Left maxillar teeth: P4, M1 Right: P3, P4, M1, M2	13–15
H2	Maxillar fragment of the left side with deciduous canine and deciduous molar in situ	3–4
H3	Maxillar + zygomatic of left side with P3, M1, M2, and left M3 & right I2-M1	10–12
H4	Left mandibular I2	
H5	Left mandibular I2	
H6	Left maxillar germ I2	3–4

The complete collection of the Gran Dolina included at the time more than 85 specimens that, with regard to the cranial material, include frontal, parietal, temporal, occipital, maxillary, zygomatic, sphenoid, and maxillary bones (see Arsuaga et al., 1999). Regarding the postcranial elements, until 2001, 43 remains of clavicle, radius, femur, vertebrae, ribs, kneecaps, metacarpals, metatarsals, and hand and foot phalanges had appeared (see Carretero et al., 1999; Lorenzo et al., 1999). This was a diverse and rich collection that permitted identification of primitive and derived features in comparison with specimens of *Homo* of similar age, older, or younger, to those of the Gran Dolina.

It was the confirmation that the hominins from Gran Dolina had a mosaic of plesiomorphies and apomorphies, which defined the species *Homo antecessor*; in particular, the possession of an intermediate mandibular morphology between that of the early *Homo* and those of the Upper Pleistocene—*H. neanderthalensis* and *H. sapiens*—plus the massive facial specimen H3, with traits of *H. sapiens* (Bermúdez de Castro and Arsuaga, 2000–01).

The consideration of *H. antecessor* as a species with a long temporal range was supported with the attribution of the specimens from Sima del Elefante, of an older age (Carbonell et al., 2008; Bermúdez de Castro, Martinón-Torres, Robles et al., 2010). If the younger Ceprano specimen is also added to the hypodigm (feasible to do, as argued by Bermúdez de Castro and

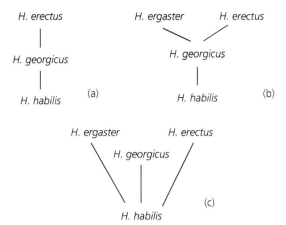

Figure 8.58 Alternative hypotheses in the evolutionary transition from *Homo habilis* to *Homo erectus*. Explanation in the text.

Arsuaga 2000–01), *Homo antecessor* will occupy a key place in human evolution. In phylogenetic terms, and in accordance with the identified apomorphies, Bermúdez de Castro, Arsuaga et al. (1997) argued that its affinities with *Homo ergaster* are considerable, which suggests that *H. antecessor* could be the last common ancestor (LCA) of the Neanderthals and modern humans, besides being the oldest protagonist of the occupation of Europe (Figure 8.58). This idea has been continuously maintained by the team members of Atapuerca.

A considerable problem in evaluating the phylogenetic relationships of *Homo antecessor* is the immature nature of most of the species' hypodigm. Bermúdez de Castro and Arsuaga (2000–01) cited the work of Manzi et al. (2001), in which the Ceprano specimen is described, favoring that it belongs to *H. antecessor*. But Manzi et al. (2001) stated in their paper that the specimen cannot be compared with any of the specimens of the Gran Dolina due to the lack of common anatomical elements, and also the frontal TD6-15 cannot serve for comparison purposes because it is a juvenile.

Box 8.35 Chronospecies in the transition to the Upper Pleistocene

In the diagram in Figure 8.58, two species, *H. heidelbergensis* and *H. rhodesiensis*, which we will examine in Chapter 9, appear interspersed between *H. antecessor* and their descendants *H. neanderthalensis* and *H. sapiens*. In the consideration made by the proponents of *H. antecessor*, these are chronospecies which do not disqualify the character of *H. antecessor* as LCA.

There are some inherent difficulties in the consideration of *H. antecessor* as LCA of Neanderthals and modern humans. First is the affinity of *Homo antecessor* with respect to *Homo erectus* sensu stricto. Emiliano Aguirre (2000) maintained that the presence of traits both from dental and facial morphology link the Gran Dolina sample with Asian specimens of *Homo erectus*. In the same way, the gracility of the ATD6-96 mandible, in addition to the reinterpretation of midfacial traits of the complete specimen, are arguments that led Eudald Carbonell and collaborators (2005) to relate the Gran Dolina TD6-level hominins with *Homo erectus* sensu stricto from China—such as the specimen Nanjing 1 from Hulu Cave (Tangshan Hill, East Central China). Thus, the origin of *H. antecessor* would be Asia. However, for *H. antecessor* to be considered the ancestor of *Homo sapiens*, which appeared in Africa (see Chapter 9), African specimens of this taxon should exist. Up to now, there are none.

Despite the evidence that we have mentioned about a possible Asian origin, the hypothesis that the migratory wave bringing the Gran Dolina hominins came from Africa is not far-fetched. If their African origin is confirmed, the original population could also be the protagonist in the subsequent evolution toward *Homo sapiens* and, if correspondent specimens could be found, they should be included in *H. antecessor* because of taxonomic priority rules. In that way there would have been (1) a transition African *H. antecessor*—*H. sapiens*—and (2) a transition European *H. antecessor*—*H. neanderthalensis*. This evolutionary model is consistent with that of Bermúdez de Castro and Arsuaga (2000–01).

8.7.6 The evolutionary transition indicated by the Atapuerca evidence

To consider that *H. antecessor* is a direct ancestor of *H. neanderthalensis* requires us to sustain an evolutionary continuity whose best indication would be an eventual transition in situ Gran Dolina–Sima de los Huesos–Neanderthals. But such an evolution is not necessarily inferred from the geographical proximity of the two sites. The continuity or discontinuity of the series Sima de los Huesos–Neanderthals will be addressed in Chapter 9. With respect to how the populations of Gran Dolina (GD) and Sima de los Huesos (SH) could relate, three main alternatives are as follows:

(1) Simple continuity: GD becomes SH in situ.
(2) Incorporation: SH substituted a GD but, in the process a hybridization episode occurred so part of GD genome is incorporated in SH.

(3) Replacement: SH for an extinct GD.

Only the first two interpretations are consistent with the hypothesis of *Homo antecessor* as ancestor of the Neanderthals, but how can we assess its degree of confidence? To some extent, the study of paleoclimatic alternatives has allowed us to define the migratory episodes of the European occupation. Can we apply the same techniques to Atapuerca to establish which is the most likely GD–SH relationship?

The age attributed to the fossils from Trinchera de Atapuerca is within the reverse Matuyama chron: above Jaramillo subchron for the Gran Dolina and between Jaramillo and Cobb subchrons in the case of Sima del Elefante. That means TD6 specimens would likely be related to the MIS 21 interglacial that, as we have seen, was a time of great importance for the occupation of Europe (see Muttoni et al., 2011). Regarding the E9 level of Sima del Elefante—and the specimen ATE9-1—this corresponds to the late glacial MIS 36 or early interglacial MIS 35. Both populations are

separated, therefore, by seven glacial episodes and their corresponding interglacials.

Although the southern latitude of Atapuerca is an argument in favor of this locality as a refuge during glacial conditions, these cycles are sufficient to accept as probable a model of colonization–extinction–recolonization, involving the disappearance of the population during episodes of extreme cold. In fact, Robin Dennell, María Martinón-Torres, and José M. Bermúdez de Castro (2011) argued that the evidence shows that occupation of Orce and Atapuerca took place in periods of warm climate, and could indicate that the Iberian Peninsula was abandoned during cold periods.

Morphological studies support the hypothesis of abandonment. The mandible ATE9-1 was provisionally placed as part of the *Homo antecessor* hypodigm in the initial description of the specimen (Carbonell et al., 2008). Such a taxonomic diagnosis speaks in favor of a population continuity between Sima del Elefante and Gran Dolina, and remained so after the initial study

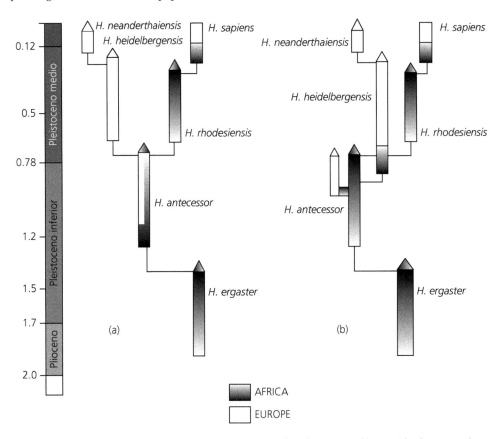

Figure 8.59 Phylogenetic role of *Homo antecessor* according to two alternative hypothesis proposed by Bermúdez de Castro and Arsuaga (2000–01). The original drawing uses the Spanish terms "Plioceno, Pleistoceno inferior y Pleistoceno medio" to refer to Pliocene, Early Pleistocene, and Medium Pleistocene. Bermúdez de Castro, J. M. & Arsuaga, J. L. (2000–2001). 1997-2001 El estatus de *Homo antecessor*. *Zephyrus*, 53–54, 5–14.

of the specimen (Bermúdez de Castro, Martinón-Torres, Prado et al., 2010). However, the subsequent comparative study of ATE9-1 (Bermúdez de Castro et al., 2011) led to removing the fossil from *H. antecessor*, and leaving it as belonging to *Homo* sp.; a decision that considers the specimens from Sima del Elefante and Gran Dolina as members of two distinct migratory processes.

The same argument for a model of colonization–extinction–recolonization is applicable in favor of the discontinuity between the population of Gran Dolina (GD) and Sima de los Huesos (SH).

An ancestor–descendant relationship between GD and SH was suggested when the taxon *H. antecessor* was first proposed (Bermúdez de Castro, Rosas et al., 1997). This remained as the preferred hypothesis in subsequent studies (Arsuaga et al., 1999; Bermúdez de

Castro, Rosas, & Nicolás, 1999; Carretero et al., 1999; Lorenzo et al., 1999; Bermúdez de Castro & Arsuaga, 2000–01). However, the comparative analysis of dental materials from Gran Dolina and Sima de los Huesos conducted by Bermúdez de Castro and collaborators (2003) suggested that the hominins of these two Atapuerca samples belong to two different paleospecies. The authors suggested as most likely an African origin for the specimens of Sima de los Huesos by a late migratory wave. It is, therefore, another episode within the dispersal that resulted in the European hominins of c. 0.5 Ma. In this way, the hypothesis of the "short chronology" is supported.

Although Bermúdez de Castro et al. (2003) stated that some genetic contact with the previous inhabitants of Europe was possible, in accordance with the hypothesis of the incorporation (2) indicated previously, the

Figure 8.60 *Homo antecessor* as terminal lineage in the occupation of Europe (Finlayson, 2005). Reprinted from Finlayson, C. (2005). Biogeography and evolution of the genus *Homo*. *Trends in Ecology & Evolution*, 20(8), 457–463, with permission from Elsevier.

truth is that the dental comparison analysis between GD and SH done by these authors has been taken as an argument against the ancestor–descendant relationship between the hominins from Gran Dolina and Sima de los Huesos (Finlayson, 2005; Mounier et al., 2009).

In the opinion of Giorgio Manzi (2011), the large glaciation MIS 16 (Günz), which covered the territories of Eastern Europe above 50° latitude under ice, was the episode that led to the disappearance of the first inhabitants of the continent. A new wave of migration, with tools of Mode 2, would have replaced them. Actually, the idea of the partial genetic absorption maintained by Bermúdez de Castro et al. (2003) and Carbonell et al. (2005) also gives the phylogenetic protagonist's role to the hominins of this new migratory wave, leaving the specimens of the Gran Dolina as members of a terminal lineage in the cul-de-sac of the Iberian Peninsula (Figure 8.59).

Nonetheless, *Homo antecessor* could still keep its role as LCA of Neanderthals and modern humans if a new hominin migratory wave that occurred after the end of the MIS 16—c. 620,000 years—during the interglacial Günz–Mindel would belong to this taxon. Up to now there are no specimens supporting that hypothesis.

CHAPTER 9

Hominin transition to Upper Pliocene

The Middle Pleistocene began 780,000 years ago, a time when the latest evolutionary events of our lineage took place, events that were complete when modern humans became the prevalent species, approximately 30,000 years ago. This range of time includes the end of the Mindel glaciation (750,000 years), the next two glaciations—Riss and Würm—and their correspondent interglacial periods. These climatic changes, although global, did not affect equally Africa, Asia, and Europe. Thus, evolutionary events had distinctive features in each continent. At the end of the Upper Pleistocene the situation was, however, shared by the whole ancient world: *Homo sapiens* was the only surviving species of its kind. The transition to our species from the ancestors of the Middle Pleistocene has been traditionally explained by two opposing hypotheses (Box 9.1). The first is known as the "Multiregional Evolution Hypothesis" or "Hybridization Hypothesis" (Multiregional Hypothesis or MH onward; Figure 9.1, left); inspired by the work of Franz Weidenreich (1943), it suggests that evolutionary changes happened contemporaneously in different regions of the world. Weidenreich's study (1943) was focused, in particular, on the Zhoukoudian specimens and their possible evolution to *Homo sapiens*, but the most solid evidence in favor of a multiregional model are the common traits detected in the last Javanese *Homo erectus*, such as those of Ngandong, and the first modern humans of Australia (Antón et al., 2011). As Milford Wolpoff, John Hawks, and Rachel Caspari (2000) argued, "multiregional" does not mean independent multiple origins, or an ancient divergence of modern human populations, or the simultaneous appearance of adaptive traits in various regions, or parallel evolution. The underlying assumption for MH is the existence of a global network of genetic exchange between evolving human populations in continuous contact (Wolpoff et al., 2000), so that the unity of the species would be preserved without divergence.

The alternative perspective, known as the "Out of Africa Hypothesis" or "Replacement Hypothesis" (Replacement Hypothesis or RH onward; Figure 9.1, right), is based, first, on molecular studies (Cann et al., 1987; Vigilant et al., 1991). It suggests that the transition from *H. erectus* (or *H. ergaster*) to *H. sapiens* occurred in a fairly localized population in East Africa. *H. erectus* grade Asian hominins would not have contributed genetically to the appearance of *H. sapiens*. Asian *H. erectus* sensu stricto remained relatively unchanged in that continent, until they were displaced by modern humans, or simply disappeared, leaving their territories to be occupied by *Homo sapiens*. A different species, *H. neanderthalensis*, inhabited Europe and the Near East. Eventually, it was also displaced, or substituted, by anatomically modern humans.

As stated earlier, the beginning of the transitional process to *Homo sapiens* is related to various species in Africa, Asia, and Europe. In addition to being relatively contemporary, all these taxa exhibit certain similar morphological features. In comparison with *Homo erectus*, their crania show:

- Greater capacity, in most instances.
- Higher cranial vault.
- Expansion of the parietal region.
- Reduction in prognathism, the frontal projection of the face.

There are, of course, differences among the diverse aforementioned taxa. Moreover, given that many of the fossils were found a long time ago and in places difficult to date with the techniques available at the time, there is a fair amount of doubt regarding their age. These circumstances account for their assignation to different species at the time of their discovery. For instance, European remains were assigned to *H. heidelbergensis* (Schoetensack, 1908), *H. erectus petraloniensis* (Murrill, 1975), and *H. swanscombensis* (Kennard, 1942). During the same period we find *H. rhodesiensis* (Woodward, 1921) and *H. helmei* in Africa, and *Homo (Javanthropus) soloensis* (Oppennoorth, 1932) in Asia, among other taxonomical proposals. But, as it became apparent that they shared certain common characteristics, more parsimonious solutions were suggested. The first,

Processes in Human Evolution. Francisco J. Ayala and Camilo J. Cela-Conde, Oxford University Press (2017).
© Francisco J. Ayala and Camilo J. Cela-Conde.
DOI 10.1093/acprof:oso/9780198739906.001.0001

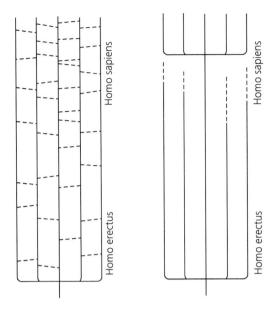

Figure 9.1 Left: Representation of the Multiregional Hypothesis, with contemporaneous *H. erectus–H. sapiens'* transitions in different continents. There were continuous hybridizations that preserved the species' unity. Right: Representation of the Replacement Hypothesis. The transition took place in Africa, beginning with a small population. Populations in other continents were eventually replaced, without hybridization.

Box 9.1. Authors that promote either MH or RH

Emiliano Aguirre, Eric Trinkaus, and Milford Wolpoff, among other authors, have advocated MH. Christopher Stringer, Ian Tattersall, and Bernard Wood, for instance, are supporters of RH. Evaluation of the two alternative hypotheses encompasses a huge number of studies. We will refer to some of them, only when they contribute to clarifying some aspects of the interpretations summarized in this book. In Chapter 10 we will examine molecular evidence in favor of each hypothesis.

historically speaking, was to place them in *H. erectus*, noting that each case was an evolved form of *H. erectus*, but not different enough to justify a new species. This taxonomic solution is part of the "single species" (*Homo erectus*) proposal put forward by Milford Wolpoff (1971a). However, Chris Stringer (Stringer, 1984, 1985; Stringer, Hublin, & Vandermeersch, 1984) and other authors opposed the existence of *H. erectus* in Europe.

The different specimens from the early Late Pleistocene were considered as predecessors of *H. sapiens*, without further taxonomic detail. Thus, given the lack of such specification, they were informally designated as "archaic" *Homo sapiens* (Stringer, 1985; Bräuer, 1989).

Because the proposal of archaic humans was initially related with human evolution in Europe, we will begin in this continent.

9.1 European archaic *Homo sapiens*

9.1.1 Mauer mandible

The Mauer mandible was the first specimen from the Mindel–Riss interglacial period to be discovered. It is also the earliest specimen of those traditionally considered as archaic *H. sapiens*. It was discovered in 1907 by a worker at the Grafenrain gravel pit, close to Mauer (Germany). It is a very complete specimen, though the left premolars and the first molar are broken (Figure 9.2).

The Grafenrain gravel pit is part of the "Mauer Sands," fluvial sedimentary terrains with abundant sandstone and gravel in two stratigraphic horizons: "upper sands" and "lower sands." A photograph taken at the time shows the place where the fossil was found, within the "lower sands," at a depth of about 25 m

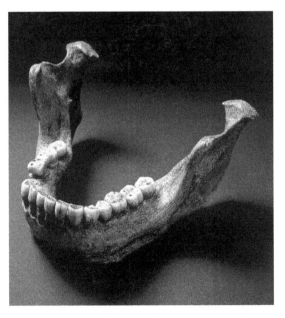

Figure 9.2 Mauer mandible, *H. heidelbergensis* (photograph from Johanson & Edgar, 1996).

Box 9.2 Searching for the ancestors

The Mauer discovery parallels the attempts to find human ancestors in the late nineteenth century and early twentieth century by followers of Darwin. A relevant effort was initiated by Otto Schoetensack, a paleontologist from Heidelberg (Germany), who directed his attention to the Grafenrain gravel pit after the discovery there in 1887 of *Elephas antiquus* fossils. The coexistence of primitive humans and *E. antiquus*—known due to the discoveries of Weimar-Taubach—led Schoetensack to teach Grafenrain's workers how to differentiate between human and ape bones, and to identify possible hominin fossils (Wagner et al., 2011). The Grafenrain gravel pit had provided a large vertebrate fossil fauna, described by Heinrich Georg Bronn (1830). In the monograph on the Mauer discovery, Schoetensack (1908) stated that the presence of *Elephas* in Granfenrain had been supported by Achilles Andreae (1884), although with some reservations about this identification. A few years before the work of Schoetensack, Hugo Möller (1900) published a paper on the emergence of humans in Europe, relating it to the "antediluvian" hunting of elephants and rhinos.

below the surface and 4.65 m below the stratigraphic level of Lettenbanck clay deposits (Wagner et al., 2011).

The Mauer mandible was found fragmented in two pieces by the symphysis and was embedded in a limestone matrix. After its cleaning and reconstruction, Schoetensack conducted a rigorous comparative analysis, which included X-rays, to observe the roots of the molars. Several features attracted his attention: the absence of a chin, the great robustness, and, in particular, the width of the ramus antero-posterior direction—traits indicating the specimen's primitiveness, in contrast to modern molars, and the canines' reduced size. Schoetensack attributed the fossil to the genus *Homo* and considered it more primitive than the newly discovered Neanderthals: a "pre-Neanderthaloid." Accordingly, he attributed it to the new species *Homo heidelbergensis*.

The age of the Mauer mandible was estimated repeatedly over the decades, usually taking as a guide the fauna found in the "lower sands," according to the taphonomic considerations of the site (Box 9.3). The origin of the sediments is attributed to the Neckar River, today far from Mauer, but its old meanderings would have reached the locality. A quiet stream would have created deposits with ideal conditions for fossilization; hence, the excellent state of the fauna fossils in both stratigraphic horizons. Because there is no direct dating, the fossil's age has been estimated from the fauna present at the site, which suggests an adaptation to mild temperatures. The fossil fauna of the Mauer's sands include 78% large herbivores (*Elephas*, *Rhinoceros*, *Bison*, *Alces*, and *Equus*), plus 2–3% bears (*Ursus*), beavers (*Castor* and *Trogontherium*), and some felines (*Leo* and *Homotherium*), beside a wolf species (*Canis mosbachensis*), another of hyena (*Hyaena arvernensis*), and animals of aquatic habitat such as *Hippopotamus* and other microfauna. The assemblage denotes a forest biotope with an interglacial climate (Howell, 1960). The question remains where to place this warm interval. Combining two different radiometric methods, uranium series and electron spin resonance, Wagner et al. (2010; see Box 9.3) obtained an age of 609,000±40,000 years, i.e., in MIS 15 (interglacial Günz–Mindel). The fossil is usually placed (Day, 1986) at the end of this interglacial.

Box 9.3 The age of the Mauer sands

Wolfgang Soergel (1933) stated that the Mauer sands originated before the Mindel glaciation, probably during the Günz–Mindel interglacial. However, it has also been proposed that the deposits correspond to a short, warm period in the Mindel glaciation (Zeuner, 1945). Due to the difficulties to relate, in geological terms, the Alpine sequence with Mauer's sediments, F. Clark Howell (1960) argued that the best option is to consider whether the Mauer fauna belongs to the Cromerian stage—which is coincident with the Alpine Günz–Mindel interglacial—or later. Consideration of the various loess depositions located above the Mauer sands led Howell (1960) to accept as probable that the fauna predates the Saale glaciation (equivalent to the Alpine Riss). This means it would be older than 350,000 years. In regard to the maximum age that would be attributed to the Mauer sands, paleomagnetic studies place the two beds and filling deposits in a normal magnetic episode, indicating that the sands correspond to the Brunhes chron and, thus, previous to the Matuyama boundary (Wagner et al., 1997). This would imply a maximum age of 780,000 years. A more precise dating of the fossil has been provided by Günther Wagner et al. (2010). According to these authors, the fauna of the "lower sands" of Mauer, and particularly the large herbivores, is similar to that of Isernia La Pineta (Italy), located in MIS 15—621,000–563,000 years by the $^{40}Ar/^{39}Ar$ method (Coltorti et al., 2005). But the faunal assemblage may also correspond to the Cromerian III, characteristic of MIS 13—533,000–478,000 years.

There are a total of 36 artifacts related to the location of the Mauer mandible. While their anthropic origin was questioned for decades, in particular because of their small size, Manfred Löscher et al. (2007) have argued that these are real tools. Some of the utensils retain a very sharp edge, disallowing the idea that they were carried by the river current; they might have been manufactured or used on the sandbar from which the fossils came.

The 600,000 years old Mauer mandible is the oldest specimen pointing to a permanent hominin occupation of Europe; an age even greater than the one established by the "short chronology" (Roebroeks & Van Kolfschoten, 1994). These figures fit well with the

model by Giorgio Manzi (2011) of a great migratory wave after the MIS16 glacial episode as being responsible for the colonization of the European continent (see Chapter 8), of which the Mauer fossil would be an example.

Table 9.1 shows the available fossils that give evidence of the process. Let's consider some of them.

9.1.2 Stenheim

In July 1933 a fossil cranium appeared in the Sigrist gravel pit, Steinheim (Germany, 20 km north of Stuttgart). It was extracted by Berckhemer Fritz, professor at Stuttgart Naturaliensammlung Württermbergischen and the expert who first described the specimen.

The Sigrist sediments come from the Murr River—a tributary of the Neckar River—and contain numerous faunal fossils. Berckhemer (1938) identified the age of the Sigrist deposits as the late Saalian glaciation (corresponding to the Alpine Riss). Raymond Vaufrey (1931) considered them to be younger, from the Saalian–Weichselian interglacial (Riss–Würm). F. Clark Howell (1960) argued that it was the Holstein interglacial (Elsterian–Saalian, i.e., Mindel–Riss), contemporary in Central Europe with the Hoxnian. The Sigrist sediments are currently attributed quite unanimously to the Mindel–Riss interglacial period. This is MIS 11 and has an estimated age of c. 400,000 years, much older than the 250,000 years figure previously attributed to the Mindel–Riss interglacial and its correlates in central and northern Europe.

The Steinheim cranium (Figure 9.3) includes the main part of the face, the upper molars, and a premolar, but the left and basal sides are very deteriorated, which has been occasionally attributed to cannibalism (Adam, 1988), although it is likely that they are distortions resulting from the sedimentary process—which also contributed numerous mineral intrusions (Haidle & Pawlik, 2010). Hermann Prossinger and collaborators (2003), using a CT scan, accomplished a virtual removal of the interior cranial encrustations, reaching the conclusion that the first examinations by Berckhemer deformed the eye socket, as well as the nasal cavity. The description of the Steinheim cranium by Hans Weinert (1936) may have been thus affected by those circumstances. The similarities of the Steinheim specimen are divided between those that refer to *Homo erectus*, such as a low cranial capacity, and those that approach the Neanderthals, such as a very developed supraorbital torus and a broad nose. Weinert (1936) described the cranium as narrow, elongated, and gracile, attributing it to a female.

Table 9.1 Specimens of European hominins between MIS 15 and MIS 7

Country	Locality	Age (Ma)	Specimens	Reference
Germany	Mauer	0.61	mandible	Wagner et al. (2010)
Spain	Sima de los Huesos (Atapuerca)	0.6	fossils	Bischoff et al. (2007)
Italy	Ceprano	0.45	cranium	Muttoni et al. (2010)
Italy	Castel di Guido	c. 0.43	partial skeleton	Mallegni et al. (1983)
England	Swanscombe	0.4	cranium	Abbate & Sagri (2011)
Germany	Bilzingsleben	0.4	cranial fragments	Abbate & Sagri (2011)
France	Caune de l'Arago	MIS 14, 13 and 12	fossils	Barsky & de Lumley (2010)
Italy	Notarchirico	MIS 13/10	femoral fragment	Manzi (2011)
Hungary	Vértesszöllös	c. 0.35 MIS 11/9	occipital fragment	Despriée et al. (2011)
Germany	Bad Canstatt	MIS 11/7	canine crown	Haidle & Pawlik (2010)
Greece	Petralona	>0.35	cranium	Stringer (1983)
Germany	Steinheim	0.3/0.25	cranium	Haidle & Pawlik (2010)
Germany	Dorn-Dürkheim 3	0.295/0.17	molar crown	Haidle & Pawlik (2010)
Germany	Reiningen	0.25/0.115	3 cranial fragments	Haidle & Pawlik (2010)
England	Boxgrove	0.24	tibia	Abbate & Sagri (2011)

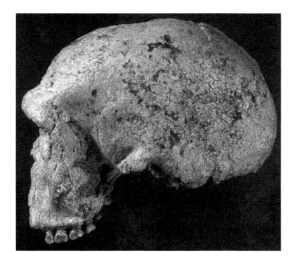

Figure 9.3 Steinheim cranium, *H. steinheimensis* (photograph from Johanson & Edgar, 1996).

However, Prossinger et al. (2003) disagreed with these observations, arguing that when the distortions are ignored the cranium shows a wide and massive appearance, with a broad face. Weinert (1936) estimated its cranial capacity at 1,070 cc, while F. Clark Howell (1960) calculated that between 1,150 and 1,175 cc would be more correct. The virtual reconstruction of Prossinger et al. (2003) reported 1,140 cc as the most probable value within the 1,200–1,100 cc range.

Fritz Berckhemer (1936) created a new species for the Steinheim specimen, *H. steinheimensis*. In his reorganization of hominin taxonomy, Bernard Campbell (1964) lowered the specimen's category to subspecies, *H. sapiens steinheimensis*.

9.1.3 Swanscombe

The Swanscombe cranium was found in successive stages. In June 1935, Alvan Marston—a professional dentist and amateur paleontologist—discovered in Barnfield Pit (Swanscombe, Kent, England), in the Thames Valley, a complete occipital. In March of the following year, Marston found in the same locality a left parietal and, improbable as it may seem, 20 years later John Wymer (1955) found the right parietal of the same individual within 30 m.

The fluvial deposits of Barnfield pit belong to the top of the Boyn Hill Terrace and have also supplied, besides the cranium, numerous tools—handaxes, flakes, and debris, but fewer cores—characteristic of Acheulean industry (Mode 2) (Howell, 1960). Wymer (1958)

has identified, in the same place as the skull, evidence of the use of fire: red and cracked flints and pieces of charred wood.

Barnfield fauna—*Elephas, Rhinoceros, Hippopotamus, Cervus, Equus,* and *Trogontherium*—is characteristic of a temperate forest environment, i.e., of an interglacial, probably the Hoxnian (Mindel–Riss) (Howell, 1960). Thus, the fauna belongs to MIS 11, as does the Steinheim fossil. Actually, there is a great similarity between the crania from Steinheim and Swanscombe, which has been pointed out numerous times. The small teeth, the higher placement of the maximum cranial width, and the shape and thickness of the cranial vault are the traits that bring these two specimens together, according to Michael Day (1986). The biggest difference is in the size of the specimens. Milford Wolpoff's comparative study (1980) stated that in all cranial measurements the Steinheim skull is smaller, in particular those from the occipital and both parietal bones. With respect to volume, Wolpoff estimated a 40% increase for the Swanscombe cranium.

The Swanscombe specimens were placed in a new species, *H. swanscombensis* (Kennard, 1942). A committee in charge of studying the first two bones, led by Le Gros Clark, suggested its classification as *Homo* cf. *sapiens* (Clark, 1938). However, Wolpoff (1971b) argued that the Swanscombe specimen's modern human-like traits are explained by the fact that it is a female exemplar. According to Wolpoff, the differences among the Swanscombe, Vértesszöllös (Hungary), Petralona (Greece), Steinheim, and Bilzingsleben (Germany) specimens are sexual dimorphisms, so that all of these specimens belong to *H. erectus.* Other authors have related the Swanscombe cranium to Neanderthals (Howell, 1960). After examining these interpretations, Day (1986) concluded that the Swanscombe specimens belong to a female transitional between *H. erectus* and *H. sapiens,* which could be placed at the base of the lateral branch leading to European Neanderthals.

9.1.4 Petralona

Few discoveries related to "archaic" humans have created more doubt and controversy than the Petralona cranium (Box 9.4).

The Petralona cave is located in the northwestern part of the Chalkidiki peninsula, south of Thessaloniki (Greece). According to Aris Poulianos (1981), it was discovered in 1959 by peasants who were looking for a spring on the slopes of the mountain. One year later, a fairly complete and very well preserved cranium was

Box 9.4 The discovery of the Petralona fossil

The discovery of the Petralona cave and the research on its fossil appear on the webpage of the Anthropological Association of Greece (http://www.petralona-cave.gr/—accessed June 10, 2016).

discovered, missing only incisors, the right zygomatic arch, and, probably, the mastoid process (Stringer & Melentis Howell, 1979).

The first description of the fossil was made the same year of its discovery (Kokkoros & Kanellis, 1960). The systematic excavation of the cave began in 1968. Aris Poulianos (1981) identified about 27 stratigraphic levels within a sedimentary deposit of 15 m, with signs of human occupation in all levels. According to Poulianos, the parts that the specimen lacked were destroyed during the previous work performed in search of the mandible and the rest of the skeleton.

The age of the Petralona fossil is also a matter of debate. It has been dated by all kinds of techniques except radiometry: faunal comparison, electron spin resonance (ESR), thermoluminescence, and paleomagnetism, with inaccurate results (for a critique on methods of direct radiation used in the cave, see Wintle & Jacobs, 1982). In any case, the Petralona cranium must be considered a discovery "on the surface," impossible to relate to any stratigraphic horizon and, therefore, to any faunal group. As indicated by Christopher Stringer, F. Clark Howell, and John Melentis (1979), it seems unlikely that the fossil would be younger than the last fauna occupying the cave, so that the minimum attributable age is c. 300,000 years old.

The Petralona cranium is of a significant size and displays traits considered to be related to both Neanderthals (Kokkoros & Kanellis, 1960) and the most primitive *Homo sapiens* (Stringer et al., 1979), and even to the advanced *erectus* (Hemmer, 1972). The Petralona specimen was one of those used by Stringer to propose the archaic *Homo sapiens*.

Other examples of ancient hominins from Greece are those from Apidima, Mani peninsula (southern Peloponnese). In 1978, two partial crania, Apidima 1 and 2, were found embedded in the sedimentary breccia of Cave A (one of the four forming the site). Katerina Harvati, Eleni Panagopoulou, and Curtis Runnels (2009)

attributed them an age characteristic of the Middle Pleistocene because of their archaic appearance. After conducting a multivariate analysis, Katerina Harvati, Christopher Stringer, and Panagiotis Karkanas (2011) concluded that Apidima 2 shows more affinities with *Homo neanderthalensis* than with *H. heidelbergensis*, indicating, based on their morphology, a probable age between 400,000 and 150,000 years old.

The specimens from Mauer, Steinheim, Swanscombe, and Petralona are the "classic" set that support the idea of a permanent hominin presence in Europe estimated, in accordance with the hypothesis of the "short chronology," at around 500,000 years. Subsequently, other more fragmentary fossils were added that contributed more evidence about the human presence on the continent.

9.1.5 Bilzingsleben

Steinrine, located 1 km south of Bilzingsleben, Artern district—in the region of Halle/Saale of what once was the German Democratic Republic—has provided hominin remains since 1969, when research by the Landesmuseum für Vorgeschichte Halle/Saale began under the supervision of Dietrich Mania. Bilzingsleben belongs to the Thuringia basin and the site containing human remains is part of a complex of travertine limestone on the terraces of the Wipper River.

In 1974, the discovery of hominin remains, including an occipital, fragments of the parietal and frontal bones, and a molar attributed to one individual, along with numerous lithic artifacts and an extensive flora and fauna, including molluscs, was published (Grimm et al., 1974; Mania & Grimm, 1974). The morphology of the hominin fossils was described and compared to the Asian specimens—*Sinanthropus* III and Sangiran 17, the Africans—OH 9, and the Europeans—Mauer, Ehringsdorf, Arago, Swanscombe, Steinheim, Petralona, and Vértesszöllös by Emanuel Vlcek (1978; Vlcek & Mania, 1977), who named the new species *Homo erectus bilzingslebenensis* (Figure 9.4). The Bilzingsleben macromammal fauna indicates an interglacial period identified as Holstein (Mindel–Riss) (Vlcek, 1978).

In November 1999, a fragment of a right mandible, Bilzingsleben E7, was discovered, increasing to 28 the cranial specimens found. Geological and radiometric studies concluded that Bilzingsleben belongs to the episode MIS 11 (400,000–350,000 years) (Vlcek et al., 2000). With the available fossils, Vlcek rebuilt two crania, Bilzingsleben I and II, whose resemblance to OH 9, *Sinanthropus* III and *Pithecanthropus* VII, was described by Emanuel Vlcek, Dieter Mania, and Ursula Mania

Figure 9.4 Bilzingsleben E7 right mandible (Vlcek et al., 2000). Vlcek, E. & Mania, D. (1977). In neuer fund von Homo erectus in Europa: Bilzingsleben (DDR). *Anthropologie*, 15, 159–169. With permission of Springer.

(2000) as "striking." The taxonomic attribution made by Vlcek et al. (2000) was to *Homo erectus*.

9.1.6 Vértesszöllös

Henry-Victor Vallois (1965) reported the discovery of cranial remains of an ancient hominin of the Mindel glaciation in the Hungarian town of Vértesszöllös (50 km west of Budapest). The specimens were described by Andor Thoma (1966), who calculated a cranial capacity for Vsz. II of 1,350 cc (i.e., similar to the occipital average of modern humans). Thoma ascribed Vsz. I and II to *Homo paleohungaricus* as a subspecies of *Homo erectus* or *H. sapiens*.

The study of Vsz. II by Milford Wolpoff (1977) reduced its cranial volume with respect to the measurements taken from Sangiran 2, 4, 12, and 17, and Peking 3, 11, and 12, placing it a bit above the upper end of *Homo erectus* variability. Vsz. II would be in the range of 1,115–1,437 cc, depending on the measurements used in the extrapolation.

Miklós Kretzoi and László Vertés (1965) calculated the age of the Vértesszöllös specimens by studying their associated fauna. The macromammals indicated a forest-steppe environment characteristic of a warm episode followed by a cooling episode in the loess complex of the upper strata. The authors attribute the human occupation of Vértesszöllös to a warm inter-stage within the Mindel glaciation, i.e., about 400,000 years ago. The associated lithic industry, with more than 500 artifacts, abounds in choppers, flakes, and quartz and flint scrapers of Clactonian technique (Kretzoi & Vertés, 1965) (Box 9.5).

The specimens from the Mindel–Riss interglacial described so far provide strong evidence about the hominin European occupation. However, to interpret the character of the final settlement and the relationships of the hominins during the European Middle and Upper Pleistocene, the best available evidence is that emerging from the localities L'Arago (France) and Sima de los Huesos (Spain).

9.1.7 Caune de l'Arago

The so-called Tautavel Man corresponds to a series of discoveries in the Caune (cave) de l'Arago, near the village of Tautavel, 19 km northeast from Perpignan (France). The cave has three stratigraphic complexes (upper, middle, and lower), bounded by the upper and lower stalagmite levels (Falguères et al., 2004). In the middle stratigraphic complex are the units I, II, III, IV, and V, each one with various levels, from A to Q (Figure 9.5). Studying the associated fauna (micromammals), Henry and Marie-Antoinette de Lumley (1973) assigned the lower strata of Caune de l'Arago to the beginning of the Riss glaciation. Later, Henry de Lumley (1979) argued in favor of an older epoch for the cave, perhaps the Mindel glaciation, of approximately 400,000 years. De Lumley and collaborators (1984) pushed back the levels M–Q to 600,000 years (MIS 14), giving an age of 450,000 years to level G, where the hominin fossils were found. The ESR

Box 9.5 Clactonian industry

The Clactonian technique is a cultural development of Mode 2 that took place in Europe. See later in Section 11.3.8 the cultural change that developed after the continued occupation of the continent.

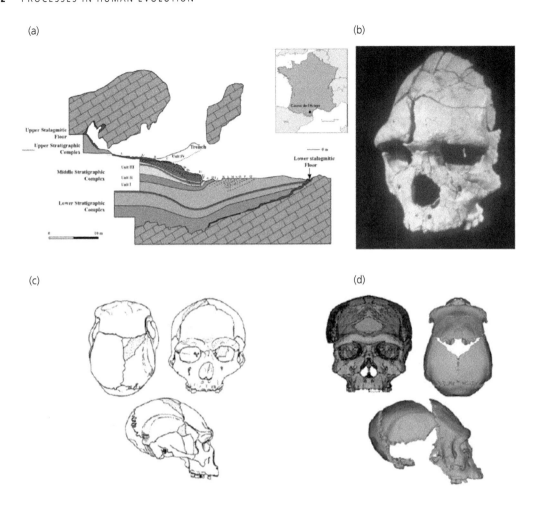

Figure 9.5 (a) Stratigraphy of Caune de l'Arago; (b) Arago XXI "Tautavel Man" (illustration by Falguères et al., 2004); (c) "Tautavel Man" reconstruction by Marie-Antoinette and Henry de Lumley in 1980; (d) digital reconstruction from X-ray images (illustrations by Guipert et al., 2004). Top left and right: Reprinted from Falguères, C., Yokoyama, Y., Shen, G., Bischoff, J. L., Ku, T.-L., & de Lumley, H. (2004). New U-series dates at the Caune de l'Arago, France. *Journal of Archaeological Science*, 31(7), 941–952, with permission from Elsevier. Bottom: Reproduced from Guipert, G., de Lumley, M.-A., & de Lumley, H. (2014). Restauration virtuelle d'Arago 21. *Comptes Rendus Palevol*, 13, 51–59, by Elsevier Masson SAS. All rights reserved.

and uranium series' studies by Fei Han (2008) on six teeth from level G confirmed the age of c. 450,000 years, although for some samples the age was 100,000 years less.

In 1969, the team directed by Henry de Lumley found in level G of unit III (middle stratigraphic complex) a mandible with five teeth, Arago II, belonging to a hominin, to which a similar mandible was later added, Arago XIII. In 1971, Henry and

Marie-Antoinette de Lumley (1971) discovered a partially deformed cranium (Arago XXI), with the face, and the zygomatic and maxillar bones. In 1980, an incomplete right parietal (Arago XLVII) appeared that fit well with the cranium. Both specimens together formed the exemplar (informally) called "Tautavel Man," which Marie-Antoinette and Henry de Lumley first reconstructed in 1981 using an occipital cast from a Swanscombe specimen and a temporal bone cast from Sangiran 17 (de Lumley et al., 1982). Later, Gaspart Guipert and collaborators (2004) made a new digital reconstruction.

In the various stratigraphic levels of the Tautavel, numerous lithic tools have been found covering a considerable period of time, 690,000–90,000 years, which allows for studying the occupation of the cave and the evolution of carving techniques, and the use of utensils from MIS 14–4 (Henry de Lumley & Barsky, 2004). Henry de Lumley and Deborah Barsky (2004) conducted a study on the different ages, MISs, climate, fauna, and occupations of the cave based on the various artifacts found in each unit. The P levels in the stratigraphic unit I, which are attributed to an occupation during a cold and dry phase of the MIS 14, contain numerous fossils of large herbivores, mainly of horses, reindeers, and bison. The hunting tools indicate a careful selection of raw materials and a high level of carving technology that belongs to Mode 2 (Barsky & Henry de Lumley, 2010). The issue of the transitional European cultures toward the Upper Paleolithic will be addressed in Section 9.4.

Once again we see that the morphology of the Tautavel specimens is intermediate, with traits of *erectus* and Neanderthals (de Lumley & de Lumley, 1973). Their resemblance to the Steinheim cranium is noteworthy, but with a larger capacity (1,100–1,200 cc). In contrast, the appearance of the cranial vault is more elongated and lower, as in the case of Asian *erectus*. The fragments of the femoral bone (Arago XLVIII, LI, and LIII) and left hip (Arago XLIV) are reminiscent of Olduvai OH 28 (Day, 1982). No wonder, then, that Michael Day (1986) argued that it was a transitional European form between *Homo erectus* and *Homo sapiens*.

9.1.8 Sima de los Huesos

Sima de los Huesos is a site on the Atapuerca Ridge (Spain), near the Gran Dolina and Sima del Elefante, mentioned in Chapter 8. It is a deep well with access through Cueva Mayor (at a distance of 500 m) or through Cueva del Silo (Figure 9.6). A complex system of passageways and caverns with various signs of human occupation, mainly from the Bronze Age, leads to a vertical shaft inside Cueva Mayor, which is 13 m deep and of difficult access, followed by a sloping tunnel of another 15 m.

A large number of fossils of hominins and other mammals have appeared in the Sima de los Huesos since late August 1976, when Trinidad Torres found a number of human specimens from the Middle Pleistocene. The sample included a mandible without the

ramus, two fragments of another mandible, 13 teeth, and two fragments of the parietal bone (Aguirre Basabe & Torres, 1976). The collection of specimens in Sima is very arduous work given the difficulties of access, which require the use of speleology techniques. Even so, the work of the team led by Emiliano Aguirre initially, and José Luis Arsuaga, José María Bermúdez de Castro, and Eudald Carbonell after the retirement of Aguirre in 1990, removed over a ton and a half of red clay with no defined stratigraphic planes, in which hominin remains were found—which were sometimes disturbed by hikers who gained access to the Sima. After the screening and selection work, fossils of at least 20 different individuals appeared, including specimens of all parts of the human skeleton (Arsuaga et al., 1997).

The dating of one of the Sima de los Huesos mandibles by gamma ray spectrometry provided an age of 300,000 years (Aguirre, 1995). Comparison with the large mammal fauna yielded an age between 525,000 and 340,000 years (Aguirre, 1995). The speleothem SRA-3 underlying the human bones (see Figure 9.7) was dated by James Bischoff et al. (2003) at more than 350,000 years using mass spectrometry (TIMS, see Box 9.6). A new analysis of six SRA-3 samples by a more precise technique (MC ICP-MS, see Box 9.6) increased the date to c. 600,000 years, with 530,000 years as the most conservative limit to minimize possible errors (Bischoff et al., 2007).

A huge number of hominin specimens come from the Sima de los Huesos. The postcranial remains discovered there account for 70% of those from the Middle Pleistocene found worldwide (Arsuaga, 1994). Furthermore, as they are hominins of a geographical, temporal, and probably even familial unity, it has been possible to conduct a paleodemographic study indicating size distribution, sexual dimorphism, and existing polymorphisms in a single population (Bermúdez de

---

Box 9.6 Mass spectrometry

Thermal-ionization mass-spectroscopy (TIMS) is an improvement of the uranium series' technique, which achieves a direct isotope analysis of high reliability in samples as small as a few centigrams. A further step has been achieved with the MC ICO-MS technique (multi-collector inductively coupled plasma mass-spectrometry), which uses an ionization source under standard atmospheric pressure: the argon or xenon plasma.

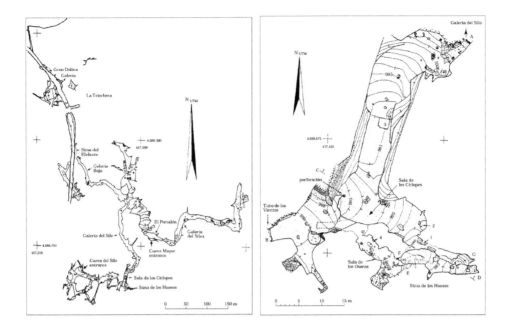

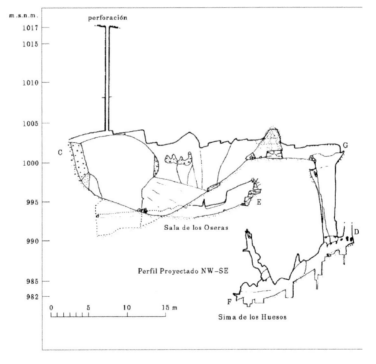

Figure 9.6 (a) General map of the Atapuerca Ridges sites. (b) Detailed map of the access to Sima de los Huesos through the Sala de los Cíclopes inside Cueva Mayor. Number 1 indicates accumulation of debris from the excavations of Trinidad Torres in 1976. Numbers 2, 3, and 4 correspond to trial pits. (c) The plane of Sima de los Huesos has been located at the level of Sala de los Cíclopes and Sala de las Oseras, although the site is actually below those chambers, as shown in the diagram. A hole drilled from the surface to facilitate removal of debris and specimens is shown in this diagram (drawings by G. E. Edelweiss compiled in Arsuaga et al., 1997). Reproduced from Arsuaga, J. L., Martínez, I., Gracia, A., Carretero, J. M., Lorenzo, C., García, N., & Ortega, A. I. (1997). Sima de los Huesos (Sierra de Atapuerca, Spain). The site. *J Hum Evol*, 33, 109–127, with permission from Elsevier.

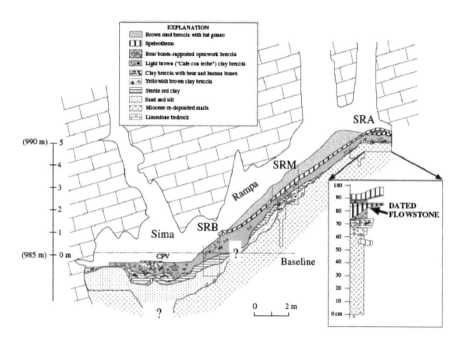

Figure 9.7 Placement of the speleothem SRA of unit 3 of Sima de los Huesos. Reprinted from Bischoff, J. L., Williams, R. W., Rosenbauer, R. J., Aramburu, A., Arsuaga, J. L., García, N., & Cuenca-Bescós, G. (2007). High-resolution U-series dates from the Sima de los Huesos hominids yields: implications for the evolution of the early Neanderthal lineage. *Journal of Archaeological Science*, 34(5), 763–770, with permission from Elsevier.

Castro, 1995). An example of the type of studies possible with the population of the Sima de los Huesos is the comparative analysis of the two segments of the dental arch: the anterior (incisors and canines) and posterior (premolars and molars). After this analysis, José María Bermúdez de Castro and María Elena Nicolás (1996) argued that the Asian *H. erectus* sample is easily distinguishable from that of the European Middle Pleistocene, which tends to group together with the African *H. erectus*.

In July 1992, three more crania were found in Sima de los Huesos: a calotte (cranium 4), an almost complete cranium (cranium 5; Figure 9.8, right), and a third fragmentary one from an infant (cranium 6) (Arsuaga et al., 1993). All these specimens have a large cranial capacity, although not of the same degree: 1,390 cc for cranium 4 and 1,125 cc for number 5.

A new fossil from the Sima de los Huesos was described in 2009, cranium 14 (Figure 9.9), attributed to an immature female of less than 18 years age, "Benjamina" (Gracia et al., 2009, 2010). After the reconstruction, cranium 14 showed some cranial deformities that were a strong indication that it had suffered a severe pathology, craniosynostosis simple:

a premature closure of the lamboid suture (Gracia et al., 2009, 2010). Ana Gracia and collaborators (2009, 2010), after analyzing the etiological factors involved in this discovery, advanced a hypothesis about the hominin living patterns in the temporal range of approximately 500,000 years.

The finding in Sima de los Huesos of a lumbar spinal column (Bonmatí et al., 2010) attributed to the

---

Box 9.7 Age at death of Sima de los Huesos sample

All the remains of Sima de los Huesos are of adult individuals close to youth, i.e., belonging to the population segment where less mortality would be expected. The explanation for the presence in such an inaccessible place of so many hominin remains is quite tentative. In an early taphonomic interpretation it was ventured that the remains could have been swept away by floods, but later some members of Atapuerca research team suggested that it could be the result of mortuary practices (Carbonell & Mosquera, 2006).

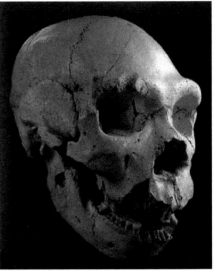

Figure 9.8 Left: Grid U-16 and T-16 (area B) in the excavation of Sima de los Huesos, campaign of July 1992; all the fossils that appeared were human with the exception of a bear rib fragment located at the top center. Right: Cranium 5 of Sima de los Huesos (Arsuaga et al., 1993). Reprinted from Arsuaga, J. L., Martínez, I., Gracia, A., Carretero, J. M., Lorenzo, C., García, N., & Ortega, A. I. (1997). Sima de los Huesos (Sierra de Atapuerca, Spain). The site. *J Hum Evol*, 33, 109–127, with permission from Elsevier.

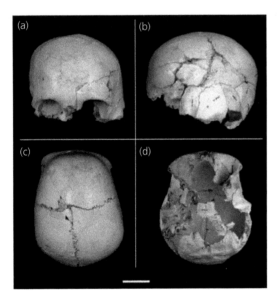

Figure 9.9 Cranium 14 from Sima de los Huesos, in frontal (a), lateral (b), superior (c), and inferior views (d); scale: 5 cm (illustration by Gracia et al., 2009). Gracia, A., Arsuaga, J. L., Martínez, I., Lorenzo, C., Carretero, J. M., Bermúdez de Castro, J. M., & Carbonell, E. (2009). Craniosynostosis in the Middle Pleistocene human Cranium 14 from the Sima de los Huesos, Atapuerca, Spain. *Proceedings of the National Academy of Sciences*, 106(16), 6573–6578.

same individual from which the complete male pelvis SH 1 (Arsuaga et al., 1999) had come, allowed the reconstruction of an exceptional and very informative lumbar–pelvic set for the hominin fossil record of the Middle Pleistocene (Figure 9.10). From the analysis of the whole set, Alejandro Bonmatí et al. (2010) deduced the presence in *Homo heidelbergensis*—the species to which the specimens from Sima de los Huesos belong—of the primitive pelvic pattern in the genus *Homo*, with sexual dimorphism and a pelvic canal resulting in difficult births.

The lumbar-pelvis set of SH 1 (Figure 9.10) indicates the presence of various degenerative diseases related to the advanced age of death of the individual (over 45 years). This is another indication, besides the aforementioned skull 14, pointing to the social assistance practices of *H. heidelbergensis*.

Another indication of behavior, but in the opposite sense of violent aggression, was deduced by Naomi Sala et al. (2015, p. 1) after the reconstruction of the cranium 17 of Sima de los Huesos. As the authors noted, the specimen "shows two clear perimortem depression fractures on the frontal bone, interpreted as being produced by two episodes of localized blunt force trauma. The type of injuries, their location, the strong similarity of the fractures in shape and size, and the different

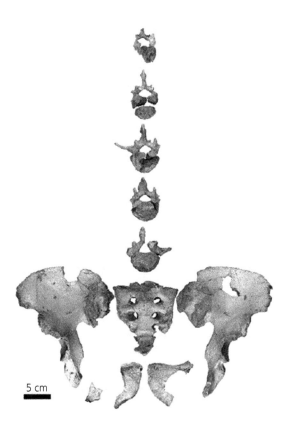

5 cm

Figure 9.10 Pelvis SH 1, after its reconstruction and adding the lumbar spine. Bonmatí, A., Gómez-Olivencia, A., Arsuaga, J.-L., Carretero, J. M., Gracia, A., Mart√ínez, I.,. . . Carbonell, E. (2010). Middle Pleistocene lower back and pelvis from an aged human individual from the Sima de los Huesos site, Spain. *Proceedings of the National Academy of Sciences*, 107, 18386–18391.

orientations and implied trajectories of the two fractures suggest they were produced with the same object in face-to-face interpersonal conflict. Given that either of the two traumatic events was likely lethal, the presence of multiple blows implies an intention to kill."

9.2 African archaic *Homo sapiens*

The transformation of the Middle Pleistocene hominins did not only take place in Europe. Authors supporting the multiregional hypothesis (MH), to which we have referred at the beginning of this chapter, point out the presence of fossils that give evidence of similar changes in both Africa and Asia. However, and as we will see later, it is not the same kind of evolution. If the evolution characteristic of Europe resulted in the emergence of *Homo neanderthalensis*—with pending doubts about the relationship of that taxon with *Homo sapiens*—a large majority of authors agree today that the origin of

modern humans took place on the African continent, where the process of change involved the transition in situ from *Homo erectus* to *Homo sapiens*. With regard to Asia, the most generally accepted position was that of a *Homo erectus* stasis that lasted approximately until 70,000 years ago, when it is replaced—or assimilated—by modern humans. But more recent studies call into question such a simple model.

The genetic evidence that we will examine in Chapters 10 and 11 supports the following scheme:

- The evolution of hominins to the forms that are characteristic of the Upper Pleistocene occurred differently in Europe, Africa, and Asia.
- *Homo neanderthalensis* and *Homo sapiens* are two different species.
- In the Middle and Upper Pleistocene, several migratory flows occurred from Africa to Asia and Europe.
- The emergence of modern humans occurred in Africa, spreading later to Asia and Europe.
- In the process of dispersal, several limited episodes of hybridization occurred between populations of *H. sapiens* and the local species.

Neither of the two models, MH and RH, is perfectly consistent with such a process of evolutionary change. Although the evidence available today supports the idea of a species of modern humans emerging in Africa and later occupying other continents, the ways in which the dispersion occurred includes some hybridization that was absent in the original Out of Africa model. However, the recovery of fossil human DNA and its comparison with current populations have revealed the relationships between the various human taxa in the Upper Pleistocene. We will return to these issues in Chapter 10.

9.3 African hominins from the Mindel–Riss interglacial period

The European specimens we have reviewed, which are only part of those available, highlight the rather ambiguous status of the hominins from the Mindel–Riss interglacial period. They share traits with *H. erectus*-grade hominins, Neanderthals, and even anatomically modern humans. But there are African remains of similar age and appearance that introduce additional complexities.

Michael Day (1973) suggested separating the Middle and Late Pleistocene African specimens in three different grades: early, intermediate, and modern. Following this proposal, Sally McBrearty and Alison Brooks

(2000), carried out an extensive classification of African specimens in each of these groups, while cautioning about the problems involved in making a classification using grades, and arguing that their proposal cannot be taken as supportive of anagenetic schemes.

Following the same criterion used by Chris Stringer to classify transitional European specimens, Günther Bräuer (1984) included in the "archaic *sapiens*" grade various African specimens from Ndutu, Broken Hill, Bodo, Hopefield, Eyasi, the Cave of Hearths, Rabat, and, with some hesitance, from Salé, Sidi Abderrahman, and Thomas. Although in the case of Europe the name "archaic *sapiens*" is inappropriate, because there the transition was to the Neanderthals—which we will review in Chapter 10—referring to African specimens using the term "archaic" is more pertinent. Nevertheless, we must bear in mind that such "archaic" fossils have little to do in phylogenetic terms with the European specimens.

To what extent is it necessary to define an intermediate grade or transitional species in Africa? G. Philip Rightmire (2009) summarized well the situation: at an age of about 0.7 Ma, African *H. erectus* showed changes that were indicative of the transition to *Homo sapiens*. The most conspicuous changes occurred in the face and cranium and, thus, specimens such as Bodo (Ethiopia), to which we will refer shortly, can be used as a reference to document the evolution toward:

- An outline of the temporal arch, proportions of the squama frontalis, and base of the skull—in part, at least—similar to the traits of modern humans.
- A cranial capacity intermediate between *H. erectus* and *H. sapiens*, close to 1,250 cc.
- A supraciliary arch divided in lateral and medial segments.
- A vertical nasal profile and the incisive canal open in the front part of the palate.

All these features are unique to *H. sapiens*. However, the "archaic" Africans retain plesiomorphies of *Homo erectus*:

- Massive facial skeleton.
- Projected front.
- Low frontal with a medial keel.
- Angular parietal torus.

The analysis by Rightmire (2009) included, as the best examples of "transitional" exemplars besides the Bodo fossil, other African specimens—Elandsfontein, to which we refer by the name of Saldanha (South

---

Box 9.8 The use of grades to explain the transition to *Homo sapiens*

Part of the taxonomic diversity that appears in the models of the transition to *Homo sapiens* is due to the habit of naming new taxa whenever new and different specimens emerge. However, the underlying question does not refer to that trend, but to the number of species necessary to explain the transition to our own species.

If no adequate answer is found, the possibility exists of accepting the multiplication of proposed species, grouping them in grades that reflect the different stages of the evolution to *H. sapiens*. That is the formula used both by Stringer (1985) for Europe, as well as by Day (1973), Bräuer (1984), and McBrearty and Brooks (2000) for Africa. Needless to say, the proposal of "archaic *sapiens*" was not generally accepted in its African version. For example, the Ndutu fossil was classified as a subspecies of *Homo erectus* by Ron Clarke (1976) after carrying out its reconstruction. Later, Clarke (1990) assigned the species to *Homo leakeyi*, arguing that *Homo erectus* originated in Asia and was confined there, and that it was not involved in the evolution to *Homo sapiens*.

Africa), Broken Hill (Zambia), and Salé (Morocco)—which show similar features, although, obviously, with some differences among them. In any case, contrasting with those closer to the morphology of modern humans are: Florisbad (South Africa), Laetoli (Tanzania), Omo I and 2, Herto and Aduma (Ethiopia), and Djebel Irhoud (Morocco). A further step is revealed in the specimens from Skhul and Qafzeh (Israel), and Klasies River (South Africa), which are essentially modern.

Rightmire's work (2009) indicates the presence of several hominin groups that manifest the evolutionary transition in Africa: from *H. erectus* to those specimens already considered *Homo sapiens*. In order not to prejudge the taxonomic considerations and phylogenetic relationships, we will simply group them separately in the following sets:

Grade I: Bodo, Elandsfontein, Broken Hill, Salé, and Eyasi.

Grade II: Florisbad, Laetoli, Omo, Herto, Aduma, and Djebel Irhoud.

Grade III: Skhul, Qafzeh, and Klasies River.

Next we will examine their characteristics.

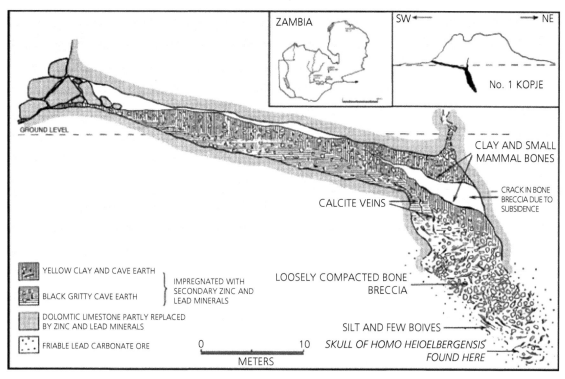

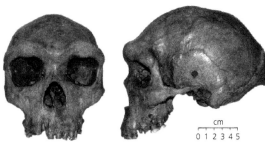

Figure 9.11 Up: No1 Kopje deposits indicating the probable place of origin of the Broken Hill cranium. Bottom: Broken Hill 1 cranium, Kabwe (Zambia). Up: Barham, L. S., Pinto Llona, A. C., & Stringer, C. (2002). Bone tools from Broken Hill (Kabwe) Cave, Zambia, and their evolutionary significance. *Before Farming*, 2, 1–12. Bottom: Rightmire, G. P. (1998). Human evolution in the Middle Pleistocene: the role of Homo heidelbergensis. *Evol. Anthrop.*, 6, 218–227.

9.3.1 African grade I

The locality of Broken Hill was discovered in 1906 during the course of mining work to obtain zinc, performed in a dolomitic outcrop, No1 Kopje, near the town of Kabwe (Zambia) (Figure 9.11, bottom and Box 9.9). The site was destroyed in the 1930s, when the deposit was transformed into an open quarry and later the well became filled with water (Barham et al., 2002). On June 17, 1921, the miners accessed, at the bottom of No1 Kopje, a deep cave in which Tom Zwigelaar and an assistant discovered a human skull. Other specimens, including a parietal bone, a maxillar, and several postcranial remains, which with the cranium corresponded to three or four individuals, appeared at the same time (Clark,

---

Box 9.9 Broken Hill?

The locality called Broken Hill in British colonial times is nowadays called Kabwe, a town near the site. The name of the country has also changed: Rhodesia under the British Crown, Zambia after its independence. The *H. rhodesiensis* taxon refers to the colonial name at the time when the fossil was discovered. As Philip Rightmire (2001) has said, the details of the discovery of the skull of Broken Hill are confusing and even contradictory. Rightmire said it is likely that the fossil was isolated when found, without a clear relationship to the other bones and lithic artifacts from the site.

1959; Clark et al., 1968). From the mine, in addition to an abundant fauna, came utensils belonging to the tradition of the African Middle Stone Age (MSA) called Still Bay.

The cranium of Broken Hill, one of the best-preserved exemplars of Middle Pleistocene hominins, was first examined and described by Arthur Smith Woodward (1921), who considered the specimen to be very modern, belonging in evolutionary terms to a more advanced grade than the Neanderthals, and it was assigned to the taxon *Homo rhodesiensis*.

The E691 Kabwe specimen is the only complete tibia available from the Middle Pleistocene. It was discovered along with the cranium Broken Hill 1 in 1921. In his detailed study, Ales Hrdlika (1930) considered that both specimens belong to the same individual. Erik Trinkaus (2009) conducted a comparative analysis between those specimens and similar fossils from the Lower and Upper Pleistocene, concluding that the hominin from Kabwe would be among those of larger size, exceeded only by the KNM-WT 15000, *Homo erectus*, Nariokotome (Kenya).

The first age attributed to the Kabwe fossils was around 40,000 years, i.e., from the Upper Pleistocene, although they were later considered as characteristic of the Middle Pleistocene, according to the fauna and lithic tools of the site (Klein, 1973). Michael Day (1986) argued that the fossils belonged to the final segment of the Middle Pleistocene. Sally MacBrearty and Alison Brooks (2000) gave them an age between 0.78 and 1.33 Ma, following studies comparing the fauna of the Kabwe site with that of Bed IV in Olduvai. However, Eric Trinkaus (2009) considered an age of at least 300,000 years to be more appropriate.

The specimens from Bodo d'ar, on the east bank of Awash River (Afar region of Ethiopia), belong to the same grade. In October 1976, Alemayehu Asfaw found the first fragment of what would later become most of a cranium associated with numerous Acheulean artifacts (Kalb et al., 1980) (Figure 9.12). Although collected on the surface, the remains were attributed to an upper unit (Upper Bodo Sand Unit, UBSU) of the member Bodo from the Wehaietu formation (Middle Awash, Ethiopia) (Kalb et al., 1980). The cranium was reconstructed from 26 initial fragments to which others were added later. The numerous fauna indicate a permanent lake or river environment, given the presence of *Hippopotamus* and *Crocodylus*, and large herbivores such as *Elephas, Giraffa, Onotragus, Homoioceras, Kobus,* and *Equus* (Kalb et al., 1980). Faunal comparisons related UBSU with the Olorgesailie formation—of an age of 0.7–0.5 Ma—and with the Olduvai Bed IV (Kalb et al., 1982), i.e., with terrains from the Middle Pleistocene. John Desmond Clark

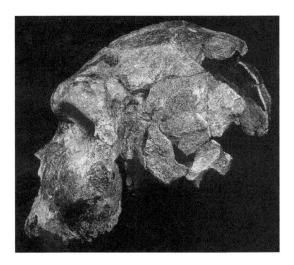

Figure 9.12 "Bodo man" (photograph from Johanson & Edgar, 1996).

and colleagues (1994), with the $^{40}Ar/^{39}Ar$ method, estimated a volcanic tuff located in unit "u" to 0.64±0.04 Ma. The specimens were found in a different unit, "t," but because of the correlation between the sediments, Clark et al. (1994) considered Bodo man to be that same age as the Günz–Mindel interglacial period. The totality of faunal studies, archaeological evidence, and radiometric analysis, led Philip Rightmire to confirm that date as the most probable (Rightmire, 1996).

A second specimen, a left parietal fragment, BOD-VP-1/1, was found in 1981, 350 m from where the first one had been discovered, probably belonging to another individual, along with a fragment of distal humerus coming from the same stratigraphic horizon (Asfaw, 1983).

Bodo cranium was first described by Glen Conroy and collaborators (1978). After a detailed analysis, G. Philip Rightmire (1996) concluded that the remains of Bodo matched those of Asian *H. erectus* in many of its metric and non-metric traits: a flat cranial vault and a great bone thickness in both the cranium and the parietal bone; robust and projected supraorbital torus; narrow and flat frontal profile; and a mid-cranial keel and bregmatic eminence. Some facial features, such as the width of the zygomatic region, bring the Bodo specimen, in particular, closer to Sangiran 17 (Java, Indonesia). But Rightmire (1996) indicated that that closeness to *H. erectus* might be misleading because it corresponds largely to primitive features. Based especially on the robust aspect, the cranial volume—which some authors estimated at up to 1,500 cc, but which he reduced to 1,300 cc (Rightmire, 1996)—and the architecture of the parietal walls, Rightmire (1996) anticipated the same conclusion that later he maintained in his 2009 review of the panorama of the Middle Pleistocene: Bodo is a transitional specimen.

The Salé specimen consists of a cranium, which lacks the face, and a left maxillar with some teeth. The specimen was found in 1971 in an open quarry near El Hamra, north of Salé (Morocco) (Jaeger, 1975). Since the first description, Jean-Jacques Jaeger (1975) indicated its clear affinities with *Homo erectus*, although the occipital area is much more bulky and rounded than would be expected, giving it a more modern appearance. This feature has been attributed to a disease: congenital torticollis (Howell, 1999; Anton, 2003; Bruner, 2004; Gracia et al., 2009), but Jean-Jacques Hublin (1985), argued that it is more logical to consider the shape of the Salé cranium as corresponding to a primitive *H. sapiens* with some retained traits of *H. erectus*.

The precise age of the Salé fossil is difficult to estimate. Jean-Jacques Hublin (1985) attributed to the site an age contemporary to the marine transgression of Anfantiana (flooding due to the increase of sea level). Andrew Millard (2008) indicated that, given the discrepancy between the different correlations of this episode with the stages of marine oxygen isotopes (MIS)—the Anfantiana transgression has been attributed to the stages 7, 9, 11, and 13—it can be assigned an age range of 525,000–190,000 years.

In 1953, Keith Jolly and Ronald Singer discovered a calotte and a mandibular fragment associated with Acheulean tools (see Straus, 1957) in the Elandsfontein ranch, located 10 miles from the city of Hopfield (South Africa) in Saldanha Bay. In the description of the exemplar, Singer (1954) related it to the Broken Hill fossil due to its appearance and measurements, although M. R. ("Maxie") Drennan (1953) qualified the fragment as "Neanderthaloid," indicating that it would not fit the Broken Hill maxillar—which lacks a mandible. Subsequently, Drennan (1954) named the species *Homo saldanensis*, which William Straus (1957) deplored, arguing that both the specimen from Broken Hill and the one from Saldanha should be classified as *Homo sapiens rhodesiensis*. In the comparison between the two fossils, Günther Bräuer (2008) estimated the cranial volume of the Saldanha fossil at 1,225 cc and, according to Haile-Selassie, Asfaw, and White (2004), the curvature of the parietal arch would bring it closer to modern humans, such as the one from Qafzeh. The age of the Saldanha specimen would be less than that of the Broken Hill specimen, i.e., less than 600,000 years old.

Three hominin cranial exemplars, one of them almost complete, Eyasi I, were found in Lake Eyasi (Tanzania) during the early twentieth-century expeditions of Kohl Larsen. Although the imprecision of its localization hindered assigning them a reliable age, Michael Mehlman (1987) granted the exemplars an age of more than 130,000 years. A new frontal bone, EH 06, associated with tools of the early MSA, was described by Manuel Domínguez-Rodrigo et al. (2008). Using the thorium content in teeth found in the same horizon, the authors dated the stratigraphic level—Gray Sands—corresponding to this fossil, placing it in the MIS 5, with a maximum of 140,000 years.

9.3.2 African grade II

The grade II of the African transitional specimens, deduced in the analyses by G. Philip Rightmire (2009), includes the fossils from Florisbad, Laetoli, Omo, Herto, Aduma, and Djebel Irhoud. This is a set that, although considered intermediate, shows accentuated features that correspond to *Homo sapiens*, to the degree that many authors consider some of these specimens to be modern humans.

The specimen Laetoli Hominid 18, found in situ in 1976 by Mary Leakey and collaborators in fluvial deposits of the Ngaloba beds, Laetoli (Tanzania) (Magori & Day, 1983), is a nearly complete cranium reconstructed from 21 fragments. It has the two temporal bones, part of the sphenoid, most of the base and part

---

Box 9.10 Kibish specimens

The deposits in the Kibish formation, of lacustrine origin, are divided into four members ranging from approximately 3,100 BP to c. 197,000 years BP (Butzer, 1971; Brown & Fuller, 2008) (Figure 9.13). Omo I and Omo II fossils came from member I, located below the KHS tuff (Kamoya's Hominid Site). By correlating Kibish members I, III, and IV with the sapropels from the eastern Mediterranean, and with various radiometric analyses of volcanic intrusions, Francis Brown and Chad Fuller (2008) attributed an age of c. 195,000 years to Omo I and Omo II. (Sapropel: unconsolidated sedimentary deposit rich in bituminous substances—*Encyclopaedia Britannica*.)

Other specimens of the Kibish formation, including a left partial tibia and a fibula fragment from the AHS site, member I—assigned to the same individual, Omo I—and a partial occipital bone of a juvenile specimen from the CHS site placed in members III or IV, i.e., of a younger age, have been described by Osbjorn Pearson and collaborators (Pearson, Fleagle et al., 2008; Pearson, Royer et al., 2008).

of the zygomatic bone, and the upper dentition. The age of the beds of Ngaloba could not be determined by radiometric or amino acid racemization but, when correlated with the Ndutu beds in Olduvai (Tanzania), a date was estimated of 120,000±30,000 years, which coincides with the age calculated by analyses of the faunal and archaeological remains (personal communication of Mary Leakey to Cassian Magori and Michael Day, 1983). Ngaloba and Ndutu tools are very similar and belong to the African MSA.

The features indicated by Cassian Magori and Michael Day (1983) in the Laetoli specimens show the continuous presence of numerous plesiomorphies in *Homo erectus*, such as low cranial vault, medial keel in the sagittal plane of the frontal bone, and a developed occipital torus. However, the cranial capacity—c. 1,200 cc, according to Magori and Day (1983)—the development of modern apomorphies in the face, the style of two separate orbital arches, and a moderate glabellar development, demonstrate the advanced nature of the

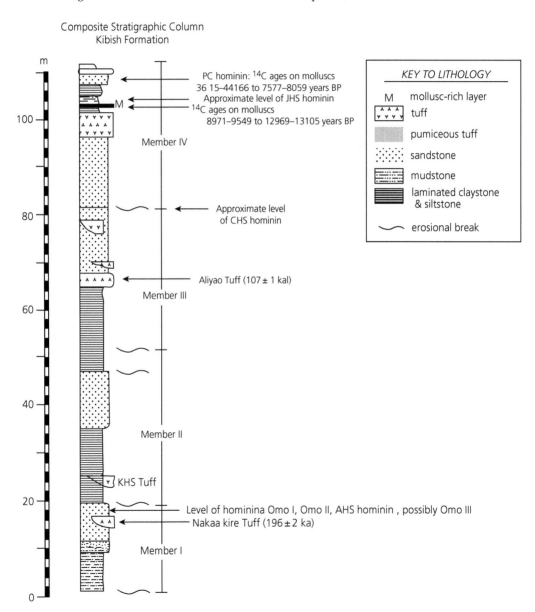

Figure 9.13 Stratigraphy of the Kibish formation (Omo region, Ethiopia). Reprinted from Brown, F. H. & Fuller, C. R. (2008). Stratigraphy and tephra of the Kibish Formation, southwestern Ethiopia. *Journal of Human Evolution*, 55(3), 366–403, with permission from Elsevier.

specimen. Cassian Magori and Michael Day (1983) recognized the differences between Laetoli Hominid 18 and the first *Homo sapiens* but, after a comparative analysis, they assigned the fossil to the latter species. It should be noted that in the comparative analysis, Magori and Day (1983) considered the specimens Kabwe, Eyasi I, Omo I, Omo II, Saldanha, Iwo Eleru, Florisbad, and Singa to be modern humans, regarding the differences between all of them as the result of mosaic evolution.

In 1967, the Kenyan team of the International Paleontological Research Expedition found a partial skeleton (Omo I), a cranium (Omo II), and some fragments of another cranium (Omo III) in member I of the Kibish formation, at Omo (Ethiopia), associated with tools of the MSA (Middle Stone Age, see later, Section 11.3.1) (Fleagle et al., 2008) (Figure 9.14).

The Omo I specimen, the most complete of the set, has a rounded cranial vault and an overall appearance similar to modern humans, though with rather robust teeth. Its height, deduced by the size of the left humerus, would be large—1.82 to 1.78 m—although the first metatarsal indicates a smaller size—1.73 to 1.62 m—(Pearson, Royer et al., 2008). Michael Day (1969b, 1972) classified it as *H. sapiens*, though perhaps of a slightly archaic kind. All the specimens from the Kibish formation, assumed to be the same age, were equally classified. But the Omo II cranium has traits that are reminiscent of *H. erectus*. Day and Stringer (1982) reinterpreted the Omo sample, keeping the classification of *Homo sapiens* for Omo I but assigning Omo II to *H. erectus*. Following a similar criterion, McBrearty and Brooks (2000) placed Omo I in group 3 (modern) and Omo II in 2 (intermediate, though McBrearty and Brooks placed

the specimens most similar to *H. erectus* in group 1). Such a classification suggests that not all the Omo specimens are of the same age. However, upon consideration of the fossil localization evidence—such as diagrams and original photographs—Brown and Fuller (2008) accepted as probable that Omo II belongs to member I of the Kibish formation, so that both Omo I and II would have the same age. Nevertheless, there are still doubts about the meaning of the differences between the crania. John Fleagle et al. (2008), after agreeing that Omo I is a modern human—and thus it should be removed from the "transitional" group II—suggested as an alternative hypotheses that either the population of Kibish member I was very variable, or that in this site *Homo erectus* and *Homo sapiens* coexisted for some time.

The Bouri formation (Middle Awash, Ethiopia), about 80 m thick, is divided into three members: Hata, Daka, and Herto (de Heinzelin et al., 1999), of which the latter belongs to the late Middle Pleistocene. Cranial remains of an immature individual and two adults associated with a complex of artifacts, which includes Acheulean tools characteristic of the MSA (White et al., 2003), appeared in 1997 in the Herto member.

The best-preserved cranium is of an adult, BOU-VP-16/1 (Figure 9.15), found in locality 16 of Bouri Vertebrate Paleontology (White et al., 2003). Despite a certain thickness in the supraciliar arch, Tim White and collaborators (2003) rejected the idea that all the morphological characters could correspond to *H. neanderthalensis*, considering that the three specimens represent an evolutionary stage immediately prior to the emergence of modern humans. Consequently, it was attributed to the subspecies *H. sapiens idaltu*, an ancestor of *H. sapiens*.

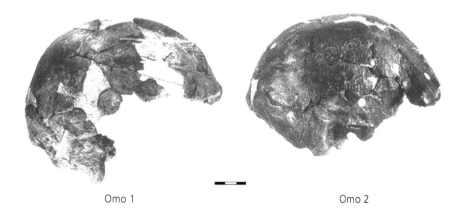

Omo 1   Omo 2

Figure 9.14 Omo's crania (Fleagle et al., 2008; original photography by Michael Day). Reprinted from Fleagle, J. G., Assefa, Z., Brown, F. H., & Shea, J. J. (2008). Paleoanthropology of the Kibish Formation, southern Ethiopia: Introduction. *Journal of Human Evolution*, 55(3), 360–365, with permission from Elsevier.

The age of the Herto specimens, obtained by relating the sediments and volcanic tuff dated using the $^{40}Ar/^{39}Ar$ method, is in the range of 160,000–154,000 years (Clark et al., 2003). An interesting question posed by J. Desmond Clark and collaborators (2003) is whether Herto specimens show signs of mortuary practices.

In the same paper in which the discovery of BOU-VP-16/1 was announced, White et al. (2003) presented the results of a multivariate analysis comparing the different crania of the following fossils: Omo 2 and Kabwe (African "archaic" *sapiens*); KNM-ER 3733, KNM-ER 3883, Sangiran, Ngandong, and Zhoukoudian (*Homo erectus*); Amud, Atapuerca, Gibraltar, La Ferrassie, La Chapelle, La Quina, Circeo Monte, Petralona, Saccopastore, Shanidar, Steinheim, and Tabun (Neanderthals sensu lato); Qafzeh 6, 9, Skhul 5, Cro-Magnon, and Predmostí 3 (ancestral modern humans); and 28

samples of current modern male humans. The results are shown in Figure 9.16.

The phenetic multivariate analysis conducted by White et al. (2003) shows graphically the relationship that exists between ancestral humans (Middle East) and modern humans on the one hand, and between African and Asian specimens of *Homo erectus* on the other. According to phenetic considerations, Neanderthals are located in the middle. The analysis by White et al. (2003) indicates that BOU-VP-16/1 is closer to modern humans than the "archaics" from Kabwe and Bodo, so that the specimens from Herto would correspond to an intermediate grade such as grade II by Rightmire (2009).

A new right parietal from the level Herto of the Bouri formation, along with four more specimens from the nearby locality of Aduma, were described in the following year by Yohannes Haile-Selassie, Berhane Asfaw,

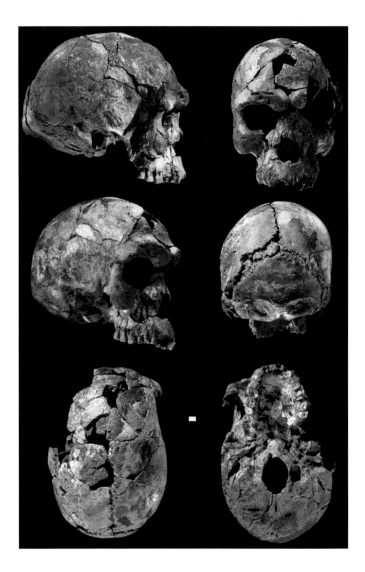

Figure 9.15 Specimen BOU-VP-16/1, the adult cranium from Herto; the white scale bar is 1 cm. Reprinted by permission from Macmillan Publishers Ltd: White, T. D., Asfaw, B., DeGusta, D., Gilbert, H., Richards, G. D., Suwa, G., & Clark Howell, F. (2003). Pleistocene Homo sapiens from Middle Awash, Ethiopia. *Nature*, 423(6941), 742–747.

and Tim White (2004). The findings were made on the surface, except the most complete fossil, ADU-VP-1/3, which was collected in situ. ADU-VP-1/3 is a partial skull whose best-preserved part is the occipital bone. The Ardu beds in the Aduma region have previously supplied lithic artifacts typical of the MSA and very similar to those from Herto (Haile-Selassie, Asfaw, & White, 2004). The tools from the Ardu beds were described by the authors as characteristic of the latter part of the MSA.

The age of the Aduma sediments has been studied by $^{40}Ar/^{39}Ar$ radiometry, uranium series, thermoluminescence, and radiocarbon dating, showing some inconsistencies; Haile-Selassie, Asfaw, and White (2004) accepted an age of 105,000–79,000 years, which places the Aduma fossils later than the Herto dating; these fossils are also more advanced in morphological terms.

Haile-Selassie, Asfaw, and White (2004) carried out a metric analysis and a morphological comparison of

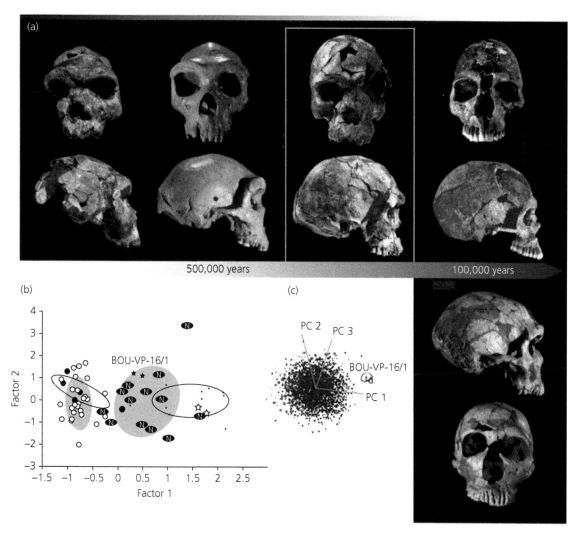

Figure 9.16 (a) Anterior and lateral views, Bodo, Kabwe, Herto BOU-VP-16/1 (boxed), Qafzeh 9, and, below, the La Ferrassie Neanderthal, all to the same scale, with approximate timeline. (b) Plot of the first two principal component scores, with the position of Herto BOU-VP-16/1 given by the fossil symbol marked "x." *Homo erectus*, includes KNM-ER 3733 and KNM-ER 3883 (open stars), and Sangiran, Ngandong, and Zhoukoudian crania (plus signs). "Neanderthals" (Amud, Atapuerca, Gibraltar, La Ferrassie, La Chapelle, La Quina, Monte Circeo, Petralona, Saccopastore, Shanidar, Steinheim, and Tabun crania) are shown by circled letter N, Omo 2 and Kabwe by filled stars, and fossil AMHS (Qafzeh 6, 9, Skhul 5, Cro-Magnon, and Predmostí 3 crania) by filled circles. Population means of 28 male modern human samples (shown by open circles) were taken from Howells (1989) and included in the principal-components analysis (PCA). (c) Plot of the first three principal components of the complete Howells (1989) dataset of 3024 modern (recent) human individuals, plus Herto BOU-VP-16/1 (the fossil symbol marked "x"). Reprinted by permission from Macmillan Publishers Ltd: White, T. D., Asfaw, B., DeGusta, D., Gilbert, H., Richards, G. D., Suwa, G., & Clark Howell, F. (2003). Pleistocene Homo sapiens from Middle Awash, Ethiopia. *Nature*, 423(6941), 742–747.

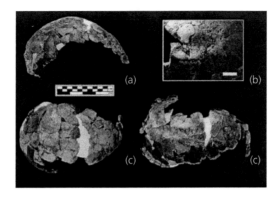

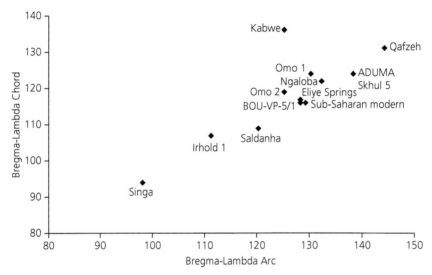

Figure 9.17 Up: Aduma cranium (ADU-VP-1/3)—(a) lateral; (b) posterior (cm scale); (c) superior; (d) inferior. Below: Bivariate comparison of Upper and Middle Pleistocene hominid crania, based on bregma-lambda arc and chord. Haile-Selassie, Y., Asfaw, B., & White, T. D. (2004). Hominid cranial remains from upper pleistocene deposits at Aduma, Middle Awash, Ethiopia. *American Journal of Physical Anthropology*, 123(1), 1–10.

the specimen ADU-VP-1/3 with other fossils from the late Middle Pleistocene and early Upper Pleistocene (see Box 9.11 and Figure 9.17). The results showed that the Aduma fossil is similar to many of the modern humans of the Upper Pleistocene of Africa and the Middle East in features such as overall size, the high vault, the rounded profile of the occipital, and the well-curved parietals. While it departs from the specific characteristics of Neanderthals, it retains some shared plesiomorphies with older specimens such as LH-18 (Haile-Selassie, Asfaw, & White, 2004). The authors' conclusion supported the inclusion of the Herto and Aduma fossils within the species *Homo sapiens*. As such, these individuals should be included in grade III.

The Irhoud Djebel cave is located 55 km southeast of the city of Safi (Morocco). During mining work to extract barite, a cranium (Irhoud 1) appeared in 1961. A second fossil, the calotte Irhoud 2, was obtained after excavations under the direction of Émile Ennouchi

(1963). Subsequent fossils, including a juvenile mandible (Irhoud 3) and some postcranial elements, complete the sample (Hublin & Tillier, 1981; Hublin, 1985). The fauna at the site—with *Gerbillus grandis* and *Alcelaphine*—indicated arid conditions that could correspond to the maximum extension of the Sahara during MIS 5, as well as to a dry episode of MIS 6 (Hublin, 1985). The ESR technique also indicated that the Djebel Irhoud sediments would have a long history of deposition in both stages, MIS 5 and 6, and the most probable temporal range for the hominin age would be in MIS 6—190,000–140,000 years—(Grün & Stringer, 1991).

Since their discovery, the Djebel Irhoud hominins were related to the Neanderthals, with the implication that the Neanderthals would have been present in North Africa, and therefore would no longer be an exclusively European taxon, extending at the most to the Middle East. However, other authors like Jean-Jacques Hublin—one of the leading specialists in the Upper Pleistocene hominins

---

Box 9.11 Comparative analysis by White et al. (2003) and Haile-Selassie, Asfaw, and White (2004)

In the analysis corresponding to Figure 9.17, the average of the 28 samples of modern human males was taken from Howells (1989). The principal component analysis only included measures of the cranial vault, so that as White et al. (2003) warned, the graph illustrates the phenetic affinities reflected by that part of the anatomy. Comparative data are by Arsuaga et al. (1997) and Howells (1989).

For the metric and morphological comparison of ADU-VP-1/3 (Figure 9.17), Haile-Selassie, Asfaw, and White (2004) used the original remains of Omo I, Omo II, Bodo, the casts of Ngaloba (LH-18), Eliye Springs (ES-11693), Kabwe, Saldanha, Florisbad, Border Cave, Singa, Djebel Irhoud, Skhul 5, and the data of Neanderthals and their predecessors by Arsuaga and collaborators (Arsuaga et al., 1997).

---

and, in particular, in Maghreb paleoanthropology—rejected that African presence of Neanderthals. The reassessment made by Hublin and collaborators (2007) indicated that the specimens from Djebel Irhoud lack the derived traits of the western Eurasian clade of Neanderthals, with whom they share some primitive traits such as a general robustness. A weak convexity of the parietal bone, the elongated temporal, and the lower shape of the occipital squama may be related to a certain degree of platycephaly, according to these authors. To Hublin et al.

(2007), these and other parallel traits place the fossils of Djebel Irhoud in relationship with the first modern humans from East Africa and the Middle East.

Florisbad is a locality with many mineral hot springs, located near Lake Sautpan, 25 miles north of Bloemfontein, Free State Province (former State of Orange), South Africa. It is a site with fauna fossils known since 1912 and has provided since then more than 30 taxa (Figure 9.18). In 1932, in one of the Florisbad closed springs, Thomas Dreyer unearthed a

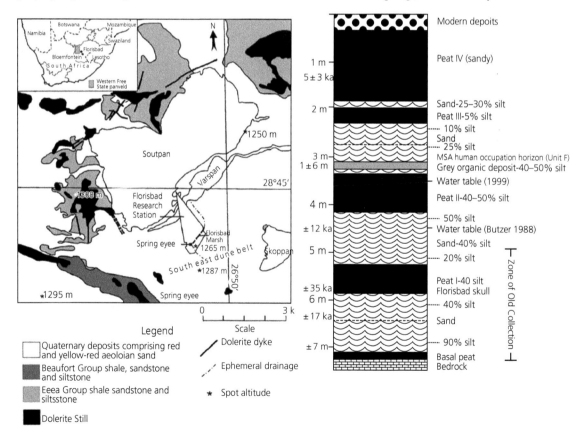

Figure 9.18 Left: Localization of Florisbad (South Africa). Right: Site stratigraphy. Reprinted from Douglas, R. M., Holmes, P. J., & Tredoux, M. (2010). New perspectives on the fossilization of faunal remains and the formation of the Florisbad archaeozoological site, South Africa. *Quaternary Science Reviews*, 29(23, 24), 3275–3285, with permission from Elsevier.

Figure 9.19 Images obtained by CT scan of the "archaic" sapiens crania from Laetoli (left), Florisbad (center), and Bodo (right). Curnoe, D. (2011). A 150-Year Conundrum: Cranial Robusticity and Its Bearing on the Origin of Aboriginal Australians. *International Journal of Evolutionary Biology*, 2011: 632484. doi:10.4061/2011/632484.

cranium described by the author as being from a very primitive form related to *Homo sapiens* (Dreyer, 1935). Dreyer (1935) classified it as *Homo helmei*, although later (Dreyer, 1936) he considered it to be a very old member of the current South African Bushmen (Figure 9.19). The campaigns of the National Museum of Bloemfonstein, started in 1981, provided an abundance of MSA tools later studied by Kathleen Kuman, Moshe Inbar, and Ron Clarke (1999). There is also evidence of occupation in unit F, such as the use of fire and butchery activities (Douglas, 2009).

The origin site of the Florisbad cranium has a depth of 7 m, with an age difficult to determine. Attempts of dating with thermoluminescence and uranium series, conducted in 1989 and 1990, were unsuccessful (Kuman et al., 1999). Using electron spin resonance and optically stimulated luminescence (ESR-OSL), a tooth associated with the cranium yielded an age of 259,000±35,000 years, a figure to which Kuman et al. (1999) gave little credence, as it did not match the Florisbad cultural sequence, established by the authors in

three horizons with an early MSA of approximately 279,000 years in the basal unit, a very retouched form of MSA of c. 157,000 years, and a slightly retouched one c. 121,000 years (Box 9.12).

9.3.3 African grade III

The grade III of Rightmire (2009) included specimens that can be considered already modern humans in sensu stricto, such as those from the Klasies River (South Africa) and those from Skhul and Qafzeh (Israel), which indicate an exit from the African continent. We will explore them in Chapter 10.

9.4 Asian archaic *Homo sapiens*

The transition in Asia from Middle Pleistocene hominins to those in the Upper Pleistocene has been discussed in the classic literature under the two opposing models cited at the beginning of this chapter. According to MH, the Asian *Homo erectus* would have become *Homo sapiens* through an evolutionary process in which the populations of different localities exchanged genes while maintaining the unity of species. By contrast, RH argued that modern humans reached Asia and displaced *Homo erectus* populations, which had existed there in a long process of stasis, with no genetic exchange.

As we have said, MH emerged from the classical interpretation of Chinese specimens (Weidenreich, 1943) and has its most significant morphological justification in the comparison of traits between the last Javanese *Homo erectus* and the first modern humans from Australia (Anton et al., 2011). Conversely, RH was proposed after genetic studies (Cann et al., 1987; Vigilant et al., 1991). To what extent can two hypotheses supported by different empirical evidence be contrasted? We will now try to answer this question. To do so, we will consider data that can shed light on how

Box 9.12 The difficulties of Florisbad dating

The doctoral thesis by Rodney Douglas (2009) tested the hypothesis that the geological events related to the Florisbad hot springs have led to misinterpretation of the sedimentation and fossilization processes, creating discrepancies that appear when obtaining the age of the stratigraphic levels. Douglas performed logarithmic transformations of the published data and obtained as the best option an age of 127,000 years for the level F of human occupation and of 420,000–250,000 years for the lower levels starting from the basal unit. The analysis of the mineral salts in the fossils and their possible origins were published a year later (Douglas et al, 2010).

the replacement in Asia of *Homo erectus* by modern humans occurred:

I. Specimens indicating the permanence of *Homo erectus* throughout the Asian Upper Pleistocene.
II. Existing morphological distance between those persistent Asian specimens of *Homo erectus* and the oldest specimens of *Homo sapiens* from the same continent.
III. Model of the evolution of *Homo erectus* in Asia.
IV. Possible dispersal routes available to modern humans for entry into Asia.
V. Genetic relationships existing between the different populations involved.

Despite what occurred in Africa, Asian early modern humans did not evolve in situ from *Homo erectus*; rather, they occupied Asia by one or more migrations from their African places of origin. This assessment is similar, then, to what happened in Europe, but with the distinction of involving the stasis of *Homo erectus* in Asia, which in contrast to the evolution of Neanderthals in Europe, took place in parallel to the African transition to modern humans.

The first three points of this general interpretive scheme—permanence of *H. erectus*, morphological

distance from *H. sapiens*, and evolution of *H. erectus*—will be examined in this chapter separately regarding Java and China. The fourth point, with the possible limited episode of hybridization that might have occurred in the Middle East and Asia between new immigrants and local populations, will be left for Chapter 10.

9.4.1 The transition in Java

The permanence of *Homo erectus*

At the beginning of the 1930s in the Solo River valley, the Ngandong site provided 12 cranial remains (Ngandong 1–8 and 11–14, with Ngandong 7 as the most complete) and two tibias (NG 9 and 10), coming from a sedimentary terrace situated 20 m over the current river bed ("terrace 20 m") (Figure 9.20). These specimens were attributed by Willem Oppennoorth (1932) to the species *Homo (javanthropus) soloensis* and are known colloquially as the "Solo Man." The age attributed to the Ngandong specimens has been a continuing controversy. The fauna and geomorphology of the levels from where the fossils supposedly came correspond to the Upper Pleistocene, but the morphological comparison with other similar specimens, as well as the suspicion that the fossils might have been

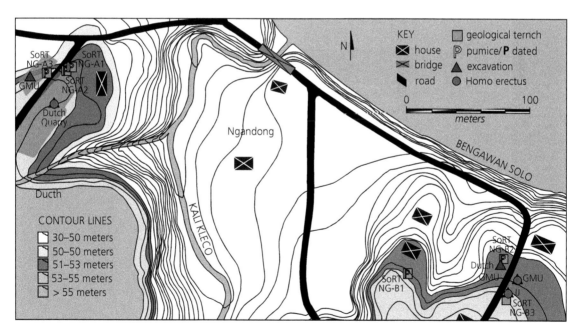

Figure 9.20 Ngandong map indicating the localities where fossils were discovered in the initial work by Oppennoorth (Dutch), Jacob (Gadjah Mada University, GMU), and the Indonesian-Japanese expedition (IJ). The square with a P identifies the places from which the samples were taken during the terrain dating work. Indriati, E., Swisher, C. C., III, Lepre, C., Quinn, R. L., Suriyanto, R. A., Hascaryo, A. T.,. . . Antón, S. C. (2011). The Age of the 20 Meter Solo River Terrace, Java, Indonesia and the Survival of Homo erectus in Asia. *PLoS One*, 6(6), e21562.

moved from their original stratum, led to their placement within the Middle Pleistocene (Santa Luca, 1980). Teuku Jacob (1975) estimated that they would have an age between 400,000 and 200,000 years.

Using the techniques of electron spin resonance and mass spectrometry of uranium series (ESR/U) on bovid teeth fossils from the same levels as the hominin specimens, Carl Swisher III and collaborators (1996) obtained an age range of 53,300±4,000 to 27,000±2000 years for Ngandong and Sambungmacan. These figures reduce the 70,000–30,000 years obtained in previous studies with the same techniques (Bartstra et al., 1988; van der Plicht et al., 1989), but do not pose a serious discrepancy. However, Rainer Grün and Alan Thorne (1997) criticized the work of Swisher et al. (1996), indicating two possible errors. First, the authors made an improper use of ESR techniques. Second, it is not correct to apply to the human specimens an age obtained by dating bovid teeth, because the upper terrains of the Solo River—such as the "terrace 20 m"— are a mixture of materials from different locations and ages (Box 9.13). Because the color of the hominin bones and the fauna fossils do not match, Grün and Thorne (1997) argued that they have different ages. However, in response to these criticisms, Swisher et al. (1997) defended their technical procedures and referred to the continuous interpretation—indicated by Gert-Jan Bartstra et al. (1988)—granting the same age to the hominins as the rest of the faunal fossils of Ngandong.

The first direct estimation of Ngandong's age by means of radiometry—the $^{40}Ar/^{39}Ar$ method—was made by Etty Indriati and collaborators (2011). On average, the age obtained was of 546.000±12.000 years, placing the specimens after the first Javanese *Homo erectus*—those from Sangiran (see Chapter 8)—but at a time much older than that proposed by Swisher et al. (1996).

The dating controversy and the difficulty of comparing the fauna of Java—subject to the isolation that an island imposes—with other continental localities, disappears if the samples in the analysis are those of the hominin fossil remains whose age we want to determine. Using the technique of gamma-ray spectroscopy, Yuji Yokoyama et al. (2008) obtained an age for the Ngandong fossils between 70,000 and 40,000 years, closer to the estimate by Swisher et al. (1996) than to that of Indriati and collaborators (2011). Without disregarding the doubts about the origins of the specimen terrains, Susan Antón (2003) noted in her study on the natural history of *Homo erectus* that the available data support a Late Pleistocene age for both Ngandong hominins and the rest of the fauna. This is the conclusion that we accept.

Box 9.13 The age of the "terrace 20 m"

Indriati and collaborators (2011) also conducted studies by ESR/U series on faunal teeth from the same "terrace 20 m," yielding an age of 143,000–73,000 years. As Indriati et al. (2011) have indicated, the discrepancy between the results produced by each technique is more serious when considering that the results are consistent within themselves. Indriati and collaborators (2011) argued that there are two possible explanations for this contradictory dating: (1) all the pumice used in these radiometric studies has been moved to more modern terrains, so their age does not match that of the hominin fossils; (2) the studies of ESR/U provided the age of some specimens that, because of hydrologic activity, appear in the same place as others of a different age, so that different fossils of the same stratigraphic horizon may correspond to different dates. Carl Swisher, Rainer Grün, and Susan Antón (the main protagonists of the current debate on the age of the site) are among the authors of the study by Indriati et al. (2011).

Besides the age of the specimens, the basic issue to interpret the evolutionary meaning of late Javanese specimens is their morphology, particularly in comparison with the rest of the island's sample. The description by Susan Antón (2003) of the Ngandong and Sambungmacan specimens indicated as the most outstanding feature—apart from the large volume—the shape of the cranium, with a moderate supraorbital constriction, but retaining the area of wider expansion toward the back—a shared trait with the oldest specimens of the island. In lateral view, the cranial vault is long and fairly low. The rear side shows a continuous torus. The glabellar torus is continuous and of a large or moderate size. Taken together these features indicate that specimens from Java maintained an overall structure that changed little except in regard to the cranial size, becoming larger in most modern fossils.

Morphological proximity *Homo erectus–Homo sapiens* in Java

The more detailed study on the existing morphological relations between the late *Homo erectus* in Java and the oldest specimens of *Homo sapiens* are centered on the specimen Sm 3.

The Sambungmacan calotte, Sm 3, is a fairly complete specimen that lacks the face. It was found in 1977 and remained as part of the inventory of an antiques' shop in Jakarta until it was secretly removed from Indonesia in 1998, appearing the following year in a natural history curiosity shop in New York (Box 9.14). Its exact origin is unknown, although Márquez and colleagues indicate with near certainty that it was found near Ngadirejo, between Chemeng and Poloyo (Márquez et al., 2001).

The comparative study of Sm 3 conducted by Eric Delson and collaborators (2001) used morphometric techniques—procrustean markings—applied to the localization corresponding to points of the glabella, bregma, lambda, inion, and opisthion, in a sample of *Homo erectus* from Indonesia, China, and Kenya, "archaic" *sapiens* from Kabwe and Petralona, and ten modern human crania. Statistical analyses (main component and canonical discrimination) showed that, in all cases, Sm 3 is in an intermediate position between the other fossil groups and the current samples. The morphological comparison of Sm 3 brought it closer to other specimens from Sambungmacan (Sm 1; Jacob, 1975) and Ngandong. The differences with other *Homo erectus* are, in particular, related with a less projected glabella in anterior direction, a more vertical supraoral plane, and a less angular occipital torus (Delson et al., 2001). The gracility of the specimen points, however, to a female, so those differences might correspond in part to sexual dimorphism.

Thanks to the excellent condition of the calotte, Douglas Broadfield and collaborators (2001) conducted a study on the Sm 3 endocranium using a computational tomography scanner (CT) in addition to traditional methods. This revealed that, although in the essential traits the specimen shares a neurological structure with other *Homo erectus* from Java and China, some features indicate a striking modernity: the degree of asymmetry between the hemispheres, with petalia in the left occipital and right frontal. The frontal lobe is more rounded and shorter, contrasting with the flat and elongated lobe of the other Javanese *Homo erectus*, such as Sangiran 17. According to Broadfield

et al. (2001), the endocranium Sm 3 shows a different morphology to those that appear in the hominin fossil record, expanding the great variability of *Homo erectus* in Indonesia.

Another remarkable specimen is the Ngawi I cranium, discovered in August 1987 by peasants in a riverbank just outside the town of Selopuro (Ngawi, Java, Indonesia) (Widianto & Zeitoun, 2003). Although its origin was attributed to the Pitu terraces (Watualang, west of Ngawi), the circumstances of the discovery make it infeasible to achieve accurate localization and dating.

Ngawi I is a nearly complete but somewhat eroded cranium, showing part of the face without maxillar or mandible. In the first description and interpretation of the fossil, Harry Widianto and Valery Zeitoun (2003) separated it from the Trinil–Sangiran group, considering it to be comparable to the Ngandong–Sambungmacan group. Authors disagreed over its best taxonomic consideration. Widianto argued that it is part of an advanced grade of *Homo erectus*, while Zeitoun preferred either to rehabilitate the name *Homo soloensis* or to attribute the fossil to a subspecies of *Homo sapiens*. However, both argued for the relevance of giving the same consideration to Ngawi I and the specimens from Ngandong–Sambungmacan. In a subsequent study, Valery Zeitoun and collaborators (among them Widianto) (2010) left open the possibility of dividing the specimens of Javanese *Homo erectus* into two different species.

The multivariate analysis by Arthur Durband (2006) of a full sample of *Homo erectus* specimens from Africa, Indonesia, and China resulted in showing how Ngawi I shared close morphological similarities with the Ngandong and Sambungmacan specimens, as Widianto and Zeitoun (2003) had concluded (Box 9.15). But, when the range of the comparisons is expanded, Ngawi I becomes very close to the African *Homo erectus* and the other Javanese specimens—including the oldest from Sangiran—while becoming more different from the Zhoukoudian specimens. According to Durband (2006), this observation implies that there was a characteristic population in Java that developed and maintained its own derived traits, thereby becoming different from the majority of Chinese *H. erectus* (with some exceptions like the Hexian exemplar).

Evolutionary meaning of the advanced specimens of Javanese *Homo erectus*

The somewhat advanced features of the specimens from Ngandong, Sambungmacan, and Ngawi raise the question of whether to uphold the idea that a taxon, *Homo erectus*, maintained such a long presence in Java without variation until the arrival of *Homo sapiens* or, on the contrary, whether it might be necessary to advance a different evolutionary model.

The large cranial capacity of some of the specimens from Ngandong allows us to suppose that they could be transitional from *Homo erectus* to *Homo sapiens*. Nevertheless, as in other cases of "archaic" specimens, there are several conflicting views. Thus, the initial idea that the Ngandong specimens were a new species was followed by the attribution of the later specimens to a variety of Neanderthal (Vallois, 1935; von Koenigswald, 1949). The examination by Franz Weidenreich (1933) brought the fossils from Solo closer to *Homo erectus* but without proposing any classification for them. After a detailed study, A. P. Santa Luca (1980) said that the Ngandong specimens were similar to those of Trinil and Sangiran and should be classified as *Homo erectus*. Differences in size and other features could be due to sexual dimorphism (Ngandong 6, considerably larger, belonged to a male, and Ngandong 7, smaller, to a female).

In order to assess the possible relationships between the *Homo erectus* from Sangiran, the more advanced Ngandong specimens, and the current Australian Aborigines—relationships seen as indicative of evolutionary continuity by supporters of MH—Arthur Durband (2007) conducted a comparative analysis of traits from the cranial base such as the foramen morphology, the placement of the squamo-timpanic fissure in the temporomandibular fossa, and the extreme expression of the post-condyloid tubercle; all these characters are mentioned by several authors, including Franz Weidenreich and Teuku Jacob. Durband's study (2007) concluded that these features, present in the population of Ngandong, seem not to appear elsewhere, which supports the idea of a discontinuity at this stage of human evolution in Australasia.

At the same time, the modernity of the frontal and parietal areas of Sm 3 could be interpreted as evidence of an ancestral relationship with *Homo sapiens*.

A similar result showed in the CT scans of the specimen Sambungmacan Sm 4 conducted by Hisao Baba and collaborators (2003). The basicranial flexion—lacking a well-developed occipital torus—and a low vault indicate, according to the authors, an independent evolution of the parietal and occipital regions. But, neither Broadfield et al. (2001) nor Baba et al. (2003) saw in the intermediate condition of Sambungmacan specimens any evidence of the phylogenetic link *Homo erectus–Homo sapiens*. By contrast, the morphology of late Javanese specimens indicate to these authors that these are substantially isolated populations. It should be recalled that the separation of glabellar torus into two arches was already present in the KNM-ER 42700 specimens from Koobi Fora.

Probably the great variation of *Homo erectus* in Java follows different aspects with both local (geo-ecological) as well as temporal, sexually dimorphic, and even individual patterns. In any case, it does not seem very likely that a large intraspecific variability should occur within one and the same population—option 1 by Márquez et al. (2001; see Box 9.16). The work of synthesis of the biometric, morphometric, and cladistic approaches by Valery Zeitoun and collaborators (2010) concluded that late Javanese specimens can be considered a separate species from *Homo erectus* (a position that corresponds to option 2 by Márquez et al., 2001), with its first proposed name being *Homo soloensis*. However, Susan Antón (2003) has pointed out that the major differences that exist between ancient and modern forms of *H. erectus* in Java are related to the increase in cranial size in the latter, affecting the height of the vault and the decrease of the post-orbital constriction. For Antón (2003), many of the features listed as distinctive in the Ngandong specimens with respect to *Homo erectus*—the morphology of the supraorbital

---

Box 9.16 Taxonomic alternatives of the Sambungmacan sample

Márquez et al. (2001) posed alternative taxonomic considerations compatible with the morphology of the Sambungmacan sample:

(1) Extends the *Homo erectus* range of variation in Indonesia and China.
(2) Indicates an evolution within the species *Homo erectus.*
(3) Supports the presence of two species in the Middle Pleistocene in Java.

Box 9.17 Coexistence in Java

The possibility of coexistence in Java of local *Homo erectus* and *Homo sapiens* has been proposed by Swisher and collaborators (1996) through the calculation made by these authors about the age of late specimens of *H. erectus* on the island. Thus, *H. erectus* would overlap in Southeast Asia with individuals of our species in a way similar to that between Neanderthals and modern humans in Europe. Swisher et al. (1996) suggested that *H. erectus* and *H. sapiens* might even have exchanged genes.

torus, occipital, mastoid, and supramastoid—are in fact differences between the entire Javanese sample and the Chinese. According to this idea, if we ignore the enlargement of the skull of the late specimens, we find that the population of Java, associated broadly with *Homo erectus*, forms a close morphological group that contrasts with that of *Homo erectus* from the Asian continent.

In any case the cranial increment shows that the widespread idea of an evolutionary stasis without evolutionary changes in this taxon in Asia is not correct, a stance previously supported by Milford Wolpoff (1984) by means of a comparative analysis of the cranial, mandibular, and dental features from a sample of 92 specimens. Even if the transition to *Homo sapiens* did not occur on the island, an evolution of the first populations of *Homo erectus* toward forms whose cranial size was larger is likely, as also happened in Europe with the Neanderthals and in Africa with modern humans. This would be another case of parallelism (see

Box 9.18 Brain expansion beginning with *Homo erectus*

The different patterns of brain expansion in *H. erectus, H. neanderthalensis*, and *H. sapiens* indicated by Emiliano Bruner (Bruner et al., 2003; Bruner, 2004) clearly show that a "cranial increment" is a trait evolved separately in different lineages. This is the essential difference between homoplasy—in a parallelism version—and homology. It is logical that such a tendency will be difficult to distinguish from homologies resulting from a common evolution, which would be a necessary issue to consider in some proposals of MH of human evolution.

Chapter 2). However, in order to explain the differences between the Asian transition process, and the European and African, we will continue to speak about a stasis of *H. erectus* in Asia.

9.4.2 The transition in China

The permanence of *Homo erectus* in China

Various Chinese sites provide evidence about the late specimens of *Homo erectus*. Let's summarize their ages and main characteristics.

Yunxian

Two crania, EV 9001 and EV 9002, come from Yunxian in Hubei province, central China, a site formed by deposits from the Han River terrace. These fossils were discovered in situ, embedded in a limestone matrix. They are Middle Pleistocene, with a large cranial capacity exceeding 1,000 cc, adult, and possibly male (Tianyuan & Etler, 1992). Their dating, though imprecise, could reach 400,000 years (Schwartz & Tattersall, 2002).

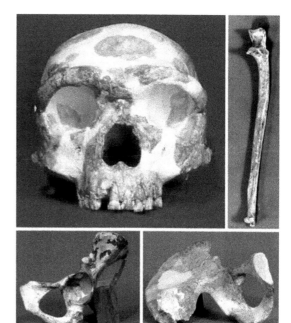

Figure 9.21 Skull, coccyx, and ulna from the Jinniushan specimen. The various elements are not on the same scale. Rosenberg, K. R., Zuné, L., & Ruff, C. B. (2006). Body size, body proportions, and encephalization in a Middle Pleistocene archaic human from northern China. *Proceedings of the National Academy of Sciences of the United States of America*, 103(10), 3552–3556.

Jinniushan

The specimen found in Jinniushan, near the city of Yinkou (Liaoning Province, northeast China), consists of a cranium with most of the upper dentition, six vertebrae, two ribs, a kneecap, a complete ulna of a very modern appearance, part of the coccyx, and several articulated bones from both hands and feet (Rosenberg et al., 2006) (Figure 9.21). The absence of duplication indicates that it is a single individual, probably a female.

The specimen was discovered by researchers from the Department of Archaeology at Peking University in 1984, but its antiquity, estimated at about 260,000 years, was established later by electron spin resonance (Tiemei et al., 1994). This age agrees with the fauna of the site and, despite subsequent studies, which yielded more recent dates by averaging the different stratigraphic levels, Karen Rosenberg, Lü Zuné, and Christopher Ruff (2006) accepted it as correct.

Dali

Dali's cranium is a specimen that lacks the lower part of the face. It was found in an alluvial stratum in Tianshuigou gravel pit, near the city of Jiefang (Shaanxi province, China) (Yongyan et al., 1979). The specimen, a male adult, shows developed supraorbital arches that do not form a continuous line, as is the case of *Homo erectus* sensu stricto (Figure 9.22). The uranium series' dating performed by Chen Tiemei, Yuan Sixun, and Gao Shijun (1984) yielded an age for Dali of 209,000±23,000 years. By paleomagnetism and stratigraphic chronology, Yin Gongming and collaborators (2002) established a range between 300,000 and 260,000 years for the stratum containing the skull.

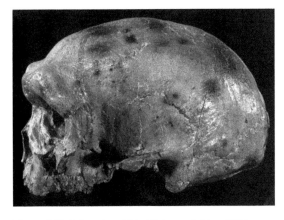

Figure 9.22 Dali specimen (China) (Johanson & Edgar, 1996).

Hexian

The Hexian site in southern China is situated on the northern side of the Wangjiashan hill, within the Taodian commune in Hexian County, north of the Yangtze River (Wu Rukang & Dong Xingren, 1982). The construction of a canal in 1973 revealed abundant animal fossils, and in the campaigns of 1980 and 1981, under the direction of Huang Wanpo, scientists from the Beijing Institute of Vertebrate Paleontology and Paleoanthropology, together with local colleagues, unearthed an almost complete calotte, two cranial fragments of another individual, and a mandibular fragment with two molars and nine isolated teeth. Biostratigraphic evidence placed the clay soils in the Middle Pleistocene.

The Hexian cranium (Figure 9.23) was described by Wu Rukang and Dong Xingren (1982) as a male juvenile with many morphological traits similar to those of Peking Man (Zhoukoudian taxon previously classified as *Sinanthropus pekinensis*). Consequently, it was assigned to *Homo erectus*. However, some advanced traits, such as a weak post-orbital constriction and a significant cranial capacity of approximately 1,025 cc, brought it closer to more modern specimens like Zhoukoudian 5 (Tiemei & Yinyun, 1991), to which we will refer later.

Hexian's age could be similar to that of the later levels of Zhoukoudian (Rukang, 1985). Xu Quinqi and You Yuzhu (1984), using faunal analysis and comparing with evidence from coral fossil deposits, granted the Hexian cranium an age of 240,000–280,000 years. Using uranium series, Chen Tiemei and Zhang Yinyun (1991) attributed an even more recent age, 190,000–150,000 years.

Nanjing

The Nanjing 1 specimen consists of three cranial fragments discovered in 1993 in the Hulu cave, located near the city of Tangshan (Nanjing, South Chima) (Tangshan Archaeological Team & Archaeology Department of Peking University, 1996). The reconstruction of the specimen carried out by Wu Liu, Yinyun Zhang Yinyun, and Wu Xinzhi (2005), estimated a cranial capacity of about 876 cc, with a low position of the cranial vault in the area of its maximum width, a remarkable bone thickness, and a development of the supraorbital, occipital, and angular tori.

The fragmentary condition of the fossil would not have allowed proceeding much further; but, in this case, as well as in Zhokoudian 5, acquiring the endocranium permitted accurate comparisons. Thus, by comparisons with endocrania from 20 specimens of *Homo erectus*, 6

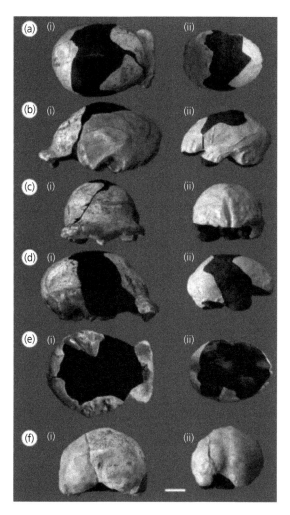

Figure 9.23 Cranium (I) and endocranial cast (II) ZKD V after reconstruction in 1973. Views: upper (a), lateral (b and d), anterior (c), basal (e), and posterior (f); scale: 4 cm. Wu, X., Schepartz, L. A., & Liu, W. (2010). A new Homo erectus (Zhoukoudian V) brain endocast from China. *Proceedings of the Royal Society of London – B*, 277(1679), DOI: 10.1098/rspb.2009.0149, by permission of the Royal Society.

Box 9.19 Nanjing I injury

Hong Shang and Erik Trinkaus (2008) indicated that the specimen Nanjing 1 shows an ectocranial injury extending through the anterior neurocranium between the temporal lines and from the supratoral sulcus to the anterior parietal bone. The authors noted that the damage cannot be attributed to neoplastic pathologies, poor diet, or infection, because of the normal appearance of the endocranium. It must have been caused by a blow or the action of fire.

---

of Neanderthal, and 38 of modern humans, whose metric data were collected using virtual three-dimensional laser reconstruction, and conducting an analysis of the main components, Wu Liu, Yinyun Zhang, and Xinzhi Wu (2005) concluded that Nanjing 1 must be classified as *Homo erectus*. However, its unique features of the upper frontal convolutions and the lack of posterior projection of the occipital lobes contribute to increasing the variation within the species.

Despite an age initially established at c. 350,000 years, Xiujie Wu and collaborators (2011) considered that Nanjing 1 might be older, up to 0.62–0.58 Ma. Being associated with cold climate fauna, Hong Shang and Erik Trinkaus (2008) accepted as valid its ascription to MIS 16, i.e., of an approximate age of 620,000 years.

Zhoukoudian V

The specimen Zhokoudian V (ZKD V) is a partial cranium we addressed in Chapter 8. It is formed by four fragments that appeared in locus H III of the site Locality I; two of the fragments were found in 1934 and the other two in 1966 (Wu, Schepartz, & Liu 2010). The cranium and the endocranial cast—nearly complete—were reconstructed in 1973 (Figure 9.23). A new endocranial reconstruction was done in 2010 (Wu, Schepartz, & Liu 2010).

ZKD V is the newest specimen of the entire crania series of Zhokoudian. The stratigraphic level from which it came was dated at 230,000 years using uranium series on mammal fossils, but a subsequent measurement, along with the use of the $^{26}Al/^{10}Be$ radiometric technique, placed it in the range of 500,000–400,000 years (Wu, Schepartz, & Liu, 2010).

The morphology of the Zhokoudian crania, including the parts of specimen V known at the time, was analyzed by Franz Weidenreich (1943), who argued that in its basic character it remained the same throughout the nearly half a million years that the fossil series spans. Weidenreich attributed all the specimens to *Sinanthropus pekinensis*, a taxon equivalent today to *Homo erectus*.

Morphological proximity of *Homo erectus*–*Homo sapiens* in China

The taxonomic consideration of the Chinese hominin specimens described in Section 9.4.2 has been an object of much discussion. The interpretations are almost as diverse as the number of studies conducted, but most authors suggest that the specimens be considered as "intermediate" between *Homo erectus* and *Homo sapiens*. To begin with, there are questions about what is understood by the term "intermediate."

The precise meaning of an intermediate morphology can be achieved through the taxonomic consideration of the cranium Zhoukoudian 5. The comparative study of the specimen, and in particular of the endocranial surface of ZKD 5 done by Xiujie Wu, Lynne Schepartz, and Wu Liu (2010), yielded a volume of 1,140 cc, corresponding to the upper range of variability of *H. erectus* and to the lower range of the Chinese sample of modern humans. The measurements of the specimens are divided between those corresponding to *Homo erectus* and those corresponding to *Homo sapiens*. The authors argued that the pattern of the meninges of ZKD V is similar to those of *Homo erectus* of Hexian, Sangiran 2 and 17, Trinil II, and Sambungmacan 3, while differing from those of KNM-WT 15000 and *Homo sapiens*. However, Xiujie Wu, Lynne Schepartz, and Wu Liu (2010) indicated the presence of "progressive" features in the Zhoukoudian specimens, which overlap with those of modern humans.

"Intermediate" means that even when a certain distinctive pattern of *Homo erectus*' features is retained, the specimens of this category have some characters that are outside that pattern and are indicative of a progressive tendency whose account is, in fact, the most controversial aspect. Thus, Li Tianyuan and Dennis Etler (1992) considered that the Yunxian sample had *sapiens*-like facial features, which depart from the typical *Homo erectus*, starting from a very early time. So much so that, with respect to the evolution to modern humans, much more primitive features are maintained in Africa—especially in male specimens. The comparative study of Yunxian and Zhoukoudian fossils by Zhang Yinyun (1998) led the author to argue that the trait-overlapping is due to distortion and damage suffered by the fossils and that, from a morphological point of view, Yunxian crania deviate from the Zhoukoudian sample, being like those of *Homo sapiens*.

Certain progressiveness in the same direction also appears in other specimens. The initial description of the Dali cranium placed it as another transitional sample but under the classification of *Homo erectus* (Yongyan et al., 1979). Therefore, based on its advanced features, Wu Xinzhi (1981) assigned it to the subspecies *Homo sapiens daliensis*. Meanwhile, the attribution of the Jinniushan specimen to *Homo sapiens* made by Chen Tiemei, Yang Quan, and Wu En (1994), will also point to the presence of modern humans in China at a very early time and, of course, their coexistence with *Homo erectus*—in Zhoukoudian, if not in Hexian—over a long period of time. However, the consideration of the Jinniushan specimen as a *Homo sapiens* is not always accepted. Frank Poirier (1987), for example, included it in *Homo erectus*.

The "intermediate" traits of the specimens of the late Middle Pleistocene in China are, to authors like Wu Xhinzi (2004), the result of a similar process to that supported by MH: a continuity with hybridization maintained for a considerable period of time, in which local people exchanged genetic material with successive migrations from Africa. However, that is not the only plausible explanation. As Glenn Conroy (1997) reminded us, the "modern" traits of the Yunxian specimens have also been identified in other African specimens of the Middle Pleistocene, so they could be considered primitive and, therefore, of no significance to the possible phylogenetic relationships between Asian *H. erectus* and *H. sapiens*. With regard to the most evolved characters—like the cranial expansion—when speaking about the transition to the Upper Pleistocene in Java, we have noted how the coincidence of environmental pressures could lead, in close species, to the development of parallelisms that reflect a parallel evolutionary convergence.

Evolutionary meaning of the most advanced specimens of *Homo erectus* in China

The similarity between the Chinese specimens with advanced morphology in respect to *Homo erectus* relative to modern humans becomes more pronounced as we approach the Upper Pleistocene. The review by Christopher Bae (2010) on the specimens of the late Middle Pleistocene in eastern Asia sought to clarify those morphological differences comparing fossils from China, Java, Vietnam, Thailand, and Korea whose ascription is in doubt (Figure 9.24).

Among the various possibilities available to classify taxonomic specimens in his study, Bae (2010) considered whether to place them in *Homo heidelbergensis*, name a new species, or maintain the somewhat fuzzy status of "archaic *sapiens*" for those fossils that cannot be included either in *Homo erectus* or *Homo sapiens*. The author opted for the latter solution.

The Chinese specimens of Dali, Jinniushan, and Xujiayao—along with those of Ngandong, Java—were included by Bae (2010) as "archaic *sapiens*." With respect to the rest, the most interesting are the fossils of Ryonggok (North Korea).

The Ryonggok cave is located near the North Korean capital, Pyongyang. Four of its stratigraphic levels have provided a set of hominin fossils representing at least five individuals. The age of the specimens is somewhat uncertain; both Christopher Norton (2000) and Christopher Bae (2010) have noted discrepancies between the dates obtained by thermoluminescence (500,000–400,000 years) and uranium series (48,000–46,000

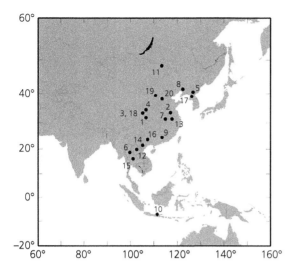

Figure 9.24 Sites with specimens considered by Christopher Bae (2010) in his study of "archaic" *sapiens* in the Far East: (1) Changyang; (2) Chaoxian; (3) Chenjiawo; (4) Dali; (5) Danyang Kunangul; (6) Dokchon Soongnisan; (7) Had Pu Dai; (8) Hexian; (9) Jinniushan; (10) Ngandong; (11) Maba; (12) Salkhit; (13) Tam Hang; (14) Tangshan; (15) Tham Kuyen, Ma U'Oi; (16) Thum Wiman Nakin; (17) Tongzi; (18) Yokpo Daehyundong, Ryonggok, Mandalli; (19) Yunxian; (20) Xujiayao; (21) Zhoukoudian. Map of Bae (2010) (in the author's original, Maba and Ngandong numbers are interchanged). Bae, C. J. (2010). The Late Middle Pleistocene Hominin Fossil Record of Eastern Asia: Synthesis and Review. *Yearbook of Physical Anthropology*, 143(51), 75–93.

years), but the study (Bae & Bae 2011) that estimated the cranial capacity between 1,450 and 1,650 cc and the presence of a chin in the associated mandible indicates that it is likely they are modern humans. The possible burial practices point in the same direction. Christopher Bae (2010) had already classified the Ryonggok crania as modern *Homo sapiens*.

As we see, the use of an "archaic" grade is convenient to help understand the intermediate state of the fossils observed in the late Middle Pleistocene throughout the ancient world. In a study of *Homo erectus* as extensive as the one conducted by Susan Antón (2003), the author used this resource to state that the specimens from

Jinniushan and Dali should be considered "archaic" *sapiens*, i.e., specimens in transition. The identical position is held by Rosenberg et al. (2006) and many other authors, now including Bae (2010; Bae & Bae, 2011).

Ian Tattersall (1986) has expressed his strong objection to distinguishing between "archaic" and "modern" specimens when discussing the same species. Such a distinction would require at least admitting two chronospecies. However, we have said on several occasions that the name "archaic" fits better for a particular grade. But, even this exception might be infeasible when the temporal coincidences indicate that we are talking about contemporary populations. The discovery of specimens in the Longlin Cave and Maludong (China) with an aspect intermediate between "archaic" and "modern" humans, but of a very recent time, highlights the need to clarify both the presence of primitive features in fossils transitioning to the Holocene, and the meaning of the arrival of modern humans to the Far East. We will return to this issue in Chapter 11.

9.5 Are the transitional species between *Homo ergaster* and *Homo sapiens* necessary?

The transformations from "archaic *sapiens*" to modern humans encompass, as we have seen, different fossils that illustrate the process of evolution to *Homo sapiens* informally referred to as grades I, II, and III, according to the guidelines of Philip G. Rightmire (2009). The convenience of using the resource of grades instead of addressing directly the issue of the species involves the retention, in almost all cases, of plesiomorphies within a process that is a mosaic evolution. However, there are abundant paleoanthropological studies that have proposed different species' names for the specimens "in transition" that are difficult to place in *Homo erectus* or *Homo sapiens*, beginning with the species that, in regard to the European process of evolution, are generally accepted: *Homo heidelbergensis*, *H. rhodesiensis*, and *H. helmei*, which would be taxa equivalent to *H.*

Box 9.20 Neanderthals in Far East Asia?

The authors who described the Ryonggok fossils assigned them to the species *Homo sapiens neanderthalensis*. However, Christopher Bae has indicated that it is common use among North Korean authors to include in the Neanderthals those "intermediate" specimens between *Homo erectus* and *Homo sapiens*. Needless to say, the morphology of the Ryonggok specimens does not closely correspond with that of *Homo neanderthalensis* (Norton, 2000).

heidelbergensis in Africa. To what extent are these species necessary to describe the phylogenetic transition to modern humans?

The answer depends on the model of evolution we accept. Within MH, *Homo heidelbergensis* would be sufficient to include all the specimens in "transition," although, being a chronospecies, it would not be necessary to take it into account. In fact, Milford Wolpoff, famous defender of the MH model, has indicated sometimes (Wolpoff et al., 1994; Wolpoff, 1999) that, strictly speaking, the taxonomy of the genus

Homo can only hold a single taxon, *Homo sapiens*, while all the others should be considered chronospecies. By contrast, RH would require adding other species.

If chronospecies are not admitted, it is obligatory to reduce the species of the genus *Homo*, or any other genus. A monophyletic lineage can be considered as a single species. But, if we also reject the idea that there has been any cladogenesis in the lineage of the genus *Homo*, then all *Homo* specimens should be included in *Homo sapiens*, which is the first defined taxon. We have not followed such a restrictive criterion in this book.

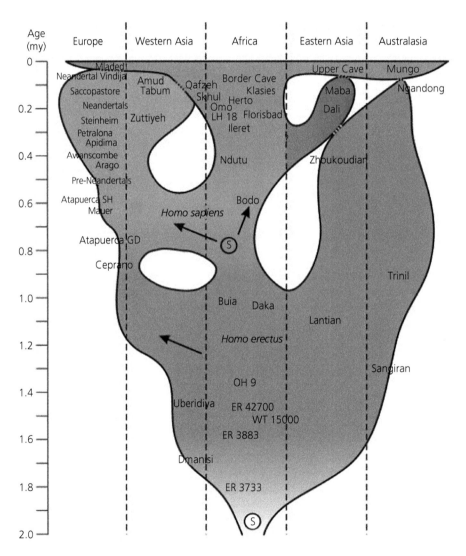

Figure 9.25 Outline of the evolution from *Homo erectus* to *Homo sapiens* proposed by Günther Bräuer (2008). Bräuer, G. (2008). The Origin of Modern Anatomy: By Speciation or Intraspecific Evolution? *Evolutionary Anthropology*, 17, 22–37.

Box 9.21 The scope of the taxon *Homo sapiens*

We have already seen that not all authors agree that the taxon *Homo sapiens* should be only applied to modern humans. Some authors reject *Homo neanderthalensis* and even *Homo erectus* and *Homo habilis* as valid taxa, separate from *Homo sapiens*. Milford Wolpoff is perhaps the leading figure of this reductionist stance on the number of species that fit into *Homo* (and in other taxa of the human lineage) (see, for example, Wolpoff et al., 1994). Milford Wolpoff (1999) supported the exclusion of *Homo habilis* from the genus as a way to elucidate the systematics of *Homo*. This proposal, as we saw, was also made previously by Bernard Wood and Mark Collard (1999b), and was reiterated by Wolpoff and Caspari (2000). While agreeing on the need to place *H. habilis* in *Australopithecus*, Wolpoff disagreed entirely with Wood and Collard on the number of species that are part of *Homo* (see Wood & Collard, 1999c).

Günther Bräuer has summarized very clearly the viability of possible intermediate species, illustrating the issue with the diagram in Figure 9.25.

The scheme by Bräuer (2008) suggests that the taxon *Homo sapiens* first appeared in Africa, later becoming the Neanderthals in Europe, while *Homo erectus* remained in Asia until being replaced by modern humans. Although he did not suggest a need to propose any intermediate species, if we admit them we have several taxonomic options (also mentioned by Bräuer, 2008), which include:

- The acceptance of a single "transitional" species, *Homo heidelbergensis*, present both in Europe and in Africa.
- The separation between a European species, *Homo heidelbergensis*, and an African *H. rhodesiensis* (which some authors call *H. helmei*).
- Although not considered by Bräuer (2008), the taxon *Homo heidelbergensis* could still include the Asian "transitional" specimens; alternatively, one or more intermediate species could be created for them—such as *Homo soloensis* (Oppennoorth, 1932); or, they could be placed in the subspecies' category, *Homo sapiens daliniensis* (Xinzhi, 1981).

The decision about which formula to choose depends, as we have said, on the accepted speciation model and on the variation and flexibility granted to the concept of species.

9.6 *Homo floresiensis*

A fossil emerging from very recent terrain on the Island of Flores represents a special case in Asian hominin evolution.

In October 2004, the journal *Nature* published a discovery made by the team of the Indonesian Centre for Archaeology in Jakarta under the direction of Mike Morwood and R. P. Soejono. At the Liang Bua site on the Island of Flores (Indonesia), near Java, with an Upper Pleistocene stratigraphic horizon, appeared a cranium, mandible, pelvis, and postcranial elements—some of them still articulated (see Figure 9.26)—of a female hominin specimen, LB1 (Brown et al., 2004). The height was nearly 1 m, and the endocranial volume was around 380 cc.

LB1 did not fit within the common framework of the process of human evolution. While there was much controversy, as we have seen in previous chapters, specialists unanimously agreed on one point: beginning from *Homo habilis*, the process of human evolution has led to increasingly larger creatures with greater cranial capacity. In fact, the increasing complexity of stone tools is often understood as a consequence of the increased brain size. Then, suddenly, this general scheme was challenged by a specimen such as LB1, similar in body and cranial volume to the smaller Pliocene *Australopithecus*, although with an age of c. 18,000 years, almost in the Holocene.

9.6.1 Liang Bua site

The Liang Bua site is a dolomitic cave located in the Wae Racang valley, Island of Flores. The first excavations of the cave took place in 1965, under the direction of the priest Theodor Verhoeven, while Raden Soejono worked in ten sectors between 1978 and 1989 (Morwood et al., 2004). Mike Morwood's team began excavations in 2001 in the areas I, III, IV, and VII (Figure 9.27a). In 2004, they continued working in the sector VII and added the adjacent sector XI (Morwood et al., 2005).

The geomorphic study of Kira Westaway, Thomas Sutikna, and collaborators (2009) identified in the cave nine major sedimentary units, numbered from the oldest to the youngest, although not all of them are present in all sectors (Figure 9.27b). The oldest showing signs of occupation—lithic artifacts—is number 3, with collapsed materials present in sectors I and IV

(a)

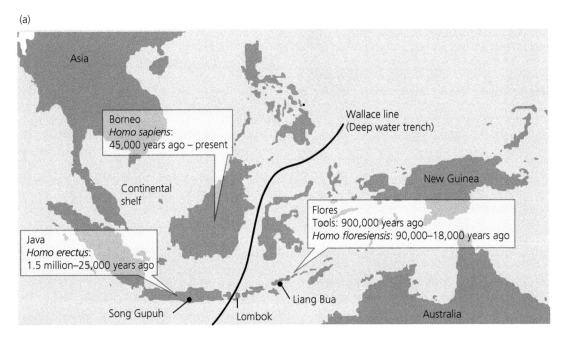

(b)

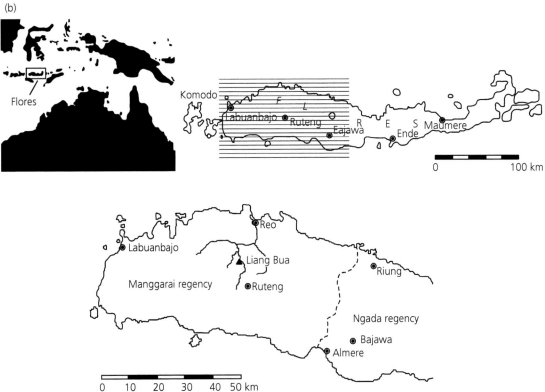

Figure 9.26 Location of the island of Flores (Indonesia) and the Ling Bua site. Up: Picture from Dalton (2005). Bottom: Pictures from Morwood et al. (2004).

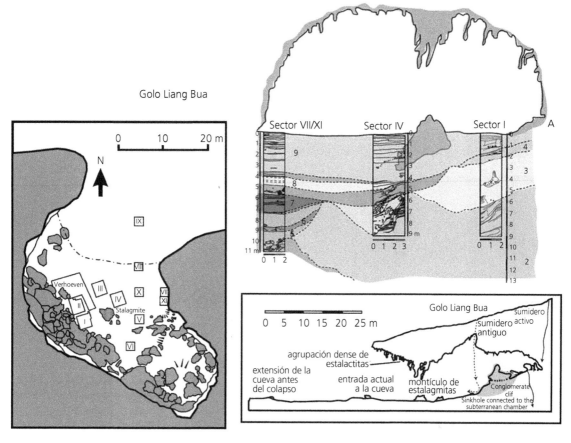

Figure 9.27 The Liang Bua cave: (a) horizontal plane; (b) sedimentary units; (c) vertical plane. Top left: Reprinted from van den Bergh, G. D., Meijer, H. J. M., Rokhus Due Awe, Morwood, M. J., Szabó, K., van den Hoek Ostende, L.W., Sutikna, T., Saptomo, E.W., Piper, P.J., & Dobney, K. M. (2009). The Liang Bua faunal remains: a 95 k.yr. sequence from Flores, East Indonesia. *Journal of Human Evolution*, 57(5), 527–537, with permission from Elsevier. Top right: Reprinted from Westaway, K. E., Sutikna, T., Saptomo, W. E., Jatmiko, Morwood, M. J., Roberts, R. G., & Hobbs, D. R. (2009). Reconstructing the geomorphic history of Liang Bua, Flores, Indonesia: a stratigraphic interpretation of the occupational environment. *Journal of Human Evolution*, 57(5), 465–483, with permission from Elsevier. Bottom right: Reprinted from Westaway, K. E., Roberts, R. G., Sutikna, T., Morwood, M. J., Drysdale, R., Zhao, J. x., & Chivas, A. R. (2009). The evolving landscape and climate of western Flores: an environmental context for the archaeological site of Liang Bua. *Journal of Human Evolution*, 57(5), 450–464, with permission from Elsevier.

(Figure 9.27c) and an age between 95,000 and 52,000 years (thermoluminescence method, TL). Unit 4, also with archaeological signs of occupation, is present in sectors I, III, and IV; its age is between 74,000 years (ESR/uranium series on a *Stegodon* tooth) and just under 38,000 years (uranium series).

LB1 comes from unit 7, which is the first with hominin remains. Michael Morwood and collaborators (2004) analyzed two samples of coal by ^{14}C from the deposits adjacent to the skeleton, obtaining an average of 18,000 years (18.7–17.9 and 18.2–17.4 Ka, respectively). It is so recent that the specimens only can be called "fossil" in a loose sense: the skeleton has not yet undergone the process of mineralization.

With regard to the sedimentary unit 7, the authors obtained by thermoluminescence an age range between 35±4 and 14±2 Ka (Morwood et al., 2004). The work by Kira Westaway, Thomas Sutikna et al. (2009) reported a range between 40,000 and 14,000 years (uranium series).

The sedimentary unit 9, with modern human remains (Morwood et al., 2005), has been dated between 11,200 and 2,620 years by the ^{14}C method (Westaway, Sutikna et al., 2009).

Thus, the Liang Bua sample reveals an occupation that continued until approximately 100,000 years ago. However, other sites on the island push the signs of anthropic activity back to 800,000 years ago (Section 9.6.5).

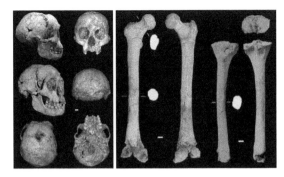

Figure 9.28 Left: Cranium of LB1. Right: Femur and tibia of LB1 (Brown et al., 2004). Scale, 1 cm in both cases (Morwood et al., 2004).

What kind of hominin is the "Hobbit of Flores," as the media baptized it?

9.6.2 LB1 morphology

Figure 9.28 shows the cranium, femur, and tibia of LB1. Figure 9.29 shows the LB1 mandibles. In the description of the specimen, Peter Brown and collaborators (2004) indicated its great morphological proximity to *Homo erectus*, leaving aside the question of size. Thus, cranial shape ratios closely follow the patterns of *H. erectus*. Viewed from behind, the parietal outline is similar to that of the cranium of *H. erectus*, although with a reduced cranial height. In addition, the LB1 face is in general similar to that of the other members of the genus *Homo*.

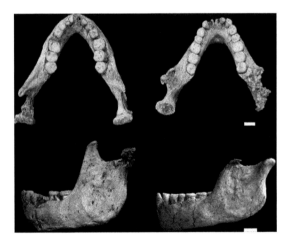

Figure 9.29 LB1 mandibles from sector VII and LB6/1 from sector XI of Liang Bua. Reprinted by permission from Macmillan Publishers Ltd: Morwood, M. J., Brown, P., Jatmiko, Sutikna, T., Wahyu Saptomo, E., Westaway, K. E.,. . . Djubiantono, T. (2005). Further evidence for small-bodied hominins from the Late Pleistocene of Flores, Indonesia. *Nature*, 437(7061), 1012–1017.

The lack of characteristic traits of the *Australopithecus* masticatory apparatus, and a femur within the wide range of variation of *H. sapiens*, are features that place LB1 within the genus *Homo*. However, its very small size and traits related allometrically with *Australopithecus* led Peter Brown and collaborators (2004) to name a new species, *Homo floresiensis*, with LB1 as holotype, and only one paratype at the time: LB2, a P3 tooth of the left mandible. Other exemplars found in sectors IV, VII, and IX of Liang Bua in 2004 completed the hypodigm available today of *Homo floresiensis* (Morwood et al., 2005). The new specimens, from at least nine individuals, confirmed the relevance, according to Mike Morwood and colleagues (2005), of the taxon *Homo floresiensis*, indicating that it is not an allometric version of a smaller size of *Homo erectus*. In the words of the authors, some morphological features, such as the humeral torsion and ulna, are not present in any other known hominin species.

9.6.3 Reasons for LB1 dwarfism

Peter Brown et al. (2004), as well as Michael Morwood et al. (2005), maintained that *Homo floresiensis* retained a close morphological relationship with *Homo erectus*, separating entirely from *Australopithecus*—a taxon with comparable body size. The explanation for the small size of the Flores hominin has been the subject of considerable debate (Balter, 2004a, 2004b) in which the different stances can be reduced to the following:

I. Isolation Hypothesis. The small size of the Flores hominin is due to the effect of natural selection and/or genetic drift under isolated island conditions, which would justify the proposal of a new species.
II. Small colonizer hypothesis. The explanation of *H. floresiensis* size is that the colonizer of the island was a small-sized hominin, *Homo habilis* or an australopith.
III. Microcephaly Hypothesis. They are pygmies (modern humans), which in some cases show microcephaly, a pathological condition.

Let's start with the latter. The study of LB1 morphology conducted by Teuku Jacob and collaborators (2006), using osteometric data and skeleton scans, led the authors to conclude that the particularities of the specimen are due to a combination of traits that are not primitive but characteristic of regional varieties present in other modern human populations—including those currently in Flores—nor are they derived, but are due to a remarkable developmental disorder. Thus, Jacob et al. (2006) indicated that dental and

---

Box 9.22 The LB1 cranium

Two primitive traits mentioned by Peter Brown and collaborators (2004) as present in LB1 and *Homo erectus*, but absent in modern humans, are (1) the deep fissure that separates the mastoid process from the tympanic petrous crest and (2) the gap between the tympanic plate and the glenoid pyramid. Jacob et al. (2006) noted that the skulls of modern humans from Australia and Tasmania quite often exhibit these two features, while the second feature is also present in two Australian Pleistocene specimens: Kow

Swamp 5 and, in reduced form, Keilor. Newly discovered fossil crania such as B:OR-15:18-001 and B:OR-14:8-005 of small bodied *Homo sapiens* from Palau (Micronesia) (Berger et al., 2008) may represent the case for modern human insular dwarfism. Lee Berger and collaborators (2008) argued that some of the traits attributed exclusively to *Homo floresiensis* might appear, according to the evidence of Palau, as correlates of the small body development in pygmy populations of *Homo sapiens*.

---

mandibular features that appear to differ from those of *Homo sapiens*, as do the rotated premolars and the absence of chin, are present in the Rampasasa pygmies living today near Liang Bua. Up to 140 cranial features of LB1 are within the range of variation of modern humans, according to Jacob et al. (2006).

In contrast to these extended characters, the study by Teuku Jacob and collaborators (2006) pointed to the presence in LB1 of individual signs of a developmental abnormality—including microcephaly—not found in any other specimens at the site. Among these is the advanced closure of the sutures on the LB1 cranium, which in some cases is difficult to identify even by a scanner, and the facial asymmetry. This was made obvious when the specular image of each side of the face was obtained (Figure 9.30).

9.6.4 The LB1 brain

The study of the cranium and brain of LB1 and the evaluation of its possible microcephalic condition have been the subject of various studies, among which are those conducted by Dean Falk.

Dean Falk and collaborators (2005) carried out a comparison of a virtual endocranial cast of LB1 with

endocrania from great apes, *Homo erectus, Homo sapiens*, a pygmy, a microcephalic human, specimen Sts 5 (*Australopithecus africanus*), and KNM-WT 17000 (*Paranthropus aethiopicus*). The morphometric, allometric, and morphological results supported the view that LB1 is not microcephalic or a pygmy; the LB1 brain/body ratio is characteristic of an australopithecine, but the shape of the endocranium resembles that of *Homo erectus* (Falk et al., 2005). Nevertheless, Falk et al. (2005) indicated the presence of derived traits in the brain of LB1 in the frontal and temporal lobes, and apomorphies in the lateral groove consistent with a large capacity for higher cognitive processing.

After examining the reconstruction by Dean Falk et al. (2005), Jochen Weber, Alfred Czarnetzki, and Carsten Pusch (2005) criticized some of its methodological aspects, such as deducing the size and proportions of the brain from an endocranial cast, and not considering the great variability of the shapes of microcephalic brains. Because Falk et al. (2005) only included one specimen in their comparisons, Weber et al. (2005) argued that the hypothesis of a pathology in LB1 could not be excluded. A similar point of view was held by Robert Martin and collaborators (2006), noting that in some cases—he included two

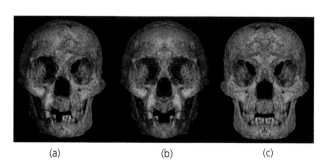

(a) (b) (c)

Figure 9.30 Facial asymmetries of LB1 indicated by comparison between the specimen (a) and the specular duplication of the right side (b) and of the left side (c). Jacob, T., Indriati, E., Soejono, R. P., Hsu, K., Frayer, D. W., Eckhardt, R. B. . . . Henneberg, M. (2006). Pygmoid Australomelanesian Homo sapiens skeletal remains from Liang Bua, Flores: Population affinities and pathological abnormalities. *PNAS*, 103, 13421–13426. Copyright (2006) National Academy of Sciences, U.S.A.

examples, one of them a woman of Lesotho—the microcephalic brain appears normal aside from its size.

In addition to direct responses to these criticisms, which appeared in the same journal that published them (*Science*), Dean Falk and collaborators (2007) conducted a new comparative analysis in ten healthy modern humans, nine microcephalics, the microcephalic woman indicated in the comment by Martin et al. (2006), and one woman with pathological dwarfism (Figure 9.31, right). Discriminant and canonical analyses performed on the sample grouped LB1 with healthy modern humans, while the two women with pathologies (358f, the women of Lesotho, and 752f, the dwarf, at the bottom right of Figure 9.31) grouped

with the microcephalics. This result supports the view of LB1 as a member of a different species, and not as a pathological modern human. A later comparative study by Dean Falk and collaborators (2009) elaborated on the same conclusions.

9.6.5. Older fossils of *Homo floresiensis*

The site of Mata Menge (So'a Basin, Central Flores) has provided the oldest known specimens of *H. floresiensis*. The haul is meager: a mandible fragment and six isolated teeth belonging to at least three small-jawed and small-toothed individuals. The Mata Menge materials are close in morphological characteristics to the

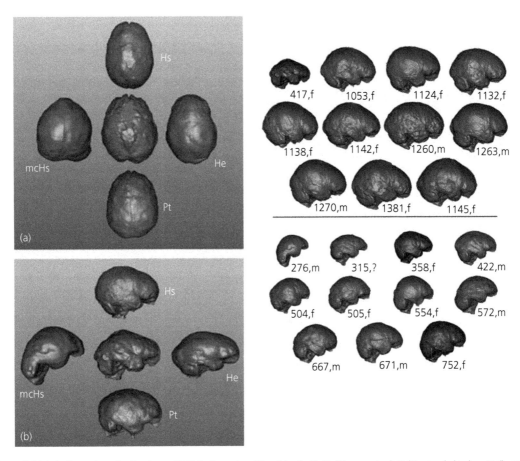

Figure 9.31 Left: Comparison of a virtual cast of LB1 (in the center of the picture) with Hs (*Homo sapiens*), Pt (*Pan troglodytes*), *mcHs* (human microcephalic), He (*Homo erectus*). (a) Dorsal view; (b) right lateral view (the images are not at the same scale). Right: Comparison in right lateral view of the virtual cast of ten normal humans (above) and nine microcephalics (below). LB1 appears at the top. The images indicate their cranial capacity and sex: f, woman; m, man. For explanation, see text. Right: Falk, D., Hildebolt, C., Smith, K., Morwood, M. J., Sutikna, T., Jatmiko,. . . Prior, F. (2007). Brain shape in human microcephalics and Homo floresiensis. *Proc Natl Acad Sci U S A*, 104 2513–2518. Copyright (2007) National Academy of Sciences, U.S.A. Left: From Falk, D., Hildebolt, C., Smith, K., Morwood, M. J., Sutikna, T., Brown, P.,. . . Prior, F. (2005). The Brain of LB1, Homo floresiensis. *Science*, 308, 242–245. Reprinted with permission from AAAS.

Box 9.23 Analysis of LB1 traits: the dwarfism issue

Several morphometric studies have aimed to analyze the condition of the LB1 traits. Debbie Argue and collaborators (2006), Adam Gordon, Lisa Nevell, and Bernard Wood (2008), and Karen Baab and Kieran McNulty (2009) agreed that the metric and non-metric results support the consideration of a different species, and not of a pathological specimen of *Homo sapiens*. However, these studies did not agree in stating the possible phylogenetic relationships of *Homo floresiensis*. The initial position of Peter Brown and collaborators (2004), in the paper reporting the discovery of LB1, was definitive with respect to the non-pathological condition of the specimen. For these authors, the dwarfism of the specimen should be attributed to the selection of small bodies in environments of low caloric supply. This is the Isolation Hypothesis we mentioned above (Section 9.6.3), which relates the decrease in size with the insular condition. Thus, scientists have interpreted in this way the presence in Madagascar of two fossil pygmy hippopotamus species—*Hippopotamus lemerlei* and *Hippopotamus madagascarensis*—and a dwarf caprid, *Myotragus balearicus*, in the island of Mallorca. In fact, in the same sedimentary unit LB1 of Liang Bua Cave are remains of a dwarf elephant, *Stegodon*.

Box 9.24 Optimal body size

The theory of optimal body size has been criticized by Pasquale Raia, Francesco Carotenuto, and Shai Meiri (2010), analyzing the relationship between the size of 4,004 mammal species and the phylogenetic tree that relates them. The authors found no specific size to serve as an evolutionary attractor but provided consistent evidence that large mammals (>10 kg) tend to reduce their size on islands.

Ling Bua fossils, with the exception of the lower first molar, more primitive, and even smaller in size (van den Bergh, Kaifu et al., 2016). Their age is 0.7 Ma, as indicated by ^{40}Ar/^{39}Ar and fission track dates (Brumm et al., 2016).

The hypothesis of an early colonization of the island by small hominins seems to be supported by the Mata Menge fossils. However, as Gerrit D. van den Bergh, Yousuke Kaifu et al. (2016) say: "The Mata Menge fossils are derived compared with *Australopithecus* and *H. habilis*, and so tend to support the view that *H. floresiensis* is a dwarfed descendent of early Asian *H. erectus*. Our findings suggest that hominins on Flores had acquired extremely small body size and other morphological traits specific to *H. floresiensis* at an unexpectedly early time" (p. 245).

9.6.6 Consequences of island life

The model that supports the relationship between isolation and body size reduction is the Optimal Body Size Theory (OST), which affirms that, taking into account

the variables of obtaining food and reproduction, for any large lineage there is a certain size that provides the optimal biological fitness (Box 9.24). In mammals, the calculation of the OST produces a low figure, 100 g (Brown et al., 1993). Interspecies' competition leads to increased body size; however, in isolation and in the absence of large predators, as is the case on islands—although the Flores Komodo dragon, whose remains were found in Liang Bua, contradicts this idea—a size tending toward a low OST can be selected.

As Eleanor Weston and Adrian Lister (2009) pointed out, a reduction of body size in mammals usually involves only a moderate decrease in brain size. This is so because the brain completes its growth during ontogenesis before the rest of the body. Nevertheless, mammalian dwarfism in the islands represents a special case. In the pygmy hippo of Madagascar, to which we have referred, the endocranial capacity is up to 30% less than that of its African ancestor with equivalent body mass. The study of Weston and Lister (2009) showed that in islands it is possible to find a reduction in endocranial capacity greater than that indicated by a general model of intraspecific proportions between body and brain. The most obvious reason for this has been expressed by Daniel Lieberman (2009): brain tissue is metabolically expensive, so that organisms with smaller brains can save energy when resources are scarce.

African *Homo erectus*' values of 60 kg weight and 991 cc of cranial volume would produce, following a general scale and taking as average the 23 kg weight for *Homo floresiensis*, a cranial volume of 704 cc. This circumstance was one of the main reasons that led Robert Martin et al. (2006) to prefer the hypothesis of a pathological condition for LB1. But, when applying the reduced ratios obtained by Weston and Lister (2009) in Madagascan hippos, the figure calculated for the cranial volume of *H. floresiensis* would be 493 cc. It is still higher than that displayed by the specimens of Liang Bua (380–430 cc) but, as Weston and Lister (2009) indicated, beginning with a specimen of African *Homo erectus* (*H. ergaster*) such as KNM-ER 3883, with a smaller size brain, it would be possible to get to figures in the range of LB1 cranial size. The same conclusion is reached when considering that the ancestor of *H. floresiensis* is *H. georgicus*, with a body mass of 40 kg (Lordkipanidze et al., 2006) and a cranial capacity of 650 cc (Lordkipanidze et al., 2007).

9.6.7 LB1 postcranial elements

Given the impossibility of retrieving genetic material from the Flores' specimens—due to climate conditions whose influence we will explain in Chapter 10—the analysis of the postcranial morphology of LB1 completes the available evidence for its taxonomic classification.

After the initial descriptions by Brown et al. (2004) and Morwood et al. (2005), Matthew Tocheri and collaborators (2007) conducted a study on the wrist of the specimen. The authors identified primitive traits shared by the "African apes + human" clade, which became derived traits in Neanderthals and modern humans. This result goes against the hypothesis that sees in LB1 a pathological *Homo sapiens*. Susan Larson and collaborators (2009)—among them Tocheri—conducted an analysis of the six specimens of *Homo floresiensis* with upper limb remains—LB1 to LB6—some reduced to fragments but others almost complete (Box 9.25). The results obtained by Larson et al. (2009) indicated the presence in the sample of a mosaic combination (primitive and derived traits), never found in modern humans, whether pathological or healthy.

With respect to the locomotor system preserved in, at least, nine specimens—LB1, LB4, LB6, LB8, LB9, LB10, LB11, LB13, and LB14 (see Figure 9.32)—William Jungers and collaborators (Jungers, Harcourt-Smith et al.,

2009; Jungers, Larson et al., 2009) carried out a study that showed a mosaic combination similar to that detected in the upper limbs, also nonexistent in modern humans. This evidence, as well as that derived from the foot anatomy, led Jungers, Harcourt-Smith et al. (2009) to consider that, although undoubtedly a biped, *Homo floresiensis* had as a hominin ancestor one more primitive than *Homo erectus*. Peter Brown and Tomoko Maeda (2009) arrived at the same conclusion after a comparative examination of the mandibular and dental apparatus of LB1, LB2, and LB6.

9.6.8 *The phylogenetic consideration of Homo floresiensis*

A question to be resolved concerns, therefore, the phylogenetic affinities of *Homo floresiensis*. The work by Peter Brown and collaborators (2004), who named the

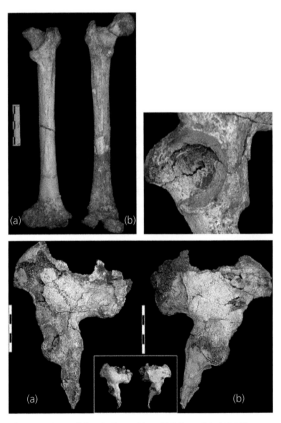

Figure 9.32 Top left: right femur (a) and left femur (b) of LB1. Top right: acetabulum of the left LB1 pelvis. Bottom, outer side (a) and internal side (b) of the coccyx bone LB1; cm scale. Left: Reprinted from Larson, S. G., Jungers, W. L., Tocheri, M. W., Orr, C. M., Morwood, M. J., Sutikna, T.,. . . Djubiantono, T. (2009). Descriptions of the lower limb skeleton of *Homo floresiensis*. *Journal of Human Evolution*, 57(5), 538–554, with permission from Elsevier.

Box 9.25 LB1 humerus

Susan Larson and collaborators (2007) studied the humerus LB1, detecting a torsion lower than in modern humans. Trenton Holliday and Robert Franciscus (2012) compared the size of the upper and lower limbs in two specimens of similar size, AL 288–1, *Australopithecus afarensis*, and LB1, *Homo floresiensis*. Both share a somewhat larger humerus and a smaller femur than what would correspond to a modern human of the same height. Both studies distance LB1 from the anatomical configuration of *Homo sapiens*.

taxon, and the subsequent study by Michael Morwood et al. (2005) expanding its hypodigm, interpreted that *H. floresiensis* had descended from an ancestral population of *Homo erectus*, basing this idea, particularly, on its cranial features; we must point out that for these authors, the taxon includes the African specimens and those from Dmanisi. However, at the end of their paper, Morwood and collaborators (2005) asserted that the genealogy of *H. floresiensis* remains uncertain. The idea of an affinity to *Homo erectus* was maintained with some doubts in the first phylogenetic tree proposed in the commentary by Marta Mirazón Lahr and Robert Foley (2004)—with a somewhat popularizing purpose—that accompanied the publication of the discovery (Figure 9.33, left). Daniel Lieberman (2009) included as a possible, but uncertain, alternative that *H. floresiensis* could be an offspring of either *Homo erectus* or *H. habilis* (Figure 9.33, right).

An alternative account of the possible phylogenetic relationships between *Homo floresiensis, Homo habilis,* and *Homo erectus* was suggested by Bernard Wood (2011). As noted before, Wood is one of the authors who prefer to remove *H. habilis* from the *Homo* genus. If the

first species of our genus was *H. erectus,* then, according to Wood (2011) it would be possible to propose an evolutionary scenario that considers the emergence of *Homo* either in continental Asia or Southeast Asia, which then migrated to Africa. This would change the model "Out of Africa" for one of "Into Africa" and, according to Wood (2011), such an undertaking would be consistent with the oldest morphology of the Liang Bua and Dmanisi specimens when compared to the Asian *Homo erectus* (sensu stricto). However, for that idea to win strong support, the presence of hominins in Flores should be previous to that of *Homo erectus* in Java; a possibility that, for now, does not have any empirical support.

Debbie Argue and collaborators (2009) conducted a cladistic analysis on 60 cranial, mandibular, and postcranial traits of a hominin sample, including *Homo floresiensis* (Table 9.2). The conclusions of the study point out that the taxon would have appeared in the Upper Pliocene or Lower Pleistocene, with two equally parsimonious cladograms for (1) a separation of *H. floresiensis* posterior to *H. rudolfensis* (represented by KNM-ER 1470), but prior to the appearance of *H.*

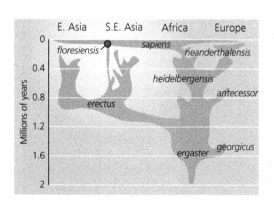

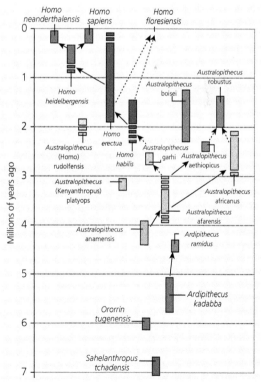

Figure 9.33 Tentative phylogenetic trees suggested by Mirazón Lahr and Foley (2004) (left) and Lieberman (2009) (right). Left: Reprinted by permission from Macmillan Publishers Ltd: Mirazón Lahr, M. & Foley, R. (2004). Human evolution writ small. *Nature*, 431(7012), 1043–1044. Right: Reprinted by permission from Macmillan Publishers Ltd: Lieberman, D. E. (2009). Palaeoanthropology: Homo floresiensis from head to toe. *Nature*, 459(7243), 41–42.

Table 9.2 Liang Bua sedimentary units, indicating the estimated ages and the dating techniques used (Westaway, Sutikna et al., 2009) (modified)

Unit	Sedimentary Classification -mean particle size (im)	Average Thickness (m)	Extent	Sedimentary structure	LBC	Clasts	Sorting[a]	Sorting[b]	Composition	Section 1.01 Dip (° and changes key mean particle (m) direction)	Variability	Fossils	Archaeology	Diagenetic/changes reworking	Summary	Location Key Sectors
1 Con-glomerate	Gravel (clast supported) matrix silty sand- 134	~6	Rear cave	Massive with some fining upward	Sharp	Y	2.0	0.8	Meta/Volcanic limestone	9NW	Variable coarse and fine	None	Stone tools	Eroded reworked	High energy alluvial deposit	CP1-3
2 Basal	Silty clay with lenses of iron oxide and sand (matrix convoluted supported)—138	4–5	Front cave	Low angle cross/non parallel convoluted bedding	?	N	-	-	-	20–45 s	Homogenous	Sterile	None	Convoluted	Silty basal suspension deposit	I, III, IV
3 Collapse	Silty clay matrix (matrix supported)- 14	2.5–3	Center cave	Massive	Sharp	Y	1.0	0.4	Limestone blocks (>4 m)	45 E	Homogenous	Bone- high densities	Stone tools (high densities)	Slumping and compaction	Clayey suspension deposit cont. collapse material	I, IV
4 Occupation	Silty clay (matrix supported)—22	1–2	Center cave	Parallel bedding layered by flowstones	Sharp	Y	0.5	0.5	Meta/Volcanic limestone	-	Homogenous	Bone: Stegodon hominins	Stone tools	-	Massive silty clay containing occupation evidence	I, III, IV
5 Channel	Clayey silt (matrix supported)-	1.5–2	East wall	Massive	Diffuse	N	-	-	Limestone	30 N	Homogenous	Bone	Stone tools hearth	-	Silty suspension deposit	VII/XI
6 Eroded / Reworked Conglomerate	Gravel clasts/clayey silt matrix (clast supported)—143—20	0.5	East wall	Massive	Sharp	Y	0.5	0.8	Meta/Volcanic	30 N	Variable	Sterile	Stone tools	Reworked matrix and clasts	Reworked conglomerate unit	VII/XI
7 Skeleton level	Clayey silt (matrix supported)- 8	1.5–2	East wall	Massive	Sharp	Y	0.5	0.8	Limestone	30 N	Homogenous	Bone— high densities	Stone tools hearth	-	Massive silty deposits cont. extensive occupation	VII/XI
8 Volcanic	Fine silts/sandy silts (matrix supported)—20	2.5–3	East wall/center cave	Laminations/loading structures/massive	Diffuse/very sharp	N	-	-	-	30 N	Variable many diff units	Sterile	None	Loading and lenses reworked	Primary and reworked volcanically derived deposits	III, IV, VII/XI
9 Younger occupation	Clayey silts/sandy silts (matrix supported)- 35	2.5–4	Front cave/rear cave	Horizontal bedding	Diffuse	Y	1.0	0.3	Limestone	-	Homogenous	Bone	Stone metal tools burials pottery beads	-	Horizontally lain silts by sheetwash	I, III, IV, VII/XI

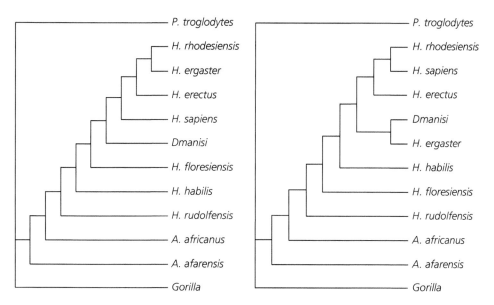

Figure 9.34 Cladograms of the same parsimonious level by Argue et al. (2009). Reprinted from Argue, D., Morwood, M. J., Sutikna, T., Jatmiko, & Saptomo, E. W. (2009). Homo floresiensis: a cladistic analysis. *Journal of Human Evolution*, 57(5), 623–639, with permission from Elsevier.

habilis (represented by KNM-ER 1813 and OH 24), or (2) a separation of *H. floresiensis* posterior to *H. habilis* (Figure 9.34). The hypotheses considering the specimens of Liang Bua as a dwarf population of *Homo erectus* or *H. sapiens* are not supported by the work of Argue et al. (2009), and, thus, the pathological explanation was rejected.

9.6.9 The industry of Flores

The stone tools associated with *Homo floresiensis* raise further controversy about the evolutionary stage, adaptation, and skills of the Flores hominins.

Human presence on the island of Flores has been known since 1957, when Theodor Verhoeven identified *Stegodon* fossils associated with lithic tools (Verhoeven, 1958). The sites with the most ancient carvings are in the Soa basin, central Flores. The sites of Mata Menge and Tangi Talo are the most studied. In Mata Menge a total of 507 artifacts appeared and were associated with *Stegodon* fossils in a fluvial level of the Ola Buia formation, to which Paul Sondaar and collaborators (1994), using paleomagnetism, granted an age close to the Matuyama–Brunhes transition. The 0.78 Ma of the transition led the authors to attribute the production of the utensils to *Homo erectus*. But, as Michael Morwood and collaborators (1998) indicated, the paleomagnetic column did not directly date the horizon containing the fossils and tools, but only the deposits located three meters above it.

The abundance of volcanoes in the area allowed taking samples at Mata Menge, above and below the

levels with lithic tools (MM1 and MM2 at the bottom of Figure 9.35b). The respective ages obtained by Morwood et al. (1998) using radiometry (fission track) were: 0.90±0.07 Ma for Tangi Talo; 0.88±0.07 Ma (MM1) and 0.80±0.07 Ma (MM2) for Mata Menge. The authors also attributed the carving to *Homo erectus*.

New excavations in Mata Menge provided over 487 artifacts in 2004–05. Among the utensils recovered, Adam Brumm and collaborators (2006) identified up to five basic types of carving technique that, in all cases, involved percussion with a core serving as a hammer. On average six flakes were extracted before discarding the worked core, with a maximum of 26 flakes (Figure 9.36).

Aside from the increase in the number of tools found and the better identification of the working process, one aspect of the research of Brumm et al. (2006) deserves to be highlighted: the authors emphasized the cultural continuity that occurred in Flores with the utensils of Mata Menge and the oldest of Liang Bua. Despite the geographical—50 km—and particularly, the temporal distance—over 700,000 years—there are striking similarities in the lithic sets of both sites, especially in relation to the chosen materials, the maximum size of the flakes, and the techniques of core reduction. The biggest differences are (1) the more common use of flakes as cores in Liang Bua in contrast to the use of pebbles in Mata Menge, (2) the abundance of bipolar reductions in Liang Bua—only one example in Mata Menge—and (3) the exclusive presence in the most modern site of heat fractures.

Table 9.3 Sample used in the cladistic analysis by Argue et al. (2009)

Specimen	Original/Cast/ Reference	Curatorial Institution	Original Site	Date	Species
Sangiran 2, Sangiran 17, Trinil	Original	Forschungsinstitut Senckenberg, Frankfurt, Senckenberg, Frankfurt, Germany (Sangiran 2); Geological Museum, Bandung, Indonesia (Sangiran 17); National Museum of Natural History, Leiden, Holland (Trinil)	Indonesia	c. 1.8 Ma to c. 50 Ka	*H. erectus*
KNM-ER 3733, KNM-ER 3883	Original	Kenya National Museum, Nairobi, Kenya	East Africa	1.8 Ma, 1.55–1.6 Ma	*H. ergaster*
02280, 02282, 02700	Casts; Rightmire et al. (2006)	Georgian State Museum, Tbilisi, Georgia	Georgia	1.8 Ma	Possible affinities: *H. erectus, H. georgicus* (Gabunia et al., 2002), *H. ergaster*
KNM-ER 1813, OH 24	Casts; Wood (1991) (KNM-ER 1813)	Australian National University (ANU), Canberra, Australia	East Africa	1.7–1.88 Ma	*H. habilis*
KNM-ER 1470	Casts; Wood (1991)	Australian National University (ANU), Canberra, Australia	East Africa	1.88 Ma	*H. rudolfensis*
Sts 5, Sts 7, Stw 505	Casts	Australian National University (ANU), Canberra, Australia	South Africa	2.8–2.3 Ma	*A. africanus*
AL444-2	Kimbel et al. (2004)		Awash River Tributary, Ethiopia	3.0±0.02 Ma	*A. afarensis*
H. floresiensis: LB1	Original	National Archaeological Research Centre, Jakarta, Indonesia	Flores, Indonesia	18 Ka (luminescence dates of 35±4 Ka and 14±2 Ka)	*H. floresiensis*
H. floresiensis post-cranial material: right humerus LB1/50; clavicle LB1/5; right scapula LB6/4; 3 carpals of LB1 left wrist	Larson et al. (2007); Tocheri et al. (2007)	National Archaeological Research Centre, Jakarta, Indonesia	Flores, Indonesia	LB1: 18 Ka (luminescence dates of 35±4 Ka and 14±2 Ka), LB6/4: 15.7–17.1 Ka	*H. floresiensis*
Kabwe	Original	National History Museum, London, UK	Zimbabwe	Unknown	*H. rhodesiensis*
H. sapiens (6 males, 5 females)	Original	Australian National University (ANU), Canberra, Australia	Indonesia (2), India (1), Africa (1), Egypt (1), "Caucasoid" (1), New Guinea (3), Polynesia (3), Japan (Ainu) (3)	Modern	*H. sapiens*
Chimpanzees (2 males, 2 females)	Original	Australian National University (ANU), Canberra, Australia; Australian Museum, Australian Museum, Sydney, Australia	Unknown	Modern	*P. troglodytes*
Gorilla (2 males, 2 females)	Original	Australian National University (ANU), Canberra, Australia; Australian Museum, Sydney, Australia	Gabon, Cameroon	Modern	*G. gorilla, G. beringei*

Contrasting with the remarkable absence of lithic utensils in Southeast Asia, Flores represents an exceptional opportunity, both for the abundance of tools and the ease of dating the terrain in which they were found in situ. Due to these advantages, it is possible to ascertain the cultural level of the hominins who colonized the island.

(a)

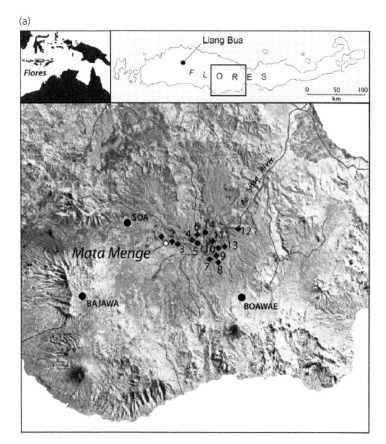

(b)

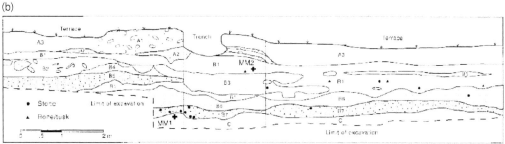

Figure 9.35 (a) Lower and Middle Pleistocene sites in Soa basin on the Island of Flores (Indonesia): (1) Kobatuwa; (2) Lembahmenge; (3) Wolo Sege; (4) Boa Lesa; (5) Ola Bula; (6) Tangi Talo; (7) Wolo Milo; (8) Wolo Keo; (9) Sagala; (10) Dozu Dhalu; (11) Kopowatu; (12) Ngamapa; (13) Pauphadhi; (14) Deko Weko. (b) Mata Menge stratigraphic levels indicating the extraction zones of the volcanic samples MM1 and MM2. Units A and B contain fossils and tools (marked with black circles in the figure). Top: Reprinted by permission from Macmillan Publishers Ltd: Brumm, A., Jensen, G. M., van den Bergh, G. D., Morwood, M. J., Kurniawan, I., Aziz, F., & Storey, M. (2010). Hominins on Flores, Indonesia, by one million years ago. *Nature*, 464, 748–752. Bottom: Reprinted by permission from Macmillan Publishers Ltd: Morwood, M. J., O'Sullivan, P. B., Aziz, A., & Raza, A. (1998). Fission-track ages of stone tools and fossils on the east Indonesian island of Flores. *Nature*, 392, 173–176.

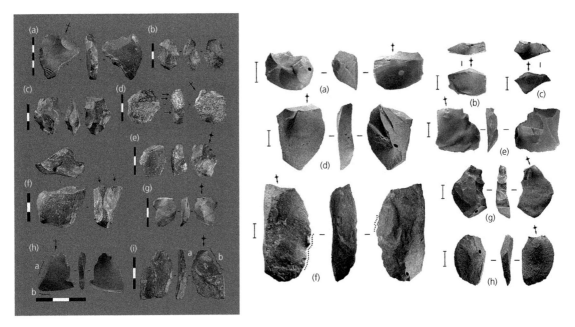

Figure 9.36 Left: Stone tools of Mata Menge (Flores): (a, b) radial bifacial flint cores; (c, d) radial bifacial cores of volcanic/metamorphic material (V/M, arrows in d indicate three carving notches); (e) retouched flakes V/M; (f) cores V/M of pebbles with carved notches indicated by arrows; (g) flint flakes; (h) flint flakes with rounded microwear marks (line *a*) and carved edge (line *b*); (i) flakes V/M with signs of abrasive polishing and smoothing (*a*) small notches and grooves (*b*). Right: Liang Bua tools. Flakes extracted from units 9 (A, B), 2 (C), and 4 (D–H). Flakes were obtained by knocking on the ventral surface of a larger flake (indicated by a point). The arrows show the axis of percussion. Left: Reprinted by permission from Macmillan Publishers Ltd: Brumm, A., Aziz, F., van den Bergh, G. D., Morwood, M. J., Moore, M. W., Kurniawan, I.,. . . Fullagar, R. (2006). Early stone technology on Flores and its implications for Homo floresiensis. *Nature*, 441(7093), 624–628. Right: Reprinted from Moore, M. W., Sutikna, T., Jatmiko, Morwood, M. J., & Brumm, A. (2009). Continuities in stone flaking technology at Liang Bua, Flores, Indonesia. *Journal of Human Evolution*, 57(5), 503–526, with permission from Elsevier.

None of the work on the industries of Flores mentioned up to now defined the lithic mode of the island's hominin carvings. This issue was addressed by Mark Moore and Adam Brumm (2007), who showed the inadequacy of the classical approaches based on African cultural traditions. When trying to relate stone artifacts that appear to the East of the Movius line with Modes 1 and 2 in Africa, West Asia, and Europe, researchers experienced the difficulty of using those categories. Moore and Brumm (2007) indicated that, as an alternative, many authors focused on the morphological differences of larger scale between the different sets from Southeast Asia, creating a dichotomy between large tools based on large pebbles, or a *core tool*, and the small, flake-based *flake tool*. By performing a chronological differentiation between these industries, it was possible to support the hypothesis that these corresponded to different hominin dispersals, concluding that even different species were involved.

However, Moore and Brumm (2007) argued that the archaeological record of Flores makes infeasible the distinction between *core tools* and *flake tools*. For these authors, both processes would be aspects of the same core-reduction sequence, which involves, first, obtaining large, flaked stones from pebbles, abandoning some of the flake-cores, and then transporting the remaining cores to another location for further carving. This sequence would give rise to two geographically separate sets: those containing discarded large flake blanks and those showing work on the flake-core. Small flakes are produced as a by-product of reducing the core, which in the second set become the final product.

9.6.10 The Liang Bua advanced utensils

The oldest deposits of Liang Bua (95,000–74,000 years) have provided, as we indicated, similar tools to those found in Mata Menge.

In addition, Liang Bua contains more advanced tools: points, perforators, blades, and microblades that probably served as fishhooks, associated with *Stegodon* remains (Morwood et al., 2004). These advanced tools

---

Box 9.26 A problem of navigation

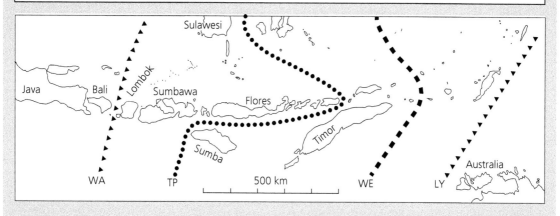

Bednarik, R. G. (1997a). The earliest evidence of ocean navigation. *The International Journal of Nautical Archaeology*, 26, 183–191. Islands of the Malay Archipelago near Flores. WA, WE, and LY are the lines of Wallace, Weber, and Lydekker; TP is the boundary of the Eurasian and Australian continental plates. All the lines represent boundaries for the fauna and flora (illustration by Robert Bednarik, 1997a).

An early occupation of the island raises the question of how the hominins could overcome the geographical barrier of the straits separating Flores from the rest of the islands of the Malay Archipelago. Although at the end of the Upper Pliocene the lowering of the sea level joined together some of the islands—Flores and Lomblen, for example—and Java was part of the Asian continent, the depth of the sea straits, indicated by dotted lines in the illustration, meant a permanent barrier to dispersals of fauna and flora. However, the crossing of the waters by humans was performed at least since the Middle Pleistocene onward in Southeast Asia, up to the occupation of Australia c. 60,000 years ago (Bednarik, 1997a). To Robert Bednarik (1997a), this fact requires rethinking the assumptions about both the technological and cognitive levels of the hominins at that time, and grants them the skill to navigate on rafts. We agree, although we should remember, as Jared Diamond (2004) noted, that the elephants—*Stegodon*—reached Flores, Timor, and the Celebes without a raft, so that humans might have done likewise.

---

are found in sedimentary units ranging from the oldest with stone tools—between 95,000 and 74,000 years—until the disappearance of *Stegodon* around 12,000 years ago (Morwood et al., 2004).

The interpretation of the Liang Bua artifacts and who were the possible makers becomes complicated by four factors:

- The coincidence of simple and more advanced tools in all the units with artifacts.
- The uneven distribution of utensils. The scarcity of tools in sector VII—only 32 were found at the same level as LB1—contrasts with the abundance of sector IV, where up to 5,500 tools per m³ appeared in the same sedimentary level as LB1 (Morwood et al., 2004).
- The diversity of the hominin remains. Those of *H. floresiensis* come from sectors IV, VII, and IX, but in sector XI adjacent to VII—within the stratigraphic sections corresponding to the Holocene—modern human fossils have been found (Morwood et al., 2005).

Box 9.27 Sequence of the Flores' utensils

If the sequence established for the Island of Flores had occurred elsewhere, the model proposed by Moore and Brumm (2007) would explain why large cores found in the islands of Southeast Asia tend to appear mainly on terraces and river banks, because these would be the sources of raw material discarded after the first manipulation. Also, this theory would explain why smaller artifacts appear in caves and shelters, being the result of the last reduction, and why the sets of large cores and small flakes are contemporary throughout the Pleistocene and Holocene, a dichotomy reflecting, in fact, two steps of the same reduction sequence.

Michael Morwood and collaborators (2004) argued that the Liang Bua hominins would have selectively hunted juvenile dwarf elephants, based on the coincidence of advanced tools and *Stegodon* fossils.

- The chronological sequence. Michael Morwood and collaborators (2004) indicated that the age of the sedimentary units with signs of occupation in the cave (3, 4, and 7) show that *H. floresiensis* was at the site perhaps since 95,000 years ago, and up to 18,000 years ago. That means, according to Morwood et al. (2004), that the presence of *H. floresiensis* would overlap with the temporal range of 55,000–35,000 years ago, corresponding to the arrival of *Homo sapiens* in the region.

It is always difficult, as we have said several times, to assign specimens to tools. However, since the remains of *Homo sapiens* found in Liang Bua correspond only to the Holocene deposits—none of those from the Pleistocene can be attributed to modern humans—Morwood et al. (2004) concluded that *H. floresiensis* was the maker of the lithic Pliocenic artifacts of the site. But, as we have seen, in the ancient levels with lithic tools there are also advanced tools.

The hypothesis that a hominin with such a small brain as *H. floresiensis* could make the tools of Liang Bua, including the most advanced, contradicts the usual models of hominization prior to the discovery of Flores. The correlation between brain development and technological progress is at odds with this attribution of tool-making—especially perforators, microblades, and fishing hooks—to a hominin with a cranial volume similar to that of *Australopithecus afarensis* (Box 9.28).

According to various studies of Dean Falk and collaborators, LB1 brain traits separated *Homo floresiensis* from hominins with the same or similar cranial volume. Would they be so important as to allow *H. floresiensis* to make the most advanced tools of Liang Bua? This is a question impossible to answer according to the evidence available today, except in hypothetical terms. Mark Moore et al. (2009) analyzed the continuity of the stone core-reduction techniques of Liang Bua, indicating that they follow the same pattern over 95,000 years. The carving sequence was maintained during the transition from the Pleistocene to the Holocene, with some small changes from 11,000 years ago, such as the use of fire, the selection of raw materials, the tendency to unifacial retouching, the polishing of flake edges, and the presence of rectangular axes (Moore et al., 2009).

According to the dates of the disappearance of *Homo floresiensis*, these recent technological advances had

---

Box 9.28 Cranium, brain, and intelligence

We must remember that cranial volumetric measurements are not suitable for assessing the cognitive potential of a species. With all possible precaution when making an extrapolation from the brain morphology to the mental processes, what seems clear is that it is better to examine the morphology of the cortex than to look to the cranial volume. Dean Falk and collaborators (2009) carried out the comparative examination of the LB1 endocranium, detailing in its cortex seven derived traits—in the occipital, prefrontal, orbitofrontal, and temporal—which indicate the presence of a substantial neurological reorganization. Neurological reorganizations have been related to key processes in human evolution (Holloway, 1983).

---

to be carried out by modern humans. But the permanence, even in those later dates, of the essential techniques of reduction in Liang Bua indicates that *Homo sapiens* did not significantly modify the carving procedures. For Moore et al. (2009), *Homo sapiens* introduced "additions," more than real "changes"—a model that could be extended to other parts of Southeast Asia. As such, the most reasonable hypotheses with respect to the carvings of Flores are:

- Colonization around 900,000–800,000 years ago by some unknown hominins.
- Existence of places of flake-stones to produce flake-cores, some discarded and others taken to places of subsequent carving into a finished tool.
- Continuity in Liang Bua, over 95,000 years, of a simple carving technique coincident with that of other sites on the island.
- Technical additions in Liang Bua, from the first occupations of the cave, which produce more complex tools than flakes (perforators, microblades, and hooks).
- New technical additions in Liang Bua within the Holocene.

Homo floresiensis would be the maker of the simple and more complex tools from the sedimentary units of the Pleistocene in Liang Bua. *Homo sapiens* would be the maker of both the simple and more complex tools of the Holocene in Liang Bua.

CHAPTER 10

Species of the Upper Pleistocene

The Upper Pleistocene began 126,000 years ago. As we saw in Chapter 9, during the Middle Pleistocene there was, in Africa and in Europe, an evolution toward hominins that shared certain features. So similar are the African, Asian, and European forms of the late Middle Pleistocene that the Multiregional Hypothesis (MH) holds that all of them are one single species, which evolved similarly in the different continents.

We prefer to separate the European species, *Homo neanderthalensis*, from the African species, *Homo sapiens*. However, exchanges will happen between them, genetic as well as cultural, which will be detailed in Chapter 11. This chapter is dedicated to the description of each one of the two species.

10.1 *Homo neanderthalensis*

Few hominins have received as much scientific and popular attention as the Neanderthals, which is one reason why few other taxa have also been subjected to such diverse interpretation. The Neanderthal story begins, as Trinkaus and Shipman (1993) described, in the Neander Valley, close to Düsseldorf (Germany). Mining at the Feldhofer calcareous caves, about 20 m above the Düssel River, led to a paleontological discovery of great importance (Box 10.1). In August 1856 the entrance to the caves was blasted in order to extract materials for construction. While cleaning the debris produced by the explosion some fossil bones appeared, including a skullcap, hipbones, ribs, and part of an arm, which seemed to belong to an animal more robust than living humans (Figure 10.1). The foreman believed these were the remains of a cave bear, and had the bones set aside for the teacher at the Elberfeld school. This teacher, Johann Fuhlrott, immediately identified them as belonging to a very primitive and robust human.

At a meeting of the British Association for the Advancement of Science, William King, professor at Queen's College, Galway (Ireland), suggested (in a footnote) the classification of the Neanderthal remains as *Homo neanderthalensis* (King, 1864). King argued, however, that the differences between this organism and humans were so considerable that it not only merited a separate species, but another genus altogether. In fact, King thought the Neander specimens exhibited a greater similarity with chimpanzees (Box 10.2).

Despite his reservations, history remembers William King as the author of the proposal of *Homo neanderthalensis*. Up to 34 different species and six genera have been proposed to accommodate Neanderthals since that first taxonomic solution (Heim, 1997). Today the only widely accepted alternatives are to regard Neanderthals as *Homo neanderthalensis* or as a subspecies of our own species, *Homo sapiens neanderthalensis* (Campbell, 1964).

10.1.1 The first Neanderthals

The Neander Valley fossils were the first to receive a taxonomic classification. However, some specimens currently included within this taxon had appeared three decades earlier, although they were not associated with Neanderthals at the time. In 1829, Phillip Charles Schmerling, a Belgian doctor and anatomist, found three crania in the Engis Cave (Belgium). The first was considerably deteriorated, and the second was from a modern human, but the third belonged to a 2- or 3-year-old Neanderthal child (Conroy, 1997). It is remarkably well preserved; its state is the finest among currently available infantile Neanderthal crania (Figure 10.2).

Another very complete Neanderthal cranium, which was not identified as such at the time, was discovered in 1848 at Forbes Quarry during the Gibraltar fortification works (*Homo calpicus*; Keith, 1911). After the Feldhofer discovery, other European Neanderthals were found in Moravia, Croatia, France, Germany, Italy, and Spain; the Neanderthal occupation of the continent excluded the northernmost countries. As

Processes in Human Evolution. Francisco J. Ayala and Camilo J. Cela-Conde, Oxford University Press (2017).
© Francisco J. Ayala and Camilo J. Cela-Conde.
DOI 10.1093/acprof:oso/9780198739906.001.0001

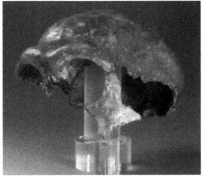

Figure 10.1 Left: A contemporary engraving of the Feldhofer cave (Bongard, 1835; picture from http://www.ateliereigenart.de/historie.htm). Right: Neander calotte and zygomatic bone, *Homo neanderthalensis*. (Photograph from Schmitz, 2003.)

Box 10.1 The circumstances of the Neander discovery

The discovery of the Neander specimens took place three years before the publication of Darwin's *Origin of Species*. Lacking the evolutionary perspective we have today, it is not surprising that contemporary explanations of the morphology of the specimens seem outlandish to us. On February 4, 1857, Hermann Schaafhausen, professor of anatomy at the nearby University of Bonn, presented the set of fossils before the Lower Rhine Medical and Natural History Society (Bonn). Later that year, on June 2, Fuhlort and Schaafhausen delivered a detailed description of the Neander specimens before the Natural History Society of Prussia, Rhineland, and Westphalia (Trinkaus & Shipman, 1993). Schaafhausen's anatomical interpretation was rigorous when compared with subsequent ones. He attributed the shape of the cranium, deformed in comparison with that of modern humans, to a natural condition, a shape that was unknown "even in the most barbarous of races," to put it in his own words. Thus, two years before the publication of Darwin's *Origin of Species*, Schaafhausen suggested the possible existence of

humans prior to Germans and Celts, in a "period at which the latest animals of the diluvium still existed." He believed that the bones belonged to some savage tribe from Northeastern Europe subsequently displaced by Germans. He denied that the deformations, both traumatic (the discovered radius exhibited an improperly healed fracture) and anatomical, could be due to pathologies such as rickets.

Despite Schaafhausen's cautions, the idea that the Neander bones belonged to a pathologically deformed human was widely spread during the years following the discovery. For instance, after examining a replica of the cranium, C. Carter Blake (1862) concluded that it corresponded to "some poor idiot or hermit" with pathological malformations that indicated an anomalous development and that had died in the cave he had used as a shelter. Trinkaus and Shipman (1993) emphasize Blake's strange statement that the Neander remains could not be attributed to a different species other than *Homo sapiens*, given that no one had yet suggested anything like it.

Box 10.2 Moral obscurity

Erik Trinkaus and Pat Shipman (1993) attributed the way in which William King interpreted the Neander cranium's morphology, tinted with that time's emphatic language, to influences from phrenology, in vogue at the time. This style was blatant regarding the "moral obscurity" of the organism whose remains he was interpreting. It should not be forgotten that Darwin himself manifested a belief in the moral inferiority of savage tribes in his *Descent of Man*, published in 1871.

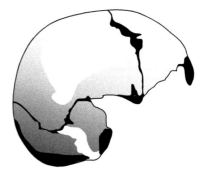

Figure 10.2 Engis infantile cranium (Belgium), *H. neanderthalensis* (picture from http://aleph0.clarku.edu/huxley/CE7/FosRem.html).

expected, the specimens were attributed to several different species (*Homo primigenius*, Schaaffhausen, 1880; *Homo transprimigenius*, Forrer, 1908; *Homo breladensis*, Marett, 1911). The Near East Neanderthals were first discovered in 1929, during a campaign led by Dorothy Garrod. The Shanidar (Iraq) Neanderthal site has been excavated by Ralph Solecki and colleagues since 1951.

In chronological terms, the oldest known specimens considered Neanderthals—if we do not consider the Sima de los Huesos exemplars as such—were retrieved at Ehringsdorf, not far from Weimar (Germany). The specimens found at Ehringsdorf between 1908 and 1927 include a cranium, a mandible, and some postcranial fragments belonging to different individuals, of which at least one is an infant (Weidenreich, 1927). The cranium restored by Weidenreich shares many common features with the Würm glaciation Neanderthal specimens, though it has a high and vertical forehead, which, in Gail Kennedy's (1980) opinion, could be the result of incorrect reconstruction. Electronic spin resonance and uranium series' dating suggest that the Ehringsdorf fossils are about 230,000 years old (Cook et al., 1982; Grün & Stringer, 1991). This is the age given by Chris Stringer and Clive Gamble (1993) for the separation of the Neanderthal and modern human lineages.

Several authors, such as Colin Groves (1989), classify the Ehringsdorf specimens as Neanderthals, but not everyone agrees with this option. Glenn Conroy (1997), to cite an example, considers the Ehringsdorf fossils as "archaic" *H. sapiens*, that is to say, as predecessors of *H. neanderthalensis*. Frank Poirier (1987) noted the proximity between the Ehringsdorf sample and the Swanscombe and Steinheim specimens, though he noticed that the Ehringsdorf specimens exhibit greater similarity with later Neanderthals.

The step from "archaic" humans to Neanderthals can be considered as an evolutionary transition between *Homo erectus* and *Homo neanderthalensis*, involving a chronospecies: *Homo heidelbergensis* (see Chapter 9). *H. heidelbergensis* could also be considered an ancestor of modern humans, i.e., the Last Common Ancestor (LCA) of our species and the Neanderthals.

There is, however, another possibility to interpret the transition to Neanderthals. This is the "accretion model" by David Dean, Jean-Jacques Hublin, Ralph Holloway, and Reinhart Ziegler (1998), which distinguishes between four different stages in this evolutionary transition (Table 10.1).

According to the available evidence, this seems the best interpretation of the process: the specimens that we call "in transition" correspond to the first two of

Table 10.1 Accretion model: stages in the anagenetic evolution of Neanderthals (according to Dean et al., 1998, as amended)

Stage	Specimens
Neanderthal 1 (early pre-Neanderthals)	Arago Mauer Petralona
Neanderthal 2 (pre-Neanderthals)	Bilzingsleben (morphologically more similar to Stage 1) Vértesszöllös (morphologically more similar to Stage 1) Atapuerca (Sima de los Huesos Site) Swanscombe Steinheim Reilingen
Neanderthal 3 (early Neanderthals)	Ehringsdorf Biache 1 La Chaise Suard Lazaret La Chaise Bourgeois-Delaunay Saccopastore (most of) Krapina (part of) (unclear) Shanidar (unclear)
Neanderthal 4 (classic Neanderthals)	Neander Spy Monte Circeo (part of) Gibraltar Forbes Quarry (unclear) La-Chapelle-of-Saints La Quina La Ferrassie Le Moustier Shanidar Amud

these stages, i.e., to the period of time between the MIS 15 to MIS 7. According to Chris Stringer (1993), such a scheme allows actually removing the taxon *Homo heidelbergensis* from the evolutionary succession as unnecessary. According to this idea, the European

---

Box 10.3 Antecedents of the accumulation model

As Aurélien Mounier, François Marchal, and Silvana Condemi (2009) indicated, the accumulation model follows a pattern defined by Jean Piveteau (1970) and later developed by Bernard Vandermeersch and his disciples (Vandermeersch, 1978; Condemi, 1989). The most important aspect of this framework is that *H. heidelbergensis* and *H. neanderthalensis* are considered chronospecies of an exclusively European lineage.

"archaic" humans would be incipient Neanderthals, which does not justify their ascription to a different species. Therefore, the accretion model by Dean et al. (1998) would establish different grades within what is a single species, *Homo neanderthalensis*, with a temporal range beginning with the specimens of *the early pre-Neanderthal* stage: Arago, Mauer, and Petralona. It might be possible to add those from Sima de los Huesos, owing to both the morphological analysis of the specimens and the last dating of the site.

It is not possible to include a complete list of Neanderthal specimens and sites because they are very numerous. Remains from the Riss–Würm interglacial period and the Würm glaciation are very abundant in southern, central, and eastern Europe, North Africa, and the Near East extending toward Siberia (Figure 10.3). Such an extended range raises the question whether it would be appropriate to recognize two different morphologies within the Neanderthal taxon, corresponding to the geographical variations: "classic" and "progressive."

10.1.2 "Classic" Neanderthals

Earnest Hooton's distinction in 1946 between "classic" Neanderthals, that is to say, those from western and southern Europe during the Würm glaciation, and "progressive" Neanderthals, including certain specimens from Near East sites, which show some similarities with anatomically modern humans, was for a time widely accepted. This viewpoint—somewhat general in the 1930s, 1940s, and 1950s (see, for example: Leakey, 1935; Paterson, 1940; Weckler, 1954)—reflects the idea that, although the Neanderthal lineage is prior to ours, it contributed genetic material by means of hybridizations to *Homo sapiens*. The "progressive" would thus be a kind of intermediate stage generated by genetic exchanges in the Middle East between ancient Neanderthals and modern humans. Later, we will address the scope of this distinction between these two types of specimens. Let's first begin with the description of the "classic" type specimens.

Neanderthal specimens exhibiting classic morphology have appeared in many places: France, with various sites (which will be next mentioned), Italy (Guattari), Croatia (Krapina and Vindija), Czechoslovakia (Ochoz, Sipka, Sala), Hungary (Subalyuk), in addition to the aforementioned Engis (Belgium), Forbes (Gibraltar), and Neander (Germany) specimens and without neglecting the later specimens of the Iberian Peninsula. The list offered is somewhat partial, but in the entire Neanderthal hypodigm the most complete and informative fossils are those from southwest France. We will focus on one of them, from the cave of La-Chapelle-aux-Saints, as an example of what its discovery and interpretation meant.

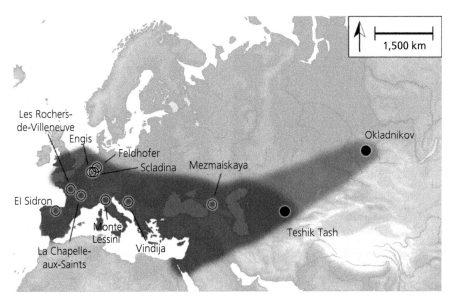

Figure 10.3 Geographical extension of the sites with Neanderthal specimens. The dark gray area corresponds to the conclusive evidence. Light gray includes the site with the specimen of dubious identification from Okladnikov (Russia) to which we'll refer in Chapter 11 (illustration by Krause et al., 2007). Reprinted by permission from Macmillan Publishers Ltd: Krause, J., Orlando, L., Serre, D., Viola, B., Prufer, K., Richards, M. P., . . . Paabo, S. (2007). Neanderthals in central Asia and Siberia. *Nature*, 449(7164), 902–904.

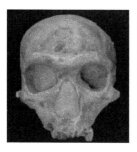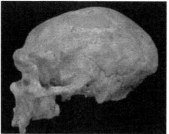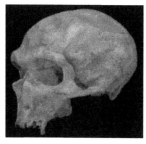

Figure 10.4 The "Old Man" from La Chapelle-aux-Saints (France) (pictures from www.mnh.si.edu/anthro/humanorigins/ha/lachap.htm).

In the early twentieth century, very complete remains of Neanderthals appeared in four French sites:
Le Moustier (1908), La Chapelle-aux-Saints (1908), La
Ferrassie (1909), and La Quina (1911). The one that immediately drew the most attention was the one from the
cave of La Chapelle-aux-Saints (Corrèze, France) (Figure 10.4). Inside the cave on August 3, 1908, the abbots
J. Bardon and L. Bouyssonie found an almost complete
skeleton buried in the ground, which was handed over
for study to Pierre Marcellin Boule, professor at the
Museum of Natural History in Paris. Boule undertook
the reconstruction and description of the specimen,
immediately causing such a stir that other specimens
of similar value, like that from La Ferrassie—also in
the hands of Boule—and La Quina—whose discovery coincided with the publication of materials from
La Chapelle—had to wait their turn to be interpreted
(Brace, 1964). So, as Loring Brace (1964) said, the specimen of La Chapelle imposed, for somewhat accidental
reasons, the image of what a Neanderthal was.

This image was adjusted to the results of the initial
studies by Marcellin Boule (1908; 1911–13). Known colloquially as the "Old Man" of La Chapelle-aux-Saints,
because it lacks almost all the teeth, the original description follows the "ape-like" patterns already indicated for the Neander specimen to which Boule
immediately related it: a being with large and divergent toes, which would force him to walk almost like
an ape, unable to fully extend the knees; with such a
curved spine that he could not quite stay upright; a face
projected forward by a short, thick neck; and a brain
qualitatively inferior to the human. This interpretation
was so remote from what was understood as a modern human, that it should have certainly influenced the
taxonomic proposal by Boule (1911–13) for the fossil
from La Chapelle and the other Neanderthals: it was a
different species from ours, one already nominated by
King as *Homo neanderthalensis* (Box 10.4).

Box 10.4 Neanderthals, according to Boule

In the review of the Neanderthal features published in
1911–13, Boule took into account the specimens from
Neander, Forbes' Quarry, The Naulette, Spy, Krapina, Le
Moustier, and La Ferrassie. All of them were considered
together with *Pithecanthropus* as a group of old forms
in contrast to the modernity of current humans. Despite
prejudices, Boule offered two arguments for the idea of
an absolute separation between the Neanderthal and
the human species. First, modern humans already existed when Neanderthals lived. Second, that the Mousterian
culture—characteristic of the Neanderthals—was suddenly replaced by the traditions of the Upper Paleolithic
when they arrived in Europe.

Almost all the attributions in the first interpretation
by Boule (1911–13) were discarded in later studies of
the specimen (see Brace, 1964). So, while a valid description of the Neanderthal features was entering the
discussion, the idea of a close proximity between modern humans and Neanderthals—a position contrary to
that of Boule—was also gaining favor, and the correct
taxonomic rank of both tended to be that of two subspecies of *Homo sapiens*.

Parallel to the discovery of the La Chapelle skeleton
was that of the remains of La Ferrassie (Dordogne,
France), as we mentioned. These are two almost complete adult skeletons (a male, La Ferrassie I, and a
female, La Ferrassie II) and several fragmentary specimens of at least half a dozen child-like creatures
found between 1909 and 1913 in a rocky shelter. Boule
(1911–13) associated them with the Neanderthal from
La Chapelle, although Michael Day (1986) argued
much later that the mandible of number I is perhaps

Box 10.5 Dental wear

In relation with the use of dentition among Neanderthals, Bernard Vandermeersch and María Dolores Garralda (1994) remarked that not all specimens show incisor wear. They noted that the Qafzeh and Skhul specimens, in the Near East, show no such trait despite living in the same environmental conditions and using the same objects as Neanderthals.

Box 10.6 The environment of the last Neanderthals

The paleoclimatic study by P. T. Tzedakis and collaborators (2007) posed the need to precisely describe the environment in which the last Neanderthals might have survived, assuming as a hypothesis the figures of 38,000, 33,000, and 24,000 years attributed alternatively to the Gorham cave. Depending on which date is used in the study, the role of climate would be different as a factor to explain their extinction. The first two dates correspond to intervals of climatic instability, with wide swings in the conditions. But 24,000 years ago there was a change in climate that led to the maximum accumulation of ice on the continent and a decrease of 20–30 m in sea level (Tzedakis et al., 2007). These changes would have transformed the Western Mediterranean into an area of relatively high temperatures. To Tzedakis and collaborators, it is possible that migration and the competition for warm territories among hominins would be a factor to consider when explaining why the Neanderthals disappeared.

---

more modern, since it has a slight chin. The conspicuous dental wear observed on many Neanderthal specimens has often been seen as a consequence of the use of dentition as tools, helping the hands during work (Box 10.5). But the possible use of teeth as work aids is a very contested hypothesis, and the question has not been definitively answered (see, Brace et al., 1981; Wallace, 1975).

The most common position in the late 1970s accepted that Neanderthals were the human species present in Europe since the Würm glaciation until their extinction approximately 30,000 years ago, which coincided with the arrival of modern humans on the continent (Figure 10.5). Work to widen a road in St. Césaire (Charente-Maritime, France) in 1979 turned up new specimens. They included the right part of a cranium and mandible, together with some postcranial remains (Lévêque & Vandermeersch, 1980). Their traits are clearly Neanderthal (Day, 1986), but their relevance lies mainly in their age. The chatelperronian tools found at the same site are very modern. Radiocarbon dating estimates the localities bearing chatelperronian tools to be between 34,000 and 31,000 years. If the skeleton is the same age, this represents one of the last Neanderthals (ApSimon, 1980). If 30,000 years is the correct date of their extinction, the specimen of St. Césaire would be one of the last Neanderthals.

However, Clive Finlayson and collaborators (2006) dissented with respect to the boundaries accepted for the life of the taxon. They did it by analyzing the Gorham cave, in Gibraltar, where several hundred Mousterian tools were found in level III. This level had previously been dated at 32,280±420 years (Barton, 2000). The stratigraphic reinterpretation of levels III and IV, including various age measurements of the terrain by mass spectrometry (AMS), yielded three dates, providing a stratigraphic sequence between 24,010±320 and 30,560±720 years ago. Given these data, the authors concluded that Neanderthals occupied

the cave close to 28,000 years ago, and possibly even as recently as 24,000 years ago. In a commentary on the paper by Finlayson and collaborators (2006), Eric Delson and Katerina Harvati (2006) indicated various places where Neanderthals would have been present at later dates than 30,000 years, when previously it was thought that the Neanderthals had disappeared.

Eric Delson and Katerina Harvati (2006) raised a question, of considerable significance, when trying fairly to understand the respective roles of Neanderthal and modern human lineages, which emerged from the acceptance of such a late date for the extinction of Neanderthals. *Homo neanderthalensis* and *Homo sapiens* would have coexisted in Europe for a considerable time, between 8000 and 10,000 years, stretching even longer if we consider a possible prior entry of modern humans into the continent.

10.1.3 The Neanderthal morphology

We have reviewed some of the main European fossils before addressing the general morphological description of the species, to illustrate how historical interpretations influenced what is considered as a Neanderthal.

Let's began with the traits that F. Clark Howell (1952) considered to be the typical features of Neanderthal morphology. Howell observed that such

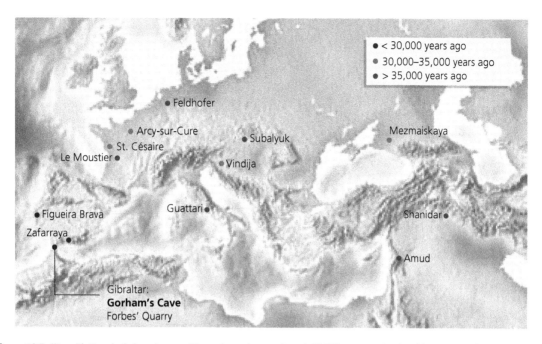

Figure 10.5 Sites with Neanderthal specimens or Mousterian tools approximately 30,000 years ago. Reprinted by permission from Macmillan Publishers Ltd: Delson, E. & Harvati, K. (2006). Return of the last Neanderthal. *Nature*, 443(7113), 762–763.

typical features are characteristic of the latest Neanderthals, survivors of a broader general population. Thus, there is not much sense in applying the term "classic" to specimens that constitute the last stage of a species. Therefore, Clark Howell used quotation marks to refer to "classic" Neanderthals, a practice we will continue here. We will do so to distinguish them from "progressive" Neanderthals.

According to Clark Howell's (1952) description, "classic" Neanderthals exhibit the following distinctive traits, among others (Figure 10.6):

- long, low, and wide cranial vault, which represents the preservation, to a lesser degree, of the typical platycephaly of *erectus*-grade specimens;
- broad facial skeleton, with a prominent zygomatic bone and large nose;
- thick semicircular and separate supraorbital tori, which do not extend laterally and are considerably pneumatized by the frontal air sinuses, which lighten these structures that appear so massive;
- absence of chin;
- strong mandible, with a retromolar diastema between M3 and the mandibular ramus;

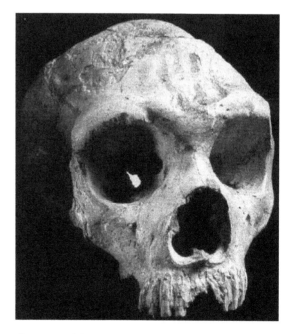

Figure 10.6 "Classic" Neanderthal; Forbes Quarry specimen, Gibraltar (photograph from http://www.msu.edu/_heslipst/contents/ANP440/neanderthalensis.htm).

- large cranial capacity, on average greater than that of *H. sapiens*: Porier (1987) attributed this high brain volume to a heavier body, while Trinkaus (1984) suggested it could be the result of a longer gestation period;
- short and massive vertebral column;
- robust and short limbs in relation to total height.

To these traits, considered typical in Neanderthals, could be added a significant facial projection and a very wide nostril (Box 10.7).

Can these traits be considered as apomorphies, derived traits of the Neanderthal taxon? In order to answer this question, Erik Trinkaus (1988) analyzed the traits usually associated with Neanderthals. For instance, Trinkaus (1988) considered the features related to the mandible and dentition, such as the notorious traits of the retromolar diastema and the absence of chin, as the result of a combination of two factors: first, the facial prognathism retained from *erectus*-grade hominins; and, second, the posteriorly placed masticatory musculature. The first factor is a plesiomorphy— the persistence of an ancestral character—while the second is a synapomorphy shared with anatomically modern humans. Neither of these traits, thus, is an exclusively Neanderthal apomorphy, although in combination could be considered a characteristic of the lineage (Trinkaus, 1988).

Trinkaus (1988) believed that the short limbs constitute a derived trait, which differentiated Neanderthals

---

Box 10.7 Neanderthal typical morphology

F. Clark Howell (1952) described the "Classic Neanderthals" as the typical specimens of this taxon, but it should be noted that their characteristic traits are not present in all specimens. The Neanderthal sample shows that there is a very wide range of variation, largely attributed by Milford Wolpoff (1980) to sexual dimorphism. The problems identifying the sex of the Neanderthal specimens were addressed by Fred Smith (1980). Naturally, the pelvis shape is sufficient evidence, but in many cases it is not available. Based on examination of the Krapina sample and of the specimens whose pelvis is available, Smith (1980) noted three cranial morphological features indicative of the gender: the mastoid process, the shape of the supraorbital torus in the glabella, and the roughness of the nuchal plane. After studying these traits, Smith concluded that sexual dimorphism would be greater in the crania of Neanderthals than in recent humans.

Box 10.8 Possible origin of the human chin

Lieberman and McCarthy (1999) described and interpreted the lack of facial projection and the prominent chin of anatomically modern humans in a very different way—as the result of the early ontogenetic reduction of the sphenoid bone. In their view, this is a favorable argument for the classification of Neanderthals and modern humans as two separate species. J. C. N. Ahern (2006) conducted a statistical comparison between modern humans and Neanderthals of nonmetric mandibular and cranial traits—considered in different studies as autapomorphies, by authors such as Rak, Stringer, and Santa Luca. The results indicated that in all but one case, the Neanderthal traits are not outside the range of variation of modern humans (Ahern, 2006).

---

from Middle Pleistocene hominins. However, he considered this trait to fall within the variability range of current humans and, consequently, suggested it should not be considered a Neanderthal autapomorphy. According to Trinkaus, the same could be said about such traits as the high encephalization and the shape of the cranium's occipital region. Hence, Trinkaus (1988) only accepted as true Neanderthal apomorphies a few traits relative to the cranium's temporal and occipitomastoid regions. This point of view is not unrelated, in our opinion, to Trinkaus' view of Neanderthals as a variety of our own species, *Homo sapiens*. If this were the case, we would expect Neanderthals and, anatomically, modern humans to share many synapomorphies. But the analysis of the ontogenetic development of the two lineages provides argument for the presence of autapomorphies. Are these distinctive of *H. neanderthalensis* or *H. sapiens*?

10.1.4 Neanderthal ontogenetic development

The great encephalization and the shape of the occipital part of the Neanderthal cranium were considered by Trinkaus (1984) to be the result of a generally more robust and heavy body. He tentatively attributed this main feature to a longer gestation period. However, the hypothetical extended gestation does not seem to be followed in Neanderthals—if we rely on the dental development—by a slower and, therefore, longer postnatal maturation. We will discuss this issue in Chapter 11.

10.1.5 The origin of Neanderthal apomorphies

If the comparison between the two hominin types of the Upper Pleistocene, Neanderthals and modern humans, shows different characters, it can be for one of three reasons:

(A) Because *Homo sapiens* has developed autapomorphies that are not present in *Homo neanderthalensis*. Neanderthals retain the primitive character for these features.
(B) The reverse case: autapomorphies of *Homo neanderthalensis*, with the primitive traits in modern humans.
(C) Both species developed apomorphies but different in each lineage, so they are different autapomorphies.

The initial descriptions (Boule, 1911–12, etc.) granted to *H. neanderthalensis* a condition similar to that of *Pithecanthropus*, which is the same as opting for hypothesis (A), but we have seen that such statements were made without the support of an entirely sound empirical base. The subsequent descriptions, which were claiming a modern character for Neanderthals, tended to indicate that the overall morphology of the "classic" specimens responded to their own autapomorphies, although without discarding the derived character from humans. This would be the hypothesis (C) of the listed alternatives.

The study of the upper part of the mandibular ramus (the coronoid process, the condylar process, and the existing gap between the two) led Yoel Rak, Avishag Ginzburg, and Eli Geffen (2002) to conclude that both European and Middle-Eastern Neanderthals differ more from early *Homo sapiens* specimens, such as Tabun II, Skhul, and Qafzeh, and also from the distant contemporary populations like those of Alaska and Australia, than the first modern humans and remote European populations differ from *Homo erectus*. In other words, the great adaptive specialization of Neanderthals would have led to fix the distinctive autapomorphies of the taxon while, at least with respect to this trait, the primitive condition remained in modern humans. This is, therefore, alternative (B).

If all three options regarding the apomorphies of Neanderthals and modern humans can be defended, which one is the most plausible? In the absence of an analysis of the trait polarities in both species, it is difficult to establish which taxon has the derived status. So a new study conducted by Erik Trinkaus (2006) determined these polarities to clarify the general direction

of the evolution that separates Neanderthals and modern humans.

Trinkaus (2006) conducted the detailed comparison that appears in Table 10.2 with the original notes of the author. In Table 10.2, Trinkaus indicates the polarities of neurocranial, facial, dental, and axial traits, as well as those typical of the upper and lower limbs in both taxa, Neanderthals and modern humans, distinguishing in the latter case between early and late specimens.

After discarding as true apomorphies some traits given as characteristic of the Neanderthals, such as the supraciliar arch and the robustness of the body—which either fall within the range of modern human variation or are a secondary consequence of the body size—Trinkaus (2006) concluded that our species would be the peculiar one, by developing its own autapomorphies. It is a return to hypothesis A, but now with empirical evidence derived from the analysis of the polarities. The results on brain development during ontogenesis by Gunz et al. (2010) speak in favor of that very idea of autapomorphies in modern humans.

The analysis of the Kebara 2 specimen conducted by Asier Gómez-Olivencia and collaborators (2009) concluded, hypothetically, that the large Neanderthal thorax would not be an autapomorphy but, on the contrary, a primitive pattern related to body mass. According to the authors, a broad thorax, indicative of a great lung capacity, also indicates the need for a large consumption of oxygen resulting from the increased energy demand of a large body. Needless to say, adaptive hypotheses relating morphology and the occupation of cold climate territories have been proposed.

10.1.6 Is it an adaptation to colder climates?

The Neanderthal morphology has been generally considered to be a consequence of adaptation to a cold climate (Brose & Wolpoff, 1971) (Box 10.9).

The climate conditions would have been extreme during the glaciations (see Figure 10.7). It is conceivable that hominins as advanced as *Homo neanderthalensis* would have had the cultural means to allow them to live under such demanding conditions. However, the absence of Neanderthal remains above 55°N latitude can be explained, according to Ian Gilligan (2007), by the hypothesis that clothes made at the Mousterian technological level would not have provided sufficient insulation from the severe cold. If the fossil Okladnikov (discussed in Chapter 11, Section 11.1.5) belongs to a Neanderthal, it would be an argument against Gilligan's suggestion. In any case,

Table 10.2 Trait polarity in Neanderthals and modern humans (Trinkaus 2006; see text)

	Neanderthals	Earliest Modern Humans	Later Modern Humans
Neocortical expansion	+	+	+
Relative parietal expansion	−	+	+
Cerebellar expansion	−	?	+
Delayed neonatal brain growth	+	+	+
Supraorbital torus absence	−	(+)	+
Sagittal keeling absence	+	+	+
Nuchal torus absence (or Miminal development)	−	+	+
Reduced or absent angular torus	+	+	+
Infrequent platycephaly	−	+	+
Prominent parietal bosses	−	+	+
Suprainiac fossa (horizontal oval form)	+	−	−
Laterally bulbous mastoid processes	−	+	+
Reduced juxtamastoid eminence	−	+	+
External occipital protuberance (esp. in males)	−	+	+
High and rounded temporal squamous portion	(−)	+	+
Labyrinthine morphology	+	−	−
Elongate foramen magnum	+	−	−

The sign + indicates that the polarity shows a large presence in the taxon; the sign − shows that it is absent. Parentheses indicate the polarity is shared in 10–90% of the taxon sample. For details and commentary see Trinkaus (2006).

Box 10.9 Cold adaptive traits

The hypothesis that Neanderthals have an anatomy adapted to severe climate was already expressed, with plenty of paleoclimatic support, by F. Clark Howell (1952). Clark Howell said that the features of the periglacial areas had important consequences on the morphology of the Neanderthals and on their distribution. But the presence of "classic" Neanderthals in the Middle East and Africa challenges the hypothesis that considers those features of Neanderthals to be an adaptation to cold climates, according to Gómez-Olivencia et al. (2009).

Simon Parfitt et al. (2010) have pointed out the presence of Lower Pleistocene lithic tools of more than 780,000 years of age in the village of Happisburgh (Norfolk, UK) located at a latitude of 52°49'N. The analysis of the Happisburgh flora and fauna points to climatic conditions during its occupation with average summer temperatures (between 16° and 18°C) that would be equal to, or slightly warmer than, southern

Great Britain today. The average winter (between 0° and −3°C), however, would be colder. It is a similar picture to that of current-day Scandinavia near the border between the temperate and boreal vegetation zones (Parfitt et al., 2010).

Jeffrey Schwartz and Ian Tattersall (1996) also classified Neanderthals in a different species from *Homo sapiens* on the grounds of certain autapomorphies of the internal nasal region. This would be an anatomical accommodation to the environmental conditions of low temperatures and low humidity. An increased surface for mucus/ciliated membranes and a very large sinusal cavity would have warmed and humidified the cold and dry air (Figure 10.8).

Jean-Louis Heim (1997) disagreed with this argument, noting that there are Neanderthals in warm and humid climates, such as the Near East. In regard to body shape, Trinkaus (1981) acknowledged that the shape of Neanderthals' lower limb bones reflect an adaptation to cold climate. However, Trinkaus also considered a complementary hypothesis: that such a trait could be related with biomechanical aspects resulting from an increase in running power, though at the price of losing speed.

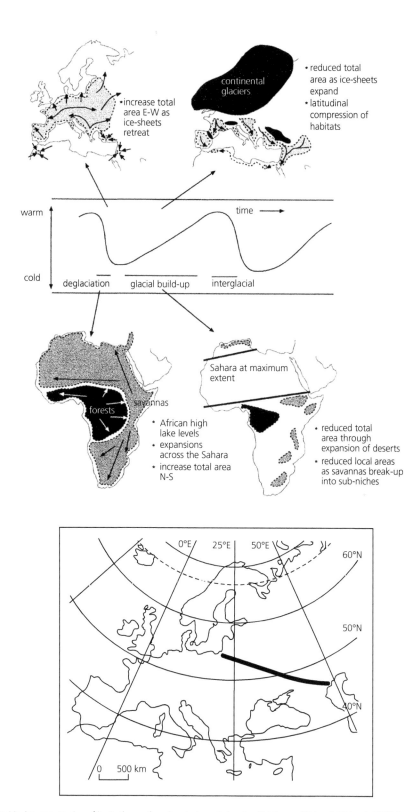

Figure 10.7 Top: Maximum extension of ice in the northern hemisphere during the Pleistocene (Ehlers & Gibbard, 2007) (repetition of Figure 8.33, above). Bottom: The gray line indicates the northern boundary of the Neanderthal occupation. Gilligan, I. (2007). Neanderthal extinction and modern human behaviour: the role of climate change and clothing. World Archaeology, 39(4), 499–514; reprinted by permission of the publisher (Taylor & Francis Ltd, http://www.tandfonline.com).

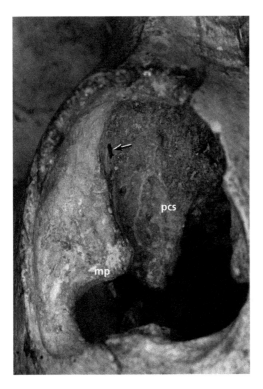

Figure 10.8 Frontal view of the nasal cavity of the cranium Gibraltar I, *H. neanderthalensis*. A cranium illustrating the internal margin with a great projection (mp), behind which there is a larger swelling inside the nasal cavity. The arrow indicates the hole in the wall of the swelling that reveals the enlarged maxillary sinus. Reprinted from Schwartz, J. H. & Tattersall, I. (1996a). Significance of some previously unrecognized apomorphies in the nasal region of *Homo neanderthalensis*. *Proceedings of the National Academy of Sciences*, 93, 10852–10854. Copyright (1996) National Academy of Sciences, U.S.A.

The explanation that Neanderthal facial apomorphies are a response to cold climates is challenged in the study by Todd Rae, Thomas Koppe, and Chris Stringer (2011). These authors hypothesized that the relationship established between Neanderthal facial features and cold climate is based on two assumptions: (1) that the increase in the craniofacial pneumatization is the result of an adaptation to low temperatures, and (2) that the Neanderthals have large, wide sinuses, even hyper-pneumatized. Rae et al. (2011) collected empirical evidence indicating otherwise: the frontal and maxillary sinuses of Arctic human and Japanese macaques are small sized. This reduction is also found experimentally in rodents subjected to extreme cold. Regarding the sinuses of Neanderthals, Rae et al. (2011) argued that the claim that their larger size does not correspond to their own body size (extra-allometric)

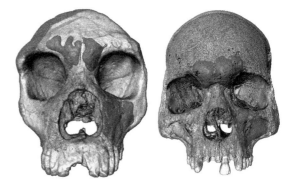

Figure 10.9 Nasal and frontal sinuses of Forbes Quarry 1 (Neanderthal) and a recent human. Although they are not at the same scale, the size of the pneumatized areas is equivalent in both specimens. Reprinted from Rae, T. C., Koppe, T., & Stringer, C. B. (2011). The Neanderthal face is not cold adapted. *J Hum Evol*, 60, 234–239, with permission from Elsevier.

is not based on any volumetric research. When this was completed, the authors obtained two-dimensional and three-dimensional values for the Neanderthal sinuses, which fell within the range of modern humans, both for Neanderthals coming from cold (Europe) and warmer places (the Orient) (Figure 10.9 and Box 10.10).

Box 10.10 The Neanderthal nose

Nathan Holton and Robert Franciscus (2008) opposed the idea that an adaptation to cold climate could be the cause of the large nasal cavity of Neanderthals, given that in modern human populations the opposite occurs. This is a paradox that has been addressed by establishing a relationship between the distance of the canines and the width of the nasal cavity. The primitive prognathism trait of Neanderthals, which correlates with both the palate and nasal fossa size, could have another explanation. Holton and Franciscus (2008) dismiss the strong correlation between intercanine distance and nasal cavity size. These authors maintain that the facial projection correlates better than the distance between canines, but warn that, by itself, prognathism cannot explain the paradox. Rae et al. (2011) pointed out, as an alternative to the adaptive and the biomechanic hypothesis, that the facial shape of Neanderthals was the result of genetic drift. They followed Susan Antón (1994) in this respect, for whom many of the cranial differences that exist between *H. neanderthalensis* and *H. sapiens* follow similar patterns to those expected if genetic drift would have intervened.

If not an adaptation to cold climates, the peculiar shape of the Neanderthal face must be explained in other terms. Rae et al. (2011) proposed a biomechanical hypothesis: the use of the anterior dentition as a working tool, already suggested by several authors (Rak, 1986; Trinkaus, 1987), and shared with current northern human populations, such as the Inuit (Spencer & Demes, 1993).

10.1.7 In search of an interpretive model

In search of an interpretive model, Steven Churchill made a cast of the human nasal cavity to study the aerodynamic flux produced by a fluid passing through them, and suggested that the very wide nose of Neanderthals could be an adaptive improvement to achieve considerable air flow while minimizing the turbulence. Churchill believed the need of a great amount of oxygen was related with Neanderthal metabolism, which would have required close to double our own energetic needs (see Holden, 1999).

A high metabolism and high body mass point to the need for a high-quality diet, such as one obtained by hunting (Box 10.11).

Neanderthal remains from the beginning of the Würm glaciation are found in rocky shelters, in caves near streams in the valleys of southern France, and in the Italian and Spanish peninsulas (Howell, 1952). This does not mean that Neanderthals did not leave these protected areas. They would have done so, according to Howell, especially during summer months, on hunting expeditions to the north, although always of a limited duration.

A migration noted by Clark Howell (1952), which took advantage of periods of low sea-level, would have gone from Italy through the Northern Adriatic into the central European corridor. Isolated by the subglacial Alps, the fauna in the center of Europe would have been a hunting resource that Neanderthals could exploit; however, the subsequent rise of sea level would have submerged those lands, as suggested by evidence showing consistent population movement (see the study by Tzedakis et al., 2007).

10.1.8 The "Neanderthal question"

Treatises on Neanderthals usually include discussion of their phyletic relations with anatomically modern humans. Three main perspectives can be distinguished. The first argues that Neanderthals are a prior and ancestral species to *H. sapiens*, such that anatomically modern humans evolved from them. This idea developed during the early twentieth century (Day, 1986), but it reappeared with Loring Brace (1964) (see Figure 10.10A). This view proposes that typical Neanderthal morphological traits disappeared as they evolved toward *H. sapiens*. Milford Wolpoff (1980) adhered to this hypothesis, though considering Neanderthals and *sapiens* as a single species.

The second perspective considers Neanderthals as a subspecies of *H. sapiens*, *H. sapiens neanderthalensis*, which may have contributed genetically to the appearance of *H. sapiens sapiens*. This hypothesis is accepted by specialists, such as Erik Trinkaus (Trinkaus & Smith, 1985). Supporters of this model not only believe that genetic exchange was possible between the two subspecies, but that fossil evidence shows that it actually took place (see Figure 10.10B).

The third hypothesis suggests that Neanderthals and anatomically modern humans were two different species, sister groups in a cladogram, about whose origin there are various explanations (see Figure 10.10C). Neanderthals are conceived as a lateral branch holding no ancestral relation to *H. sapiens*. In fact, the latter might have had an earlier origin than *H. neanderthalensis*. This is Stringer and Gamble's (1993) view, which adds that anatomically modern humans displaced Neanderthal populations wherever they overlapped.

The late presence of classic Neanderthals, such as those found in southern France and the Iberian Peninsula, is an argument against the first hypothesis, that there was a "Neanderthal phase" prior to modern humans and ancestral to our species. It could be said in

---

Box 10.11 The Neanderthal diet

Neanderthal hunting activities and diet have been the subject of highly prejudiced speculations, as are many other characteristics related to our sister group. The work of Elspeth Ready (2010) on the influence of the historical moment on the perception of Neanderthal subsistence, illustrates the various hypotheses, some of them farfetched, which have been proposed. However, there are abundant empirical studies with a proliferation of radioisotope analyses capable of indicating the original diet. One example is the analysis by Michael Richards et al. (2000) of the presence of C^{13} and N^{15} isotopes in the bone collagen of two Neanderthal specimens and other fossils from the fauna of Vindija (Croatia), which confirmed that Neanderthals got most of their protein from animal sources through hunting, not scavenging.

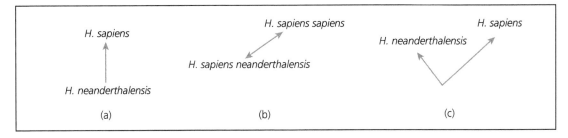

Figure 10.10 Relational models between Neanderthal and modern humans. (a) Neanderthal is a species that evolved by anagenesis into modern humans. (b) Neanderthals and modern humans are two subspecies that exchanged genes. (c) Neanderthals and humans are sister groups, two species which appeared by cladogenesis (although with prior chronospecies, maybe in both cases).

Box 10.12 Subspecies or population?

The question of whether the Neanderthals are a subspecies is more a formal one and is almost the same as the hypothesis that argues that they are a population variety. Trinkaus and Howells (1980) argued that the term subspecies is unnecessary and suggested that it is better to speak of the evolution of local populations, some of which are those considered as Neanderthals.

---

response that this evolutionary transition took place locally, maybe in the Near East, while the Neanderthal forms remained unchanged at other places. However, the presence of modern human specimens such as the Cro-Magnon, contemporary if not previous to the Neanderthal of St. Nazaire (France), weakens the argument.

Box 10.13 MH and RH, once again

As we might recall, MH advocates a gradual transition from archaic to modern populations, which happened in numerous locations through a process in which Neanderthals contributed to the gene pool of modern humans. RH suggests that there was a sudden replacement of the earlier populations of *Homo neanderthalensis* by *Homo sapiens*, without crossbreeding. The approach we have followed is to accept the taxon *Homo neanderthalensis*, for the reasons of genetic distance that will become clear in Chapter 11. Nevertheless, the Neanderthals maintained relationships with *Homo sapiens* that, as we shall see, are far from the radical interpretation of RH.

The "Neanderthal question," therefore, comes down to whether Neanderthals and modern humans are one or two species. This decision has much to do with the two evolutionary scenarios: MH and RH (Box 10.13). In the second case, Neanderthals are considered as a different species (as in Figure 10.10C).

10.1.9 The oriental contact

A good way to test MH and RH is to study in detail those localities in which Neanderthal and modern human specimens appear. Let´s begin with the Near and Middle East.

The easternmost territories occupied by Neanderthals extend from Israel to Iraq (see Figure 10.11). Sites in this region have not only yielded "classic" Neanderthals, similar to the Europeans, but two different kinds of Neanderthal specimens have been found in nearby sites, or even in the same one. The specimens found in Tabun (Mount Carmel, Israel), Shanidar (Iraq), Amud (Amud Valley, Israel), and Kebara (Mount Carmel, Israel) are similar to "classic" Neanderthals. A different kind of specimen, exhibiting greater morphological resemblance to *H. sapiens*, has appeared at Skhul (Mount Carmel, Israel), Qafzeh (close to Nazareth), and Ksar Akil (Lebanon). The question is whether these specimens are highly variable members of a single species ("classic" and "progressive" Neanderthals) or whether they belong to two different species (*H. neanderthalensis* and *H. sapiens*). In the second case, it is necessary to clarify what was the evolutionary relationship between the two species.

In order to assess the hypotheses regarding the evolution of the Near East specimens, it is necessary to settle two issues. First, the morphological relationships among different exemplars must be determined by comparative studies. Second, the correlation of the sedimentary sequences must be established by dating

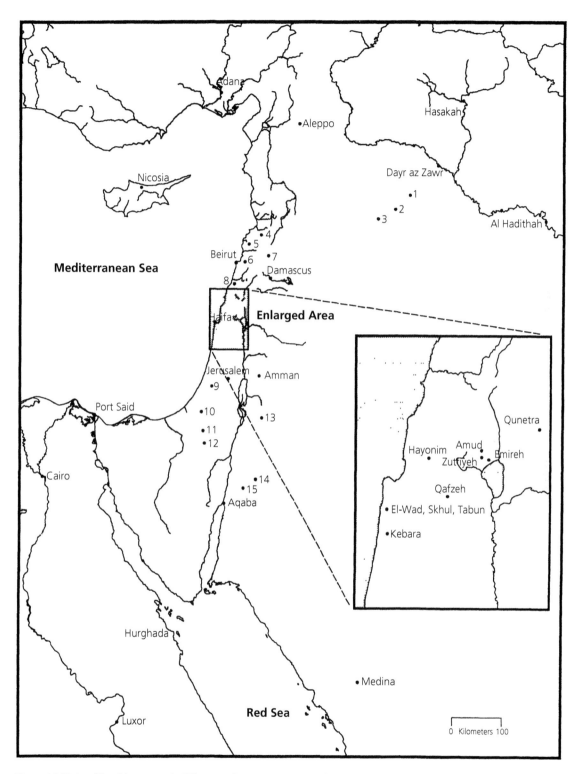

Figure 10.11 Localities of the Near and Middle East with sites in which "classic" and "progressive" Neanderthals are found. The enumerations correspond to the localities not considered here: 1, El Kowm; 2, Douara; 3, Jerf Ajla; 4, Keoue; 5, Nahr Ibrahim; 6, Ksar Akil; 7, Yabrud; 8, Adlun; 9, Shukbah; 10, Fara II; 11, Rosh Ein Mor; 12, Ain Aqev; 13, Ain Difla; 14, Tor Faraj; 15, Tor Sabiha (Bar-Yosef et al., 1992).

the sites in which indications of both species (or populations) appear. We will first address the morphology and distribution of the various types of Upper Pleistocene hominins found in the Near and Middle East.

Between 1929 and 1934, the joint expedition, led by Dorothy Garrod, of the British School of Archaeology of Jerusalem and the American School of Prehistoric Research, carried out a 5-year excavation plan in the Skhul, Tabun, and el-Wad caves, on the hills of Mount Carmel, close to Wadi el-Mugharah (Valley of the Caves). Two of the many caves in these slanted calcareous terrains are very near each other: the large Tabun cave (Mugharet el-Tabun, "Cave of the Oven") and the smaller Skhul cave (Mugharet es-Skhul, "Cave of the Kids") have provided hominin specimens, as well as Acheulean and Mousterian tools. A nearly complete skeleton of an adult female (Tabun I), a mandible (Tabun II), and some other postcranial remains were found at the Tabun cave (Garrod & Bate, 1937). The Skhul cave, deteriorated and reduced to a rocky shelter, contained a very broad collection of human remains, which appeared to have been buried deliberately. This collection included six adult specimens (Skhul II–VII) (one of them partial), one skeleton belonging to a child (Skhul I), remains of lower limbs (Skhul III and VIII), and a mandible together with a femoral fragment, both of them infantile (Skhul X) (McCown & Keith, 1939) (Figure 10.12).

The Qafzeh cave is close to Nazareth. The initial excavations carried out between 1933 and 1935 by René

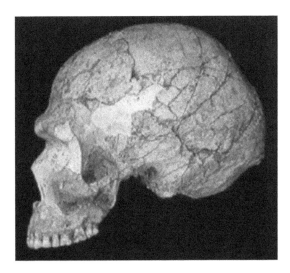

Figure 10.12 Skhul V cranium (photograph from http://humanorigins.si.edu/evidence/human-fossils/fossils/skh%C5%ABl-v) (object's negative number: NHB2014-03453). Donald E. Hurlbert, Smithsonian.

Neuville, from the Institute of Human Paleontology in Paris, and Moshe Stekelis, from the Hebrew University, led to the discovery of several fossil hominins (Qafzeh 1–7), with tools of Mode 3, Mousterian, close to the cave's entrance, though their description was not published at the time. In the preliminary study of *Homo* VI, Vallois and Vandermeersch (1975) maintained that many of its traits coincided with those of modern humans. It is a young adult cranium of male aspect, coming from level L, and has a very larger volume (1,568 cc). Since 1966, the team led by Bernard Vandermeersch has discovered the remains of about 14 individuals, at least three of which seem to have been buried intentionally. In some cases, such as the Qafzeh 11 infantile specimen, the remains appeared accompanied in the same tomb by objects that could have a ritual meaning, e.g., a deer cranium in Qafzeh 11 (Vandermeersch, 1981).

At Wadi Amud, north of Tiberias and about 50 km northeast of Haifa, the Amud cave has been excavated since 1959 by the Tokio University Scientific Expedition to Western Asia, under the direction of Hishasi Suzuki. Amud contains two stratigraphic levels: A and B. The specimen known as Amud I, an incomplete cranium, a mandible in excellent condition, and some postcranial remains (femur and tibia) were found in level B in 1969 (Suzuki & Takai, 1970).

The Mugharet el-Kebara cave, on the western hillside of Mount Carmel, has been excavated since 1982 by Ofer Bar-Yosef and Bernad Vandermeersch (Arensburg et al., 1985; Bar-Yosef & Vandermeersch, 1991, 1993; see the summary of the work done between 1982 and 1990 in Bar-Yosef et al., 1992), with the main objective of identifying the behavioral patterns of hominins discovered in Israel. The research was a great success. In 1983, unit XII of Kebara, of approximately 60,000 years according to thermoluminescence studies (Bar-Yosef et al., 1992), yielded one of the most complete Neanderthal specimens to date: a male adult lacking the cranium but not the mandible (Arensburg et al., 1985). Some authors (Bar-Yosef & Vandermeersch, 1993; Conroy, 1997) have tentatively attributed the absence of the cranium to funerary practices. The excellent conservation of this specimen—it was buried—has allowed a revision of the theories concerning the Neanderthal capability for speech, based on the specimen's hyoid bone. The Kebara hyoid, together with the one recently discovered in Dikika, Ethiopia (Chapter 4, Box 4.18) plus another one from Atapuerca's Sima de los Huesos, are the only hominin hyoids known, other than those from modern humans.

Figure 10.13 Burial 4 of Shanidar. Each segment on the scale bar is 5 cm. Solecki, R. (1971). Shanidar: The First Flower People. New York, NY: Knopf.

Box 10.14 Funerary offerings at Shanidar?

Some aspects of the Shanidar specimens, like the possible deliberate burial and even the presence of associated floral offerings, suggest the existence of funerary practices at the site. The evolution of symbolism is not part of this book; this question, therefore, remains pending.

Leaving aside the Teshik-Tash site (Uzbekistan), the Shanidar Valley, in the Zagros Mountains of Northern Iraq, has revealed indications of the westernmost Neanderthal occupation. The Shanidar cave has yielded the greatest number of Neanderthal remains in the entire East. Between 1951 and 1960 the team led by Ralph Solecki retrieved up to 28 burials with remains of nine individuals (Figure 10.13).

The site and its contents were described by Ralph Solecki (1953), T. Dale Stewart (1958), and, more extensively, by Erik Trinkaus (1983). According to Trinkaus, the Shanidar I–VI fossils correspond to adults, VII to a youngster, while Shanidar VIII and the one known as

"the Shanidar child" are infantile specimens that had not reached the age of one. Shanidar I, a skeleton, includes the best-preserved cranium of the entire sample.

10.1.10 One or two species in the Near East?

The morphological study of the Upper Pleistocene fossils in the Near East revealed the existence of conspicuous differences among the specimens from nearby sites. McCown and Keith (1939) suggested that the Tabun I female was similar to European Neanderthals, with separate and large supraorbital tori, large face and nose, and no mental protuberance. However, the specimen was small, and its cranial capacity was very low for a Neanderthal (1,270 cc). In contrast, the Tabun II mandible showed indications of a chin.

The Skhul specimens were somewhat different to those from Tabun. Skhul V, the best-preserved cranium, has a high and short cranial vault with vertical lateral walls (see Figure 10.12). The specimen exhibits marked supraorbital tori, in contrast with Skhul IV, which lacks them, although they are not of the typical Neanderthal shape (Stringer and Gamble, 1993). To a greater or lesser extent, the Skhul specimens also show a mental protuberance—which is incipient in Skhul V—and lack a midfacial projection. Overall, the Skhul specimens appear more modern than those from Tabun. Even so, McCown and Keith (1939) considered that they were all highly variable members of the same population, and attributed the anatomical differences to a gradual evolutionary trend.

The idea of a single species at Tabun and Skhul was supported by the widespread notion at the time that they were contemporary caves. If they were quite close and contemporary, it would seem unlikely that two species or even two subspecies were present. When Tabun was estimated to be older (Higgs, 1961), the differences between the specimens collected at the caves were interpreted differently. The Skhul specimens were considered younger and more modern-looking than those from Tabun. These samples seemed to be an excellent testimony of the gradual evolution from Neanderthal forms to modern humans.

The remaining sites in the Near and Middle East can be classified consonantly with the two populations from Skhul and Tabun, as "classic" and "progressive" Neanderthals (Figure 10.14). The morphology of the Qafzeh specimens is similar to that of Skhul exemplars. Some of the crania within the Qafzeh sample exhibit small supraorbital tori (Qafzeh 9), whereas others exhibit protruding supraorbital tori but a shape different from that of classic Neanderthals (Qafzeh 6).

"Classic" Neanderthals, similar to European specimens

Close-to-modern-humans specimens; "progresive" Neanderthals

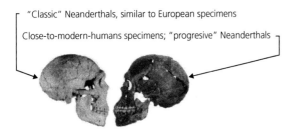

Figure 10.14 The two forms of "Neanderthal" from the Upper Pleistocene in the Near and Middle East (illustration by Bar-Yosef & Vandermeersch, 1993; see text for the cranial interpretation).

Qafzeh 9 exhibits modern traits, such as the elevated vault and the supraorbital torus, together with some primitive ones: prognathism and large teeth. In spite of this, the Qafzeh exemplars have been classified together with those from Skhul as part of the same species, considered to be different from the one in Tabun.

In contrast to the more modern Skhul and Qazfeh specimens, the Amud exemplar is similar to those from Tabun and Shanidar. In fact, materials from the latter, and other sites, were used to complete the missing parts during the reconstruction of the cranium. This shows some modern traits, such as an incipient mental protuberance or a modest supraorbital torus, but Suzuki and Takai (1970), as well as Howells (1974), included it in the Near East Neanderthal population found at Tabun and Shanidar. This idea has been widely accepted.

The affinities shown by the Kebara specimen cannot be doubted either. Despite the absence of the cranium, its remaining traits are reminiscent of the Tabun Neanderthals (Bar-Yosef & Vandermeersch, 1991). Arensburg (1989) preferred not to use the term "Neanderthal" and referred to the Kebara specimen as "Mousterian," without a taxonomical label. However, Arensburg (1989) associated some of the specimen's traits, such as the pelvis, with those of European and Shanidar specimens.

Regarding Shanidar, the tendency is to include its specimens in the category of Near East classic-like Neanderthals. The Shanidar I cranium, which exhibits such traits as a low and long vault, marked supraorbital torus, facial prognathism, and occipital torus, constitutes a very typical Neanderthal example. The mandible, which is complete, shows the typical Neanderthal traits: robusticity and lack of mental protuberance.

We have mentioned that MH suggests that Neanderthals were a subspecies, *H. sapiens neanderthalensis*, and could have contributed with their genetic material to the origin of *H. sapiens sapiens*. If this scenario is

---

Box 10.15 Populations of the Near East

The general view in regard to the Eastern Mousterian specimens is to assume that there were two populations. But some authors, such as Milford Wolpoff (1991) continued to support McCown and Keith's (1939) initial idea that all Near East human remains constitute a highly variable single population. Nevertheless, the division between the Tabun, Kebara, Amud, and Shanidar samples on one side, and those from Skhul and Qafzeh on the other, is evident. Whether this difference has phylogenetic significance is a different matter. Joseph Weckler (1954) challenged the idea of "progressive" Neanderthals as a result of a cross between Neanderthals and ancestral modern humans— that he supported—with an alternative hypothesis. This would consider the "progressive" as a taxon previous to the "classic" Neanderthals and the current modern humans. The "progressive" Neanderthal would thus be the common ancestor. Weckler mentioned McCown and Keith (1939), and Clark Howell (1951), as supporters of this hypothesis.

---

accepted, then the transition of "classic" Neanderthals (Tabun, and so on) to "progressive" forms (similar to Skhul specimens) could be excellent evidence in favor of the gradual evolution that transformed Neanderthals into anatomically modern humans. The hypothesis requires that the ages of the caves coincide with the evolutionary sequence, that is to say, that the remains of the Neanderthal-looking specimens are also older than the modern-looking exemplars. Thus, we need to turn to the chronological sequence of the Near East sites.

10.1.11 The ages of the Near East caves

We will refer to the estimates obtained by the early studies of the caves, prior to the application of methods such as electronic spin resonance (ESR) or thermoluminescence (TL), as "initial" ages.

Tabun includes a series of sedimentary levels, A through G. Tools have been found at levels Tabun B and C, whereas human remains appeared in sediments 1 and 2 of level C. The initial age of Dorothy Garrod and Dorothy Bate (1937) assigned levels C and D to the Riss–Würm interglacial period (between 275,000 and 125,000 years). Later, Garrod (1962) considered that level C was younger, toward the second half of the Würm glaciation (less than 100,000 years). Radiocarbon dating estimated the minimum age of Tabun at 51,000 years (Farrand, 1979). But these ages are too old for the method of carbon isotopes to yield reliable results.

In regard to Skhul, Theodore McCown identified three stratigraphic levels: A, B, and C. The specimens were buried in level B, whereas A, above it, contained stone tools. The fauna and the tools at Skhul correspond well to those at Tabun level C; so, Garrod and Bate (1937) assigned it the same initial age, the Riss–Würm interglacial period. Subsequent macrofaunal studies suggested that Skhul was at least 10,000 years younger than Tabun C (Higgs, 1961), which allowed an evolutive sequence for the specimens found at both caves to be established. If Tabun was prior to Skhul, then the evolutionary sequence classic–progressive specimens from both caves fitted nicely. The cave with other modern-looking remains, Qafzeh, also seemed to be younger than Tabun. Qafzeh includes up to 24 sedimentary levels, with human remains (which in some instances are difficult to associate stratigraphically) found in level XVII. Amino acid racemization studies estimate that that level is 32,000–39,000 years (Farrand, 1979). This age was subsequently raised, by the same technique, to 39,000–78,000 years (Masters, 1982).

Hence, the chronological scenario suggested that the Tabun, Amud, and Kebara set were older than the Skhul and Qazfeh caves. Sites that yielded similar remains to classic Neanderthals were older than those containing modern-looking specimens. McCown and Keith's (1939) gradual evolution hypothesis was supported by such a chronology. However, that hypothesis became difficult to sustain after Bar-Yosef and Vandermeersch (1981) reassessed the situation. These authors carried out a comparative study of the stratigraphic relations, paleoclimatic data, and microfaunal and cultural sequences. They concluded that Qazfeh was about as old as Tabun D, close to 100,000 years old; this questioned the evolution of "classics" toward "progressives."

The application of more modern and precise dating techniques—ESR, TL—toward the end of the 1980s has confirmed Bar-Yosef and Vandermeersch's results. The TL study of 20 burnt flint-stones from Qafzeh yielded an average of 92±5 years, with a range between 85 and 105 (Valladas et al., 1988). In regard to Skhul, the study of mammal teeth with ESR estimated its age to be between 80,000 and 100,000 years—much older than previously thought (Stringer et al., 1989). TL assigned to Skhul level A, believed to be the most modern, an even greater age, about 120,000 years (Mercier et al., 1993). Given that one of the sites containing "classic"

Neanderthals, Kebara, was estimated by means of TL to about 60,000 years (Valladas et al., 1987), it seemed clear that the chronological sequence that assumed that "classic" Neanderthals sites were old and "progressive" Neanderthal sites were young could no longer be upheld. The broader sedimentary sequence at Tabun was estimated by ESR from 200,000 years, for the oldest level, to 90,000 for the youngest (Grün et al., 1991). Level C, where the Tabun I female specimen was found, was estimated by Grün and colleagues (1991) to be between 102±17 and 119±11 years old, a similar age to the Skhul specimens.

Despite the dates granted to the specimens from Tabun, the "Tabun C2" industry could be older than the estimated dates of the Tabun specimens. By means of thermoluminescence, similar tools have been dated in Tabun C and Hayonim E at 150,000–170,000 years (Bar-Yosef, 2000). The question is, who manufactured the instruments, given the lack of fossil specimens?

These dates, which completely upset the assumed archaeological sequences of Mount Carmel, were criticized by many authors. But the very precise technique of uranium series' mass spectrometry applied to samples of bovid teeth from Tabun, Skhul, and Qafzeh, confirmed the new scenario (McDermott et al., 1993). Tabun level C was estimated to be about 100,000 years (three samples yielded 97.8±0.4, 191.7±1.4, and 105.4±2.6 years). The Skhul specimens are difficult to date using this technique, owing to the presence of two different faunas in level B. McDermott and colleagues (1993) suggested that there are human specimens in both, but they also expressed the need for further detailed studies to confirm this point. In any case, the coincidence of different dating techniques (ESR, TL, and uranium series) in the attribution of a greater (but similar) age to Tabun and Skhul seems conclusive, taking into account that the results rendered by these techniques are very similar.

Chris Stringer and Clive Gamble (1993) provided an illustrative summary of how the chronological sequence of the Middle East sites changed with the application of modern dating techniques. The ages of the localizations are diverse: Tabun is the oldest site, with about 120,000 years, then come Qafzeh and Skhul, and Kebara and Amud are around 50,000 years old.

New analysis with ESR and uranium series' methods, conducted by Rainer Grün et al. (2005), calculated for Skhul a range between 135,000 and 100,000 years, consistent with the thermoluminescence data—119,000±18,000 years (Mercier et al., 1993). To Grün et al. (2005), Tabun, Skhul, and Qafzeh can be considered contemporary.

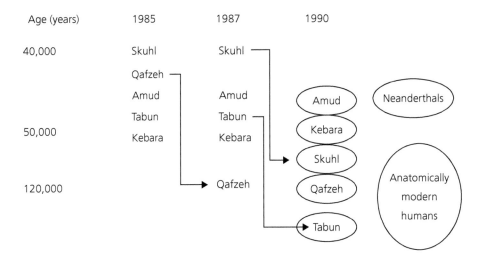

Figure 10.15 Chronological sequence of Upper-Pleistocene sites in the Near East, according to Stringer and Gamble (1993).

A sequence like that of the right column in Figure 10.15 leads to the assumption that creatures similar to the "classic" Neanderthal and those first understood as "progressive," were successively occupying these sites, alternately and without overlapping.

Bernard Vandermeersch (1996) considered that cultural practices shared by the two populations present in the Near East since the beginning of the Riss–Würm interglacial period exclude the hypothesis that there was an alternated occupation of the area. If both populations were present at the same time, then cohabitation can be understood either as:

(1) corresponding to two separate species, with no possibility of hybridization; or
(2) as the contact between varieties of a single species, which would allow genetic exchange.

Vandermeersch noted that the available data do not justify denying that such exchange occurred, though the possibility must be tentative because of the great difficulties involved in its demonstration. A tentative way of doing so would be to identify specimens that were morphologically intermediate between the two populations living in the area. However, this kind of evidence is controversial because of the broad variation range among Neanderthals, as well as among our own species.

10.1.12 Modern humans in the Near East

The most convincing way to explain a scenario in which creatures of an aspect similar to the classic Neanderthals appeared first, and were then followed by others looking like modern humans, is, in our view, that which F. Clark Howell posed in 1959: that the "progressive" specimens of the Middle East actually belong to our species. Clark Howell (1959) called them Proto-Cro-Magnon. That idea implied, as Vandermeersch (1989) pointed out, that in the first part of the Würm glaciation there were two different populations in the area. Stringer and Gamble (1993) led the argument in favor of the presence of two species in the Middle East: *Homo neanderthalensis*, to which the specimens belong that are reminiscent of those from La Ferrassie; and *Homo sapiens*, for those with a modern look.

The debate about the coexistence or succession of Neanderthals and modern humans in the Middle East has led to a precise dating sequence of the levels in which human remains are found (Bar-Yosef, 2000). But, the idea of occupation alternatives in the Levant could change considerably if the remains found belong to levels different from those previously assumed. Henry Schwarcz, John Simpson, and Chris Stringer (1998) followed that argument when they carried out a reinterpretation of Tabun C1— the nearly complete skeleton of a female found by Garrod and Bate (1937). Schwarcz and collaborators described the placement of the specimen as being so close to the surface of level C that possibly it could really correspond to level B. Having been buried, it would be artificially included in the lower stratigraphic level (something already suspected by Dorothy Garrod; see Bar-Yosef, 2000). Using gamma-ray spectrometry, Schwarcz et al. (1998) attributed 34,000±5000 years to the Tabun C1 specimen, suggesting that the latest dates of Neanderthal taxon existence, contrary to what is often believed, are similar in both the Middle East and Europe. However, Ofer Bar-Yosef (2000) declined to consider the results obtained by gamma-ray spectrometry due to its unreliability.

In summary, one might say that the development of the controversy about the presence of one or two taxa in the specimens of the Middle East is too broad to reflect here in more detail than that already mentioned. On the other hand, it follows closely the previous idea about whether Neanderthals and modern humans belong to the same species. This is a matter that fortunately can be evaluated more precisely thanks to the recovery of the Neanderthal genome and one which we will address in detail in Chapter 11.

10.2 *Homo sapiens*

Carolus Linnaeus (1758) placed humans in the taxon *Homo sapiens*, which is the starting point for investigating modern humans from a taxonomic perspective

(Chapter 2). As we have said throughout this book, our species belongs to the order Primata, superfamily Hominidea, family Hominidae, tribe Hominini, genus *Homo*, and species *Homo sapiens* (Figure 10.16).

10.2.1 What is a modern human?

The question about what is meant by modern human seems unnecessary, or belated, as we have been using that expression continuously in this book. But, now we would like to recapitulate systematically what has been said up to now on this matter.

If we apply to *Homo sapiens* the criteria that we have used with other hominin species, a starting point would be to identify the holotype and paratypes that constitute the hypodigm of modern humans; this is not necessary or possible. It is not necessary because we know who are the members of our species. Linnaeus did not define a holotype for *H. sapiens*, nor is this a matter of concern for any current anthropologist. Yet,

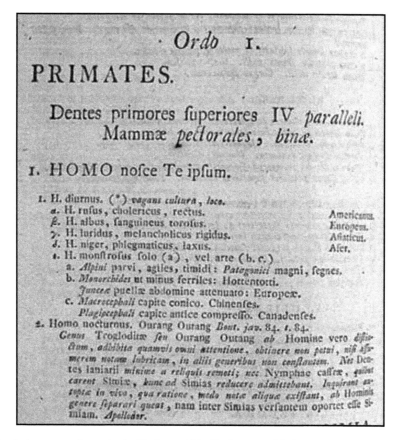

Figure 10.16 Order of primates, according to the classification established by Linnaeus. Within the genus *Homo*, Linnaeus (1758) distinguished between *Homo diurnus*, which today we call *Homo sapiens*, and *Homo nocturnus* (orangutan). Such a classification, including up to seven different human races, is not acceptable today, although Linnaeus' taxonomy remains the source of our current taxonomic system.

on reflection, we realize that it is far from easy to list the apomorphies that identify our species. Moreover, listing the paratypes is impossible; it would be necessary to include millions of specimens.

The relevant questions concerning our species refer to the apomorphies that identify it, and, in contrast to taxonomic considerations, they are not obvious.

Excluded from *H. sapiens* are other species that have coexisted with modern humans, such as the Neanderthals, but also *Homo floresiensis* and, as we will see in Chapter 11, the Denisova hominins. Nevertheless, if we include in *H. sapiens* only modern humans and their ancestors, which are not ancestors of *H. neanderthalensis* as well, then we need to define the autapomorphies of *H. sapiens*, the derived traits not shared with any other hominin.

It seems easy to identify the traits characteristic of modern humans, because we have an immense number of exemplars. But their abundance entails enormous intraspecific variation, which handicaps any effort to establish quantitative measurements. Consider, for example, the paradox faced by Chris Stringer, Jean-Jacques Hublin, and Bernard Vandermeersch (1984) seeking to specify the cranial measurements of *H. sapiens*. If statistical bounds are used that would distinguish modern humans and their direct ancestors from other taxa, those bounds would exclude many living and ancestral humans from the taxon.

The observation of modern humans, even when reduced to a minimal sample of those living now, makes apparent their distinctive apomorphies, anatomical as well as functional; but, "distinctive" with respect to

---

Box 10.16 The aborigines, after Haeckel

Haeckel (1905) maintained that "the lower races—such as the Veddahs or Australian Negroes—are physiologically nearer to the mammals, apes and dogs, than to the civilized European," (from Milford Wolpoff & Rachel Caspari, 2000). Naturally, the question is not to decide who is human and who is not among current individuals. The ethnocentrism of the past, which classified the so-called inferior races as arboreal primates, was discarded a long time ago. However, the question remaining is how to identify the human taxon so that it can be determined whether or not certain fossils belong to *Homo sapiens*. Sometimes this is not so obvious.

Box 10.17 The naked ape

Certain traits that seem distinctive in comparison with other primates, such as the absence of hair from much of our body—we are the "naked ape"—occur in other mammals, such as mole rats and whales. However, these mammals are not bipedal, nor do they have an advanced technology or literature and art. A more meaningful comparison is that made with organisms closer to us.

---

which other species? We have a large brain, bipedal gait, use clothing as well as an advanced technology, which includes airplanes and computers, and a well-developed culture, which includes art, literature, legal codes, and political institutions. All these traits distinguish us from the other apes.

But bipedal gait, even though it distinguishes *H. sapiens* from other apes, was already present in *H. erectus* and, with some distinctive traits, in the previous members of the entire human lineage. More generally, direct comparison of modern humans with chimps and other close relatives is likely to yield plesiomorphic primitive features inherited from ancestral hominins, in addition to autapomorphies that would characterize *H. sapiens*.

In order to identify valid human autapomorphies, we need to compare the traits of modern humans with those of our sister taxon, *H. neanderthalensis*—assuming, of course, that this taxon is a different species.

10.2.2 Morphological differences between *Homo neanderthalensis* and *Homo sapiens*

Table 10.3 shows the morphological and functional differences between modern humans and Neanderthals according to Bernard Wood (2005). In Chapter 2 we have given another, somewhat different, list provided by Sean Carroll (2003). Wood's list has the advantage of including several traits related to what looks like a very conspicuous element when evaluating the phenotype of one of the human beings around us: the face and the cranium. However, in Chapter 11, we will see how the consideration of the utensils of Neanderthals and modern humans might be closer than Wood (2005) indicates.

However, the consideration of the size and shape of the cranium and face leads to a paradox similar to the one indicated earlier. William Howells (1973, 1989) conducted a comprehensive study of the variation of

Table 10.3 Main morphological and functional differences among modern humans and Neanderthals (Wood, 2005)

Morphology	Modern humans	Neanderthals
Brain size (absolute)	Large	Very large
Brow ridges	Weak	Thick and arched
Nose and mid-face	Flat	Projecting
Cranial vault	Straight sides	Bulging sides
Occipital region	Round	Bulging
Incisors	Small	Large
Chest	Narrow	Broad
Hip	Small and narrow	Large and wide
Limb bones	Straight	Curved
Limb joints	Small	Large
Thumb	Short	Long
Development of bones and teeth	Slow	Fast
Behavior*		
Lithic utensils	Small	Larger and cruder and specialized
Complex utensils	Yes	No
Bone made utensils	Yes	No
Personal decoration	Yes, and well developed	No

*We consider the behavioral and cognitive differences between Neanderthals and modern humans in Chapter 11.

Box 10.18 Apomorphies and climates related to *H. sapiens*

When studying the apomorphies of *H. sapiens*, Wood and Richmond (2000) noted that they seem to fit best the modern humans from hot, arid climates. This is hardly surprising since modern humans evolved in tropical Africa and their earlier expansion was through tropical or subtropical lands.

---

57 cranial measurements in 28 human groups, concluding that the variation among these groups is similar to that which separates all human crania from the small sample of *H. neanderthalensis* crania, also included in the study. In other words, to distinguish modern humans from Neanderthals based on these traits leads to as many differences as those shown when comparing various groups of current humans.

Human phenotypic variation is immense. Why? The answer seems obvious: we have colonized all the continents—except Antarctica—adapting to very diverse climates since tens, if not hundreds, of thousands of years. The result can be no other than the great variability of phenotypes, whose genetic basis we will return to later. To understand its scope and meaning we are going to first examine how the origin and dispersal of our species took place.

10.2.3 The African origin: the coalescence of the current mtDNA

The origin of anatomically modern humans is a somewhat controversial issue. Over the past four decades two hypotheses have prevailed, incompatible with one another at least in their extreme formulation, which we have described and discussed in Chapter 9 and have since then used: the Multiregional Hypothesis (MH) and the Replacement (or From Africa) Hypothesis (RH) (Figure 10.17).

As we shall see in Chapter 11, the molecular evidence favors the RH model of the origin of modern humans, although not in its extreme form. This argues, first, that the transition from archaic to modern *Homo*

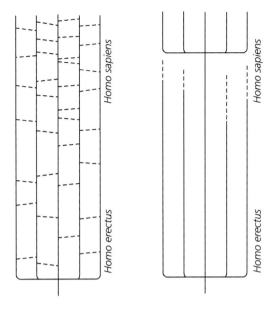

Homo sapiens

Homo erectus

Homo sapiens

Homo erectus

Figure 10.17 Multiregional Hypothesis (MH, left) and Replacement Hypothesis (RH, right). As indicated in Chapter 9 (Figure 9.1 is repeated here for reader convenience), for MH to be viable, there must be continuous migrations to ensure genetic exchange between populations of different continents; RH rejects that exchange.

sapiens was associated with an episode of a very narrow population bottleneck at the African origin (we will see in Chapter 11 that the second assertion of the extreme RH, that of a complete absence of hybridization, conflicts with molecular data). The population bottleneck was a drastic reduction in the number of our immediate ancestors, to the point where very few individuals (even only two) would, in the most extreme case, be the ancestors of all modern humans. The most radical version of RH, called the "Noah's Ark" or the "mitochondrial Eve" (Brown, 1980; Lowenstein, 1986), argues that all modern humans are descended from a single woman. Let's see, using the coalescence theory, why this is not feasible.

The coalescence theory examines the genealogical relations between genes (Griffiths 1980; Hudson 1990). According to this theory, all genes (alleles) present in extant populations must have descended from a single gene, to which they coalesce (Box 10.19).

The arguments in favor of the Noah's Ark hypothesis come from the confusion between gene genealogies and individuals' genealogy. Genes gradually coalesce into fewer and fewer ancestors and, ultimately, into a single ancestor. Individuals, by contrast, expand from any individual, with a factor ×2 for each ancestral generation. As we go back many generations, more subjects appear as ancestors (Box 10.20).

In Chapter 11 we will explore the genetic relationship existing between Neanderthals and modern humans, but we can say in advance that the mtDNA analysis of 100 ethnically heterogeneous individuals done in the 1980s indicated that the mtDNA sequences of modern humans coalesce into an ancestral sequence existing in Africa about 200,000 years ago (Cann et al., 1987; Stoneking et al., 1990; Vigilant et al., 1991). The legitimate conclusion of the mtDNA analysis is that the mitochondrial Eve is the matrilineal ancestor of modern humans. Everyone has a single matrilineal ancestor in any given generation. Everyone inherits the mtDNA from the mother and, in turn, from the maternal grandmother, and from the maternal lineage great-grandmother, and so on. But every person also inherits other genes from the other three great-grandmothers and from their four great-grandfathers.

The mtDNA we inherit from the mitochondrial Eve represents a small fraction of our total DNA. The rest of DNA has been received from other individuals, contemporary or not, of the mitochondrial Eve. The coalescence of the mtDNA of modern humans into a single ancestor is a feature that necessarily occurs for any one gene or genetic trait. As one proceeds back in time, at any gene locus (or DNA segment) all 2N genes of a species with N individuals derive from fewer and fewer ancestral genes, eventually converging into a single gene ancestor to all 2N descendants. But the ancestral genes for different gene loci occur in different generations and, of course, different individuals. The genome of each living human individual derives from many ancestors. The converse of this is the non-intuitive inference that any human who lived a few thousand generations ago and who has living descendants is an ancestor of all living individuals (Rohde et al., 2004; Hein, 2004), although he/she would have contributed different genes to different living individuals (Box 10.21).

10.2.4 The genealogical tree of mtDNA

The reconstruction of the mtDNA genealogical tree places its roots, i.e., the origin of ancestral mtDNA, in Africa (Cann et al., 1987; Stoneking et al., 1990; Vigilant et al., 1991; Ruvolo et al., 1993; Horai et al., 1995).

Box 10.19 Coalescence

The phylogeny of individual genes is star-like, with the most recent common ancestor at the vertex of the star as shown in the Box Figure (Slatkin & Hudson, 1991).

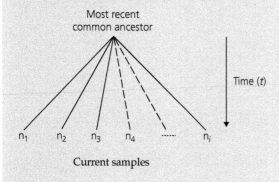

Current samples

The theory was first formulated for neutral or nearly neutral genes, which are genes that do not modify the welfare of the organism. In a random mating population at equilibrium, the mean coalescence time of neutral genes is given by:

$$T = 4N_e[1-(1/i)],$$

where T is the number of generations to coalescence, N_e is the effective size of the population, i is the number of sampled genes, and the variance is large (Kingman, 1982a, 1982b; Tajima, 1983; Tavaré, 1984; Takahata & Nei, 1985). For any two genes ($i = 2$), the mean coalescence time reduces to $T \approx 2N_e$ generations; for a large number of genes, the mean coalescence time is $T \approx 4N_e$. Thus, in a population with $N_e = 1$ million individuals, genes are expected to converge to their one ancestor 4 million generations earlier.

The coalescence equation can be used in the opposite direction, so that we can estimate population size if the coalescence time is known. In order to determine the time to the coalescent (the "most recent common ancestor" or MRCA), we need to know the rate of neutral mutation. This can sometimes be determined by the number of neutral substitutions between the genes of two species, of which the time of divergence is known. Under the assumptions of the coalescence theory, and ignoring the possibility of multiple hits at individual sites, the number of neutral polymorphisms that we observe in a sample of multiple genes will have a Poisson distribution with a mean that depends on the neutral mutation rate, the time elapsed, and the number of lineages examined. The expected number of polymorphisms is:

$$\lambda = \mu t \sum n_i l_i$$

where μ is the neutral mutation rate, t is the time since the MRCA, n_i is the number of lineages sampled at the ith locus, and l_i is the number of neutral sites at the ith locus. Solving for t and replacing λ with S, the observed number of polymorphisms:

$$t = S/\mu \sum n_i l_i$$

If S is assumed to have a Poisson distribution, the 95% confidence intervals can be estimated. When examining genes that are not located in the autosomes, one needs to take into account that the relative effective population size of the autosomes: X chromosomes: non-recombining Y: mtDNA is 4:3:1:1. Accordingly, the estimated N based on mtDNA, for example, is N_f, the estimated effective population size of females, so that $N_e = 2N_f$.

The molecular evidence used in the "Noah's Ark" model is derived from the study of the genealogies of the mitochondrial DNA (mtDNA). Approximately 100–200 thousand years ago, they coalesced into one haploid mtDNA, ancestor of all current mtDNAs.

Early mtDNA studies focused on the control region, which represents less than 7% of all mitochondrial genetic information and does not have a coding role. A study of the complete mtDNA (16,500 nucleotides in length) from 53 individuals confirmed the same African origin (Ingman et al., 2000). The mtDNA evidence would not be conclusive by itself, given that mtDNA is based on a minimal amount of deoxyribonucleic acid;

each of the two nuclear genomes of a human consists of about 3 billion nucleotides, which is about 250,000 times more than the mtDNA. But in the 1990s, analysis of chromosome DNA microsatellites (Goldstein et al., 1995) and of a large sample of nuclear genes distributed throughout the entire human genome (Cavalli-Sforza et al., 1994) also yielded genealogical trees rooted in Africa.

Box 10.20 Genetic and family relationship

The difference between ancestral genes and ancestral relatives can be illustrated by an analogy. The Ayala surname is shared by many people who live in Spain, Mexico, the Philippines, and other countries. Historians have concluded that all Ayalas descend from Don Lope Sánchez de Ayala, the grandson of Don Vela, vassal of King Alfonso VI, who established the dominion of Ayala in the year 1085, in the present province of Álava, in the País Vasco (Marqués de Lozoya, 1972; Luengas Otaola, 1974). Don Lope is the Adam from whom all who bear the name of Ayala descend in the patrilineal line, but current Ayalas also descend from many other men and women who lived in the eleventh century, as well as before and after that time.

Box 10.21 How much remains in us from the mitochondrial Eve?

The mtDNA that remains from the mitochondrial Eve is 1/400,000 of the DNA of any modern human. The rest of the genetic material comes from other individuals contemporary to the mitochondrial Eve and their ancestors. Many women contemporary of the one carrying that mtDNA molecule have left descendants among the current human population, to which they have transmitted nuclear genes.

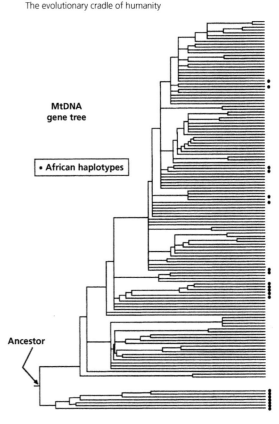

The evolutionary cradle of humanity

MtDNA gene tree

• African haplotypes

Ancestor

Figure 10.18 Mitochondrial DNA phylogeny of modern humans. The dots mark mtDNAs from native Africans; the others are from native European, Asians, Australians, and New Guineans (from Avise, 2006, redrawn after Cann et al., 1987).

In the mtDNA tree, ancestral African populations are set apart from all non-African populations, which are located on a single branch emerging from the multi-branched African tree. The most profound divergence of non-African populations in the genealogical trees is calculated at about 156,000 years ago (with a possible error of tens of thousands of years). The time estimates to the most recent common ancestor of modern humans vary from one to another study; for example, Ingman et al. (2000) set the time at 175,000 years. In any case, the first divergence between African and non-African populations would mark the earliest possible point in time at which modern humans would have dispersed from Africa to the rest of the world.

In summary, the weight of the molecular evidence supports the hypothesis of a recent African origin

Box 10.22 Genealogy of the RRM2P4 pseudogene

mtDNA phylogenies point to Africa as the place of origin of the most recent common ancestor (MRCA) of modern humans (see Figure 10.18). Autosomal and X-linked loci also typically point to Africa. One important exception is the X-linked ribonucleotide reductase M2 polypeptide pseudogene 4 (RRM2P4), which yields a tree with a cenancestor (most recent common ancestor) in East Asia (Garrigan et al., 2005; Garrigan and Hammer, 2006). A total of 13 polymorphic sites were observed in 2,385 nucleotides sequenced from 42 individuals: 12 from Asia and 10 from each of Africa, Europe, and America. Coalescent simulations yield Asia as the most likely origin of the MRCA, with a probability of $P = 0.92$, while $P = 0.05$ for Africa and smaller probabilities for Europe or America.

Box 10.23 Genetic distances

The results obtained in the genetic distance calculation are especially remarkable in studies such as that by Cavalli-Sforza et al. (1994), because the analyses carried out sought to determine the history of human populations and not the ancestral line of individual genes. However, despite the results obtained by these authors, there seems not to be definitive reasons to completely exclude the hypothesis that different genes might have different population origins. The average distance reflects the average—and, therefore, predominant—genetic contribution coming from several ancestral populations, so the results would also be consistent with a model in which the African replacement coincides with some regional continuity. Given that evidence of regional genetic continuity is nonexistent or very scarce, many authors agreed at the time with RH; however, see Li et al. (1993), Ayala et al. (1994), and Deeb et al. (1994).

---

of modern humans. Ethnic differentiation between modern populations would be recent in phylogenetic terms: the result of a divergent evolution between populations that have been separated only for the last 50,000–100,000 years. This conclusion is consistent, furthermore, with the extensive studies of genetic polymorphism that refer to a large number of genes in populations around the world.

Luigi Luca Cavalli-Sforza and colleagues (1994) pointed out discrepancies between the calculated bifurcation time between African and non-African populations based on nuclear genes (about 100,000 years ago) and mtDNA (close to 200,000 years ago). Divergence time estimates in such studies show great variation, largely due to the limited dataset they are based on. It is unsurprising, therefore, that mtDNA polymorphism coalescence has been estimated at 143,000 years by Horai and colleagues (1995) and at 298,000 years by Ruvolo and colleagues (1993), with confidence intervals ranging from 129,000 to 536,000 years. The appearance of differences by a factor of two between the various estimates of mitochondrial and nuclear DNA should not be a concern, given the high degree of uncertainty in the calculations. On the contrary, what is surprising is that some molecular scientists assert precise dates inferred from their analyses. In any case, from the analysis of 30 polymorphisms of nuclear SSR (microsatellite), Goldstein

and collaborators (1995) estimated the divergence between African and non-African populations at 156,000 years, which is an intermediate date between those obtained by means of nuclear genes and mitochondrial DNA.

10.2.5 Size of the ancestral human populations

The idea of an actual "mitochondrial Eve" as a single woman who could have produced, by herself, the entire human diversity is, as we have seen, the result of a misunderstanding. But, how large would the population have been at the time when our species appeared?

The number of ancestral individuals per generation can be determined by the genetic coalescence theory, which studies highly polymorphic genes, such as those of the Major Histocompatibility Complex (MHC) of the immune system, which protects us against pathogens and parasitic infections. In humans, the MHC genes are known as Human Leukocyte *Antigen* (HLA) (Box 10.24).

The amino acid composition varies from one to another MHC molecule; this variation is responsible for the enormous polymorphism characteristic of MHC molecules and of the genes encoding them. In human populations, as in other mammals, there are a large number of genetic variants (alleles) in any of the

---

Box 10.24 HLA genes

The HLA genes encode molecules with a critical role in tissue compatibility and in the defense against pathogens and parasites. These genes are distributed into two distinct groups, class I and class II, and are separated by several dozen other genes that have functions primarily related to the immune response. MHC molecules bind to protein fragments (antigens) on the surface of certain cells, which present the fragments to lymphocytes called T cells. When the T cells bearing the corresponding receptors for a particular antigen combine with it, the proliferation of these T cells is specifically stimulated and thus initiates the immune response by secreting specific antibodies. The recognition of the protein fragments is performed in a specialized region on the surface of the MHC molecules called PBR (peptide binding region), which has about 50 amino acids (Bjorkman et al., 1987a, 1987b; Brown et al., 1993).

different MHC loci; these alleles can differ from each other in as many as 100 nucleotides (Klein & Figueroa, 1986; Marsh & Bodmer, 1991; World Health Organization, 1992; O'Huigin et al., 1993; Bontrop, 1994; Bontrop et al., 1995; McDevitt, 1995).

MHC polymorphisms refer us to ancient times, with genetic lineages that can be traced back millions of years in primates (Lawlor et al., 1988; Mayer et al., 1988; Fan et al., 1989; Gyllenstein & Erlich, 1989; Gyllenstein et al., 1990; Ayala et al., 1994; Bergström & Gyllenstein, 1995; Ayala & Escalante, 1996) and in rodents (Arden & Klein, 1982; Figueroa et al., 1988; McConnell et al., 1988).

The method used to reconstruct the genealogy of the immune system genes can be illustrated by a simple example. Figure 10.19 shows a comparison among four *DRB1* gene sequences, one of the 100 genes that constitute the HLA in mammals and other vertebrates (Klein, 1986; Kaufman et al., 1995). The upper two sequences of Figure 10.19 are from humans and the two below from chimpanzees, each with 270 nucleotides. The differences between the various pairs of sequences are given in Table 10.2. The most similar are the human gene *Hs*1103* and chimpanzee gene *Pt*0309*, associated with each other in Figure 10.19 to indicate that they share a recent common ancestor. The other two genes, *Hs*0302* and *Pt*0302*, are also more similar to each other than with respect to *Hs*1103* or *Pt*0309*. The two pairs were linked to indicate that the four genes derived from a common ancestral gene. The length of the branches is proportional to the number of nucleotide changes that have occurred in each branch, as inferred from Table 10.4.

A remarkable property of these four *DRB1* genes is that the two human sequences have more differences

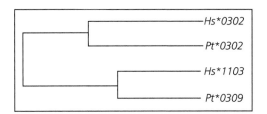

Figure 10.20 Genealogical tree of four *DRB1* genes showing a representation of their evolutionary divergence patterns. *Hs* symbolizes *Homo sapiens* and *Pt* symbolizes *Pan troglodytes*. The length of the branches is proportional to the difference in nucleotides between the genes (see Table 10.3).

Table 10.4 Number of nucleotide differences among four *DRB1* genes, two human (*Hs*) and two chimpanzees (*Pt*)

	*Hs*1103*	*Pt*0309*	*Pt*0302*
*Hs*0302*	18	18	12
*Hs*1103*		9	20
*Pt*0309*			20

between them than between each of them and a chimpanzee gene. From this it follows that the lineages of these two human genes diverged from each earlier than 7 Ma, the approximate time when the separation between humans and chimpanzees took place (Figure 10.20). The ancient origin of lineages of *DRB1* genes is the property that makes them particularly suitable for the study of the history of ancient human populations.

Figure 10.21 represents the genealogy of 119 *DRB1* genes, of which 59 are from humans, 40 from apes, and 20 from Old World monkeys (Table 10.4). In this genealogy the length of each branch is proportional to the number of nucleotide substitutions that have occurred in every lineage. For example, the bottom of Figure 10.21 shows that of the two macaque genes, *Mm*0301* and *Mm*0302*, the first has changed more than the second.

The associations of some human genes are very ancient. For example, the nine genes at the top of Figure 10.21, closely related, constitute a gene cluster formed before their relationship with any of the six genes that appear immediately beneath them—among which one is from a drill and four are from macaques. The divergence between hominoids (apes and humans) and Old World monkeys took place around 35 Ma, at the limit between the Eocene and Oligocene, so the relationships depicted in Figure 10.21 show that

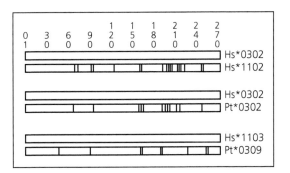

Figure 10.19 Nucleotide differences in four *DRB1* gene sequences. The upper two are of humans, and the bottom two of chimpanzees. Nucleotides that are different in each comparison are indicated by vertical stripes in the lower sequence.

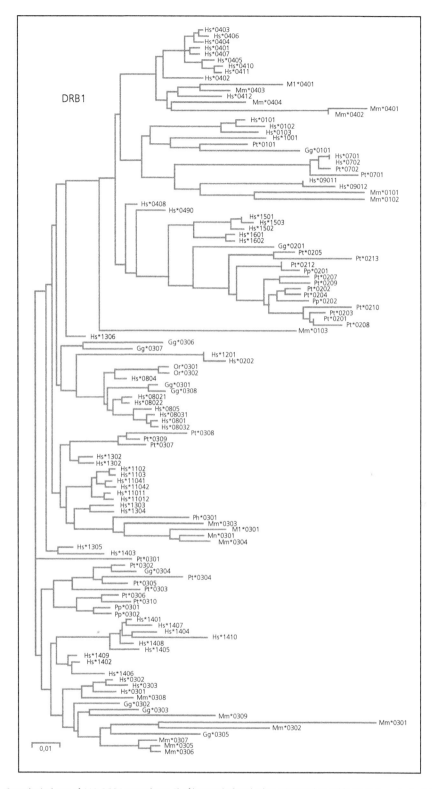

Figure 10.21 Genealogical tree of 119 *DRB1* genes (exon 2) of human (*Hs*) and other primates (see Table 10.4). From Ayala, F. J. (1995b). The Myth of Eve: Molecular Biology and Human Origins. *Science*, 270, 1930–1936. Reprinted with permission from AAAS.

Box 10.25 *DRB1* as a molecular clock

The *DRB1* gene has a substitution rate of 1.06×10^{-9} nucleotides per locus per year. This ratio is obtained by the "minimum" method (Ayala & Escalante 1996; see also Satta et al., 1991).

---

at that time several lineages of human *DRB1* genes already existed which have descendants among current humans.

The age of the human *DRB1* lineages can be calculated by calibrating the molecular clock (Box 10.25). Figure 10.22 shows a genealogy of the 59 human *DRB1* genes, constructed using the average rate of evolution to determine the length of the branches.

The timescale indicated at the bottom of Figure 10.22 specifies three stages in the evolution of hominoids and hominins: the divergence of the orangutan lineage c. 15 Ma, the divergence of humans and chimpanzees c. 7 Ma, and the appearance *H. erectus* c. 1.8 Ma.

These benchmarks are useful for calculating the number of existing genetic lineages at a given time, which is done by counting how many are intersected by the vertical line corresponding to the time indicated on the abscissa. For example, there were 32 human genetic lineages around 7 Ma, which means in turn that the other 27 human lineages appeared after the hominins diverged from the African apes. The genealogy of all human genes coalesces at about 60 Ma, which means that the human *DRB1* genes present today in our species began to diverge at that time (Box 10.26).

The conclusions reached by the application of the theory of coalescence establish that the verifiable age of polymorphisms requires that the *effective average size* of the ancestral population of modern humans from 60 million years ago to the present—that is, the number of individuals per generation having direct descendants now among the living humans—to be about a hundred thousand individuals (Box 10.27). This estimate refers to the average effective population number; the "effective" population number includes only those individuals able to reproduce at any given time (children and the elderly are not included).

The results offered by the coalescence theory have been confirmed by computational experiments. In these studies, populations are created with a given number of individuals that reproduce according to the rules of human populations and maintain that

reproduction process for as many thousands of generations as desired. These studies confirm that for the present humanity to have about 59 *DRB1* alleles of this gene it would be required during the last 2 million years—since the origin of *H. erectus* at least—that the human populations from which we descend consisted, on average, of at least 100,000 reproducing individuals, similar to the number obtained by the coalescence theory (see, in particular: Ayala, 1995b; Ayala & Escalante, 1996).

The ancestral population number at the time of the emergence of our species obtained by studying the polymorphism of human *DRB1* genes can also be calculated using other techniques:

* Using mtDNA. If the coalescence took place around 200,000 years ago, the average number of ancestors from which we descend would be approximately 10,000 (Ayala 1995b; Wills, 1995). Other estimates indicate that the coalescence of mtDNA took place between 143,000 and 298,000 years ago (Ruvolo et al., 1993; Horai et al., 1995), or even earlier, between 622,000 and 889,000 years ago (Wills, 1995). These figures correspond to an ancestral population average of 31,100–44,450 individuals.

* Using the chromosome Y (Box 10.28). The coalescence theory indicates that the origin of modern humans goes back to a chromosome Y from about 270,000 years ago, with a confidence interval that extends from zero to 800,000 years ago. If we assume 20 years per generation, the coalescence of the *ZFY* gene in the Y chromosome leads to an effective population size of 13,500 individuals with an upper confidence limit of 95% establishing a population of 40,000 individuals. But if we take into account the standard deviation of the coalescence median, the upper limit of 95% rises to 80,000 individuals (Ayala 1995b).

The ancestral Adam from whom all living men have inherited their Y chromosome was not, however, our only male ancestor in his own or any other generation, in the same way that the woman from whom all modern humans have inherited their mtDNA was not the single woman of her generation ancestral to modern humans. The rest of our genes, other than *ZFY*, and mtDNA come from many other different male and female ancestors (Ayala, 1995a, 1995b). See Box 10.20.

10.2.6 Founder effect

The speciation theories based on the "founder effect" suggest that the emergence of a new species occurs

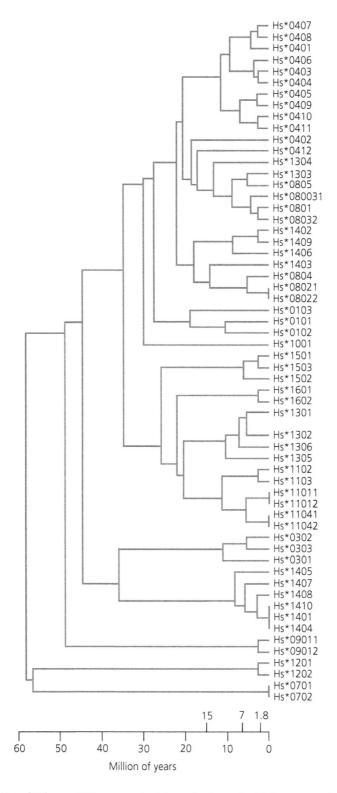

Figure 10.22 Genealogical tree of 59 human *DRB1* genes, constructed according to a method that assumes equal evolutionary rates along the branches (Ayala, 1995a). From Ayala, F. J. (1995b). The Myth of Eve: Molecular Biology and Human Origins. *Science*, 270, 1930–1936. Reprinted with permission from AAAS.

Box 10.26 *DRB1* lineages

If 32 *DRB1* human lineages of this gene have persisted since 7 Ma, we can infer that at any given moment since that time there could not have been less than 16 individuals. The minimum number of people must have been much higher, because the probability that each of the 16 individuals were carriers of two different genes, different also with respect to the genes of all the others, is virtually zero. Beyond these hypothetical considerations we can get accurate estimates of the size of populations by mathematical theories formulated in recent years, including the coalescence theory (see Ayala et al., 1994; Ayala, 1995b; Ayala & Escalante, 1996; in addition to the data provided by DNA sequencing, e.g., Meyer et al., 2012).

Box 10.27 Population size

The average we refer to when calculating the size of a population is not the arithmetic mean, but the *harmonic* mean, which is consistent with numbers much above 100,000 individuals in any one generation, but is not consistent with numbers much below the average in any one generation.

Census numbers, i.e., the total number of the population, would be at least two to three times greater than the effective number. The census number of ancestral human populations would be, then, more than 200,000 individuals on average.

The number of ancestors of an individual in a past generation is 2^n, where n is the number of generations. The value of 2^n is 1,024 for 10 generations, more than 1 million in the case of 20 generations and more than a billion for 30 generations. But, in fact, when we are dealing with many generations, the numbers of ancestors are much lower than those given by the calculations, because many of the same ancestors are likely to appear repeatedly in the ancestral genealogies of different individuals.

Box 10.28 The Y chromosome

The Y chromosome is the genetic counterpart of mtDNA in that it is inherited only from fathers to sons. There are regions on chromosome Y that are not homologous to chromosome X and thus are transmitted only through the paternal line in the same way that the mtDNA is transmitted only through the maternal line. A DNA fragment of 729 nucleotides of the *ZFY* gene (probably involved in testicle or sperm maturation) found on chromosome Y was sequenced in 38 men representative of major ethnic groups, without finding any variation (Dorit et al., 1995; see also: Hammer, 1995; Whitfield et al., 1995).

Box 10.29 Is the founder effect common?

The importance of speciation by founder effect is a matter subject to debate. Certain experiments have led to the conclusion that this form of speciation is less likely than some authors propose (Moya & Ayala, 1989; Galiana et al., 1993).

---

habitat such as an island, a lake, or a previously uninhabited territory. If the population prospers, the gene pool can be very different from the original due to the so-called sampling errors, so that a "gene revolution" takes place during the adjustment process of the new gene pool (Mayr, 1954, 1963, 1982).

The founder effect was, according to some authors (e.g., Brown, 1980; Cann et al., 1987; Stoneking et al., 1990; Vigilant et al., 1991), the circumstance that favored the evolution of the distinctive features of modern humans, who descend from very few individuals.

It is possible to consider that one or more bottlenecks occurred in human evolution. The conclusion of both theoretical studies and computer simulation models indicates that, if there ever was a population bottleneck, it never reduced the total to less than 5,000–10,000 individuals (Ayala & Escalante, 1996). This number is consistent with the lower estimations of mtDNA polymorphisms and *ZFY* gene studies. On the contrary, it can be shown that for millions of years the populations from which we descend had, on average, at least 100,000 individuals throughout history (Ayala et al., 1994; Ayala, 1995b).

precisely as a result of a founder or "bottleneck" event, so that the new population is established from very few individuals, or even just one (one fertilized female) (Box 10.29). This phenomenon can occur when a population is suffering a drastic reduction due to biological or physical causes (climate, for example) or, more often, when the "founders" colonize a new

10.2.7 The imprint of the first modern humans

Nuclear evidence supports the hypothesis of a recent African origin (c. 150,000–100,000 years) for modern humans, but with some interesting nuances that are necessary to clarify. We will begin with the imprint that the emergence episode of the modern humans left in the current populations. Is there any group today that we could qualify as close to the first *Homo sapiens*?

Studies of single nucleotide polymorphisms (SNPs) in the hypervariable region of human mtDNA show that human mtDNA is geographically structured. It may be classified into groups of related haplotypes (L0 to L6, plus M and N macrohaplogroups). M and N macrohaplogroups persisted in non-African populations after the migration of modern humans out of Africa, while macrohaplogroup L is limited to sub-Saharan Africa and has been divided into L0–L6 haplogroups (Gonder et al., 2007) (Figure 10.23).

In order to establish the evolutionary sequence of the sub-Saharan macrohaplogroup L, Toomas Kivisild and collaborators (2006) conducted an analysis of the complete mtDNA of 277 individuals belonging to haplogroups L0–L5. The most parsimonious cladogram obtained shows that the L0D, corresponding to the Khoe-San people, is the ancestral subhaplogroup with respect to the rest of Africa (Figure 10.24 and Box 10.30).

Recent analysis of single nucleotide polymorphisms (SNP) of nuclear DNA supports the sub-Saharan origin

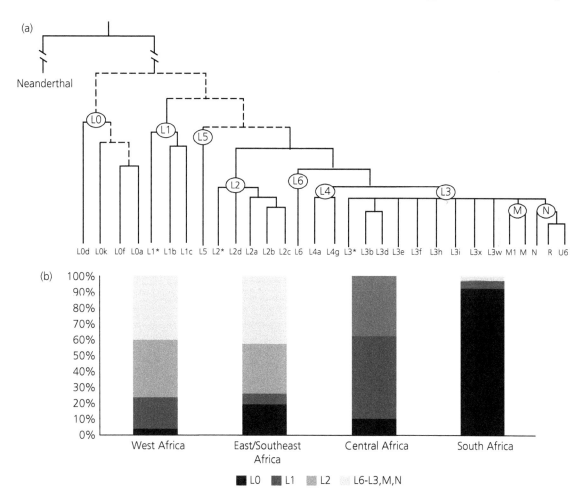

Figure 10.23 (a) Relationships among different mtDNA haplogroup lineages. Dashed lines indicate previously unresolved relationships. (b) Relative frequencies of haplogroups L0, L1, L5, L2, L3, M, and N in different regions of Africa. Gonder, M. K., Mortensen, H. M., Reed, F. A., de Sousa, A., & Tishkoff, S. A. (2007). Whole-mtDNA Genome Sequence Analysis of Ancient African Lineages. *Molecular Biology and Evolution*, 24(3), 757–768, by permission of Oxford University Press.

Box 10.30 Khoe-San ethnicity

Khoe-San is the name recommended by the San Coun-cil to replace the more commonly used Khoisan. In both cases it is the group of South African populations Khoek-hoe and San who speak a particular language, "Clicks," different to any other in the world but shared with other African hunter-gatherers like the Hadza and Sandawe. The Khoekhoe include the Nama groups (province of Nor-thern Cape), Koranna (Kimberley and Free State), and Gri-qua (Western Cape, Eastern Cape, Northern Cape, Free State, and Kwa-Zulu-Natal). The San include the Khomani San groups (Kalahari region), Khwe and !Xun (Platfon-tein, Kimberley) (http://www.iwgia.org/regions/africa/south-africa/894-update-2011-south-africa, accessed June 12, 2016).

Box 10.31 The South African sample

The choice of the sample made by Schuster et al. (2010) acquired a symbolic meaning when he performed the complete sequencing of the Archbishop Desmond Tutu genome, from the Bantu ethnic, a well-known figure of the Anglican Church, whose role during the struggle against apartheid in South Africa reached a global dimension. The authors hypothetically attributed the genetic proximity between ethnic groups alien to hunter-gatherers to the possible adaptive selection occurring when the agricul-tural lifestyle emerged.

of modern humans (Jakobsson et al., 2008; Li, Absher, et al., 2008). The full genome comparisons carried out by Stephan Schuster et al. (2010) have further clarified the relationships between the Khoe-San and other eth-nic groups. Schuster et al. (2010) obtained the complete

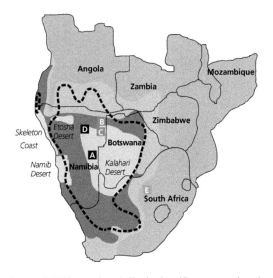

Figure 10.24 The area bounded by the dotted line corresponds to the geographical area of the Khoe-San language (Click language). The dark gray area is that of the Niger–Congo languages. The light gray area indicates arid climate zones and the lighter gray area desert climate zones. The letters correspond to subjects about whom the genome study was conducted: (A) Bushman of Tuu language; (B–D) Bushman of Juu Juu language—the Juu and Tuu languages belong to the Khoe-San group; (E) Bantu. Reprinted by permission from Macmillan Publishers Ltd: Schuster, S. C., Miller, W., Ratan, A., Tomsho, L. P., Giardine, B., Kasson, L. R., . . . Hayes, V. M. (2010). Complete Khoisan and Bantu genomes from southern Africa. *Nature*, 463(7283), 943–947.

nuclear DNA of a San Bushman and a Bantu—another South African ethnic group whose language, Bantu, does not belong to the group of Click languages—along with protein-coding regions of another three groups of Khoe-San hunter-gatherers from the Kala-hari (Box 10.31).

The San Bushmen show on average greater genetic differences among the various groups than those ex-isting, for example, between a European and an Asian (Schuster et al., 2010).

The details of Khoe-San genetic variation have been offered by Carina Schlebusch et al. (2012) by genotyp-ing c. 2.3 million of SNPs in 220 South Africans. The re-sults of the study indicated that the divergence between the Khoe-San and other modern African humans took place more than 100,000 years ago—that is to say, near the very beginning of the emergence of *Homo sapiens*—although the genetic distribution of the modern-day Khoe-San goes back only about 35,000 years.

According to the distribution of specific genetic vari-ations by Schlebusch et al. (2012), the regions which most likely have undergone positive selection in mod-ern humans would be those containing the genes:

ROR2, involved in the regulation of bone and cartil-age development.

SPTLC1, related to hereditary sensory neuropathy.

SULF2 regulates the development of cartilage, and its anomalies distort brain growth.

RUNX2, regulator of the fontanelle closure, with alleles associated with cleidocranial dysplasia (Schlebusch et al., 2012).

SDCCAG8, involved in microcephaly.

LRAT has been linked to Alzheimer's disease.

The first two of these regions are those with higher evidence of positive selection. Of the remaining four, the one containing *RUNX2* is the largest (c. 900 kb)

Box 10.32 *RUNX2*

Richard Green and collaborators (2010) pointed to *RUNX2* as a candidate gene for positive selection in modern humans, but not in the Neanderthals (see Chapter 11).

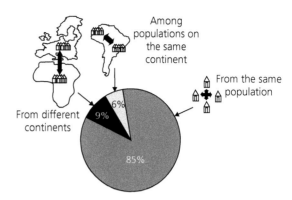

Figure 10.25 Genetic diversity in human living populations. Most (85%) of human genetic variation can be found within a single village. Populations from other villages of the same continent contribute an additional 6%, and those from different continents an additional 9% of humankind's total genetic variation (Ayala, 2007, p. 103).

(Box 10.32). All of them have some connection with brain development.

10.2.8 Human genetic diversity

Modern human traits are shared by all current humans and reflect both the unity of the species and the relatively recent origin of the differences appearing in *Homo sapiens*. The characteristics, sometimes called "racial," which roughly identify human populations from different regions of the world, evolved separately after dispersion into the different continents. Distinctive regional traits include, as it is widely known, skin pigmentation, hair type, facial configuration, including the eyes, and size and structure of the body. These traits have evolved partly due to natural selection, including sexual selection and cultural preferences—as Darwin emphasized in *The Descent of Man* (1871)—but also because of random processes such as genetic drift (including the founder effect, which, as we have seen, is due to the small number of the colonizers of different regions, and also to the small size of various populations over time—Carroll, 2003; Stringer, 2012).

The recent gradual dispersal of modern humans, mostly since less than 60,000 years ago, is consistent with studies of human genetic polymorphism. When analyzing the distribution of the genetic diversity of human populations in their geographic components, the results are remarkable, because they are totally the opposite of what intuition tells us. Studies of genetic diversity in living human populations have revealed information consistent with a recent origin of all living human populations, as proposed by the Out-of-Africa hypothesis. When the genetic diversity of human populations is mapped out geographically, it is found that 85% of it is present in any local population, this is to say, in any village or city of any continent (although the genes contributing to this 85% vary from one population to another). From 5 to 6% additional genetic variation is found when local populations on the same continent are compared, and an additional 10% when populations from different continents are compared (Barbujani et al., 1997; Jorde et al., 1997; Kaessmann et al., 1999) (Figure 10.25).

The calculations about extended genetic diversity seem to stumble upon the difficulty of the concept. Thus, 85% of the current genetic diversity could be present in any population, provided that a small number of immigrants are included that are not weighted highly in terms of what we might call social impact, but significant enough to indicate that diversity has arrived in the locality. In that case, the extended genetic diversity could be compatible with a "geographical grouping," able to couple some continents—or all of them—with "races." Africa would be, for example, the "black continent," Asia, the "yellow," etc.

The work of Noah Rosenberg and collaborators (2002) could be interpreted as supporting the hypothesis of the "geographical grouping." Using data from Howard Cann et al. (2002) (see Box 10.33), the authors studied 377 autosomal satellites from 1,056 subjects of 52 populations from different regions of the world.

The results of Rosenberg et al. (2002) indicate that differences between individuals of the same population represent 93–95% of the whole genetic variation and the differences between groups are only 3–5%. Of 4,119 alleles analyzed, 46.7% of them appear in all

Box 10.33 CEPH diversity panel

Howard Cann et al. (2002) made available to researchers the "CEPH Diversity Panel" composed of 1,064 cell lines from the CEPH (Centre d'Étude du Polymorphisme Humain) (Jean Dausset Foundation, Paris). The genetic data correspond to individuals unrelated to each other coming from the entire world.

regions, and only 7.4% are exclusive to one region. Such figures seem to support the idea of a large distinction of genetic diversity.

But the computational algorithm *Structure* used by the authors gave a different result. To apply the algorithm, the first step is to choose a number "K" of clusters (inferred groups). After indicating the desired number of clusters, the *Structure* program assigns each individual of the sample, without any a priori information about their origin, to one or more of the chosen K clusters (inferred groups). Thus, if K = 2, the program assigns the various subjects to the inferred two groups, which were separated by relatively significant genetic distances corresponding to very large groups of geographical regions (see Table 10.5). As clusters are added, these groups are divided to include those subjects that are closer in genetic terms. For K = 5, Rosenberg et al. (2002) concluded that the inferred groups corresponded largely to the major geographical regions of Africa, Europe, Central Asia, East Asia, and America.

It is not easy to get an intuitive idea of the results reported by Rosenberg et al. (2002). But, at the K = 5 level, they seem to have identified genetic discontinuities between continents. In other words, one could understand—in popular terms not used by the authors—that despite the great dispersion that characterizes humans from the genetic point of view, it is still possible to detect groupings that would be somehow equivalent to the attribution of "races" to "continents."

However, David Serre and Svante Pääbo (2004) have pointed out some weaknesses in the analysis of Rosenberg et al. (2002). Thus, when increasing the number of clusters to K = 6, the new inferred group did not correspond to any geographical region but, for the most part, to the Kalash from Northern Pakistan (Rosenberg et al., 2002 had already indicated that anomalous grouping would occur when K = 6 or greater). It is not clear, then, what number of clusters would best represent the genetic diversity of humans.

Serre and Pääbo (2004) argued that the key to the possible genetic discontinuities detected between

Table 10.5 Number of *DRB1* genes in primate species

Symbol	Scientific name (common name)	Genes
	Primates	
Hs	*Homo sapiens* (human)	59
Pp	*Pan paniscus* (bonobo, or pygmy chimpanzee)	4
Pt	*Pan troglodytes* (chimpanzee)	24
Gg	*Gorilla gorilla* (gorilla)	10
Or	*Pongo pygmaeus* (orangutan)	2
	Old World Monkeys	
Mm	*Macaca mulatta* (rhesus macaque)	16
Mn	*Macaca nemestrina* (southern pig-tailed macaque)	1
Ml	*Mandrillus leucophaeus* (drill)	2
Ph	*Papio hamadryas* (hamadryas baboon)	1

geographic regions might be in an imbalance introduced when obtaining the samples of individuals for analysis. To eliminate this possible source of error, Serre and Pääbo used three different samples, including in each five subjects from each population of the 52 from the CEPH Diversity Panel. By introducing this correction of the possible geographical imbalances, Serre and Pääbo (2004) found that after the application of *Structure* with K = 4 inferred groups (A, B, C, and D), most of the subjects of the sample were very mixed among all these clusters, so it was not possible to assign clusters to continents:

- Subjects from African populations showed "mixture" of the inferred groups A, B, and C.
- Subjects from Eurasia and Oceania showed mixture of A and B.
- Subjects from America showed mixture of B and D.

The most interesting result obtained by Serre and Pääbo (2004) has to do with the emergence of geographic gradients in relation to the ancestry coefficients (see Box 10.34). Thus, in regard to group B (with individuals belonging to populations in Africa, Eurasia,

Table 10.6 Assignment of groups of genetic proximity by the program *Structure* (see text)

K	Inferred groups							
2	Africa + Europe + Central Asia	/	East Asia + America					
3	Africa	/	Europe + Central Asia	/	East Asia + America			
4	Africa	/	Europe + Central Asia	/	East Asia	/	America	
5	Africa	/	Europe	/	Central Asia	/	East Asia	/	America

> **Box 10.34 Individuals and groups in *Structure***
>
> By "mixture" it should not be understood here as a mix of subjects, but that they were placed by *Structure* in one or more of the inferred groups K. An individual is placed in a group when their genetic characteristics are very close to those of the other group members. This condition is expressed by saying that it has a high ancestry coefficient.

Oceania, and America), the ancestry varies depending on the origin of the subject (Figure 10.26).

Regarding Eurasia, the ancestry expressed by the vertical Z dimension in Figure 10.26 (genetic overlapping between members of population B) is much higher in the population of East Asia than in the European. The population of Oceania is close to the East Asian ancestry. Melanesia is closer to that of Southeast Asia. In America, the population of Mexico is approaching the ancestry coefficients of the Melanesians, whereas that of Brazil is much less so. In Africa, the ancestry has a north–south gradient, although small, with higher coefficients in the north.

The work of Rosenberg et al. (2002) and Serre and Pääbo (2004) used a larger number of subjects than any other study to determine the extent of genetic diversity. Even though, if we talk about samples including ten subjects in each of the 52 populations analyzed, the possibilities of introducing statistical bias are very high.

We are, in fact, facing a bottleneck phenomenon in each tested population. Also, the 52 populations available in the CEPH Panel of Diversity do not cover the entire planet; the panel does not have, for example, North American subjects. Despite such limitations, the fact is that we have strong indications about how human genetic diversity is distributed. This evidence shows that the geographical and ethnic groups in the style of the "black continent" lack sufficient scientific support.

10.2.9 Adaptive phenotypic changes

Considering how easy it is to distinguish a Congolese from a Swede or a Japanese, it seems surprising that the genetic differentiation between humans from different continents involves only 10% of the genes. This fact has, however, an explanation. Ethnic differentiations based on both skin color and other obvious morphological traits are caused by a small number of genes that have a high adaptive value with respect to different climates. That is the case with the presence of melanin—dark skin—in the tropics, which protects the skin from ultraviolet rays and therefore from cancer, in contrast with its absence in the Nordic countries where the sun's rays are much more limited. In these countries a greater protection from the sun by a melanin filter would limit the synthesis of vitamin D in the inner layer of the dermis.

Nina Jablonski and George Chaplin (2000) carried out a study on how the potential for the synthesis of previtamin D_3 and the annual average UVMED on the

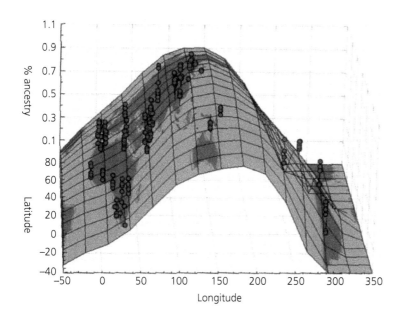

Figure 10.26 Geographical representation of the ancestry coefficients corresponding to the subjects of inferred population B (see text). The circles represent individuals defined by their geographical origin (x-axis longitude, y-axis latitude) and ancestry coefficient (z-axis) (illustration by Serre and Pääbo, 2004). Serre, D. & Pääbo, S. (2004). Evidence for Gradients of Human Genetic Diversity Within and Among Continents. *Genome Research*, 14, 1679–1685.

Box 10.35 Reflectance index

The reflectance index is a measure of the relationship between the electromagnetic power of light falling on a unit of unprotected body surface and the reflected electromagnetic power. The index expresses the amount of light energy—essentially ultraviolet rays—absorbed by the body per area unit. The increase in reflectance is associated with a large presence of melanin, as in the dark skin of tropical indigenous populations. The UVMED dose is defined as the amount of ultraviolet radiation necessary to produce a faint redness in minimally pigmented skin. UVMED averages for different parts of the world have been established with the help of measurements made by the Nimbus-7 satellite between 1978 and 1993 using the NASA Total Ozone Mapping Spectrometer (Jablonski & Chaplin, 2000).

planet correlate, producing the graph shown in Figure 10.27. The gray colors in the graph indicate the distribution of the various potentials, from highest to lowest, in relation to the distance from the Equator in different latitudes. The area without dots located on the Equator is the one with a sufficient capacity to synthesize, with only sporadic exposure throughout the year, the amount of vitamin D_3 required for the human body. Inside this area the darkest gray indicates the highest value of UVMED. Toward the north and south, the weak dotted area indicates that there will be, on average, insufficient vitamin synthesis during at least one month of the year. The boldly dotted area (northern and southern regions) indicates that the synthesis will be insufficient for the entire year.

To the extent that the graph in Figure 10.27 corresponds roughly to the darker or lighter pigmentation of indigenous peoples, Jablonski and Chaplin (2000) concluded that the results of their study suggest that skin pigmentation is somewhat labile, adapting to local conditions in short periods—short in terms of geological time. Therefore, human groups might have gone through successive periods of pigmentation and depigmentation when moving from a region with a certain average UVMED to another with a different average. Although we must not forget that in prehistoric times, cultural solutions—wearing clothes, taking shelter—would have contributed to offsetting any excess exposure to solar radiation (Jablonski, 2004).

The verification of the genetic diversity scope of humanity and its geographical expression, as well as explanations about the adaptive nature of the main ethnic differences, highlight an interesting issue. The enormous mobility of individuals of the modern human species is a very powerful factor in favor of miscegenation. In zoological terms, it would be

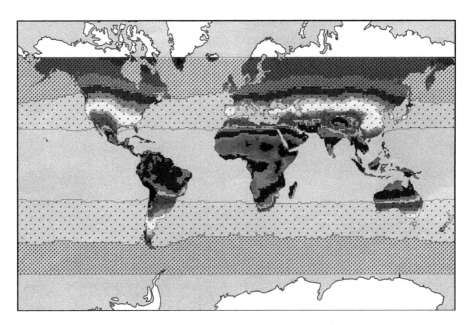

Figure 10.27 Capacity to synthesize vitamin D by human skin with low pigmentation related to the annual average of UVMED. To see the explanation of the gray color code, see text (illustration by Jablonski & Chaplin, 2000). Reprinted from Jablonski, N. & Chaplin, G. (2000). The evolution of human skin coloration. *Journal of Human Evolution*, 39(1), 57–106, with permission from Elsevier.

difficult to understand that widespread populations in such diverse and distant geographical areas would not have initiated speciation processes—the separation into different species. In anthropological terms the tendency is the opposite: the human species has a great mobility and hence it has such a strong tendency toward miscegenation that the number of different alleles present almost everywhere on the planet is becoming larger and larger.

10.2.10 The fossil record of the first modern humans

As you might recall, grade III of the transition from "archaic" to modern humans proposed by Philip Rightmire (1990) is comprised of the specimens already considered as *Homo sapiens* in a strict sense. In Chapter 9 we examined the first two grades (I and II) of this process, and when addressing the characteristics of the specimens of grade II, we referred to creatures with characteristics close to modern humans, which appeared in the sites of Skhul and Qafzeh (Israel) (Box 10.36); commonly these are considered as modern humans. They are dated at approximately 100,000 years (Valladas et al., 1988; Stringer et al., 1989; McDermott et al., 1993), although some authors grant them a more recent age, 78,000–39,000 years (Masters, 1982), or older, up to 120,000 years (Mercier et al., 1993). The date of c. 100,000 years is consistent with the most plausible interpretation of the first migration of modern humans out of Africa and their arrival in the Near East.

10.2.11 The Klasies River: the first *Homo sapiens*

The oldest specimens of *Homo sapiens* identified as such come from the southernmost extreme of South Africa, near Cape Town. Different sites like the Border cave (de Villiers, 1973), the Klasies River mouth (Singer & Wymer, 1982), the Equus cave (Grine & Klein, 1985), the Die Kelders cave (Grine et al., 1991), the Witkrans cave (McCrossin, 1992), and Hoedjiespunt (Berger & Parkington, 1995) provide the most important samples available today of the emergence of modern humans. Among the localities mentioned, the Klasies River, located between Plettenberg Bay and Cape St. Francis, is the best known (Figure 10.28).

In 1967–68 the excavations at the main site of the Klasies River began, leading to the first description of the sediments of the locality, the fossil fauna, and the different signs of human occupation, including tools from the Middle Stone Age (MSA; see Box 10.37), and various ornaments such as perforated shells (Singer & Wymer, 1982) (Figure 10.29). Already in these first studies a very ancient presence, older than 100,000 years, was attributed to this locality. Since 1984, Hillary Deacon directed the work at the site, establishing more precise stratigraphic sequences and dating.

The main site is located on the coast, on the sheltered side of a 40 m cliff ending in a pebble beach, a major source of cores for lithic carvings. Despite having several caves located at various altitudes above sea level (1, 1A, 1B, and 2), the main site forms a single unit of deposition and not a series of isolated shelters (Rightmire & Deacon, 2001). Human occupation began after the subsidence of the waters, in the interglacial MIS 5c, which had flooded the locality, so human presence cannot be older than 120,000 years.

The schematic chart and stratigraphy of the Klasies River main site is shown in Figures 10.30 and 10.31. Since this is a single structure of deposition, the layers could be correlated between different caves with the exception of 1B (Deacon & Geleijnse, 1988).

The oldest deposits are those of pebble gravel that remain in the back of cave 1, indicating the highest level of the seawater, but there are no fossils. Above this level is the LBS member, with human fossils and utensils from MSA I, whose most likely age is 120,000–115,000 years (MIS 5e) (Deacon & Geleijnse, 1988). Within the Upper Member, the cave 1A contains evidence of the use of fire.

Box 10.36 Skhul and Qafzeh

Although belonging to the taxon *Homo sapiens*, the reason why we have examined the Skhul and Qafzeh specimens in Chapter 9 is that, as we saw there, these exemplars were taken for a long time to be a particular type of *Homo neanderthalensis*—that of the "progressive Neanderthals." We refer the reader to Chapter 9 for the discussion about the taxonomic proposal, as well as for the details on the fossil traits.

Box 10.37 African MSA

The different stages of the African MSA, the role of the tools and decorative objects of this cultural tradition, their potential makers and the relationship that may exist with the culture of early modern humans who entered Europe, are issues to be addressed in Chapter 11.

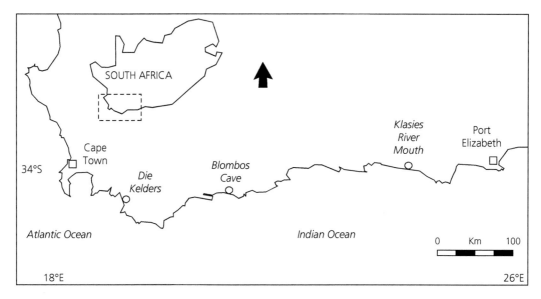

Figure 10.28 Location of the South African sites of Die Kelders, Blombos, and the Klasies River. Reprinted from Grine, F. E., Henshilwood, C. S., & Sealy, J. C. (2000). Human remains from Blombos Cave, South Africa: (1997–1998 excavations). *Journal of Human Evolution*, 38(6), 755–765, with permission from Elsevier.

From the level LBS of the Klasies River come two fragmentary maxillas, as reported by Hilary Deacon and Vera Geleijnse (1988), but most of the human remains of the Klasies River have appeared in the level SAS (Figure 10.32 and Box 10.38). The mandibles are the best represented element of a sample formed by KRM 13400, 16424, 21776, and 41815. The differences in size are considerable—a fact that is interpreted as indicative of significant sexual dimorphism (Rightmire & Deacon, 2001).

Also from level SAS come a frontal fragment and another of a temporal bone (Singer & Wymer, 1982). With regard to postcranial elements, this level of the Klasies River has provided very few exemplars. Worth mentioning are a fragment of a radius (Singer & Wymer, 1982), a fragment of an ulna (Churchill et al., 1996), and three foot metatarsals (Rightmire et al., 2006). The study by Philip Rightmire et al. (2006) of these remains revealed a coincidence in certain characters with the specimens of Qafzeh and Skhul from Israel, although the individuals from Klasies would have had a considerable height. Another postcranial trait also falls outside the range of the oldest modern humans in the Middle East. Rightmire et al. (2006) pointed out that, given the small information provided by the available specimens, it is difficult to draw firm conclusions about the phylogenetic affinities between the Klasies hominins and other Pleistocene populations.

Colin Groves and Alan Thorne (2000) have indicated that various hypotheses on the affinities of the Klasies River specimens can be asserted. These specimens could be related to:

(1) Current South African natives (Khoe-San Bushmen).
(2) Current Africans, in general.
(3) Modern humans, in general.
(4) "Archaic" *sapiens* related to the populations of Skhul and Qafzeh (Israel), and Djebel Irhoud (Morocco).
(5) Populations even more archaic, like those of Florisbad, Saldanha (South Africa), Kabwe (Zambia), Tighenif (Algeria), and Casablanca (Morocco), and their contemporaries in Europe.
(6) None of them, constituting a separate group not related to any other.

In order to test these hypotheses, Groves and Thorne (2000) conducted canonical analyses of discriminant functions for the most likely placement of the Klasies River sample (Box 10.39).

With the results of their analysis (Box 10.39), Groves and Thorne (2000) concluded that the population of the Klasies River must be regarded as akin to *Homo sapiens* in general terms (hypothesis 3) or as a separate group (hypothesis 6). Between these alternatives, the authors chose the first option. In other words, the Klasies River hominins would be modern humans.

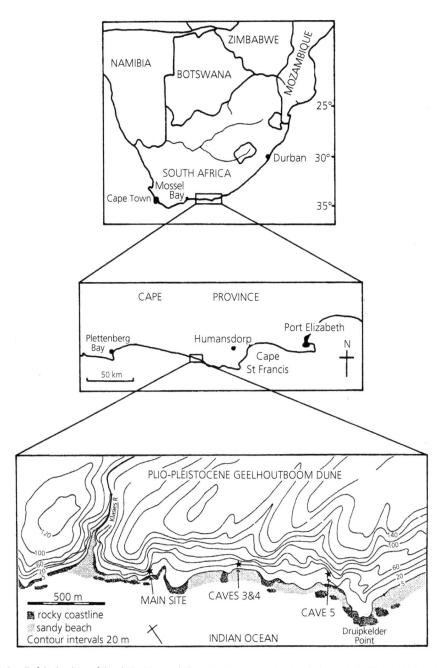

Figure 10.29 Detail of the localities of the Klasies River, with the main site. Deacon, H. J. & Geleijnse, V. B. (1988). The Stratigraphy and Sedimentology of the Main Site Sequence, Klasies River, South Africa. *The South African Archaeological Bulletin*, 43(147), 5–14.

Other localities in the Cape of Good Hope have provided human remains, but even more scarce than in the Klasies River mouth. During the campaigns of 1997–98, four teeth were found in the Blombos cave, two of them deciduous (Grine et al., 2000). The level corresponding to the MSA of Die Kelders 1 (DK1) provided additional specimens whose age has been estimated between 60,000 and 80,000 years (Grine, 2000).

In the opinion of G. Philip Rightmire and Hilary Deacon (2001), two of the molars found in DK1 fall

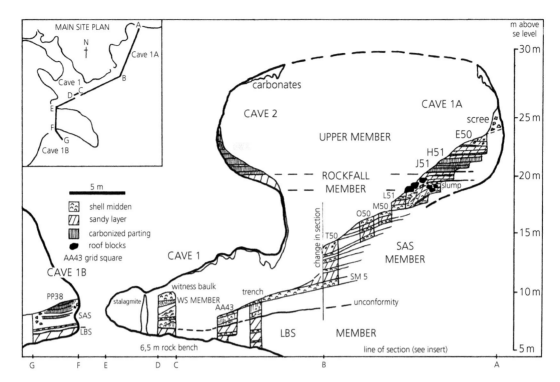

Figure 10.30 Diagrammatic section through the main site depository of the Klasies River, caves 1A, 1, 1B, and 2. Deacon, H. J. & Geleijnse, V. B. (1988). The Stratigraphy and Sedimentology of the Main Site Sequence, Klasies River, South Africa. *The South African Archaeological Bulletin*, 43(147), 5–14.

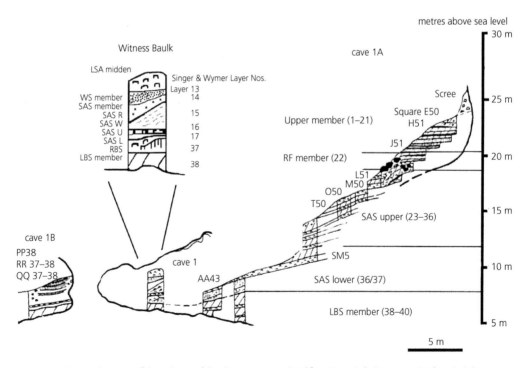

Figure 10.31 Members and sections of the main site of the Klasies River. Reprinted from Wurz, S., le Roux, N. J., Gardner, S., & Deacon, H. J. (2003). Discriminating between the end products of the earlier Middle Stone Age sub-stages at Klasies River using biplot methodology. *Journal of Archaeological Science*, 30, 1107–1126, with permission from Elsevier.

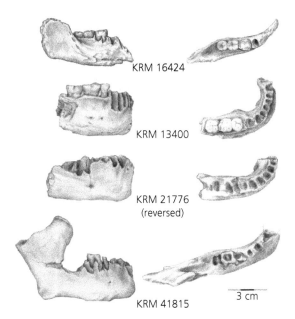

KRM 16424

KRM 13400

KRM 21776
(reversed)

KRM 41815

3 cm

Figure 10.32 Mandibles from the Klasies River in lateral and occlusal views. Reprinted from Royer, D. F., Lockwood, C. A., Scott, J. E., & Grine, F. E. (2009). Size variation in early human mandibles and molars from Klasies River, South Africa: Comparison with other middle and late Pleistocene assemblages and with modern humans. *American Journal of Physical Anthropology*, 140(2), 312–323, with permission from Elsevier.

Box 10.38 The size of the mandibles from the Klasies River

Yin Man Lam, Osbjorn Pearson, and C. M. Smith (1996) pointed out that the high degree of sexual dimorphism in the mandibular sample of the Klasies River is due to the very small size of KRM 16424, attributed to a female. The rest of the specimens, much larger, are considered male. But, if any of these larger pieces did correspond to a female, the dimorphism of the site's population would have been overvalued.

---

completely within the range of current South Africans. However, the analysis by Frederick Grine (2000) provided the main reason for the lack of certainty about their taxonomic status. The sample is comprised of 27 specimens—24 isolated teeth, a mandibular fragment (AP 6276), and two phalanges (AP 6267 and 6268)—coming from up to ten individuals, many of them infants. Their dental morphology looks modern

---

Box 10.39 Sample contained in the analysis of Klasies River (*Groves & Thorne, 2000*)

The analysis by Groves and Thorne (2000) included the frontal KRM 16425 and the mandibles KRM 13400, 14695, 16424, 21776, and 41815 of the Klasies River, which were compared to 5 specimens from the Middle Pleistocene of Europe and Africa, 3 from Skhul/Qafzeh, 13 of proto-Bushmen from South Africa, 20 from Western Africa, and 20 from the Upper Paleolithic of North Africa and Europe. The conclusions of Groves and Thorne (2000) were: the analysis barely discriminated the very fragmented frontal KRM 16425 from the others, excluding only hypothesis 2 and to some extent hypothesis 5. Two mandibular analyses were more indicative. In the first analysis were excluded only hypotheses 1, 2 (probably), and 4, while in the second analysis were excluded hypotheses 2 (probably), 4, and 5. Using non-metric comparisons, Groves and Thorne excluded hypotheses 1, 4, and 5.

Box 10.40 LSA level

Several adult human bones and a partial child's skeleton, accepted as "fully modern" without any doubt, appeared in the LSA level of DK1, Die Kelders (Rightmire, 1979b). But their age is almost contemporary: 2,000 years old at the maximum (Grine et al., 1991).

---

but, as indicated by Grine (2000), their morphological variants shared with current humans are mostly plesiomorphies—primitive features—that do not permit a determination of phylogenetic affinities or confirmation of their possible membership in *H. sapiens*.

Then, why the near unanimous consideration of South Africa and, in particular, the localities of the Cape of Good Hope, as the most ancient birthplace of all modern humans identified with certainty?

From the evidence available in the Klasies River, Hilary Deacon (1989) calculated the presence of modern humans in South Africa at approximately 100,000 years ago. They would have been able to cope with the environmental degradation, maintaining their populations despite the decline in the availability of resources caused by the cooling climate. According to Deacon

(1989), it was their social and cognitive mechanisms, similar to those of our species, with the use of artifacts as symbols, that permitted the adaptive success of the Klasies River, and placed at the site some of the oldest evidence for the emergence of modern humans. This is an issue that will be discussed more extensively in Chapter 11.

10.2.12 The specimen from Hofmeyr

A very well preserved cranium, although with a history of damage after its recovery, was discovered in 1954 in a dry channel of the Vlekpoort River, near the town of Hofmeyr (Eastern Cape, South Africa) (Grine et al., 2007) (Box 10.41). At the time of the discovery no more fossils or associated tools appeared and, as the channel disappeared under the mud due to the construction of a dam downstream, the original location became inaccessible.

Figure 10.33 shows the state of the Hofmeyr cranium at the time of its discovery and today. Frederick Grine et al. (2007) attributed the damage done to the fossil in the 1960s to the extraction of a sample for C^{14} dating, which produced no results.

Frederick Grine and collaborators (2007) determined a more precise age of the cranium by the luminescence

method, obtaining a figure of 36.2±3.3 Ka. This date shows the importance of the specimen, since it belongs to a temporal range coinciding with the appearance of the modern cultures—Upper Paleolithic in Eurasia and Late Stone Age (LSA), in sub-Saharan Africa—and the entry of modern humans in Europe.

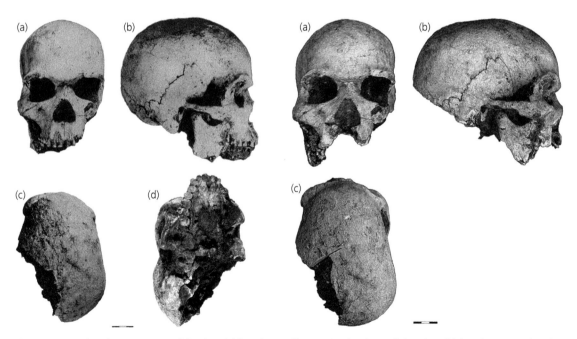

Figure 10.33 Hofmeyr's cranium in 1965 (left, a frontal, b lateral, c top, d bottom views) and 1993 (right, a frontal, b lateral, c top views). Scales in cm in both cases. Reprinted from Grine, F. E., Gunz, P., Betti-Nash, L., Neubauer, S., & Morris, A. G. (2010). Reconstruction of the late Pleistocene human skull from Hofmeyr, South Africa. *Journal of Human Evolution*, 59(1), 1–15, with permission from Elsevier.

The specimen from Hofmeyr is, both in age and morphology, *Homo sapiens*. Its appearance is, however, primitive when compared with current African features such as the prominent glabella, the continuous and somewhat thick supraorbital torus, the narrow, high, and thin malar bone, a broad front wall in the maxillar, and molars with comparatively large crowns (Grine et al., 2007). However, the authors who conducted the reinterpretation of 2007 indicated that it has no Neanderthal apomorphies.

The phenetic affinities of the Hofmeyr cranium were studied by Frederick Grine and collaborators by a multivariate analysis of linear measurements and the coordinates of 19 three-dimensional points, compared with those of modern humans from North Africa (Mesolithic), sub-Saharan Africa, Western Eurasia, Oceania, and East Asia/New World, and two Neanderthals, four Upper Paleolithic modern humans, and one modern human from the Levant, also in the Upper Paleolithic. The dendrogram in Figure 10.34 shows the phenetic relationships among all these samples calculated using the UPGMA ("Unweighted Pair Group Method with arithmetic Average") method, a clustering algorithm used to establish phenotypic similarities.

The phenogram in Figure 10.34 reveals a closer morphological affinity between the Hofmeyr cranium and the specimens of the Eurasian Upper Paleolithic (EUP in the diagram) than with the Africans from the

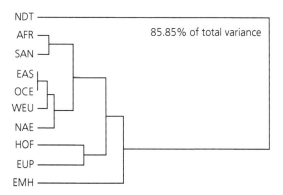

Figure 10.34 UPGMA tree of phenetic affinities among Neanderthal samples (NDT), sub-Saharian Africa (AFR), South Africa (SAN), East Asia/New World (EAS), Oceania (OCE), Western Eurasia I (WEU), North Africa (NAE), the Hofmeyr specimen (HOF), Eurasian Upper Paleolithic (EUP), and the cranium Skhul 5, modern human of the Upper Pleistocene (EMH). From Grine, F. E., Bailey, R. M., Harvati, K., Nathan, R. P., Morris, A. G., Henderson, G. M., . . . Pike, A. W. G. (2007). Late Pleistocene Human Skull from Hofmeyr, South Africa, and Modern Human Origins. *Science*, 315(5809), 226–229. Reprinted with permission from AAAS.

Mesolithic (NAE) or the oldest populations in South Africa (the sample SAN was taken from indigenous Khoe-San). These results support, according to Grine et al. (2007), the hypothesis that the first modern humans migrated to Eurasia, but it contradicts what was stated before about the San Bushmen being the closest group to our ancestor, in accordance with the molecular methods.

10.2.13 Other older African specimens of *Homo sapiens*

With an indeterminate age, but close to the emergence of our species, are two crania (Irhoud 1 and 2) and a juvenile mandible (Irhoud 3) from Djebel Irhoud, Morocco, classified as Neanderthals by its discoverer (Ennouchi, 1963), although later attributed to *Homo sapiens sapiens* (Hublin & Tillier, 1981). Also from Morocco are two sets of cranial, maxillar, and mandibular fragments from Dar-es-Soltan (Ferembach, 1976), to which McBrearty and Brook (2000) granted an age between 60,000 and 90,000 years, and the mandibles found in Temara (Vallois, 1960) and Zouhrah (Close, 1984), both with more than 40,000 years, an age increased to c. 130,000 by McBrearty and Brooks (2000). A partial juvenile skeleton appeared in Taramsa (Egypt) (Vermeersch et al., 1998). Cranial and mandibular fragments of three individuals came from Soleb (Sudan) (Giorgini, 1971).

Some important fossils from Omo (Ethiopia) are a partial skeleton (Omo I) with an age about 130,000 years, and an even older cranium (Omo II)—but of dubious attribution to *H. sapiens*—found, respectively, in the members KHS and PHS of the Kibish formation

---

Box 10.42 A prolonged coexistence?

In a paper devoted to analyzing the origin of the anatomy of modern humans, Günther Bräuer (2008) indicated that the evidence accumulated from the mid-1970s—including the discovery of fossils such as the crania from Ndutu, Laetoli, and Bodo—showed that "archaic" humans of very old aspect would have remained in Africa until recently, with ages between 40,000 and 30,000 years. Therefore, if it is accepted that modern humans, i.e., *Homo sapiens*, appeared at least 100,000 years ago, to what extent were the "archaic" and "modern" specimens, which coexisted for such a long time, related? In Chapter 11 we will suggest a possible answer.

(Day, M. H. 1969b). McDougall, Brown, and Fleagle (2005) using ^{40}Ar/^{39}Ar on a tuff in Member I below the hominin levels obtained a maximum age of 198 ± 14 Ka (weighted mean age 196 ± 2 Ka). Maxime Aubert et al. (2012), using direct U-series determination on the Omo I fossil, confirmed the older age. From Tanzania came several molars found in Mumba Rock Shelter, 110,000–130,000 years old (Bräuer & Mehlman, 1988), and from Kabua (Kenya) the cranial and mandibular fragments of two individuals of uncertain age (Whitworth, 1966).

But the fossil evidence about the specimens of our own species could go even further back in time. A cranium (KNM-ER 3884) and a femur (KNM-ER 999) from Koobi Fora (Kenya) are considered by Bräuer et al. (1997) as the oldest specimens available of beings close to modern humans, with a respective age of 270,000 and 300,000 years, obtained by uranium series (Bräuer et al., 1997). The 640,000 years attributed to the Bodo specimens (Clark et al., 1994) would push even further back the origin of our species, if the "Bodo Man" is considered a human of modern appearance. If this condition would be granted to the specimen from Danakil (Abbate et al., 1998), we would have gone back to an age around 1 million years. These dates are very far from the range of the evidence provided by the molecular methods.

CHAPTER 11

Neanderthals and modern humans

similarities and differences

No morphological, taphonomic, geological, or cultural evidence currently available serves to resolve completely the controversy over the origin of modern humans posed by the alternative models of the Multiregional Hypothesis (MH) versus the Replacement Hypothesis (RH). In other words, to estimate the evolutionary distance between Neanderthals and modern humans is not so easy. In this chapter we will return to this matter, using a new kind of evidence: that obtained by sequencing fossil DNA. This evidence allows very accurate determination of the genetic distance between our species and the Neanderthals. But beyond genetics, how different or similar are Neanderthals and modern humans? What do they share and what separates them in physical, as well as mental and behavioral, terms? In particular, to what extent would Neanderthals have achieved the "modern mind," allowing the symbolic insights characteristic of the Cro-Magnon?

The acquisition of the modern mind is a process involving genetic, brain, and cognitive pattern changes, forming a set in which it is difficult to tell which are the causes and which are the effects. An extremely adaptive model would argue that the correct sequence of events is: genetic change–brain change–behavioral change. It is unnecessary to resort to Lamarckian proposals to understand that it is not always so. The present mechanism of exaptation (the use of an inherited capacity for functions other than those that led to its appearance) can lead to a course of action that takes advantage of existing capabilities in a different and crucial way without genetic or brain changes. But, as it is necessary to follow an explanatory order, the comparison between *Homo neanderthalensis* and *Homo sapiens* will be carried out by the following sequence of similarities and differences in (1) genetics, (2) brain, and (3) cognition (referring only to those related to symbolic behavior). A hint of our conclusions is shown in Table 11.1.

The particularities of the phylogenetic history of Neanderthals versus modern humans can be summarized in one sentence: the concurrent emergence of two different but very similar species on two separate continents. Let's consider their distinctive features in detail.

11.1 Genetic distance between *Homo neanderthalensis* and *Homo sapiens*

The genetic differences between Neanderthals and modern humans have been accurately documented thanks to the retrieval of ancient DNA. It began with the mitochondrial genetic material—mitochondrial DNA or mtDNA. Mitochondria are cellular organelles that act as suppliers of most of the energy needed by the cell and possess their own genetic material, mtDNA, separate from the DNA in the cell nucleus. The mtDNA encodes 13 proteins, two ribosomal genes, and 22 tRNA—RNA of transference—essential for energy production (Kivisild et al., 2006). The mtDNA is inherited through the maternal lineage, transmitted through the ovule.

According to Lynn Margulis (1970), the origin of mitochondria goes back to the appearance of the first eukaryotic cells (with nuclei), as a symbiosis between some prokaryote cells (without nuclei) and some bacteria (with nuclei). Accordingly, the mitochondrial DNA may be considered a descendant of those symbiotic prokaryotes. In fact, its circular shape (Figure 11.1) is typical of prokaryotic (including bacterial) DNA.

Human mtDNA comprises around 16,000 nucleotides, a very small number compared to the 3 billion nucleotides in each nuclear genome, where most genes reside. The mtDNA sequences that were recovered in the early days of DNA sequencing technology belonged to a non-coding control region whose database contains 1,210 pairs (bp) and exhibits a great variability, usually referred to with the acronym HVR, "hypervariable region."

Processes in Human Evolution. Francisco J. Ayala and Camilo J. Cela-Conde, Oxford University Press (2017).
© Francisco J. Ayala and Camilo J. Cela-Conde.
DOI 10.1093/acprof:oso/9780198739906.001.0001

Table 11.1 Similarities and differences between *Homo neanderthalensis* and *Homo sapiens*

Genetics

	Genetic changes shared since the hominin and panin divergence		Positive selection in *H. sapiens*
H. neanderthalensis	88%		
H. sapiens			brain genes (*MEF2A*)
Brain			
	whole developmental pattern		time and mode of brain development
H. neanderthalensis	common		different
H. sapiens			
Cognition (symbolic behavior)			
	geometric markers	decorative objects	figurative representations
H. neanderthalensis	yes	yes	uncertain
H. sapiens	yes	yes	yes

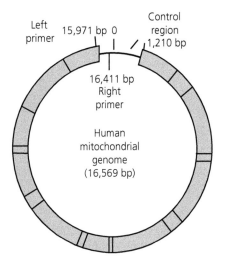

Figure 11.1 A schematic of mitochondrial DNA, or mtDNA. At the top is the non-coding control region (hypervariable region, or HVR). It contains 7.5% of all the mitochondrion's genetic information.

11.1.1 Ancient mtDNA retrieval

In 1984, Allan Wilson and collaborators managed to retrieve ancient DNA for the first time from the quagga, an extinct relative of horses (Higuchi et al., 1984). Shortly thereafter, Svante Pääbo (1984, 1985) retrieved human mtDNA from a 2,500-year-old Egyptian mummy. By the year 2000, mtDNA had been cloned from up to 18 fossil or extinct organisms (quagga,

marsupial wolf, sabre-toothed cat, moa, mammoth, cave bear, blue antelope, giant ground sloth, auroch, mastodon, New Zealand coot, South Island piopio11, Steller's sea cow, Neanderthal, *Aptornis defossor*, Shasta ground sloth, pig-footed bandicoot, moa-nalo, and *Myotragus balearicus*) (Hofreiter et al., 2001). The DNA of these organisms was cloned by using the technique known as polymerase chain reaction (PCR), which allows millions of identical copies from a single DNA molecule to be obtained (Pääbo et al., 2004). According to Svante Pääbo and collaborators (2004), PCR allowed production of a virtually unlimited number of copies from a very few original DNA samples, or even just one. The same DNA sequence can be amplified several times from the same ancient DNA sample and studied in a scientifically rigorous manner.

But retrieved DNA remains from the fossil record are in fact very rare. Obtaining useful genetic material from a fossil depends on the conservation of the specimens, which is a function of their age and the conditions of the site. Recent specimens and a dry and cold environment offer the greatest possibilities for successful cloning. One difficulty is the possible contamination of the material to be analyzed, which requires that the samples be handled with great care. Even so, contamination frequently happens.

11.1.2 Neanderthal mtDNA

The first Neanderthal mtDNA to be obtained came from the Feldhofer cranium (Neander, Germany)

Box 11.1 Protocol for fossil mtDNA retrieval

A very precise experimental protocol specifies the procedures that must be followed to retrieve fossil genetic material (Table 11.2). The protocol includes, among other precautions, analyzing the samples in two independent laboratories so that contaminating mtDNA can be detected. For a detailed description of the technique and its limitations, see O'Rourke et al. (2000), Hofreiter et al. (2001), and Pääbo et al. (2004); for a review of the different techniques involved see Cipollari et al. (2005).

(Krings et al., 1997). The same 379 bp-long sequence was obtained by the two laboratories that carried out the analysis, and the recovered mtDNA was compared with the same sequence corresponding to a current *Homo sapiens* reference population (Anderson et al., 1981), yielding 27 differences. Comparison of the mtDNA segment between individuals from different living human populations revealed an average of between five and six differences. The results obtained by Krings and colleagues (1997) indicated that the difference between Neanderthal and modern human mtDNA was three times larger than that among living humans. The comparison of humans and chimpanzees renders twice the number of differences between Neanderthals and modern humans. Krings et al. (1997) concluded that these results support the two species hypothesis.

The second recovering of Neanderthal mtDNA was carried out with material from a rib belonging to a 29,000-year-old juvenile specimen found in the Mezmaiskaya cave (northern Caucasus) (Ovchinnikov et al., 2000). A 345 bp-long sequence was obtained, corresponding to the same mtDNA segment as that of the Feldhofer specimen. The comparison of the Mezmaiskaya sequence with the Anderson et al. (1981) reference and the Feldhofer sequence revealed 22 and 12 differences, respectively. The two Neanderthals share 19 substitutions relative to the reference sequence.

Fossil nuclear DNA is usually degraded and mixed with microbial contaminants. DNA is rapidly degraded by enzymes such as lysosomal nucleases. In addition, bacteria, fungi, and insects feed on, and degrade, DNA macromolecules (Eglinton & Logan, 1991). Nuclear DNA cloning has already been achieved by sequencing

Table 11.2 Criteria for obtaining and sequencing ancient DNA (after Pääbo et al., 2004)

---

1. Cloning of amplification products and sequencing of multiple clones.
   This serves to detect heterogeneity in the amplification products, due to contamination, DNA damage, or jumping PCR.
2. Extraction controls and PCR controls.
   At least one extraction blank that does not contain any tissue but is otherwise treated identically should be done. During each PCR blank, PCR controls should be performed to differentiate between contamination that occurs during the extraction and during the preparation of the PCR.
3. Repeated amplifications from the same or several extracts.
   This serves two purposes. First, it allows detection of sporadic contaminants. Second, it allows detection of consistent changes due to miscoding DNA lesions in extracts with extremely low numbers of template molecules.
4. Quantization of the number of amplifiable DNA molecules.
   This shows whether consistent changes occur or not. If consistent changes can be excluded (roughly for extracts containing >1,000 template molecules), a single amplification is sufficient. Quantitation has to be performed for each primer pair used, as the number of amplifiable molecules varies dramatically with the length of the amplified fragment, the sensitivity of the specific primer pair used, and the base composition of the amplified fragment.
5. Inverse correlation between amplification efficiency and length of amplification.
   As ancient DNA is fragmented, the amplification efficiency should be inversely correlated with the length of amplification.
6. Biochemical assays of macromolecular preservation.
   Poor biochemical preservation indicates that a sample is highly unlikely to contain DNA.
   Good biochemical preservation can support the authenticity of an ancient DNA sequence.
7. Exclusion of nuclear insertions of mtDNA.
   It is highly unlikely that several different primer pairs all select for a particular nuclear insertion. Therefore, substitutions in the overlapping part of different amplification products are a warning that nuclear insertions of mtDNA may have been amplified. A lack of diversity in population studies can also be taken as an indication that nuclear insertions may have confounded the results is warranted.
8. Reproduction in a second laboratory.
   This serves a similar purpose as 2 and 3, i.e., to detect contamination of chemicals or samples during handling in the laboratory. This is not warranted in each and every study, but rather when novel or unexpected results are obtained. Note that contaminants that are already on a sample before arrival in the laboratory will be faithfully reproduced in a second laboratory.

Box 11.2 mtDNA and nuclear DNA

In his commentary on the recovery of the Mezmaiskaya specimen, Matthias Höss (2000) warned that these studies indicate that none of the examined Neanderthals contributed mtDNA to the human lineage; but this doesn't imply that they did not contribute nuclear DNA. To know to what extent they would have done so, it would be necessary to recover nuclear DNA, which was achieved later. The reason why retrieval studies of ancient genetic material initially was limited to the mtDNA analysis is because fossil nuclear DNA is normally degraded and mixed with microbial contaminants (see text).

---

a single gene locus through the direct use of PCR on the material to be sequenced (Huynen et al., 2003). Using a different technique, known as the metagenomic approach, Noonan and colleagues (2005) were able to clone a 26,000 bp-long DNA sequence from two 40,000-year-old extinct cave bears. The procedure involved obtaining the ancient DNA first, composed of a mixture of genome fragments from the ancient organisms and sequences derived from other organisms in the environment. Thereafter, the metagenomic approach was used, in which all genome sequences are anonymously cloned into a single library. To avoid likely contamination, the operation must be carried out in a laboratory into which modern carnivore DNA has never been introduced. The next step was to sequence

the library, which gave way to chimeric inserts due to the presence of DNA from different organisms. Finally, the inserts were compared with GenBank nucleotide, protein, and environmental sequences, which include those of bears. These precedents suggest that nuclear DNA from the Mezmaiskaya specimen might be retrieved in the future, given its excellent preservation, as well as from other Neanderthal fossils.

11.1.3 Evolutionary consequences of the mtDNA comparison from the Neander and Mezmaiskaya specimens

Igor Ovchinnikov and colleagues (2000) arrived at several conclusions based on the comparison of the two Neanderthal mtDNA sequences with each other and with those of modern humans. The genetic differentiation between the two Neanderthal sequences is greater than among 300 individuals in a sample from current Caucasoid and Mongoloid populations, but comparable to a similar sample from Africans. The Feldhofer and Mezmaiskaya sites are far apart and thus may be taken as representative of Neanderthals as a whole (Figure 11.2).

A second conclusion was that "these data provide further support for the hypothesis of a very low gene flow between the Neanderthals and modern humans. In particular, these data reduce the likelihood that Neanderthals contained enough mtDNA sequence diversity to encompass modern human diversity" (p. 492). Third, based on their results, Ovchinnikov et al. (2000) estimated the age of the common ancestor of western

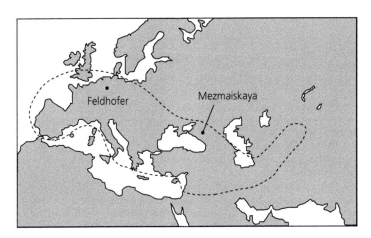

Figure 11.2 The Feldhofer and Mezmaiskaya sites, sources of the two first Neanderthal specimens whose mtDNA was cloned. The dashed line encompasses the region where Neanderthal specimens have been found (picture from http://www.nature.com/nature/journal/v404/n6777/fig_tab/404490a0_F1.html#figure-title).

(Feldhofer) and eastern (Mezmaiskaya) Neanderthals to be between 151,000 and 325,000 years. The divergence between Neanderthals and modern humans would have taken place between 365,000 and 853,000 years (Ovchinnikov et al., 2000). Ward and Stringer (1997) had estimated that date as no less than 500,000 years, based on the Feldhofer results of Krings et al. (1997), which falls within that interval.

Supporters of MH for the emergence of modern humans also use ancient DNA retrieval to support their model. Following the publication of Ovchinnikov et al. (2000), Gregory Adcock and colleagues (2001) obtained mtDNA from a very early modern human, Lake Mungo 3 (LM3) (Australia), described in 1974 by Bowler and Thorne (1976) (Figure 11.3).

The comparison of the mtDNA from LM3 with the modern human reference sequence yielded 13 differences. This is large since, on the grounds of its morphology and the tools found at the site, there is no doubt that LM3 is a modern human. The authors concluded that the mtDNA of LM3 eventually disappeared from the human lineage. If this were so, it could also be that the same might have happened in other cases, such as the Neanderthals. If the absence of the genetic peculiarities of LM3 in current populations does not justify excluding the Lake Mungo specimen from the modern human lineage, then Neanderthals need not be excluded either on the grounds of the results from the cloning of their mtDNA. This argument has been used by the supporters of MH to invalidate the evidence of the separation deduced from Neanderthal mtDNA.

The controversy regarding the significance of the cloned material from Lake Mungo illustrates some possibilities and limits afforded by molecular tests for the understanding of human evolution. The specialists in mtDNA cloning were quick to criticize the work by Adcock et al. (2001) for three reasons. First, the findings were not replicated by an independent laboratory (Cooper et al., 2001; Gillespie, 2002). The second difficulty is the specimen's doubtful age (Cooper et al., 2001). The third line of criticism embraces the same arguments leveled to criticize the cloning procedures used with Cro-Magnon specimens (see Box 11.3).

The dating techniques used by Alan Thorne et al. (1999) were: electron spin resonance (ESR) on dental enamel; U-series on calcite crust covering the skeleton; and optically stimulated luminescence (OSL) on the sediment surrounding the skeleton. The latter technique examines individual grains of minerals, which absorb radiation from the sediment. The mineral releases this energy when erosion exposes it to sunlight. The energy still retained by the mineral indicates when it was last exposed to sunlight. But Gillespie (2002, p. 466) noted, "whether LM3 remains turn out to be

> **Box 11.3 The age of the Lake Mungo specimen**
>
> The age of the LM3 specimen from Lake Mungo was initially established, on the basis of geomorphological criteria and stratigraphic association with Mungo 1, as between 28,000 and 32,000 years (Bowler & Thorne, 1976). However, by a combination of different dating systems (see below), Thorne and colleagues (1999) determined the age for LM3 as 62,000 years. This date was soon criticized. It turned out to be 20,000 years older than the lower level of the Mungo Unit from which LM3 was recovered, reliably dated at 43,000 years; see Bowler and Magee (2000) and Gillespie and Roberts (2000), who argued that, given that LM3 had been buried—and thus deposited in an inferior sedimentary level—it had to be younger than 43,000 years.

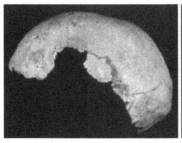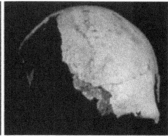

Figure 11.3 LM3 from Lake Mungo (Australia), *Homo sapiens* (Bowller & Thorne, 1976): left, lateral view; middle, superior view; right, frontal view (photographs from http://www-personal.une.edu.au/~pbrown3/LM3.html).

40, 50, or 60 ka does not seem to matter for the human origins debate, with *modern* people living in Africa >100 ka there is plenty of time to reach Europe or Australia by any of those dates." An earlier estimate for LM3 could run against the credibility of the mtDNA study carried out by Gregory Adcock et al. (2001), because "DNA is not expected to survive for this length of time outside of cold environments" (Cooper et al., 2001, p. 1655).

Another line of criticism of the study by Adcock et al. (2001) used the same arguments that led to criticizing the Cro-Magnon specimens' recovery (see Box 11.4). Endogenous mtDNA and contaminating current mtDNA cannot be distinguished in early modern human remains, a circumstance that affects LM3 (Pääbo et al., 2004; Serre et al., 2004). However, it is not completely fair to apply this caution to the Lake Mungo specimen because, in this case, it highlights precisely the differences regarding the reference sequence. They could not be produced by contaminating current mtDNA, which lacks the peculiarities of LM3. Pääbo et al. (2004) have argued that it is virtually impossible to exclude all modern human DNA sequences as possible sources of contamination, "including excavators, museum personnel, or laboratory researchers." That exception applies entirely, in our view, to the specimen from Lake Mungo.

The fairest criticism to the interpretation of LM3 as supporting MH has to do with the distance of the specimen in regard to modern humans and Neanderthals. Caramelli and colleagues (2003) carried out a comparative analysis by means of multidimensional scaling (MDS), which graphically shows the genetic distances between sequences. The analyzed sample included current humans (60 Europeans and 20 non-Europeans), Neanderthals (4), Cro-Magnons (2), and the LM3 specimen. The results (Figure 11.4.) provide a clear idea of the situation. Even though some mitochondrial peculiarities of the Lake Mungo specimen are absent in modern populations, it falls within the variation range of *Homo sapiens*, whereas Neanderthals do not. A similar argument was advanced by Cooper et al. (2001).

11.1.4 The multiplication of Neanderthal mtDNA studies

After the recovery of the Mezmaiskaya specimen, and until 2005, eight additional specimens were sequenced: Feldhofer 2 (Germany) (Schmitz et al., 2002), Vindija 75 (Croatia) (Krings et al., 2000), Vindija 77 and 80 (Croatia) (Serre et al., 2004), Engis 2 (Belgium) (Serre et al., 2004), a specimen from La Chapelle-aux-Saints (France) (Serre et al., 2004), RdV 1 from Les Rochers-de-Villeneuve (France) (Beauval et al., 2005), and El Sidrón (Spain) (Lalueza-Fox et al., 2005). The comparisons between the obtained and reference sequences have produced very similar results to the ones we have already discussed. But some of the studies merit further attention.

Matthias Krings et al. (2000) obtained a 357 bp HVRI sequence and a 288 bp HVRII sequence from

---

Box 11.4 Cro-Magnon mtDNA

David Caramelli and colleagues (2003) obtained mtDNA from two Italian Cro-Magnon specimens. Their results suggested that the specimens were well within the variation range of current humans. However, their work was criticized on the grounds of a methodological issue. If the mtDNA that is being cloned is very similar to that of current humans, it is not possible to determine which part of the obtained results corresponds to endogenous material and which is due to contaminations (Nordborg, 1998; Trinkaus & Zilhao, 2002; Pääbo et al., 2004; Serre et al., 2004). In Pääbo's words, "Cro-Magnon DNA is so similar to modern human DNA that there is no way to say whether what has been seen is real" (cited by Abbott, 2003, p. 468).

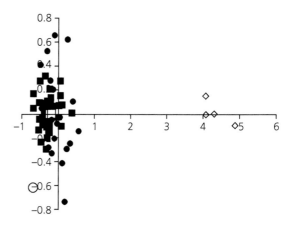

Figure 11.4 Multidimensional scaling (MDS) of HVRI sequences of 60 modern Europeans (filled squares), 20 modern non-Europeans (filled circles), 4 Neanderthals (open diamonds), the Australian Lake Mungo 3 (open circle), and 2 Cro-Magnons (open squares); the axes have different scales (picture from Caramelli et al., 2003).

the Vindija 75 specimen. Comparison with the Anderson reference sequence produced very similar results to the two previous cases: it is "highly unlikely that a Neanderthal mtDNA lineage will be found that is sufficiently divergent to represent an ancestral lineage of modern European mtDNAs" (p. 145). The authors analyzed Neanderthal genetic diversity, based on the Feldhofer, Mezmaiskaya, and Vindija 75 specimens. When compared to African apes and current humans, Neanderthals are more similar to modern humans than to apes in having a low species-wide mtDNA diversity (Krings et al., 2000) (Table 11.3).

According to Krings et al. (2000, p. 145–146): "If the Neandertals, similar to humans, had a diversity lower than that of the great apes, in spite of inhabiting a region much larger than the apes, this may indicate that they also had expanded from a small population." A phylogenetic tree corresponding to these three Neanderthal specimens is illustrated in Figure 11.5.

The study carried out by David Serre et al. (2004) is especially interesting, because it takes into account the two main problems noted by the supporters of MH:

(1) possible Neanderthal contribution to the gene pool of modern humans might have been erased by genetic drift (Nordborg, 1998) (Box 11.5); and

(2) if some Neanderthals carried mtDNA sequences similar to contemporaneous humans, such sequences may be erroneously regarded as modern contaminants (Trinkaus, 2001).

Serre et al. (2004) retrieved mtDNA from four Neanderthals and five early modern humans. In order to minimize the problem of contamination of current genetic material on early modern humans, the authors concentrated on a region that contains two particular substitutions found in previously studied Neanderthals. Their results showed that "All four Neanderthals yielded 'Neandertal-like' mtDNA sequences, whereas none of the five early modern humans contained such mtDNA sequences, even though they were as well preserved as the Neandertals" (p. 313). Serre and colleagues used these results to elaborate a statistical model based on coalescence theory, which "excludes any large genetic contribution by Neandertals to early modern humans but does not rule out the possibility of a smaller contribution" (p. 0313) (Box 11.6).

In addition to serving as a synthesis of ancient mtDNA studies, the work performed by Lalueza-Fox and colleagues (2005) offers new information on the

Table 11.3 Mitochondrial DNA genetic diversity of African apes, modern humans and Neanderthals

Gorillas	Chimpanzees	Modern humans	Neanderthals
18.57±5.26%	14.82±5.70%	3.43±1.22%	3.73%

Data taken from Krings et al. (2000).
The numbers indicate substitutions for every 100 nucleotide sites. The sequences of current humans and apes correspond to 50,000 randomly chosen among 5,530 humans, 359 chimpanzees, and 28 gorillas. The Neanderthal sequences are from the three Feldhofer, Mezmaiskaya, and Vindija 75 specimens. The probability that the Neanderthal sample's diversity corresponds to the highest diversity in the complete population is 50%.

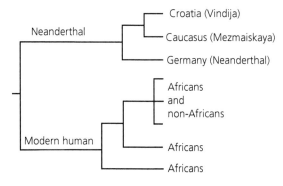

Figure 11.5 Phylogenetic tree with current humans and Neanderthal specimens from Feldhofer, Mezmaiskaya, and Vindija 75 (figure from Hofreiter et al., 2001).

Box 11.5 Genetic drift

As we said in Chapter 1, genetic drift is a phenomenon leading to the disappearance of part of the allele variation in a population, after an occasional bottleneck has reduced it to a few individuals. Only the alleles from the survivors are preserved, while the others are eliminated.

Box 11.6 Genetic contribution among species

The methodology used by Serre et al. (2004) and their conclusions were criticized by Alan Cooper, Alexei Drummond, and Eske Willerslev (2004). Later we will address to what extent the comparison of nuclear DNA clarifies the possible genetic cross between Neanderthals and modern humans.

episode that saw the replacement of *H. heidelbergensis* by *H. neanderthalensis*. The two explanatory hypotheses for the substitution of the former chronospecies by the latter assume:

(1) a gradual emergence of distinctive Neanderthal features through chronospecies continuity; and
(2) emergence of Neanderthals as the result of a clearly defined speciation event, occurring around 250,000–300,000 years.

Lalueza-Fox et al. (2005) argued that "the present genetic data support the latter hypothesis that *H. neanderthalensis* emerged as a distinct biological entity after a speciation event, c. 250,000 years. This event not only coincides with the TMRCA [time to the most recent common ancestor] estimates of the Neandertal mtDNA variation but also with the appearance in Europe of the cultural Mode 3 industry and a decrease in the morphological variation observed in *H. heidelbergensis*" (p. 1080).

There is little doubt that the use of molecular methods represents a significant advance for the understanding of the evolutionary relationships of Upper Pleistocene hominins. The mtDNA offered only indications, although pointing insistently to a lack of evidence of any gene flow between Neanderthals and modern humans (Serre et al., 2004). The sequences of Neanderthal mtDNA recovered up to 2005 were not present in modern humans. This can be for two reasons. First,

---

Box 11.7 Chronospecies replacement

Aside from the relevance that the original dating of Lalueza-Fox et al. (2005) could have for the history of the Neanderthal lineage, their work suggests a possibility of estimating with relative accuracy when the replacement of chronospecies took place. As we have said on several occasions, it is generally quite difficult to pinpoint the time at which one chronospecies was replaced by another. The genetic data and the estimation of the different dental growth patterns in *H. heidelbergensis* and *H. neanderthalensis* (Rozzi & Bermúdez de Castro, 2004) would allow, exceptionally, establishing a precise frontier between these two chronospecies. However, the abundance paradox could blur a scheme that for the moment seems clear due to the lack of intermediate specimens. The sample of *H. heidelbergensis* in Rozzi and Bermúdez de Castro's study (2004) included only specimens from Sima de los Huesos (Atapuerca, Spain).

Neanderthal mtDNA might have evolved in a separate lineage to that of anatomically modern humans, a scenario that favors RH and considers the Neanderthals and modern humans as two different species. Second, it could be that Neanderthal mtDNA disappeared in modern humans due to genetic drift. As several authors have noted (Kahn & Gibbons, 1997; Höss, 2000; Cooper et al., 2001; Paunovic et al., 2001; Beauval et al., 2005), some genetic exchange between Neanderthals and modern humans cannot be discarded on the grounds of evidence available after retrieving the full Neanderthal mtDNA. David Serre and collaborators (2004) stated that, assuming that human populations have consisted of 10,000 reproductive individuals per generation, it is possible to exclude that the amount of DNA coming from Neanderthal ancestors would have been greater than 25% of the total (Figure 11.6).

11.1.5 Okladnikov: did Neanderthals reach Siberia?

The geographical boundaries of the Neanderthal presence are established by an accurate or probable identification of fossils that fit their morphological characteristics. That can only be achieved if there are sufficiently complete specimens so that their attribution can be fulfilled. Since dubious evidence often appears, the boundaries of the Neanderthal presence were established without using the doubtful evidence. Thus, the child skeleton found in the Teshik-Tash cave in Uzbekistan was considered evidence of the occupation most distant from Europe (Debetz, 1940) (see Figure 11.7, which reproduces the same figure included in Chapter 10).

Johannes Krause and collaborators (2007) conducted a mtDNA analysis of the Teshik-Tash specimen and some uninformative remains—teeth and postcranial fragments—found in the cave of Okladnikov, part of the Altai region of southern Siberia. These latter materials, dated between 43,700 and 37,750 years ago, do not have a firm attribution: they can be considered as Neanderthals, modern humans, or even *Homo erectus* (Krause et al., 2007).

Krause et al. (2007) retrieved mtDNA from the left femur of Teshik-Tash and from three fragmentary bones of Okladnikov. The sequences obtained from the Teshik-Tash specimen largely correspond to those of modern humans, although two of them agreed with previously identified Neanderthal sequences. Regarding the Okladnikov specimen, the results were very similar: only a small part of the sequences was Neanderthal mtDNA. But these coincidences were enough

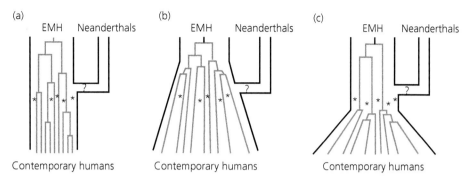

Figure 11.6 (a) Under the assumption of a constant effective population size of 10,000 for modern humans, contemporary mtDNAs trace back to approximately five mtDNA lineages 25,000 years ago. The modern human fossils represent five additional samples from around the time of putative admixture (stars). The contemporary and early modern human (EMH) samples reject a Neanderthal contribution of 25% or more to modern humans about 30,000 years. (b) Under the more realistic scenario of an expansion of the human population during and after the colonization of Europe, a smaller Neanderthal contribution can be excluded because the number of ancestors of the current human gene pool was larger than 30,000 years. The contribution that can be excluded would depend on when and how the expansion occurred. (c) Under the scenario that population size was constant before a putative merging with the Neanderthal population and expanded only thereafter, the Neanderthal contribution could have been larger, but similarly depends on how the expansion occurred (figure and explanation by David Serre et al., 2004).

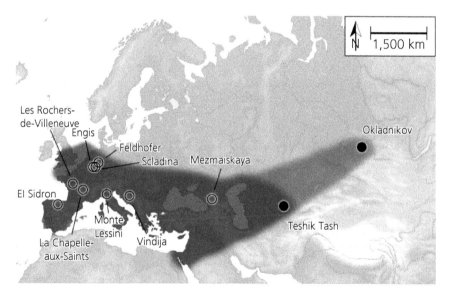

Figure 11.7 Geographic extension of sites with Neanderthal specimens, including the caves of Okladnikov and Teshik-Tash. Reprinted by permission from Macmillan Publishers Ltd: Krause, J., Orlando, L., Serre, D., Viola, B., Prufer, K., Richards, M. P., . . . Paabo, S. (2007). Neanderthals in central Asia and Siberia. *Nature*, 449(7164), 902–904.

for Krause et al. (2007) to highlight the fact that some of the remains contained genetic material of *Homo neanderthalensis*. A study of the remains with a positive identification, and the subsequent statistical analyses of the results of comparing them with both the available Neanderthal mtDNA samples, and with current and ancient modern humans, was carried out by Krause et al. (2007), reaching the conclusion that both the specimen from Teshik-Tash and the one from Okladnikov belonged to a population related to European Neanderthals; a result already pointed to by the Teshik-Tash morphological studies and Okladnikow teeth. In turn, the Teshik-Tash specimen seems to be closer to European Neanderthals than to the Okladnikov fossils, but their differences are not large enough to conclude that their populations lived apart for a long period of time. This observation supports the hypothesis of a late colonization of Central Asia by the Neanderthals, perhaps taking advantage of an episode of relatively mild climate 125,000 years ago (Krause et al., 2007).

11.1.6 mtDNA background: populations and species

The results of ancient mtDNA studies contrast with the anatomical analysis of some late Neanderthal specimens, which have been interpreted as evidence of hybridization. This is the case of the almost complete skeleton of a Neanderthal child discovered in 1998 in the late site (24,500 years) of Abrigo do Lagar Velho (Portugal) (Duarte et al., 1999) (Figure 11.8). Morphological examination of the specimen (Lagar Velho 1) reveals, in the opinion of the discoverers, mosaic evolution. Some traits, such as the incipient chin, the width of the molar crowns, and the relative size of the thumb's proximal and distal phalanxes, are close to traits considered "derived" (rather than "primitive" or

Figure 11.8 Lapedo child's burial from the Abrigo do Lagar Velho (Portugal) (picture from http://artsci.wustl.edu/~anthro/blurb/LagarVelho1.jpg).

ancient) in anatomically modern humans. Other traits, such as the size of the incisors or the proportions of the femur, are characteristically Neanderthal. Trinkaus et al. (1999) concluded that Neanderthal and anatomically modern human populations interbred, at least in the Iberian Peninsula.

Solid evidence of such interbreeding would speak against the hypothesis of a replacement of one population by another without genetic exchange. But the authors who interpreted the Lagar Velho finding advance an additional argument. Trinkaus et al. (1999) believe that such genetic exchange was made possible by the similarities in Neanderthal and modern human behavioral traits. In their opinion, the most important implications of Lagar Velho 1 are not taxonomic or phylogenetic, but behavioral. These authors argue that the morphological mosaic demonstrates that when modern humans dispersed throughout the Iberian Peninsula they encountered Neanderthal populations. Both groups would have recognized each other as humans with similar behavioral strategies, social systems, communication structures, and adaptive strategies. As Zilhao and Trinkaus (2002) said about the two populations that they believe coincided at Lapedo: "The fact that, anatomically, the child is principally a modern human, suggests an imbalanced interaction, with Neandertal populations being essentially absorbed and genetically swamped, with particular features characteristic of such populations being still present a few millennia after contact but disappearing subsequently. In such a scenario of absorption as is suggested by the biological data, it is not surprising, particularly given the much faster pace of change permitted by Lamarckian mechanisms of cultural transmission, that no specifically Mousterian traits of culture are to be found in the archeological record of Portugal ca. 25 000 years ago" (p. 547).

The perspective taken by Zilhão and Trinkaus (2002) is not, however, the only way to interpret the Lapedo specimen. The interpretation of anatomical traits is often controversial, more so when the specimen is infantile, as is the case with Lagar Velho 1. Tattersall and Schwartz (1999) argued that it was an anatomically modern human (see Box 11.8).

Anatomical analyses cannot settle whether the infantile specimen from Lagar Velho was a Neanderthal or a modern human. But we must not forget William Goodwin's warning: it is possible that Neanderthals and humans interbred but produced sterile offspring (Chang, 2000). If the Lapedo child was actually a hybrid of a Neanderthal and a modern human, this still would not prove that these were populations of the same species. To verify that—according to the

---

Box 11.8 Lapedo's taphonomy

In their study on the Lapedo specimen, Lagar Velho 1, Ian Tattersall and Jeffrey Schwartz (1999, p. 7119) stated: "The archaeological context of Lagar Velho is that of a typical Gravettian burial, with no sign of Mousterian cultural influence, and the specimen itself lacks not only derived Neanderthal characters but any suggestion of Neanderthal morphology. The probability must thus remain that this is simply a chunky Gravettian child, a descendant of the modern invaders who had evicted the Neanderthals from Iberia several millennia earlier." Mousterian culture is common in Neanderthal sites, although it is also present in some of the Near East modern human sites. Gravettian is characteristic of modern humans that arrived at Europe.

Box 11.9 A remote African origin for the Neanderthals?

If Neanderthals were the descendants of immigrants to Europe, arriving from Africa and not Asia, the argument we are using will not change much. The issue is to understand the possible causes of genetic separation reflected in the mtDNA and what these differences can tell us about the human evolutionary transition from the Middle to the Upper Pleistocene.

---

definition of biological species—it should also be demonstrated that the Lagar Velho 1 specimen was fertile, or to find other specimens with their own indicative characteristics of being hybrid.

Anatomical or cultural analyses cannot prove beyond reasonable doubt whether the Lagar Velho child was a Neanderthal or a modern human. Throughout this book we have seen too many examples of how different starting presuppositions lead to the interpretation of empirical evidence in contradictory ways. Population genetic considerations, such as the one indicated by Goodwin—reproductive isolation—are not decisive when it comes to the fossil record. The effort to retrieve mtDNA opened a new perspective but, as we have seen, the evidence obtained is not conclusive. Before proceeding to the next step, the recovery of nuclear DNA, we will address the evidence provided by the mtDNA retrieved from a Siberian specimen.

For reasons related to the scientific method, and as the philosopher of science Karl Popper (1963) revealed, hypotheses like "Neanderthal mtDNA is not present in modern humans," are falsifiable merely by offering an example in which it does appear. Thus, such hypotheses can only be held as provisionally true. Some future experiment might eventually demonstrate that they are false.

Yet, the accumulation of experiments indicating consistently the separation between Neanderthal and modern human mtDNA increasingly supported the hypothesis of two separate groups with no gene flow between them. But, are we talking about two populations or two species? Incidentally, what is the extent of

the differences between one or the other response with regard to the understanding of human evolution?

Let us assume that Neanderthals are descendants from Asian populations after their migration to Europe. (If they were successors of some new immigrants to Europe, but from Africa and not Asia, this scenario would not change much except for the timing of the episodes.) Let us also assume that modern humans are descendants of an African population. We will refer to the Neanderthal population arriving at Europe as Np, and the modern human population that evolved in Africa as Mhp. The two alternative hypotheses are: H1, that there was no genetic flow between Np and Mhp (replacement); H2, that such flow did in fact occur (hybridization) (Figure 11.9).

How can we test these hypotheses? It can be done, to a certain extent, by genetic comparison between Neanderthals and modern humans. In this respect, mtDNA studies support H1, but, for the reasons stated above, such support is always provisional.

Let us now assume that, in spite of the provisional nature of the argument, we are convinced that there was no such genetic flow. The next question is why this happened. Np and Mhp might not have shared genetic flux because they evolved to become two different species after their geographic separation. But it could also be the case that, while belonging to the same species, there was no genetic flow because the populations remained isolated in different places. Such flux could not have taken place without physical contact between them. Thus, the possible alternatives are:

(1) Np and Mhp constitute two different species, *Homo neanderthalensis* and *Homo sapiens*.
(2) The subspecies hypothesis: *H. sapiens neanderthalensis* and *H. sapiens sapiens*.
(3) Pn and Pmh are just two populations of the same taxon.

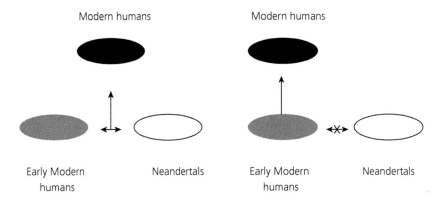

Figure 11.9 Alternatives to the possible Neanderthal genetic contribution to modern humans. Left : Early modern humans and Neanderthals interbred and both populations contributed their DNA to modern humans. Right: Early modern humans and Neanderthals did not interbreed, so only early modern humans contributed their DNA to modern humans. That hybridization did not occur does not necessarily imply that Neanderthals and modern humans were different species (see text).

Choosing one of these alternatives is a decision that cannot be made strictly in terms of ancient DNA retrieval, for the moment. It depends on other factors, especially in common paleontological practice. These practices impose the use of the concept of biological species to name fossil taxa and, thus, throughout this book we have referred to *Orrorin tugenensis, Australopithecus afarensis, Paranthropus robustus*, and *Homo erectus*, among many others. But, as we know, the technical means available today do not allow testing the criteria relative to biological species (whether they can produce fertile progeny) for these fossil "species."

The case of Neanderthals is similar to that of any other fossil group. If we compare Neanderthals with modern humans, mtDNA studies support the hypothesis of two different populations, Np and Mhp, which did not exchange genetic material. Serre and colleagues' (2004) estimation of a flow below 25% between both populations does not mean that there was that amount of genetic exchange. It only establishes that, if indeed there was genetic exchange, it could not have been above that limit. Negative evidence suggests, thus, there was either none or a very small exchange. This fact would not indicate whether they were two populations or two species.

Why do discussions such as the one concerning the Lagar Velho 1 specimen become heated? As Geoffrey Clark (1997a, 1997b) noted, this might be a paradigmatic problem. The hybridization and replacement hypotheses are grounded on radically different preconceptions about humans' remote past. Hence, the variables that researchers measure and the methods they employ to do so, such as mtDNA comparisons, cannot resolve this issue. The debate has to do more with implicit concepts about what a human being is than with the origins of our own species. For reasons that have to do with the concept of biological species and with currently available molecular evidence, we favor that Neanderthals and modern humans are separate species. Placing Neanderthals in another species does not imply viewing them pejoratively. Neanderthals are no longer reduced to the brutish image they were associated with for a time (see Tattersall, 1998). Molecular methods do not provide information about what sort of humans the Neanderthals were, or about their apomorphies. The archaeological record is the best source of information as to what Neanderthals were.

11.1.7 High-throughput DNA sequencing

As we have seen, Neanderthal mtDNA sequencing became possible, despite the difficulties inherent in the poor preservation of DNA macromolecules after death, thanks to the technique of polymerase chain reaction (PCR). But this technique, although it provided spectacular results, is slow and expensive. Shotgun sequencing technique, although an improvement, could not resolve the problems arising from the need to restore badly damaged sequences. Hence the need to replace these initial methods by one of better performance. Thus, the technique of primer extension capture (PEC), or high-throughput sequencing, was created (Briggs et al., 2009).

Box 11.10 Technique 454

T454 was initially used by Adrian Briggs et al. (2007) in a study that detected the deterioration patterns of ancient DNA in a Neanderthal, a mammoth, and a cave bear. Briggs et al. (2007) indicated that the results suggested the possibility of obtaining reliable genomic sequences of Pleistocene organisms. The webpage http://scott.sherrillmix.com/blog/tag/pyrosequencing/ (accessed June 12, 2016) explains how the pyrosequencing technique is applied.

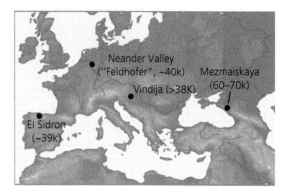

Figure 11.10 Origin and age of Neanderthal samples used by Briggs and collaborators (2009) (K = thousand years). From Briggs, A. W., Good, J. M., Green, R. E., Krause, J., Maricic, T., Stenzel, U., Lalueza-Fox, C., Rudan, P., Brajkovic, D., & Kucan, Z. (2009). Targeted retrieval and analysis of five Neandertal mtDNA genomes. *Science*, 325(5938), 318–321. Reprinted with permission from AAAS.

PEC permits identification of short sequences of nitrogenous bases from very degraded DNA samples contaminated by the presence of microbial DNA, a common occurrence when speaking of fossils. It was introduced in 2009 by Adrian Briggs and collaborators— Svante Pääbo among them—researchers who were all members of the Max Planck Institute for Evolutionary Anthropology in Leipzig (MPIEA), directed by Pääbo.

The PEC method is based on a pyrosequencing procedure called "technique 454"—T454, from now on (Box 11.10)—a resource obtained thanks to the collaboration between the MPIEA and one US company, the 454 Life Sciences Corporation. Using luminescence, this technology allows the determination of DNA sequences on a large scale, and is applicable to whole genomes. T454 can determine a sequence of 20 million bases in 4.5 hours, which speeds the retrieval process and reduces costs significantly (Briggs et al., 2009).

High-throughput sequencing was used by the team of Adrian Briggs and collaborators (2009) to obtain the whole mtDNA of five Neanderthal specimens: the Vindija Vi 33.25 (Croatia), Feldhofer I and II (Germany), El Sidrón 1253 (Asturias, Spain), and Mezmaiskaya (Russia), covering a wide geographic range, as well as a considerable time span (Figure 11.10).

The results obtained by Briggs et al. (2009) indicated that the mtDNA diversity of Neanderthals between 70,000 and 38,000 years ago was about one-third of the genetic diversity of contemporary humans. Analyzing the evolution of mtDNA proteins, the authors concluded that their data suggested that the long-term Neanderthal effective population size was smaller than that of modern humans or of great apes (Briggs et al., 2009). However, they pointed out an alternative hypothesis: that the found mtDNA diversity was typical of only the last Neanderthals, after they had

Box 11.11 Protein evolution

In the cladogram of Figure 11.11 the protein evolution ratio is indicated under each branch, a ratio that is higher in the case of Neanderthals than in modern humans. The ratio is obtained as the amino acid substitution rate divided by the rate of neutral substitution (changes in the nitrogenous bases which do not lead to an amino acid change).

experienced a considerable reduction in their numbers due to the direct or indirect influence of the expansion of modern humans out of Africa.

The cladogram drawn by Briggs et al. (2009) from the recovery of the entire mtDNA sequence is shown in Figure 11.11.

11.1.8 A new species? The Denisova specimen

In a summary of the 12 Neanderthal mtDNA studies conducted up to 2009, some general conclusions could be drawn:

• The Neanderthal mtDNA falls outside the range of the current human variation.
• The likelihood of a Neanderthal genetic contribution to the modern human genome is 0.1–25%.
• The common ancestor of Neanderthals and modern humans is c. 500,000 years.

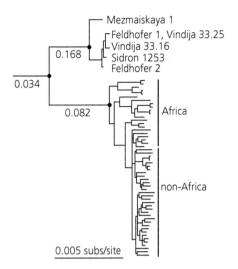

Figure 11.11 Phylogeny of the mtDNA of Neanderthals and closer relatives. From Briggs, A. W., Good, J. M., Green, R. E., Krause, J., Maricic, T., Stenzel, U., Lalueza-Fox, C., Rudan, P., Brajkovic, D., & Kucan, Z. (2009). Targeted retrieval and analysis of five Neandertal mtDNA genomes. *Science*, 325(5938), 318–321. Reprinted with permission from AAAS.

In 2010, the use of the PEC method for recovering fossil DNA allowed Johannes Krause and collaborators to take a step forward in the analysis of the complete mtDNA of the Denisova specimen.

From the Denisova cave, located in the Altai Mountains (Russia), comes the first fossil of the Asian Upper Pleistocene (besides the Flores hominin), which might be alien to the taxa of both Neanderthals and modern humans. It is a very incomplete specimen, limited to a partial distal phalanx of the fifth finger of a hominin (Box 11.12), but the cave contains evidence of a hominin episodic occupation since 125,000 years ago, with an early Mousterian industry in the stratigraphic levels 21 and 22 (Shunkov, 2005).

With respect to morphology, the Denisova example is not very informative. However, the analysis of its genome makes it possible to determine its relationships to other *Homo* taxa, going beyond the interpretations reached after the mtDNA retrieval from the Denisova specimen—incidentally, a female. Comparison with known samples of Neanderthals and modern humans, current as well as from the Pleistocene, led Johannes Krause and collaborators (2010) to conclude that the Denisova specimen is from an unknown hominin type that can be considered a sister group of the whole "Neanderthal + modern human" lineage, from which it separated about 1 million years ago. Consequently, it

---

Box 11.12 The Denisova phalanx

The partial phalanx corresponds to level 11, whose age is 48,000–30,000 years (Krause et al., 2010). Although the specimen is not sufficiently indicative to be assigned to any taxon, its hominin status is well established. The industry of the Denisova level 11 is transitional between the Middle and Upper Paleolithic techniques (Green et al., 2010).

---

follows that the presence at such high latitudes of the Denisova specimen corresponds, according to Krause et al. (2010), to a migration from Africa separate from those of the Neanderthals and modern human ancestors, although the authors argue that the cave stratigraphy suggests that the individual under study could have lived close in time and place to one or the other. The corresponding cladogram is shown in Figure 11.12.

In a commentary on the conclusions of Krause et al. (2010), María Martinón-Torres, Robin Dennell, and José María Bermúdez de Castro (2011) cautioned about the need for greater empirical evidence before supporting a phylogenetic model of the relationship among the Denisova specimen, Neanderthals and modern humans, in particular with regard to the exit of the *Homo neanderthalensis* ancestors from Africa, which Krause et al. (2010) fixed at around 500,000–300,000 years ago. To Martinón-Torres et al. (2011), the hypothesis of a hominin dispersal into Asia nearly a million years ago and, therefore, of an Asian origin for the Denisova population, cannot be excluded at this time.

11.1.9 Retrieval of Neanderthal nuclear DNA

In 2001 the complete sequencing of the human genome was achieved (Venter et al., 2001). From that moment there was renewed hope that someday it could be compared with the Neanderthal genome, to discover the genetic distance between the two taxa and, therefore, clarify whether there is one or two species. Although it seemed very difficult to achieve the same feat with the Neanderthal genome, the development of the high-throughput PEC sequencing technique allowed the picture to be completed.

The biggest problem to be solved in order to decipher the Neanderthal nuclear genome continued to be the contamination of material, despite the improvement in sequencing techniques. So, the team of Richard Green and collaborators (2006) conducted a preliminary

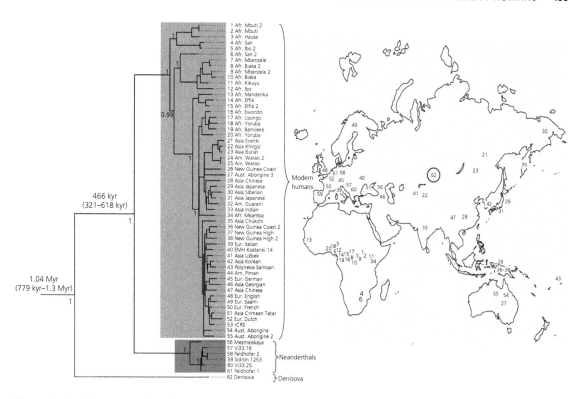

Figure 11.12 Phylogenetic tree of complete mtDNAs. The phylogeny was estimated using 54 present-day and one Pleistocene modern human mtDNA, 6 Neanderthals, and the Denisova hominin. The tree is rooted with a chimpanzee and a bonobo mtDNA. Posterior probabilities are given for each major node. The map shows the geographical origin of the mtDNAs (24, 25, 32, and 44 are in the Americas). Reprinted by permission from Macmillan Publishers Ltd: Krause, J., Fu, Q., Good, J. M., Viola, B., Shunkov, M. V., Derevianko, A. P., & Paabo, S. (2010). The complete mitochondrial DNA genome of an unknown hominin from southern Siberia. *Nature*, 464(7290), 894–897.

study to determine which sites would be freer of such contaminants, especially from modern human DNA. To do this they conducted amino acid conservation tests on more than 70 specimens—bones and teeth—from different parts of Europe and West Asia. The vast majority of the tested samples had either a very low amino acid content, or high rates of racemization. Furthermore, six of the specimens showed a sufficient presence of non-degraded amino acids. Green et al. (2006) extracted the genetic material from samples of these six specimens and performed the amplification of the mtDNA hypervariable region by PCR, obtaining, of course, Neanderthal material, as well as modern contaminations. But, since the sequences of the hypervariable region are different in Neanderthals, and we know them, Green et al. (2006) were able to establish the ratio of the presence of mtDNA Neanderthal-like and mtDNA modern human-like, showing the level of contamination. The differences found between the various samples were very important (Figure 11.13).

As shown in Figure 11.13, only 1% of the recovered mtDNA was Neanderthal-like in the samples from France, Russia, and Uzbekistan. A sample from Croatia (VI-77) yielded 5%; another from Spain (El Sidrón, Asturias), 75%. But the sample from Vi-80 (Vindija, Croatia) contained about 99% of Neanderthal mtDNA in the recovery of a 63 bp segment and of c. 94% in a segment of 119 bp. Thus, a sample with contamination levels of less than 6% was identified.

The Vindija cave, located in the Hrvatsko Zagorje region, contains 13 sedimentary levels (units A to M) ranging from 25,000 to 125,000 years. The majority of Neanderthal material discovered was found in Unit G (Jankovic et al., 2006). The condition of preservation of the specimens is very diverse, and their state probably depends on water intrusions that have occurred in different places (Green et al., 2006).

In 1928, excavation in the cave began. The specimen Vi-80 (now Vi33.16) was discovered in 1980 by Mirko Malez et al. (1980). This fossil comes from unit G3, including Mousterian utensils of an age of 32,010±2120 years, obtained by C¹⁴ accelerator mass spectrometry (Green et al., 2006).

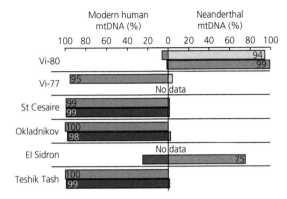

Figure 11.13 Percentage of human mtDNA (left) and Neanderthals mtDNA (right) in the retrieval made of Neanderthal specimen samples from Croatia (Vi-77 and Vi-80), France (St. Césaire), Russia (Okladnikov), Spain (El Sidrón), and Uzbekistan (Teshik-Tash). Each specimen has two horizontal columns corresponding to the two retrieved series. See text for interpretation. Reprinted by permission from Macmillan Publishers Ltd: Green, R. E., Krause, J., Ptak, S. E., Briggs, A. W., Ronan, M. T., Simons, J. F., . . . Paabo, S. (2006). Analysis of one million base pairs of Neanderthal DNA. *Nature*, 444(7117), 330–336.

The high-throughput PEC technique led to the retrieval from Vi-80 of 254,953 nuclear DNA sequences with about 1 million bp (0.03 of the total) (Green et al., 2006), and was the first major step toward obtaining the Neanderthal genome. Despite being a small part of the nuclear genetic material, Green and collaborators could make comparisons with the complete genomes of modern humans, chimpanzees, and mice, the set of redundant GenBank sequences, and the sequence set of GenBank environmental samples. The results allowed estimation of how the detected differences had been produced (Figure 11.14). Thirty-five million base pairs separate the genome of modern humans from chimpanzees.

From the genetic material recovered by Green et al. (2006), James Noonan et al. (2006) identified 65,250 base pairs of Neanderthal origin that enabled the divergence time between *H. neanderthalensis* and *H. sapiens* to be calculated. In the authors' words, "On average the Neanderthal genomic sequence we obtained and the reference human genome sequence share a most recent common ancestor ~706,000 years ago (. . .) the human and Neanderthal ancestral populations split ~370,000 years ago, before the emergence of anatomically modern humans" (Noonan et al., 2006, p. 1113).

Richard Green and collaborators (2006) also reached very interesting conclusions for use in evaluating the evolutionary process that led to Neanderthals and

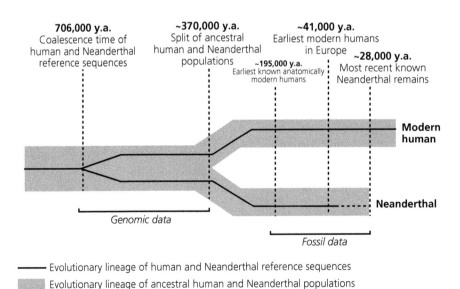

Figure 11.14 Divergence estimates for human and Neanderthal genomic sequences and ancestral human and Neanderthal populations. The branch lengths are schematic and not to scale (y.a. = years ago). From Noonan, J. P., Coop, G., Kudaravalli, S., Smith, D., Krause, J., Alessi, J., . . . Rubin, E. M. (2006). Sequencing and Analysis of Neanderthal Genomic DNA. *Science*, 314(5802), 1113–1118. Reprinted with permission from AAAS.

modern humans. How should we consider these two lineages? Were they populations of the same species that were separated over a long period of time, or were they two different species?

To decide this, Green et al. (2006) calculated the presence in the Vindija sample of the ancestral alleles shared with chimpanzees, and derived alleles that are the same as in modern humans, particularly in loci in which modern humans have a single nucleotide polymorphism (SNP) (Box 11.13).

The result was that c. 30% of derived alleles are shared by Neanderthals and our species. A very high rate, which led Green et al. (2006) to conclude that some gene flow, should have occurred between the Neanderthal lineage and the modern humans after their separation. In other words, that *H. neanderthalensis* and *H. sapiens* would have interbred, producing fertile offspring after the two species had diverged to a great extent, although not enough to make impossible the production of fertile hybrid offspring. Some of these would have continued breeding with *H. sapiens* for several generations, so that the proportion of Neanderthal DNA in the descendants would become only a small fraction of the DNA of current humans. Considerations of the Neanderthal X chromosome led Green et al. (2006) to argue that the genetic contribution was by modern human males to the Neanderthal genome.

Have we arrived, then, at the conclusion that since it is likely that these two lineages interbred, it is possible to say—contradicting everything said up to now—that Neanderthals and modern humans belong to the same species? Before giving an answer, let's address the retrieval of the complete Neanderthal genome.

11.1.10 The Neanderthal genome

In the 2009 meeting of the American Association for the Advancement of Science, the Max Planck Institute for Evolutionary Anthropology and the 454 Life Sciences Corporation announced the first draft of the Neanderthal genome, corresponding to 63% or 3,300 millions of bases, with the entire genome to be published soon. This promise was fulfilled in May 2010, when the paper by Richard Green and collaborators (2010) appeared presenting the results.

The entire retrieval of the Neanderthal genome was achieved by applying PEC techniques of high-throughput sequencing to three bones from the Vindija cave. The selection of the material was made from 89 Neanderthal bones coming from 19 sites, and a total of 201 DNA extractions. Only three bone fragments were selected:

- Vi33.16 (the specimen previously known as Vi.80, mentioned earlier) coming from unit G3, with an age of 32,010±2130 years.
- VI33.25, from unit I, deeper and older, although of an age still undetermined, whose mtDNA hasn't been sequenced yet (Green et al., 2008).
- Vi33.26, coming from unit G (unknown sub-unit), of 44,450±550 years of age.

In order to determine with certainty whether the fragments belonged to one or more individuals, Green and collaborators (2010) first compared known mtDNA sequences from Vi.33.25 and Vi33.13, which differed in ten positions—they belonged, therefore, to two different individuals. When performing for the first time the sequencing of Vi33.26, the result obtained was identical to that available for Vi33.16. In order to verify whether they came from the same individual, the authors conducted an allele comparison (Green et al., 2010; supplementary information). After analyzing the results, the authors attributed the exemplars to different individuals, although Vi33.16 and Vi33.26 could have had family relationships. The virtual absence of DNA fragments characteristic of the chromosome Y (see Box 11.14) indicated that they were females in all three cases.

After comparison with modern human and chimpanzee genomes, the results of Green et al. (2010) led to very significant conclusions. The substitutions and suppressions of nitrogen bases with respect to the chimpanzee appear in Table 11.4 and Figure 11.15.

The main part of substitutions and suppressions found in the Neanderthal genome (87.9 and 87.3%, respectively) occurred prior to the separation from the modern human lineage. Derived traits in this lineage and primitive traits in Neanderthals could be encoded by only 78 nucleotides affecting the protein structure, which, according to Green and collaborators (2010), indicates that only a few changes occurred at the level

---

Box 11.13 Single nucleotide polymorphism (SNP)

A single nucleotide polymorphism, SNP, occurs when alleles corresponding to a certain locus differ in a single nitrogen base (although the term is applied when only a few bases are involved). The largest part of the genetic variation in current humans (around 90%) corresponds to SNPs.

Box 11.14 Analyzing Vi33.16

It was a surprise that Vi33.16 was a female, because in the previous analysis of 1 million bp, Green et al. (2006) had determined that the specimen belonged to a male.

The recovery of the entire genetic material of the three chosen fragments yielded, as expected, contamination problems related with bacterial and human DNA, as well as the presence of analytic artifacts. Richard Green and collaborators (2010) followed procedures to eliminate these sources of error. Bacterial contamination—between 95 and 99% of the material obtained previously by traditional methods—was eliminated by selecting restriction enzymes that affect the microbial DNA but did not have any effect on primate sequences. Artifacts (misidentification of nitrogenous bases) came mostly from deamination of the cytosine nitrogen base, which loses its amino group. This is a normal process in which the deamination of 5-methyl cytosine produces thymine (guanine also becomes adenine by the same process), and it occurs in greater numbers around the 5' end

of the DNA strand. Forty percent of cytosines placed in the first position of this end become thymines by deamination. To detect these artifacts, the authors compared by mapping (alignment) the DNA sequences of a modern human, chimpanzee, orangutan, macaque, and Neanderthal. The deamination error ratio was calculated by comparison of the differences of the base ratios of C/T and G/A between the mtDNA previously known in the analyzed specimens and the newly obtained mtDNA. Artifacts by deamination were modest (rate of 0.1%), but much greater than those due to the technique in use (whose rate was limited to 0.001%). The modern human DNA contamination was calculated in three ways: mtDNA, nuclear DNA, and the presence of the chromosome Y. In the latter case, it was fortunate that they were Neanderthal females in all three samples. Contamination coming from current males reached 0.6% (95% confidence interval to 1.53%). A similar rate (0.7%) was found for modern human nuclear DNA contamination.

Table 11.4 Substitutions and suppressions of bases when comparing the modern human and Neanderthal genomes with that of the chimpanzee (Green et al., 2010)

	Substitutions	Suppressions
Modern human genome	10,535,445	479,863
Neanderthal genome	3,202,190	69,029
%	30	14

of amino acid substitutions during the hundreds of thousands of years Neanderthal and modern human lineages were separate.

11.1.11 Hybridization traces

As we know, MH maintains that modern human and local populations hybridized in every continent continuously, while RH denies the occurrence of any interbreeding. Studies on ancient genetic material (Green et al., 2006, 2010; Reich et al., 2011; Fu et al., 2015) show that neither of these happened; the principal event was the migration from Africa of new modern human populations along with sporadic and isolated episodes of hybridization, with an impact reflected in current DNA.

The genetic relationships between *Homo neanderthalensis* and *Homo sapiens* began to be known thanks to

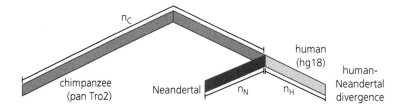

Figure 11.15 Nucleotide substitutions inferred to have occurred on the evolutionary lineages leading to the Neanderthals, the human, and the chimpanzee genomes. In dark gray are substitutions on the Neanderthal lineage (n_N), in light gray the human lineage (n_h), and in medium gray the combined lineage from the common ancestor of these to the chimpanzee (n_c). UCSC hg18 = reference human genome; *panTro2* = reference chimpanzee genome. From Green, R. E., Krause, J., Briggs, A. W., Maricic, T., Stenzel, U., Kircher, M., . . . Paabo, S. (2010). A Draft Sequence of the Neandertal Genome. *Science*, 328(5979), 710–722. Reprinted with permission from AAAS.

the analysis of Richard Green and collaborators (2006) made on 1 million base pairs of the Neanderthal genome, which we mentioned earlier. SNP comparisons of the entire Neanderthal genome and five individuals of different ethnic groups of modern humans from West Africa (Yoruba), South Africa (San), France, China (Han), and New Guinea (Papua), with the chimpanzee and orangutan genomes as a contrast to determine the ancestral character, allowed clarification of the existing genetic distances between the various populations (Green et al., 2010).

One of the most surprising results of the work of Green et al. (2010) is the discovery that Neanderthals are further from Africans than from European and Asian individuals. Between 2 and 4% of the human genome of non-Africans or North Sahara Africans comes from the Neanderthal genome, but this genetic trace does not exist in sub-Saharan populations. The most-parsimonious explanation for this fact suggests that there was gene flow between Neanderthals and modern humans after they left Africa, but before the differentiation of non-African populations, i.e., at a date close to 80,000 years ago. The most likely place for this gene flow would be the Middle East (Green et al., 2010) (Figure 11.16).

The retrieval of the Denisova specimen genome (Reich et al., 2011) verified that, in the case of the Australia and New Guinea aboriginals, in addition to the c. 2.5% DNA from Neanderthal, they have nearly 3% from the Denisovans. This second episode of hybridization most likely occurred due to Denisovan migrations to New Guinea and Australia (Rasmussen et al., 2011; Skoglund & Jakobsson, 2011).

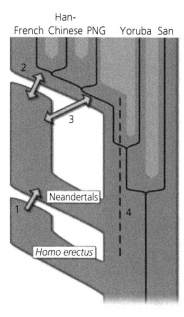

Figure 11.16 Possible interpretations about the genetic material flow from Neanderthals to modern humans. The explanation indicated by arrow 3 is the most-parsimonious one, involving a flow occurring after the exit of modern humans from Africa and before the separation of non-African populations. Green et al. (2010) also considered, but with less probability, the hypotheses: (1) gene flow from archaic hominin populations (represented in the figure by *Homo erectus*) that would have contributed to Neanderthals with unexpected large segments of divergence when compared to the modern human genome; and (2) gene flow from Neanderthals to Europeans and Asians. Reverse flows, although undetected as yet, can also be considered as hypothetical; hence the two-way arrows (Green et al., 2010). From Green, R. E., Krause, J., Briggs, A. W., Maricic, T., Stenzel, U., Kircher, M., . . . Paabo, S. (2010). A Draft Sequence of the Neandertal Genome. *Science*, 328(5979), 710–722. Reprinted with permission from AAAS.

11.1.12 High-quality Neanderthal genome

Earlier in this chapter (Section 11.1.2) we discussed the high fragility of the DNA molecule, subjected after the death of any organism to very rapid degradation processes due to the intervention of bacteria, fungi, and insects. This degradation is very sensitive to the moisture and heat of the site, such that remains from cold and dry places can potentially provide the greatest amounts of ancient DNA preserved in conditions that allow sequencing. Kay Prüfer et al. (2014) analyzed the quality and quantity of the ancient DNA present in the toe phalanx found in 2010 in the cave of Altai (Siberia), whose mtDNA analysis related it closely to the Neanderthal child specimen 1 from Mezmaiskaya (Caucasus). Prüfer et al. (2014) analyzed five DNA libraries that provided 52-fold sequence coverage of the genome (see Box 11.15), and calculated that in present-day human DNA contamination among the fragments sequenced was around 1%.

The high-quality genome of the Neanderthal specimen of Altai (a woman) allowed researchers to precisely measure the scope of the Neanderthal gene flow into modern humans: "the proportion of Neanderthal-derived DNA in people outside Africa is 1.5–2.1%" (Prüfer et al., 2014, p. 45). When adding the Denisovan gene flow estimation in mainland Asia (c. 0.2%) and Papua–Australia (c. 3.18%), and that of the Neanderthal gene flow into Denisovans (a minimum of 0.5%), it can be stated that "interbreeding, albeit of low magnitude, occurred among many hominin groups in the Late Pleistocene" (Prüfer et al., 2014, p. 43).

Box 11.15 Sequencing coverage

The high or low coverage of genome sequences is expressed by the "N"-fold genomic coverage, where "N" is a number indicating the depth of the coverage. This depth is "the number of times a reference base is represented within a set of sequencing reads (. . .). For example, a genome with 30× coverage [30-fold] will have an average of 30 reads spanning any given position within the genome" (Illumina Technical Note: www.illumina.com/content/dam/illumina-marketing/documents/products/technotes/hiseq-x-30x-coverage-technical-note-770–2014-042.pdf, accessed June 12, 2016).

A low number "N" indicates low coverage. For example, the coverage of the three Neanderthal specimens from Vindija (Croatia), whose nuclear DNA was recovered by Green et al. (2010), yielded a 1.3-fold genome coverage. The Neanderthal Mezmaiskaya infant was sequenced to 0.5-fold genomic coverage (Prüfer et al., 2014). Matthias Meyer et al. (2012) obtained a 30-fold coverage (high-coverage) for the "Denisovan" specimen of Altai. Obviously, the 52-fold coverage of Neanderthal genome of Prüfer et al. (2014) was also high coverage.

An interesting aspect of the study of Kay Prüfer et al. (2014) is the inference of the population size of these hominin groups over time. Using coalescence estimates of the two chromosomal copies of the genome, the authors detected a reduction shared by the populations of Neanderthal, Denisovan, and current humans, which occurred prior to 1.0 Ma ago (Figure 11.17), and pointed out that, "subsequently, the population ancestral to present-day humans increased in size, whereas the Altai and Denisovan ancestral populations decreased further in size" (Prüfer et al., 2014, p. 45).

After obtaining DNA from a Cro-Magnon 42,000–37,000 years old from the site of Pestera cu Oase (Romania) (Fu et al., 2015) it was possible to calculate that this particular modern human has 6–9% of Neanderthal DNA in his genome. The interpretation given by the authors of the large difference in the extent of gene flow compared to other estimates suggests a hybridization that happened in Europe in a very recent time, i.e., a few generations before the analyzed specimen (four to six generations, or about 200 years, before the birth of this fossil specimen). Perhaps the most important aspect of this finding is the detection of a hybridization occurring long after that of c. 85 Ka in the Middle East, indicated by Green et al. (2010). An earlier work

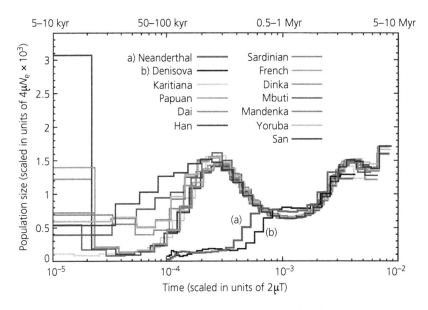

Figure 11.17 Inference of population size. Authors' note: The analysis assumes that the Neanderthal and Denisova remains are of the same age, whereas archaeological evidence and the branch shortening indicate that the Neanderthal bone is older than the Denisovan bone. However, because the exact difference in ages is not known, it is not possible to determine whether the reduction in population size experienced by both archaic groups—but not by modern humans—coincided in time. Reprinted by permission from Macmillan Publishers Ltd: Prüfer, K., Racimo, F., Patterson, N., Jay, F., Sankararaman, S., Sawyer, S., . . . Paabo, S. (2014). The complete genome sequence of a Neanderthal from the Altai Mountains. *Nature*, 505(7481), 43–49.

by Ekaterina Khrameeva and collaborators (2014), who compared the genomes of different populations of Neanderthals and modern humans, provided evidence for a late interbreeding with Neanderthals in the Old Continent. The comparative study by Khrameeva et al. (2014) showed that much of the common genome in both species is related to genes controlling the metabolism of fat. Thus, the characteristics of the genes that catabolize lipids in the Europeans could have been inherited from a crossing with Neanderthals in this continent. However, current Asians—who also retain the Neanderthal inheritance—lack this abundance of fat-controlling genes. The key to a possible explanation can be given in terms of functionality. Both Khrameeva's paper and the commentary emphasized the possibility that, as Neanderthals were well adapted to cold, it is possible that a significant portion of their features would correspond to this adaptation—as is assumed by many, but not all, anthropologists. Hence, perhaps the first modern humans entering Europe took advantage of the Neanderthal heritage to adapt to the extreme European winter climate. Unfortunately, the Oase Cro-Magnon does not serve as evidence, either in favor or against the functional hypothesis (Fu et al., 2015). While seeking to explain why this specimen shares the same alleles with current Europeans and Asians, Quiaomei Fu et al. suggested that, nevertheless, the population of Oase might not have contributed substantially to the current genome of Europeans. It is likely that it disappeared without leaving descendants.

11.1.13 African hybridization

We finished Chapter 10 saying that "archaic" humans could have coexisted for some time with the first *Homo sapiens*. If so, what might have been their relationship? Could some hybridization also have occurred?

Michael Hammer et al. (2011) provided one possible clue on the relationships between these different fossil forms. Hammer et al. (2011) tested the degree of plausibility for the hypothesis of absolute genetic separation between the populations of these two groups, namely, of the "archaics," indicating an evolution from *Homo erectus*, and modern humans. As indicated by Hammer et al. (2011), despite the evidence of interbreeding between populations of Neanderthals, Denisovans, and

---

Box 11.16. DNA from Sima de los Huesos

The sequence of old genetic materials from hominins collected in Sima de los Huesos (Altapuerca, Spain) is an excellent example of the wonderful possibilities of this kind of information, but also of how difficult it may be to interpret it. Matthias Meyer et al. (2013) obtained a nearly complete mtDNA sequence from samples of Femur XII belonging to a Sima hominin dated c. 430,000 years. After comparing it with other mtDNA sequences, they concluded that its closest sequence come from Denisovan mtDNA. According to Meyer et al. (2013, p. 403): "The fact that the Sima de los Huesos mtDNA shares a common ancestor with Denisovan rather than Neanderthal mtDNAs is unexpected in light of the fact that the Sima de los Huesos fossils carry Neanderthal-derived features (for example, in their dental, mandibular, midfacial, supraorbital and occipital morphology)." The issue was resolved later when Meyer et al. (2016) obtained a partial sequence of the nuclear genome from Sima de los Huesos hominins and concluded that these were more closely related to Neanderthals than to Denisovans, to the extent that the Sima honinins could be considered "early Neanderthals" (p. 506). They explained the discrepancy between the nuclear and mtDNA asserting that "given that the SH hominins are early Neanderthals (or closely related to these), and assuming that the mtDNA they carried was typical of early Neanderthals, an additional possibility that appears reasonable is that the mtDNAs seen in Late Pleistocene Neanderthals were acquired by them later, presumably because of gene flow from Africa" (Meyer et al., 2016, p. 506).

The wane of ancestral mtDNA is relatively likely because the mtDNA is inherited, as well known, from the mother (see Box 10.20). Similarly, the chromosome Y is inherited from the father. The analysis of about 120 kb of exome-captured Y-chromosome DNA from a Neanderthal individual from El Sidrón site (Spain) revealed that none of the sequences obtained is present in modern humans (Méndez et al., 2016). Given that Neanderthals and humans hybridized, it could be the case that the DNA came from an extinct Y-chromosome lineage. But there is a possible alternative. As asserted by Fernando L. Méndez et. al. (2016, p. 732), "Although the Neandertal Y chromosome (and mtDNA) might have simply drifted out of the modern human gene pool, it is also possible that genetic incompatibilities contributed to their loss." In this case, the absence of Neanderthal Y-chromosome DNA in modern humans could be due to its possible incompatibilities with other human DNA, which might explain why the hybridization events between the two species were quite rare.

modern humans, the possibility of hybridization in Africa had not been addressed.

The study of genetic material from three current sub-Saharan populations: Bianka, San, and Mandenka, the first two hunter-gatherers and the latter farmers, led the authors to conclude that an introgression of a small amount of genetic material (about 2%) from the "archaic" populations occurred 35,000 years ago (Hammer et al., 2011). The protagonists of this partial hybridization would have been, according to Hammer et al. (2011), hominins that split from the lineage leading to modern humans about 700,000 years ago, i.e., in the early Middle Pleistocene, and which survived for a very long time.

Although these findings cannot be extrapolated unequivocally to all "transitional" fossils included in the paradox of Bräuer (2008), the work of Hammer et al.

Box 11.17 The archaic introgression

The study of Hammer et al. (2011) pointed to the loci *4qMB179*, *18qMB60*, and *13qMB179* as the best candidates to indicate the archaic introgression, with a haplotype showing different frequencies in populations of Central, East, West, and South Africa (Figure 11.18).

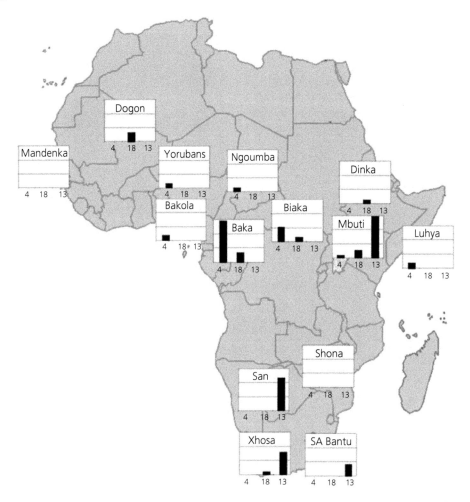

Figure 11.18 Frequency of introgressive variants within three sequenced regions in an expanded sample of 502 sub-Saharan Africans. The filled bar represents the frequency of a variant marking the divergent haplotype at 4qMB179 (left), 18qMB60 (center), and 13qMB179 (right) in each of 14 population samples. Each horizontal line on the bar charts represents a frequency of 5%. Hammer, M. F., Woerner, A. E., Mendez, F. L., Watkins, J. C., & Wall, J. D. (2011). Genetic evidence for archaic admixture in Africa. *Proceedings of the National Academy of Sciences*, 108(37), 15123–15128.

(2011) provided a compelling model to explain how relationships could occur between archaic and modern populations in the process of the transition and consolidation of *Homo sapiens*.

The problematic preservation of nucleic acids in warm-weather sites, such as in Africa, has not allowed so far the recovery of genetic material from populations involved in the evolution of African toward modern humans; not even from the last part of the transitional process. With these deficiencies in mind, Philipp Gunz and collaborators (2009) conducted a comparative analysis of neurocranial geometry in current modern humans, early modern humans, Upper Paleolithic *Homo sapiens*, Neanderthals, and "archaic" humans (see distribution shown in Table 11.5).

The results of the analysis by Gunz et al. (2009) indicated that, among all the groups, the early modern humans—left column in Table 11.5—contain greater

Table 11.5 Sample of the fossils included in the neurocranial geometry analysis conducted by Gunz et al. (2009)

First MH	Upper Paleolithic MH	Neanderthals	Archaic Sapiens
Brno 2	Jebel Irhoud 1	Amud 1	Dali
Combe-Capelle	Jebel Irhoud 2	Atapuerca SH5	Kabwe 1
Cro-Magnon 1	Ngaloba	Guattari 1	KNM-ER 3733
Cro-Magnon 3	Omo 2	La Chapelle-aux-Saints	Ngandong 7
Dolní Vestonice 2	Qafzeh 6	La Ferrassie 1	Ngandong 14
Fish Hoek	Qafzeh 9	La Quina 5	Petralona
Grotte des Enfants 4	Skhul I 5	Le Moustier 1	Sangiran 17
Mladec1		Spy 1	Trinil 2
Mladec5		Spy 2	Zhoukoudian 1
Mladec6		Tabun C1	Zhoukoudian 11
Oberkassel 1			Zhoukoudian 12
Oberkassel 2			
Pavlov 1			
Predmostí 3			
Predmostí 4			
Zhoukoudian Upper Cave 103			

MH means "modern humans." The population of modern humans incorporated Holocene specimens as well as subfossils such as Hohlenstein 1, Hohlenstein 2, Kaufertsberg, Wahlwies, Wadjak 1, Cohuna, Kow Swamp 5, and Paderborn 1.

variation in the measures taken (average of the Procrustean distance among several hundred neuronal landmarks). The morphology of these specimens differs from that of the "archaic" (Figure 11.19).

11.1.14 Asiatic hybridization

Apart from considerations regarding the alternative models—MH vs. RH—deduced from the analysis by Gunz and collaborators (2009), the results indicate that most likely early modern humans were already divided into different populations within Africa; therefore, migration patterns become complex depending upon the origin of the different migrants. A similar point of view was expressed by Jean-Jacques Hublin and collaborators (2007) after examining fossil evidence from Djebel Irhoud (Morocco).

According to genetic studies on mtDNA and the diversity of the Y chromosome, the presence of *Homo sapiens* in Australasia began around 70,000–50,000 years ago, by a migration from Africa that could have followed the northern or southern routes, or even both, rapidly occupying different locations (Kong et al., 2010; Stoneking & Delfin, 2010). However, fossils corresponding to those dates are difficult to classify or present dating problems, while specimens of an age beyond doubt—usually classified as *Homo sapiens*—are much younger, about 16,000 years (Curnoe et al., 2012).

The Longlin and Maludong specimens represent a different case, of great interest to the history of the Asian occupation, studied by Darren Curnoe and collaborators (2012).

In 1979, geologists looking for oil accidentally found human remains in a cave near Dc'e (Guangxi Province, China) within a block of fine sediments. After transportation to Kunming (Yunnan Province), a mandibular fragment and some postcranial elements were extracted. In 2010, the Curnoe team isolated from the block a partial cranium (Figure 11.20, left) and some postcranial bones. Although the stratigraphical origin of the specimen cannot be established, the radiocarbon analysis using sediments extracted from the endocranium yielded an age of 11,510±255 years (Curnoe et al., 2012).

Maludong cave near Mwngzi (Yunnan Province, China) was first excavated in 1989, with numerous cranial, mandibular, dental, and postcranial exemplars recovered. In 2008 a re-excavation of the cave completed the extraction of the materials, established the stratigraphy—with 11 aggregates—and obtained samples for dating the site. The age estimated by radiocarbon dating for the levels with human remains

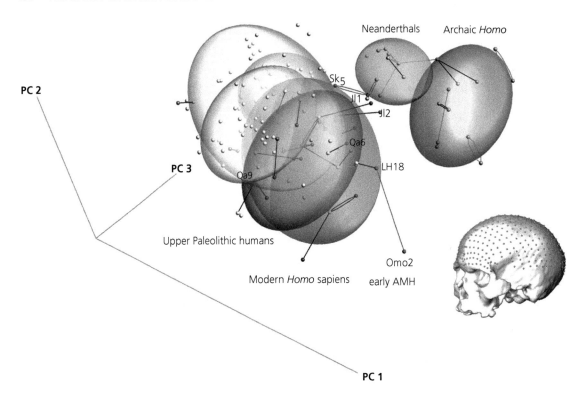

Figure 11.19 Shape space of modern humans and archaic forms. Two-dimensional projection of the first three principal components of the neurocranial shape. The drawn skull is an example (Mladec 1) for the full set of landmarks measured in each neurocranium. The graph offers two different kinds of information: (1) position of each specimen in the first three principal components, and (2) nearest neighbor relations according to the full Procrustean neurocranium shape distance (using all dimensions, not just the first three components). Connections between the nearest neighbors from the same group are shown in their own gray tones; connections between the nearest neighbors of different groups are drawn as black lines. Ellipsoids for recent humans are based on their geographical origin: Africa, Asia, Australia, and Europe. Gunz, P., Bookstein, F. L., Mitteroecker, P., Stadlmayr, A., Seidler, H., & Weber, G. W. (2009). Early modern human diversity suggests subdivided population structure and a complex out-of-Africa scenario. *Proceedings of the National Academy of Sciences U S A*, 106, 6094–6098.

was between 13,990±165 and 13,890±140 years (Curnoe et al., 2012).

Given such a recent age, the fossils of Longlin and Maludong could be considered as modern humans, adding to the many samples of this type found in the Far East (Box 11.18). However, their morphological analysis reveals some important discrepancies. The specimens Longlin 1 and Maludong 1704 have an important glabellar development that, in the first case, separates it from modern humans by the absence of the dividing groove between the medial and lateral parts, and, in the second case—which has a groove—shows a very developed supraorbital torus that is present in the first *Homo sapiens* but is very rare today (Curnoe et al., 2012). The facial skeleton LL1 shows a marked alveolar prognathism, which sets it apart from the first *Homo sapiens*. The mandibles of LL1, as well as of the

specimen from Maludong, contain traits that diverge from modern humans, such as the small chin and the retromolar space.

Darren Curnoe and collaborators (2012) maintained that a combination of apomorphies characteristic of modern humans and plesiomorphies shared with "archaic" fossils, such as those of Longlin and Maludong, is unusual in Eurasia. Although some African exemplars like those from the Klasies River and Hofmeyr (South Africa), Iwo Eleru (Nigeria), Nazl et Khater (Egypt), and Dar-es-Soltane and Témara (Morocco)—much older than the Chinese specimens—show a mosaic that also combines "modern" and "archaic" traits. The comparative analysis conducted by Curnoe et al. (2012)—including specimens from Skhul, Qafzeh (Israel), and Pestera cu Oase (Romania)—rejected the simplest hypothesis: that the Longling and Maludong

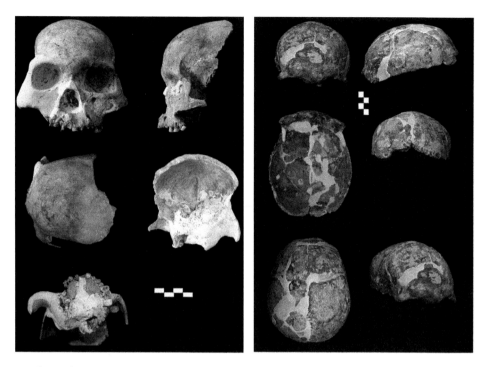

Figure 11.20 Left: Partial cranium Longlin 1. Right: Calotte Maludong 1704. In both cases the scale is in cm. Curnoe, D., Xueping, J., Herries, A. I. R., Kanning, B., Taçon, P. S. C., Zhende, B., . . . Rogers, N. (2012). Human Remains from the Pleistocene-Holocene Transition of Southwest China Suggest a Complex Evolutionary History for East Asians. *PLoS One*, 7, e31918.

Box 11.18 Cranial and dental traits of Longlin and Maludong specimens

The entire list given by Curnoe et al. (2012) of the characteristics that relate the Longlin and Maludong fossils with *Homo sapiens* includes, as affinities, certain traits of the neurocranium, viscerocranium, mandibles, and dentition. However, in all these anatomic elements differences appear that extend to the cranial base.

---

fossils correspond to very robust individuals that were a part of an epipaleolithic population of high variability. As the most plausible explanation, Curnoe et al. (2012) stated that they could be:

(1) Surviving members of an archaic population (i.e., that of Dar-es-Soltane and Témara), which might be reflected in other Chinese fossils such as the mandibular fragment from Zhirendong (China) or Salkiht (Mongolia).

(2) Specimens of a *Homo sapiens'* population that retained an ample number of ancestral polymorphisms, as a result of a hypothetical population subdivision in Africa prior to the later migrations to Eurasia, according to the model of Philipp Gunz and collaborators (2009).

The analysis by Curnoe et al. (2012) concluded that, particularly in the second case, the Longlin and Maludong fossils could belong to a previously unknown population whose genetic contribution to Asian modern humans would be difficult to establish. The authors attempted to retrieve ancient DNA from the studied samples without success, so they did not venture any hypotheses in this regard. In any case, after what we have seen with respect to isolated episodes of hybridization in Africa and the Middle East, this possibility cannot be discarded for now; thus, in addition to the two alternatives of Curnoe et al. (2012) for the Longlin and Maludong specimens, it could also be:

(3) Hybrids of the local population (*Homo erectus*) and migrants from Africa who exited a long time ago (modern humans).

11.1.15 A model adjusted for the origin of modern humans

The hybridization processes indicated between modern humans, the African "archaic" humans, Neanderthals, Denisovans, and other potential Asian cases—even when sporadic and limited—oblige a rethinking of the hypotheses about the origin and dispersal of *Homo sapiens*.

After what has been established by MH, as well as RH, the intermediate alternative of the Assimilation Hypothesis (AH) is a respectful model for both the African origin of modern humans and the localized hybridization episodes, whose scheme is shown in Figure 11.21.

Given that hybridization processes occurred both in the Near East and Asia, and in fact also in the place of origin of modern humans, Africa, we could conclude that proposed intermediate taxa between *Homo erectus* and *Homo sapiens*, such as *H. heidelbergensis*, *H. rhodesiensis*, or others, could be no more than the reflection of particular hybridization episodes for which fossil evidence has been discovered. If one had to assign a species to all those different populations in geographic and morphological terms, the transition from *Homo erectus* to *Homo sapiens* and *Homo neanderthalensis* would require the number of species to be multiplied exaggeratedly. Hence, it is often more practical to refer to the *Homo sapiens'* transition in terms of successive grades. But, even if it is the case that retrieval of fossil genetic material significantly complicates this transition model, at the same time it allows detection of some genetic changes related to apomorphies characteristic of our species.

11.1.16 Positive selection of genes

Changes in the genetic material do not only become reflected in differences at the protein level. As we saw when discussing the neutral selection and molecular clock (see Chapter 1, Section 1.2.18), the nucleotide differences found in the same position in a sequence indicate lineages that have taken separate paths. For understanding the evolutionary process of the hominin lineage in its last stage, single nucleotide polymorphism (SNP) comparisons are the most relevant analysis.

Neanderthals are in the range of SNP variation of modern humans with respect to many regions of the genome; thus, both lineages share the alleles present in modern humans. But, in order to determine a positive selection in modern humans, Green et al. (2010) focused on those SNPs that appeared as derived for our species, while Neanderthals retained the plesiomorphic condition (shared with the chimpanzee genome). That is, changes in the SNPs that emerged after the separation of the Neanderthal and modern human lineages through a process of positive selection in the latter.

Green et al. (2010) identified a total of 212 regions that may have experienced positive selection in *Homo sapiens*. Because there is a direct relationship between the strength of the positive selection and the length of the affected sequence, the authors focused on 20 extensive regions with SNPs showing positive selection in order to identify the genes encoded for those sets of nucleotides. Five regions did not contain encoded genes. The other 15 regions had between 1 and 12 genes. The longest, with a length of 336,000 bp and almost no derived SNPs in the Neanderthal genome, is localized in chromosome 2 and contains the *THADA* gene. Some of the alterations of this gene cause diabetes type II. Green et al. (2010) posed a hypothesis of the possible role of *THADA* in the regulation of energy metabolism in the first modern humans.

Table 11.6 shows several genes associated with disorders affecting cognitive capabilities and which belong to some of the 20 regions with higher probability of having been subjected to positive selection (Box 11.19).

Naturally, positive selection is meaningless if it leads to the selection of defective genes. In fact, what happens is that, for reasons having to do with the need to guide medical research into the correlation between diseases and genes, the first identified function is always the altered one.

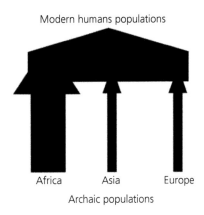

Modern humans populations

Africa   Asia   Europe

Archaic populations

Figure 11.21 Assimilation Hypothesis graph, representing an evolution and dispersal of modern humans compatible both with an African origin and limited episodes of hybridization.

Table 11.6 Genes belonging to regions with SNPs indicating a probable positive selection in modern humans, and whose defective alleles are associated with cognitive disorders (Green et al., 2010)

Chromosome	Region	Genes	Associated disorder
7	68,662,946–69,274,862	AUTS2	autism
21	37,580,123–37,789,088	DYRK1A	Down syndrome
10	83,336,607–83,714,543	NRG3	schizophrenia
6	45,440,283–45,705,503	RUNX2	cleidocranial dysplasia
7	121,763,417–122,282,663	CADPS2	autism

Box 11.19 Positive selection

The evolutionary model of Tim Crow on the origin of schizophrenia, sees this disorder as a collateral and unwanted result of the positive selection of genes that in their non-defective version are related with language (see, for example: Crow, 1993, 1995a, 1995b, 1996, 1997).

Many of the difficulties in completely understanding the meaning of positive selection derive from knowing only the defective and not the normal allele functions of genes. As Schwartz (2011) said, it makes no sense to argue that the primitive version of *RUNX2* matches the alterations of that gene, which produces defects in cranial development such as cleidocranial dysplasia. Neanderthals, the bearers of the "original" version, show a phenotype opposite to that of modern humans with a hypoplasia that prevents a full dental and bone development. Neanderthals had large and robust epiphysis in the long bones, a high and curved supraorbital arch, often substantial diaphysis diameter, and sometimes taurodontism—an extended pulp chamber. Schwartz's stance (2011) stresses the need to understand the genetic relationships between Neanderthals and modern humans, including the positive selection of genes such as *RUNX2*, while leaving aside orthogenetic evolution hypotheses. By "orthogenesis" we mean a gradual process in which it is possible to distinguish between lower and higher stages of evolution. The author emphasized the highly derived nature of many Neanderthal traits.

Until now, functions performed by the genes of Table 11.6 in their non-defective version are unknown. It is not even known the extent to which they are expressed in brain tissue, much less what meaning they could have for cognitive functions. In fact, the development of mental capacity is one of the most intricate issues, but also the most interesting that can be addressed within the hominin phylogeny. However, this is an issue we will have to leave out of this book for reasons of space.

Having evidence such as that provided by the work of Green et al. (2010) on the positive selection of modern human genes, is a good beginning for going beyond speculation. Xiling Liu et al. (2012) took a step forward by using microarrays and RNA-sequencing. Xiling Liu et al. (2012), compared gene expression changes taking place during postnatal brain development in the cerebellum and in five different modules of the prefrontal cortex of 23 humans, 12 chimpanzees, and 26 rhesus macaques. The results indicated that "the most prominent human-specific expression change affects genes associated with synaptic functions and represents an extreme shift in the timing of synaptic development in the prefrontal cortex, but not the cerebellum. Consequently, peak expression of synaptic genes in the prefrontal cortex is shifted from <1 year in chimpanzees and macaques to 5 years in humans" (Liu et al., 2012).

Of the PFC modules studied by Xiking Liu and collaborators (2012), the 1, 2, and 5 show a delay in the development of synaptic connections that cannot be explained by the developmental rate differences among species. This delay is not present in the PFC modules 2 and 4, or in the human cerebellum (Figure 11.22).

Within each module, the synchronized expression of genes implies coordinated regulation by shared mechanisms, such as transcription factors. Among the five PFC modules, associations between transcription factors and their predicted target genes were particularly pronounced in module1. Within this module the coordinated delay in expression of genes in the human PFC "could be linked to a similar delay in the expression of known regulators of activity-dependent synaptic development programs, such as MEF2A" (Xiling Liu et al., 2012, p. 618) (Box 11.20).

The work by Xiling Liu et al. (2012) indicates that "the module 1 expression profiles reflect a human-specific shift in the timing of cortical synaptogenesis, which coincides with a shift in the synaptic development program mediated by *MEF2A*" (p. 617). The possibility of linking an evolutionary event, such as the delay of synaptic development in humans, with its possible genetic correlate (that there are 12 times

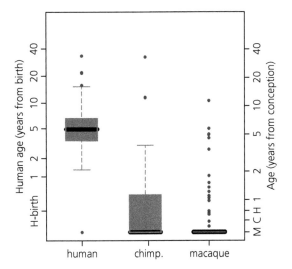

Figure 11.22 Distribution of ages at which expression of PFC module 1 genes reaches its maximum. The left and right y-axes show ages in years from birth for humans and from estimated conception time, respectively. H, C, and M indicate birth age for humans, chimpanzees, and macaques, respectively. Liu, X., Somel, M., Tang, L., Yan, Z., Jiang, X., Guo, S., . . . Khaitovich, P. (2012). Extension of cortical synaptic development distinguishes humans from chimpanzees and macaques. *Genome Research*, 22(4), 611–622.

Box 11.20 *MEF2A* gene

The gene *MEF2A* (myocyte enhancer factor 2A, location 15q26) encoded a DNA-binding transcription factor. According to the information given by the NCBI (The National Center for Biotechnology Information), *MEF2A* encoded protein "is involved in several cellular processes, including muscle development, neuronal differentiation, cell growth control, and apoptosis" (http://www.ncbi. nlm.nih.gov/gene/4205#). The study by Liu et al. (2012) suggests that *MEF2A* plays multiple roles in neuronal development, including neuronal survival, dendritic differentiation, synaptic density of hippocampal neurons, spine density in nucleus accumbens, and both synapse weakening/elimination and synaptic strengthening.

---

more developmental gene expression changes in humans than in chimpanzees, which are, moreover, centered in a specific region—*MEF2A*) opens a window to the interpretation of the phylogeny leading to the human brain. Xiling Liu and colleagues (2012), echoing Mochida and Walsh (2001), notice that the full spectrum of cognitive differences between humans

---

Box 11.21 Direct genetic comparison

Xiling Liu and collaborators (2012) raised the question: if the genetic region that might be related to synaptic development in the PFC has been identified, the next step would be to "identify the exact genetic basis of the human-specific developmental delay with subsequent comparison to the genomes of Neanderthals (. . .) and Denisovans (. . .)." (p. 619). However, the direct comparison of the *MEF2A* regions has not been carried out, as far as we know, between the three genomes.

---

and apes cannot be explained by size alone. Changes in development, plus the delay of the massive formation of synapses in humans until 5 years of age, would be, to Xiling Liu et al. (2012), a potential mechanism for the emergence of human-specific cognitive skills.

11.2 Brain distance between Neanderthals and modern humans

The findings of Xiling Liu and collaborators (2012) meant a new source of evidence for the comparison between the Neanderthal and the modern human brain. If size is similar in both species in allometric terms, a difference in ontogenetic maturation could be the mechanism accounting for the different cognitive capabilities of *Homo neanderthalensis* and *Homo sapiens*. But, before beginning to consider the developmental patterns of the two species, it is better to compare the detectable similarities and differences in the anatomy of their respective brains.

11.2.1 Brain expansion

The increase in brain capacity is, as we know, one of the more relevant apomorphies of the genus *Homo*. However, the patterns of brain expansion differ between the various species of the genus—Neanderthals and modern humans, and their ancestors of the Hominini tribe. Neanderthals and modern humans share an extra-allometric development of the frontal lobe (Bruner & Holloway, 2010) and differ in the lateral expansion (Bruner, 2010), as well as—hypothetically—in the visual cortex (Pearce et al., 2013). The study conducted by Emiliano Bruner and Ralph Holloway (2010) compared the dimensions of the hemispheric length, the maximum endocranial width, and the frontal width. The result showed that "within the genus *Homo*, the most encephalized taxa (Neandertals and modern humans) show relatively wider frontal lobes than either *Homo erectus* or australopithecines." (Bruner & Holloway, 2010, p. 138) The frontal relative widening makes the endocranium appear more "squared" (see Figure 11.23, bottom). As the authors maintain, "Taking into account the contrast between the intraspecific patterns and the between-species differences, the relative widening of the anterior fossa can be interpreted as a definite evolutionary character instead of a passive consequence of brain size increase" (Bruner & Holloway, 2010, p. 138).

The functional interpretation made by Bruner and Holloway (2010) of this extra-allometric change states that: "This expansion is most likely associated with correspondent increments of the underlying neural mass, or at least with a geometrical reallocation of the frontal cortical volumes. Although different structural changes of the cranial architecture can be related to such variations, the widening of the frontal areas is nonetheless particularly interesting when some neural functions (like language or working memory, decision processing, etc.) and related fronto-parietal cortico-cortical connections are taken into account" (Bruner & Holloway, 2010, p. 138).

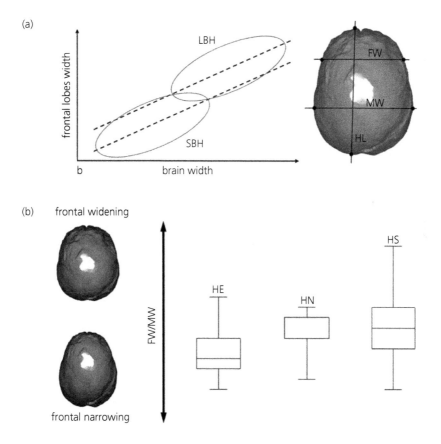

Figure 11.23 (a) A discrete morphological change in the frontal widening of the large-brained human species (LBH) (left) is favored by using a bivariate approach to the relationships between (right) hemispheric length (HL), maximum endocranial width (MW), and frontal width (FW). (b) Variation and distribution of the ratio between frontal width and maximum endocranial width for the three human groups (HE 1/4 *H. erectus* sensu lato; HN 1/4 Neanderthals; HS 1/4 modern humans). Box and whiskers' plots show the non-parametric descriptors (maximum and minimum, interquartile, and median). Differences between HE and the other two groups are significant (all pictures, Bruner & Holloway, 2010). Reprinted from Bruner, E. & Holloway, R. L. (2010). A bivariate approach to the widening of the frontal lobes in the genus Homo. *Journal of Human Evolution,* 58(2), 138–146, with permission from Elsevier.

Emiliano Bruner (2010) compared the anatomic configuration of the frontal lobe with the endocrania of three fossils: (1) the Salé specimen (Morocco), "generally referred to as *Homo erectus* (*sensu lato*) (. . .) this endocast could well represent the basic endocranial morphology"; (2) the cranium from La Chapelle-aux-Saints, which "represents one of the best-known 'classic' Neanderthals"; and (3) the skull from Vatte di Zambana, which "represents an anatomically modern human from northern Italy" (quotations by Bruner, 2010, p. 578). In these fossils, Bruner chose nine two-dimensional landmarks (see Figure 11.24). Landmarks were sampled in three dimensions using a Microscribe 3DX (Immersion) on both hemispheres.

The results of the analysis by Bruner (2010) indicate that the Vatte di Zambrana fossils, as well as that from La Chapelle-aux-Saints, display lateral widening of the parietal areas, when compared with the Salé specimen. The author understands that this result could be extrapolated to the whole set of modern humans and Neanderthals. However, the lateral extension is not equivalent in both species. *Homo neanderthalensis* shows a midsagittal flattening, while the cranium of *Homo sapiens* displays in that area a bulging aspect. This is an important difference because, as Bruner (2010) reminds us, the upper parietal areas are involved in a neural network formed together with the occipital and frontal lobes (Figure 11.25). The newest perspective in cognitive science understands human cognition as the result of a co-activation of different brain regions, which are distant from each other but are part of a functionally integrated network (Sporns et al., 2004). For obvious reasons, it is impossible to detect functionally active networks in endocrania. But, there are corticocortical connections from the inferior parietal lobe to the dorsolateral prefrontal surface, on the one hand, and from the upper parietal lobe to the dorsomedial prefrontal cortex, on the other (Bruner, 2010, quotes the work by Wise et al., 1997). This observation leads to the conclusion that evolutionary changes in these interconnected regions may have had a significant effect on the cognitive functions controlled by the network. Bruner (2010) mentions the working memory, relating this to cognitive abilities such as "reordering of elements," "internal mental images," "internal mental representation," "imagined world," "thought experiment," "orienting attention," and "abstract representation." The parietofrontal connection is also part of the first network to be identified as being involved in tasks of esthetic preference (Cela-Conde et al., 2013; Cela-Conde & Ayala, 2014), whose importance in the "modern mind" is beyond doubt, and to which we will return later.

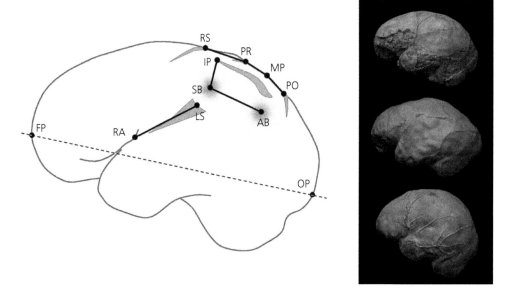

Figure 11.24 Left: Configuration used in the Bruner (2010) analysis. It includes two baseline points (FP, frontal pole; OP, occipital pole) and nine parietal landmarks, four on the midsagittal profile (RS, Rolandic sulcus; PR, post- Rolandic sulcus; PO, parieto-occipital sulcus; MP, midparietal point), three on the lateral surface (IP, anterior edge of the intraparietal sulcus; SB, supramarginal boss; AB, angular boss), and two localizing the lateral sulcus (RA, anterior edge of the Rolandic sulcus; LS, posterior edge of the lateral sulcus). Right: Endocasts used in the comparison—Salé (top), La Chapelle-aux-Saints (middle), and Vatte di Zamabana (bottom). (Bruner, 2010).

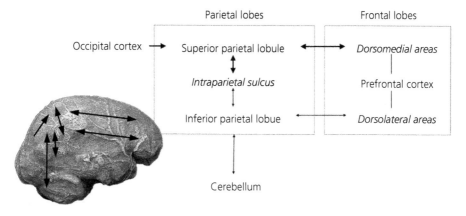

Figure 11.25 Upper parietal network (Bruner, 2010).

Box 11.22 Brain and metabolism

The existence of high metabolic requirements is an important consideration when assessing how natural selection favors an expanding cortex. Katherine Milton (1988) suggested that the only way to meet the brain's metabolic demands for the genus *Homo* was a dietary change toward richer nutrients, particularly meat. The intestine of modern humans is relatively smaller than the intestine of other primates. It would seem strange that throughout evolution, one organ—the brain—required a greater number of nutrients, while the intestines responded by reducing their size. Milton (1988) believed that this is explained by the fact that the human digestive system is specialized, with a relatively long small intestine when compared to the considerable length of the colon of apes.

Aiello and Wheeler's (1995) study of the relation between brain size and the length of the intestines led to similar conclusions. Herbivorous primates, with larger intestines, have relatively smaller brains than frugivorous primates. Aiello and Wheeler's explanation is known as the "expensive tissue" hypothesis. Two different animal species with a similar metabolic rate have had to "choose" between intestinal and cerebral tissues, given that both have very high energetic requirements. Herbivorous diets require very large intestines for digestion, so this expensive system would constitute a barrier for high encephalization.

The differences between the higher parietal cortex in Neanderthals and humans could be due to differences in ontogenetic development, as pointed out by Xiling Liu and collaborators (2012). If so, it is necessary to establish the ontogenetic comparative patterns between *H. neanderthalensis* and *H. sapiens*.

Herman Pontzer and collaborators (2016) used doubly labeled water measurements of total energy expenditure (TEE) in humans and apes to test the hypothesis that the human lineage has experienced an acceleration in metabolic rate, providing energy for larger brains. Their results indicate that "human TEE exceeded that of chimpanzees and bonobos, gorillas and orangutans by approximately 400, 635 and 820 kcal day^{-1}, respectively, readily accommodating the cost of humans' greater brain size and reproductive output. Much of the increase in TEE is attributable to humans' greater basal metabolic rate (kcal day^{-1}), indicating increased organ metabolic activity. Humans also had the greatest body fat percentage. An increased metabolic rate, along with changes in energy allocation, was crucial in the evolution of human brain size and life history" (Pontzer et al., 2016, p. 390).

11.2.2 The ontogenetic development of Neanderthals

When in Chapter 10 we spoke about traits derived from Neanderthals, we left open the question of their ontogenesis. There are two competing hypotheses about the ontogenetic development of Neanderthals when compared with that of modern humans. The first argues that Neanderthals reached adulthood more rapidly, which is the same as saying that their childhood was more brief than that of modern humans. The second argues that Neanderthals and modern humans followed the same temporal pattern of ontogenetic development, with equivalent childhoods.

There are two main indicators that can be used to calibrate the age of death of a hominin fossil: its dental development and its brain development.

11.2.3 Dental development

The ontogenetic condition of fossil specimens usually is inferred from the developmental patterns of teeth, the episodes of deciduous teeth substitution, and the emergence of the third molar, the most conspicuous characteristics indicating the end of childhood (Box 11.23).

Studies on the growth of dentition have established a model of interpretation, generally accepted, indicating that ontogeny was faster among hominins during the Pliocene and the Lower Pleistocene than in subsequent specimens (Bromage & Dean, 1985; Dean et al., 2001; Lacruz & Ramírez Rozzi, 2010). The acquisition of the modern pattern—a prolonged childhood—was identified as being close to the emergence

---

Box 11.23 Immature fossils

The dental maturation process is not, however, so simple; analysis can be performed even on the microscopic growth that takes place every day during childhood and adolescence. Markus Bastir and Antonio Rosas (2004), by a comparative study of various dental ages in chimpanzees and modern humans, have shown how for each age the average size (indicating growth) and the average shape (indicating the degree of maturation) may pose the paradox of high levels of maturity linked to lower rates of growth. Most identifications as "immature" fossil specimens come from studies of dental development, either in terms of growth or maturity, although there are other signs such as suture ossification within the cranial morphology.

Alistair R. Evans et al. (2016) have identified a default pattern of tooth sizes for all lower primary postcanine teeth. In Evans et al.'s (2016) words, "This configuration is also equivalent to a morphogenetic gradient, finally pointing to a mechanism that can generate this gradient. The pattern of tooth size remains constant with absolute size in australopiths (including *Ardipithecus, Australopithecus* and *Paranthropus*). However, in species of *Homo*, including modern humans, there is a tight link between tooth proportions and absolute size such that a single developmental parameter can explain both the relative and absolute sizes of primary postcanine teeth" (p. 477).

---

of *Homo sapiens*, thanks to a study by Tanya Smith and collaborators (2007) carried out by using microtomography on a juvenile *Homo sapiens* from Irhoud Djebel (Morocco) of 160,000 years. Its developmental pattern (degree of eruption, developmental stage, and time of crown formation) is, according to the analysis made by Smith et al. (2007), closer to modern humans than to the early species of *Homo*. However, the study by microtomography of the immature mandible H5 from the TD6 level of Gran Dolina (Atapuerca) by José María Bermúdez de Castro et al. (Bermúdez de Castro, Martinón-Torres, Prado et al., 2010), contradicted that hypothesis. Comparing the data obtained with the timings of enamel development and with the rates of the root extension in chimpanzees, ancient hominins, and modern humans, Bermúdez de Castro, Martinón-Torres, Prado et al. (2010) estimated that the growth period for the first upper and lower molar—emerged just before the death of H5—would have been between 5.3 and 6.6 years old, i.e., within the variation range of modern human populations. These authors noted that, due to the fact that the eruption time of M1 is a strong indicator of the life history, it is arguable—at least as a working hypothesis—that the Gran Dolina hominins, with just under a million years of antiquity, had a childhood that falls into the normal patterns of modern humans. The view of Bermúdez de Castro, Martinón-Torres, Prado et al. (2010) implies that the appearance of this important feature of biological development, and the consequent increase of the brain size associated with a long childhood, would have preceded the development of the neocortical areas leading to the cognitive abilities considered unique to *Homo sapiens*.

The Neanderthal case is somewhat complex and difficult to interpret. Jennifer Thompson and Andrew Nelson (2000) proposed that, compared to modern humans, Neanderthals had either slower growth or, if their ontogeny took an equivalent time, their teeth development was more rapid. The work of Fernando Ramírez Rozzi and José María Bermúdez de Castro (2004) agreed with that interpretation, highlighting the fact that both *H. antecessor* and *H. heidelbergensis*, and even *H. neanderthalensis*, had a faster dental development than modern humans. However, the opposite view was defended by Roberto Macchiarelli and collaborators (2006) after the study using computer tomography of two molars from the site of La Chaise-de-Vouthon (Charente, France). The results indicated an age for the eruption of M1 and a developmental pattern similar to those of modern humans (Macchiarelli et al., 2006).

The discussion about the patterns of dental development in Neanderthals and modern humans gained new perspectives due to the work of Tanya Smith and collaborators (2010), who applied virtual histological techniques using a synchrotron micro-CT scanner (Figure 11.26). The study confirmed that, in most cases, dental Neanderthal crowns grew rapidly, producing a quicker maturation of the denture. If that accelerated ontogeny affected traits other than dental, it would open issues of great interest, such as the growth and development of the brain.

The different patterns of the Neanderthal and modern human brain development were studied by Phillipp Gunz and collaborators (2010) using tomography (CT) to virtually compare 58 crania of modern humans and 9 Neanderthals, confirming the hypothesis of a similar brain size in both species at the time of birth. However, the type of development differs. The "globularization-phase," which for humans, in comparison with chimpanzees, leads to a cranium and brain of globular shape, does not occur in the case of Neanderthals (Gunz et al., 2010). The work of Gunz et al. (2010) established that the derived character is the development pattern of modern humans, and that changes toward globularization occur mainly in the frontal and parietal bones shortly after birth, when the skull bones are thin and not fully ossified.

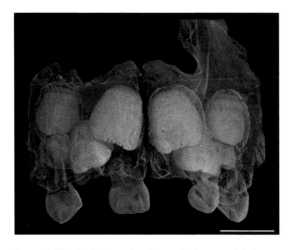

Figure 11.26 Synchrotron micro-CT scan (31.3-µm voxel size) showing central incisors, lateral incisors, canines, and third premolars, not yet erupted (the four lower teeth are deciduous) (scale: 10 mm). Smith, T. M., Tafforeau, P., Reid, D. J., Pouech, J., Lazzari, V., Zermeno, J. P., . . . Hublin, J.-J. (2010). Dental evidence for ontogenetic differences between modern humans and Neanderthals. *Proceedings of the National Academy of Sciences*, 107(49), 20923–20928.

11.2.4 Recreating birth

Gunz and collaborators (2010) conducted a computer simulation of what a human head would be if, from its shape at the moment of birth it followed the Neanderthal developmental pattern. The result produced a look of "striking resemblance" to an adult Neanderthal, as Gunz and collaborators remarked. So, it would be the process of ontogenesis, and not the neonate starting point, which mainly produces the final shape (Box 11.24).

The consequences of a similar brain size at birth and the diverse developmental patterns between Neanderthals and humans have been studied by Marcia Ponce de León and collaborators (2008).

Thanks to virtual reconstructions of a Neanderthal newborn (Figure 11.27) from the Mezmaiskaya cave (Russia) and of two infant skeletons from Dederiyeh (Syria), Ponce de León et al. (2008) concluded that the Neanderthal brain size at birth was similar to modern humans. As the total period of ontogenesis is the same for Neanderthals and modern humans, the higher rate of growth in Neanderthals does not imply an earlier maturation; after an initial stage of equal size it leads to a larger brain size.

Following Leigh and Blomquist (2007), Ponce de León et al. (2008) maintained the hypothesis that the high energy demands of the postnatal brain growth in Neanderthals would have led to a peculiar kind of life history in which mothers would have assumed the costs of their child's brain maturation. This is one of the few empirical evidences that can be given about the social life of Neanderthals. Another feature, related to the patrilocal mating pattern—the couple remains with the male group—was concluded by Carlos Lalueza-Fox and collaborators (2010) by the genetic study of 12 specimens of Neanderthals from El Sidrón (Spain).

Box 11.24 The particular ontogeny of Neanderthals

Fernando Ramírez Rozzi and José María Bermúdez de Castro (2004) concluded that, as the increase in brain size correlates with a longer period of teeth development during the evolution of hominins, the fact that Neanderthals accelerated their dental growth rate while having a very large brain is an autapomorphic reversion, pointing firmly to their status as a species different from modern humans.

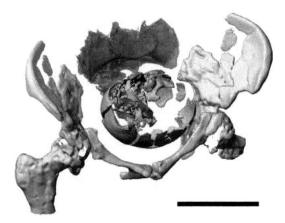

Figure 11.27 Reconstruction of Neanderthal birth. The pelvic reconstruction combines elements of the Tabun 1 specimen (ilium, ischium, and pubis) with their mirror-imaged counterparts (transparent), and replacement parts from a modern human female pelvis (sacrum and ischial spines). The Mezmaiskaya skull is visualized during its passage through the pelvic inlet (scale bar: 10 cm). Ponce de León, M. S., Golovanova, L., Doronichev, V., Romanova, G., Akazawa, T., Kondo, O., . . . Zollikofer, C. P. E. (2008). Neanderthal brain size at birth provides insights into the evolution of human life history. *Proceedings of the National Academy of Sciences*, 105(37), 13764–13768. Copyright (2008) National Academy of Sciences, U.S.A.

11.3 Cognitive distance between Neanderthals and modern humans

The genetic distance between Neanderthals and modern humans shows that, even in the case of two different species that appeared in different continents, their evolutionary proximity is so great that they were able to interbreed. Investigations of the distance between the brains of Neanderthals and modern humans are inconclusive for establishing their respective cognitive capabilities. As we have seen, the brains of both species have an equivalent expansion in the frontal region, but there are differences in the parietal regions due to the dissimilar patterns of ontogenetic development. Those differences might have developed into two different minds. But if so, what was the scope of the "Neanderthal mind"? To answer that question it is necessary to explore the evidence of how the "modern mind" emerged.

11.3.1 The emergence of the "modern mind": South Africa

In Chapter 7, when discussing cultural traditions, we arrived at the African Middle Stone Age (MSA) stage

and its transition to the Late Stone Age (LSA). We left pending some questions such as who were the makers of the MSA advanced techniques—Still Bay (SB) and Howieson's Poort (HP)—and which relationships were maintained between these industries and their associated decorative objects and the cultural manifestations of the European Upper Palaeolithic. Let's begin with the first question. In Chapter 10, we mentioned that, according to Terrence Deacon (1989), social and cognitive mechanisms similar to our own species, which allowed the use of artifacts as symbols, furthered the adaptive success of the South African populations that made SB and HP, such as those of the Klasies River. Let's now review the evolutionary steps so we can later compare the modern human achievements in South Africa with those of the Neanderthals in Europe.

As we saw in Chapter 10, South Africa provided evidence on the origin and dispersals of the first *Homo sapiens*. Various sites from the most southerly part of South Africa, near Cape Town, such as the Border cave (de Villiers, 1973), the Klasies River mouth (Singer & Wymer, 1982), the Equus cave (Grine & Klein, 1985), the Die Kelders cave, the Blombos cave (Henshilwood et al., 2001), Sibudu (Backwell et al., 2008), Hofmeyr (Grine et al., 2007), and Hoedjiespunt (Berger & Parkington, 1995), among others, have provided the most important samples of the emergence of modern humans. The fossils from these sites are normally of lesser significance and of dubious dating when compared with those coming from other African sites. But the importance of the association between fossil specimens and archeological remains in South Africa lies in the fact that the Khoe-San hunter-gatherers, the oldest living identified ethnicity, are found there. In Chapter 10 we mentioned that their separation from the rest of the human populations occurred at least 100,000 years ago (Schlebusch et al., 2012).

The evidence of an earlier division and dispersion of populations of *H. sapiens* reveals that the transit from cultural Mode 3 to Mode 4 cannot be deduced by comparison between the technological level of Neanderthals and Cro-Magnon in the range of c. 40 Ka. It must be found in the cultural development of *Homo sapiens* in Africa, which took place at a time (100–200 Ka) when the appearance and early evolution of our species occurred. A review of African cultural evolution of that period reveals the meaning of the proposed model of gradual change by McBrearty and Brooks (2000) to which we referred in Chapter 7. Jayne Wilkins (2013) pointed out the general characteristics of the human evolution related to the development of MSA in South Africa. For this author the earlier MSA is "generally attributed to a group of hominins that are variably described as late archaic *Homo sapiens*, or *H. helmei* [meanwhile] by ~195–150 Ka, anatomically modern human fossils are known from East Africa (. . .) and modern *H. sapiens* are responsible for the later MSA" (Wilkins, 2013, p. 6).

In a review article that overviews how the South African Pleistocene *Homo* fossil record correlates with the Stone Age sequence, Gerrit Dusseldorp, Marlize Lombard, and Sarah Wurz (2013) have argued about the basic problem in establishing the correlation between fossils and tools: few South African hominin fossils can be placed between c. 200 Ka and 110 Ka, i.e., during the probable dates of transformation from Mid-Pleistocene *Homo* to modern *H. sapiens* in the region. After these dates, the Klasies River and other South African sites have offered specimens belonging to our species with an estimated age of c. 110 Ka and 40 Ka (see Chapter 10). The set of South African fossils associated with MSA technocomplexes mainly belong to MIS 5 and MIS 4, and can be called "transitional." The morphology of the fossils is modern, but the process of gracilization, leading to the form and dimensions of contemporary populations, was not yet completed (Dusseldorp et al., 2013). As Dusseldorp et al. (2013, p. 5) have said: "On the whole, the fossil record from this period suggests that South Africa was occupied by populations showing a wide range of anatomical variation."

Between the end of MSA and the beginning of LSA, two complete fossils of modern morphology are available in South Africa: the Hofmeyr skull and the jaw of Bushman Rock Shelter. Both specimens are attributed to MIS 3, with an age for the child's jaw of Bushman Rock Shelter—assigned tentatively to site levels 16 or 17—of c. 29.5 Ka (Protsch & de Villiers, 1974). The Hofmeyr skull has been dated at 36.2±3.3 Ka by thermoluminiscence and uranium-series (see Chapter 10).

The anatomical connection, the presence of an advanced technology of tool-making, and the evidence of the emergence of what could be called modern behavior, in cognitive terms that goes beyond the technological level, make South Africa a site of great value for understanding the last steps of human evolution. In this respect, the Blombos site becomes important. Its human remains provide little information; during the 1997–98 campaigns, four teeth were found in the Blombos cave, two of them deciduous teeth. With respect to the crown diameter, some of the four teeth belong to the modern human range. But the peculiarity of Blombos is linked to the presence of ocher pieces.

The presence of red ocher—haematite (iron oxide)— is very common in all South African sites of the Late MSA, and stones with signs of use have striations, which are attributed widely to the acquirement of powder for pigment-making. In the absence of polychrome on the walls of caves, it is possible to infer that the use of the pigment is related to other symbolic behavior (Figure 11.28). As argued by Christopher Henshilwood and collaborators (2001), one intuitive conclusion, shared by most archaeologists, is that MSA ocher was used for body-paint/cosmetics and possibly the decoration of organic artifacts. But, in the absence of empirical evidence, that hypothesis is entirely speculative. Blombos' value lies in the contribution of evidence linking ocher and symbolism.

The Blombos cave is located near the Indian Ocean, 25 km west of Still Bay town and 300 km east of Cape Town. The site is located 100 m from the coast and at an elevation of 34.5 m above sea level. In Blombos MSA levels, more than 8,000 pieces of ocher have been found, many with signs of use (Henshilwood et al., 2002), among which the most outstanding are the geometric engravings. Sixteen new engraved ocher pieces

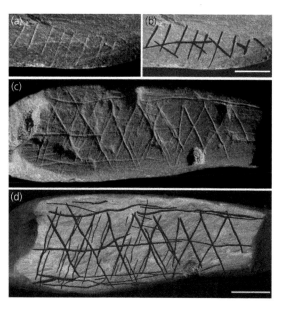

Figure 11.28 Engraved ochers from Blombos Cave. (A) SAM-AA 8937, flat piece of shale-like ochre. (B) Tracing of lines verified as engraved by study under magnification (scale bar: 5 mm). (C) SAM-AA 8938, rectangular slab of ocherous shale. (D) Tracing of lines verified as engraved by study under magnification, superimposed on flat-bed scan of engraved surface (scale bar: 10 mm). From Henshilwood, C. S., d'Errico, F., Yates, R., Jacobs, Z., Tribolo, C., Duller, G. A. T., . . . Wintle, A. G. (2002). Emergence of Modern Human Behavior: Middle Stone Age Engravings from South Africa. *Science*, 295(5558), 1278–1280. Reprinted with permission from AAAS.

from the Still Bay levels of Blombos were discovered and later analyzed, which indicated the existence of various engraving patterns on ochers of the M1, M2, and M3 phases (Henshilwood et al., 2009).

11.3.2 The way out of Africa for *Homo sapiens*

The process leading to modern humans, from their emergence and development of Mode 4 in Africa, to their entry into Europe, and to the "artistic explosion" that appears in the caves of southern France and northern Spain, is controversial. Both the presence of various populations in South Africa in the crucial environment for the evolution of modern humans (Dusseldorp et al., 2013), as well as the close relationship between Eurasian and South African fossils of the MIS 3 (Late MSA and Early African LSA) (Grine et al., 2007), make it difficult to compare the anatomy of the various specimens. Hence, the need to search for genetic markers associated with this dispersal.

As we saw in Chapter 10, recent analysis of SNPs in nuclear DNA supported the sub-Saharan origin of modern humans (Jakobsson et al., 2008; Li, Ruan, & Durbin, 2008, Schlebusch et al., 2012). Recent studies point to the San Bushmen as the best candidates available when identifying the closest modern humans to the first *Homo sapiens*. The mtDNA haplogroup analysis (see Chapter 10, Secion 10.2.7) of current populations has identified the expansion in Africa of haplogroups L2 and L3 around 85,000 years ago (Forster & Matsumura, 2005), and a possible way out of Africa after the volcanic supereruption of Toba in Sumatra (Toba, hereinafter; the eruption occurred 74,000 years ago—Soares et al., 2012). However, these dates do not match the archaeological evidence, which indicates the presence of modern humans outside Africa in the Near East (Skuhl and Qafzeh), c. 100,000 years ago (Mellars, 2006a, for example).

Genetic and archaeological data would be compatible if one or a few more exits had occurred from Africa, as proposed by Marta Mirazón Lahr and Robert Foley (1994). Genetic studies (mtDNA and Y chromosome) of the current populations point firmly to a limited genetic diversity in Eurasia, something that is incompatible with multiple exits from Africa over extended periods of time (Mellars 2006b). However, archaeological and genetic data might fit together with a hypothesis of a very ancient first exit c. 100,000 years ago, whose genetic trace has not been retained in the current populations—perhaps due to its lack of success in the medium or long term—and one other later exit c. 65,000 years ago (Mellars 2006b). The discovery in the Fuyan cave in Daoxian (southern China) of 47 human teeth dated more than 80,000 years old, and with an inferred maximum age of 120,000 years (Liu et al., 2015), supports an ancient exit.

The scope of an early pre-Toba exit of modern humans from Africa would place the two models in confrontation. The first model, by Michel Petraglia and collaborators (2007, 2010), maintains that the available archeological and genetic evidence supports the older exit from Africa. The second, by Paul Mellars and collaborators (2013), denies the existence of this evidence. An important discrepancy points to the type of dispersal. According to David Bulbeck (2007), the exit of modern humans from Africa happened rapidly and efficiently by taking advantage of the best coastal areas for dispersal (coastal model). Mellars et al. (2013) consider that the coastal model adjusts well to the Southeast Asia colonization reflected in the genetics of the current population. However, Petraglia et al. (2010) argue that the coastal model does not explain the

---

Box 11.25 Traces of the first modern humans in Europe

By means of genome-wide analysis of the DNA from 51 modern humans, dated between 45,000 and 7,000 years ago, Fu et al. (2016) established the genetic track present in the European populations since the first modern humans arrived. As stated by Fu et al. (2016, p. 200), "Modern humans arrived in Europe ~45,000 years ago, but little is known about their genetic composition before the start of farming ~8,500 years ago. (. . .). [From 45,000 to 7,000 years ago], the proportion of Neanderthal DNA decreased from 3–6% to around 2%, consistent with natural selection against Neanderthal variants in modern humans. Whereas there is no evidence of the earliest modern humans in Europe contributing to the genetic composition of present-day Europeans, all individuals between ~37,000 and ~14,000 years ago descended from a single founder population which forms part of the ancestry of present-day Europeans. An ~35,000-year-old individual from northwest Europe represents an early branch of this founder population which was then displaced across a broad region, before reappearing in southwest Europe at the height of the last Ice Age ~19,000 years ago. During the major warming period after ~14,000 years ago, a genetic component related to present-day Near Easterners became widespread in Europe."

diversity of the occupation after the exit from Africa, making a very complex process of dispersal necessary.

In fact, both models (Petraglia et al., 2010; Mellars et al., 2013) accept the coastal model, although giving it a different significance. The advantage of the idea of a coastal dispersal is that it provides a convincing explanation for how modern humans exited Africa by passing through areas of "high marine productivity." These areas would have ensured plentiful resources—which would also explain the success of the late industries of the African MSA, Howieson's Poort and Still Bay.

As Jane Balme et al. (2009) have indicated, the occupation by hominins of diverse ecosystems with hostile environments and depressed fauna could only have occurred with the use of complex systems of exchange and communication, including language. Although it is difficult to verify this hypothesis, what these authors argue is equivalent to an acceptance that Southeast Asian settlers in that time range possessed the modern mind. There is archeological evidence of such a dispersal. The microlithic industry of the Late African MSA shows a significant similarity to that of the subcontinental Indian and Southeast Asian sites. Even more important for documentation of the modern mind is the coincidence between the engravings in ocher pieces and ostrich eggshells (Blombos and Diepkloof), and those of Patne (India), or the perforated beads that are found in Africa and Asia (Figure 11.29).

11.3.3 The encounter of species

Regardless of the dispersion model accepted, the fact of a departure from Africa by itself shows that early modern humans had the opportunity to meet with local inhabitants in both the Near and Middle East, as well as in Asia. This would be an "encounter" between lineages that diverged from each other long before the c. 100,000 years attributed to the exit of the first modern humans out of Africa. Of these encounters, the most important is with the Neanderthals, both in the Near East and Europe.

The cultural trajectory of Neanderthals and modern humans followed curiously parallel lines. This is surprising because the evolution of the Neanderthal culture took place in Europe, with ramifications in the Near and Middle East, while the cultural development of modern humans toward the LSA has been documented in South Africa. Even so, by the time the two population protagonists of the Upper Paleolithic cultural leap met, some of their technological and mental traits coincided to a great degree. One wonders whether, in addition to genetic hybridization, there

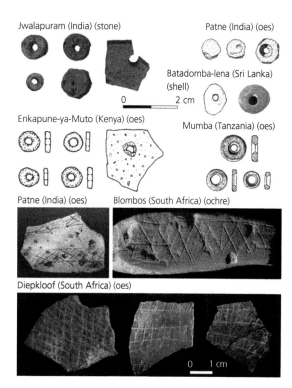

Figure 11.29 Ornamental beads and geometric symbols from later MSA sites in South and East Africa and early microlithic sites in India and South Asia (oes, ostrich eggshell). Mellars, P., Gori, K. C., Carr, M., Soares, P. A., & Richards, M. B. (2013). Genetic and archaeological perspectives on the initial modern human colonization of southern Asia. *Proceedings of the National Academy of Sciences*, 110(26), 10699–10704.

Box 11.26 Out of Africa

The existence of a surprisingly precise correlation between the genetic diversity of a population and the geographical distance between this population and East Africa—indicated by Prugnolle et al. (2005)—is consistent with the idea of a gradual geographical dispersal from the African origin. A different issue is how to document the symbolic behavior of dispersing populations. We will discuss this matter in Section 11.3.3.

---

was also a kind of "cultural hybridization" between the two species that led to the sharing of more than a few traits of what we call the modern mind.

We have noted earlier in this chapter genetic (Section 11.1), and brain (Section 11.2) differences between Neanderthals and modern humans. Let's now focus on the cognitive distance, i.e., the issue of symbolism. Following a widely accepted criterion, Erella Hovers, Bernard Vandermeersch, and Ofer Bar-Yosef (1997) considered that there are two types of detectable symbolic behaviors in the fossil record: burials and the production of non-utilitarian objects. Let's start with the first.

11.3.4 The tomb issue

In Chapter 7 we indicated the presence of Mousterian industry in one Near East Neanderthal site—Tabun—as well as in sites of modern humans—Skuhl and Qafzeh—around 100,000 years ago. We left a question pending: if both species shared the same technological level, would this mean that Neanderthals had reached a cognitive level equivalent to modern humans? Some authors arguing in favor of high cognitive capacities in Neanderthals went beyond lithic culture shared at the Near East. They presented other kinds of items, which, in their opinion, were indications of Neanderthal esthetic, religious, symbolic, and even maybe linguistic, capacities.

The possibility that Neanderthals buried their dead is the best basis on which to attribute transcendental thought to them. Voluntary burial is indicative of respect and appreciation, as well as a way to hide the body from scavengers. This may also imply concern about death, about what lies beyond death, and the meaning of existence. The argument for religiousness is convincing when burial is accompanied by some sort of ritual.

Neanderthal burials have been located in four areas: Southern France, Northern Balkans, the Near East (Israel and Syria), and Central Asia (Iraq, Caucasus, and Uzbekistan). In most cases these burials seem to be deliberate. Hence, the "old man" from La Chapelle-aux-Saints appeared in a rectangular hole dug in the ground of a cave that could not be attributed to natural processes (Bouyssonie et al., 1908). In regard to La Ferrassie and Shanidar, the possible evidence of the existence of tombs led Michael Day (1986) to remark, in a technical and unspeculative treatise, that these exemplify the first intentional Neanderthal burials that have been reliably determined. Eric Trinkaus' (1983) taphonomic considerations point in the same direction. The abundance and excellent state of Neanderthal remains at those sites, together with the presence of infantile remains, are a proof that the bodies were out of the reach of scavengers. Given that there is no way natural forces could produce those burials, Trinkaus believes the most reasonable option is to accept that the remains were intentionally deposited in tombs. However, William Noble and Ian Davidson (1996) argued that, at least in the case of Shanidar (Iraq), it is probable that the cave's ceiling collapsed while its inhabitants were sleeping.

Some of the aforementioned remains are not only buried intentionally, but they are accompanied by evidence of rituals. This is the case of the Kebara skeleton (Israel), which, despite being excellently preserved—it even includes the hyoid bone—is lacking the cranium. Everything suggests that the absence of the cranium is due to deliberate action carried out many months after the individual died (Bar-Yosef & Vandermeersch, 1993). It is difficult to imagine a different taphonomic explanation. Bar-Yosef and Vandermeersch (1993) wondered about the reasons for such an action, suggesting that the answer might lie in a religious ritual.

A Neanderthal tomb with an infantile specimen was found in the Dederiyeh cave (Syria), 400 km north of Damascus. Takeru Akazawa and colleagues (1995) interpreted the burial as an indication of the existence of a ritual. The reason behind this argument is the posture in which the specimen was deposited in the tomb. The excellently preserved skeleton was found with extended arms and flexed legs. Mousterian lithic industry also turned up in the cave, which Akazawa et al. (1995) associated with that from Kebara and Tabun B, though there were few tools at the burial level. An almost rectangular limestone rock was placed on the skeleton's cranium and a small triangular piece of flint appeared where the heart had once been. Although Akazawa and colleagues (1995) did not elaborate an interpretation of these findings, they implicitly suggest that these objects had ritual significance.

The Shanidar IV specimen is one of the most frequent references in relation to ritual behaviors. The discovery of substantial amounts of pollen at the tomb was interpreted as evidence of an intentional floral offering (Leroi-Gourhan, 1975). If this were the case, it would represent the beginning of a custom that lasts today. It must not be forgotten either that two of the Shanidar crania, I and V, show a deformation that was attributed to esthetic or cultural motives. However, Chris Stringer and Eric Trinkaus (1981) indicated that the specimens had been reconstructed incorrectly and that the shape of the first one was due to pathological circumstances. In his study of the Shanidar IV burial, Ralph Solecki (975) argued that there is no evidence of an intentional deposit of flowers at the burial; the pollen must have been deposited there in a natural way by the wind. Supporting the notion of an unintentional presence, Robert Gargett (1989) suggested that the pollen could have been introduced simply by the boots of the workers at the cave's excavation. Paul Mellars (1996) believes that the accidental presence of objects at French burial sites, such as La Ferrassie o Le Moustier, is inevitable: the tombs were opened at places in which faunal remains and Mousterian utensils were abundant.

The Teshik-Tash site (Uzbekistan), located on high and precipitous terrain, contains an infantile burial associated with wild goat crania. According to Hallam Movius (1953), the horns formed a circle around the tomb. This would support a symbolic purpose and a ritual content associated with the burial. Currently, however, even those who favor Neanderthals as individuals with remarkable cognitive capacities are quite skeptical about the presumed intentional arrangement of the crania (Trinkaus & Shipman, 1993; Akazawa et al., 1995; Mellars, 1996).

Neanderthal burials can be interpreted as a functional response to the need of disposing of the bodies, even if only for hygienic reasons. But they could also be understood as the reflection of transcendent thinking, beyond the simple human motivation of preserving the bodies of deceased loved ones. According to Mellars (1996, p. 381), "we must assume that the act of deliberate burial implies the existence of some kind of strong social or emotional bonds within Neanderthal societies." However, Mellars believes that there is no evidence of rituals or other symbolic elements in those tombs. The appearance of such evidence would demonstrate that Neanderthals were capable of religious thinking. Similarly, Gargett (1989) argued that the evidence of Neanderthal burials is much more solid than the evidence of offerings or rituals. Julien Riel-Salvatore and Geoffrey A. Clark (2001) have noted that applying Gargett's criterion to the early Upper Paleolithic would also lead to doubting the intentionality of the first modern human burials. They believe that there is a continuity, regarding the tombs, between the Middle and early Upper Paleolithic archaeological records. True differences do not appear until the late phase of the Upper Paleolithic (20–10 Ka).

However, Neanderthal burials contrast sharply with the burials made by modern humans, living approximately at the same time. The differences are especially illustrative in the Near East. The only intentional, and potentially symbolic, funerary Middle Paleolithic objects are the bovid and pig remains found in burials at Qafzeh and Skuhl (Mellars, 1996). Both appeared in modern human sites. Taking into account that humans

---

Box 11.27 Tombs outside caves

William Noble and Ian Davidson (1996) stressed that Neanderthal burials have not been found outside caves. In contrast, there are examples of very early human tombs in open terrains at places such as Lake Mungo (Australia), Dolni Vestonice (Czech Republic), and Sungir (Russia). In Noble and Davidson's (1996) view, the appearance of a Neanderthal tomb outside the caves would be the best proof that this is an intentional burial. For now, known tombs provide no conclusive clues about Neanderthal self-awareness, not to speak of their religion.

---

and Neanderthals living at those sites shared the same Mousterian tradition, this is a significant difference. It not only has to do with the manufacture of objects, but with much more subtle aspects, which are associated with mental processes like symbolism, esthetics, or religious beliefs.

11.3.5 Non-utilitarian objects

The second type of symbolic behavior indicated by Erella Hovers, Bernard Vandermeersch, and Ofer Bar-Yosef (1997) refers to the production of non-utilitarian objects. The analysis of such objects is, in fact, the core of discussions about the meaning and scope of symbolism.

When dealing with the evolution of symbolic behavior it is necessary to understand what exactly is a symbol. It is common to accept that an artifact should lack practical use as a tool in order to have a symbolic character. There is an almost unanimous agreement that drawings, engravings, and paintings are symbolic objects, but beyond that the consensus dissolves. As one of the most prolific authors on the evolution of symbolism, Robert Bednarik (1997b, p. 147), has maintained: "Previous reviews of pre-Upper Palaeolithic traces of symbolling behaviour and their critiques have usually considered perceived evidence of several types: the cognitive aspects of lithic typology, human burials, intentional deposition of human and other animal remains, language ability, and 'art.'" Although these are functional features, all these behaviors provide empirical evidences that can be found in the fossil and archaeological records. However, in the same work, Bednarik warns that "past discussions have shown that, of all of these types of evidence, only the last mentioned can provide reasonably unambiguous information" (Bednarik, 1997b, p. 147).

Bednarik and most experts on the evolution of symbolisim include in the "art" concept the precedent of what today is understood as artistic creation, i.e., the so-called palaeoart. This includes non-utilitarian objects of anthropic collection or manipulation, objects whose presence could be traced back to more than 2 million years. For example, a jasperita stone with holes produced naturally, but which might remind one of a human face, was found in 1925 in Makapansgat (South Africa) on terrain of c. 3 Ma (Oakley, 1981). While this piece does not show traces of anthropic manipulation, the review by James B. Harrod in 2014 of the possible oldest symbolic artifacts granted that status to the stone. It was included in the list of various samples of Oldowan symbolic behavior shown in Table 11.7.

According to Harrod (2014, p. 147) "Considering the seven stone artifacts from the Oldowan of Olduvai Gorge and Koobi Fora, the strongest candidates for palaeoart, in chronological order, appear to be the FxJj1 'broken core' with inner rhomboid shape, ~1.87 million years ago, the FLK North grooved and pecked cobble, ~1.80 million years ago, and the MNK Main subspheroid, ~1.5–1.6 million years ago." The reason why these objects are considered candidates to represent the oldest symbolic behavior is because they meet some (or all) of the following conditions: they are non-utilitarian pieces; there are no indications of anthropogenic manipulation, or of curation (see Box 11.28); they are geometric artifacts; they are exotic or accidentally decorated tools (Harrod, 2014).

Determining the degree to which a very ancient object fulfills the necessary conditions to be considered palaeoart is not easy. For example, Harrod (2014, p. 147) maintained that "if the FLK North grooved and pecked cobble could be re-examined and confirmed as non-utilitarian and the markings to one degree or another intentional, this would be the earliest evidence for intentional hominin-made palaeoart mark making. It might be categorized as the earliest petroglyph, one that combines both geometric shapes and glyph-like motifs. Further, the apparent two sets of four and two incised dots ('cupules'), if confirmed to have been intentional, could be interpreted as the earliest evidence of symbolic representation of numerosity (comparison of arithmetic sets and larger and smaller size shapes) in human evolution." Such an abundance of doubts and conditionals highlights the essential question about the hypothetical symbolic meaning of such artifacts: the attribution of intention. To confirm or reject that these pieces fall into the category of palaeoart, it

Table 11.7 Putative candidates for Oldowan symbolic behavior (Harrod, 2014; modified—the original table includes mortuary practice and spoken language not included here)

Symbolic behavior	Site and date	Artifact
Curation of exotic object + exotic tool + geometric shape	FxJj1 (KBS), Koobi Fora, ~1.87 Ma, Classic Oldowan	Broken/irregular chopper ("rhomboid-in-a core") on basalt pebble core, curated
Rock art, glyph-like motifs + geometric shaped artifact (numerosity?)	FLK North 1, Upper Bed I, ~1.80 Ma, Olduvai Gorge, Classic Oldowan	Artificially grooved and pecked phonolite cobble (incised circular groove, cupules)
	MNK Main, Middle Bed II, ~1.5–1.6 Ma, Olduvai Gorge, Dev. Oldowan/Early Acheulian 6.6 kg (14.5 lb)	Subspheroid with apparent natural marking (dot and undulating line), possibly framed by hexagonal flaking
Geometric shaped artifact	FLK North Clay—Root Casts, Lower Bed II, ~1.74–1.80 Ma	Anvil, hexagonal block, quartzite (battered or flaked shape?)
	FLK North Sandy Congl. 6, Middle Bed II, ~1.6–1.66/1.74 Ma, Developed Oldowan A	Anvil, hexagonal block, hornblende gneiss (battered or flaked shape?)
Exotic tool, non-iconic ("exaggerated anvil pit") + geometric shaped (concentric circles) or rock art (cupule)	FLK North 1, Upper Bed I, ~1.80 Ma, Olduvai Gorge, Classic Oldowan	Pitted anvil, conical block, pecked pit 29 × 17 mm, depth 9 mm
	FLK North Sandy Conglomerate 6, Middle Bed II, ~1.6–1.66/1.74 Ma, Developed Oldowan A	"Unusual anvil," with pit 35 × 24 mm, depth 5 mm in center of one face
Pigment use	BK, Upper Bed II, c. 1.48 Ma, Olduvai Gorge, Dev. Oldowan B/Early Acheulian	Two (2) lumps of non-local "red ocher," but reanalyzed as local red volcanic tuff (no evidence of use)

Box 11.28 Curation

Curation is the act of organizing and preserving an art and artifact collection. It is also called "manuport," which alludes to objects transported by anthropic intervention. The fact of their transportation evidences an interest in them.

is necessary to specify the intentions of those who collected or manipulated them, which requires the use of indirect evidence in entirely speculative terms.

In an attempt to eliminate speculation from palaeo-art analysis, Robert Bednarik (2008) put forward the need to distinguish between two different types of symbolism in the icons that appear in the archaeological record:

(1) A simple symbolism based on the recognition of an object as an icon "in the sense that an organism capable of cognitively perceiving visual ambiguity detects at least some meaning without any cultural faculties coming into play" (Bednarik, 2008, p. 85). This would be the symbolism characteristic of the appreciation of geometric or colorful natural shapes; a capacity shared with other animals.
(2) An advanced symbolism, "requiring the link between referent and referrer to be negotiated culturally." This type of symbolism implies "the cultural and intentional creation of features prompting visual responses to a signifier" (Bednarik, 2008, p. 85, both quotations).

According to the classification of Bednarik (2008), all that can be said of the oldest samples of palaeoart—as indicated by Harrod (2014)—is that they would belong to the simple symbolism category. They would be the expression of "natural" cognitive capabilities, in the sense that they are not subject to cultural modification. The "symbolic" icons, qualified as such by their geometric or colorful shape, fall into this category. For example, the red and pink quartzite biface found in Sima de los Huesos (Atapuerca, Spain) and interpreted by Marie-Hélène Moncel and collaborators (2012) as typical of the origins of symbolic behavior, belongs to this category. To be able to speak of advanced symbolism it is necessary to go beyond a preference for certain shapes and colors by adding the link referent–referrer that is characteristic of a specific cultural tradition. Neither domes (rounded concavities) nor the use of pigments are subject to this requirement. Hence, Bednarik (2008) reduces the evidence of advanced symbolism in its early manifestations to geometric engravings and perforated shells.

Bednarik's distinction between simple and advanced symbolism serves to guide the analysis of the first icons with anthropic modifications. However, the modern mind includes much more advanced symbolic icons, with the relevant example of figurative polychromes from the caves of the Upper Paleolithic. Therefore, in order to compare the cognitive capabilities of Neanderthals and modern humans, it is convenient to increase the number of degrees of symbolic development. Here we consider four degrees:

- Symbolism Mode 1 (SM1, natural icons requiring an attribution of intention to be considered symbols).
- Symbolism Mode 2 (SM2, modified icons such as geometric engravings in stone or bone).
- Symbolism Mode 3 (SM3, a new degree of symbolism that appears with personal ornaments).
- Symbolism Mode 4 (SM4, figurative representations).

In the comparison between Neanderthals and modern humans it is unnecessary to address the SM1; both species have ample capacity to detect symbolic conditions (shapes and colors) in natural, non-manipulated objects. We will proceed with SM2.

11.3.6 Symbolism Mode 2 (SM2)

The most important icons of SM2 are modified geometric engravings made on stone or bone. Leaving aside the question of whether the geometric shapes of bifaces might have been appreciated as symbols (which, therefore, would belong in SM1), the oldest objects with engravings appear in Acheulean sites (Mode 2). Some of the pieces from the Bilzingsleben site (Germany), which has yielded *H. erectus* specimens

---

Box 11.29 Modes of symbolism

It is obvious that the naming of symbolic modes we propose follows the well-known classification of cultural modes. However, we do not intend to assert here that there is a close correspondence between cultural and symbolic modes. The clearest relationship between them is the existing difficulty to assign both cultural traditions and symbolic degrees to certain hominin species. Beyond that, the absolute disparity in quantitative terms in the archaeological record between cultural artifacts in the broader sense (tools) and symbolic objects, makes it difficult to establish links between them.

Figure 11.30 Geometrical marks performed on an elephant metatarsal from the Bilzingsleben Acheulean site (Germany, c. 350,000 years) (illustration from Bednarik, 1995).

and has been dated to around 350,000 years, show geometrically arranged lines (Bednarick, 1995; an example is Figure 11.30).

The geometric motifs reappear in various isolated objects from sites with carvings of Mode 3 (Mousterian). An artifact found in a Mousterian site at Quneitra (Golan Heights, between Israel and Syria) is an example of this. It is an approximately triangular plate of flint cortex, about 7.2 cm high (Goren-Inbar, 1990) (Figure 11.31). The site's age, estimated by means of electronic spin resonance applied on bovid dental enamel, is close to 54,000 years (Ziaei et al., 1990). The Quneitra fragment shows a set of marks that were described as the earliest sample of a representation in the form of an engraving (Goren-Inbar, 1990). The microscopic examination of the piece revealed a set of four concentric semicircles carefully carved and surrounded by angular lines that roughly follow their form, together with other vertical lines on the right-hand side. A sinuous line traces the shape of the broken right side of the flint (Marshack, 1996).

Marshack's (1996) interpretation of the Quneitra fragment was based on how fast and how many incisions were made in order to produce the different lines on the flint plate. According to Marshack, the vertical lines indicate a clear intention of covering the entire available surface. Some of them were made by means of a long blow followed by a shorter one in order to reach the edge of the plate. The sinuous line of the plate's right-hand border shows that the intention was to mark the trajectory of the flint's side (Figure 11.32).

No human remains have been found in Quneitra, making it impossible to directly assign the manipulation of the piece to Neanderthals or modern humans that, as we know, met in the Near East.

An identification of the populations associated with the site was made by Maya Oron and Naama Goren-Inbar (2014) using a computational process called GSI (Geographic Information System), which carried out a spatial analysis of lithic and paleontological assemblages. The conclusion of Oron and Goren-Inbar (2014, p. 201) asserts that "the results of this study augment our knowledge of Middle Paleolithic sites by attesting to different patterns of activity areas that are more complex and dynamic than those previously discerned. This conclusion supports a general division of activities, similar to those seen for example in Amud Cave (. . .) or at Kebara Cave (. . .), but with spatial patterns pertaining specifically to an open-air site of short term duration, as opposed to trends of continuity which are typical for cave use." Both Amud and Kebara are sites of undoubted Neanderthal occupation.

The importance of the marks left on the Bilzingsleben bones and the Levantine artifacts resides in the fact that they show that geometrical designs were created by hominins throughout a long span of time. Some explanations for the origin of the earliest iconic representations suggest they could be randomly produced

Figure 11.31 Quneitra artifact (Syria, c. 54,000 years) (illustration from Goran-Inbar, 1990).

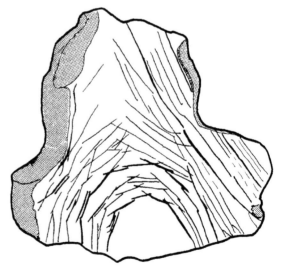

Figure 11.32 Schematic representation of the lines on the Quneitra fragment (illustration from Marshack, 1996).

marks. Only accidentally are they reminiscent of a figure, and only later are they recognized as such. Neither the Bilzingsleben nor the Quneitra signs correspond to any stylized image of animals or human beings. As Bednarik (1995) noted, the random distribution of marks left by natural processes can produce geometrical patterns. But in the case of the Bilzingsleben objects, the relation between the spatial distribution of the marks and the available space on the bone argue against the absence of intentionality (Bednarik, 1995). This is even truer of the Quneitra plate, which in no way could be considered a personal decorative object, such as a pendant or a piece of a necklace (Box 11.30). The concentric circle motif did not appear again until the European Upper Paleolithic, 25,000 years later, and only occasionally. The Quneitra artifact is unique in that respect, unparalleled at the time.

The cognitive complexity necessary for manipulating a fragment such as the Quneitra plate must not be interpreted as artistic capability. The visual-motor coordination required to carry out the hand movements for engraving the lines on the plate is also a necessary prerequisite to create Acheulean bifaces and, to a certain extent, the most primitive Oldowan instruments. What distinguishes the Bilzingsleben bone engravings and the Quneitra stone engravings from other manufactured objects is, as we said earlier, the absence of any kind of useful value, the fact that their only purpose seems to be symbolic.

The Quneitra piece, likely of Neanderthal manufacture, is also reminiscent of some stones from Mousterian sites (with Levallois' technique), which are attributed to modern humans. Erella Hovers, Bernard Vandermeersch, and Ofer Bar-Yosef (1997) reported the finding in the Qafzeh cave (Near East) of a stone fragment with geometric markings (Figure 11.33). After analysis of the piece, the authors suggested that "the Qafzeh object was intentionally engraved and is probably of non-utilitarian character." The Qafzeh layers XXIV–XVII include remains identified as modern human. The engraved piece comes from layer XVII, dated at c. 100 Ka and was found near the Qafzeh 8 tomb (Hovers et al., 1997).

Comparing the pieces from Quneitra and Qazfeh reveals that the cognitive capacity necessary for manipulating such fragments is similar. Even the first piece, most likely manufactured by Neanderthals, seems to have a greater degree of complexity than the second, which is attributed to modern humans. It can be concluded that, besides the Mousterian technology (Mode 2), both species shared the Near East symbolic mode SM2. But the Mousterian levels of Qazfeh and Skuhl—both with fossils of modern humans—also contain perforated shells (Hovers et al., 1997; Vanhaeren et al., 2006), often associated with the manufacture of necklaces, that is to say, decorative pieces for personal use that would be part of SM3. The simultaneous presence of objects from SM2 and SM3, which can provide clues about the transition of one symbolic level to another, is in itself characteristic of some sites of the latter stages of the South African MSA.

Box 11.30 The intentionality of the Quneitra piece

Marshack (1996) believes that the engraving process indicates a surprising cognitive complexity of the author of the lines on the Quneitra artifact. Far from being a random carving, it shows an intentional trend to center the semicircles that had to be kept in mind while the stone was being turned for making the marks. According to Marshack, the required technique was complex and required precise coordination of hand movements, under the supervision of the visual system and following a preconceived plan. Very sophisticated cognitive capacities are necessary to carry out a designed plan by means of hand coordination. Marshack talked about a true "gestalt" to produce the lines of the engraving in accordance with the shape and size of the flint.

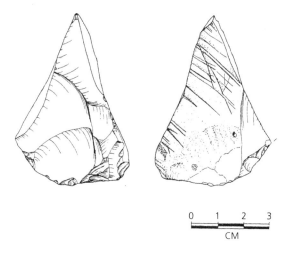

Figure 11.33 The incised artifact of Qafzeh: left, flaking surface; right, cortical surface (Hovers et al., 1997, modified). Hovers, E., Vandermeersch, B., & Bar-Yosef, O. (1997). A Middle Palaeolithic engraved artefact from Qafzeh Cave, Israel. *Rock Art Research*, 14, 79–87.

11.3.7 The transition to Symbolism Mode 3 (SM3)

As we have mentioned several times, the final stages of cultural evolution within the MSA correspond in South Africa to the traditions of Still Bay (SB) and Howieson's Poort (HP). We have also warned that both SB and HP show different innovations, such as lanceolate points and composite weapons, which previously were associated only with the technology of the Upper Paleolithic. The South African site of the Blombos cave includes the presence of engravings, pigments, and perforated beads.

The various patterns identified by Christopher Henshilwood and collaborators (2009) of the marks on the ocher pieces from the Blombos MSA phases M1, M2, and M3, document for the first time the existence of a changing cultural tradition of engraving processes (Henshilwood et al., 2009). As we might recall, Robert Bednarik's criteria for advanced symbolism specially included the use of traditions determined by culture.

Thus, the engravings of Blombos should be considered characteristic of the symbolism of SM3. Without using the classification we propose, Christopher Henshilwood, Francesco d'Errico, and Ian Watts (2009) pointed out, "The fact that they were created, that most of them are deliberate and were made with representational intent, strongly suggests they functioned as artefacts within a society where behavior was mediated by symbols" (p. 45).

Besides geometric engravings, the MSA levels of the Blombos cave contain perforated shells of *Nassarius krausianus* (Figure 11.34). A total of 41 pieces, 39 of them from the levels of the M1 phase dated at c. 75 ka (Henshilwood et al., 2004) and the remaining two pieces from the M2 levels with a provisional dating of c. 78 Ka (d'Errrico et al., 2005). The *N. krausianus* perforated shells from the most technologically advanced levels that belong to the African LSA have been found in both the Blombos cave and Die Kelders.

As we saw in Chapter 10, the population in the Klasies River site (South Africa), which exhibits the final

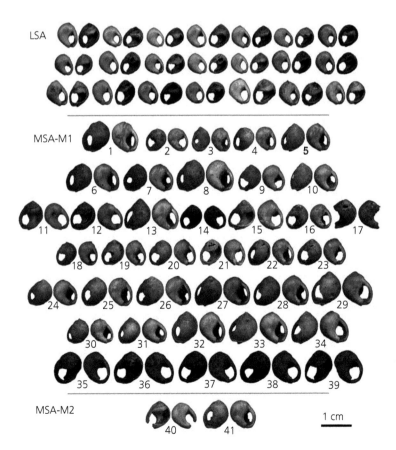

Figure 11.34 Perforated *Nassarius kraussianus* beads from LSA layers and MSA Phases M1 and M2 at the Blombos cave. Reprinted from d'Errico, F., Henshilwood, C., Vanhaeren, M., & van Niekerk, K. (2005). Nassarius kraussianus shell beads from Blombos Cave: evidence for symbolic behaviour in the Middle Stone Age. *Journal of Human Evolution*, 48(1), 3–24, with permission from Elsevier.

Box 11.31 Intentionality in the perforated shells

The holes in the shells can be caused naturally, without anthropic manipulation (see examples in Anderson, 2009). Such shells, already perforated, could have been collected and used for decorative purposes, so the most important question is to determine their character in order to obtain evidence about their use. Francesco d'Errico et al. (2005, p. 16) have indicated that "Microscopic analysis of MSA tick shells reveals a distinct use-wear, absent on LSA beads and natural shells, consisting of facets that flatten the outer lip or create a concave surface on the lip close to the anterior canal." The traces of pigment detected by d'Errico et al. (2009) on the shells found in France (Grotte des Pigeons) and North Africa (Rhafas, Contrebandiers, and Ifri n'Ammar) add evidence about their personal use for decorative purposes.

stages of South African MSA, can most likely be assigned to *Homo sapiens*. In addition, the Blombos cave shows a sequence of cultural changes associated with the final stages of the MSA and the emergence of the LSA, which documents the presence of SM3, since c. 75 Ka. This date indicates that the cognitive capacity of our own species, i.e., what we have described as the "modern mind," was achieved in Africa tens of thousands of years before the arrival of modern humans in Europe.

In short, we can say that the cognitive capacity of our own species, identified today as the distinctively human process of thought, and the evidence gathered in South Africa, identify the following features as characteristic of the modern mind:

• Achieved by *Homo sapiens* around 75 Ka.
• Corresponds to the symbolic stage we call SM3.
• Includes the use of personal decorative elements.

SM4, which adds figurative elements to the above symbolic representations, also has an ancient presence in Africa. From the Apollo cave (Namibia) come seven pieces of quartzite with paintings depicting animals (Wendt, 1976). However, its dating would be close to, or even later than, the Chauvet cave (France) (Anderson, 2009).

With regard to Europe, SM4 has been found at dates close to the arrival of Cro-Magnons on the continent. The process of transition from SM3 to SM4 can be very complex, but let's not anticipate events.

11.3.8 Population encounters in Europe

Cultural traditions commonly attributed to modern humans entering Europe are, cited in order of antiquity, the industries Aurignacian, Gravettian, Solutrean, and Magdalenian. The Aurignacian culture was defined by Edouard Lartet (1860) in accordance with the tools found at the site of Aurignac (French Pyrenees), but is also assigned to similar industry sets from large parts of Eastern, Western, and Central Europe, and also to some of the existing technocomplexes in parts of the Middle East (Mellars, 2006a). The technological level of Aurignacian contrasts with the Mousterian Mode 3 of Neanderthals. The most obvious difference is the use of microbladelet technology, representing a major difference from the most advanced tools up to that time. If Châtelperronian points, as well as bladelets, fulfill the same purpose, to serve as projectiles, the Aurignacian microliths are "serially, laterally hafted along the shaft of projectiles, not mounted at their extremities" (Bon, 2006, p. 141).

The need to compare the technocomplexes of Mode 3 and Aurignacian led to distinguishing the latter as belonging to a homogeneous tradition, following the initial descriptions, such as those by Abbe Breuil (1913). However, technological analysis determined the presence of different forms of tool production in the Aurignacian. For the initial carvings, in contrast with Mode 3, different names were proposed—"Classic Aurignacian," "Aurignacian I," "Proto-Aurignacian," "Early Aurignacian," "Pre-Aurignacian," "Archaic Aurignacian," "Initial Aurignacian"—and applied to techniques that, in some instances, are quite similar. Some authors have even suggested that between the initial carvings of Mode 4 and the most advanced Aurignacian there are very few differences (Nejman, 2008). However, François Bon (2006) has argued that, from a technological point of view, two different systems can be distinguished, which the author called "Archaic (or Proto) Aurignacian" and "Early Aurignacian." We will maintain the same distinction.

The difference indicated by Bon (2006) consists in that only one *chaîne opératoire* is required to obtain Archaic Aurignacian tools, meanwhile two distinct *chaînes opératoires* are required to obtain blades and bladelets of the Early Aurignacian. As William Banks, Francesco d'Errico, and Joao Zilhão (2013, p. 41) have stated: "For the Proto-Aurignacian, blades and bladelets were produced from unidirectional prismatic cores within a single, continuous reduction sequence (. . .). During the Early Aurignacian, blades and bladelets were produced via two distinct core reduction strategies. Blades continued to be produced from prismatic cores, were robust, and were typically heavily retouched on their lateral edges. Carinated 'scrapers' served as specialized cores whose reduction yielded short, straight, or curved bladelets that were typically left unretouched. The Early Aurignacian is also characterized by the appearance of split-based bone points."

According to the available archaeological evidence, modern humans would have followed two different routes into Europe: the Danube route indicated by the Aurignacian technology, and the Mediterranean route, where the Proto-Aurignacian industry appears (Mellars, 2005) (Figure 11.35).

The oldest documented presence of Archaic or Proto-Aurignacian is found before the cold Heinrich Event 4 (HE4) (c. 40 Ka) in northern Spain (El Castillo, Cantabria, level 18, 41–38 Ka and l'Arbreda, Catalonia, level 11, 41–39 Ka), and northern Italia (Paina 38.6–37.9 and Fumane 36.8–32.1 Ka) (Kozlowski & Otte, 2000), southern France (Isturitz 37.18±4.2 Ka) (Szmidt et al., 2010), and Moravia (Brno-Bohunice, c. 48 Ka) (Hoffecker, 2009). The oldest evidence of the Early Aurignacian would be of almost 34 Ka in France (Castanet

Lower, Combe Saunière VIII, Flageolet I XI, Pataud 11 and 13, Roc de Combe 7c, and Tuto de Camalhot 70e8) and Germany (Geissenklösterle IIIa and Wildscheuer III) (Banks et al., 2013). The lowest strata of Hohle Fels cave, which ranged between 36 and 33 Ka, contains ivory sculptures in addition to facies of Early Aurignacian (Conard, 2003). Banks et al. (2013, Table 2) give radiocarbon calibrated dates associated with the Proto-Aurignacian and Early Aurignacian sites in Europe (see Figure 11.36).

The biggest obstacle to give meaning to the Archaic and Early Aurignacian cultures is the absence of associated fossils. Therefore, as stated by Joao Zilhao (2006), these industries can be attributed to both Neanderthals and modern humans. We have, then, a problem for the identification of the "modern mind." If we associate it to the technological level of cultures immediately preceding the advanced Aurignacian, Neanderthals could also have had that cognitive capacity. But, if we relate the modern mind with realistic representations of figurative cave art, these only appear at the end of the Aurignacian; early modern humans would then have lacked such a capacity. To determine to what

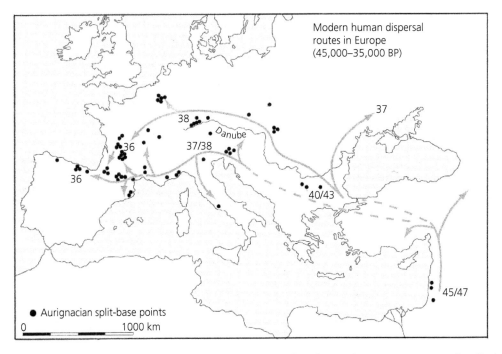

Figure 11.35 Apparent dispersal routes of the earliest anatomically and behaviorally modern populations across Europe, as reflected in the archeological data. Figures indicate the earliest radiocarbon dates for these technologies in different areas, expressed in thousands of radiocarbon years (between 2,000 and 4,000 years less than calendrical ages). Mellars, P. (2005). The Impossible Coincidence. A Single-Species Model for the Origins of Modern Human Behavior in Europe. *Evolutionary Anthropology*, 14, 12–27.

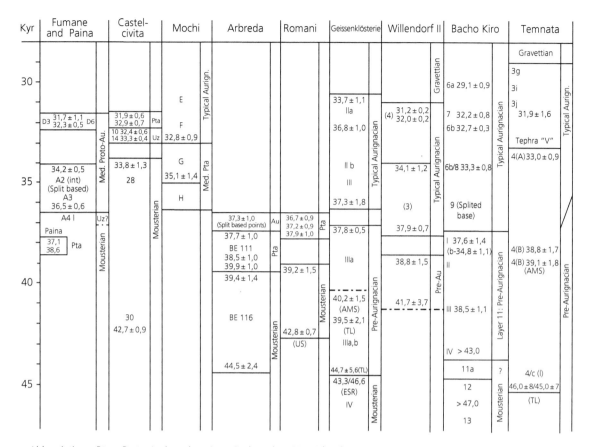

Abbreviations: Pta = Proto-Aurignacian, Au = Aurignacian, Uz = Uluzzian.

Figure 11.36 Some of the most important Aurignacian sequences of Mediterranean Europe, Central Europe, and the Balkans. Kozlowski, J. K. & Otte, M. (2000). The Formation of the Aurignacian in Europe. *Journal of Anthropological Research*, 56(4), 513–534.

extent there are comparable cognitive capabilities, the best procedure is to compare the so-called transitional industries.

11.3.9 Transitional industries

According to Ivor Jankovic and collaborators (2006, p. 461), the so-called transitional industries "include the Châtelperronian of France and northern Spain, Szeletian and Jankovichian of central and parts of eastern Europe, Uluzzian of Italy (Tuscany, Calabria, southern Adriatic part, Uluzzo Bay, etc.), Streletskian of eastern Europe, Jerzmanowician of eastern Germany and Poland, Althmulian of southern Germany, Bohunician of Czech Republic, Brynzeny and Kostenki Szeletian of Russia and several other unnamed or site-specific assemblages from Poland, Slovakia,

Czech Republic, Romania, etc." They are called "transitional" because they contain elements of the Middle Paleolithic (Mousterian and Micoquian), absent in the Early Aurignacian, such as curved-backed points and foliate points (Kozlowski & Otte, 2000), but also tools that are considered characteristic of the Upper Paleolithic, such as carinated scrapers or bone points. David Brose and Milford Wolpoff (1971) provide a long list of Upper Paleolithic utensils found in Middle Paleolithic contexts.

The problem of the transitional industries appears when we need to assign them to a species. As was indicated by Jankovic and collaborators (2011, p. 304) "even if we accept the earliest Aurignacian as an industrial complex that has its origins outside this area (...) (which is far from proven) and attribute it to anatomically modern newcomers (for which there are no known hominin/ industrial associations) we are left with the problem of who is responsible for these Initial Upper Paleolithic assemblages." The absence of fossil remains associated with almost every transitional technocomplex generally prevents the association of hominin/industry, and of confirming who were the architects of this cultural change. However, two sites with Châtelperronian culture, Saint-Césaire (c. 36 Ka) (Lévêque & Vandermeersch, 1980; Mercier et al., 1991) and Arcy-sur-Cure (c. 34 Ka) (Hublin et al., 1996), contain in the same stratigraphic-level fossils of *H. neanderthalensis* (questioned by: Bar-Yosef & Bordes, 2010; Higham et al., 2010). This coincidence has been at times enough to attribute all transitional industries to Neanderthals (Allsworth-Jones, 1986; Stringer & Gamble, 1993; Mellars, 1996); a consideration sustained in some revisions of specialists (Churchill & Smith, 2000; Francesco d'Errico, 2003).

The general assignment of transitional industries to the Neanderthals encounters the problem of the morphology of fossil specimens recently found at Uluzzian levels. In the Grotta di Fumane (Lessini Mountains, North Italy) several human teeth have been found: Fumane, 1, 4, 5, deciduous; Fumane 6, adult. Stefano Benazzi et al. (2014) classified Fumane 1 as clearly Neanderthal and Fumane 5 as supporting Neanderthal affinity. Both specimens come from the Mousterian levels of Fumane. At the same time, Fumane 6, of the Uluzzian levels, does not show morphological features useful for taxonomic discrimination (Benazzi et al., 2014). The Fumane fossil specimens, therefore, did not seem to contradict the general attribution of Uluzzian to Neanderthals. However, a later analysis of the mtDNA of Fumane 2 indicated that it fits within the range of variation of modern humans (Benazzi et al., 2015). In the same article, a tooth from Riparo Bombrini (western Ligurian Alps, Italy), a lower left lateral deciduous incisor, is attributed to a modern human because of the relatively thick enamel. Moreover, new analysis by Stefano Benazzi and collaborators (2011) of

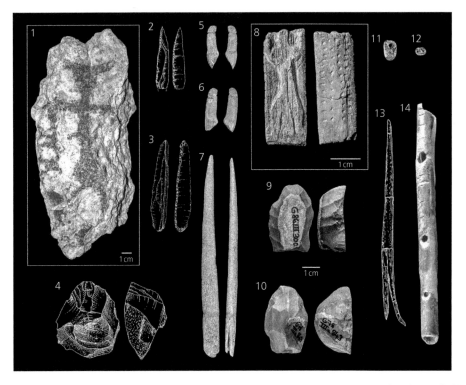

Figure 11.37 Protoaurignacian (Uluzzian) finds from Grotta di Fumane (1–7) are compared to early Aurignacian finds from Geißenklösterle (8–14). (1) Therianthropic painting on limestone block; (2, 3) points with abrupt retouch; (4, 9, 10) carinated scrapers; (5, 6) grooved cervid teeth; (7, 13) split-based bone points; (8) therianthropic relief carved from ivory; (11, 12) personal ornaments made from ivory; (14) bone flute made from a swan radius. Conard, N. J. & Bolus, M. (2015). Chronicling modern human's arrival in Europe. *Science*, 348(6236), 754–756 (Panels 1–7, courtesy of Marco Peresani; Panels 9–14, courtesy of University of Tübingen).

two deciduous molars from the Uluzzian levels (EIII) of the Grotta del Cavallo (Apulia, southern Italy), one initially classified as a Neanderthal, leads to different conclusions. By means of morphometric methods based on microtomographic data, Benazzi et al. (2011) stated that the Cavallo specimens can be attributed to modern humans. Also, in the EIII level of the Grotta del Cavallo appeared several marine shells (*Dentalium* sp., *Nuculana* sp., and *Cyclope neritea*) snapped or pierced to be transformed into beads.

If the Uluzzian technocomplex, very ancient, is the production of modern humans, we find ourselves with the possibility of establishing plausible dates for the entry into Europe of *H. sapiens*. The Grotta di Fumane (Lessini Mountains, North Italy) contains levels of the late Mousterian (A11, A5), Uluzzian (A4, A3), and Proto-Aurignacian (A2, A1 up to D3) technocomplex (Benazzi et al., 2014) (Figure 11.37). Fumane Mousterian levels were dated by calibrated radiocarbon between 45.4–41.7 (A11) and 38.875±1.497 Ka (A5), while the Uluzzian level (A4) received 37.8–36.9 Ka (Peresani et al., 2008). Applying a development of radiocarbon dating (acid–base–oxidation-stepped combustion—ABOx-SC—and acid–base–acid—ABA—pretreatments for removing contaminants, then accelerator mass spectrometry—AMS), Thomas Higham and collaborators (2009) increased the age of the fossils of Cavallo. The age of the Proto-Aurignacian A2 level would be 41.20–40.45 Ka, i.e., prior to the Campanian Ignimbrite eruption. The latest Mousterian occupation (A5) would be 43.58–42.98 Ka, and the Uluzzian levels should be found between that date and 41.20–40.45 Ka. The analysis by Katerina Douka and collaborators (2014) pushed back even further the age of the Uluzzian. By an integrated synthesis of new radiocarbon results and a Bayesian statistical approach from four stratified Uluzzian cave sequences in Italy and Greece (Cavallo, Fumane, Castelcivita, and Klissoura 1), Douka et al. (2014) concluded that the Uluzzian arrived in Italy and Greece shortly before 45 Ka. Its final stages are c. 39.5 Ka, coinciding with the Campanian Ignimbrite eruption. Fumane dates agree with that of the Grotta del Cavallo. Benazzi et al. (2011) dated the Cavallo shells by AMS radiocarbon at an age of 45.01–43.38 Ka.

The latest scenario presents, therefore, the arrival of Uluzzian technocomplexes—i.e., of modern humans—in Italy and Greece, with the modern mind necessary to use personal ornaments (beads), shortly before 45 Ka; a date old enough to match the Châtelperronian levels of Neanderthals in France and northern Spain. Additionally, beads and ocher pigments also appear at the Châtelperronian sites. In addition to Châtelperronian

tools constructed in situ, the Grotte du Renne (Arcy-sur-Cure, France)—a site inhabited by Neanderthals—has yielded a series of up to 36 objects such as carved ivory pieces and perforated bones, the sole purpose of which must have been decorative (Hublin et al., 1996). Since 1949 Leroi-Gourhan carried out studies that revealed important differences between the engraving techniques used to produce the Arcy-sur-Cure Châtelperronian artifacts and the latest Aurignacian utensils that were found in the most modern strata of the same cave (Leroi-Gourhan, 1958, 1961). Hence, the Châtelperronian (Neanderthal) and Aurignacian (modern human) cultures were different.

11.3.10 The hypothesis of SM3 acculturation

One of the main aspects that characterize the modern mind is, as we have seen, the common use by *Homo sapiens* of ocher pigment and decorative objects. Such use is shown by objects coming from both South Africa and the European Aurignacian sites. However, the presence in the Châtelperronian levels of Arcy-sur-Cure of significant amounts of red and black pigments, some of them shaped by grinding to obtain crayons (d'Errico et al., 2009), indicate that Neanderthals would also have had the cognitive level associated with the use of pigments. This is also the case of a set of perforated shells and bones, whose obvious use is to serve as pendants, coming from the same site of Grotte du Renne. In both respects, Neanderthals would have reached the SM3 level, i.e., the "modern mind."

Many authors, headed by Trinkaus, Howells, and Zilhão, believe that Neanderthal cognitive abilities were as complex as those currently characteristic of our own species. The main reason to maintain this idea relies on tools and decorative objects coming from sites attributed to Neanderthals. If they were the makers, we would have evidence of the emergence, even in *H. neanderthalensis*, of the modern mind in both technological and symbolic terms. But, even if this is accepted, there would still be doubt as to whether the modern mind had been achieved independently in Neanderthals and modern humans, or rather, whether it would be a process of acculturation, of imitation of the pieces introduced in Europe by the Cro-Magnons.

Hublin and colleagues (1996) interpreted the Arcy-sur-Cure artifacts as the result of trading process rather than the result of technical imitation of modern human technology. Francesco d'Errico and colleagues (1998) arrived at a different conclusion: those objects were the result of an independent and characteristically Neanderthal cultural development, which had managed to cross the threshold of the symbolism inherent in decorative objects. There is no reason to assume that the biological differences between Neanderthals and modern humans necessarily translated into differences between their intellectual capacities. Paul Bahn (1998) also believed the Arcy-sur-Cure objects merited attributing Neanderthals a sophisticated and modern symbolic behavior.

Randall White (2001, p. 3) has offered an alternative interpretation of the decorative objects from the Grotte du Renne: "It seems implausible that [...] Neanderthals and Cro-Magnons independently and simultaneously invented personal ornaments manufactured from the same raw materials and using precisely the same techniques." Consequently, he argues that the Châtelperronian ornaments from the Grotte du Renne are Aurignacian and were produced by modern humans. The question whether the authors of the Châtelperronian culture were Neanderthals, modern humans, or both, has sparked numerous discussions. The evidence from Saint-Césaire (France), with both Middle and Upper Paleolithic strata, allowed in situ studies of the association of specimens and tools, as well as the cultural transition (Mercier et al., 1991). Norbert Mercier et al. (1991) used thermoluminescence to estimate the age of the Neanderthal specimens found in levels with Châtelperronian industry. Their results suggest they were 36,300±2700 years old. Mercier et al. (1991) argued that there was contact between Neanderthals from Western Europe and the first modern humans that arrived there. They also noted something we have said on several occasions: the straightforward identification of cultures with taxa is not possible.

Arcy-sur-Cure suggests Neanderthals were possibly capable of producing decorative objects; other sites provide evidences of cultural sharing. Ivor Karavanic and Fred Smith (1998) documented the presence of two contemporary sites at Hrvatsko Zagorje (Croatia), which are close to each other. The Vindija cave has yielded Neanderthals, while Velika Pécina has only produced remains of anatomically modern humans. The authors believed that the coincidences exhibited by the tools from both sites are due to imitation or even commercial exchange. These Croatian sites do not include ornaments, but they provide remarkable indications of cultural exchange. This is corroborated beyond a doubt by *Homo neanderthalensis* and *Homo sapiens* coincident at Palestine caves. Although the shared Near East Mousterian culture could be interpreted as the maximum horizon Neanderthals could reach, the Arcy-sur-Cure objects, assuming they were constructed or used by Neanderthals, suggest this was not the case. They seem to support the notion that Neanderthals appreciated pendants enough to identify them as "beautiful objects." At least in this sense, they would have achieved the "modern mind."

The hypothesis that Neanderthal decorative elements found in the Châtelperronian deposits are imitations of Aurignacian objects made by modern humans, implies that both cultures were contemporary or that the Aurignacian culture was older. Joao Zilhão et al. (2006) have investigated the sequence of sediments and the archeological association of the Grotte des Féees at Châtelperron (France) and reject the Châtelperronian–Aurignacian contemporaneity: They assert that "its stratification is poor and unclear, the bone assemblage is carnivore accumulated, the putative interstratified Aurignacian lens in level B4 is made up for the most part of Châtelperronian material, the upper part of the sequence is entirely disturbed, and the few Aurignacian items in levels B4-5 represent isolated intrusions into otherwise *in situ* Châtelperronian deposits" (Zilhão et al., 2006). Their conclusion is that "as elsewhere in southwestern Europe, this evidence confirms that the Aurignacian postdates the Châtelperronian and that the latter's cultural innovations are better explained as the Neanderthals' independent development of behavioral modernity" (Zilhão et al., 2006, p. 12643). This hypothesis deserves attention, but in order to be accepted, similar studies should be carried out at places other than the Grotte des Feés.

Any chronological table of the cultural sequences reveals the difficulties we are encountering. Direct correspondences are usually drawn between cultural manifestations and species, associating Mousterian with Neanderthals and Aurignacian with modern humans. Hence, it seems clear that attributing or not to Neanderthals sufficient cognitive capacities for esthetic experience is heavily influenced by a given author's point of view about the Mousterian evidence. Those who argue that Neanderthals and *Homo sapiens* belong to different species tend to reject the presence of the modern mind in the former's contrivances, and vice versa.

11.3.11 The step to Symbolism Mode 4 (SM4): from Sulawesi to Cantabria

The appearance of the highest symbolic capabilities, those of SM4, has taken a new turn, due to the discovery of figurative representations in caves such as Wallacea (Southeast Asia) and the Cantabrian Mountains (Southwest Europe), which are so widely separated.

Box 11.32 Sulawesi tools

Between 2007 and 2012, excavations were carried out in the Walanae Basin (South Sulawesi) seeking to ascertain the significance of the "Cabenge Industry" tools found there in the late 1940s. At Talepu, 3 km southeast of Cabenge, the trench T2, as many as 270 stone artifacts appeared between the surface and 4.2 m depth associated with fossils of megafauna (*Bubalus* sp., *Stegodon* and *Celebochoerus*) (van den Bergh, Li et al., 2016). Most of the T2 industries are medium- to large-sized flakes found *in situ*, recovered from deposits dated from before 200,000 to c. 100,000 years ago. As the authors said, "Our findings at Talepu attest to the presence of early tool-makers on Sulawesi by the late Middle Pleistocene, but the absence of Pleistocene human fossils on the island precludes a definitive answer as to which hominin species was first to make landfall. With regard to potential island colonizers, there are at least three candidates in the region: the known and inferred distributions of *H. floresiensis* on Flores (. . .), *H. erectus* on the southern margin of Sunda (present-day Java) (~1.5 million years ago to ~140 ka) (. . .), and 'Denisovans', whose geographic range may have extended into Wallacea" (van den Bergh, Li et al., 2016, p. 210).

Since the 1950s, approximately 90 dolomitic caves with paintings (van Heekeren, 1952) have been located in the Maros and Pangkep regions on the Sulawesi Island (Indonesia). Maxine Aubert and collaborators (2014), using uranium-series' disequilibrium dating, calculated the age of 19 coralloid speleothems associated with 12 hand stencils and two figurative paintings of animals from seven sites in Maros (Figure 11.38). Most of the samples were obtained "*in situ* so as to produce a continuous profile microstratigraphic extending from the outer surface of the coralloid through the pigment layer and into the underlying rock face" (Aubert et al., 2014). Thus, the authors were able to obtain the minimum and maximum age of the drawings, although the underlying rock provides temporal ranges of more than 140 Ka. However, "in some cases, hand stencils and paintings made over coralloids that were then continued to grow, providing an opportunity to obtain both minimum and maximum ages for the art" (Aubert et al., 2014, p. 224). The samples LL1.1 and LL1.2, collected in 2013 at Lompoa Leang, associated with hand stencils, correspond to that condition and yield

a maximum age of 39 Ka and a minimum age of 27 Ka for the artistic motif.

Several icons are particularly interesting in the work of Aubert et al. (2014). The first are hand stencils from the caves of Leang Timpunseg and Leang Jarie, with a dating for the deposited speleothem samples of minimum ages of 40.70 Ka and 39.67 Ka, respectively. The second is the figure of an undetermined animal from the Leang Barugayya 2 cave of 44.99 Ka. Finally, in Leang Timpunseng, appearing close to a hand stencil whose speleothem has been dated, was a realistic figure in which a babirusa can be recognized, and is associated with a coral speleothem of 36.90 Ka. Thus, it is possible to conclude that in Sulawesi, symbols in the form of hand stencils and figurative representations of animals appeared progressively, during a span of 8,000 years around the central date of 40,000 years. These dates place modern humans who occupied Wallacea with the symbolism SM4, at a time very close to the SM4 iconography produced in the caves of southern France and northern Spain.

Northern Spain cave dating studies are comparable to that of Aubert et al. (2014) in Sulawesi. Alistair Pike and collaborators (2012), using uranium-series' disequilibrium dating, estimated the age of the calcite speleothems deposited on paintings and engravings in 11 caves in Asturias and Cantabria (Spain). The minimum age was thus assessed, although a maximum age could also be estimated, in cases where the motifs were deposited on a flowstone, or on engraved flowstone. The results obtained by Pike et al. (2012) appear in Table 11.8.

As shown in Table 11.8, the samples O-69 and O-87 correspond to speleothems located below and above the same artistic motif (large red disk) in the cave of El Castillo (Cantabria, Spain) with ages of c. 35.72 and c. 34.25 Ka, respectively (Figure 11.39). Thus, a very precise dating was obtained of the moment when the icon was made, which turns out to be slightly more modern than those in Sulawesi. But, the sample O-83 from El Castillo also provided a minimum age very close to the central completion date of the SM4 symbols of Maros.

The set of dated icons from Maros and Cantabria reveals an astonishing similarity in artistic motif (hand stencils and realistic representation of animals), as well as in age (c. 40 Ka), corresponding to places as distant as Sulawesi and Spain. This includes a detail, in the latter case, which raises more questions about the origin of SM4. In the conclusion to their paper, Alistair Pike and collaborators (2012, p. 1412) stated the following: "If the earliest cave paintings appeared in the region shortly before 40.8 Ka, this would,

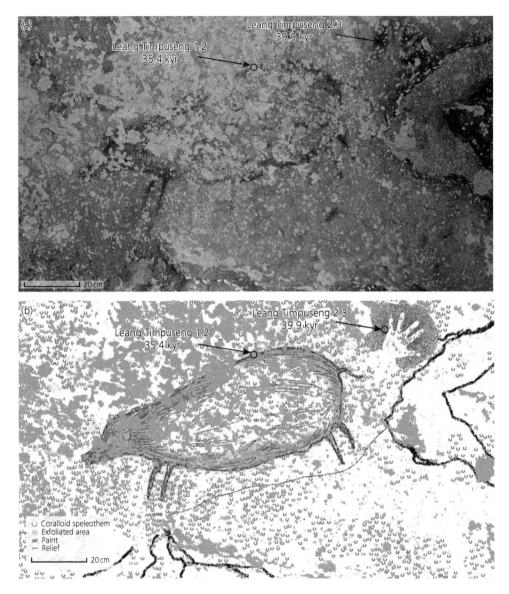

Figure 11.38 Photograph (a) and tracing (b) showing the locations of the dated coralloid speleothems and associated paintings: a hand stencil and a large naturalistic depiction of an animal (Leang Timpuseng cave). Although the animal figure is badly deteriorated and obscured by coralloids, the authors interpret it as a female babirusa. A painted line below the babirusa—not clearly visible in (a) but illustrated in (b)—seems to represent the ground surface on which the animal is standing or walking. The rock art panel is located on the ceiling about 8 m from the cave entrance and 4 m above the current cave floor. Reprinted by permission from Macmillan Publishers Ltd: Aubert, M., Brumm, A., Ramli, M., Sutikna, T., Saptomo, E. W., Hakim, B., . . . Dosseto, A. (2014). Pleistocene cave art from Sulawesi, Indonesia. *Nature*, 514(7521), 223–227.

assuming that the Proto-Aurignacian cultural complex was made exclusively by *Homo sapiens*, support the notion that cave art coincided with their arrival in western Europe ~41.5 Ka and that the exploration and decorating of caves was part of their cultural package.

However, because the 40.8 Ka date for the disk is a minimum age, it cannot be ruled out that the earliest paintings were symbolic expressions of the Neanderthals, which were present in Cantabrian Spain until at least 42 Ka."

Table 11.8 Results of U-series disequilibrium dating (Pike et al., 2012; modified)

Sample	Site	Description	Corrected age (Ka)
Minimum ages			
O-53	Altamira	Overlays red spotted outline horse of *Techo de los Polícromos* chamber	22.11±0.13
O-80	El Castillo	Overlays black outline drawing of indeterminate animal of corridor of *Techo de las Manos*	22.88±0.27
O-58	El Castillo	Overlays red stippled negative hand stencil of *Techo de las Manos*	24.34±0.12
O-21	Tito Bustillo	Red pigment associated with anthropomorphic figure of *Galería de los Antropomorfos*	30.8±5.6 29.65±0.55 (different correction)
O-69	El Castillo	Large red disk of *Galería de los Discos*	34.25±0.17
O-50	Altamira	Large red claviform-like symbol of *Techo de los Polícromos*	36.16±0.61
O-82	El Castillo	Sample overlays red negative hand stencil, and underlies yellow outline bison of *Panel de las Manos*	37.63±0.34
O-83	El Castillo	Overlays large red stippled disk of *Panel de las Manos*	41.40±0.57
Maximum ages			
O-87	El Castillo	Underlies large red disk of *Galería de los Discos* (same panel as O-69)	35.72±0.26
O-48	Tito Bustillo	Underlies red anthropomorph figure of *Galería de los Antropomorfos* (see also O-21)	36.2±1.5 35.54±0.39 (different correction)

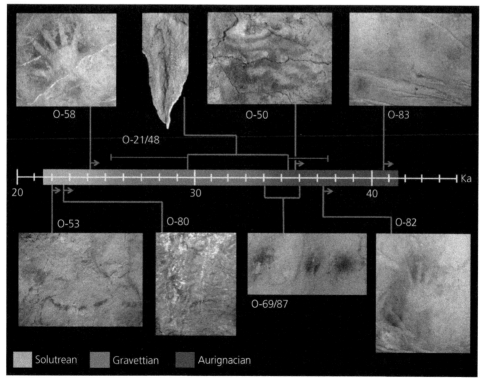

Figure 11.39 Time line of the cave art dated by Pike et al. (2012). A single arrow represents a minimum age, but, where two dates are indicated, both maximum and minimum ages have been obtained. From Pike, A. W. G., Hoffmann, D. L., García-Diez, M., Pettitt, P. B., Alcolea, J., De Balbín, R., . . . Zilhão, J. (2012). U-Series Dating of Paleolithic Art in 11 Caves in Spain. *Science*, 336(6087), 1409–1413. Reprinted with permission from AAAS.

In the classification we use here, the modern mind was already expressed in the SM3 symbolism. The decorative objects and the use of ocher that were, in all probability, used by Neanderthals belong in this level. But if hand stencils are granted the distinctive status of SM4 icons, then Neanderthals may have also reached SM4, equaling the cognitive level of modern humans entering Europe. Although it remains speculative, Pike et al. (2012) argue that Neanderthals could have been the authors of the oldest icons of Cantabria. In this case the acculturation—the cultural sharing— would be the most-parsimonious hypothesis for the origin of this leap forward by the Neanderthals. It is very unlikely, even hypothetically, that modern humans and Neanderthals made independently, and isolated from each other, the same icons of SM3 and SM4, simultaneously and in places as far apart as Sulawesi and Cantabria.

Glossary

Acheulean A lower *Paleolithic* lithic technology. It evolved from the *Oldowan* and is characterized by handaxes and cleavers. Widespread in Africa, Europe and parts of Asia from around 1.5 million to 150,000 years ago. Associated with *Homo erectus* and archaic *Homo sapiens*.

adapiform A group of diverse *fossil* primate mammals that lived during the *Eocene*. They are considered by some authors as the first true Primates.

adaptation A structural or functional characteristic of an organism that allows it to cope better with its environment; the evolutionary process by which organisms become adapted to their environment.

adaptive radiation Diversification of a group of related species associated with the colonization of novel ecological niches.

adaptive value A measure of the reproductive efficiency of an organism (or *genotype*) compared with other organisms (or genotypes); also called *selective value*.

allele Each of the two or more different forms of a *gene*, such as the alleles for *A, B,* and *O* at the *ABO* blood-group gene locus.

allometry Relative growth relationships between two parts of an organism or between two species.

allopatric Geographically separated populations or species (see also *sympatric*).

altruism Behavior that benefits other individuals at the expense of the one who performs the action.

amino acid The building blocks of *proteins*. Several hundred are known, but only 20 are normally found in proteins.

anagenesis The evolutionary change of a single lineage in the course of time (see also *cladogenesis*).

analogy (adj. analogous) Resemblance in function but not in structure, due to independent evolutionary origin; e.g., the wings of a bird and of an insect.

anatomically modern human Human beings that appeared about 150,000 years ago and which share conspicuous morphological traits with current humans, despite cultural and perhaps symbolic differences.

anthropoid A *primate* belonging to one of the following superfamilies: Ceboidea (New World monkeys), Cercopithecoidea (*Old World monkeys*), or *Hominoidea* (lesser apes, great apes and humans).

antibody A *protein*, synthesized by the immune system of a higher organism, that binds specifically to the foreign molecule (*antigen*) that induced its synthesis.

antigen See *antibody*.

anvil A hard surface used as a base when extracting *flakes* from a *core*. It is usually a flat stone, although chimpanzees have been observed using roots as anvils to open nuts.

ape A member of the *primate* group that includes the gibbons, orangutans, gorillas and chimpanzees; gibbons are often called the lesser apes and the others great apes.

apomorphy A trait that has appeared after the node where a certain *clade* originated. It is also known as a *derived trait*. A distinction is made between *synapomorphy* and *autapomorphy*.

archaic *Homo sapiens* The name given to *fossil* humans that lived in Africa, Europe and Asia from about 400,000 to 200,000 or 100,000 years ago with features intermediate between *Homo erectus* and Late Pleistocene species (*Homo neanderthalensis* and modern *Homo sapiens*), and which may represent separate species.

Ardipithecus The *genus* including late Miocene *hominin* remains found in Middle Awash region of Ethiopia.

argon-argon A variation of the *potassium/argon* dating method.

arthropod A member of the animal *phylum* Arthropoda (insects, arachnids, millipedes, crustaceans, etc.).

articulation The joint between two bones.

assemblage A group of objects found together in an archaeological setting.

Aurignacian More developed cultural level that was traditionally assigned to Cro-Magnon tools.

Australopithecus An extinct *genus* of the tribe *Hominini*. Species commonly assigned to this *genus* include *A. anamensis, A. afarensis, A. bahrelghazali, A. africanus* and *A. garhi*. We have included *A. platyops* too. Some authors include *Paranthropus* species in the *Australopithecus genus*.

autapomorphy A *derived trait* found only in a single *clade* of a cladogram.

autosome A *chromosome* other than a sex chromosome.

base pair Two nitrogenous bases that pair by hydrogen-bonding in double-stranded *DNA* or *RNA*.

bed A geologic layer.

biface A flat stone tool produced by extracting *flakes* from both sides of a *core* until an edge is obtained along the whole perimeter. The most common form of bifacial tools is *handaxes*.

biogeography The geographic distribution of plants and animals.

biostratigraphy The study of the sequence of appearance and disappearance of *fossil* species throughout a series of deposits.

biota All plants and animals of a given region or time.

bipedalism A mode of locomotion involving a vertical position of the body and walking by use of only the hindlimbs.

blade A tool created by striking a long and narrow flake from a core of stone.

bladelet A small *blade*.

bottleneck A period when a *population* becomes reduced to only a small number of individuals.

brachiation A mode of arboreal locomotion involving swinging alternate forelimbs to move from branch to branch.

brain case Refers to the set of bones that surround the brain; also known as *cranium*.

breccia Cave sediments that have been calcified by filtering lime solutions.

brow ridge A ridge of bone that arches above the eye sockets, the main function of which is to protect the ocular cavities.

burin A piercing tool commonly used to engrave materials such as antler, ivory or bone.

calcaneus The largest of the *hominin* tarsal bones.

calcareous Composed of, or containing lime (calcium carbonate).

calotte The roof of the skull.

canine Each of the two lower and two upper large teeth at the corners of the mouth, between the *incisors* and *premolars*, and whose function is to pierce food.

Carbon 14 (^{14}C) See *radiocarbon dating*.

carpal bone Each of the eight bones that constitute the wrist. They are placed in two rows; the proximal one articulates with the *radius* and *ulna* while the distal one articulates with the *metacarpals*.

catarrhine A member of the *primate* infraorder Catarrhini (*Old World monkeys*, apes and humans).

category Each of the levels in the Linnaean classification system (*kingdom, phylum, class, order, family, genus, species*, and their intermediates).

ceboid A member of the *primate* superfamily Ceboidea (New World monkeys).

Cenozoic The era from 65 Ma to the present, also called the Age of Mammals. It is divided into the Tertiary and Quaternary periods.

cercopithecoid A member of the *primate* superfamily Cercopithecoidea (*Old World monkeys*).

cerebral cortex The outer layer of gray matter of the cerebral hemispheres, comprising layers of nerve cells and their interconnections.

character A trait or feature of an organism.

Châtelperronian A tool industry that exhibits certain Upper-*Paleolithic* features. Such instruments are found in Western Europe and they are thought to have been developed by late *Neanderthals*.

chin The anterior projection of the *mandible* at its midline that begins below the alveolar bone of the central *incisors* and extends down and out to form a raised inverted T.

chopper A stone tool that has been flaked irregularly to produce a cutting edge on one side.

chromosome A thread-shaped structure visible in the cell's *nucleus* during cell division. Chromosomes contain most of the hereditary material or genes.

chronospecies A species that gradually changes through time such that the original organisms and resulting ones are too different to be classified within the same species, although there is no clear cutting point along the lineage at which they can be differentiated.

chronostratigraphy The study of the temporal aspect of sediments aiming to establish a temporal reference for all elements related with rocks and *fossils* found at a site.

clade A complete group of organisms derived from a common ancestor; or the branches that separate in a cladistic event.

cladistic event See cladogenesis.

cladistics A classification system proposed by Hennig based on phylogenetic hypotheses and common ancestry that only admits speciation by means of *cladogenesis*.

cladogenesis An evolutionary process whereby a species gives rise to two different ones, after which it disappears. In a *cladogram*, the divergence point is known as a node.

cladogram The graphic representation of the branching relations between *species*, genera, families, and so on, which are represented as *clades*.

class A category formed by a set of *orders*.

clavicle Each of the two collarbones that articulate with the scapulae and sternum.

cleaver A large bifacial stone tool with a transverse sharp edge on one side.

coalescence The convergence of the phylogenies of several species into their most recent common ancestor.

codon A group of three adjacent *nucleotides* in an *mRNA* molecule that code either for a specific *amino acid* or for *polypeptide* chain termination during *protein* synthesis.

cognitive capacity A technical concept related to what is informally named "mind."

conspecific Belonging to the same species.

continental drift The slow movement of the continents and their crustal plates over the Earth's surface.

convergence The parallel development of the same feature in unrelated organisms.

core What remains of a stone after *flakes* have been removed from it.

cranium The set of bones that constitute the skull except for the *mandible* (also known as the *brain case*).

Cretaceous The last geological period of the Mesozoic era, about 145-65 Ma.

Cro-Magnon The earliest anatomically modern human *populations* in Europe.

cuboid The tarsal bone placed on the lateral side of the foot. It articulates with the *calcaneous*, the lateral *cuneiform* and the fourth and fifth *metatarsals*.

cuneiform Each of three footbones (medial, intermediate, and lateral cuneiforms) that link the *navicular* and the medial *metatarsals*.

dating methods See *radiocarbon dating, paleomagnetism, potassium/argon* and *argon-argon, fission track, uranium series, thermoluminescence, electronic spin resonance*.

deciduous dentition The first set of teeth to appear, before the *mandible* is capable of accommodating the *permanent dentition*.

deletion A chromosomal mutation due to the loss of a chromosomal segment.

Denisovans Old *Homo* species found in the Denisova Cave in the Altai Mountains (Russia), contemporary of Neanderthals and archaic humans, and which may have hybridized with both of them.

dental arcade Shape of the lower and upper rows of teeth.

derived trait See *apomorphy*.

diastema The gap between *incisors* and *canines*. It is very marked in rodents and archaic *primates*.

digit A toe or finger.

diploid A cell, tissue, or organism having two *chromosome* sets.

DNA Deoxyribonucleic acid; *nucleic acid* composed of units consisting of a deoxyribose sugar, a phosphate group, and the nitrogen bases adenine, guanine, cytosine, and thymine. The self-replicating genetic material of all living cells; it is made up of a double helix of two complementary strands of *nucleotides*.

dominant A character that is manifest in the *phenotype* of heterozygous individuals.

duplication A chromosomal mutation characterized by two copies of a *chromosome* segment.

effective population size The number of reproducing individuals in a population.

electronic spin resonance A dating method based in determining the amount of electrons trapped in defects of crystal lattices, which are caused by the decay of radioactive elements, through the measurement of their absorption of microwave radiation.

electrophoresis A technique for separating molecules based on their differential mobility in an electric field.

enamel A hard substance that forms a layer around the crown dentine of teeth.

encephalization The increase of brain size relative to body size.

endemic Applied to species restricted to a certain region or part of a region; in epidemiology, applied to diseases that are constantly present at relatively low levels in a particular *population*.

endocast or endocranial cast A natural or artificial mold of the inner surface of the *cranium*.

enzyme A biochemical catalyst based on specialized *protein* molecules that speeds up biochemical processes.

Eocene The second epoch of the Tertiary period, about 56.5⊠35.4 Ma, during which the second wave of *primates* appeared.

epigenesis Cellular processes that modify the expression of genes without changing their DNA sequence.

epithelium A tissue consisting of one or more layers of tightly bound cells that covers the external and internal surfaces of the body.

epoch A subdivision of a geological period.

era The largest division of geological time, including one or more periods.

eukaryote An organism whose cells contain a distinct *nucleus* as well as *mitochondria* and other *organelles*.

exon The *DNA* of a eukaryotic transcription unit whose transcript becomes a part of the *mRNA* produced by splicing out introns.

extant Living; the opposite of extinct.

family The category composed by a set of *genera*.

fault A fracture of the earth's crust, across which there has been observable displacement.

fauna The animals of a region, country, special environment, or period.

femoral head Rounded upper part of the *femur* that forms the joint with the hip.

femoral neck Section of the *femur* that extends medially separating the *femoral* head from the shaft.

femur The proximal bone of the lower limbs. It is the largest bone in the human body. It is usually divided into three sections: the upper end, which includes the *femoral head* and *femoral neck*, the femoral shaft, and the lower or condylar end.

fibula The smaller of the two distal hindlimb bones.

Fission-track dating A method used for dating volcanic rocks associated with *fossil* remains. The age of the rocks is estimated as a function of the amount of uranium in them and the density of the damage trails left by the spontaneous fission of uranium (^{238}U).

fitness The reproductive contribution of an organism or *genotype* to future generations.

flake Long, small, and cutting stone fragment obtained from cobbles by percussion or pressure.

flora The plants of a region, country, particular environment, or period.

folivore An animal whose diet is composed mainly of leaves.

forager One who collects wild animal or plant food.

foramen magnum The large opening in the back of the skull through which the spinal chord passes to join the brainstem; where the vertebral column connects with the *cranium*.

formation In geology, a fundamental unit of stratigraphic classification. Often formations are given geographic names; for example, the Hadar Formation.

fossil Any preserved remains or traces of past life, more than about 10,000 years old, embedded in rock either as mineralized remains or impressions, casts, or tracks.

frugivore An animal whose diet is composed mainly of fruit.

gamete A mature reproductive cell capable of fusing with a similar cell of opposite sex to give a *zygote*; also called a *sex cell*.

gene A genetic unit found on a specific *locus* of a *chromosome*. It consists of a sequence of *DNA* that codes for an *enzyme* or a *protein*, or that regulates activity of other genes.

genetic drift Chance fluctuations in gene frequency observed especially in small *populations*.

genome The genetic content of a cell; in *eukaryotes*, it sometimes refers to only one complete (*haploid*) *chromosome* set.

genotype The genetic information of an individual.

genotyping Procedure to obtain the genome of an individual and/or a species.

genus The category formed by a set of closely related *species*.

glabella A prominence above the nose in the midline of the external surface of the frontal bone.

glacial period A period of cold climate, during which time a certain amount of oceanic water is deposited on the continents as glaciers.

Gorilla The *genus* to which gorillas belong.

gracile Slender or light in build, often used to characterize the *Australopithecus species*, *A. afarensis* and *A. africanus*.

grade Group of organisms that do not necessarily form a *clade* but share a set of characteristics that set them apart from others.

great apes Chimpanzees and gorillas (in Africa) and orangutans (in Asia).

habitat The natural home or environment of a plant or animal.

half-life Statistical average time in which half of a certain amount of a radioactive isotope disappears.

hammerstone An unaltered stone used for hammering. It is commonly considered as the simplest stone tool.

handaxe The most common *biface* stone tool found in Acheulean archaeological sites. It is usually oval or teardrop-shaped.

haploid Of cells, such as *gametes*, that in *eukaryotes* have half as many *chromosome* sets as the somatic cells.

Hardy-Weinberg law Describes the genetic equilibrium in a population stating that genotypes, the genetic constitution of individual organisms, exist in certain frequencies that are a simple function of the allelic frequencies.

heterozygote An organism with two different *alleles* at a certain locus.

high-throughput genetic sequencing See PEC.

holotype The example of a certain organism which is used for its classification. Ideally it should be typical of its *taxon*, though in *fossil* taxa, the holotype is often only a partial specimen.

hominin An individual belonging to the tribe *Hominini*.

Hominini A tribe composed of current humans and their direct and lateral ancestors that are not also ancestral to chimpanzees.

hominoid An individual belonging to the superfamily *Hominoidea*.

Hominoidea The superfamily composed of lesser apes, great apes, and humans, as well as their direct and lateral ancestors that are not also ancestral to *Old World monkeys*.

Homo The genus to which the human species belongs. The species *H. habilis*, *H. naledi*, *H. ergaster*, *H. antecessor*, *H. erectus*, *H. heidelbergensis*, *H. neanderthalensis*, and *H. sapiens* are usually included within this genus. *H. rudolfensis* is also included by some authors.

homologous (noun, homology) A trait that is shared by two taxa, which inherited it from their closest common ancestor. Can also be used to refer to *chromosomes*.

homoplasy A similar trait in two or more taxa that is not due to inheritance from their closest common ancestor, but rather has evolved independently in the different taxa.

homozygote A cell or organism having the same *allele* at a given locus on *homologous chromosomes* (adj., *homozygous*).

humerus The arm's largest bone. Its proximal end articulates with the *scapula* at the shoulder joint, and at the distal end it articulates with the *radius* and the *ulna*.

hunter-gatherer One who lives by hunting and scavenging wild animals, gathering plants and, in some places, collecting shellfish and fishing, often moving in small groups (bands) from place to place.

hybridization Genetic admixing between two *species*.

hybridization hypothesis The proposition that the appearance of modern humans involved the admixing of modern humans with *Neanderthals* or other early *species*.

Hylobates The genus including gibbons and siamangs.

hyoid bone A u-shaped bone in the neck lying just above the larynx and below the tongue.

hypodigm The whole available set of fossil material belonging to a given species.

inbreeding Mating between relatives.

incisors Frontal teeth whose primary function is to cut food. All *hominoids* have four upper and four lower incisors.

insectivore An animal that feeds mostly on insects.

interglacial period A warm interval between two glaciations.

introgression Presence of some amount of genetic material coming from a different species.

intron A length of *DNA* within a functional gene in *eukaryotes*, separating two segments of coding DNA (*exons*).

inversion A chromosomal mutation characterized by the reversal of a *chromosome* segment.

K/Ar See *potassium/argon*.

Kenyanthropus The *hominin genus* introduced in 2001 to accommodate a *cranium* and other findings from Lomekwi (northern Kenya), which have been attributed to the species *Kenyanthropus platyops*.

kin selection A form of altruism related to parental care.

kingdom The category of classification formed by a set of phyla; plants, animals, and fungi are kingdoms.

knuckle-walking A kind of quadrupedal locomotion typical of gorillas and chimpanzees, in which the weight of the upper body leans on the finger knuckles, bent into the palm of the hands and placed on the floor.

lesser apes The Asian apes, gibbons and syamang.

Levallois technique A tool-making technology involving significant pre-shaping of the *core*.

Levantine corridor A migration and settlement area between Africa and Eurasia on the eastern coast of the Mediterranean Sea.

lithic traditions The use of stone tools by *Homo* species.

locus A *gene*'s specific place on a *chromosome*; sometimes used to refer to the gene itself (plural: *loci*).

LSA Name given to the African Late Stone Age that is an evolution of the *MSA* It is close to the European Later *Paleolithic*.

magnetostratigraphy A dating method based on the polarity inversions of Earth's magnetic field. Calibration using other techniques has allowed charting many such episodes, which can constrain the estimated age of *fossils*.

mandible The lower jaw.

mandibular symphysis The midline separating the left and right halves of the *mandible*.

maxilla A facial bone that forms the floor of the orbital cavity, the lateral wall of the nasal cavity, and the support for the upper teeth.

member In geology, a rock unit which is a subdivision of a broader *formation*.

mentum See *chin* (also known as a mental protuberance).

messenger RNA (mRNA) An RNA molecule whose *nucleotide* sequence is translated into an *amino acid* sequence on ribosomes during *polypeptide* synthesis.

metacarpal Each of a set of five bones in each hand that form the palm and the thumb's first bone. They articulate proximally with the wrist's carpal bones and distally with the fingers' *phalanges*.

metatarsal Each of a set of five bones in each foot that form the instep.

Miocene The fourth epoch of the Tertiary period, about 23.3-5.2 Ma.

mitochondria Organelles in a eukaryotic cell that are involved in energy metabolism; each mitochondrion has its own small circular genome.

mitochondrial DNA (mtDNA) Genetic information contained in *mitochondria* in the form of a single circular strand which is inherited only through the maternal line.

mitochondrion An organelle in eukaryotic cell that is involved in energy metabolism; each mitochondrion has its own small circular genome (plural; *mitochondria*).

modern mind *Cognitive capacity* of *Homo sapiens*.

molar Each of the teeth with large occlusal surfaces located at the back of the jaws for grinding and crushing food. Humans have three molars on each side of the jaw.

molecular clock The estimated regularity of changes in *DNA* and *proteins* through time, which can be used to estimate the timing of evolutionary episodes.

monophyletic Describing a group of organisms, including their common ancestral stock and all its descendants.

Mousterian A middle-*Paleolithic* stone culture characterized by small and precise instruments, such as small *handaxes*, side-scrapers, and triangular points. It is commonly associated with *Neanderthals*.

MSA Name given to the African Middle Stone Age that is an evolution of the Acheulian in Africa.

mtDNA See *mitochondrial DNA*

multiregional hypothesis An hypothesis suggesting that the appearance of modern humans occurred by an independent evolution from earlier *hominins* in different geographical regions.

mutant An *allele* different from the wild type; or an individual carrying such an allele.

mutation An inheritable modification of genetic material.

nasal Of or relating to the nose.

natural selection The differential reproduction of alternative *genotypes* due to variable *fitness*.

navicular A footbone located on the medial side of the foot that articulates posteriorly with the *talus*, laterally with the *cuboid* and distally with the three *cuneiform* bones.

Neandertal Common designation for *Homo neanderthalensis*, a *hominin species* closely related to *Homo sapiens* which inhabited southern Europe and the Levant during the Würm glaciation. They are generally associated with the *Mousterian* material culture.

Neolithic The New Stone Age, usually associated with the beginnings of agriculture, pottery, and settlements in the Old World. In parts of western Asia, farming began as early as 10,000 years ago (although without pottery).

neoteny The persistence of a juvenile character in adulthood.

niche The place of an organism in its environment, including the resources it exploits and its association with other organisms.

nitrogen base An organic compound composed of a ring containing nitrogen. Used here to refer to each of the complementary molecules that keep the two *DNA* strands together transversally or form *RNA* strands (the bases are adenine, cytosine, guanine, thymine, and uracil).

node The point where two *clades* diverge in a cladogram.

nomad One who continually moves from place to place to find food.

nucleic acid See *DNA* and *RNA*.

nucleotide A nucleic acid unit, composed of a sugar molecule, a nitrogen base, and a phosphate group. A set of three nucleotides constitutes a *triplet*, or *codon,* and each triplet codes for an *amino acid* or represents a stop signal during protein synthesis.

nucleus A membrane-enclosed *organelle* of *eukaryotes* that contains the *chromosomes*.

occipital torus A protuberance found on the posterior part of the *cranium*.

Oldowan The oldest known *hominin* lithic cultural tradition, usually associated with the appearance of *Homo habilis*. The oldest Oldowan remains are about 2.5 million years old.

Old World monkeys Relating to monkeys in all geographical areas except South and Central America, in the superfamily Cercopithecoidea.

Oligocene The third epoch of the Tertiary period, lasting about 35.4-23.3 Ma, during which there was a great reduction of the *primate order*.

omnivore An eater of both animal and plant food.

ontogenetic Relative to *ontogeny*.

ontogeny Development of an organism after conception.

orbit The bony socket for the eye.

order The taxonomic category formed by a set of *families* and a division of a *class*.

organelle A functional membrane-enclosed body inside cells (e.g., a *nucleus* or a *mitochondrion*).

Orrorin The *genus* including late Miocene *hominin* remains found in the Tugen Hills region of Kenya.

orthognathic With little facial projection, opposite to *prognathic*.

orthograde The upright position of the body.

orthologous genes or chromosomes Referring to *genes* or *chromosomes* of different *species* which are similar because they derive from a common ancestor.

out of Africa See *replacement hypothesis*.

ovum A female *gamete*.

palate The roof of the mouth. There are two different parts: a hard bony anterior one, and a soft posterior one.

Paleocene The first epoch of the Tertiary period, about 65-56.5 Ma, during which archaic, or first wave, *primates* appeared.

Paleolithic The Old Stone Age, the first and longest part of the stone age that began some 2.6 Ma in Africa with the first recognizable stone tools belonging to the *Oldowan* industrial tradition and ended some 12,000-10,000 years ago.

paleomagnetism An important dating method based on the history of changes in the Earth's magnetic polarity throughout a stratigraphic column.

Pan The *genus* to which chimpanzees and bonobos belong.

Paranthropus Extinct *hominin genus* including Pliocene and Pleistocene robust individuals. The species *P. boisei, P. robustus* and *P. aethiopicus* are usually assigned to this genus.

paraphyletic A group of organisms including some, but not all, of the descendants of the group's common ancestor.

paratype A set of specimens in the type series other than the *holotype*.

parsimony In evolution, the principle proposing that evolution has followed the most economical route, involving the assumption that closely related species (those that diverged more recently) will consistently have fewer differences than species that diverged longer ago.

patella The sesamoid bone located at the knee joint.

PCR (polymerase chain reaction) Technique for cloning genetic material which allows obtaining millions of identical copies from a single DNA molecule.

PEC (primer extension capture) Also known as high-throughput sequencing. Technique that permits identification of short sequences of nitrogenous bases from very degraded DNA samples contaminated by the presence of microbial DNA.

pelvis The structure composed by two pelvic bones and the sacrum. Each pelvic bone is composed of three bones, the ilium (the upper blade), ischium (the lower part), and pubis (the ventral part).

permanent dentition The second generation of teeth that appear once the *deciduous dentition* has been lost.

phalanges One of the set of 14 finger bones in each hand and foot. There are five proximal phalanges that articulate proximally with the five *metacarpals* in the hand, and the five *metatarsals* in the foot. There are four middle phalanges in each hand and foot (absent in the thumb and the toe). There are five distal phalanges in each hand and foot.

phenetics A system of classification of organisms based principally on the similarity of morphological traits, also known as numerical *taxonomy*.

phenotype An organism's observable traits.

phyletic Applied to a group of *species* with a common ancestor; a line of direct descent.

phylogenesis The process of evolution and differentiation of organisms.

phylogenetic tree The graphic representation of the evolutionary relations among living and extinct organisms.

phylogeny The evolutionary history of a group of living or extinct organisms.

phylum The category formed by a set of *classes* (plural: *phyla*).

Pithecanthropus A *genus* created by Eugène Dubois to include the specimens discovered by his expedition to Java. Those specimens are currently included within *Homo erectus*.

Platyrrhine A member of the *primate* infraorder Platyrrhini (*New World monkeys*).

Pleistocene The first epoch of the Quaternary period, which lasted from about 1.64 million to 10,000 years ago, and saw the radiation of the *genus Homo*.

plesiadapiform A group of diverse *fossil* primate-like mammals that lived during the *Paleocene* and early *Eocene*. Traditionally, they have been considered as a suborder of Primates, but this view has gradually lost support in favor of their consideration as a different *order* of mammals.

plesiomorphy A trait that is already present in the ancestral group of the taxon being studied.

Pliocene The final epoch of the Tertiary period, about 5.2-1.64 Ma.

polymorphism The existence of alternative allelic forms at a *locus* within a *population*. Thus, in humans, there is a polymorphism for the ABO blood groups.

polypeptide A chain of *amino acids* covalently bound by peptide linkages.

polyphyletic The grouping of organisms derived from at least two different ancestral stocks.

Pongo The *genus* to which orangutans belong.

population A set of individuals belonging to the same *species* that constitute an effective reproductive community.

postcranial Referring to any skeletal element except those forming the *cranium*.

potassium/argon A dating method that estimates the age of volcanic terrains based on the amount of radioactive potassium (^{40}K) remaining in a sample.

premolar Each of the teeth located between the canines and the molars. Humans have two premolars in each half of each jaw.

primate The *order* to which the human *species* belongs, together with prosimians, tarsoids and the rest of anthropoids.

primitive trait A feature exhibited by an organism or set of organisms that appeared before the node representing the origin of their *clade*, opposite of *derived trait* or *apomorphy*.

prognathic Describing the outward facial projection beyond the vertical plane that passes through the orbital cavities, opposite to *orthognatic*.

prosimian Any *primate* in the suborder Prosimii (lemurs, lorises, and tarsiers).

protein A molecule composed of one or more *polypeptide* subunits and possessing a characteristic three-dimensional shape imposed by the sequence of its component *amino acid* residues.

Protoaurignacian Cultural level preceding *Aurignacian*.

quadrupedal An arboreal or terrestrial mode of locomotion that involves the use of the four limbs, forelimbs and hindlimbs.

Quaternary The period of the Cenozoic era that began about 1.64 Ma.

radioactivity The emission of ionizing radiation from an unstable chemical isotope.

radiocarbon dating Radiometric dating method for organic materials based on the rate of decay of ^{14}C to ^{14}N.

radiometric dating Dating methods, such as *radiocarbon dating*, based on the measurement of radioactive decay.

radius A bone of the forearm, somewhat shorter than the *ulna*, but broader at its distal side.

ramus An ascending backward projection from each side of the *mandible*'s body (plural: *rami*).

recessive An allele, or the corresponding trait, that is manifest only in *homozygotes*.

replacement hypothesis The proposal that the modern humans that had dispersed from Africa did not admix with earlier *populations* living in the territories they colonized.

RNA Ribonucleic acid *nucleic acid* composed of units consisting of a ribose sugar, a phosphate group, and the nitrogen bases adenine, guanine, cytosine, and uracil.

robust Heavily built *fossil* specimen, especially the face and jaw, normally with large masticatory apparatus.

sacrum The continuation of the vertebral column into the pelvis; composed of several fused vertebrae.

sagittal crest A bony protuberance along the skull's superior midline for the attachment of large muscles, observed in some robust *hominins*.

Sahelanthropus The *genus* including late Miocene *hominin* remains found in Chad.

savanna Subtropical or tropical grassland with scattered trees and shrubs and a pronounced dry season.

scapula The shoulder blade.

sedimentary rock A rock formed by the accumulation and hardening of rock particles (sediments) derived from existing rocks and/or organic debris and deposited by agents such as wind, water, and ice at the Earth's surface; the source of fossils.

selection See *natural selection*.

selective value See *adaptive value*.

sensu lato Latin expression meaning "in a broad sense."

sensu stricto Latin expression meaning "in a strict sense"

sex cell See *gamete*.

sex chromosome A chromosome that differs between the two sexes and is involved in sex determination (see *autosome*).

sexual dimorphism A set of traits that distinguish male and female individuals from the same *species*.

shared derived character A feature shared by descendants from an ancestral stock that was not present in the remote common ancestor.

simian Any member of the *primate* suborder Anthropoidea (monkeys, apes, and humans); a higher primate.

single-species hypothesis The notion that all *hominin* specimens after *Homo habilis* can be adequately accommodated within a single *species*.

single nucleotide polymorphism See *SNP*.

sister group Each one of the clades that separates at a node (see *clade, cladogenesis,* and *cladogram*).

skull Set formed by the cranial bones and the *mandible*.

SNPs Single nucleotide polymorphisms. Alleles corresponding to a certain locus that differ in a single nitrogen base.

soft-hammer A tool-making technique in which the hammer is made of a softer material than the *core*. It allows a greater precision in the production of *flakes*.

speciation The process of evolution of a new *species*.

species The basic unit of Linnaean classification, always expressed by two Latin names (such as *Homo sapiens*), the first of which specifies the *genus*; defined as groups of interbreeding natural *populations* that are reproductively isolated from other such groups.

stratocladistics Method of evaluating competing phylogenetic hypotheses using a parsimony criterion that takes into account the stratigraphic (temporal) sequence of the fossil taxa, in addition to their morphological characters.

stratigraphy In geology, the study of a site's strata; usually represented by stratigraphic columns.

stratum A layer of sedimentary terrain that corresponds to a specific sedimentary period and is differentiable from other layers, above or below it.

supraorbital torus A bar of bone extending over the superior margins of the *orbits*.

suspensory behavior A locomotion or posture that involves hanging from branches or clinging to them.

suture In anatomy, the junction of two parts immovably connected.

symbol A word, behavior, or object that conveys meaning.

symbolism The capacity of creating and interpreting symbols.

sympatric Referring to species that share the same territory (see also, *allopatric*).

synapomorphy A *derived trait* that is shared by two or more *clades*.

systematics The discipline that studies the classification of organisms and their evolutionary relationships.

talus A footbone that links the leg and the rest of the foot. Proximally, it articulates with the *tibia*; anteriorly, with the *navicular*; and inferiorly, with the *calcaneus*.

taphonomy A discipline that studies the processes involved in fossilization.

tarsal bone One of the set of seven ankle bones in each foot: *calcaneus, talus, navicular, cuboid,* and three *cuneiforms.*

taxon A defined unit in the classification of organisms. For example, *Homo* is a taxon of the *genus* category; *Homo sapiens* is a taxon of the *species* category (plural: *taxa*).

taxonomy The rules and procedures used in the classification of organisms.

temporal bone One of a pair of bones that form part of the side wall and base of the human *brain case* and cover the auditory ossicles in the middle ear.

Tertiary The first period of the Cenozoic era, from 65⊠1.64 Ma.

tetrapod A four-footed animal: any amphibian, reptile, bird, or mammal.

thalamus A large mass of gray matter deep in the cerebral hemispheres in the middle of the forebrain.

thermoluminescence A dating method based on the calculation of the amount of electrons trapped in defects of crystal lattices, caused by the decay of radioactive elements; it measures the intensity of the light emitted by electrons escaping as the sample is heated to a high temperature.

tibia The largest of the two distal leg bones.

triplet In genetics, set of three contiguous *DNA* or *RNA nucleotides* that specifies a particular *amino acid* in a *protein*, or indicates its end signal (also called *codon*).

tuff Rock formed by the cementing or compression of volcanic ashes.

type specimen An individual exemplar used as the reference for naming a new *species*.

ulna A bone of the forearm.

uranium series A dating method based on the radioactive decay of uranium in a geological sample.

vertebrate An animal with a backbone or vertebral column.

Villafranchian Fauna of the early Pleistocene, after the Italian village of Villafranca.

Wernicke's area The region of the human brain involved in the comprehension of speech, lying in the upper part of the temporal cortex and extending into the parietal cortex in the left cerebral hemisphere.

Y The designation of the shape of the lower molars with five cusps, typical of *hominoids*.

zygomatic arch The arch of bone that extends along the front or side of the skull beneath the *orbit*.

zygomatic bone The lateral facial bone below the eye that in mammals forms part of the *zygomatic arch* and part of the *orbit*.

zygote The diploid cell formed by the union of egg and sperm nuclei in the cell.

References

Abbate, E. & Sagri, M. (2011). Early to Middle Pleistocene *Homo* dispersals from Africa to Eurasia: geological, climatic and environmental constraints. *Quaternary International, 33*, 271–296.

Abbate, E., Albianelli, A., Azzaroli, A., Benvenutti, M., Tesfamariam, B., Bruni, P., . . . Villa, I. (1998). A one-million-year-old *Homo* cranium from the Danakil (Afar) Depression of Eritrea. *Nature, 393*, 458–460.

Abbott, A. (2003). Anthropologists cast doubt on human DNA evidence. *Nature, 423*, 468.

Adam, K. D. (1988). Der Urmensch von Steinheim an der Murr und seine Umwelt. Ein Lebensbild aus der Zeit vor einer Viertelmillion Jahre. *Jahrbuch des Römisch-Germanischen Zentralmuseums, 35*, 3–23.

Adcock, G. J., Dennis, E. S., Easteal, S., Huttley, G. A., Jermiin, L. S., Peacock, W. J., & Thorne, A. (2001). Mitochondrial DNA sequences in ancient Australians: implications for modern human origins. *Proceedings of the National Academy of Sciences, USA, 98*, 537–542.

Agassiz, L. (1840). *Etudes Sur Les Glaciers*. Neuchâtel, Switzerland: Jent & Gassmann.

Agnew, N. & Demas, M. (1998). Preserving the Laetoli Footprints. *Scientific American, 279*, 26–37.

Aguirre, E. (1970). Identificación de "Paranthropus" en Makapansgat. In X. C. N. d. Arqueología (Ed.), *Crónica* (pp. 97–124). Madrid: Crónica.

Aguirre, E. (1995). Los yacimientos de Atapuerca. *Investigación y ciencia, 229*, 42–51.

Aguirre, E. (2000). *Evolución humana. Debates actuales y vías abiertas*. Madrid: Real Academia de Ciencias Exactas, Físicas y Naturales.

Aguirre, E., Basabe, J. M., & Torres, T. (1976). Los fósiles humanos de Atapuerca (Burgos). *Zephyrus, 26–27*, 489–511.

Aguirre, E. & de Lumley, M.-A. (1977). Fossil men from atapuerca, Spain: their bearing on human evolution in the Middle Pleistocene. *Journal of Human Evolution, 6*, 681–688.

Aguirre, E., de Lumley, M.-A., Basabe, J. M., & Botella, M. (1980). Affinities between the mandibles from Atapuerca and L'Arago, and some East African fossil hominids. In R. E. F. E. Leakey & B. A. Ogot (Eds.), *Proceedings of the 8th Panafrican Congress of Prehistory and Quaternary Studies* (pp. 171–174). Nairobi: The International Louis Leakey Memorial Institute for African Prehistory.

Agustí, J. (2000). Un escenario para la evolución de los homínidos del Mioceno. In J. Agustí (Ed.), *Antes de Lucy. El agujero negro de la evolución humana*. Barcelona: Tusquets.

Agustí, J. & Lordkipanidze, D. (2011). How "African" was the early human dispersal out of Africa? *Quaternary Science Reviews, 30*(11–12), 1338–1342.

Agustí, J., Köhler, M., Moyà-Solà, S., Cabrera, L., Garcés, M., & Parés, J. M. (1996). Can Llobateres: the pattern and timing of the Vallesian hominoid radiation reconsidered. *Journal of Human Evolution, 31*, 143–155.

Ahern, J. C. M. (1998). Underestimating intraspecific variation: the problem with excluding Sts 19 from *Australopithecus africanus*. *American Journal of Physical Anthropology, 105*, 461–480.

Ahern, J. C. M. (2006). Non-metric variation in recent humans as a model for understanding Neanderthal-early modern human differences: just how "unique" are Neanderthal unique traits? In K. Harvati & T. Harrison (Eds.), *Neanderthals Revisited: New Approaches and Perspectives* (pp. 255–268). New York, NY: Springer.

Aiello, L. C. (1992). Allometry and the analysis of size and shape in human evolution. *Journal of Human Evolution, 22*, 127–147.

Aiello, L. C. (1994). Variable but singular. *Nature, 368*, 399–400.

Aiello, L. C. & Andrews, P. (2000). The Australopithecines in review. *Human Evolution, 15*, 17–38.

Aiello, L. C. & Collard, M. (2001). Our newest older ancestor? *Nature, 410*, 526–527.

Aiello, L. C., Collard, M., Thackeray, J. F., & Wood, B. (2000). Assessing exact randomization-based methods for determining the taxonomic significance of variability in the human fossil record. *South African Journal of Science, 96*, 179–183.

Aiello, L. C. & Wheeler, P. (1995). The expensive tissue hypothesis: the brain and the digestive system in human and primate evolution. *Current Anthropology, 36*, 199–221.

Aigner, J. S. & Laughlin, W. S. (1973). The dating of Lantian man and his significance for analyzing trends in human evolution. *American Journal of Physical Anthropology, 39*, 97–110.

Akazawa, T., Muhesen, M., Dodo, Y., Kondo, O., & Mizoguchi, Y. (1995). Neanderthal infant burial. *Nature, 377*, 585–586.

Alemseged, Z., Coppens, Y., & Geraads, D. (2002). Hominid cranium from Omo: description and taxonomy of Omo-323-1976-896. *American Journal of Physical Anthropology, 117*, 103–112.

Alemseged, Z., Spoor, F., Kimbel, W. H., Bobe, R., Geraads, D., Reed, D., & Wynn, J. G. (2006). A juvenile early hominin skeleton from Dikika, Ethiopia. *Nature, 443*(7109), 296–301.

Alemseged, Z., Wynn, J. G., Kimbel, W. H., Reed, D., Geraads, D., & Bobe, R. (2005). A new hominin from the Basal Member of the Hadar Formation, Dikika, Ethiopia, and its geological context. *Journal of Human Evolution, 49*, 499–514.

Alexeev, V. P. (1986). *The Origin of the Human Race*. Moscow: Progress Publishers.

Alimen, M. H. (1975). Les "Isthmes" hispano-marocain et Siculo-Tunisien au temps Acheuléens. *L'Anthropologie, 79*, 399–436.

Allsworth-Jones, P. (1986). *The Szeletian and the Transition from Middle to Upper Paleolithic in Central Europe*. Oxford: Oxford University Press.

Almécija, S., Moyà-Solà, S., & Alba, D. M. (2010). Early origin for human-like precision grasping: a comparative study of pollical distal phalanges in fossil hominins. *PLoS One, 5*(7), e11727.

Ambrose, S. H. (2001). Paleolithic technology and human evolution. *Science, 291*, 1748–1753.

An, Z. S. & Ho, C. K. (1989). New magnetostratigraphic dates of Lantian *Homo erectus*. *Quaternary Research, 32*, 213–221.

Anderson, H. (2009). *Beginnings of Art: 100,000–28,000 BP. A Neural Approach*. (Ph.D. dissertation), University of East Anglia (UK).

Anderson, S., Bankier, A. T., Barrell, B. G., de Bruijn, M. H., Coulson, A. R., Drouin, J., . . . Young, I. G. (1981). Sequence and organization of the human mitochondrial genome. *Nature, 290*, 457–465

Andreae, A. (1884). *Der Diluvialsand von Hangenbieten im Unter-Elsass, seine geolog. u. paläontolog. Verhältnisse u. Vergleich seiner Fauna mit der recenten Fauna des Elsass*. Straßburg: Abh. z. geolog. Specialkarte von Elsass-Lothringen,Bd. IV. Heft II.

Andrews, P. (1981). Species diversity and diet in monkeys and apes during the Miocene. In C. B. Stringer (Ed.), *Aspects of Human Evolution* (pp. 25–61). London: Taylor & Francis.

Andrews, P. (1983). Small mammal faunal diversity at Olduvai Gorge, Tanzania. In J. Clutton-Brock & C. Grigson (Eds.), *Animals and Archaeology: I. Hunters and Their Prey* (pp. 77–85). Oxford: BAR Int. Series 163.

Andrews, P. (1992). Evolution and environment in the Hominoidea. *Nature, 360*, 641–646.

Andrews, P. (1995). Ecological apes and ancestors. *Nature, 376*, 555–556.

Andrews, P. & Cronin, J. E. (1982). The relationships of *Sivapithecus* and *Ramapithecus* and the evolution of the orang-utan. *Nature, 297*, 541–546.

Andrews, P. & Martin, L. B. (1987). Cladistic relationships of extant and fossil hominoids. *Journal of Human Evolution, 16*, 101–118.

Antón, S. C. (1994). Mechanical and other perspectives on Neandertal craniofacial morphology. In R. S. Corruccini & R. L. Ciochon (Eds.), *Integrative Paths to the Past* (pp. 677–695). Englewood Cliffs, NJ: Prentice-Hall.

Antón, S. C. (2002). Evolutionary significance of cranial variation in Asian *Homo erectus*. *American Journal of Physical Anthropology, 118*, 301–323.

Antón, S. C. (2003). Natural history of *Homo erectus*. *Yearbook of Physical Antropology, 46*, 126–170.

Antón, S. C. (2004). The face of Olduvai Hominid 12. *Journal of Human Evolution, 46*, 337–347.

Antón, S. C., Carter-Menn, H., & DeLeon, V. B. (2011). Modern human origins: continuity, replacement, and masticatory robusticity in Australasia. *Journal of Human Evolution, 60*(1), 70–82.

Antón, S. C., Leonard, W. R., & Robertson, M. L. (2002). An ecomorphological model of the initial hominid dispersal from Africa. *Journal of Human Evolution, 43*, 773–785.

ApSimon, A. M. (1980). The last Neanderthal in France? *Nature, 287*, 271–272.

Arambourg, C. (1954). L'hominine fossile de Ternifine (Algérie). *Comptes-Rendus de l'Académie de Scenices de Paris, 239*, 893–895.

Arambourg, C. (1955). A recent discovery in human paleontology: *Atlanthropus* of ternifine (Algeria). *American Journal of Physical Anthropology, 13*, 191–202.

Arambourg, C. & Coppens, Y. (1968). Découverte d'un Australopithécien nouveau dans les gisements de l'Omo (Ethiopie). *South African Journal of Science, 64*, 58–59.

Arden, B. & Klein, J. (1982). Biochemical comparison of major histocompatibility complex molecules from different subspecies of *Mus musculus*: evidence for trans-specific evolution of alleles. *Proceedings of the National Academy of Sciences, USA, 79*, 2342–2346.

Arensburg, B. (1989). New skeletal evidence concerning the anatomy of Middle Paleolithic populations in the Middle East: the Kebara skeleton. In P. Mellars & C. Stringer (Eds.), *The Human Revolution: Behavioural and Biological Perspectives on the Origins of Modern Humans* (pp. 165–171). Princeton, NJ: Princeton University Press.

Arensburg, B., Bar-Yosef, O., Chech, M., Goldberg, P., Laville, H., Meignen, L., . . . Vandermeersch, B. (1985). Une sépulture néandertalienne dans la grotte de Kébara (Israël). *Comptes-Rendus de l'Académie des Scenices de Paris (Série D), 300*, 227–230.

Argue, D., Donlon, S., Groves, C., & Wright, R. (2006). *Homo floresiensis*: microcephalic, pygmoid, *Australopithecus*, or *Homo*? *Journal of Human Evolution, 51*, 360–374.

Argue, D., Morwood, M. J., Sutikna, T., Jatmiko, & Saptomo, E. W. (2009). *Homo floresiensis*: a cladistic analysis. *Journal of Human Evolution, 57*(5), 623–639.

Aronson, J. L., Hailemichael, M., & Savin, S. M. (2008). Hominid environments at Hadar from paleosol studies in a framework of Ethiopian climate change. *Journal of Human Evolution*, 55(4), 532–550.

Aronson, J. L., Schmitt, T. J., Walter, R. C., Taieb, M., Tiercelin, J. J., Johanson, D. C., . . . Nairn, A. E. M. (1977). New geochronologic and palaeomagnetic data for the hominid-bearing Hadar Formation of Ethiopia. *Nature*, 267, 323–327.

Arribas, A. & Palmqvist, P. (1999). On the Ecological Connection between sabre-tooths and hominids: faunal dispersal events in the Lower Pleistocene and a review of the evidence for the first human arrival in Europe. *Journal of Archaeological Science*, 26, 571–585.

Arsuaga, J. L. (1994). Los hombres fósiles de la sierra de Atapuerca. *Mundo científico*, 143, 167–168.

Arsuaga, J. L., Lorenzo, C., Carretero, J. M., Gracia, A., Martinez, I., Garcia, N., . . . Carbonell, E. (1999). A complete human pelvis from the Middle Pleistocene of Spain. *Nature*, 399(6733), 255–258.

Arsuaga, J. L., Martínez, I., Gracia, A., Carretero, J. M., & Carbonell, E. (1993). Three new human skulls from the Sima de los Huesos Middle Pleistocene site in Sierra de Atapuerca, Spain. *Nature*, 362, 534–537.

Arsuaga, J. L., Martínez, I., Gracia, A., Carretero, J. M., Lorenzo, C., García, N., & Ortega, A. I. (1997). Sima de los Huesos (Sierra de Atapuerca, Spain). The site. *Journal of Human Evolution*, 33, 109–127.

Arzarello, M., Marcolini, F., Pavia, G., Pavia, M., Petronio, C., Petrucci, M., . . . Sardella, R. (2007). Evidence of earliest human occurrence in Europe: the site of Pirro Nord (Southern Italy). *Naturwissenschaften*, 94(2), 107–112.

Arzarello, M. & Peretto, C. (2010). Out of Africa: the first evidence of Italian peninsula occupation. *Quaternary International*, 223–224(0), 65–70.

Ascenzi, A., Biddittu, I., Cassoli, P. F., Segre, A. G., & Segre-Naldini, E. (1996). A calvarium of late *Homo erectus* from Ceprano, Italy. *Journal of Human Evolution*, 31, 409–423.

Ascenzi, A., Mallegni, F., Manzi, G., Segre, A. G., & Segre Naldini, E. (2000). A re-appraisal of Ceprano calvaria affinities with *Homo erectus*, after the new reconstruction. *Journal of Human Evolution*, 39, 443–450.

Asfaw, B. (1983). A new hominid parietal from Bodo, Middle Awash Valley, Ethiopia. *American Journal of Physical Anthropology*, 61, 387.

Asfaw, B. (1987). The Belohdelie frontal: new evidence of early hominid cranial morphology from the Afar of Ethiopia. *Journal of Human Evolution*, 16, 611–624.

Asfaw, B., Beyene, Y., Suwa, G., Walter, R. C., White, T. D., WoldeGabriel, G., & Yemane, T. (1992). The earliest Acheulan from Konso-Gardula. *Nature*, 360, 732–735.

Asfaw, B., Gilbert, W. H., Beyene, Y., Hart, W. K., Renne, P. R., WoldeGabriel, G., . . . White, T. D. (2002). Remains of *Homo erectus* from Bouri, Middle Awash, Ethiopia. *Nature*, 416, 317–320.

Asfaw, B., White, T., Lovejoy, O., Latimer, B., Simpson, S., & Suwa, G. (1999). *Australopithecus garhi*: a new species of early hominid from Ethiopia. *Science*, 284, 629–635.

Ashton, N., Lewis, S. G., Parfitt, S., & White, M. (2006). Riparian landscapes and human habitat preferences during the Hoxnian (MIS 11) Interglacial. *Journal of Quaternary Science*, 21(5), 497–505.

Aubert, M., Brumm, A., Ramli, M., Sutikna, T., Saptomo, E. W., Hakim, B., . . . Dosseto, A. (2014). Pleistocene cave art from Sulawesi, Indonesia. *Nature*, 514(7521), 223–227.

Avise, J. C. (2006). *Evolutionary Pathways in Nature. A Phylogenetic Approach*. Cambridge: Cambridge University Press.

Ayala, F. J. (1970). Invalidation of principle of competitive exclusion defended. *Nature*, 227(5253), 89–90.

Ayala, F. J. (1995a). Genes and origins. *Journal of Molecular Evolution*, 41, 683–688.

Ayala, F. J. (1995b). The myth of eve: molecular biology and human origins. *Science*, 270, 1930–1936.

Ayala, F.J. (2007). *Darwin's Gift to Science and Religion*. Washington, DC: Joseph Henry Press.

Ayala, F. J., Capó, M. S., Cela-Conde, C. J., & Nadal, M. (2007). Genetics and the human lineage. In A. Fagot-Largeault, J. M. Torres, & S. Rahman (Eds.), *The Influence of Genetics on Contemporary Thinking* (pp. 60–80). Dordrecht: Springer.

Ayala, F. J. & Escalante, A. (1996). The evolution of human populations: a molecular perspective. *Molecular Phylogenetics and Evolution*, 5, 188–201.

Ayala, F. J., Escalante, A., O'Huigin, C., & Klein, J. (1994). Molecular genetics of speciation and human origins. *Proceedings of the National Academy of Sciences*, 91, 6787–6794.

Azanza, B., Alberdi, M. T., & Prado, J. L. (2000). Large mammal turnover pulses correlated with latest Neogene glacial trends in the northwestern Mediterranean region. In M. B. Hart (Ed.), *Climates: Past and Present* (pp. 161–170). London: Geological Society.

Baab, K. L. (2007). *Cranial Shape Variation in Homo erectus*. (Ph.D. dissertation), City University of New York., New York, NY.

Baab, K. L. (2008). The taxonomic implications of cranial shape variation in *Homo erectus*. *Journal of Human Evolution*, 54(6), 827–847.

Baab, K. L. & McNulty, K. P. (2009). Size, shape, and asymmetry in fossil hominins: the status of the LB1 cranium based on 3D morphometric analyses. *Journal of Human Evolution*, 57(5), 608–622.

Baba, H., Aziz, F., Kaifu, Y., Suwa, G., Kono, R. T., & Jacob, T. (2003). *Homo erectus* Calvarium from the Pleistocene of Java. *Science*, 299(5611), 1384–1388.

Backwell, L. & D'Errico, F. (2001). Evidence of termite foraging by Swartkrans early hominids. *Proceedings of the National Academy of Sciences*, 98, 1358–1363.

Backwell, L. & d'Errico, F. (2008). Early hominid bone tools from Drimolen, South Africa. *Journal of Archaeological Science*, 35(11), 2880–2894.

Backwell, L., d'Errico, F., & Wadley, L. (2008). Middle Stone Age bone tools from the Howiesons Poort layers, Sibudu Cave, South Africa. *Journal of Archaeological Science*, 35(6), 1566–1580.

Bae, C. J. (2010). The late Middle Pleistocene hominin fossil record of eastern Asia: synthesis and review. *Yearbook of Physical Anthropology, 53,* 75–93.

Bae, C. J. & Bae, K. (2011). The nature of the Early to Late Paleolithic transition in Korea: current perspectives. *Quaternary International, 281,* 26–35.

Baena, J., Lordkipanidze, D., Cuartero, F., Ferring, R., Zhvania, D., Martín, D., . . . Rubio, D. (2010). Technical and technological complexity in the beginning: the study of Dmanisi lithic assemblage. *Quaternary International, 223–224*(0), 45–53.

Bahain, J.-J., Falguères, C., Voinchet, P., Duval, M., Dolo, J.-M., Despriée, J., . . . Tissoux, H. (2007). Electron spin resonance (ESR) dating of some European Late Lower Pleistocene sites. *Quaternaire, 18,* 175–186.

Bahn, P. G. (1998). Neanderthals emancipated. *Nature, 394,* 719–720.

Bailey, G. N. & King, G. C. P. (2011). Dynamic landscapes and human dispersal patterns: tectonics, coastlines, and the reconstruction of human habitats. *Quaternary Science Reviews, 30*(11–12), 1533–1553.

Bailey, W. J., Hayasaka, K., Skinner, C. G., Kehoe, S., Sieu, L., Slightom, J. L., & Goodman, M. (1992). Reexamination of the African hominoid trichotomy with additional sequences from the primate beta-globin gene cluster. *Molecular Phylogenetics and Evolution, 1,* 97–135.

Balme, J., Davidson, I., McDonald, J., Stern, N., & Veth, P. (2009). Symbolic behaviour and the peopling of the southern arc route to Australia. *Quaternary International, 202,* 59–68.

Balter, M. (2004a). Skeptic to take possession of flores hominid bones. *Science, 306,* 1450.

Balter, M. (2004b). Skeptics question whether flores hominid is a new species. *Science, 306,* 1116.

Banks, W. E., d'Errico, F., & Zilhão, J. (2013). Human–climate interaction during the Early Upper Paleolithic: testing the hypothesis of an adaptive shift between the Proto-Aurignacian and the Early Aurignacian. *Journal of Human Evolution, 64*(1), 39–55.

Bar-Yosef, O. (2000). The Middle and Early Upper Paleolithic in southwest Asia and neighboring regions. In O. Bar-Yosef & D. Pilbeam (Eds.), *The Geography of Neandertals and Modern Humans in Europe and the Greater Mediterranean* (pp. 107–156). Cambridge, MA: Peabody Museum of Archaeology and Ethnology, Harvard University.

Bar-Yosef, O. & Belfer-Cohen, A. (2001). From Africa to Eurasia—early dispersals. *Quaternary International, 75*(1), 19–28.

Bar-Yosef, O. & Belmaker, M. (2011). Early and Middle Pleistocene faunal and hominins dispersals through Southwestern Asia. *Quaternary Science Reviews, 30*(11–12), 1318–1337.

Bar-Yosef, O. & Bordes, J.-G. (2010). Who were the makers of the Châtelperronian culture? *Journal of Human Evolution, 59*(5), 586–593.

Bar-Yosef, O. & Vandermeersch, B. (1981). Notes concerning the possible age of the Mousterian layers in Qafzeh Cave. In P. Sanlaville & J. Cauvin (Eds.), *Préhistoire du Levant* (pp. 281–285). Paris: CNRS.

Bar-Yosef, O. & Vandermeersch, B. (1991). *Le Squelette Mousterien de Kebara.* Paris: CNRS.

Bar-Yosef, O. & Vandermeersch, B. (1993). Modern humans in the levant. *Scientific American, 268*(4), 94–100.

Bar-Yosef, O., Vandermeersch, B., Arensburg, B., Belfer-Cohen, A., Goldberg, P., Laville, H., . . . Weiner, S. (1992). The excavations in Kebara Cave, Mt. Carmel. *Current Anthropology, 33,* 497–534.

Barbetti, M. (1986). Traces of fire in the archaeological record, before one million years ago? *Journal of Human Evolution, 15*(8), 771–781.

Barbetti, M., Clark, J. D., Williams, F. M., & Williams, M. A. J. (1980). Palaeomagnetism and the search for very ancient fireplaces in Africa. *Anthropologie, 18,* 299–304.

Barbujani, G., Magagni, A., Minch, E., & Cavalli-Sforza, L. L. (1997). An apportionment of human DNA diversity. *Proceedings of the National Academy of Sciences, 94,* 4516–4451.

Barham, L. S., Pinto Llona, A. C., & Stringer, C. (2002). Bone tools from Broken Hill (Kabwe) Cave, Zambia, and their evolutionary significance. *Before Farming, 2,* 1–12.

Barluenga, M., Stölting, K. N., Salzburger, W., Muschick, M., & Meyer, A. (2006). Sympatric speciation in Nicaraguan crater lake cichlid fish. *Nature, 439*(7077), 719–723.

Barsky, D. & de Lumley, H. (2010). Early European Mode 2 and the stone industry from the Caune de l'Arago's archeostratigraphical levels "P". *Quaternary International, 223–224*(0), 71–86.

Barton, R. N. E. (2000). Mousterian hearths and shellfish: Late Neanderthal activities in Gibraltar. In C. B. Stringer, R. N. E. Barton, & J. C. Finlayson (Eds.), *Neanderthals on the Edge: 150th Anniversary Conference of the Forbes' Quarry Discovery, Gibraltar* (pp. 211–220). Oxford: Oxbow Books.

Bartstra, G.-J., Soegondho, S., & van der Wijk, A. (1988). Ngandong man: age and artifacts. *Journal of Human Evolution, 17,* 325–337.

Basell, L. S. (2008). Middle Stone Age (MSA) site distributions in eastern Africa and their relationship to Quaternary environmental change, refugia and the evolution of *Homo sapiens. Quaternary Science Reviews, 27*(27–28), 2484–2498.

Bastir, M. & Rosas, A. (2004). Comparative ontogeny in humans and chimpanzees: similarities, differences and paradoxes in postnatal growth and development of the skull. *Annals of Anatomy – Anatomischer Anzeiger, 186*(5–6), 503–509.

Beaulieu, C. (1998). *Australopithecus bahrelghazali* et faune associée. *Bipedia, 176*(16.2), 1–7.

Beauval, C., Maureille, B., Lacrampe-Cuyaubère, F., Serre, D., Peressinotto, D., Bordes, J. G., . . . Trinkaus, E. (2005). A late Neandertal femur from Les Rochers-de-Villeneuve, France. *Proceedings of the National Academy of Sciences, 192,* 7085–7090.

Bednarik, R. G. (1995). Concept-mediated marking in the Lower Paleolithic. *Current Anthropology, 36,* 605–634.

Bednarik, R. G. (1997a). The earliest evidence of ocean navigation. *The International Journal of Nautical Archaeology, 26*, 183–191.

Bednarik, R. G. (1997b). The global evidence of early human symboling behaviour. *Human Evolution, 12*, 147–168.

Bednarik, R. G. (2008). The origins of symbolism *Signs, 2*, 82–113.

Begun, D. R. (2007). How to identify (as opposed to define) a homoplasy: examples from fossil and living great apes. *Journal of Human Evolution, 52*(5), 559–572.

Begun, D. R., Ward, C. V., & Rose, M. D. (Eds.). (1997). *Function, Phylogeny, and Fossils: Miocene Hominoid Evolution and Adaptations*. New York, NY: Springer.

Behrensmeyer, A. K. & Laporte, L. F. (1981). Footprints of a Pleistocene hominid in northern Kenya. *Nature, 289*, 167–169.

Behrensmeyer, A. K., Todd, N. E., Potts, R., & McBrinn, G. E. (1997). Late Pliocene faunal turnover in the Turkana Basin, Kenya and Ethiopia. *Science, 278*, 1589–1594.

Belmaker, M., Tchernov, E., Condemi, S., & Bar-Yosef, O. (2002). New evidence for hominid presence in the Lower Pleistocene of the Southern Levant. *Journal of Human Evolution, 43*(1), 43–56.

Benazzi, S., Bailey, S. E., Peresani, M., Mannino, M. A., Romandini, M., Richards, M. P., & Hublin, J.-J. (2014). Middle Paleolithic and Uluzzian human remains from Fumane Cave, Italy. *Journal of Human Evolution, 70*(0), 61–68.

Benazzi, S., Douka, K., Fornai, C., Bauer, C. C., Kullmer, O., Svoboda, J., . . . Weber, G. W. (2011). Early dispersal of modern humans in Europe and implications for Neanderthal behaviour. *Nature, 479*(7374), 525–528.

Benazzi, S., Slon, V., Talamo, S., Negrino, F., Peresani, M., Bailey, S. E., . . . Hublin, J. J. (2015). The makers of the Protoaurignacian and implications for Neandertal extinction. *Science, 348*(6236), 793–796.

Bennett, M. R., Harris, J. W. K., Richmond, B. G., Braun, D. R., Mbua, E., Kiura, P., . . . Gonzalez, S. (2009). Early hominin foot morphology based on 1.5-million-year-old footprints from Ileret, Kenya. *Science, 323*(5918), 1197–1201.

Berckhemer, F. (1936). Der Urmenscheschädel von Steinheim. *Zeitschrift für Morphologie und Anthropologie, 35*, 463–516.

Berckhemer, F. (1938). Vorweisung des Steinheimer-Schädels in Original. *Verhandlungen der Deutschen Gesellschaft für Rassen- forschung, 9*, 190–192.

Berezikov, E., Cuppen, E., & Plasterk, R. H. A. (2006). Approaches to microRNA discovery. *Nature Genetics, 38*, S2–S7.

Berge, C. (1991). Quelle est la signification fonctionelle du pelvis très large de *Australopithecus afarensis* (AL 288–1)? In Y. Coppens & B. Senut (Eds.), *Origine(s) de la bipédie chez les hominidés* (pp. 113–119). Paris: Editions du CNRS.

Berge, C., Penin, X., & Pelle, E. (2006). New interpretation of Laetoli footprints using an experimental approach and Procrustes analysis: preliminary results. *Comptes Rendus Palevol, 5*, 561–569.

Berger, L. R., Churchill, S. E., De Klerk, B., & Quinn, R. L. (2008). Small-bodied humans from Palau, Micronesia. *PLoS One, 3*(3), e1780.

Berger, L. R. & Clarke, R. J. (1996). The load of the Taung Child. *Nature, 379*, 778–779.

Berger, L. R., de Ruiter, D. J., Churchill, S. E., Schmid, P., Carlson, K. J., Dirks, P. H. G. M., & Kibii, J. M. (2010). *Australopithecus sediba*: a new species of *Homo*-like australopith from South Africa. *Science, 328*(5975), 195–204.

Berger, L. R., Hawks, J., de Ruiter, D. J., Churchill, S. E., Schmid, P., Delezene, L. K., . . . Zipfel, B. (2015). *Homo naledi*, a new species of the genus *Homo* from the Dinaledi Chamber, South Africa. *eLife, 4*, http://dx.doi.org/10.7554/eLife.09560.09001.

Berger, L. R., Keyser, A. W., & Tobias, P. V. (1993). Brief Communication: Gladysvale: first early hominid site in South Africa Since 1948. *American Journal of Physical Anthropology, 92*, 107–111.

Berger, L. R., Lacruz, R., & de Ruiter, D. J. (2002). Revised age estimates of Australopithecus-bearing deposits at Sterkfontein, South Africa. *American Journal of Physical Anthropology, 119*, 192–197.

Berger, L. R. & Parkington, J. E. (1995). A new Pleistocene hominid-bearing locality at Hoedjiespunt, South Africa. *American Journal of Physical Anthropology, 98*, 601–609.

Bergström, T. & Gyllenstein, U. (1995). Evolution of Mhc class II polymorphism: the rise and fall of class II gene function in primates. *Immunological Reviews, 143*, 13–32.

Bermúdez de Castro, J. M. (1995). Nuevos datos sobre la biología del hombre de Atapuerca. *Fronteras de la ciencia y la tecnología, 9*, 52–55.

Bermúdez de Castro, J. M. & Arsuaga, J. L. (2000–2001). 1997–2001 El estatus de *Homo* antecessor. *Zephyrus, 53–54*, 5–14.

Bermúdez de Castro, J. M. & Nicolás, M. E. (1996). Changes in the lower premolar-size sequence during hominid evolution. Phylogenetic implications. *Human Evolution, 11*, 107–112.

Bermúdez de Castro, J. M., Arsuaga, J. L., Carbonell, E., Rosas, A., Martínez, I., & Mosquera, M. (1997). A hominid from the lower Pleistocene of Atapuerca, Spain: possible ancestor to Neandertals and modern Humans. *Science, 276*, 1392–1395.

Bermúdez de Castro, J. M., Carbonell, E., Cáceres, I., Díez, J. C., Fernández-Jalvo, Y., Mosquera, M., . . . van der Made, J. (1999). The TD6 (Aurora stratum) hominid site. Final remarks and new questions. *Journal of Human Evolution, 37*(3–4), 695–700.

Bermúdez de Castro, J. M., Martinón-Torres, M., Gómez-Robles, A., Prado-Simón, L., Martín-Francés, L., Lapresa, M., . . . Carbonell, E. (2011). Early Pleistocene human mandible from Sima del Elefante (TE) cave site in Sierra de Atapuerca (Spain): a comparative morphological study. *Journal of Human Evolution, 61*(1), 12–25.

Bermúdez de Castro, J. M., Martinón-Torres, M., Prado, L., Gómez-Robles, A., Rosell, J., López-Polín, L., . . . Carbonell, E. (2010). New immature hominin fossil from European Lower Pleistocene shows the earliest evidence of a modern human dental development pattern. *Proceedings of the National Academy of Sciences, USA, 107*, 11739–11744.

Bermúdez de Castro, J. M., Martinón-Torres, M., Robles, A. G., Prado, L., & Carbonell, E. (2010). New human evidence of the Early Pleistocene settlement of Europe, from Sima del Elefante site (Sierra de Atapuerca, Burgos, Spain). *Quaternary International, 223–224* (0), 431–433.

Bermúdez de Castro, J. M., Martinón-Torres, M., Sarmiento, S., & Lozano, M. (2003). Gran Dolina-TD6 versus Sima de los Huesos dental samples from Atapuerca: evidence of discontinuity in the European Pleistocene population? *Journal of Archaeological Science, 30*(11), 1421–1428.

Bermúdez de Castro, J. M., Pérez-González, A., Martinón-Torres, M., Gómez-Robles, A., Rosell, J., Prado, L., . . . Carbonell, E. (2008). A new early Pleistocene hominin mandible from Atapuerca-TD6, Spain. *Journal of Human Evolution, 55*(4), 729–735.

Bermúdez de Castro, J. M., Rosas, A., Arsuaga, J. L., Martínez, I., Carbonell, E., & Mosquera, M. (1997). *Homo antecessor*, una nueva especie del Pleistoceno inferior de Atapuerca. *Mundo científico, 181,* 649.

Bermúdez de Castro, J. M., Rosas, A., & Nicolás, M. E. (1999). Dental remains from Atapuerca-TD6 (Gran Dolina site, Burgos, Spain). *Journal of Human Evolution, 37*(3–4), 523–566.

Betzler, C. & Ring, U. (1995). Sedimentology of the Malawi Rift: facies and stratigraphy of the Chiwondo Beds, northern Malawi. *Journal of Human Evolution, 28*(1), 23–35.

Bever, M. R. (2001). An overview of Alaskan Late Pleistocene archaeology: historical themes and current perspectives. *Journal of World Prehistory, 15,* 125–191.

Bigazzi, G., Balestrieri, M. L., Norelli, P., Oddone, M., & Tecle, T., M,. (2004). Fission-track dating of a tephra layer in the Alat formation of the Dandiero Group (Danakil Depression, Eritrea). *Riv. It. Paleont. Strat., 110 (supplement),* 45–49.

Bilsborough, A. & Wood, B. (1988). Cranial morphometry of early hominids: facial region. *American Journal of Physical Anthropology, 76,* 61–86.

Binford, L. R. (1981). *Bones: Ancient Men and Modern Myths.* New York, NY: Academic Press.

Bischoff, J. L., Shamp, D. D., Aramburu, A., Arsuaga, J. L., Carbonell, E., & Bermúdez de Castro, J. M. (2003). The Sima de los Hueso hominids date to beyond U/Th equilibrium (>350 kyr) and perhaps to 400–500 kyr: new radiometric dates. *Journal of Archaeological Science, 30,* 275–280.

Bischoff, J. L., Williams, R. W., Rosenbauer, R. J., Aramburu, A., Arsuaga, J. L., García, N., & Cuenca-Bescós, G. (2007). High-resolution U-series dates from the Sima de los Huesos hominids yields: implications for the evolution of the early Neanderthal lineage. *Journal of Archaeological Science, 34*(5), 763–770.

Bjorkman, P. J., Saper, M. A., Samraoui, B., Bennett, W. S., Strominger, J. L., & Wiley, D. C. (1987a). The foreign antigen binding site and T cell recognition regions of class I histocompatibility antigens. *Nature, 329*(6139), 512–518.

Bjorkman, P. J., Saper, M. A., Samraoui, B., Bennett, W. S., Strominger, J. L., & Wiley, D. C. (1987b). Structure of the human class I histocompatibility antigen, HLA-A2. *Nature, 329*(6139), 506–512.

Black, D. (1927). On a lower molar hominid tooth from the Chou Kou Tien deposit. *Paleontol. sin., Ser D., 7,* 1–29.

Black, D. (1930). On an adolescent skull of *Sinanthropus pekinensis* in comparison with an adult skull of the same species and with other hominid skulls, recent and fossil. *Paleontol. sin., Ser D. vol. VII fascicle ii,* 1–145.

Black, D. (1931). Evidences for the use of fire by *Sinanthropus*. *Bulletin of the Geological Society of China, 11,* 107–198.

Black, D., Teilhard de Chardin, P., Young, C. C., & Pei, W. C. (1933). Fossil man in China. The Choukoutien cave deposits with a synopsis of our present knowledge of the late Cenozoic of China. *Mem. Geol. Surv. China Ser. A 11,* 1–158.

Blake, C. C. (1862). On the cranium of the most ancient races of man. *Geologist, June,* 206.

Blanc, A. C. (1940). *Industrie mousteriane et paleolithiche superiori nelle dune fossile e nelle grotte litoranee di Capo Palinuro* (Vol. 10). Roma: Sc. Fis. Mat. e nat.

Blumenschine, R. J. (1987). Characteristics of an early hominid scavenging niche. *Current Anthropology, 28,* 383–408.

Blumenschine, R. J., Peters, C. R., Masao, F. T., Clarke, R. J., Deino, A. L., Hay, R. L., . . . Ebert, J. I. (2003). Late Pliocene *Homo* and hominid land use from Western Olduvai Gorge, Tanzania. *Science, 299,* 1217–1221.

Boaz, N. T. & Howell, F. C. (1977). A gracile hominid from Upper Member G of the Shungura Formation, Ethiopia. *American Journal of Physical Anthropology., 46,* 93–10

Boaz, N. T., Ciochon, R. L., Xu, Q., & Liu, J. (2004). Mapping and taphonomic analysis of the *Homo erectus* loci at Locality 1 Zhoukoudian, China. *Journal of Human Evolution, 46*(5), 519–549.

Bobe, R. & Behrensmeyer, A. K. (2004). The expansion of grassland ecosystems in Africa in relation to mammalian evolution and the origin of the genus *Homo*. *Palaeogeography, Palaeoclimatology, Palaeoecology, 207,* 399–420.

Boesch, C., Head, J., & Robbins, M. M. (2009). Complex tool sets for honey extraction among chimpanzees in Loango National Park, Gabon. *Journal of Human Evolution, 56,* 560–569.

Boesch, C. & Tomasello, M. (1998). Chimpanzee and human cultures. *Current Anthropology, 39,* 591–595.

Boivin, P., Barbet, P., Boeuf, O., Devouard, B., Besson, J.-C., Hénot, J.-M., . . . Charles, L. (2010). Geological setting of the lower Pleistocene fossil deposits of Chilhac (Haute-Loire, France). *Quaternary International, 223–224*(0), 107–115.

Bon, F. (2006). A brief overview of Aurignacian cultures in the context of the industries of the transition from the Middle to the Upper Paleolithic. In O. Bar Yosef & J. Zilhão (Eds.), *Towards a Definition of the Aurignacian* (pp. 133–144). Lisboa: American School of Prehistoric Research/Instituto Português de Arqueologia.

Bonmatí, A., Gómez-Olivencia, A., Arsuaga, J.-L., Carretero, J. M., Gracia, A., Mart√ínez, I., . . . Carbonell, E. (2010). Middle Pleistocene lower back and pelvis from an aged human individual from the Sima de los Huesos site,

Spain. *Proceedings of the National Academy of Sciences, 107,* 18386–18391.

Bontrop, R. E. (1994). Nonhuman primate Mhc-DQA and DQB second exon nucleotide sequences: a compilation. *Immunogenetics, 39,* 81–92.

Bontrop, R. E., Otting, N., Slierendregt, B. L., & Lanchbury, J. S. (1995). Evolution of major histocompatibility complex polymorphisms and T-cell receptor diversity in primates. *Immunological Reviews, 143,* 33–62.

Booth, L. (2011). An evaluation of *Homo habilis* sensu lato variability through a comparative analysis of the coefficient of variation of three hominid species. *Totem: The University of Western Ontario Journal of Anthropology, 18,* Article 14.

Bordes, F. (1953). Essai de classification des industries «moustériennes». *Bulletin de la Société Préhistorique Française, 1953,* 457–466.

Bordes, F. (1979). *Typologie du Paléolithique ancien et moyen* (Vol. 1). Paris: CNRS.

Boule, M. (1908). L'Homme fossile de La Chapelle-aux-Saints (Corrèze). *Comptes-Rendus de l'Académie des Scenices de Paris, 147,* 1349–1352.

Boule, M. (1911–1913). L'Homme fossile de La Chapelle-aux-Saints. *Extraites Annales de Paléontologie, (1911) VI:111–72 (1–64); (1912) VII:21–192 (65–208); (1913) VIII: 1–70 (209–78).*

Bouyssonie, A., Bouyssonie, J., & Bardon, L. (1908). Découverte d'un squelette humain mousterian à la bouffia de la Chapelle-aux-Saints (Corrèze). *L'Anthropologie, 19,* 513–519.

Bowers, E. (2006). A new model for the origin of bipedality. *Human Evolution, 21*(3), 241–250.

Bowler, J. M. & Magee, J. (2000). Redating Australia's oldest human remains: a sceptics view. *Journal of Human Evolution, 38,* 719–726.

Bowler, J. M. & Thorne, A. G. (1976). Human remains from Lake Mungo. In R. L. Kirk & A. G. Thorne (Eds.), *The Origin of the Australians* (pp. 127–138). Canberra: Australian Institute of Aboriginal Studies.

Brace, C. L. (1964). The fate of the 'Classic' Neanderthals: a consideration of hominid catastrophism. *Current Anthropology, 5,* 3–43.

Brace, C. L. (1965). *The Stages of Human Evolution* (References are from the third edition, 1988 ed.). Englewood Cliffs, NJ: Prentice-Hall.

Brace, C. L., Mahler, P. E., & Rosen, R. B. (1973). Tooth measurements and the rejection of the taxon 'Homo habilis'. *Yearb. Phys. Anthropol. 1972, 16,* 50–58.

Brace, C. L., Ryan, A. S., & Smith, B. H. (1981). Tooth wear in La Ferrassie man: comment. *Current Anthropology, 22,* 426–430.

Brain, C. K. (1958). The transvaal ape-man bearing cave deposits. *Transvaal Museum Memoir, 11,* 1–125.

Brain, C. K. (1970). New finds at the Swartranks site. *Nature, 225,* 1112–1119.

Brain, C. K. (1982). *The Swartranks site: stratigraphy of the fossil hominids and a reconstruction of the environment of early Homo.*

Paper presented at the Congrès International de Paléontologie Humaine. Premier Congrés. Prétirage. Tome 2, Nice.

Brain, C. K. (1993). Structure and stratigraphy of the Swartkrans cave in the light of the new excavations. In C. K. Brain (Ed.), *Swartkrans: A Cave's Chronicle of Early Man* (pp. 23–34). Pretoria: Transvaal Museum.

Brain, C. K. & Sillen, A. (1988). Evidence from the Swartranks cave for the earliest use of fire. *Nature, 336,* 464–466.

Bramble, D. M. & Lieberman, D. E. (2004). Endurance running and the evolution of Homo. *Nature, 432*(7015), 345–352.

Brantingham, P. J. (1998). Hominid-carnivore coevolution and invasion of the predatory guild. *Journal of Anthropological Archaeology, 17*(4), 327–353.

Bräuer, G. (1984). A craniological approach to the origin of the anatomically modern *Homo sapiens* in Africa and its implications for the appearance of modern humans. In F. H. Smith & F. Spencer (Eds.), *The Origins of Modern Humans* (pp. 327–410). New York, NY: Alan R. Liss.

Bräuer, G. (1989). The evolution of modern humans: a comparison of the African and non-African evidence. In P. Mellars & C. Stringer (Eds.), *The Human Revolution: Behavioural and Biological Perspectives on the Origins of Modern Humans* (pp. 123–154). Princeton, NJ: Princeton University Press.

Bräuer, G. (2008). The origin of modern anatomy: by speciation or intraspecific evolution? *Evolutionary Anthropology, 17,* 22–37.

Bräuer, G. & Mehlman, M. J. (1988). Hominid molars from a Middle Stone Age level at Mumba Rock Shelter, Tanzania. *American Journal of Physical Anthropology, 75,* 69–76.

Bräuer, G. & Schultz, M. (1996). The morphological affinities of the Plio-Pleistocene mandible from Dmanisi, Georgia. *Journal of Human Evolution, 39,* 445–481.

Bräuer, G., Yokoyama, Y., Falguères, C., & Mbua, E. (1997). Modern human origins backdated. *Nature, 386,* 337.

Breuil, H. (1913). *Les subdivisions du Paléolithique supérieur et leur signification.* Paper presented at the Congrès International d'Anthropologie et d'Archéologie préhistoriques. Compte-rendu de la 14ème session, Genève 1912, tome 1, Genève.

Briggs, A. W., Good, J. M., Green, R. E., Krause, J., Maricic, T., Stenzel, U., & Pääbo, S. (2009). Primer Extension Capture: targeted Sequence Retrieval from Heavily Degraded DNA Sources. *JoVE, 31,* http://www.jove.com/index/Details.stp?ID=1573.

Briggs, A. W., Stenzel, U., Johnson, P. L. F., Green, R. E., Kelso, J., Prüfer, K., . . . Pääbo, S. (2007). Patterns of damage in genomic DNA sequences from a Neandertal. *Proceedings of the National Academy of Sciences, 104*(37), 14616–14621.

Broadfield, D. C., Holloway, R. L., Mowbray, K., Silvers, A., Yuan, M. S., & Márquez, S. (2001). Endocast of Sambungmacan 3 (Sm 3): a new *Homo erectus* from Indonesia. *The Anatomical Record, 262,* 369–379.

Broecker, W. S. & van Donk, J. (1970). Insolation changes, ice volumes, and the O18 record in deep-sea cores. *Reviews of Geophysics, 8*(1), 169–198.

Bromage, T. G. & Dean, M. C. (1985). Re-evaluation of the age at death of immature fossil hominids. *Nature*, 317(6037), 525–527.

Bromage, T. G., Schrenk, F., & Zonneveld, F. W. (1995). Paleoanthropology of the Malawi Rift: an early hominid mandible from the Chiwondo Beds, northern Malawi. *Journal of Human Evolution*, 28(1), 71–108.

Bronn, H. G. (1830). *Gaea Heidelbergensis, oder mineralogische Beschreibungen der Gegend von Heidelberg*. Heidelberg, Leipzig: Gross.

Brooks, A. S., Helgren, D. M., Cramer, J. S., Franklin, A., Hornyak, W., Keating, J. M., . . . Yellen, J. E. (1995). Dating and context of three Middle Stone Age sites with bone points in the Upper Semliki Valley, Zaire. *Science*, 268, 553–556.

Broom, R. (1938). The Pleistocene anthropoid apes of South Africa. *Nature*, 142, 377–379.

Broom, R. (1949). Another new type of fossil ape-man (*Paranthropus crassidens*). *Nature*, 163, 57

Broom, R. (1950). The genera and species of the South African fossil ape-man. *American Journal of Physical Anthropology*, 8, 1–13.

Broom, R. & Robinson, J. T. (1949). A new type of fossil man. *Nature*, 164, 322–323.

Broom, R. & Robinson, J. T. (1950). Man contemporaneous with the Swartranks ape-man. *American Journal of Physical Anthropology*, 8, 151–155.

Broom, R. & Robinson, J. T. (1952). Swartranks ape-man *Paranthropus crassidens*. *Transvaal Museum Memoir*, 6, 1–123.

Brose, D. S. & Wolpoff, M. H. (1971). Early Upper Paleolithic Man and late Middle Paleolithic tools. *American Anthropologist*, 73, 1156–1194.

Brown, B., Brown, F. H., & Walker, A. (2001). New hominids from the Lake Turkana Basin, Kenya. *Journal of Human Evolution*, 41(1), 29–44.

Brown, F., Harris, J., Leakey, R. E. F., & Walker, A. (1985). Early *Homo erectus* skeleton from west Lake Turkana, Kenya. *Nature*, 316, 788–792.

Brown, F. H. & Fuller, C. R. (2008). Stratigraphy and tephra of the Kibish Formation, southwestern Ethiopia. *Journal of Human Evolution*, 55(3), 366–403.

Brown, F. H., McDougall, I., Davies, T., & Maier, R. (1985). An integrated Plio-Pleistocene chronology for the Turkana Basin. In E. Delson (Ed.), *Ancestors: The Hard Evidence* (pp. 82–90). New York, NY: Alan R. Liss.

Brown, J. H., Marquet, P. A., & Taper, M. L. (1993). Evolution of body size: consequences of an energetic definition of fitness. *The American Naturalist*, 142, 573–584.

Brown, P. & Maeda, T. (2009). Liang Bua *Homo floresiensis* mandibles and mandibular teeth: a contribution to the comparative morphology of a new hominin species. *Journal of Human Evolution*, 57(5), 571–596.

Brown, P., Sutikna, T., Morwood, M. J., Soejono, R. P., Jatmiko, Wayhu Saptomo, E., & Awe Due, R. (2004). A new small-bodied hominin from the Late Pleistocene of Flores, Indonesia. *Nature*, 431(7012), 1055–1061.

Brown, W. M. (1980). Polymorphism in mitochondrial DNA of humans as revealed by restriction endonuclease analysis. *Proceedings of the National Academy of Sciences, USA*, 77, 4605–3609.

Bruce, E. J. & Ayala, F. J. (1979). Phylogenetic relationships between man and the apes: electrophoretic evidence. *Evolution*, 33, 1040–1056.

Brumm, A., Aziz, F., van den Bergh, G. D., Morwood, M. J., Moore, M. W., Kurniawan, I., . . . Fullagar, R. (2006). Early stone technology on Flores and its implications for *Homo floresiensis*. *Nature*, 441(7093), 624–628.

Brumm, A., Moore, M. W., van den Bergh, G. D., Kurniawan, I., Morwood, M. J., & Aziz, F. (2010). Stone technology at the Middle Pleistocene site of Mata Menge, Flores, Indonesia. *Journal of Archaeological Science*, 37(3), 451–473.

Brumm, A., van den Bergh, G. D., Storey, M., Kurniawan, I., Alloway, B. V., Setiawan, R., . . . Morwood, M. J. (2016). Age and context of the oldest known hominin fossils from Flores. *Nature*, 534(7606), 249–253.

Bruner, E. (2004). Geometric morphometrics and paleoneurology: brain shape evolution in the genus Homo. *Journal of Human Evolution*, 47, 279–303.

Bruner, E. (2010). Morphological differences in the parietal lobes within the human genus. *Current Anthropology*, 51(Supplement 1), S77–S88.

Bruner, E. & Holloway, R. L. (2010). A bivariate approach to the widening of the frontal lobes in the genus Homo. *Journal of Human Evolution*, 58(2), 138–146.

Bruner, E., Manzi, G., & Arsuaga, J. L. (2003). Encephalization and allometric trajectories in the genus *Homo*: evidence from the Neandertal and modern lineages. *Proceedings of the National Academy of Sciences*, 100, 15335–15340.

Brunet, M. (2008). *Origine et histoire des hominidés. Nouveaux paradigmes*. Paris: Fayard.

Brunet, M. (2010). Two new Mio-Pliocene Chadian hominids enlighten Charles Darwin's 1871 prediction. *Philosophical Transactions of the Royal Society B: Biological Sciences*, 365(1556), 3315–3321.

Brunet, M., Beauvilain, A., Coppens, Y., Heintz, E., Moutaye, A. H. E., & Pilbeam, D. (1995). The first australopithecine 2,500 kilometres west of the Rift Valley (Chad). *Nature*, 378, 273–275.

Brunet, M., Beauvilain, A., Coppens, Y., Heintz, E., Moutaye, A. H. E., & Pilbeam, D. (1996). *Australopithecus bahrelghazali*, une nouvelle espèce d'Hominidé ancien de la région de Koro Toro (Tchad). *Comptes-Rendus de l'Académie des Scenices de Paris, t. 322, série II a*, 907–913.

Brunet, M., Guy, F., Pilbeam, D., Lieberman, D. E., Likius, A., Mackaye, H. T., . . . Vignaud, P. (2005). New material of the earliest hominid from the Upper Miocene of Chad. *Nature*, 434(7034), 752–755.

Brunet, M., Guy, F., Pilbeam, D., Mackaye, H. T., Likius, A., Ahounta, D., . . . Zollikofer, C. (2002). A new hominid from the Upper Miocene of Chad, Central Africa. *Nature*, 418, 145–151.

Bulbeck, D. (2007). Where river meets sea: a parsimonious model for *Homo sapiens* colonization of the Indian Ocean rim and Sahul. *Current Anthropology*, *48*(2), 315–321.

Bunn, H. T. (1981). Archaeological evidence for meat-eating by Plio-Pleistocene hominids from Koobi Fora and Olduvai Gorge. *Nature*, *291*, 574–577.

Bush, M. E. (1980). The thumb of *Australopithecus afarensis*. *American Journal of Physical Anthropology*, 52, 210.

Bush, M. E., Lovejoy, C. O., Johanson, D. C., & Coppens, Y. (1982). Hominid carpal, metacarpal and phalangeal bones recovered from the Hadar formation: 1974–1977 Collections. *American Journal of Physical Anthropology*, 57, 651–667.

Butler, D. (2001). The battle of Tugen Hills. *Nature*, *410*, 508–509.

Butzer, K. W. (1971). *Recent History of an Ethiopian Delta*. Chicago, IL: Chicago University Press.

Butzer, K. W. (1974). Paleoecology of South African Australopithecines: Taung Revisited. *Current Anthropology*, *15*(4), 367–382.

Bye, B. A., Brown, F. H., Cerling, T. E., & McDougall, I. (1987). Increased age estimate for the Lower Paleolithic hominid site at Olorgesailie, Kenya. *Nature*, *329*, 237–239.

Callaway, E. (2015). Oldest stone tools raise questions about their creators. *Nature*, *520*, 421.

Campbell, B. (1964). Quantitative taxonomy and human evolution. In S. L. Washburn (Ed.), *Classification and Human Evolution* (pp. 50–74). London: Methuen and Co.

Cande, S. & Kent, D. V. (1995). Revised calibration of the geomagnetic polarity timescale for the Late Cretaceous and Cenozoic. *Journal of Geophysical Research*, *100*, 6093–6095.

Cann, H. M., de Toma, C., Cazes, L., Legrand, M.-F., Morel, V., Piouffre, L., . . . Cavalli-Sforza, L. L. (2002). A human genome diversity cell line panel. *Science*, *296*(5566), 261–262.

Cann, R. L., Stoneking, M., & Wilson, A. C. (1987). Mitochondrial DNA and human evolution. *Nature*, *325*, 31–36.

Caramelli, D., Lalueza-Fox, C., Vernesi, C., Martina Lari, S., Casoli, A., Mallegni, F., . . . Bertorelle, G. (2003). Evidence for a genetic discontinuity between Neandertals and 24,000-year-old anatomically modern Europeans. *Proceedings of the National Academy of Sciences*, *100*, 6593–6597.

Carbonell, E. & Mosquera, M. (2006). The emergence of a symbolic behaviour: the sepulchral pit of Sima de los Huesos, Sierra de Atapuerca, Burgos, Spain. *Comptes Rendus – Palevol*, *5*, 155–160.

Carbonell, E., Bermudez de Castro, J. M., Arsuaga, J. L., Allue, E., Bastir, M., Benito, A., . . . Verges, J. M. (2005). An Early Pleistocene hominin mandible from Atapuerca-TD6, Spain. *Proceedings of the National Academy of Sciences*, *102*(16), 5674–5678.

Carbonell, E., Bermúdez de Castro, J. M., Arsuaga, J. L., Díez, J. C., Rosas, A., Cuenca-Bescós, G., . . . Rodríguez, X. P. (1995). Lower Pleistocene hominids and artifacts from Atapuerca-TD6 (Spain). *Science*, *269*, 826–830.

Carbonell, E., Bermudez de Castro, J. M., Pares, J. M., Perez-Gonzalez, A., Cuenca-Bescos, G., Olle, A., . . . Arsuaga, J. L. (2008). The first hominin in Europe. *Nature*, *452*(7186), 465–469.

Carbonell, E., Mosquera, M., & Rodríguez, X. P. (2007). The emergence of technology: a cultural step or long-term evolution? *Comptes Rendus Palevol*, *6*(3), 231–233.

Carbonell, E., Mosquera, M., Rodríguez, X. P., & Sala, R. (1996). The first human settlement of Europe. *Journal of Anthropological Research*, *52*(1), 107–114.

Carbonell, E., Sala Ramos, R., Rodríguez, X. P., Mosquera, M., Ollé, A., Vergés, J. M., . . . Bermúdez de Castro, J. M. (2010). Early hominid dispersals: a technological hypothesis for 'out of Africa'. *Quaternary International*, *223–224*(0), 36–44.

Carlson, K. J., Stout, D., Jashashvili, T., de Ruiter, D. J., Tafforeau, P., Carlson, K., & Berger, L. R. (2011). The endocast of MH1, Australopithecus sediba. *Science*, *333*(6048), 1402–1407.

Carney, J., Hill, A., Miller, J. A., & Walker, A. (1971). Late Australopithecine from Baringo District, Kenya. *Nature*, *230*, 509–514.

Carrara, C., Frezzotti, M., & Giraudi, C. (1995). Stratigrafia plio-quaternaria. In C. Carrara (Ed.), *Lazio Meridionale, Sintesi delle ricerche geologiche multidisciplinari* (pp. 62–85). La Spezia, Italy: ENEA, Marine Environment Research Centre.

Carretero, J. M., Lorenzo, C., & Arsuaga, J. L. (1999). Axial and appendicular skeleton of Homo antecessor. *Journal of Human Evolution*, *37*(3–4), 459–499.

Carroll, S. B. (2003). Genetics and the making of *Homo sapiens*. Nature, *422*, 849–857.

Cavalli-Sforza, L. L., Menozzi, P., & Piazza, A. (1994). *The History and Geography of Human Genes*. Princeton, NJ: Princeton University Press.

Cela-Conde, C. J. (1998). The problem of hominoid systematics, and some suggestions for solving it. *South African Journal of Sciences*, *94*, 255–262.

Cela-Conde, C. J. (2001). Hominid taxon and the systematics of Hominoidea. In P. V. Tobias, M. A. Raath, J. Moggi-Cecchi, & G. A. Doyle (Eds.), *Humanity from African Naissance to Coming Millennia* (pp. 271–278). Firenze: Firenze University Press & Witwatersrand University Press.

Cela-Conde, C. J. & Ayala, F. J. (2001). *Senderos de la evolución humana*. Madrid: Alianza Editorial.

Cela-Conde, C. J. & Nadal, M. (2012). Taxonomical uses of the species concept in the human lineage. *Human Origins Research*, *2*(1), 1–9.

Cela-Conde, C. J., García-Prieto, J., Ramasco, J. J., Mirasso, C., Bajo, R., Munar, E., . . . Maestú, F. (2013). Dynamics of brain networks in the aesthetic appreciation. *Proceedings of the National Academy of Sciences, USA*, *110*(Supplement 2), 10454–10461.

Cela-Conde, C. J., Gomila, A., Munar, E., & Nadal, M. (2008). Taking Wittgenstein seriously. Indicators of the evolution of language. In D. M. A. Smith, K. Smith, & R. Ferrer i Cancho (Eds.), *The Evolution of Language: Proceedings of the*

7th International Conference (EVOLANG7) (pp. 407–408). Barcelona: World Scientific.

Cela-Conde, C. J. & Ayala, F. (2014). Brain keys in the appreciation of beauty: a tale of two worlds. *Rendiconti Lincei, 25,* 1–8.

Cela-Conde, C. J. & Ayala, F. J. (2007). *Human Evolution. Trails from the Past.* New York, NY: Oxford University Press.

Cerling, T. E., Brown, F. H., Cerling, B. W., Curtis, G. H., & Drake, R. E. (1979). Preliminary correlations between the Koobi Fora and Shungura Formations, East Africa. *Nature, 279,* 118–121.

Cerling, T. E., Mbua, E., Kirera, F. M., Manthi, F. K., Grine, F. E., Leakey, M. G., . . . Uno, K. T. (2011). Diet of *Paranthropus boisei* in the early Pleistocene of East Africa. *Proceedings of the National Academy of Sciences, 108*(23), 9337–9341.

Chaline, J., Durand, A., Marchand, D., Dambricourt Malassé, A., & Deshayes, M. J. (1996). Chromosomes and the origins of Apes and Australopithecines. *Human Evolution, 11,* 43–60.

Chaline, J., Dutrillaux, B., Couturier, J., Durand, A., & Marchand, D. (1991). Un modèle chomosomique et paléobiogéographique d'évolution des primates supérieurs. *Geobios, 24,* 105–110.

Chamberlain, A. T. (1987). *A Taxonomic Review and Phylogenetic Analysis of Homo habilis.* (Ph.D. dissertation), University of Liverpool.

Chamberlain, A. T. & Wood, B. (1987). Early hominid phylogeny. *Journal of Human Evolution, 16,* 119–133.

Chang, K. (2000). A breed apart. DNA tests: humans not descended from Neanderthals. *ABCNews-Science, 29/3/2000.*

Charrié-Duhaut, A., Porraz, G., Cartwright, C. R., Igreja, M., Connan, J., Poggenpoel, C., & Texier, P.-J. (2013). First molecular identification of a hafting adhesive in the Late Howiesons Poort at Diepkloof Rock Shelter (Western Cape, South Africa). *Journal of Archaeological Science, 40*(9), 3506–3518.

Chauhan, P. R. (2010). Comment on 'Lower and Early Middle Pleistocene Acheulian in the Indian sub-continent' by Gaillard et al. (2009) (Quaternary International). *Quaternary International, 223–224,* 248–259.

Chiarelli, B. (1962). Comparative morphometric analysis of primate chromosomes of the anthropoid apes and man. *Caryologia, 15,* 99–121.

Choi, Y. & Mango, S. E. (2014). Hunting for Darwin's gemmules and Lamarck's fluid: transgenerational signaling and histone methylation. *Biochimica et Biophysica Acta (BBA) – Gene Regulatory Mechanisms, 1839*(12), 1440–1453.

Churchill, S. E., Pearson, O. M., Grine, F. E., Trinkaus, E., & Holliday, T. W. (1996). Morphological affinities of the proximal ulna from Klasies River main site: archaic or modern? *Journal of Human Evolution, 31*(3), 213–237.

Churchill, S. E. & Smith, F. H. (2000). Makers of the early Aurignacian of Europe. *American Journal of Physical Anthropology, 113*(Suppl. 31), 61–115.

Ciochon, R. L. (1983). Hominoid cladistic and the ancestry of modern apes and humans. In R. L. Ciochon & R. S. Corruccini (Eds.), *New Interpretations of Ape and Human Ancestry* (pp. 783–837). New York, NY: Plenum Press.

Ciochon, R. L. (2009). The mystery ape of Pleistocene Asia. *Nature, 459*(7249), 910–911.

Ciochon, R. L. & Bettis III, E. A. (2009). Asian *Homo erectus* converges in time. *Nature, 458*(7235), 153–154.

Cipollari, M., Galderisi, U., & di Bernardo, G. (2005). Ancient DNA as a multidisciplinary experience. *Journal of Cellular Physiology, 202,* 315–322.

Clark, G. (1970). *Aspects of Prehistory.* Berkeley, CA: University of California Press.

Clark, G. A. (1969). *World Prehistory: A New Synthesis.* Cambridge: Cambridge University Press.

Clark, G. A. (1997a). The Middle-Upper Paleolithic transition in Europe: an American Perspective. *Norwegian Archaelogical Review, 30,* 25–53.

Clark, G. A. (1997b). Through a glass darkly. In G. A. Clark & C. M. Willermet (Eds.), *Conceptual Issues in Modern Human Origins* (Vol. 30, pp. 60–76). New York, NY: Aldine de Gruyter.

Clark, J. D. (1959). Further excavations at Broken Hill, Northern Rhodesia. *Journal of the Royal Anthropological Institute, 89,* 201–232.

Clark, J. D. (1995). Introduction to research on the Chiwondo Beds, northern Malawi. *Journal of Human Evolution, 28,* 3–5.

Clark, J. D., Beyene, Y., WoldeGabriel, G., Hart, W. K., Renne, P. R., Gilbert, H., . . . White, T. D. (2003). Stratigraphic, chronological and behavioural contexts of Pleistocene *Homo sapiens* from Middle Awash, Ethiopia. *Nature, 423*(6941), 747–752.

Clark, J. D., Brothwell, D. R., Powers, R., & Oakley, K. P. (1968). Rhodesian man: notes on a new femur fragment. *Man, 3,* 105–111.

Clark, J. D. & Harris, J. W. K. (1985). Fire and its roles in early hominid lifeways. *African Archaeological Reviews, 3,* 3–27.

Clark, J. D., Heinzelin, J., Schick, K. D., Hart, W. K., White, T. D., WoldeGabriel, G., . . . Selassie, Y. H. (1994). African *Homo erectus*: old radiometric ages and young Oldowan Assemblages in the Middle Awash Valley, Ethiopia. *Science, 264,* 1907–1920.

Clark, J. D. & Kurashina, H. (1979). Hominid occupation of the East-Central Highlands of Ethiopia in the Plio-Pleistocene. *Nature, 282,* 33–39.

Clark, W. L. G. (1938). The endocranial cast of the Swanscombe bones. *J. R. anthrop. Inst., 68,* 61–67.

Clarke, R. J. (1976). New cranium of *Homo erectus* from Lake Ndutu, Tanzania. *Nature, 262,* 485–487.

Clarke, R. J. (1977). *The Cranium of the Swartkrans Hominid SK 847 and Its Relevance to Human Origins.* (Ph.D. dissertation), University of Witwatersrand., Johannesburgh.

Clarke, R. J. (1988). A new *Australopithecus* cranium from Sterkfontein and its bearing on the ancestry of *Paranthropus.* In F. E. Grine (Ed.), *Evolutionary History of the Robust Australopithecines* (pp. 285–292). New York, NY: Aldine de Gruyter.

Clarke, R. J. (1990). The Ndutu cranium and the origin of Homo sapiens. *Journal of Human Evolution, 19*, 699–736.

Clarke, R. J. (1998). First ever discovery of a well-preserved skull and associated skeleton of *Australopithecus*. *South African Journal of Science, 94*, 460–463.

Clarke, R. J. (1999). Discovery of complete arm and hand of the 3.3 million-year-old *Australopithecus* skeleton from Sterkfontein. *South African Journal of Science, 95*, 477–480.

Clarke, R. J. (2000). A corrected reconstruction and interpretation of the *Homo erectus* calvaria from Ceprano, Italy. *Journal of Human Evolution, 39*, 433–442.

Clarke, R. J., Howell, F. C., & Brain, C. K. (1970). More evidence of an advanced hominid at Swartranks. *Nature, 225*, 1219–1222.

Clarke, R. J., Partridge, T. C., Granger, D. E., & Caffe, M. W. (2003). Dating the Sterkfontein fossils. *Science, 301*, 5956–5597.

Clarke, R. J. & Tobias, P. V. (1995). Sterkfontein member 2 foot bones of the oldest South African hominid. *Science, 269*, 521–524.

Close, A. (1984). Current research and recent radiocarbon dates from North Africa, II. *Journal of African History, 25*, 1–24.

Coe, R. S., Singer, B. S., Pringle, M. S., & Zhao, X. (2004). Matuyama-Brunhes reversal and Kamikatsura event on Maui: paleomagnetic directions, 40Ar/39Ar ages and implications. *Earth and Planetary Science Letters, 222*, 667–684.

Cohen, M. N. & Armelagos, G. J. (1984). *Paleopathology at the Origins of Agriculture*. Orlando: Academic Press.

Collard, M. & Aiello, L. C. (2000). From forelimbs to two legs. *Nature, 404*, 339–340.

Collard, M. & Wood, B. (1999). Grades among the African Early Hominids. In T. G. Bromage & F. Schrenk (Eds.), *African Biogeography. Climate Change & Human Evolution* (pp. 316–327). New York, NY: Oxford University Press.

Coltorti, M., Feraud, G., Marzoli, A., Peretto, C., Ton-That, T., Voinchet, P., . . . Thun Hohenstein, U. (2005). New 40Ar/39Ar, stratigraphic and palaeoclimatic data on the Isernia La Pineta Lower Palaeolithic site, Molise, Italy. *Quaternary International, 131*, 11–22.

Conard, N. J. (2003). Palaeolithic ivory sculptures from southwestern Germany and the origins of figurative art. *Nature, 426*(6968), 830–832.

Conard, N. J. & Bolus, M. (2015). Chronicling modern human's arrival in Europe. *Science, 348*(6236), 754–756.

Condemi, S. (1989). Décalage dans l'apparition des traits néanderthaliens sur le crâne cérébral chez les fossiles du Riss-Würm. In G. Giacobini (Ed.), *Hominidae* (pp. 357–362). Milano: Jaca Book.

Conroy, G. C. (1997). *Reconstructing Human Origins: A Modern Synthesis*. New York, NY: W.W. Norton & Company.

Conroy, G. C., Jolly, C. J., Cramer, D., & Kalb, J. E. (1978). Newly discovered fossil hominid skull from the Afar Depression, Ethiopia. *Nature, 275*, 339–406.

Conroy, G. C., Weber, G. W., Seidler, H., Tobias, P. V., Kane, A., & Brunsden, B. (1998). Endocranial capacity in an early hominid cranium from Sterkfontein, South Africa. *Science, 280*(5370), 1730–1731.

Constantino, P. & Wood, B. (2004). *Paranthropous* paleobiology. *Miscelánea en homenaje a Emiliano Aguirre. VOl. III: Paleoantropología, Madrid. Museo Arqueológico Regional*, 137–151.

Cook, J., Stringer, C., Currant, A., Schwarcz, H., & Wintle, A. (1982). A review of the chronology of the European middle Pleistocene hominid record. *Yearbook of Physical Anthropology, 25*, 19–65.

Cooke, H. B. S. (1964). Pleistocene mammal faunas of Africa, with particular reference to South Africa. In F. C. Howell & F. Bourlière (Eds.), *African Ecology and Human Evoltuion* (pp. 65–116). London: Methuen and Co.

Cooke, H. B. S. (1976). Suidae from Plio-Pleistocene strata of the Rudolph Basin. In Y. Coppens, F. Clark Howell, G. L. Isaac, & R. E. F. Leakey (Eds.), *Earliest Man and Environement in the Lake Rudolf Basin* (pp. 251–263). Chicago, IL: Chicago University Press.

Cooke, H. B. S. (1978). Suid evolution and correlation of African hominid localities: an alternative taxonomy. *Science, 201*, 460–463.

Cooper, A., Drummond, A. J., & Willerslev, E. (2004). Ancient DNA: would the real Neandertal please stand up? *Current Biology, 14*, R431–R433.

Cooper, A., Rambaut, A., Macaulay, V., Willerslev, E., Hansen, A. L., & Stringer, C. (2001). Human origins and ancient human DNA. *Science, 292*, 1655–1656.

Coppens, Y. (1972). Tentative de zonation du Pliocene et du Pléistocène d'Afrique par les grands Mammifères. *Comptes rendus des séances de l'Academie des Sciences*(274, Série D), 181–184.

Coppens, Y. (1978a). Evolution of the hominids and of their environment during the Plio-Pleistocene in the lower Omo Valley, Ethiopia. In W. W. Bishop (Ed.), *Geological Background to Fossil Man* (pp. 499–506). Edinburgh: Scottish Acad. Press.

Coppens, Y. (1978b). Les Hominidés du Pliocene et du Pléistocène d'Ethiopie; Chronologie, systématique, environnement. In F. Singer-Polignac (Ed.), *Les origines humaines et les époques de l'intelligence* (pp. 79–106). Paris: Masson.

Coppens, Y. (1980). The differences between *Australopithecus* and *Homo*: preliminary conclusions from the Omo research expedition's studies. In L. K. Koniggson (Ed.), *Current Argument on Early Man* (pp. 207–225). Stockholm: Pergamon Press.

Coppens, Y. (1983a). Les plus anciens fossiles d'hominidés. *Pontifical Academy of Science. Scripta Varia, 50*, 1–9.

Coppens, Y. (1983b). Systématique, phylogénie, environnement et culture des australopithèques, hypothèses et synthèse. *Bulletin et Memoires de la Societé d'Anthropologie. Paris, 10*, 273–284.

Coppens, Y. (1991). L'évolution des hominidés, de leur locomotion et de leurs environnements. In Y. Coppens & B. Senut (Eds.), *Origine(s) de la bipédie chez les hominidés* (pp. 295–301). Paris: Editions du CNRS.

Coppens, Y. (1994). East side story: the origin of mankind. *Scientific American, 270,* 62–69.

Coppens, Y. & Howell, F. C. (1976). Mammalian faunas of the Omo group: distributional and biostratigraphical aspects. In Y. Coppens, F. C. Howell, G. L. Isaac, & R. E. F. Leakey (Eds.), *Earliest Man and Environments in the Lake Rudolf Basin* (pp. 177–192). Chicago, IL: Chicago University Press.

Cracraft, J. (1974). Phylogenetic models and classification. *Systematic Zoology, 23,* 71–90.

Crawford, T., McKee, J., Kuykendall, K., Latham, A., & Conroy, G. C. (2004). Recent paleoanthropological excavations of in situ deposits at Makapansgat, South Africa—a first report. *Collegium Antropologicum, Suppl. 2,* 43–57.

Crochet, J.-Y., Welcomme, J.-L., Ivorra, J., Ruffet, G., Boulbes, N., Capdevila, R., . . . Pickford, M. (2009). Une nouvelle faune de vertébrés continentaux, associée à des artefacts dans le Pléistocène inférieur de l'Hérault (Sud de la France), vers 1,57 Ma. *Comptes Rendus Palevol, 8,* 725–736.

Crompton, R., H., Vereecke, E. E., & Thorpe, S. K. (2008). Locomotion and posture from the common hominoid ancestor to fully modern hominins, with special reference to the last common panin/hominin ancestor. *Journal of Anatomy, 212,* 501–543.

Crompton, R. H. & Pataky, T. C. (2009). Stepping out. *Science, 323*(5918), 1174–1175.

Crow, T. (1995). A Darwinian approach to the origins of psychosis. *British Journal of Psychiatry, 167,* 12–25.

Crow, T. J. (1993). Sexual selection, Machiavellian intelligence, and the origins of psychosis. *The Lancet, 342,* 594–598.

Crow, T. J. (1995). A theory of the evolutionary origins of psychosis. *European Neuropsychopharmacology, 5*(Supplement 1), 59–63.

Crow, T. J. (1996). Language and psychosis: common evolutionary origins. *Endeavour, 20,* 105–109.

Crow, T. J. (1997). Is schizophrenia the price that *Homo sapiens* pays for language? *Schizophrenia Research, 28,* 127–141.

Curnoe, D. (2008). On the Number of Ancestral Human Species. *Encyclopedia of Life Sciences.* Chichester: John Wiley & Sons.

Curnoe, D. (2010). A review of early *Homo* in southern Africa focusing on cranial, mandibular and dental remains, with the description of a new species (*Homo gautengensis* sp. nov.). *HOMO – Journal of Comparative Human Biology, 61*(3), 151–177.

Curnoe, D. (2011). A 150-year conundrum: cranial robusticity and its bearing on the origin of aboriginal Australians. *International Journal of Evolutionary Biology, 2011.*

Curnoe, D., Grün, R., Taylor, L., & Thackeray, F. (2001). Direct ESR dating of a Pliocene hominin from Swartkrans. *Journal of Human Evolution, 40,* 379–391.

Curnoe, D. & Thorne, A. (2003). Number of ancestral human species: a molecular perspective. *Homo, 53,* 201–224.

Curnoe, D. & Tobias, P. V. (2006). Description, new reconstruction, comparative anatomy, and classification of the Sterkfontein Stw 53 cranium, with discussions about the taxonomy of other southern African early *Homo* remains. *Journal of Human Evolution, 50*(1), 36–77.

Curnoe, D., Xueping, J., Herries, A. I. R., Kanning, B., Taçon, P. S. C., Zhende, B., . . . Rogers, N. (2012). Human remains from the Pleistocene-Holocene transition of southwest China suggest a complex evolutionary history for East Asians. *PLoS One, 7,* e31918.

Czarnetzki, A., Frangenberg, K.-W., & Pusch, C. M. (2007). Zwei neue Schädelfragmente der frühesten Vertreter der Gattung *Homo* aus Sarstedt, LDKR. Hildesheim. In H. Thieme (Ed.), *Die Schöninger Speere. Mensch und Jagd vor 400.000 Jahren.* (pp. 229–234). Sttutgart: Theiss.

Czarnetzki, A., Frangenberg, O., Frangenberg, K.-W., Gaudzinski, S., & Rohde, P. (2002). Die Neandertaler-Fundstätte bei Sarstedt, Landkreis Hildesheim. *Geologie, Archäologie und Anthropologie, 53,* 23–45.

Czarnetzki, A., Gaudzinski, S., & Pusch, C. M. (2001). Hominid skull fragments from Late Pleistocene layers in Leine Valley (Sarstedt, District of Hildesheim, Germany). *Journal of Human Evolution, 41*(2), 133–140.

d'Errico, F. (2003). The invisible frontier. A multiple species model for the origin of behavioral modernity. *Evolutionary Anthropology, 12,* 188–202.

d'Errico, F. & Backwell, L. (2009). Assessing the function of early hominin bone tools. *Journal of Archaeological Science, 36*(8), 1764–1773. 5

d'Errico, F. & Backwell, L. R. (2003). Possible evidence of bone tool shaping by Swartkrans early hominids. *Journal of Archaeological Science, 30*(12), 1559–1576.

d'Errico, F., Henshilwood, C., Vanhaeren, M., & van Niekerk, K. (2005). *Nassarius kraussianus* shell beads from Blombos Cave: evidence for symbolic behaviour in the Middle Stone Age. *Journal of Human Evolution, 48*(1), 3–24.

d'Errico, F., Zilhão, J., Julien, M., Baffier, D., & Pelegrin, J. (1998). Neanderthal acculturation in western Europe? A critical review of the evidence and its interpretation. *Current Anthropology, 39*(S1), S1–S44.

d'Errico, F., Vanhaeren, M., Henshilwood, C., Lawson, G., Maureille, B., Gambier, D., . . . van Niekerk, K. (2009). From the origin of language to the diversification of languages. What can archaeology and palaeoanthropology say? In F. d'Errico & J.-M. Hombert (Eds.), *Becpoming Eloquent* (pp. 13–67). Amsterdam: John Benjamins.

Dainton, M. (2001). Did our ancestors knuckle-walk? *Nature, 410,* 324–325.

Dalton, R. (2005). Looking for the ancestors. *Nature, 434,* 432–434.

Dart, R. (1925). *Australopithecus africanus*: the man-ape of South Africa. *Nature, 115,* 195–199.

Dart, R. (1948). The Makapansgat proto-human *Australopithecus prometheus. American Journal of Physical Anthropology, 6,* 259–284.

Dart, R. (1949a). Innominate fragments of *Australopithecus prometeus. American Journal of Physical Anthropology, 7,* 301–334.

Dart, R. (1949b). The predatory implemental technique of *Australopithecus. American Journal of Physical Anthropology, 7,* 1–16.

Dart, R. (1953). The predatory transition from ape to man. *International Anthropological and Linguistic Review, 1,* 201–218.

Dart, R. (1957). *The Osteodontokeratic Culture of Australopithecus Pprometheus.* Pretoria: Transvaal Museum.

Darwin, C. (1859). *On the Origin of Species by Means of Natural Selection.* London: John Murray.

Darwin, C. (1871). *The Descent of Man, and Selection in Relation to Sex.* London: John Murray.

Davidson, I. & McGrew, W. C. (2005). Stone tools and the uniqueness of human culture. *Journal of the Royal Anthropological Institute, 11,* 793–817.

Davis, P. R. (1964). Hominid fossils from Bed I, Olduvai Gorge, Tanganyika: a tibia and a fibula. *Nature, 3*(201), 967–970.

Day, M. H. (1969a). Femoral fragment of a robust australopithecine from Olduvai Gorge, Tanzania. *Nature, 221,* 230–233.

Day, M. H. (1969b). Omo human skeletal remains. *Nature, 223,* 1234–1239.

Day, M. H. (1971). Postcranial remains of *Homo erectus* from Bed IV, Olduvai Gorge, Tanzania. *Nature, 232,* 383–387.

Day, M. H. (1972). The Omo human skeletal remains. In F. Bordes (Ed.), *The Origin of Homo Sapiens* (pp. 31–35). Paris: Unesco.

Day, M. H. (1973). *The Development of Homo sapiens Darwin Centenary Symposium on the Origin of Man* (pp. 87–95). Rome: Accadémia Nazionale dei Lincei.

Day, M. H. (1982). The *Homo erectus* pelvis: punctuation or gradualism? *Cong. Int. Paléont. hum. I, Nice, Prétirage,* 411–421.

Day, M. H. (1984). The postcranial remains of *Homo erectus* from Africa, Asia and possibly Europe. In P. Andrews & J. L. Franzen (Eds.), *The Early Evolution of Man with Special Emphasis on Southeast Asia and Africa* (Vol. 69, pp. 113–121). Frankfurt: Cour. Forsch.-Inst. Senckenberg.

Day, M. H. (1986). *Guide to Fossil Man* (Fourth *Edition).* Chicago, IL: University of Chicago Press.

Day, M. H. (1992). Posture and childbirth. In S. Jones, R. Martin, & D. Pilbeam (Eds.), *The Cambridge Encyclopedia of Human Evolution* (pp. 88). Cambridge: Cambridge University Press.

Day, M. H. (1995). Remarkable delay. *Nature, 376,* 111.

Day, M. H. & Napier, J. R. (1964). Hominid fossils from Bed I, Olduvai Gorge, Tanganyika: fossil foot bones. *Nature, 201,* 969–970.

Day, M. H. & Stringer, C. B. (1982). A reconsideration of the Omo Kibish remains and the erectus-sapiens transition. In H. de Lumley (Ed.), *L'Homo erectus et la Place de l'Homme de Tautavel parmi les Hominidés Fossiles. Vol. 2* (Vol. Prétirage, pp. 814–846). Nice: Louis-Jean Scientific and Literary Publications.

Day, M. H. & Wickens, E. H. (1980). Laetoli Pliocene hominid footprints and bipedalism. *Nature, 286,* 385–387.

Day, M. H. & Wood, B. (1968). Functional affinities of the Olduvai Hominid 8 talus. *Man, 3,* 440–445.

de Heinzelin, J., Clark, J. D., Schick, K. D., & Gilbert, W. H. (Eds.). (2000). *The Acheulean and the Plio-Pleistocene Deposits of the Middle Awash Valley, Ethiopia.* Tervuren, Belgium: Royal Museum of Central Africa.

de Heinzelin, J., Clark, J. D., White, T., Hart, W., Renne, P., WoldeGabriel, G., . . . Vrba, E. (1999). Environment and behavior of 2.5-million-year-old Bouri hominids. *Science, 284,* 625–629.

de la Torre, I. (2011). The Early Stone Age lithic assemblages of Gadeb (Ethiopia) and the Developed Oldowan/early Acheulean in East Africa. *Journal of Human Evolution, 60*(6), 768–812.

de la Torre, I., Mora, R., & Martínez-Moreno, J. (2008). The early Acheulean in Peninj (Lake Natron, Tanzania). *Journal of Anthropological Archaeology, 27*(2), 244–264.

de Lumley, H. (1979). *L'homme de Tautavel* (Vol. 36).

de Lumley, H. & Barsky, D. (2004). Évolution des caractères technologiques et typologiques des industries lithiques dans la stratigraphie de la Caune de l'Arago. *L'Anthropologie, 108,* 185–237.

de Lumley, H. & de Lumley, M.-A. (1971). Découverte de restes humains anténéandertaliens datés du début de Riss à la Caune de l'Arago à Tautavel (Pyrénées-Orientales). *Comptes-Rendus de l'Académie des Scenices de Paris, 272,* 1729–1742.

de Lumley, H. & de Lumley, M. A. (1973). Pre-neanderthal human remains from Arago cave in South-eastern France. *Yearbook of Physical Antropology, 17,* 162–168.

de Lumley, H., de Lumley, M. A., & David, R. (1982). *Découverte et reconstruction de l'Homme de Tautavel.* Paper presented at the 1er congrès de paléontologie humaine, Nice.

de Lumley, H., Fournier, A., Park, Y. C., Yokoyama, Y., & Demouy, A. (1984). Stratigraphie du remplissage pléistocène moyen de la Caune de l'Arago à Tautavel. Étude de Huit carottages effectués de 1981 à 1983. *L'Anthropologie, 88,* 5–18.

de Lumley, M. A. & Sonakia, A. (1985). Première découverte d'un *Homo erectus* sur le continent indien à Hathnora, dans la moyenne vallée de la Narmada. *L'Antrhopologie, 89,* 13–61.

De Queiroz, K. & Donoghue, M. J. (1988). Phylogenetic systematics and the species problem. *Cladistics, 4,* 317–338.

de Ruiter, D. J., Pickering, R., Steininger, C. M., Kramers, J. D., Hancox, P. J., Churchill, S., E., . . . Backwell, L. (2009). New *Australopithecus robustus* fossils and associated U-Pb dates from Cooper's Cave (Gauteng, South Africa). *Journal of Human Evolution, 56,* 497–513.

de Ruiter, D. J., Sponheimer, M., & Lee-Thorp, J. A. (2008). Indications of habitat association of Australopithecus robustus in the Bloubank Valley, South Africa. *Journal of Human Evolution, 55*(5), 1015–1030.

de Ruiter, D. J., Steininger, C. M., & Berger, L. R. (2006). A cranial base of Australopithecus robustus from the hanging remnant of Swartkrans, South Africa. *American Journal of Physical Anthropology, 130,* 435–444.

de Villiers, H. (1973). Human skeletal remains from Border Cave, Ingwavumu District, KwaZulu, South Africa. *Annals of the Transvaal Museum, 28,* 229–256.

de Vos, J. & Sondaar, P. (1994). Dating hominid sites in Indonesia. *Science, 266*, 1726–1727.

Deacon, H. J. (1989). Late Pleistocene Palaeoecology and archaeology in the southern Cape, South Africa. In P. Mellars & C. Stringer (Eds.), *The Human Revolution: Behavioural and Biological Perspectives on the Origins of Modern Humans* (pp. 547–564). Princeton, NJ: Princeton University Press.

Deacon, H. J. (2001). Guide to Klasies River. academic.sun.ac.za/archaeology/KR**guide**2001.PDF.

Deacon, H. J. & Geleijnse, V. B. (1988). The stratigraphy and sedimentology of the main site sequence, Klasies River, South Africa. *South African Archaeological Bulletin, 43*(147), 5–14.

Dean, C., Leakey, M. G., Reid, R., Schrenk, F., Schwartzk, G. T., Stringer, C., & Walker, A. (2001). Growth processes in teeth distinguish modern humans from Homo erectus and earlier hominins. *Nature, 414*, 628–631.

Dean, D. & Delson, E. (1995). *Homo* at the gates of Europe. *Nature, 373*, 472–473.

Dean, D., Hublin, J. J., Holloway, R., & Ziegler, R. (1998). On the phylogenetic position of the pre-Neandertal specimen from Reilingen, Germany. *Journal of Human Evolution, 34*, 485–508.

Dean, M. C., Stringer, C. B., & Bromage, T. G. (1986). Age at death of the Neanderthal child from Devil's Tower, Gibraltar and the implications for studies of general growth and development in Neanderthals. *American Journal of Physical Anthropology, 70*(3), 301–309.

Dean, M. C. & Wood, B. (1988). Basicranial anatomy of Plio-Pleistocene hominids from East and South Africa. *American Journal of Physical Anthropology, 59*, 157–174.

Debetz, G. (1940). The anthropological features of the human skeleton from the cave of Teshik-Tash [in Russian]. *Trudy Uzbekist. Fil. Akad. Nauk., 1*, 46–49.

Deeb, S. S., Jorgensen, A. L., Battisti, L., Iwasaki, L., & Motulsky, A. G. (1994). Sequence divergence of the red and green visual pigments in the great apes and man. *Proceedings of the National Academy of Sciences, 91*, 7262–7266.

Deinard, A. & Kidd, K. (1999). Evolution of a HOXB6 intergenic region within the great apes and humans. *Journal of Human Evolution, 36*, 687–703.

Deino, A. & McBrearty, S. (2002). 40Ar/39Ar chronology for the Kapthurin Formation, Baringo, Kenya. *Journal of Human Evolution, 42*, 185–210.

Deino, A. L. (2011). 40Ar/39Ar dating of Laetoli, Tanzania. In T. Harrison (Ed.), *Paleontology and Geology of Laetoli: Human Evolution in Context* (Vol. 77–97). Amsterdam: Springer.

Deino, A. L. & Hill, A. (2002). 40Ar/39Ar dating of Chemeron Formation strata encompassing the site of hominid KNM-BC 1, Tugen Hills, Kenya. *Journal of Human Evolution, 42*(1–2), 141–151.

Deino, A. L. & Potts, R. (1990). Single-crystal 40Ar/39Ar dating of the Olorgesailie Formation, southern Kenya Rift. *Journal of Geophysical Research, B, Solid Earth and Planets, 95*, 8453–8470.

Deino, A. L., Scott, G. R., Saylor, B., Alene, M., Angelini, J. D., & Haile-Selassie, Y. (2010). 40Ar/39Ar dating, paleomag-netism, and tephrochemistry of Pliocene strata of the hominid-bearing Woranso-Mille area, west-central Afar Rift, Ethiopia. *Journal of Human Evolution, 58*(2), 111–126.

Deloison, Y. (1991). Les Australopithèques marchaient-ils comme nous? In Y. Coppens & B. Senut (Eds.), *Origine(s) de la bipédie chez les hominidés* (pp. 177–186). Paris: Editions du CNRS.

Deloison, Y. (1995). Le pied des premiers hominidés. *La Recherche, 281*, 52–55.

Delson, E. (1984). Cercopithecid biochronology of the African Plio–Pleistocene: correlations among eastern and southern hominid-bearing localities. *Courier Forschungsinstitut Senckenberg, 69*, 199–218.

Delson, E. (1986). Human phylogeny revised again. *Nature, 322*, 496–497.

Delson, E. (1988). Chronology of South African Australopith Site Units. In F. E. Grine (Ed.), *Evolutionary History of the "Robust" Australopithecines* (pp. 317–324). New York, NY: Aldine de Gruyter.

Delson, E. (1997). One skull does not a species make. *Nature, 389*, 445–446.

Delson, E., Eldredge, N., & Tattersall, I. (1977). Reconstruction of hominid phylogeny: a testable framework based on cladistic analysis. *Journal of Human Evolution, 6*, 263–278.

Delson, E. & Harvati, K. (2006). Return of the last Neanderthal. *Nature, 443*(7113), 762–763.

Delson, E., Harvati, K., Reddy, D., Marcus, L. F., Mowbray, K., Sawyer, G. J., . . . Márquez, S. (2001). The Sambungmacan 3 Homo erectus calvaria: a comparative morphometric and morphological analysis. *The Anatomical Record, 262*, 380–397.

deMenocal, P. B. (2004). African climate change and faunal evolution during the Pliocene-Pleistocene. *Earth and Planetary Science Letters (Frontiers), 220*, 3–24.

den Boer, P. J. (1986). The present status of the competitive exclusion principle. *Trends in Ecology & Evolution, 1*(1), 25–28.

Dennell, R. & Roebroeks, W. (1996). The earliest colonization of Europe: the short chronology revisited. *Antiquity, 70*, 535–542.

Dennell, R., Martinón-Torres, M., & Bermudéz de Castro, J. M. (2011). Hominin variability, climatic instability and population demography in Middle Pleistocene Europe. *Quaternary Science Reviews, 30*, 1511–1524.

Derricourt, R. (2005). Getting "out of Africa": sea crossings, land crossings and culture in the hominin migrations. *Journal of World Prehistory, 19*, 119–132.

DeSilva, J. M. (2011). A shift toward birthing relatively large infants early in human evolution. *Proceedings of the National Academy of Sciences, 108*(3), 1022–1027.

DeSilva, J. M. & Throckmorton, Z. J. (2010). Lucy's flat feet: the relationship between the ankle and rearfoot arching in early hominins. *PLoS One, 5*(12). DOI: 10.1371/journal.pone.0014432.

Despriée, J., Voinchet, P., Tissoux, H., Bahain, J.-J., Falguères, C., Courcimault, G., . . . Abdessadok, S. (2011). Lower and

Middle Pleistocene human settlements recorded in fluvial deposits of the middle Loire River Basin, Centre Region, France. *Quaternary Science Reviews*, *30*(11–12), 1474–1485.

Diamond, J. (2004). The astonishing micropygmies. *Science*, *306*(5704), 2047–2048.

Díez Martín, F. (2003). La aplicación de los "modos tecnológicos" en el análisis de las industrias paleolíticas. Reflexiones desde la perspectiva europea. *SPAL. Revista de prehistoria y arqueología de la Universidad de Sevilla*, *12*, 35–51.

Dirks, P. H. G. M., Berger, L. R., Roberts, E. M., Kramers, J. D., Hawks, J., Randolph-Quinney, P. S., . . . Tucker, S. (2015). Geological and taphonomic context for the new hominin species *Homo naledi* from the Dinaledi Chamber, South Africa. *eLife*, *4*, http://dx.doi.org/10.7554/eLife.09561.09001.

Dirks, P. H. G. M., Kibii, J. M., Kuhn, B. F., Steininger, C., Churchill, S. E., Kramers, J. D., . . . Berger, L. R. (2010). Geological setting and age of *Australopithecus sediba* from Southern Africa. *Science*, *328*(5975), 205–208.

Ditchfield, P. & Harrison, T. (2011). Sedimentology, lithostratigraphy and depositional history of the Laetoli area. In T. Harrison (Ed.), *Paleontology and Geology of Laetoli: Human Evolution in Context* (pp. 47–76). Dordrecht: Springer.

Dobzhansky, T. (1935). A critique of the species concept in biology. *Philosophy of Science*, *2*, 344–355.

Dobzhansky, T. (1937). *Genetics and the Origin of Species*. New York, NY: Columbia University Press.

Domínguez-Rodrigo, M., Barba, R., & Egeland, C. P. (2007). *Deconstructing Olduvai: A Taphonomic Study of the Bed I Sites*. Dordrecht: Springer.

Domínguez-Rodrigo, M., Mabulla, A., Luque, L., Thompson, J. W., Rink, J., Bushozi, P., . . . Alcala, L. (2008). A new archaic *Homo sapiens* fossil from Lake Eyasi, Tanzania. *Journal of Human Evolution*, *54*(6), 899–903.

Dominguez-Rodrigo, M., Pickering, T. R., Almecija, S., Heaton, J. L., Baquedano, E., Mabulla, A., & Uribelarrea, D. (2015). Earliest modern human-like hand bone from a new >1.84-million-year-old site at Olduvai in Tanzania. *Nature Communications*, *6*.

Dominguez-Rodrigo, M., Serrallonga, J., Juan-Tresserras, J., Alcala, L., & Luque, L. (2001). Woodworking activities by early humans: a plant residue analysis on Acheulian stone tools from Peninj (Tanzania). *Journal of Human Evolution*, *40*, 289–299.

Dorit, R. L., Akashi, H., & Gilbert, W. (1995). Absence of polymorphism at the ZFY Locus on the human Y chromosome. *Science*, *268*, 1183–1185.

Doronichev, V. & Golovanova, L. (2003). Bifacial tools in the Lower and Middle Paleolithic of Caucasus and their contexts. In M. Soressi & H. L. Dibble (Eds.), *Multiple Approaches to the Study of Bifacial Technologies* (pp. 77–107). Philadelphia, PA: University of Pennsylvania Museum of Archaeology and Anthropology.

Douglas, R. M. (2009). *A New Perspective on the Geohydrological and Surface Processes Controlling the Depositional Environment at the Florisbad Archaeozoological Site*. (Ph.D. dissertation), Free State University, Bloemfontein.

Douglas, R. M., Holmes, P. J., & Tredoux, M. (2010). New perspectives on the fossilization of faunal remains and the formation of the Florisbad archaeozoological site, South Africa. *Quaternary Science Reviews*, *29*(23,Äì24), 3275–3285.

Douka, K., Higham, T. F. G., Wood, R., Boscato, P., Gambassini, P., Karkanas, P., . . . Ronchitelli, A. M. (2014). On the chronology of the Uluzzian. *Journal of Human Evolution*, *68*(0), 1–13.

Drapeau, M. S. M., Ward, C. V., Kimbel, W. H., Johanson, D. C., & Rak, Y. (2005). Associated cranial and forelimb remains attributed to *Australopithecus afarensis* from Hadar, Ethiopia. *Journal of Human Evolution*, *48*(6), 593–642.

Drennan, M. R. (1953). The Saldanha skull and its associations. *Nature*, *172*, 791–793.

Drennan, M. R. (1954). The special features and status of the Saldanha skull. *American Journal of Physical Anthropology*, *13*, 625–634.

Dreyer, T. F. (1935). A human skull from Florisbad, Orange Free State, with a note on the endocranial cast, by C.U. Ariëns Kappers. *Kon. Ned. Akad. Wet.*, *38*, 118–128.

Dreyer, T. F. (1936). The endocranial cast of the Florisbad skull. *Soologiese Navorsing van die Nasional Museum Bloemfontein*, *1*, 21–23.

Duarte, C., Maurício, J., Pettitt, P. B., Souto, P., Trinkaus, E., van der Plicht, H., & Zilhão, J. (1999). The early Upper Paleolithic human skeleton from the Abrigo do Lagar Velho (Portugal) and modern human emergence in Iberia. *Proceedings of the National Academy of Sciences*, *96*, 7604–7609.

Dubois, E. (1894). *Pithecanthropus erectus. Eine menschanähnliche Übergangsform aus Java*. Batavia: Landsdruckerei.

Dubois, E. (1935). On the gibbon-like appearance of *Pithecanthropus erectus*. *Proceedings, Royal Academy of Amsterdam*, *38*.

Durband, A. C. (2006). Craniometric variation within the Pleistocene of Java: the Ngawi 1 cranium. *Human Evolution*, *21*(3), 193–201.

Durband, A. C. (2007). The view from down under: a test of the multiregional hypothesis of modern human origins using the basicranial evidence from Australasia. *Collegium Antropologicum*, *31*, 651–659.

Dusseldorp, G., Lombard, M., & Wurz, S. (2013). Pleistocene *Homo* and the updated Stone Age sequence of South Africa. *South African Journal of Science*, *5/6*, 1–7.

Duval, M., Moreno, D., Shao, Q., Voinchet, P., Falguères, C., Bahain, J., . . . Martínez, K. (2011). Datación por ESR del yacimiento arqueológico del Pleistoceno inferior de Vallparadís (Terrassa, Cataluña, España). *Trabajos de Prehistoria*, *68*, 7–24.

Duyfjes, J. (1936). Zur geologic und stratigraphie des Kendenggebietes zwischen Trinil und Soerabaja (Java). De Ingenieur in Nederlandsch-Indië, IV. *Mijbouw & Geologic, De Mijningenieur Jaargang III*, *8*, 136–149.

Eberz, G. W., Williams, F. M., & Williams, M. A. J. (1988). Plio-Pleistocene volcanism and sedimentary facies changes at Gadeb prehistoric site, Ethiopia. *Geologische Rundschau*, *77*, 513–527.

Eckhardt, R. B. (1977). Hominid origins: the Lothagam mandible. *Current Anthropology, 18,* 356.

Eglinton, G. & Logan, G. A. (1991). Molecular preservation. *Philosophical Transactions of the Royal Society B: Biological Sciences, 333,* 315–328.

Ehlers, J. & Gibbard, P. L. (2007). The extent and chronology of Cenozoic Global Glaciation. *Quaternary International, 164–165,* 6–20.

Elango, N., Thomas, J. W., Program, N. C. S., & Yi, S. V. (2006). Variable molecular clocks in hominoids. *Proceedings of the National Academy of Sciences, 103,* 1370–1375

Elias, S. A. (2007). Introduction. In S. A. Elias (Ed.), *Encyclopedia of Quaternary Science* (pp. 10–18). Amsterdam: Elsevier.

Enard, W., Przeworski, M., Fisher, S. E., Lai, C. S. L., Wiebe, V., Kitano, T., . . . Pääbo, S. (2002). Molecular evolution of *FOXP2,* a gene involved in speech and language. *Nature, 418,* 869–872.

Ennouchi, E. (1963). Les Néanderthaliens du Jebel Irhoud (Maroc). *Comptes-Rendus de l'Académie des Scenices de Paris, 256,* 2459–2460.

Etler, D. A., Crummett, T. L., & Wolpoff, M. H. (2001). Longuppo: Early *Homo* colonizer or Late Pliocene *Lufengpithecus* survivor in South China? *Human Evolution, 16,* 1–12.

Evans, A. R., Daly, E. S., Catlett, K. K., Paul, K. S., King, S. J., Skinner, M. M., . . . Jernvall, J. (2016). A simple rule governs the evolution and development of hominin tooth size. *Nature, 530(7591),* 477–480.

Evernden, J. F. & Curtiss, G. H. (1965). Potassium argon dating of Late Cenozoic rocks in East Africa and Italy. *Current Anthropology, 6,* 348–385.

Ewer, R. F. (1956). The dating of the Australopithecinae: faunal evidence. *South African Archaeological Bulletin, 11(42),* 41–45.

Facorellis, Y., Kyparissi, N., & Maniatis, Y. (2001). The cave of Theopetra, Kalambaka: radiocarbon evidence for 50,000 years of human presence. *Radiocarbon, 43,* 1029–1048.

Falguères, C. & Yokoyama, Y. (2001). Datation des hominidés fossiles en Asie. In F. Sémah, C. Falguères, D. Grimaud-Hervé, & A.-M. Sémah (Eds.), *Origine des peuplements et chronologie des cultures Paléolithiques dans le sud-est Asiatique* (pp. 55–64). Paris: Semenanjung.

Falguères, C., Bahain, J.-J., Duval, M., Shao, Q., Han, F., Lebon, M., . . . Garcia, T. (2010). A 300–600 ka ESR/U-series chronology of Acheulian sites in Western Europe. *Quaternary International, 223–224(0),* 293–298.

Falguères, C., Bahain, J.-J., Pérez-González, A., Mercier, N., Santonja, M., & Dolo, J.-M. (2006). The Lower Acheulian site of Ambrona, Soria (Spain): ages derived from a combined ESR/U-series model. *Journal of Archaeological Science, 33(2),* 149–157.

Falguères, C., Yokoyama, Y., Shen, G., Bischoff, J. L., Ku, T.-L., & de Lumley, H. (2004). New U-series dates at the Caune de l'Arago, France. *Journal of Archaeological Science, 31(7),* 941–952.

Falk, D. (2009). The natural endocast of Taung (*Australopithecus africanus*): insights from the unpublished papers of Raymond Arthur Dart. *Yearbook of Physical Anthropology, 52,* 49–65.

Falk, D., Hildebolt, C., Smith, K., Morwood, M. J., Sutikna, T., Brown, P., . . . Prior, F. (2005). The Brain of LB1, *Homo floresiensis. Science, 308,* 242–245.

Falk, D., Hildebolt, C., Smith, K., Morwood, M. J., Sutikna, T., Jatmiko, . . . Prior, F. (2007). Brain shape in human microcephalics and *Homo floresiensis. Proceedings of the National Academy of Sciences,* USA, 104 2513–2518.

Falk, D., Hildebolt, C., Smith, K., Morwood, M. J., Sutikna, T., Jatmiko, . . . Prior, F. (2009). LB1's virtual endocast, microcephaly, and hominin brain evolution. *Journal of Human Evolution, 57(5),* 597–607.

Fan, W., Kasahara, M., Gutknecht, J., Klein, D., Mayer, W. E., Jonker, M., & Klein, J. (1989). Shared class II MHC polymorphism between humans and chimpanzees. *Human Immunology, 26,* 107–121.

Farrand, W. R. (1979). Chronology and paleoenvironment of Levantine prehistoric sites as seen from sediment studies. *Journal of Archaeological Science, 6,* 369–392.

Feibel, C. S., Brown, F. H., & McDougall, I. (1989). Stratigraphic context of fossil hominids from the Omo group deposits: northern Turkana Basin, Kenya and Ethiopia. *American Journal of Physical Anthropology, 78,* 595–622.

Ferembach, D. (1976). Les reste humaines de la Grotte de Dar-es-Soltane 2 (Maroc) Campagne 1975. *Bull. Mém. Soc. Anthrop., Paris, 3,* 183–193.

Fernández García, M. G. (2001). *Análisis morfométricos de cráneos del Museo de Antropología de la Escuela de Medicina Legal.* Madrid: Universidad Complutense. Facultad de Ciencias Biológicas. http://eprints.ucm.es/tesis/ghi/ucm-t25040.pdf.

Fernández-Jalvo, Y., Denys, C., Andrews, P., Williams, T., Dauphin, Y., & Humphrey, L. (1998). Taphonomy and palaeoecology of Olduvai Bed-I (Pleistocene, Tanzania). *Journal of Human Evolution, 34,* 137–172.

Ferring, R., Oms, O., Agustí, J., Berna, F., Nioradze, M., Shelia, T., . . . Lordkipanidze, D. (2011). Earliest human occupations at Dmanisi (Georgian Caucasus) dated to 1.85–1.78 Ma. *Proceedings of the National Academy of Sciences, 108,* 10432–10436.

Figueroa, F., Gunther, E., & Klein, J. (1988). MHC polymorphism predating speciation. *Nature, 335,* 265–267.

FInarelli, J. A. & Clyde, W. C. (2004). Reassessing hominoid phylogeny: evaluating congruence in the morphological and temporal data. *Paleobiology, 30,* 614–651.

Finlayson, C. (2005). Biogeography and evolution of the genus Homo. *Trends in Ecology & Evolution, 20(8),* 457–463.

Finlayson, C., Giles Pacheco, F., Rodriguez-Vidal, J., Fa, D. A., Maria Gutierrez Lopez, J., Santiago Perez, A., . . . Sakamoto, T. (2006). Late survival of Neanderthals at the southernmost extreme of Europe. *Nature, 443,* 850–853.

Fischman, J. (1994). Putting our oldest ancestors in their proper place. *Science, 265,* 2011–2012.

Fitch, F. J. & Miller, J. A. (1970). Radioisotopic age determinations of Lake Rudolf artifact site. *Nature, 231,* 241–245.

Fitch, F. J., Hooker, P. J., & Miller, J. A. (1976). 40Ar/39Ar dating of the KBS Tuff in Koobi Fora Formation, East Rudolf, Kenya. *Nature*, *263*, 740–744.

Fleagle, J. G. (1992). Primate locomotion and posture. In S. Jones, R. Martin, & D. Pilbeam (Eds.), *The Cambridge Encyclopedia of Human Evolution* (pp. 75–79). Cambridge: Cambridge University Press.

Fleagle, J. G., Assefa, Z., Brown, F. H., & Shea, J. J. (2008). Paleoanthropology of the Kibish Formation, southern Ethiopia: introduction. *Journal of Human Evolution*, *55*(3), 360–365.

Flemming, N., Bailey, G., Courtillot, V., King, G., Lambeck, K., Ryerson, F., & Vita-Finzi, C. (2003). Coastal and marine palaeo-environments and human dispersal points across the Africa-Eurasia boundary. In C. A. Brebbia & T. Gambin (Eds.), *The Maritime and Underwater Heritage* (pp. 61–74). Southampton: Wessex Institute of Technology.

Foley, R. A., Maíllo-Fernández, J. M., & Mirazón Lahr, M. (2013). The Middle Stone Age of the Central Sahara: biogeographical opportunities and technological strategies in later human evolution. *Quaternary International*, *300*(0), 153–170.

Formicola, V. & Giannecchini, M. (1999). Evolutionary trends of stature in Upper Paleolithic and Mesolithic Europe. *Journal of Human Evolution*, *36*, 319–333.

Forrer, R. (1908). *Urgeschichte des Europäers von der Menschwerdung bis zum Anbruch der Geschichte*. Stuttgart: Spemann.

Forster, P. & Matsumura, S. (2005). Did early humans go north or south? *Science*, *308*(5724), 965–966.

Franciscus, R. G. (2009). When did the modern human pattern of childbirth arise? New insights from an old Neandertal pelvis. *Proceedings of the National Academy of Sciences*, *106*(23), 9125–9126.

Fu, Q., Hajdinjak, M., Moldovan, O. T., Constantin, S., Mallick, S., Skoglund, P., . . . Pääbo, S. (2015). An early modern human from Romania with a recent Neanderthal ancestor. *Nature*, *524*, 216–219.

Fu, Q., Posth, C., Hajdinjak, M., Petr, M., Mallick, S., Fernandes, D., . . . Reich, D. (2016). The genetic history of Ice Age Europe. *Nature*, *534*(7606), 200–205.

Gabunia, L., de Lumley, M. A., Vekua, A., Lordkipanidze, D., & de Lumley, H. (2002). Découverte d'un nouvel hominidé à Dmanissi (Transcaucasie, Géorgie). *C. R. Palevol.*, *I*, 243–253.

Gabunia, L. & Vekua, A. (1995). A Plio-Pleistocene hominid from Dmanisi, East Georgia, Caucasus. *Nature*, *373*, 509–512.

Gabunia, L., Vekua, A., & Lordkipanidze, D. (2000a). The environmental contexts of early human occupation of Georgia (Transcaucasia). *Journal of Human Evolution*, *38*(6), 785–802.

Gabunia, L., Vekua, A., & Lordkipanidze, D. (2000b). Taxonomy of the Dmanisi Crania (Response). *Science*, *289*(5476), 56.

Gabunia, L., Vekua, A., Lordkipanidze, D., Swisher III, C. C., Ferring, R., Justus, A., . . . Mouskhelishvili, A. (2000). Earliest Pleistocene hominid cranial remains from Dmanisi, Republic of Georgia: taxonomy, geological setting, and age. *Science*, *288*, 1019–1025.

Galiana, A., Moya, A., & Ayala, F. J. (1993). Founder-flush speciation in *Drosophila pseudoobscura*: a large-scale experiment. *Evolution*, *47*, 432–444.

Galik, K., Senut, B., Pickford, M., Gommery, D., Treil, J., Kuperavage, A. J., & Eckhardt, R. B. (2004). External and internal morphology of the BAR 1002'00 Orrorin tugenensis femur. *Science*, *305*(5689), 1450–1453.

Ganopolski, A. & Rahmstorf, S. (2001). Rapid changes of glacial climate simulated in a coupled climate model. *Nature*, *409*, 153–158.

Gantt, D. G. (1979). Comparative enamel histology of primate teeth. *Journal of Dental Research*, *58*, 1002–1003.

Gargett, R. (1989). Grave shortcomings: the evidence for Neandertal burial (Commentary). *Current Anthropology*, *30*, 157–180.

Garrigan, D. & Hammer, M. F. (2006). Reconstructing human origins in the genomic era. *Nature Reviews Genetics*, *7*(9), 669–680.

Garrigan, D., Mobasher, Z., Severson, T., Wilder, J. A., & Hammer, M. F. (2005). Evidence for archaic Asian ancestry on the human X chromosome. *Molecular Biology and Evolution*, *22*(2), 189–192.

Garrod, D. A. E. (1962). The Middle Palaeolithic of the Near East and the problem of Mount Carmel man. *Journal of the Royal Anthropological Institute*, *92*, 232–259.

Garrod, D. A. E. & Bate, D. M. A. (1937). *The Stone Age of Mount Carmel. Vol I. Excavations at the Wady el-Mughara*. Oxford: Clarendon Press.

Gathogo, P. N. & Brown, F. H. (2006). Revised stratigraphy of Area 123, Koobi Fora, Kenya, and new age estimates of its fossil mammals, including hominins. *Journal of Human Evolution*, *51*(5), 471–479.

Gause, G. F. (1934). *The Struggle for Existence*. Baltimore, MD: Williams & Wilkins.

Gee, H. (1995). Uprooting the human family tree. *Nature*, *373*, 15.

Gee, H. (1996). Box of bones 'clinches' identity of Piltdown paleontology hoaxer. *Nature*, *381*, 261–262.

Ghiselin, M. T. (1987). Species concepts, individuality, and objectivity. *Biology and Philosophy*, *2*, 127–143.

Gibbons, A. (1998). Old, old skull has a new look. *Science*, *280*, 1525.

Gibbons, A. (2004). Oldest human femur wades into controversy. *Science*, *305*(5692), 1885.

Gibbons, A. (2009). A new kind of ancestor: *Ardipithecus* unveiled. *Science*, *326*, 10–14.

Gibbs, S., Collard, M., & Wood, B. (2000). Soft-tissue characters in higher primate phylogenetics. *Proceedings of the National Academy of Sciences*, *97*, 1130–1132.

Gilbert, W. H., White, T. D., & Asfaw, B. (2003). *Homo erectus*, *Homo ergaster*, *Homo "cepranensis"*, and the Daka cranium. *Journal of Human Evolution*, *45*, 255–259.

Gillespie, R. (2002). Dating the first Australians. *Radiocarbon*, *44*, 455–472.

Gillespie, R. & Roberts, R. G. (2000). On the reliability of age estimates for human remains at Lake Mungo. *Journal of Human Evolution*, *38*, 727–730.

Gilligan, I. (2007). Neanderthal extinction and modern human behaviour: the role of climate change and clothing. *World Archaeology*, *39*(4), 499–514.

Giorgini, M. S. (1971). *Soleb II. Les Nécropoles*. Firenze: Sansoni.

Glazko, G. & Nei, M. (2003). Estimation of divergence times for major lineages of primate species. *Molecular Biology and Evolution*, *20*, 424–434.

Glazko, G., Veeramachaneni, G. G., Nei, M., & Makalowski, W. (2005). Eighty percent of proteins are different between humans and chimpanzees. *Gene*, *346*, 215–219.

Gleadow, A. J. W. (1980). Fission track age of the KBS Tuff and associated hominid remains in northern Kenya. *Nature*, *284*, 225–230.

Goldstein, D. B., Ruiz-Linares, A., Cavalli-Sforza, L. L., & Feldman, M. W. (1995). Microsatellite loci, genetic distances, and human evolution. *Proceedings of the National Academy of Sciences*, *92*, 6723–6727.

Gómez-Olivencia, A., Carretero, J. M., Lorenzo, C., Arsuaga, J. L., Bermúdez de Castro, J. M., & Carbonell, E. (2010). The costal skeleton of Homo antecessor: preliminary results. *Journal of Human Evolution*, *59*, 620–640.

Gómez-Olivencia, A., Eaves-Johnson, K. L., Franciscus, R. G., Carretero, J. M., & Arsuaga, J. L. (2009). Kebara 2: new insights regarding the most complete Neandertal thorax. *Journal of Human Evolution*, *57*(1), 75–90.

Gonder, M. K., Mortensen, H. M., Reed, F. A., de Sousa, A., & Tishkoff, S. A. (2007). Whole-mtDNA genome sequence analysis of ancient African lineages. *Molecular Biology and Evolution*, *24*(3), 757–768.

Gongming, Y., Hua, Z., Jinhui, Y., & Yanchou, L. (2002). Choronology of the stratum containing the skull of the Dali Man. *Chinese Science Bulletin*, *47*, 1302–1307.

Gonzalez-Jose, R., Escapa, I., Neves, W. A., Cuneo, R., & Pucciarelli, H. M. (2008). Cladistic analysis of continuous modularized traits provides phylogenetic signals in Homo evolution. *Nature*, *453*(7196), 775–778.

Goodall, J. M. (1964). Tool-using and aimed throwing in a community of free-living chimpanzees. *Nature*, *201*, 1264–1266.

Goodman, M. (1962). Evolution of the immunologic species specificity of human serum proteins. *Human Biology*, *34*, 104–150.

Goodman, M. (1963). Man's place in the phylogeny of the primates as reflected in serum proteins. In S. L. Washburn (Ed.), *Classification and Human Evolution* (pp. 204–234). Chicago, IL: Aldine.

Goodman, M. (1975). Protein sequence and immunological specifity. In W. P. Luckett & F. S. Szalay (Eds.), *Philogeny of the Primates* (pp. 219–248). New York, NY: Plenum Press.

Goodman, M. (1976). Towards a genealogical description of the primates. In M. Goodman & R. E. Tashian (Eds.), *Molecular Anthropology* (pp. 321–353). New York, NY: Plenum Press.

Goodman, M., Bailey, W. J., Hayasaka, K., Stanhope, M. J., Slightom, J., & Czelusniak, J. (1994). Molecular evidence on primate phylogeny from DNA sequences. *American Journal of Physical Anthropology*, *94*, 3–24.

Goodman, M., Porter, C. A., Czelusniak, J., Page, S. L., Schneider, H., Shoshani, J., . . . Groves, C. P. (1998). Toward a phylogenetic classification of primates based on DNA evidence complemented by fossil evidence. *Molecular Phylogenetics and Evolution*, *9*, 585–598.

Goodman, M., Poulik, E., & Poulik, M. D. (1960). Variations in the serum specifities of higher primates detected by two-dimensional starch-gel electrophoresis. *Nature*, *188*, 78–79.

Goodwin, A. J. H. & Van Riet Lowe, C. (1929). *The Stone Age Cultures of South Africa* (Vol. 27). Cape Town: Annals of the South African Museum.

Gordon, A. D., Nevell, L., & Wood, B. (2008). The *Homo floresiensis* cranium (LB1): size, scaling, and early Homo affinities. *Proceedings of the National Academy of Sciences*, *105*(12), 4650–4655.

Goren-Inbar, N. (1990). The lithic assemblages. In N. Goren-Inbar (Ed.), *Quneitra: A Mousterian Site on the Golan Heights* (pp. 61–149). Jerusalén: Monographs of the Institute of Archaeology, Hebrew University, 31.

Goren-Inbar, N., Alperson, N., Kislev, M. E., Simchoni, O., Melamed, Y., Ben-Nun, A., & Werker, E. (2004). Evidence of hominin control of fire at Gesher Benot Ya`aqov, Israel. *Science*, *304*(5671), 725–727.

Gracia, A., Arsuaga, J. L., Martínez, I., Lorenzo, C., Carretero, J. M., Bermúdez de Castro, J. M., & Carbonell, E. (2009). Craniosynostosis in the Middle Pleistocene human Cranium 14 from the Sima de los Huesos, Atapuerca, Spain. *Proceedings of the National Academy of Sciences*, *106*(16), 6573–6578.

Gracia, A., Martínez-Lage, J., Arsuaga, J.-L., Martínez, I., Lorenzo, C., & Pérez-Espejo, M.-Á. (2010). The earliest evidence of true lambdoid craniosynostosis: the case of "Benjamina", a *Homo heidelbergensis* child. *Child's Nervous System*, *26*(6), 723–727.

Graves, R. R., Lupo, A. C., McCarthy, R. C., Wescott, D. J., & Cunningham, D. L. (2010). Just how strapping was KNM-WT 15000? *Journal of Human Evolution*, *59*(5), 542–554.

Green, D. J. & Alemseged, Z. (2012). *Australopithecus afarensis* scapular ontogeny, function, and the role of climbing in human evolution. *Science*, *338*(6106), 514–517.

Green, R. E., Krause, J., Briggs, A. W., Maricic, T., Stenzel, U., Kircher, M., . . . Paabo, S. (2010). A draft sequence of the Neandertal genome. *Science*, *328*(5979), 710–722.

Green, R. E., Krause, J., Ptak, S. E., Briggs, A. W., Ronan, M. T., Simons, J. F., . . . Paabo, S. (2006). Analysis of one million base pairs of Neanderthal DNA. *Nature*, *444*(7117), 330–336.

Green, R. E., Malaspinas, A.-S., Krause, J., Briggs, A. W., Johnson, P. L. F., Uhler, C., . . . Pääbo, S. (2008). A complete Neandertal mitochondrial genome sequence determined by high-throughput sequencing. *Cell*, *134*(3), 416–426.

Greenfield, L. O. (1983). Toward the resolution of discrepancies between phenetic and paleontological data bearing on the question of human origins. In R. L. Ciochon & R. S. Corruccini (Eds.), *New Interpretations of Ape and Human Ancestry* (pp. 659–703). New York, NY: Plenum Press.

Griffiths, R. C. (1980). Lines of descent in the diffusion approximation of neutral Wright-Fisher models. *Theoretical Population Biology, 17*, 37–50.

Grimm, H., Mania, D., & Toepfer, V. (1974). Ein neuer Hominidenfund in Europa-Nachtrag zum Vorbericht fiber Bilzingsleben. *Kr. Artern. Z. Archäol., 8*, 175–176.

Grine, F. E. (1981). Trophic differences between gracile and robust *Australopithecus*: a scanning electron microscope analysis of occlusal events. *South Africa Journal of Science, 77*, 203–230.

Grine, F. E. (1985). Australopithecine evolution: the decidous dental evidence. In E. Delson (Ed.), *Ancestors: The Hard Evidence* (pp. 153–167). New York, NY: Alan R. Liss.

Grine, F. E. (1987). The diet of South African australopithecines based on a study of dental microwear. *L'Anthropologie, 91*, 467–482.

Grine, F. E. (1988). Evolutionary history of the 'robust' australopithecines: a Summary and historical perspective. In F. E. Grine (Ed.), *Evolutionary History of the "Robust" Australopithecines* (pp. 509–520). New York, NY: Aldine de Gruyter.

Grine, F. E. (2000). Middle Stone Age human fossils from Die Kelders Cave 1, Western Cape Province, South Africa. *Journal of Human Evolution, 38*, 129–145.

Grine, F. E., Bailey, R. M., Harvati, K., Nathan, R. P., Morris, A. G., Henderson, G. M., ... Pike, A. W. G. (2007). Late Pleistocene human skull from Hofmeyr, South Africa, and modern human origins. *Science, 315*(5809), 226–229.

Grine, F. E., Gunz, P., Betti-Nash, L., Neubauer, S., & Morris, A. G. (2010). Reconstruction of the late Pleistocene human skull from Hofmeyr, South Africa. *Journal of Human Evolution, 59*(1), 1–15.

Grine, F. E., Henshilwood, C. S., & Sealy, J. C. (2000). Human remains from Blombos Cave, South Africa: (1997–1998 excavations). *Journal of Human Evolution, 38*(6), 755–765.

Grine, F. E., Jungers, W. L., & Schultz, J. (1996). Phenetic affinities among early *Homo* crania from East and South Africa. *Journal of Human Evolution, 30*, 189–225.

Grine, F. E., Klein, R.-G., & Volman, T. P. (1991). Dating, archaeology and human fossils from the Middle Stone Age levels of Die Kelders, South Africa. *Journal of Human Evolution, 21*, 363–395.

Grine, F. E. & Klein, R. G. (1985). Pleistocene and Holocene human remains from Equus cave, South Africa. *Anthropology, 8*, 55–98.

Grine, F. E. & Martin, L. B. (1988). Enamel thickness and development in *Australopithecus* and *Paranthropus*. In F. E. Grine (Ed.), *Evolutionary History of the "Robust" Australopithecines* (pp. 3–42). New York, NY: Aldine de Gruyter.

Grine, F. E., Smith, H. F., Heesy, C. P., & Smith, E. J. (2009). Phenetic affinities of Plio-Pleistocene *Homo* fossils from South Africa: molar cusp proportions. In F. E. Grine, J. Fleagle, & R. E. Leakey (Eds.), *The First Humans. Origin and Early Evolution of the Genus Homo* (pp. 49–62). New York, NY: Springer. Vertebrate Paleobiology and Paleoanthropology Series.

Groves, C. P. (1986). Systematics of the Great Apes. In D. R. Swindler & J. Erwin (Eds.), *Comparative Primate Biology* (pp. 187–217). New York, NY: Alan R. Liss.

Groves, C. P. (1989). *A Theory of Human and Primate Evolution*. Oxford: Clarendon Press.

Groves, C. P. & Mazák, V. (1975). An approach to the taxonomy of the hominidae: gracile Villafranchian hominids of Africa. *Casopis pro Nineralogii Geologii, 20*, 225–247.

Groves, C. P. & Paterson, J. D. (1991). Testing hominoid phylogeny with the PHYLIP programs. *Journal of Human Evolution, 20*, 167–183.

Groves, C. P. & Thorne, A. (2000). The affinities of the Klasies River Motuh remains. *Perspectives in Human Biology, 5*, 43–53.

Grün, R., Huang, P.-H., Wu, X., Stringer, C. B., Thorne, A. G., & McCulloch, M. (1997). ESR analysis of teeth from the palaeoanthropological site of Zhoukoudian, China. *Journal of Human Evolution, 32*(1), 83–91.

Grün, R., Stringer, C., McDermott, F., Nathan, R., Porat, N., Robertson, S., ... McCulloch, M. (2005). U-series and ESR analyses of bones and teeth relating to the human burials from Skhul. *Journal of Human Evolution, 49*, 316–334.

Grün, R., Stringer, C., & Schwarcz, H. (1991). ESR dating of teeth from Garrod's Tabun cave collection. *Journal of Human Evolution, 20*, 231–248.

Grün, R. & Stringer, C. B. (1991). Electron spin resonance dating and the evolution of modern humans. *Archeometry, 33*, 153–199.

Grün, R. & Thorne, A. (1997). Dating the Ngandong humans. *Science, 276*, 1575.

Guanjun, S., Teh-Lung, K., Hai, C., Edwards, R. L., Zhenxin, Y., & Qian, W. (2001). High-precision U-series dating of Locality 1 at Zhoukoudian, China. *Journal of Human Evolution, 41*, 679–688.

Guanjun, S., Xing, G., Bin, G., & Granger, D. E. (2009). Age of Zhoukoudian Homo erectus determined with 26Al/10Be burial dating. *Nature, 458*(7235), 198–200.

Guipert, G., de Lumley, M.-A., de Lumley, H., & Mafart, B. (2004). Three-dimensional imagery: a new look at the Tautavel Man. In S. Wien (Ed.), *Computer Applications and Quantitative Methods in Archaeology* (pp. 100–102). Wien: BAR International series 1227.

Gunz, P., Bookstein, F. L., Mitteroecker, P., Stadlmayr, A., Seidler, H., & Weber, G. W. (2009). Early modern human diversity suggests subdivided population structure and a complex out-of-Africa scenario. *Proceedings of the National Academy of Sciences, USA, 106*, 6094–6098.

Gunz, P., Neubauer, S., Maureille, B., & Hublin, J.-J. (2010). Brain development after birth differs between Neanderthals and modern humans. *Current Biology, 20*(21), R921–R922.

Guy, F., Lieberman, D. E., Pilbeam, D., Ponce de León, M., Likius, A., Mackaye, H. T., ... Brunet, M. (2005). Morphological affinities of the Sahelanthropus tchadensis (Late Miocene hominid from Chad) cranium. *Proceed-*

ings of the National Academy of Sciences, USA, (102), 18836–18841.

Gyllenstein, U. B. & Erlich, H. A. (1989). Ancient roots for polymorphism at the HLA-DQ locus in primates. *Proceedings of the National Academy of Sciences, USA, 86*, 9986–9990.

Gyllenstein, U. B., Lashkari, D., & Erlich, H. A. (1990). Allelic diversification at the class H DQB locus of the mammalian major histocompatibility complex. *Proceedings of the National Academy of Sciences, USA, 87*, 1835–1839.

Haeckel, E. (1868). *Natürliche Schopfungsgeschichte*. Berlin: Reimer.

Haeckel, E. (1905). *The Wonders of Life*. New York, NY: Harper.

Haeusler, M. & McHenry, H. M. (2004). Body proportions of Homo habilis reviewed. *Journal of Human Evolution, 46*(4), 433–465.

Haeusler, M. & McHenry, H. M. (2007). Evolutionary reversals of limb proportions in early hominids? Evidence from KNM-ER 3735 (*Homo habilis*). *Journal of Human Evolution, 53*, 383–405.

Haidle, M. N. l. & Pawlik, A. F. (2010). The earliest settlement of Germany: is there anything out there? *Quaternary International, 223–224* (0), 143–153.

Haile-Selassie, Y. (2001). Late Miocene hominids from the Middle Awash, Ethiopia. *Nature, 412*, 178–181.

Haile-Selassie, Y., Asfaw, B., & White, T. D. (2004). Hominid cranial remains from Upper Pleistocene deposits at Aduma, Middle Awash, Ethiopia. *American Journal of Physical Anthropology, 123*(1), 1–10.

Haile-Selassie, Y., Gibert, L., Melillo, S. M., Ryan, T. M., Alene, M., Deino, A., . . . Saylor, B. Z. (2015). New species from Ethiopia further expands Middle Pliocene hominin diversity. *Nature, 521*(7553), 483–488.

Haile-Selassie, Y., Latimer, B. M., Alene, M., Deino, A. L., Gibert, L., Melillo, S. M., . . . Lovejoy, C. O. (2010). An early *Australopithecus afarensis* postcranium from Woranso-Mille, Ethiopia. *Proceedings of the National Academy of Sciences, 107*(27), 12121–12126.

Haile-Selassie, Y., Melillo, S. M., & Su, D. F. (2016). The Pliocene hominin diversity conundrum: do more fossils mean less clarity? *Proceedings of the National Academy of Sciences, USA, 113*(23), 6364–6371.

Haile-Selassie, Y., Saylor, B. Z., Deino, A., Alene, M., & Latimer, B. M. (2010). New hominid fossils from Woranso-Mille (Central Afar, Ethiopia) and taxonomy of early Australopithecus. *American Journal of Physical Anthropology, 141*(3), 406–417.

Haile-Selassie, Y., Saylor, B. Z., Deino, A., Levin, N. E., Alene, M., & Latimer, B. M. (2012). A new hominin foot from Ethiopia shows multiple Pliocene bipedal adaptations. *Nature, 483*(7391), 565–569.

Haile-Selassie, Y., Suwa, G., & White, T. D. (2004). Late Miocene Teeth from Middle Awash, Ethiopia, and early hominid dental evolution. *Science, 303*, 1503–1505.

Hammer, M. F. (1995). A recent common ancestry for human Y chromosomes. *Nature, 378*, 376–378.

Hammer, M. F., Woerner, A. E., Mendez, F. L., Watkins, J. C., & Wall, J. D. (2011). Genetic evidence for archaic admixture in Africa. *Proceedings of the National Academy of Sciences, 108*(37), 15123–15128.

Han, F. (2008). Effect of deposits alteration on dating the animal teeth from Caune de l'Arago site by combined Electron Spin Resonance (ESR) and uranium series methods. *Annali dell'Università degli Studi di Ferrara. Museologia Scientifica e Naturalistica, Volume speciale 2008*, 77–80.

Harcourt-Smith, W. E. H. & Aiello, L. C. (2004). Fossils, feet and the evolution of human bipedal locomotion. *Journal of Anatomy, 204*(5), 403–416.

Harris, J. H. & Leakey, M. G. (2003). Introduction. In J. H. Harris & M. G. Leakey (Eds.), *Geology and Vertebrate Paleontology of the Early Pliocene Site of Kanapoi, Northern Kenya* (Vol. Contributions in Science Number 498, pp. 1–7). Los Angeles, CA: Natural History Museum of Los Angeles County.

Harris, J. M. (1985). Age and paleoecology of the Upper Laetoli Beds, Laetoli, Tanzania. In E. Delson (Ed.), *Ancestors: The Hard Evidence* (pp. 76–81). New York, NY: Alan R. Liss.

Harris, J. M. & White, T. D. (1979). Evolution of the Plio-Pleistocene African Suidae. *Transactions of the American Philosophical Society, 69*, 1–128.

Harrison, T. (2010). Apes among the tangled branches of human origins. *Science, 327*(5965), 532–534.

Harrison, T. & Kweka, A. (2011). Paleontological Localities on the Eyasi Plateau, Including Laetoli. In T. Harrison (ed.), Paleontology and Geology of Laetoli: Human evolution in Context (pp. 17–45). Volume 1: Geology, Geochronology, Paleoecology and Paleoenvironment.

Harrison, T. (2011a). Hominins from the Upper Laetolil and Upper Ndolanya Beds, Laetoli. In T. Harrison (Ed.), *Paleontology and Geology of Laetoli: Human Evolution in Context* (pp. 141–188). Dordrecht: Springer.

Harrison, T. (2011b). The Laetoli hominins and associated fauna. In T. Harrison (Ed.), *Paleontology and Geology of Laetoli: Human Evolution in Context* (pp. 1–15). Dordrecht: Springer.

Harrod, J. B. (2014). Palaeoart at two million years ago? A review of the evidence. *Ars, 3*, 135–155.

Hartwig-Scherer, S. & Martin, R. D. (1991). Was "Lucy" more human than her "child"? Observations on early hominid post-cranial skeletons. *Journal of Human Evolution, 21*, 439–449.

Harvati, K., Frost, S. R., & McNulty, K. P. (2004). Neanderthal taxonomy reconsidered: implications of 3D primate models of intra- and interspecific differences. *Proceedings of the National Academy of Sciences, USA, 101*, 1147–1152.

Harvati, K., Panagopoulou, E., Karhanas, P., Athanassiou, A., & Frost, S. (2008). Preliminary results od the Aliakmon paleolithic/anthropological survey, Greece, 2004–2005. In A. Darlas & D. Mihailovic (Eds.), *The Paleolithic of the Balkans* (pp. 15–20). Oxford: Archaeopress.

Harvati, K., Panagopoulou, E., & Runnels, C. (2009). The paleoanthropology of Greece. *Evolutionary Anthropology, 18*, 131–143.

Harvati, K., Stringer, C., & Karkanas, P. (2011). Multivariate analysis and classification of the Apidima 2 cranium from

Mani, Southern Greece. *Journal of Human Evolution, 60*(2), 246–250.

Hattori, M., Fujiyama, A., Taylor, T. D., Watanabe, H., Yada, T., Park, H.-S., ... Yaspo, M.-L. (2000). The DNA sequence of human chromosome 21. *Nature, 405*(6784), 311–319.

Häusler, M. & Schmid, P. (1995). Comparison of the pelvis of Sts 14 and AL 288-1: implications for birth and sexual dimorphism in australopithecines. *Journal of Human Evolution, 29*, 363–383.

Hay, R. L. (1963). Stratigraphy of Beds I through IV, Olduvai Gorge, Tanganyika. *Science, 139*, 829–833.

Hay, R. L. (1971). *Geologic Background of Beds I and II. Stratigraphic Summary in Olduvai Gorge* (Vol. 3). Cambridge: Cambridge University Press.

Hay, R. L. & Leakey, M. D. (1982). Fossil footprints of Laetoli. *Scientific American Presents, February 1982*, 50–57.

Hedenström, A. (1995). Lifting the Taung's Child. *Nature, 378*, 670.

Hedges, D. J., Callinan, P. A., Cordaux, R., Xing, J., Barnes, E., & Batzer, M. A. (2004). Differential *Alu* mobilization and polymorphism among the human and chimpanzee lineages. *Genome Research, 14*, 1068–1075.

Heim, J. L. (1997). Lo que nos dice la nariz. *Mundo científico, 177*, 526–534.

Hein, J. (2004). Human evolution: pedigrees for all humanity. *Nature, 431*, 518–519.

Hemmer, H. (1972). Notes sur la position phylétique de l'homme de Petralona. *Anthropologie, 76*, 155–162.

Hennig, W. (1950). *Grundzüge einer Theorie der phylogenetischen Systematik*. Berlin: Aufbau.

Hennig, W. (1966). Phylogenetic systematics. *Annual Review of Entomology, 10*, 97–116.

Henshilwood, C. S. & Dubreuil, B. (2012). Style, symbolism, and complex technology: the Middle Stone Age in Southern Africa. A Response to Shea. *Current Anthropology, 53*(1), 132–133.

Henshilwood, C. S., d'Errico, F., & Watts, I. (2009). Engraved ochres from the Middle Stone Age levels at Blombos Cave, South Africa. *Journal of Human Evolution, 57*(1), 27–47.

Henshilwood, C. S., D'Errico, F., Marean, C. W., Milo, R. G., & Yates, R. (2001). An early bone tool industry from the Middle Stone Age at Blombos Cave, South Africa: implications for the origins of modern human behaviour, symbolism and language. *Journal of Human Evolution, 41*, 631–678.

Henshilwood, C., d'Errico, F., Vanhaeren, M., van Niekerk, K., & Jacobs, Z. (2004). Middle Stone Age shell beads from South Africa. *Science, 304*, 404.

Henshilwood, C. S., d'Errico, F., Yates, R., Jacobs, Z., Tribolo, C., Duller, G. A. T., ... Wintle, A. G. (2002). Emergence of modern human behavior: Middle Stone Age engravings from South Africa. *Sciencexpress*, http://www.sciencexpress.org / 10 January 2002 / Page 4/ 10.1126/science.1067575.

Herries, A. I. R. & Shaw, J. (2011). Palaeomagnetic analysis of the Sterkfontein palaeocave deposits: implications for the age of the hominin fossils and stone tool industries. *Journal of Human Evolution, 60*(5), 523–539.

Herrmann, E., Call, J., Hernandez-Lloreda, M. V., Hare, B., & Tomasello, M. (2007). Humans have evolved specialized skills of social cognition: the cultural intelligence hypothesis. *Science, 317*(5843), 1360–1366.

Higgs, E. S. (1961). Some Pleistocene faunas of the Mediterranean coastal areas. *Proceedings of the Prehistoric Society, 27*, 144–154.

Higham, T., Brock, F., Peresani, M., Broglio, A., Wood, R., & Douka, K. (2009). Problems with radiocarbon dating the Middle to Upper Palaeolithic transition in Italy. *Quaternary Science Reviews, 28*(13–14), 1257–1267.

Higham, T., Jacobi, R., Julien, M., David, F., Basell, L., Wood, R., ... Ramsey, C. B. (2010). Chronology of the Grotte du Renne (France) and implications for the context of ornaments and human remains within the Châtelperronian. *Proceedings of the National Academy of Sciences, 107*, 20234–20239.

Higuchi, R., Bowman, B., Freiberger, M., Ryder, O. A., & Wilson, A. C. (1984). DNA sequences from the quagga, an extinct member of the horse family. *Nature, 312*, 282–284.

Hill, A. (1999). The Baringo Basin, Kenya: from Bill Bishop to BPRP. In P. Andrews & P. Banham (Eds.), *Late Cenozoic Environments and Hominid Evolution: A Tribute to Bill Bishop* (pp. 85–97). London: Geological Society of London.

Hill, A. (2002). Paleoanthropological research in the Tugen Hills, Kenya. *Journal of Human Evolution, 42*, 1–10.

Hill, A., Drake, R., Tauxe, L., Monaghan, M., Barry, J. C., Behrensmeyer, A. K., ... Pilbeam, D. (1985). Neogene palaeontology and geochronology of the Baringo Basin, Kenya. *Journal of Human Evolution, 14*, 759–773.

Hill, A., Ward, S., & Brown, B. (1992). Anatomy and age of the Lothagam mandible. *Journal of Human Evolution, 22*, 439–451.

Hill, A., Ward, S., Deino, A., Curtis, G., & Drake, R. (1992). Earliest *Homo*. Nature, *355*, 719–722.

Hoffecker, J. F. (2009). The spread of modern humans in Europe. *Proceedings of the National Academy of Sciences, USA, 106*, 16040–16045.

Hofreiter, M., Serre, D., Poinar, H. N., Kuch, M., & Pääbo, S. (2001). Ancient DNA. *Nature Genetics, 2*, 353–359.

Holden, C. (1999). A new look into Neandertals' noses. *Science, 285*(5424), 31–33.

Holliday, T. W. & Franciscus, R. G. (2012). Humeral length allometry in African hominids (sensu lato) with special reference to A.L. 288-1 and Liang Bua 1. *PaleoAnthropology, 2012*, 1–12.

Holloway, R. L. (1973). New endocranial values for the East African early hominids. *Nature, 243*, 97–99.

Holloway, R. L. (1978). Problems of brain cast interpretation and African hominid evolution. In C. G. Jolly (Ed.), *Early Hominids of Africa* (pp. 379–401). London: Duckworth Press.

Holloway, R. L. (1980). The O.H. 7 (Olduvai Gorge, Tanzania) hominid partial brain endocast revisited. *American Journal of Physical Anthropology, 53*(2), 267–274.

Holloway, R. L. (1983). Cerebral brain endocast pattern of *Australopithecus afarensis*. Nature, *303*, 420–422.

Holton, N. E. & Franciscus, R. G. (2008). The paradox of a wide nasal aperture in cold-adapted Neandertals: a causal assessment. *Journal of Human Evolution*, *55*(6), 942–951.

Hooton, E. (1946). *Up from the Ape* (2nd *ed.*). New York, NY: The Macmillan Co.

Horai, S., Hayasaka, K., Kondo, R., Tsugane, K., & Takahata, N. (1995). Recent African origin of modern humans revealed by complete sequences of hominoid mitochondrial DNAs. *Proceedings of the National Academy of Sciences*, *92*, 532–536.

Höss, M. (2000). Neanderthal population genetics. *Nature*, *404*, 453–454.

Hou, Y., Potts, R., Yuan, B., Guo, Z., Deino, A., Wang, W., . . . Huang, W. (2000). Mid-Pleistocene Acheulan-like stone technology of the Bose Basin, South China. *Science*, *287*, 1622–1626.

Hovers, E., Vandermeersch, B., & Bar-Yosef, O. (1997). A Middle Palaeolithic engraved artefact from Qafzeh Cave, Israel. *Rock Art Research*, *14*, 79–87.

Howell, F. C. (1951). The place of Neanderthal man in human evolution. *American Journal of Physical Anthropology*, *9*, 379–416.

Howell, F. C. (1952). Pleistocene glacial ecology and the evolution of 'classic Neanderthal' man. *Southwestern Journal of Anthropology*, *8*, 377–410.

Howell, F. C. (1959). Upper Pleistocene stratigraphy and early man in the Levant. *Proceedings of the American Philosophical Society*, *103*, 1–65.

Howell, F. C. (1960). European and northwest African Middle Pleistocene Hominids. *Current Anthropology*, *1*, 195–232.

Howell, F. C. (1978). Hominidae. In V. J. Maglio & H. B. S. Cooke (Eds.), *Evolution of African Mammals* (pp. 154–248). Cambridge, MA: Harvard University Press.

Howell, F. C. (1999). Paleo-demes, species clades, and extinctions in the Pleistocene hominin record. *Journal of Anthropological Research*, *55*(2), 191–243.

Howell, F. C., Haesaerts, P., & de Heinzelin, J. (1987). Depositional environments, archeological occurences and hominids from Members E and F of the Shungura Formation (Omo basin, Ethiopia). *Journal of Human Evolution*, *1987*, 665–700.

Howells, W. W. (1973). *Cranial Variation in Man: A Study by Multivariate Analysis of Pattern of Differences Among Recent Human Populations*. Cambridge, MA: Harvard University Press.

Howells, W. W. (1974). Neanderthals: names, hypotheses, and scientific method. *American Anthropologist*, *76*, 24–38.

Howells, W. W. (1980). *Homo erectus*-who, when and where: a survey. *Yearbook of Physical Anthropology*, *23*, 1–23.

Howells, W. W. (1989). Skull shapes and the map. Craniometric analyses in the dispersion of modern *Homo*. *Papers of the Peabody Museum of Archaeology and Ethnology*, *79*, 1–189.

Howells, W. W. (1989). *Skull Shapes and the Map: Craniometric Analysis of Modern Homo*. Cambridge, MA: Harvard University Press.

Hrdlicka, A. (1930). The skeletal remains of early man. *Smithsonian Miscellaneous Collections*, *83*, 1–379.

Hsieh, W. P., Chu, T. M., Wolfinger, R. D., & Gibson, G. (2003). Mixed-model reanalysis of primate data suggests tissue and species biases in oligonucleotide-based gene expression profiles. *Genetics*, *165*, 747–757.

Hu, C. Z. (1973). Ape-man teeth from Yuanmou, Yunnan. *Acta Geologica Sinica*, *1*, 65–72.

Huang, W. & Hou, Y. (1997). Archaeological evidence for the first human colonisation of East Asia. *Indo-Pacific Prehistory Association Bulletin*, *16*, 3.12.

Huang, W., Ciochon, R., Gu, Y., Larick, R., Fang, Q., Schwarcz, H., . . . Rink, W. (1995). Early *Homo* and associated artifacts from Asia. *Nature*, *378*, 275–278.

Hublin, J. J. (1985). Human fossils from the North African Middle Pleistocene and the origin of *Homo sapiens*. In E. Delson (Ed.), *Ancestors: The Hard Evidence* (pp. 283–288). New York, NY: Alan R. Liss.

Hublin, J. J. (1986). Some comments on the diagnostic features of *Homo erectus*. *Anthropos*, *23*, 175–187.

Hublin, J.J., Harvati, K., Gunz, P., & Ben-Ncer, A. (2007). *A re-assessment of the Jebel Irhoud (Morocco) Mousterian adult cranial remains*. Paper presented at the PaleoAnthropology Society 2007 Meetings.

Hublin, J. J., Spoor, F., Braun, M., Zonneveld, F., & Condemi, S. (1996). A late Neanderthal associated with Upper Palaeolithic artefacts. *Nature*, *381*, 224–226.

Hublin, J. J. & Tillier, A. M. (1981). The Mousterian Juvenile Mandible from Irhoud (Morocco): A Phylogenetic Interpretation. In C. B. Stringer (Ed.), *Aspects of Human Evolution* (pp. 167–185). London: Taylor & Francis.

Hudson, R. R. (1990). Gene genealogies and the coalescent process. *Oxford Surveys in Evolutionary Biology*, *7*, 1–44.

Huffman, O. F. (2001). Geological context and age of the Perning/Mojokerto *Homo erectus*, East Java. *Journal of Human Evolution*, *40*, 353–362.

Huffman, O. F., Shipman, P., Hertler, C., de Vos, J., & Aziz, F. (2005). Historical evidence of the 1936 Mojokerto skull discovery, East Java. *Journal of Human Evolution*, *48*(4), 321–363.

Huffman, O. F., Zaim, Y., Kappelman, J., Ruez Jr., D. R., de Vos, J., Rizal, Y., . . . Hertler, C. (2006). Relocation of the 1936 Mojokerto skull discovery site near Perning, East Java. *Journal of Human Evolution*, *50*, 431–451.

Hughes, A. R. & Tobias, P. V. (1977). A fossil skull probably of the genus *Homo* from Sterkfontein, Transvaal. *Nature*, *265*, 310–312.

Hull, D. (1977). The ontological status of species as evolutionary units. In R. Butts & J. Hintikka (Eds.), *Foundational Problems in the Special Sciences* (pp. 91–102). Dordrecht: Reidel Publishing Company.

Hurford, A. J., Gleadow, A. J. W., & Naeser, C. W. (1976). Fission-track dating of pumice from the KBS Tuff, East Rudolf, Kenya. *Nature*, *263*, 738–740.

Huxley, J. (1942). *Evolution: The Modern Synthesis*. London: George Allen and Unwin.

Huxley, J. (1958). Evolutionary processes and taxonomy with special reference to grades. *Upssala Univ. Arssks.*, *6*, 21–38.

Huynen, L., Millar, C., Scofield, R. P., & Lambert, D. M. (2003). Nuclear DNA sequences detect species limits in ancient moa. *Nature, 425*, 175–178.

Hyodo, M., Nakaya, H., Urabe, A., Saegusa, H., Shunrong, X., Jiyun, Y., & Xuepin, J. (2002). Paleomagnetic dates of hominid remains from Yuanmou, China, and other Asian sites. *Journal of Human Evolution, 43*(1), 27–41.

Hyodo, M., Watanabe, N., Sunata, W., & Susanto, E. E. (1993). Magnetostratigraphy of hominid fossil bearing formations in Sangiran and Mojokerto, Java. *Anthropological Science, 101*, 157–186.

Igreja, M. & Porraz, G. (2013). Functional insights into the innovative Early Howiesons Poort technology at Diepkloof Rock Shelter (Western Cape, South Africa). *Journal of Archaeological Science, 40*(9), 3475–3491.

Indriati, E. & Antón, S. C. (2010). The calvaria of Sangiran 38, Sendangbusik, Sangiran Dome, Java. *HOMO - Journal of Comparative Human Biology, 61*(4), 225–243.

Indriati, E., Swisher, C. C., III, Lepre, C., Quinn, R. L., Suriyanto, R. A., Hascaryo, A. T., . . . Antón, S. C. (2011). The age of the 20 meter Solo River Terrace, Java, Indonesia and the Survival of *Homo erectus* in Asia. *PLoS One, 6*(6), e21562.

Ingman, M., Kaessmann, H., Pääbo, S., & Gyllensten, U. (2000). Mitochondrial genome variation and the origin of modern humans. *Nature, 408*, 708–713.

Isaac, G. L. (1969). Studies of early cultures in East Africa. *World Archaeology, 1*, 1–28.

Isaac, G. L. (1975). Stratigraphy and cultural patterns in East Africa during the middle ranges of Pleistocene time. In K. W. Butzer & G. L. Isaac (Eds.), *After the Australopithecines* (pp. 495–542). The Hague: Mouton.

Isaac, G. L. (1978). The food-sharing behavior of proto-human hominids. *Scientific American, 238*, 90–106.

Isaac, G. L. (1984). The archaeology of human origins: studies of the Lower Pleistocene in Africa. In F. Wendorf & A. Close (Eds.), *Advances in World Archaelogy 3* (pp. 1–87). New York, NY: Academic Press.

Ishida, H., Pickford, M., Nakaya, H., & Nakano, Y. (1984). Fossil anthropoids from Nachola and Samburu Hills, Samburu District, Kenya. *African Study Monograph, supplement 2*, 73–85.

Jablonski, N. (2004). The evolution of human skin and skin color. *Annual Review of Anthropology, 33*, 585–623.

Jablonski, N. & Chaplin, G. (2000). The evolution of human skin coloration. *Journal of Human Evolution, 39*(1), 57–106.

Jacob, T. (1975). The pithecanthropines of Indonesia. *Bulletins et Mémoires de la Société d'anthropologie de Paris, XIII° Série, 2*, 243–256.

Jacob, T., Indriati, E., Soejono, R. P., Hsu, K., Frayer, D. W., Eckhardt, R. B., . . . Henneberg, M. (2006). Pygmoid Australomelanesian *Homo sapiens* skeletal remains from Liang Bua, Flores: population affinities and pathological abnormalities. *Proceedings of the National Academy of Sciences, 103*, 13421–13426.

Jacobs, Z., Roberts, R. G., Galbraith, R. F., Deacon, H. J., Grun, R., Mackay, A., . . . Wadley, L. (2008). Ages for the Middle Stone Age of southern Africa: implications for human behavior and dispersal. *Science, 322*(5902), 733–735.

Jaeger, J. J. (1975). Decouverte d'un Crâne d'Hominidé dans le Pleistocene Moyen du Maroc. In C. i. C. n. 218 (Ed.), *Problémes actuels de Paléontologie* (pp. 897–902). Paris: CNRS.

Jakobsson, M., Scholz, S. W., Scheet, P., Gibbs, J. R., VanLiere, J. M., Fung, H.-C., . . . Singleton, A. B. (2008). Genotype, haplotype and copy-number variation in worldwide human populations. *Nature, 451*(7181), 998–1003.

James, S. R., Dennell, R. W., Gilbert, A. S., Lewis, H. T., Gowlett, J. A. J., Lynch, T. F., . . . Stahl, A. B. (1989). Hominid use of fire in the Lower and Middle Pleistocene: a review of the evidence [and Comments and Replies]. *Current Anthropology, 30*(1), 1–26.

Jankovic, I., Karavanic, I., Ahern, J. C. M., Brajkovic, D., Lenardic, J. M., & Smith, F. H. (2006). Vindija cave and the modern human peopling of Europe. *Collegium Antropologicum, 30*, 457–466.

Jankovic, I., Karavanic, I., Ahern, J. C. M., Brajkovic, D., Lenardic, J. M., & Smith, F. H. (2011). Archaeological, paleontological and genomic perspectives on late European Neandertals at Vindija Cave, Croatia. In S. Condemi & G.-C. Weniger (Eds.), *Continuity and Discontinuity in the Peopling of Europe: One Hundred Fifty Years of Neanderthal Study* (pp. 299–313). Dordrecht: Springer Science+Business Media B.V.

Jiménez-Arenas, J. M., Palmqvist, P., & Pérez-Claros, J. A. (2011). A probabilistic approach to the craniometric variability of the genus *Homo* and inferences on the taxonomic affinities of the first human population dispersing out of Africa. *Quaternary International, 243*, 219–230.

Johanson, D. C. (1976). Ethiopia yields first 'family' of early man. *National Geographic, 150*, 790–811.

Johanson, D. C. & Coppens, Y. (1976). A preliminary anatomical diagnosis of the first Plio-Pleistocene hominid discoveries in the central Afar, Ethiopia. *American Journal of Physical Anthropology, 45*, 217–234.

Johanson, D. C. & Edey, M. A. (1981). *Lucy: The Beginnings of Humankind*. New York, NY: Simon and Schuster.

Johanson, D. C. & Edgar, B. (1996). *From Lucy to Language*. New York, NY: Simon and Schuster.

Johanson, D. C., Masao, F. T., Eck, G. G., White, T. D., Walter, R. C., Kimbel, W. H., . . . Suwa, G. (1987). New partial skeleton of *Homo habilis* from Olduvai Gorge, Tanzania. *Nature, 327*, 205–209.

Johanson, D. C. & Taieb, M. (1976). Plio-Pleistocene hominid discoveries in Hadar, Ethiopia. *Nature, 260*, 293–297.

Johanson, D. C. & White, T. D. (1979). A systematic assessment of early African Hominids. *Science, 202*, 322–330.

Johanson, D. C., White, T., & Coppens, Y. (1978). A new species of the genus *Australopithecus* (Primates: Hominidae) from the Pliocene of Eastern Africa. *Kirtlandia, 28*, 1–14.

Jolly, C. J. (1970). The seed-eaters: a new model of hominid differentiation. *Man (New Series), 5*, 5–26.

Jolly, C.J. (2001). A proper study of mankind: analogies from the papionin monkeys and their implications for human evolution. *Yearbook of Physical Antropology, 44*, 177–204.

Jorde, L. B., Rogers, A. R., Bamshad, M., Watkins, W. S., Krakowiak, P., Sung, S., . . . Harpending, H. C. (1997). Microsatellite diversity and the demographic history of modern humans. *Proceedings of the National Academy of Sciences, 94*, 3100–3103.

Jouffroy, F. (1991). La 'main sans talon' du primate bipède. In Y. Coppens & B. Senut (Eds.), *Origine(s) de la bipédie chez les hominidés* (pp. 21–35). Paris: Editions du CNRS.

Jukang, W. (1980). Palaeanthropology in the New China. In L. K. Königsson (Ed.), *Current Argument on Early Man* (pp. 182–206). Oxford: Pergamon Press.

Jukang, W. & Tzekuei, C. (1959). New discovery of *Sinanthropus* mandible from Choukoutien. *Vertebr. Palastiat, 3*, 160–172.

Jungers, W. L. (1994). Ape and hominid limb lenght. *Nature, 369*, 194.

Jungers, W. L., Harcourt-Smith, W. E. H., Wunderlich, R. E., Tocheri, M. W., Larson, S. G., Sutikna, T., . . . Morwood, M. J. (2009). The foot of Homo floresiensis. *Nature, 459*(7243), 81–84.

Jungers, W. L., Larson, S. G., Harcourt-Smith, W., Morwood, M. J., Sutikna, T., Due Awe, R., & Djubiantono, T. (2009). Descriptions of the lower limb skeleton of Homo floresiensis. *Journal of Human Evolution, 57*(5), 538–554.

Kaessmann, H., Heissig, F., von Haeseler, A., & Pääbo, S. (1999). DNA sequence variation in a non-coding region of low recombination on the human X chromosome. *Nature Genetics, 22*, 78–81.

Kahn, P. & Gibbons, A. (1997). DNA from an extinct human. *Science, 277*, 176–178.

Kaifu, Y. (2006). Advanced dental reduction in Javanese Homo erectus. *Anthropological Science, 114*, 35–41.

Kaifu, Y., Zaim, Y., Baba, H., Kurniawan, I., Kubo, D., Rizal, Y., . . . Aziz, F. (2011). New reconstruction and morphological description of a Homo erectus cranium: Skull IX (Tjg-1993.05) from Sangiran, Central Java. *Journal of Human Evolution, 61*(3), 270–294.

Kalb, J. E., Jolly, C. J., Mebrate, A., Tebedge, S., Smart, C., Oswald, E. B., . . . Kana, B. (1982). Fossil mammals and artefacts from the Middle Awash Valley, Ethiopia. *Nature, 298*, 25–29.

Kalb, J. E., Wood, C. B., Smart, C., Oswald, E. B., Mabrete, A., Tebedge, S., & Whitehead, P. (1980). Preliminary geology and palaeontology of the Bodo D'ar hominid Site, Afar, Ethiopia. *Palaeogeography, Palaeoclimatology, Palaeoecology, 30*(0), 107–120.

Kappelman, J. (1997). They might be giants. *Nature, 387*, 126–127.

Karavanic, I. & Smith, F. H. (1998). The Middle/Upper Paleolithic interface and the relationship of Neanderthals and early modern humans in Hrvatsko Zagorje, Croatia. *Journal of Human Evolution, 34*, 223–248.

Katoh, S., Nagaoka, S., WoldeGabriel, G., Renne, P., Snow, M. G., Beyene, Y., & Suwa, G. (2000). Chronostratigraphy and correlation of the Plio-Pleistocene tephra layers of the Konso Formation, southern Main Ethiopian Rift, Ethiopia. *Quaternary Science Reviews, 19*(13), 1305–1317.

Kaufman, J., Völk, H., & Wall, H. J. (1995). A 'minimal essential Mhc' and an 'unrecognized Mhc': two extremes in selection for polymorphism. *Immunological Reviews, 143*, 63–88.

Kay, R. F. (1981). The nut-crackers: a new theory of the adaptations of the Ramapithecine. *American Journal of Physical Anthropology, 55*, 141–151.

Kay, R. F. & Grine, F. E. (1985). Tooth morphology, wear and diet in *Australopithecus* and *Paranthropus* from southern Africa. In F. E. Grine (Ed.), *Evolutionary History of the 'Robust' Australopithecines* (pp. 427–447). New York, NY: Aldine de Gruyter.

Keith, A. (1903). The extent to which the posterior segments of the body have been transmuted and suppressed in the evolution of man and allied primates. *Journal of Anatomy and Physiology, 37*, 271–273.

Keith, A. (1911). The early history of the Gibraltar cranium. *Nature, 87*, 314.

Kennard, A. S. (1942). Faunas of the high terrace at Swanscombe. *Proc. geol. Ass. Lnd., 53*, 105.

Kennedy, G. E. (1980). *Paleoanthropology*. New York, NY: McGraw-Hill.

Kennedy, G. E. (1991). On the autapomorphic traits of *Homo erectus*. *Journal of Human Evolution, 20*, 375–412.

Kennedy, G. E. (1999). Is "*Homo rudolfensis*" a valid species? *Journal of Human Evolution, 36*, 119–121.

Kennedy, K. A. R., Sonakia, A., Chiment, J., & Verma, K. K. (1991). Is the Narmada hominid an Indian *Homo erectus*? *American Journal of Physical Anthropology, 86*, 475–496.

Keyser, A. W. (2000). The Drimolen skull: the most complete australopithecine cranium and mandible to date. *South African Journal of Science, 96*, 189–193.

Keyser, A. W., Menter, C. G., Moggi-Cecchi, J., Pickering, T. R., & Berger, L. R. (2000). Drimolen: a new hominid-bearing site in Gauteng, South Africa. *South African Journal of Science, 96*, 193–197.

Khaitovich, P., Hellmann, I., Enard, W., Nowick, K., Leinweber, M., Franz, H., . . . Paabo, S. (2005). Parallel patterns of evolution in the genomes and transcriptomes of humans and chimpanzees. *Science, 309*(5742), 1850–1854.

Khrameeva, E. E., Bozek, K., He, L., Yan, Z., Jiang, X., Wei, Y., . . . Khaitovich, P. (2014). Neanderthal ancestry drives evolution of lipid catabolism in contemporary Europeans. *Nature Communications, 5*, 1–8.

Kibii, J. M., Churchill, S. E., Schmid, P., Carlson, K. J., Reed, N. D., de Ruiter, D. J., & Berger, L. R. (2011). A partial pelvis of Australopithecus sediba. *Science, 333*(6048), 1407–1411.

Kidder, J. H. & Durband, A. C. (2004). A re-evaluation of the metric diversity within *Homo erectus*. *Journal of Human Evolution, 46*, 299–315.

Kimbel, W. H., Johanson, D. C., & Rak, Y. (1994). The first skull and other new discoveries of *Australopithecus afarensis* at Hadar, Ethiopia. *Nature, 368*, 449–451.

Kimbel, W. H., Johanson, D. C., & Rak, Y. (1997). Systematic assessment of a maxilla of *Homo* From Hadar, Ethiopia. *American Journal of Physical Anthropology, 103,* 235–262.

Kimbel, W. H., Rak, T., & Johanson, D. C. (2004). *The Skull of Australopithecus afarensis.* New York, NY: Oxford University Press.

Kimbel, W. H., Lockwood, C. A., Ward, C. V., Leakey, M. G., Rak, Y., & Johanson, D. C. (2006). Was *Australopithecus anamensis* ancestral to *A. afarensis*? A case of anagenesis in the hominin fossil record. *Journal of Human Evolution, 51*(2), 134–152.

Kimbel, W. H. & Rak, Y. (1993). The importance of species in paleoanthropology and an argument for the phylogenetic concept of the species category. In W. H. Kimbel & L. B. LB Martin (Eds.), *Species, Species Concepts, and Primate Evolution* (pp. 461–484). New York, NY: Plenum.

Kimbel, W. H., Walter, R. C., Johanson, D. C., Reed, K. E., Aronson, J. L., Assefa, Z., . . . Smith, P. E. (1996). Late Pliocene *Homo* and Oldowan Tools from the Hadar Formation (Kada Hadar Member), Ethiopia. *Journal of Human Evolution, 31,* 549–561.

Kimbel, W. H. & White, T. D. (1988). Variation, sexual dimorphism and the taxonomy of *Australopithecus*. In F. E. Grine (Ed.), *Evolutionary History of the "Robust" Australopithecines* (pp. 175–192). New York, NY: Aldine de Gruyter.

Kimbel, W. H., White, T. D., & Johanson, D. C. (1984). Cranial morphology of *Australopithecus afarensis*: a comparative study based on a composite reconstruction of the adult skull. *American Journal of Physical Anthropology, 64,* 337–388.

Kimbel, W. H., White, T. D., & Johanson, D. C. (1988). Implications of KNM-WT 17000 for the evolution of "Robust" *Australopithecus*. In F. E. Grine (Ed.), *Evolutionary History of the "Robust" Australopithecines* (pp. 259–268). New York, NY: Aldine de Gruyter.

King, W. (1864). The reputed fossil man of the Neanderthal. *Quarterly Journal of Science, 1,* 88–97.

Kingman, J. F. C. (1982a). The coalescent. *Stoch. Proc. Appl., 13,* 235–248.

Kingman, J. F. C. (1982b). On the genealogy of large populations. *J. Appl. Probab., 19,* 27–43.

Kingston, J. D., Marino, B. D., & Hill, A. (1994). Isotopic evidence for Neogene hominid paleoenvironments in the Kenya Rift Valley. *Science, 264,* 955–959.

Kivell, T. L. & Begun, D. R. (2006). Frequency and timing of scaphoid-centrale fusion in hominoids. *Journal of Human Evolution, 55,* 321–340.

Kivell, T. L., Deane, A. S., Tocheri, M. W., Orr, C. M., Schmid, P., Hawks, J., . . . Churchill, S. E. (2015). The hand of *Homo naledi*. *Nature Communication, 6.* DOI:10.1038/ncomms9431.

Kivell, T. L., Kibii, J. M., Churchill, S. E., Schmid, P., & Berger, L. R. (2011). Australopithecus sediba hand demonstrates mosaic evolution of locomotor and manipulative abilities. *Science, 333*(6048), 1411–1417.

Kivell, T. L. & Schmitt, D. (2009). Independent evolution of knuckle-walking in African apes shows that humans did not evolve from a knuckle-walking ancestor. *Proceedings of the National Academy of Sciences, USA, 106,* 14241–14246.

Kivisild, T., Shen, P., Wall, D. P., Do, B., Sung, R., Davis, K., . . . Oefner, P. J. (2006). The role of selection in the evolution of human mitochondrial genomes. *Genetics, 172*(1), 373–387.

Klein, J. (1986). *The Natural History of the Major Histocompatibility Complex.* New York, NY: Wiley.

Klein, J. & Figueroa, F. (1986). The evolution of class I MHC genes. *Immunology Today, 7*(2), 41–44.

Klein, R. G. (1973). Geological antiquity of Rhodesian man. *Nature, 244,* 311–312.

Klein, R. G. (2008). Out of Africa and the evolution of human behavior. *Evolutionary Anthropology, 17,* 267–281.

Klein, R. G. (2016). Issues in human evolution. *Proceedings of the National Academy of Sciences, 113*(23), 6345–6347.

Klein, R. G., Avery, G., Cruz-Uribe, K., Halkett, D., Parkington, J. E., Steele, T., . . . Yates, R. (2004). The Ysterfontein 1 Middle Stone Age site, South Africa, and early human exploitation of coastal resources. *Proceedings of the National Academy of Sciences, USA, 101*(16), 5708–5715.

Kokkoros, P. & Kanellis, A. (1960). Découverte d'un crane d'homme paléolithique dans la peninsule Chalcidique. *Anthropologie, 64,* 132–147.

Kong, Q.-P., Yao, Y.-G., Sun, C., Bandelt, H.-J., Zhu, C.-L., & Zhang, Y.-P. (2010). Phylogeny of east Asian mitochondrial DNA lineages inferred from complete sequences. *American Journal of Human Genetics, 73,* 671–676.

Koulakovska, L., Usik, V., & Haesaerts, P. (2010). Early Paleolithic of Korolevo site (Transcarpathia, Ukraine). *Quaternary International, 223–224*(0), 116–130.

Kozlowski, J. K. & Otte, M. (2000). The formation of the Aurignacian in Europe. *Journal of Anthropological Research, 56*(4), 513–534.

Kramer, A. (1986). Hominid-pongid distinctiveness in the Miocene-Pliocene fossil record: the Lothagam mandible. *American Journal of Physical Anthropology, 70,* 457–473.

Kramer, A., Donnelly, S. M., Kidder, J. H., Ousley, S. D., & Olah, S. M. (1995). Craniometric variation in large-bodied hominoids: testing the single-species hypothesis for *Homo habilis*. *Journal of Human Evolution, 29,* 443–462.

Kramer, P. A. & Eck, G. G. (2000). Locomotor energetics and leg length in hominid bipedality. *Journal of Human Evolution, 38,* 651–666.

Kramer, P. A. & Sylvester, A. D. (2009). Bipedal form and locomotor function: understanding the effects of size and shape on velocity and energetics. *PaleoAnthropology, 2009,* 238–251.

Krause, J., Fu, Q., Good, J. M., Viola, B., Shunkov, M. V., Derevianko, A. P., & Paabo, S. (2010). The complete mitochondrial DNA genome of an unknown hominin from southern Siberia. *Nature, 464*(7290), 894–897.

Krause, J., Orlando, L., Serre, D., Viola, B., Prufer, K., Richards, M. P., . . . Paabo, S. (2007). Neanderthals in central Asia and Siberia. *Nature, 449*(7164), 902–904.

Kretzoi, M. & Vertés, L. (1965). Upper Biharian (Intermindel) pebble-industry occupation site in western Hungary. *Current Anthropology, 6*(1), 74–87.

Krings, M., Capelli, C., Tschentscher, F., Geisert, H., Meyer, S., von Haeseler, A., . . . Pääbo, S. (2000). A view of Neandertal genetic diversity. *Nature Genetics*, *26*, 144–146.

Krings, M., Stone, A., Schmitz, R. W., Krainitzki, H., Stoneking, M., & Pääbo, S. (1997). Neandertal DNA sequences and the origin of modern humans. *Cell*, *90*, 19–30.

Kroll, E. M. (1994). Behavioral implications of Plio-Pleistocene archaeological site structure. *Journal of Human Evolution*, *27*, 1–3.

Kuhn, S. L. (2010). Was Anatolia a bridge or a barrier to early hominin dispersals? *Quaternary International*, *223–224*(0), 434–435.

Kullmer, O., Sandrock, O., Abel, R., Schrenk, F., Bromage, T. G., & Juwayeyi, Y. M. (1999). The first *Paranthropus* from the Malawi Rift. *Journal of Human Evolution*, *37*, 121–127.

Kuman, K. (2007). The Earlier Stone Age in South Africa: site context and the influence of cave studies In T. R. Pickering, K. Schick, & N. Toth (Eds.), *Breathing Life into Fossils* (pp. 181–198). Gosport, IN: Stone Age Institute Press.

Kuman, K. & Clarke, R. (2000). Stratigraphy, artefact industries and hominid associations for Sterkfontein, Member 5. *Journal of Human Evolution*, *38*, 827–847.

Kuman, K., Inbar, M., & Clarke, R. J. (1999). Palaeoenvironments and cultural sequence of the Florisbad Middle Stone Age hominid site, South Africa. *Journal of Archaeological Science*, *26*(12), 1409–1425.

Kummer, B. (1991). Biomecanical foundations of the development of human bipedalism. In Y. Coppens & B. Senut (Eds.), *Origine(s) de la bipédie chez les hominidés* (pp. 1–8). Paris: Editions du CNRS.

Kurtén, B. (1959). New evidence on the age of Pekin Man. *Vertebr. Palastiat.*, *3*, 173–175.

Lacruz, R. S. & Ramirez Rozzi, F. V. (2010). Molar crown development in Australopithecus afarensis. *Journal of Human Evolution*, *58*(2), 201–206. 7

Laitman, J. T. & Tattersall, I. (2001). *Homo erectus newyorkensis*: an Indonesian fossil rediscovered in Manhattan sheds light on the middle phase of human evolution. *The Anatomical Record*, *262*(4), 341–343.

Lalueza-Fox, C., Rosas, A., Estalrrich, A., Gigli, E., Campos, P. F., García-Tabernero, A., . . . de la Rasilla, M. (2010). Genetic evidence for patrilocal mating behavior among Neandertal groups. *Proceedings of the National Academy of Sciences*, *6*:8431, 1–9.

Lalueza-Fox, C., Sampietro, M. L., Caramelli, D., Puder, Y., Lari, M., Calafell, C., . . . Rosas, A. (2005). Neandertal evolutionary genetics: mitochondrial DNA data from the Iberian Peninsula. *Molecular Biology and Evolution*, *22*, 1077–1081.

Lam, Y. M., Pearson, O. M., & Smith, C. M. (1996). Chin morphology and sexual dimorphism in the fossil hominid mandible sample from Klasies River Mouth. *American Journal of Physical Anthropology*, *100*, 545–557.

Lamarck, J. B. (1809). *Philosophie zoologique*. Paris: Dentu.

Landeck, G. (2008). Migration of early humans to Central Europe before the Middle Pleistocene? – New archaeological evidence from Germany. 1–23. Retrieved from Ur-/Frühgeschichte und Archäologie Mittelalter website.

Landeck, G. (2010). Further evidence of a lower pleistocene arrival of early humans in northern Europe— the Untermassfeld site (Germany). *Collegium. Antropologicum*, *34*, 1229–1238.

Larick, R., Ciochon, R. L., Zaim, Y., Sudijono, Suminto, Rizal, Y., . . . Heizler, M. (2001). Early Pleistocene 40Ar/39Ar ages for Bapang Formation hominins, Central Jawa, Indonesia. *Proceedings of the National Academy of Sciences*, *98*(9), 4866–4871.

Larson, S. G., Jungers, W. L., Morwood, M. J., Sutikna, T., Jatmiko, Saptomo, E. W., . . . Djubiantono, T. (2007). *Homo floresiensis* and the evolution of the hominin shoulder. *Journal of Human Evolution*, *53*(6), 718–731.

Larson, S. G., Jungers, W. L., Tocheri, M. W., Orr, C. M., Morwood, M. J., Sutikna, T., . . . Djubiantono, T. (2009). Descriptions of the upper limb skeleton of Homo floresiensis. *Journal of Human Evolution*, *57*(5), 555–570.

Lartet, E. (1860). Mémoire sur la station humaine d'Aurignac. *ASNZ*, *15*, 177–253.

Lartet, E. & Christy, H. (1865–1875). *Reliquiae Aquitanicae*. London: Williams & Norgate.

Latham, A. G., Herries, A. I. R., Sinclair, A. G. M., & Kuykendall, K. (2002). Re-examination of the lower stratigraphy in the classic section, limeworks site, Makapansgat, South Africa. *Human Evolution*, *17*, 207–214.

Latimer, B. (1991). Locomotor Adaptations in *Australopithecus afarensis*: the issue of arboreality. In Y. Coppens & B. Senut (Eds.), *Origine(s) de la bipédie chez les hominidés* (pp. 169–176). Paris: Editions du CNRS.

Latimer, B. & Lovejoy, C. O. (1989). The calcaneus of *Australopithecus afarensis* and its implications in the evolution of bipedality. *American Journal of Physical Anthropology*, *78*, 369–386.

Latimer, B., Ohman, J. C., & Lovejoy, C. O. (1987). Talocrural joint in African hominoids: implications for *Australopithecus afarensis*. American Journal of Physical Anthropology, *74*, 155–175.

Lawlor, D. A., Ward, F. E., Ennies, P. D., Jackson, A. P., & Parham, P. (1988). HLA-A and B polymorphisms predate the divergence of humans and chimpanzees. *Nature*, *355*, 268–271.

Le Fur, S., Fara, E., Mackaye, H. T., Vignaud, P., & Brunet, M. (2009). The mammal assemblage of the hominid site TM266 (Late Miocene, Chad Basin): ecological structure and paleoenvironmental implications. *Naturwissenschaften*. *96*(5), 565–574.

Le Gros Clark, W. E. (1955). *The Fossil Evidence for Human Evolution*. Chicago, IL: University of Chicago Press.

Le Gros Clark, W. E. (1964a). The evolution of man. *Discovery*, *25*, 49.

Le Gros Clark, W. E. (1964b). *The Fossil Evidence for Human Evolution*. 2nd *edition*. Chicago, IL: University of Chicago Press.

Le Gros Clark, W. E. & Leakey, L. S. B. (1951). The Miocene Hominoidea. In B. Museum (Ed.), *Fossil Mammals of Africa* (Vol. I, pp. 1–117). London: British Museum.

Leakey, L. S. B. (1928). The Oldoway skull. *Nature*, *121*, 499.

Leakey, L. S. B. (1935). *Adam's Ancestors*. New York, NY: Longmans, Green & Co.

Leakey, L. S. B. (1951). *Olduvai Gorge. A Report on the Evolution of the Handaxe Culture in Beds I–IV*. Cambridge: Cambridge University Press.

Leakey, L. S. B. (1958). Recent discoveries at Olduvai Gorge, Tanganyika. *Nature, 181*, 1099–1103.

Leakey, L. S. B. (1959). A new fossil skull from Olduvai. *Nature, 184*, 491–493.

Leakey, L. S. B. (1961a). The juvenile mandible from Olduvai Gorge, Tanganyika. *Nature, 191*, 417–418.

Leakey, L. S. B. (1961b). New finds at Olduvai Gorge, Tanganyika. *Nature, 189*, 649–650.

Leakey, L. S. B. (1967). *Olduvai Gorge 1951–61. Volume I.* Cambridge: Cambridge University Press.

Leakey, L. S. B., Boswell, P. G. H., Reck, H., Soloman, J. D., & Hopwood, A. T. (1933). The Oldoway human skeleton. *Nature, 131*, 397.

Leakey, L. S. B., Evernden, J. F., & Curtiss, G. H. (1961). Age of Bed I, Olduvay Gorge, Tanganyika. *Nature, 191*, 478–479.

Leakey, L. S. B. & Leakey, M. (1964). Recent discoveries of fossil hominids in Tanganyika: at Olduvai and near Lake Natron. *Nature, 202*, 5–7.

Leakey, L. S. B., Tobias, P. V., & Napier, J. R. (1964). A new species of the genus *Homo* from Olduvai. *Nature, 202*, 7–9.

Leakey, M. D. (1966). A review of the Oldowan culture from Olduvai Gorge, Tanzania. *Nature, 212*, 577–581.

Leakey, M. D. (1969). Recent discoveries of hominid remains at Olduvai Gorge, Tanzania. *Nature, 223*, 754–756.

Leakey, M. D. (1971). *Olduvai Gorge, 3. Excavations in Beds I and II 1960–1963*. Cambridge: Cambridge University Press.

Leakey, M. D. (1975). Cultural patterns in the Olduvai sequence. In K. W. Butzer & G. L. Isaac (Eds.), *After the Australopithecines* (pp. 477–493). The Hague: Mouton.

Leakey, M. D. (1981). Tracks and tools. *Philosophical Transactions of the Royal Society B: Biological Sciences, 292*, 95–102.

Leakey, M. D., Clarke, R. J., & Leakey, L. S. B. (1971). New hominid skull from Bed I, Olduvai Gorge, Tanzania. *Nature, 232*, 308–312.

Leakey, M. D., Curtis, G. H., Drake, R. E., Jackes, M. K., & White, T. D. (1976). Fossil hominids from the Laetolil Beds. *Nature, 262*, 460–466.

Leakey, M. D. & Hay, R. L. (1979). Pliocene footprints in the Laetoli Beds at Laetoli, northern Tanzania. *Nature, 278*, 317–323.

Leakey, M. D. & Hay, R. L. (1982). The chronological positions of the fossil hominids of Tanzania. In H. d. Lumley (Ed.), *L'Homo erectus andla place de l'homme de Tautavel parmi les hominidés fossiles* (pp. 753–765). Paris: CNRS.

Leakey, M. G., Feibel, C. S., McDougall, I., & Walker, A. (1995). New four-million-year-old hominid species from Kanapoi and Allia Bay, Kenya. *Nature, 376*, 565–572.

Leakey, M. G., Feibel, C. S., McDougall, I., Ward, C., & Walker, A. (1998). New specimens and confirmation of an early age for *Australopithecus anamensis*. *Nature, 393*, 62–66.

Leakey, M. G., Spoor, F., Brown, F. H., Gathogo, P. N., Kiarie, C., Leakey, L. N., & McDougall, I. (2001). New hominin genus from eastern Africa shows diverse middle Pliocene lineages. *Nature, 410*, 433–440.

Leakey, M. G., Spoor, F., Dean, M. C., Feibel, C. S., Anton, S. C., Kiarie, C., & Leakey, L. N. (2012). New fossils from Koobi Fora in northern Kenya confirm taxonomic diversity in early Homo. *Nature, 488*, 201–204.

Leakey, R. E. F. (1970). New hominid remains and early artefacts from northern Kenya. *Nature, 226*, 223–224.

Leakey, R. E. F. (1973a). Evidence for an advanced Plio-Pleistocene hominid from East Rudolf, Kenya. *Nature, 242*, 447–450.

Leakey, R. E. F. (1973b). Skull 1470. *National Geographic, 143*, 819–829.

Leakey, R. E. F. (1974). Further evidence of Lower Pleistocene hominids from East Rudolf, North Kenya, 1973. *Nature, 248*, 653–656.

Leakey, R. E. F. (1976). New hominid fossil from the Koobi Fora Formation, North Kenya. *Nature, 261*, 574–576.

Leakey, R. E. F. (1981). *The Making of Mankind*. Londres: The Rainbird Publishing Group Limited.

Leakey, R. E. F., Mungai, J. M., & Walker, A. C. (1971). New australopithecines from East Rudolf, Kenya. *American Journal of Physical Anthropology, 35*, 175–186.

Leakey, R. E. F. & Walker, A. (1985). Further hominids from the Plio-Pleistocene of Koobi Fora. *American Journal of Physical Anthropology, 67*, 135.

Leakey, R. E. F., Walker, A., Ward, C. V., & Grausz, H. M. (1989). A partial skeleton of a gracile hominid from the Upper Burgi member of the Koobi Fora Formation, East Turkana, Kenya. In G. Giacobini (Ed.), *Proceedings of the 2nd International Congress of Paleoanthropology* (pp. 167–173). Milan: Jaca Books.

Leakey, R. E. F. & Walker, A. C. (1976). *Australopithecus*, *Homo erectus* and the single species hypothesis. *Nature, 261*, 572–574.

Leakey, R. E. F. & Wood, B. (1973). New evidence of the genus *Homo* from East Rudolf, Kenya. II. *American Journal of Physical Anthropology, 39*, 355–368.

Lebatard, A.-E., Bourlès, D. L., Duringer, P., Jolivet, M., Braucher, R., Carcaillet, J., . . . Brunet, M. (2008). Cosmogenic nuclide dating of *Sahelanthropus tchadensis* and *Australopithecus bahrelghazali*: Mio-Pliocene hominids from Chad. *Proceedings of the National Academy of Sciences, 105*(9), 3226–3231.

Lee-Thorp, J. A., Sponheimer, M., Passey, B. H., de Ruiter, D. J., & Cerling, T. E. (2010). Stable isotopes in fossil hominin tooth enamel suggest a fundamental dietary shift in the Pliocene. *Philosophical Transactions of the Royal Society B: Biological Sciences, 365*(1556), 3389–3396.

Lee-Thorp, J. A. & van der Merwe, N. J. (1993). Stable carbon isotope studies of Swartkrans fossil assemblages. In C. K. Brain (Ed.), *Swartkrans: A Cave's Chronicle of Early Man (Transvaal Museum Monograph)* (pp. 243–250). Pretoria: Transvaal Museum.

Lee-Thorp, J. A., van der Merwe, N. J., & Brain, C. K. (1994). Diet of *Australopithecus robustus* at Swartkrans from stable carbon isotopic analysis. *Journal of Human Evolution, 27*, 361–372.

Leigh, S. R. & Blomquist, G. E. (2007). Life history. In C. Campbell, A. Fuentes, K. C. MacKinnon, M. Panger, & S. K. Bearder (Eds.), *Primates in Perspective* (pp. 396–407). New York, NY: Oxford University Press.

Leonard, W. R. & Robertson, M. L. (2000). Ecological correlates of home range variation in primates: implications for hominid evolution. In S. Boinski & P. Garber (Eds.), *On the Move: How and Why Animals Travel in Groups* (pp. 628–648). Chicago, IL: Chicago University Pres.

Lepre, C. J. & Kent, D. V. (2010). New magnetostratigraphy for the Olduvai Subchron in the Koobi Fora Formation, northwest Kenya, with implications for early *Homo*. *Earth and Planetary Science Letters, 290*, 362–374.

Lepre, C. J., Roche, H., Kent, D. V., Harmand, S., Quinn, R. L., Brugal, J.-P., . . . Feibel, C. S. (2011). An earlier origin for the Acheulian. *Nature, 477*(7362), 82–85.

Leroi-Gourhan, A. (1958). Étude des restes humains fossiles provenant des grottes d'Arcy-sur-Cure (Yonne). *Annales de Paléontologie, 44*, 87–148.

Leroi-Gourhan, A. (1961). Les fouilles d'Arcy-sur-Cure (Yonne). *Gallia Préhistoire, 4*, 3–16.

Leroi-Gourhan, A. (1964). *Le Geste et la Parole*. Paris: Albin Michel.

Leroi-Gourhan, A. (1975). The flowers found with Shanidar IV, a Neanderthal Burial in Iraq. *Science, 190*, 562–564.

Lévêque, F. & Vandermeersch, B. (1980). Les découvertes de restes humains dans un niveau castelperronien à Saint-Césaire (Charente-Maritime). *Comptes rendus de l'Académie des Sciences de Paris, 291D*, 187–189.

Lewin, R. (1986). New fossil upsets human family. *Science, 233*, 720–721.

Lewin, R. (1987). *Bones of Contention. Controversies in the Search for Human Origins*. New York, NY: Simon and Schuster.

Lewis, O. J. (1972). The evolution of the hallucial tarsometatarsal joint in the anthropoidea. *American Journal of Physical Anthropology, 37*, 13–34.

Lewis, O. J. (1980). The joints of the evolving foot. Part III. The fossil evidence. *Journal of Anatomy, 131*, 275–298.

Li, H., Ruan, J. & Durbin, R. (2008). Mapping short DNA sequencing reads and calling variants using mapping quality scores. *Genome Research, 18*(11), 1851–1858.

Li., H., Yang., X, Heller, F., & Li, H. (2008). High resolution magnetostratigraphy and deposition cycles in the Nihewan Basin (North China) and their significance for stone artifact dating. *Quaternary Research, 69*(2), 250–262.

Li, J. Z., Absher, D. M., Tang, H., Southwick, A. M., Casto, A. M., Ramachandran, S., . . . Myers, R. M. (2008). Worldwide human relationships inferred from genome-wide patterns of variation. *Science, 319*(5866), 1100–1104.

Li, W.-H., Xiong, W., Liu, S. A.-W., & Chan, L. (1993). Nucleotide diversity in humans and evidence for the absence of a severe bottleneck during human evolution. In C. F. Sing & C. L. Hanis (Eds.), *Genetics of Cellular, Individual,* *Family, and Population Variability* (pp. 253–261). New York, NY: Oxford University Press.

Lieberman, D. E. (2001). Another face in our family tree. *Nature, 410*, 419–420.

Lieberman, D. E. (2009). Palaeoanthropology: Homo floresiensis from head to toe. *Nature, 459*(7243), 41–42.

Lieberman, D. E. & McCarthy, R. C. (1999). The ontogeny of cranial base angulation in humans and chimpanzees and its implications for reconstructing pharyngeal dimensions. *Journal of Human Evolution, 36*, 487–517.

Lieberman, D. E., Pilbeam, D. R., & Wood, B. (1988). A probabilistic approach to the problem of sexual dimorphism in *Homo habilis*: a comparison of KNM-ER 1470 and KNM-ER 1813. *Journal of Human Evolution, 17*(5), 503–511.

Lieberman, D. E., Raichlen, D. A., Pontzer, H., Bramble, D. M., & Cutright-Smith, E. (2006). The human gluteus maximus and its role in running. *Journal of Experimental Biology, 209*(11), 2143–2155.

Lieberman, D. E., Wood, B., & Pilbeam, D. R. (1996). Homoplasy and early *Homo*: an analysis of the evolutionary relationships of *H. habilis sensu stricto* and *H. rudolfensis*. *Journal of Human Evolution, 30*, 97–120.

Limondin-Lozouet, N., Nicoud, E., Antoine, P., Auguste, P., Bahain, J.-J., Dabkowski, J., . . . Mercier, N. (2010). Oldest evidence of Acheulean occupation in the Upper Seine valley (France) from an MIS 11 tufa at La Celle. *Quaternary International, 223–224*(0), 299–311.

Linnaeus, C. (1758). *Systema Naturae per Naturae Regna Tria, Secundum Classes, Ordines, Genera, Species cum Characteribus, Synonymis, Locis.* (10ª y definitiva ed. en 1758 ed.). Stockholm: Laurentii Sylvii.

Liu, W., Martinon-Torres, M., Cai, Y.-j., Xing, S., Tong, H.-w., Pei, S.-w., . . . Wu, X.-j. (2015). The earliest unequivocally modern humans in southern China. *Nature, 526*(7575), 696–699.

Liu, W., Yinyun, Z., & Xinzhi, W. (2005). A middle Pleistocene human cranium from Tangshan, Nanjing of southeast China: a comparison with *Homo erectus* from Eurasia and Africa based on new reconstruction. *American Journal of Physical Anthropology, 25*, 253–262.

Liu, X., Somel, M., Tang, L., Yan, Z, Jiang, X., Guo, S., ... Khaitovich, P. (2012). Extension of cortical synaptic development distinguishes humans from chimpanzees and macaques. *Genome Research, 22*(4), 611–622.

Lockwood, C. A. & Kimbel, W. H. (1999). Endocranial capacity of early hominids. *Science, 283*(5398), 9b.

Lockwood, C. A. & Tobias, P. V. (1999). A large male hominin cranium from Sterkfontein, South Africa, and the status of *Australopithecus africanus*. *Journal of Human Evolution, 36*, 637–685.

Lockwood, C. A. & Tobias, P. V. (2002). Morphology and affinities of new hominin cranial remains from Member 4 of the Sterkfontein Formation, Gauteng Province, South Africa. *Journal of Human Evolution, 42*, 389–450.

Longa, V. M. (2009). En el origen. Técnica y creatividad en la prehistoria. *Ludus Vitalis, 17*, 227–231.

Lonsdorf, E. V., Ross, S. R., Linick, S. A., Milstein, M. S., & Melber, T. N. (2009). An experimental, comparative investigation of tool use in chimpanzees and gorillas. *Animal Behaviour, 77*(5), 1119–1126.

Lordkipanidze, D., Jashashvili, T., Vekua, A., de Leon, M. S. P., Zollikofer, C. P. E., Rightmire, G. P., . . . Rook, L. (2007). Postcranial evidence from early Homo from Dmanisi, Georgia. *Nature, 449*(7160), 305–310.

Lordkipanidze, D., Vekua, A., Ferring, R., Rightmire, G., Zollikofer, C., Ponce de León, M., . . . Tappen, M. (2006). A fourth hominin skull from Dmanisi, Georgia. *Anatomical Record Part A, Discoveries in Molecular, Cellular, and Evolutionary Biology, 288*, 1146–1157.

Lordkipanidze, D., Vekua, A., Ferring, R., Rightmire, G. P., Agusti, J., Kiladze, G., . . . Zollikofer, C. P. E. (2005). The earliest toothless hominin skull. *Nature, 434*, 717–718.

Lorenzo, C., Arsuaga, J. L., & Carretero, J. M. (1999). Hand and foot remains from the Gran Dolina Early Pleistocene site (Sierra de Atapuerca, Spain). *Journal of Human Evolution, 37*(3–4), 501–522.

Löscher, M., Eibner, C., & Wegner, D. (2007). Alte und neue Funde von Steinwerkzeugen aus den Mauerer Sanden. In G. A. Wagner, H. Rieder, L. Zöller, & E. Mick (Eds.), *Homo heidelbergensis – Schlüsselfund der Menschheitsgeschichte* (pp. 267–279). Stuttgart: Theiss.

Louys, J. & Turner, A. (2011). Environment, preferred habitats and potential refugia for Pleistocene Homo in Southeast Asia. *Comptes Rendus Palevol, 11*, 203–211.

Lovejoy, C. O. (1975). Biomechanical perspectives on the lower limb of early hominids. In R. H. Tuttle (Ed.), *Primate Functional Morphology and Evolution* (pp. 291–326). The Hague: Mouton.

Lovejoy, C. O. (1981). The origin of man. *Science, 211*, 341–350.

Lovejoy, C. O., Heiple, K. G., & Meindl, R. S. (2001). Did our ancestors knuckle-walk? *Nature, 410*, 325–326.

Lovejoy, C. O., Latimer, B., Suwa, G., Asfaw, B., & White, T. D. (2009). Combining prehension and propulsion: the Foot of Ardipithecus ramidus. *Science, 326*(5949), 72; 72e01–e08.

Lovejoy, C. O., Simpson, S. W., White, T. D., Asfaw, B., & Suwa, G. (2009). Careful Climbing in the Miocene: the forelimbs of Ardipithecus ramidus and humans are primitive. *Science, 326*(5949), 70; 70e01–70e08.

Lovejoy, C. O., Suwa, G., Simpson, S. W., Matternes, J. H., & White, T. D. (2009). The great divides: Ardipithecus ramidus reveals the postcrania of our last common ancestors with African apes. *Science, 326*(5949), 73; 100–106.

Lovejoy, C. O., Suwa, G., Spurlock, L., Asfaw, B., & White, T. D. (2009). The pelvis and femur of Ardipithecus ramidus: the emergence of upright walking. *Science, 326*(5949), 71e01–71e06.

Lowenstein, J. M. (1986). Where there's smoke. *Pacific Discovery, 39*, 30–32.

Lozoya, Marqués de (Juan de Contreras y Lopez de Ayala) (1972). *Introducción a la Biografía del Canciller Ayala*. Bilbao: Editorial Vizcaina.

Luengas Otaola, V. G. (1974). *Introducción a la Historia de la muy noble y muy leal Tierra de Ayala*. Bilbao: Editorial Vizcaina.

Macchiarelli, R., Bondioli, L., Chech, M., Coppa, A., Fiore, I., Russom, R., . . . Rook, L. (2004). The late early Pleistocene human remains from Buia, Danakil depression, Eritrea. *Rivista Italiana di Paleontologia e Stratigrafia, 110*, 133–144.

Macchiarelli, R., Bondioli, L., Debenath, A., Mazurier, A., Tournepiche, J.-F., Birch, W., & Dean, M. C. (2006). How Neanderthal molar teeth grew. *Nature, 444*(7120), 748–751.

MacLatchy, L. (1998). The controversy continues. *Evolutionary Anthropology, 6*, 147–150.

MacRae, A. (1998, 2004). Radiometric dating and the geological time scale. Circular reasoning or reliable tools? http://www.talkorigins.org/faqs/dating.html.

Madella, M., Jones, M. K., Goldberg, P., Goren, Y., & Hovers, E. (2002). The exploitation of plant remains by Neandertals in Amud cave (Israel): the evidence from phytolith studies. *Journal of Archaeological Science, 29*, 703–719.

Maglio, V. J. (1972). Vertebrate faunas and chronology of hominid-bearing sediments east of Lake Rudolf, Kenya. *Nature, 239*, 379–385.

Magori, C. C. & Day, M. H. (1983). Laetoli Hominid 18: an early Homo sapiens skull. *Journal of Human Evolution, 12*(8), 747–753.

Malez, M., Smith, F. H., Radovcic, J., & Rukavina, D. (1980). Upper Pleistocene hominids from Vindija, Croatia, Yugoslavia. *Current Athropology, 21*, 365–367.

Mallegni, F., Carnieri, E., Bisconti, M., Tartarelli, G., Ricci, S., Biddittu, I., & Segre, A. (2003). *Homo cepranensis sp. nov.* and the evolution of African-European Middle Pleistocene hominids. *C. R. Palevol, 2*, 153–159.

Mallegni, F., Mariani-Costantini, R., Fornaciari, G., Longo, E. T., Giacobini, G., & Radmilli, A. M. (1983). New European fossil hominid material from an Acheulean site near Rome (Castel di Guido). *American Journal of Physical Anthropology, 62*(3), 263–274.

Mallory, A. C. & Vaucheret, H. (2006). Functions of microRNAs and related small RNAs in plants. *Nature Genetics, Supplement, 38*, S31–S36.

Mania, D. & Grimm, H. (1974). Bilzingsleben, Kr. Artern–eine palöökologisch aufschlussreiche Fundstelle des Altpaläolithikums mit Hominiden-Fund. *Biologische Rundschau, 12*, 361–364.

Manzi, G. (2004). Human evolution at the Matuyama-Brunhes boundary. *Evolutionary Anthropology, 13*(1), 11–24.

Manzi, G. (2011). Before the emergence of *Homo sapiens*: overview on the Early-to-Middle Pleistocene Fossil Record (with a Proposal about *Homo heidelbergensis* at the subspecific level). *International Journal of Evolutionary Biology, 2011*. DOI :10.4061/2011/582678.

Manzi, G., Bruner, E., & Passarello, P. (2003). The one-million-year-old Homo cranium from Bouri (Ethiopia): a reconsideration of its H. erectus affinities. *Journal of Human Evolution, 44*(6), 731–736.

Manzi, G., Magri, D., Milli, S., Palombo, M. R., Margari, V., Celiberti, V., . . . Biddittu, I. (2010). The new chronology of the Ceprano calvarium (Italy). *Journal of Human Evolution, 59*(5), 580–585.

Manzi, G., Mallegni, F., & Ascenzi, A. (2001). A cranium for the earliest Europeans: phylogenetic position of the hominid from Ceprano, Italy. *Proceedings of the National Academy of Sciences, USA, 98*(17), 10011–10016.

Marett, R. (1911). Pleistocene man in Jersey. *Archaeologia, 62,* 449–480.

Margulis, L. (1970). *Origin of Eukaryotic Cells.* New Haven, CO: Yale University Press.

Marqués-Bonet, T., Cáceres, M., Bertranpetit, J., Preuss, T. M., Thomas, J. W., & Navarro, A. (2004). Chromosomal rearrangements and the genomic distribution of gene-expression divergence in humans and chimpanzees. *Trends in Genetics, 20,* 524–529.

Márquez, S., Mowbray, K., Sawyer, G. J., Jacob, T., & Silvers, A. (2001). New fossil hominid calvaria from Indonesia—Sambungmacan 3. *The Anatomical Record, 262*(4), 344–368.

Marsh, S. G. E. & Bodmer, J. G. (1991). HLA class II nucleotide sequences. *Immunogenetics, 33,* 321–334.

Marshack, A. (1996). A middle paleolithic symbolic composition from the Golan Heights: the earliest known depictive image. *Current Anthropology, 37,* 357–365.

Martin, L. B. (1985). Significance of enamel thickness in hominoid evolution. *Nature, 314,* 260–263.

Martin, R. D. (1990). *Primate Origins and Evolution. A Phylogenetic Reconstruction.* London: Chapman and Hall.

Martin, R. D., MacLarnon, A. M., Phillips, J. L., Dussubieux, L., Williams, P. R., & Dobyns, W. B. (2006). Comment on "The Brain of LB1, *Homo floresiensis*". *Science, 312*(5776), 999.

Martínez-Navarro, B. & Palmqvist, P. (1995). Presence of the African machairodont *Megantereon whitei* (Broom, 1937) (Felidae, Carnivora, Mamalia) in the Lower Pleistocene site of Venta Micena (Orce, Granada, Spain), with some considerations on the origin, evolution and dispersal of the genus. *Journal of Archaeological Science, 22,* 569–582.

Martínez-Navarro, B., Rook, L., Segid, A., Yosieph, D., Ferretti, M. P., Shoshani, J., . . . Libsekal, Y. (2004). The large fossil mammals from Buia (Eritrea). *Riv. It. Paleont. Strat., 110 (supplement),* 61–88.

Martinón-Torres, M., Dennell, R., & Bermúdez de Castro, J. M. (2011). The Denisova hominin need not be an out of Africa story. *Journal of Human Evolution, 60,* 251–255.

Martyn, J. (1967). Pleistocene deposits and new fossil localities in Kenya. *Nature, 215,* 476–479.

Masters, P. (1982). An amino acid racemization chronology for Tabun. In A. Ronen (Ed.), *The Transition from the Lower to Middle Paleolithic and the Origin of Modern Man, 151* (pp. 43–54). Oxford: Br. arch. Rep. Int. Series.

Mayer, W. E., Jonker, M., Klein, D., Ivanyi, P., van Seventer, G., & Klein, J. (1988). Nucleotide sequences of chimpanzee MHC class I alleles: evidence for trans-species mode of evolution. *The EMBO Journal, 7*(9), 2765–2774.

Mayr, E. (1942). *Systematics and the Origin of Species from the Viewpoint of a Zoologist.* New York, NY: Columbia University Press.

Mayr, E. (1944). On the concepts and terminology of vertical subspecies and species. *Common Problems of Genetics, Paleontology and Systematics Bulletin, 2,* 11–16.

Mayr, E. (1950). Taxonomic categories in fossil hominids. *Cold Spring Harbor Symposia on Quantitative Biology, 15,* 109–117.

Mayr, E. (1954). Notes on nomenclature and classification. *Systematic Zoology, 3,* 86–89.

Mayr, E. (1957). Difficulties and importance of the biological species concept. In E. Mayr (Ed.), *The Species Problem* (pp. 371–388). Washington, DC: American Assotiation for the Advancement of Science.

Mayr, E. (1963). *Animal Species and Evolution.* Cambridge: Belnakp Press.

Mayr, E. (1969). *Principles of Systematic Zoology.* New York, NY: McGraw-Hill; 6th. edition, New Delhi, Tata McGraw-Hill.

Mayr, E. (1970). *Population, Species, and Evolution.* Cambridge: Harvard University Press.

Mayr, E. (1982). Processes of speciation in animals. In C. Barigozzi (Ed.), *Mechanisms of Speciation* (pp. 1–19). New York, NY: Liss.

Mazza, P. A., Martini, F., Sala, B., Magi, M., Colombini, M. P., Giachi, G., . . . Ribecchini, E. (2006). A new Palaeolithic discovery: tar-hafted stone tools in a European Mid-Pleistocene bone-bearing bed. *Journal of Archeological Science, 33,* 1310–1318.

McBrearty, S. & Brooks, A. S. (2000). The revolution that wasn't: a new interpretation of the origin of modern human behavior. *Journal of Human Evolution, 39,* 453–563.

McBrearty, S. & Jablonski, N. G. (2005). First fossil chimpanzee. *Nature, 437*(7055), 105–108.

McBrearty, S. & Tryon, C. (2006). From Acheulean to Middle Stone Age in the Kapthurin Formation, Kenya. In E. Hover & S. L. Kuhn (Eds.), *Transitions Before the Transition* (pp. 257–277). New York, NY: Springer.

McClean, R. G. & Kean, W. F. (1993). Contributions of wood ash magnetism to archaeomagnetic properties of fire pits and hearths. *Earth and Planetary Science Letters, 119*(3), 387–394.

McConnell, T. J., Talbot, W. S., McIndoe, R. A., & Wakeland, E. K. (1988). The origin of MHC class II gene polymorphism within the genus *Mus. Nature, 332,* 651–654.

McCown, T. D. & Keith, A. (1939). *The Stone Age of Mount Carmel. Vol 2, The Fossil Human Remains from the Levallois Mousterian.* Oxford: Clarendon Press.

McCrossin, M. L. (1992). Human molars from late Pleistocene deposits of Witkrnas Cave, Gaap Escarpment, Kalahari margin. *Human Evolution, 7,* 1–10.

McDermott, F., Grün, R., Stringer, C. B., & Hawkesworth, C. J. (1993). Mass-spectrometric U-series dates for Israeli Neanderthal/early modern hominid sites. *Nature, 363,* 252–255.

McDevitt, H. (1995). Evolution of Mhc Class II allelic diversity. *Immunological Reviews*, *143*, 113–122.

McDougall, I. (1985). Potassium-Argon and Argon⁴⁰/Argon³⁹ dating of the hominid bearing sequence at Koobi Fora, Lake Turkana, Northern Kenya. *Geological Society of America Bulletin*, *96*, 159–175.

McDougall, I. & Brown, F. H. (2006). Precise 40Ar/39Ar geochronology for the upper Koobi Fora Formation, Turkana Basin, northern Kenya. *Journal of the Geological Society*, *163*, 205–220.

McDougall, I., Maier, R., Sutherland-Hawkes, P., & Gleadow, A. J. W. (1980). K-Ar age estimate for the KBS Tuff, East Turkana, Kenya. *Nature*, *284*, 230–234.

McFadden, P. L., Brock, A., & Partridge, T. C. (1979). Palaeomagnetism and the age of the Makapansgat hominid site. *Earth and Planetary Science Letters*, *44(3)*, 373–382.

McHenry, H. & Corruccini, R. (1980). Late tertiary hominoids and human origins. *Nature*, *285*, 397–398.

McNulty, K. P. (2010). Apes and tricksters: the evolution and diversification of humans' closest relatives. *Evolution: Education and Outreach*, *3*, 322–332.

McPherron, S. P., Alemseged, Z., Marean, C. W., Wynn, J. G., Reed, D., Geraads, D., . . . Bearat, H. A. (2010). Evidence for stone-tool-assisted consumption of animal tissues before 3.39 million years ago at Dikika, Ethiopia. *Nature*, *466(7308)*, 857–860.

Mehlman, M. J. (1987). Provenience, age and associations of archaic *Homo sapiens* crania from Lake Eyasi, Tanzania. *Journal of Archaeological Science*, *14(2)*, 133–162.

Meier, R. & Willmann, R. (2000). The Hennigian species concept. In Q. D. Wheeler & R. Meier (Eds.), *Species Concept and Phylogenetic Theory* (pp. 30–43). New York, NY: Columbia University Press.

Mein, P. (1975). Resultats du Groupe de Travail des Vertebres. In J. Senes (Ed.), *Report on Activity of the RCMNS Working Groups (1971–1975)* (pp. 78–81). Bratislava: Veda.

Mellars, P. (1996). *The Neanderthal Legacy*. Princeton, NJ: Princeton University Press.

Mellars, P. (2005). The impossible coincidence. a single-species model for the origins of modern human behavior in Europe. *Evolutionary Anthropology*, *14*, 12–27.

Mellars, P. (2006a). Going east: new genetic and archaeological perspectives on the modern human colonization of Eurasia. *Science*, *313(5788)*, 796–800.

Mellars, P. (2006b). Why did modern human populations disperse from Africa ca. 60,000 years ago? A new model. *Proceedings of the National Academy of Sciences*, *103(25)*, 9381–9386.

Mellars, P., Gori, K. C., Carr, M., Soares, P. A., & Richards, M. B. (2013). Genetic and archaeological perspectives on the initial modern human colonization of southern Asia. *Proceedings of the National Academy of Sciences*, *110(26)*, 10699–10704.

Mendel, G. (1866). Versuche über Plflanzen-Hybriden. Verhandlungen des Naturforschenden Vereines, Abhandlungen, Brünn. 3, 209–220.

Méndez, Fernando L., Poznik, G. D., Castellano, S., & Bustamante, Carlos D. (2016). The divergence of Neandertal and modern human Y chromosomes. *American Journal of Human Genetics*, *98(4)*, 728–734.

Mercader, J., Barton, H., Gillespie, J., Harris, J., Kuhn, S., Tyler, R., & Boesch, C. (2007). 4,300-Year-old chimpanzee sites and the origins of percussive stone technology. *Proceedings of the National Academy of Sciences*, *104(9)*, 3043–3048.

Mercader, J., & Panger, M., & Boesch, C. (2002). Excavation of a Chimpanzee Stone Tool Site in the African Rainforest. *Science*, *296*, 1452–1455.

Mercier, N., Valladas, H., Bar-Yosef, O., Vandermeersch, B., Stringer, C., & Joron, J. L. (1993). Thermoluminescence date for the Mousterian burial site of Es-Skhul, Mt. Carmel. *Journal of Archaeological Science*, *20*, 169–174.

Mercier, N., Valladas, H., Joron, J. L., Reyss, J. L., Lévêque, F., & Vandermeersch, B. (1991). Thermoluminescence dating of the late Neanderthal remains from Saint-Césaire. *Nature*, *351*, 737–739.

Meyer, M., Arsuaga, J.-L., de Filippo, C., Nagel, S., Aximu-Petri, A., Nickel, B., . . . Pääbo, S. (2016). Nuclear DNA sequences from the Middle Pleistocene Sima de los Huesos hominins. *Nature*, *531(7595)*, 504–507.

Meyer, M., Fu, Q., Aximu-Petri, A., Glocke, I., Nickel, B., Arsuaga, J.-L., . . . Paabo, S. (2013). A mitochondrial genome sequence of a hominin from Sima de los Huesos. *Nature*, *505*, 403–406.

Meyer, M., Kircher, M., Gansauge, M.-T., Li, H., Racimo, F., Mallick, S., . . . Pääbo, S. (2012). A high-coverage genome sequence from an archaic Denisovan individual. *Science*, *338(6104)*, 222–226.

Mgeladze, A., Lordkipanidze, D., Moncel, M.-H., Despriee, J., Chagelishvili, R., Nioradze, M., & Nioradze, G. (2011). Hominin occupations at the Dmanisi site, Georgia, Southern Caucasus: raw materials and technical behaviours of Europe's first hominins. *Journal of Human Evolution*, *60(5)*, 571–596.

Mietto, P., Avanzini, M., & Rolandi, G. (2003). Palaeontology: human footprints in Pleistocene volcanic ash. *Nature*, *422(6928)*, 133–133.

Milankovitch, M. (1941). *Kanon der Erdbestrahlung und seine Anwendung auf das Eiszeitenproblem*. Belgrado: Académie Royale Serbe.

Millard, A. R. (2008). A critique of the chronometric evidence for hominid fossils: I. Africa and the Near East 500–50 ka. *Journal of Human Evolution*, *54(6)*, 848–874.

Miller, J. M. A. (2000). Craniofacial variation in *Homo habilis*: an analysis of the evidence for multiple species. *American Journal of Physical Anthropology*, *112*, 103–128.

Milton, K. (1988). Foraging behaviour and the evolution of primate intelligence. In R. Byrne & A. Whiten (Eds.), *Machiavellian Intelligence* (pp. 285–305). Oxford: Clarendon Press.

Mirazón Lahr, M. (2010). Saharan corridors and their role in the evolutionary geography of "out of Africa I". In

J. G. Fleagle, J. J. Shea, F. E. Greene, A. L. Baden, & R. E. Leakey (Eds.), *The First Hominin Colonization of Eurasia, Vertebrate Paleobiology and Paleoanthropology* (pp. 27–46). Berlin: Springer.

Mirazón Lahr, M. & Foley, R. (1994). Multiple dispersals and modern human origins. *Evolutionary Anthropology, 3*, 48–58.

Mirazón Lahr, M. & Foley, R. (2004). Human evolution writ small. *Nature, 431*(7012), 1043–1044.

Mishra, S., Gaillard, C., Deo, S., Singh, M., Abbas, R., & Agrawal, N. (2010). Large Flake Acheulian in India: implications for understanding lower Pleistocene human dispersals. *Quaternary International, 223–224*(0), 271–272.

Mochida, G. H. & Walsh, C. A. (2001). Molecular genetics of human microcephaly. *Current Opinion in Neurology, 14*, 151–156.

Moggi-Cecchi, J., Tobias, P. V., & Beynon, A. D. (1998). The mixed dentition and associated skull fragments of a juvenile fossil hominid from Sterkfontein, South Africa. *American Journal of Physical Anthropology, 106*, 425–465.

Möller, H. (1900). Über *Elephas antiquus Falc.* u. *Rhinoceros Merckii* als Jagdtiere des altdiluvialen Menschen in Thüringen und über das erste Auftreten des Menschen in Europa. In G. Brandes (Ed.), *Zeitschrift für Naturwissenschaflen* (pp. 41–70). Stuttgart: Seliweizerbart'sche Verlagshandlung

Moncel, M. H., Chiotti, L., Gaillard, C., Onoratini, G., & Pleurdeau, D. (2012). Non-utilitarian lithic objects from the European Paleolithic. *Archaeology, Ethnology and Anthropology of Eurasia, 40*(1), 24–40.

Moore, M. W. & Brumm, A. (2007). Stone artifacts and hominins in island Southeast Asia: new insights from Flores, eastern Indonesia. *Journal of Human Evolution, 52*, 85–102.

Moore, M. W., Sutikna, T., Jatmiko, Morwood, M. J., & Brumm, A. (2009). Continuities in stone flaking technology at Liang Bua, Flores, Indonesia. *Journal of Human Evolution, 57*(5), 503–526.

Morell, V. (1995). African origins: west side story. *Science, 270*, 1117.

Morgan, L. E. & Renne, P. R. (2008). Diachronous dawn of Africa's Middle Stone Age: new 40Ar/39Ar ages from the Ethiopian Rift. *Geology, 36*(12), 967–970.

Mortillet, G. (1897). *Formation de la Nation Française*. Paris: Plon-Nourrit.

Morwood, M. J., Brown, P., Jatmiko, Sutikna, T., Wahyu Saptomo, E., Westaway, K. E., . . . Djubiantono, T. (2005). Further evidence for small-bodied hominins from the Late Pleistocene of Flores, Indonesia. *Nature, 437*(7061), 1012–1017.

Morwood, M. J., O'Sullivan, P. B., Aziz, A., & Raza, A. (1998). Fission-track ages of stone tools and fossils on the east Indonesian island of Flores. *Nature, 392*, 173–176.

Morwood, M. J., Soejono, R. P., Roberts, R. G., Sutikna, T., Turney, C. S. M., Westaway, K. E., . . . Fifield, L. K. (2004). Archaeology and age of a new hominin from Flores in eastern Indonesia. *Nature, 431*(7012), 1087–1091.

Morwood, M. J., Sullivan, O. P., Susanto, E. E., & Aziz, F. (2003). Revised age for Mojokerto 1, an early *Homo erectus* cranium from East Java. *Australian Archaeology, 57*, 1–4.

Morwood, M. J., Sutikna, T., Saptomo, E. W., Jatmiko, Hobbs, D. R., & Westaway, K. E. (2009). Preface: research at Liang Bua, Flores, Indonesia. *Journal of Human Evolution, 57*(5), 437–449.

Mounier, A., Marchal, F., & Condemi, S. (2009). Is *Homo heidelbergensis* a distinct species? New insight on the Mauer mandible. *Journal of Human Evolution, 56*(3), 219–246.

Mountain, E. D. (1966). Footprints in calcareous sandstone at Nahoon Point. *South African Journal of Science, 62*, 103–111.

Movius, H. L. (1948). The Lower Paleolithic cultures of southern and eastern Asia. *Transactions of the American Philosophical Society, 38*, 330–420.

Movius, H. L. (1953). The Mousterian cave of Teshik-Tash, Southeastern Uzbekistan, Central Asia. *Bulletin of the American School of Prehistoric Research, 17*, 11–71.

Moya, A. & Ayala, F. J. (1989). Fertility interactions in *Drosophila*: theoretical model and experimental tests. *Journal of Evolutionary Biology, 2*, 1–12.

Mturi, A. A. (1987). The archaeological sites of Lake Natron. *Lac Natron, Sciences Geologique, 40*, 209–215.

Mulcahy, N. J. & Call, J. (2006). Apes save tools for future use. *Science, 312*(5776), 1038–1040.

Murdock, M. (2006). These apes were made for walking: the pelves of *Australopithecus afarensis* and *Australopithecus africanus*. *Journal of Creation, 20*, 104–112.

Murrill, R. I. (1975). A comparison of the Rodhesian and Petralona upper jaws in relation to other Pleistocene hominids. *Zeitschrift für Morphologie und Anthropologie, 66*, 176–187.

Muttoni, G., Scardia, G., & Kent, D. V. (2010). Human migration into Europe during the late Early Pleistocene climate transition. *Palaeogeography, Palaeoclimatology, Palaeoecology, 296*(1–2), 79–93.

Muttoni, G., Scardia, G., Kent, D. V., Morsiani, E., Tremolada, F., Cremaschi, M., & Peretto, C. (2011). First dated human occupation of Italy at ~0.85 Ma during the late Early Pleistocene climate transition. *Earth and Planetary Science Letters, 307*, 241–252.

Muttoni, G., Scardia, G., Kent, D. V., Swisher, C. C., & Manzi, G. (2009). Pleistocene magnetochronology of early hominin sites at Ceprano and Fontana Ranuccio, Italy. *Earth and Planetary Science Letters, 286*, 255–268.

Napier, J. (1962a). The evolution of the hand. *Scientific American 207*, 56–65.

Napier, J. R. (1962b). Fossil hand bones from Olduvai Gorge. *Nature, 196*, 409–411.

Napier, J. R. (1963). Brachiation and brachiators. *Symposia of the Zoological Society of London, 10*, 183–195.

Napier, J. R. (1967). The antiquity of human walking. *Scientific American, 216*, 56–66.

Napier, J. R. (1980). *Hands*. New York, NY: Pantheon Books.

Napier, J. R. & Walker, A. C. (1967). Vertical clinging and leaping – a newly recognised category of locomotor behaviour of primates. *Folia primatologia, 6*, 204–219.

Nature. (2006). Making the paper: Zeresenay Alemseged. *Nature, 443*(7109), xiii.

Navarro, A. & Barton, N. H. (2003). Accumulating postzygotic isolation genes in parapatry: a new twist on chromosomal speciation. *Evolution, 57*, 447–459.

Nejman, L. (2008). A reinterpretation of early Upper Palaeolithic assemblages from Stránská Skála: the differences in lithic economy between the Aurignacian and the Bohunician assemblages. *Přehled Výzkumů, 49*, 24–45.

Nelson, G. J. (1973). Classification as an expression of phylogenetic relationships. *Systematic Zoology, 22*, 344–359.

Nevell, L. & Wood, B. (2008). Cranial base evolution within the hominin clade. *Journal of Anatomy, 212*, 455–468.

Noble, W. & Davidson, I. (1996). *Human Evolution, Language and Mind.* Cambridge: Cambridge University Press.

Noll, M. P. & Petraglia, M. D. (2003). Acheulean bifaces and early human behavioral patterns in East Africa and South India. In M. Soressi & H. L. Dibble (Eds.), *Multiple Approaches to the Study of Bifacial Technologies* (pp. 31–53). Philadelphia, PA: University of Pennsylvania Museum of Archaeology and Anthropology.

Nomade, S., Muttoni, G., Guillou, H., Robin, E., & Scardia, G. (2011). First 40Ar/39Ar age of the Ceprano man (central Italy). *Quaternary Geochronology, 6*(5), 453–457.

Noonan, J. P., Coop, G., Kudaravalli, S., Smith, D., Krause, J., Alessi, J., . . . Rubin, E. M. (2006). Sequencing and analysis of Neanderthal genomic DNA. *Science, 314*(5802), 1113–1118.

Noonan, J. P., Hofreiter, M., Smith, D., Priest, J. R., Rohland, N., Rabeder, G., . . . Rubin, E. M. (2005). Genomic sequencing of Pleistocene cave bears. *Science, 309*, 597–600.

Nordborg, M. (1998). On the probability of Neandertal ancestry. *American Journal of Human Genetics, 63*, 1237–1240.

Norton, C. J. (2000). The current state of Korean paleoanthropology. *Journal of Human Evolution, 38*(6), 803–825.

O'Regan, H. J., Turner, A., Bishop, L. C., Elton, S., & Lamb, A. L. (2011). Hominins without fellow travellers? First appearances and inferred dispersals of Afro-Eurasian large mammals in the Plio-Pleistocene. *Quaternary Science Reviews, 30*(11–12), 1343–1352.

O'Huigin, C., Bontrop, R., & Klein, J. (1993). Nonhuman primate Mhc-DRB sequences: a compilation. *Immunogenetics, 38*, 165–183.

O'Rourke, D. H., Geoffrey Hayes, M., & Carlyle, S. W. (2000). Ancient DNA studies in physical anthropology. *Annual Review of Anthropology, 29*, 17–42.

Oakley, K. P. (1956). The earliest fire-makers. *Antiquity, 30*, 102–107.

Oakley, K. P. (1981). Emergence of higher thought 3.0–0.2 Ma B.P. *Philosophical Transactions of the Royal Society B: Biological Sciences, B292*, 205–211.

Ohman, J., Wood, C., Wood, B., Crompton, R., Günther, M., Yu, L., . . . Wang, W. (2002). Stature-at-death of KNM-WT 15000. *Human Evolution, 17*(3), 129–141.

Ohta, S., Nishiaki, Y., Abe, Y., Mizoguchi, Y., Kondo, O., Dodo, Y., . . . Haydal, J. (1995). Neanderthal infant burial from the Dederiyeh cave in Syria. *Paléorient, 21*, 77–86.

Olson, T. R. (1985). Cranial morphology and systematics of the Hadar Formation hominids and *"Australopithecus" africanus.* In E. Delson (Ed.), *Ancestors: The Hard Evidence* (pp. 102–119). New York, NY: Alan R. Liss.

Oms, O., Parés, J. M., Marténez-Navarro, B., Agusté, J., Toro, I., Martínez-Fernández, G., & Turq, A. (2000). Early human occupation of Western Europe: paleomagnetic dates for two paleolithic sites in Spain. *Proceedings of the National Academy of Sciences, 97*(19), 10666–10670.

Oppennoorth, W. F. F. (1932). *Homo (Javanthropus) soloensis,* een plistocene Mensch von Java. *Wet. Meded. Dienst. Mijnb. Ned.-Oast. Indië, 20*, 49–75 Volume 2011, Article ID 582678, 11 pages doi:10.4061/2011/582678.

Oron, M. & Goren-Inbar, N. (2014). Mousterian intra-site spatial patterning at Quneitra, Golan Heights. *Quaternary International, 331*(0), 186–202.

Otte, M. (1996). *Le paléolithique inférieur er moyen en Europe.* Paris: Armand Collin.

Otte, M. (2003). The pitfalls of using bifaces as cultural markers. In M. Soressi & H. L. Dibble (Eds.), *Multiple Approaches to the Study of Bifacial Technologies* (pp. 183–192). Philadelphia, PA: University of Pennsylvania Museum of Archaeology and Anthropology.

Otte, M. (2010). Before Levallois. *Quaternary International, 223–224*(0), 273–280.

Ovchinnikov, I. V., Götherström, A., Romanova, G. P., Kharitonov, V. M., Liden, K., & Goodwin, W. (2000). Molecular analysis of Neanderthal DNA from the northern Caucasus. *Nature, 404*, 490–493.

Owen, R. (1843). *Lectures on the Comparative Anatomy and Physiology of the Invertebrate Animals, Delivered at the Royal College of Surgeons in 1843.* London: Longman, Brown, Green & Longmans.

Oxnard, C. & Lisowski, F. P. (1980). Functional articulation of some hominoid foot bones: implications for the Olduvai (hominid 8) foot. *American Journal of Physical Anthropology, 52*, 107–117.

Pääbo, S. (1984). Über den Nachweiss von DNA in altägyptischen Mumien. *Das Latertum, 30*, 213–218.

Pääbo, S. (1985). Molecular cloning of ancient Egyptian mummy DNA. *Nature, 314*, 644–645.

Pääbo, S., Poinar, H., Serre, D., Jaenicke-Després, V., Hebler, J., Rohland, N., . . . Hofreiter, M. (2004). Genetic analyses from ancient DNA. *Annual Review of Genetics, 38*, 645–679.

Panger, M. A., Brooks, A. S., Richmond, B. G., & Wood, B. (2002). Older than the Oldowan? Rethinking the emergence of hominin tool use. *Evolutionary Anthropology, 11*, 235–245.

Parés, J. M. & Pérez-González, A. (1995). Paleomagnetic age for hominid fossils at Atapuerca archaeological site, Spain. *Science, 269*, 830–832.

Parés, J.M. & Pérez-González, A. (1999). Magnetochronology and stratigraphy at Gran Dolina section, Atapuerca (Burgos, Spain). *Journal of Human Evolution, 37*, 325–347.

Parés, J. M., Pérez-González, A., Rosas, A., Benito, A., Bermúdez de Castro, J. M., Carbonell, E., & Huguet, R. (2006). Matuyama-age lithic tools from the Sima del Elefante site, Atapuerca (northern Spain). *Journal of Human Evolution, 50*(2), 163–169.

Parfitt, S. A., Ashton, N. M., Lewis, S. G., Abel, R. L., Coope, G. R., Field, M. H., . . . Stringer, C. B. (2010). Early Pleistocene human occupation at the edge of the boreal zone in northwest Europe. *Nature, 466*(7303), 229–233.

Parfitt, S. A., Barendregt, R. W., Breda, M., Candy, I., Collins, M. J., Coope, G. R., . . . Stuart, A. J. (2005). The earliest record of human activity in northern Europe. *Nature, 438*, 1008–1012.

Partridge, T. C. (1975). *Stratigraphic, geomorphological and paleo-environmental studies of the Makapansgat Limeworks and Sterkfontein hominid sites: a progress report on research carried out between 1965 and 1975.* Paper presented at the IIIrd Sci. Congress, Cape Town, May–June 1975.

Partridge, T. C. (1978). Re-appraisal of lithostratigraphy of Sterkfontein hominid site. *Nature, 275*, 282–287.

Partridge, T. C. (1979). Re-appraisal of lithostratigraphy of Makapansgat Limeworks hominid site. *Nature, 279*, 484–488.

Partridge, T. C. (1982). The chronological positions of the fossil hominids of Southern Africa. In M. A. De Lumley (Ed.), *L'Homo erectus et la place de l'homme de Tautavel parmi les hominids fossils.* (Vol. 2, pp. 617–675). Nice: Premier Congres International de Paleontologie Humaine.

Partridge, T. C. (2000). Hominid-bearing cave and tufa deposits. In T. C. Partridge & R. R. Maud (Eds.), *The Cenozoic of southern Africa* (pp. 100–125). New York, NY: Oxford University Press.

Partridge, T. C., Granger, D. E., Caffee, M. W., & Clarke, R. J. (2003). Lower Pliocene hominid remains from Sterkfontein. *Science, 300*(5619), 607–612.

Partridge, T. C., Shaw, J., Heslop, D., & Clarke, R. (1999). The new hominid skeleton from Sterkfontein, South Africa: age and preliminary assessment. *Journal of Quaternary Science, 14*, 293–298.

Partridge, T. C. & Watt, I. B. (1991). The stratigraphy of the Sterkfontein hominid deposit and its relationship to the underground cave system. *Palaeontologia Africana, 28*, 35–40.

Patel, B. A. & Carlson, K. J. (2007). Bone density spatial patterns in the distal radius reflect habitual hand postures adopted by quadrupedal primates. *Journal of Human Evolution, 52*(2), 130–141.

Patel, B. A., Susman, R. L., Rossie, J. B., & Hill, A. (2009). Terrestrial adaptations in the hands of Equatorius africanus revisited. *Journal of Human Evolution, 57*(6), 763–772.

Paterson, T. T. (1940). Geology and early man. *Nature, 146*, 12–16, 49–52.

Patnaik, R., Chauhan, P. R., Rao, M. R., Blackwell, B. A. B., Skinner, A. R., Sahni, A., . . . Khan, H. S. (2009). New geochronological, paleoclimatological, and archaeological data from the Narmada Valley hominin locality, central India. *Journal of Human Evolution, 56*, 114–133.

Patterson, B., Behrensmeyer, A. K., & Sill, W. D. (1970). Geology and fauna of a new Pliocene locality in north-western Kenya. *Nature, 226*, 918–921.

Patterson, B. & Howells, W. (1967). Hominid humeral fragment from early Pleistocene of northwestern Kenya. *Science, 156*, 64–66.

Paunovic, M., Krings, M., Capelli, C., Tshentscher, F., Geisert, H., Meyer, S., . . . Pääbo, S. (2001). *The Vindija hominids: a view of Neandertal genetic diversity.* Paper presented at the Paleoanthropology Society. Abstracts for the 2001 Meeting.

Pearce, E., Stringer, C., & Dunbar, R. I. M. (2013). New insights into differences in brain organization between Neanderthals and anatomically modern humans. *Proceedings of the Royal Society B: Biological Sciences, 280*(1758).

Pearson, O. M., Fleagle, J. G., Grine, F. E., & Royer, D. F. (2008). Further new hominin fossils from the Kibish Formation, southwestern Ethiopia. *Journal of Human Evolution, 55*(3), 444–447.

Pearson, O. M., Royer, D. F., Grine, F. E., & Fleagle, J. G. (2008). A description of the Omo I postcranial skeleton, including newly discovered fossils. *Journal of Human Evolution, 55*(3), 421–437.

Pei, W. C. (1929). An account of the discovery of an adult *Sinanthropus* skull in the Chou Kou Tien deposit. *Bulletin of the Geological Society of China, 8*, 203–250.

Pei, W. Z. & Zhang, S. S. (1985). A study on the lithic artifacts of *Sinanthropus*. *Palaeontologica Sinica New Series D, 12*, 1–277.

Pei-Hua, H., Si-Zhao, J., Zi-Cheng, P., Ren-You, L., Zhong-Jia, L., Zhao-Rong, W., . . . Z.-X., Y. (1993). ESR dating of tooth enamel: comparison with U-Series, FT and TL dating at the Peking Man Site. *Applied Radiation and Isotopes, 44*, 239–242.

Penck, A. & Brückner, E. (1909). *Die Alpen im Eiszeitalter.* Leipzig: Tauchnitz.

Peresani, M., Cremaschi, M., Ferraro, F., Falguères, C., Bahain, J.-J., Gruppioni, G., . . . Dolo, J.-M. (2008). Age of the final Middle Palaeolithic and Uluzzian levels at Fumane Cave, Northern Italy, using 14C, ESR, 234U/230Th and thermoluminescence methods. *Journal of Archaeological Science, 35*(11), 2986–2996.

Perry, G. H., Verrelli, B. C., & Stone, A. C. (2005). Comparative analyses reveal a complex history of molecular evolution for human MYH16. *Molecular Biology and Evolution, 22*, 379–382.

Petraglia, M., Korisettar, R., Boivin, N., Clarkson, C., Ditchfield, P., Jones, S., . . . White, K. (2007). Middle Paleolithic assemblages from the Indian subcontinent before and after the Toba super-eruption. *Science, 317*, 114–116.

Petraglia, M. D., Haslam, M., Fuller, D. Q., Boivin, N., & Clarkson, C. (2010). Out of Africa: new hypotheses and evidence for the dispersal of Homo sapiens along the Indian Ocean rim. *Annals of Human Biology, 37*, 288–311.

Pickering, R., Dirks, P. H. G. M., Jinnah, Z., de Ruiter, D. J., Churchil, S. E., Herries, A. I. R., . . . Berger, L. R. (2011). *Australopithecus sediba* at 1.977 Ma and implications for the origins of the genus Homo. *Science, 333*(6048), 1421–1423.

Pickering, R., Hancox, P. J., Lee-Thorp, J. A., R., G., Mortimer, G. E., McCulloch, M., & Berger, L. R. (2007). Stratigraphy,

U-Th chronology, and paleoenvironments at Gladysvale Cave: insights into the climatic control of South African hominin-bearing cave deposits. *Journal of Human Evolution, 53*, 602–619.

Pickering, R. & Kramers, J. D. (2010). Re-appraisal of the stratigraphy and determination of new U-Pb dates for the Sterkfontein hominin site, South Africa. *Journal of Human Evolution, 59*(1), 70–86.

Pickering, T. R., Clarke, R. J., & Heaton, J. L. (2004). The context of Stw 573, an early hominid skull and skeleton from Sterkfontein Member 2: taphonomy and paleoenvironment. *Journal of Human Evolution, 46*(3), 277–295.

Pickering, T. R., White, T. D., & Toth, N. (2000). Cutmarks on a plio-pleistocene hominid from Sterkfontein, South Africa. *American Journal of Physical Anthropology, 111*, 570–584.

Pickford, M. (1975). Late Miocene sediments and fossils from the Northern Kenya Rift Valley. *Nature, 256*, 279–284.

Pike, A. W. G., Hoffmann, D. L., García-Diez, M., Pettitt, P. B., Alcolea, J., De Balbín, R., . . . Zilhão, J. (2012). U-Series dating of Paleolithic art in 11 caves in Spain. *Science, 336*(6087), 1409–1413.

Pilbeam, D. R. (1978). Rethinking human origins. *Discovery, 13*, 2–9.

Piveteau, J. (1970). Les grottes de La Chaise (Charente). Paléontologie Humaine I. L'homme de l'abri Suard. *Ann. Paléont. (Vert.), 56*, 175–225.

Platnick, N. I. (1979). Philosophy and the transformation of cladistics. *Systematic Zoology, 28*, 537–546.

Plavcan, J. M. & Cope, D. A. (2001). Metric variation and species recognition in the fossil record. *Evolutionary Anthropology, 10*, 204–222.

Plavcan, J. M., Lockwood, C. A., Kimbel, W. H., Lague, M. R., & Harmon, E. H. (2005). Sexual dimorphism in *Australopithecus afarensis* revisited: how strong is the case for a human-like pattern of dimorphism? *Journal of Human Evolution, 48*, 313–320.

Plummer, T. (2004). Flaked stones and old bones: biological and cultural evolution at the dawn of technology. *Yearbook of Physical Anthropology, 47*, 118–164.

Poirier, F. E. (1987). *Understanding Human Evolution (2nd. edition, 1990 ed.)*. Englewood Cliffs, NJ: Prentice-Hall.

Pollard, K. S., Salama, S. R., Lambert, N., Lambot, M.-A., Coppens, S., Pedersen, J. S., . . . Haussler, D. (2006). An RNA gene expressed during cortical development evolved rapidly in humans. *Nature*.

Ponce de León, M. S., Golovanova, L., Doronichev, V., Romanova, G., Akazawa, T., Kondo, O., . . . Zollikofer, C. P. E. (2008). Neanderthal brain size at birth provides insights into the evolution of human life history. *Proceedings of the National Academy of Sciences, 105*(37), 13764–13768.

Pontzer, H., Brown, M. H., Raichlen, D. A., Dunsworth, H., Hare, B., Walker, K., . . . Ross, S. R. (2016). Metabolic acceleration and the evolution of human brain size and life history. *Nature, 533*(7603), 390–392.

Pontzer, H., Rolian, C., Rightmire, G. P., Jashashvili, T., Ponce de León, M. S., Lordkipanidze, D., & Zollikofer, C. P. E.

(2010). Locomotor anatomy and biomechanics of the Dmanisi hominins. *Journal of Human Evolution, 58*(6), 492–504.

Pope, G. G. (1988). Recent advances in far eastern paleoanthropology. *Annual Review of Anthropology, 17*, 43–77.

Popper, K. R. (1963). *Conjectures and Refutations: The Growth of Scientific Knowledge*. London: Routledge.

Porraz, G., Parkington, J. E., Rigaud, J.-P., Miller, C. E., Poggenpoel, C., Tribolo, C., . . . Texier, P.-J. (2013). The MSA sequence of Diepkloof and the history of southern African Late Pleistocene populations. *Journal of Archaeological Science, 40*(9), 3542–3552.

Porraz, G., Texier, P.-J., Archer, W., Piboule, M., Rigaud, J.-P., & Tribolo, C. (2013). Technological successions in the Middle Stone Age sequence of Diepkloof Rock Shelter, Western Cape, South Africa. *Journal of Archaeological Science, 40*(9), 3376–3400.

Potts, R. (1992). The hominid way of life. In S. Jones, R. Martin, & D. Pilbeam (Eds.), *The Cambridge Encyclopedia of Human Evolution* (pp. 325–334). Cambridge: Cambridge University Press.

Potts, R., Behrensmeyer, A. K., Deino, A., Ditchfield, P., & Clark, J. (2004). Small mid-Pleistocene hominin associated with East African Acheulean technology. *Science, 305*(5680), 75–78.

Poulianos, A. N. (1981). Pre-sapiens man in Greece. *Current Anthropology, 22*, 287–288.

Prat, S., Brugal, J. P., Roche, H., & Texier, P. J. (2003). Nouvelles découvertes de dents d'hominidés dans le membre Kaitio de la formation de Nachukui (1,65–1,9 Ma), Ouest du lac Turkana (Kenya). *C.R. Paleovol, 2*, 685–693.

Prat, S., Brugal, J. P., Tiercelin, J. J., Barrat, J. A., Bohn, M., Delagnes, A., . . . Roche, H. (2005). First occurrence of early *Homo* in the Nachukui Formation (West Turkana, Kenya) at 2.3–2.4 Myr. *Journal of Human Evolution, 49*, 230–240.

Prentice, M. L. & Denton, G. H. (1988). The deep-sea oxygen isotope record, the global ice sheet system and hominid evolution. In F. E. Grine (Ed.), *Evolutionary History of the "Robust" Australopithecines* (pp. 383–403). New York, NY: Aldine de Gruyter.

Prossinger, H., Seidler, H., Wicke, L., Weaver, D., Recheis, W., Stringer, C., & Müller, G. (2003). Electronic removal of encrustations inside the Steinheim cranium reveals paranasal sinus features and deformations, and provides a revised endocranial volume estimate. *The Anatomical Record, 273B*, 132–143.

Protsch, R. (1974). The age and stratigraphic position of Olduvai Hominid I. *Journal of Human Evolution, 3*, 379–385.

Protsch, R. & de Villiers, H. (1974). Bushman rock shelter, Ohrigstad, Eastern Transvaal, South Africa. *Journal of Human Evolution, 3*(5), 387–396.

Prüfer, K., Racimo, F., Patterson, N., Jay, F., Sankararaman, S., Sawyer, S., . . . Paabo, S. (2014). The complete genome sequence of a Neanderthal from the Altai Mountains. *Nature, 505*(7481), 43–49.

Prugnolle, F., Manica, A., & Balloux, F. (2005). Geography predicts neutral genetic diversity of human populations. *Current Biology, 15*(5), R159–R160.

Quinqi, X. & Yuzhu, Y. (1984). Hexian fauna: correlation with deep sea sediments. *Acta Anthropol. sin., 1,* 180–190.

Rae, T. C., Koppe, T., & Stringer, C. B. (2011). The Neanderthal face is not cold adapted. *Journal of Human Evolution, 60,* 234–239.

Raia, P., Carotenuto, F., & Meiri, S. (2010). One size does not fit all: no evidence for an optimal body size on islands. *Global Ecology and Biogeography, 19,* 475–484.

Raichlen, D. A., Gordon, A. D., Harcourt-Smith, W. E. H., Foster, A. D., & Haas, W. R. J. (2010). Laetoli footprints preserve earliest direct evidence of human-like bipedal biomechanics. *PLoS One, 5*(3), e9769.

Rak, Y. (1983). *The Australopithecine Face.* New York, NY: Academic Press.

Rak, Y. (1985). Systematic and functional implications of the facial morphology of *Australopithecus* and early *Homo.* In E. Delson (Ed.), *Ancestors: The Hard Evidence* (pp. 168–170). New York, NY: Alan R. Liss.

Rak, Y. (1986). The Neanderthal: a new look at an old face. *Journal of Human Evolution, 15,* 151–164.

Rak, Y. (1988). On variation in the masticatory system of *Australopithecus boisei.* In F. E. Grine (Ed.), *Evolutionary History of the "Robust" Australopithecines* (pp. 193–198). New York, NY: Aldine de Gruyter.

Rak, Y., Ginzburg, A., & Geffen, E. (2002). Does *Homo neanderthalensis* play a role in modern human ancestry? The mandibular evidence. *American Journal of Physical Anthropology, 119,* 199–204.

Ramirez Rozzi, F. (1998). Can enamel microstructure be used to establish the presence of different species of Plio-Pleistocene hominids from Omo, Ethiopia? *Journal of Human Evolution, 35,* 543–576.

Ramirez Rozzi, F. & Bermúdez de Castro, J. M. (2004). Surprisingly rapid growth in Neanderthals. *Nature, 428,* 936–939.

Rasmussen, M., Guo, X., Wang, Y., Lohmueller, K. E., Rasmussen, S., Albrechtsen, A., . . . Willerslev, E. (2011). An Aboriginal Australian genome reveals separate human dispersals into Asia. *Science, 334*(6052), 94–98.

Rayner, R. J., Moon, B. P., & Masters, J. C. (1993). The Makapansgat australopithecine environments. *Journal of Human Evolution, 24,* 219–231.

Reader, J. (1981). *Missing Links. The Hunt for Earliest Man.* London: Collins.

Ready, E. (2010). Neandertal man the hunter: a history of Neandertal subsistence. *Explorations in Anthropology, 10,* 58–80.

Reck, H. (1914). Erste vorläufige Mitteilungen über den Fund eines fossilen Menschenskeletts aus Zentralafrika. *Mammalia, 555,* 85, 92.

Reed, K. E. (2008). Paleoecological patterns at the Hadar hominin site, Afar Regional State, Ethiopia. *Journal of Human Evolution, 54,* 743–768.

Reich, D., Patterson, N., Kircher, M., Delfin, F., Nandineni, M. R., Pugach, I., . . . Stoneking, M. (2011). Denisova admixture and the first modern human dispersals into Southeast Asia and Oceania. *American Journal of Human Genetics, 89*(4), 516–528.

Renne, P. R., WoldeGabriel, G., Hart, W. K., Heiken, G., & White, T. D. (1999). Chronostratigraphy of the Miocene-Pliocene Sagantole Formation, Middle Awash Valley, Afar Rift, Ethiopia. *Geological Society of America Bulletin, 111,* 869–885.

Reno, P. L., McCollum, M. A., Meindl, R. S., & Lovejoy, C. O. (2010). An enlarged postcranial sample confirms *Australopithecus afarensis* dimorphism was similar to modern humans. *Philosophical Transactions of the Royal Society B: Biological Sciences, 365*(1556), 3355–3363.

Reno, P. L., Meindl, R. S., McCollum, M. A., & Lovejoy, C. O. (2003). Sexual dimorphism in *Australopithecus afarensis* was similar to that of modern humans. *Proceedings of the National Academy of Sciences, 100,* 9404–9409.

Richards, G. D. & Jabbour, R. S. (2011). Foramen magnum ontogeny in *Homo sapiens*: a functional matrix perspective. *The Anatomical Record, 294*(2), 199–216.

Richards, M. P., Pettitt, P. B., Trinkaus, E., Smith, F. H., Paunovic, M., & Karavanic, I. (2000). Neanderthal diet at Vindija and Neanderthal predation: the evidence from stable isotopes. *Proceedings of the National Academy of Sciences, USA, 97,* 7663–7666.

Richmond, B. G. & Jungers, W. L. (2008). *Orrorin tugenensis* femoral morphology and the evolution of hominin bipedalism. *Science, 319*(5870), 1662–1665.

Richmond, B. G. & Strait, D. S. (2000). Evidence that humans evolved from a knuckle-walking ancestor. *Nature, 404,* 382–385.

Riel-Salvatore, J. & Clark, G. A. (2001). Grave markers. Middle and Early Upper Paleolithic burials and the use of chronotypology in contemporary Paleolithic research. *Current Anthropology, 42,* 449–479.

Rightmire, G. P. (1979a). Cranial remains of *Homo erectus* from Beds II and IV, Olduvai Gorge, Tanzania. *American Journal of Physical Anthropology, 51,* 99–115.

Rightmire, G. P. (1979b). Human skeletal remains from Die Kelders, Cape. *Ann. S. Afr. Mu., 78,* 333.

Rightmire, G. P. (1980). Middle Pleistocene hominids from Olduvai Gorge, northern Tanzania. *American Journal of Physical Anthropology, 53,* 225–241.

Rightmire, G. P. (1990). *The Evolution of Homo Erectus. Comparative Anatomical Studies of an Extinct Human Species.* Cambridge: Cambridge University Press.

Rightmire, G. P. (1993). Variation among early *Homo* crania from Olduvai Gorge and the Koobi Fora region. *American Journal of Physical Anthropology, 90,* 1–33.

Rightmire, G. P. (1996). The human cranium from Bodo, Ethiopia: evidence for speciation in the Middle Pleistocene? *Journal of Human Evolution, 31*(1), 21–39.

Rightmire, G. P. (1998). Human evolution in the Middle Pleistocene: the role of *Homo heidelbergensis.* Evol. Anthrop., 6, 218–227.

Rightmire, G. P. (2001). Patterns of hominid evolution and dispersal in the Middle Pleistocene. *Quaternary International, 75,* 77–84.

Rightmire, G. P. (2009). Middle and later Pleistocene hominins in Africa and Southwest Asia. *Proceedings of the National Academy of Sciences*, 106(38), 16046–16050.

Rightmire, G. P. & Deacon, H. J. (2001). New human teeth from Middle Stone Age deposits at Klasies River, South Africa. *Journal of Human Evolution*, 41, 535–544.

Rightmire, G. P., Deacon, H. J., Schwartz, J. H., & Tattersall, I. (2006). Human foot bones from Klasies River main site, South Africa. *Journal of Human Evolution*, 50(1), 96–103.

Rissman, E. F. & Adli, M. (2014). Minireview: transgenerational epigenetic inheritance: focus on endocrine disrupting compounds. *Endocrinology*, 155(8), 2770–2780.

Robbins, L. M. (1987). Hominid footprints from Site G. In M. D. Leakey & J. M. Harris (Eds.), *Laetoli: A Pliocene Site in Northern Tanzania* (pp. 407–502). Oxford: Clarendon Press.

Roberts, D. & Berger, L. R. (1997). Last interglacial (c. 117 kyr) human footprints from South Africa. *South African Journal of Science*, 93, 349–350.

Robinson, J. T. (1954a). The genera and species of the australopithecinae. *American Journal of Physical Anthropology*, 12, 181–200.

Robinson, J. T. (1954b). Prehominid dentition and hominid evolution. *Evolution*, 8, 324–334.

Robinson, J. T. (1962). The origin and adaptive radiation of the australopithecines. In G. Kurth (Ed.), *Evolution und Hominization* (pp. 150–175). Stuttgart: Stuttgarter Verlagskontor Gmbh.

Robinson, J. T. (1968). The origin and adaptive radiation of the australopithecines. In G. Kurth (Ed.), *Evolution und Hominisation* (pp. 150–175). Sttutgart: Fischer.

Robinson, J. T. & Manson, R. J. (1957). Occurence of stone artefacts with *Australopithecus* at Sterkfontein. *Nature*, 180, 521–524.

Roche, H., Delagnes, A., Brugal, J.-P., Feibel, C., Kibunjia, M., Mourre, V., & Texier, J.-P. (1999). Early hominid stone tool production and technical skill 2.34 Myr ago in West Turkana, Kenya. *Nature*, 399, 57–60.

Roebroeks, W. & Soressi, M. (2016). Neandertals revised. *Proceedings of the National Academy of Sciences, USA*, 113(23), 6372–6379.

Roebroeks, W. & Van Kolfschoten, T. (1994). The earliest occupation of Europe: a short chronology. *Antiquity*, 68, 489–503.

Rohde, D.L.T., Olson, S., and Chang, J.T. (2004). Modelling the recent common ancestry of all living humans. *Nature* 431, 562–566.

Rohde, R. A. & Muller, R. A. (2005). Cycles in fossil diversity. *Nature*, 434(7030), 208–210.

Rolian, C., Lieberman, D. E., & Hallgrìmsson, B. (2010). The coevolution of human hands and feet. *Evolution*. DOI: 10.1111/j.1558-5646.2010.00944.x.

Rolland, N. (2004). Was the emergence of home bases and domestic fire a punctuated event? A review of the Middle Pleistocene record in Eurasia. *Asian Perspectives*, 43, 248–280.

Ron, H. & Levi, S. (2001). When did hominids first leave africa? New high-resolution magnetostratigraphy from the Erk-el-Ahmar Formation, Israel. *Geology*, 29, 887–890.

Ronen, A. (2006). The oldest human groups in the Levant. *Comptes Rendus Palevol*, 5, 345–351.

Rook, L. (2000). The taxonomy of Dmanisi crania: crucial consequences for hominid taxonomy.

Rosas, A. & Bermúdez de Castro, J. M. (1998). On the taxonomic affinities of the Dmanisi mandible (Georgia). *American Journal of Physical Anthropology*, 107, 145–162.

Rosenberg, K. R. & Trevathan, W. (2001). The evolution of human birth. *Scientific American*, 285, 72–77.

Rosenberg, K. R., Zuné, L., & Ruff, C. B. (2006). Body size, body proportions, and encephalization in a Middle Pleistocene archaic human from northern China. *Proceedings of the National Academy of Sciences, USA*, 103(10), 3552–3556.

Rosenberg, N. A., Pritchard, J. K., Weber, J. L., Cann, H. M., Kidd, K. K., Zhivotovsky, L. A., & Feldman, M. W. (2002). Genetic structure of human populations. *Science*, 298(5602), 2381–2385.

Rosselló i Verger, V. (1970). Clima y morfología pleistocena en el litoral mediterráneo español. In S. d. Publicaciones (Ed.), *Papeles del Departamento de Geografía* (pp. 79–101). Murcia: Universidad de Murcia.

Rosselló-Mora, R. (2003). Opinion: the Species problem, can we achieve a universal concept? *Systematic and Applied Microbiology*, 26(3), 323–326.

Royer, D. F., Lockwood, C. A., Scott, J. E., & Grine, F. E. (2009). Size variation in early human mandibles and molars from Klasies River, South Africa: comparison with other middle and late Pleistocene assemblages and with modern humans. *American Journal of Physical Anthropology*, 140(2), 312–323.

Ruff, C. B. (1995). Biomechanics of the hip and birth in early *Homo. American Journal of Physical Anthropology*, 98, 527–554.

Ruff, C. B. (2009). Relative limb strength and locomotion in *Homo habilis. American Journal of Physical Anthropology*, 138, 90–100.

Ruff, C. B. (2010). Body size and body shape in early hominins–implications of the Gona pelvis. *Journal of Human Evolution*, 58, 166–178.

Ruff, C. B., Trinkaus, E., & Holliday, T. W. (1997). Body mass and encephalization in Pleistocene *Homo. Nature*, 387, 173–176.

Rukang, W. (1964). A newly discovered mandible of the Sinanthropus type-*Sinanthropus lantianensis. Scientia Sinica*, 13, 891–911.

Rukang, W. (1966). The hominid skull of Lantian, Shensi. *Vertebr. Palasiat.*, 10, 1–16.

Rukang, W. (1980). Palaeanthropology in the New China. In L. K. Königsson (Ed.), *Current Argument on Early Man* (pp. 182–206). Oxford: Pergamon Press.

Rukang, W. (1985). New Chinese *Homo erectus* and recent work at Zhoukoudian. In E. Delson (Ed.), *Ancestors: The Hard Evidence* (pp. 245–248). New York, NY: Alan R. Liss.

Rukang, W. (1988). The reconstruction of the fossil human skull from Jinniushan, Yinkou, Liaoning Province and its main features. *Acta Anthropologica Sinica, 7*, 97–101.

Rukang, W. & Xingren, D. (1982). Preliminary study of *Homo erectus* remains from Hexian, Anhui. *Acta Anthropologica Sinica, 1*, 2–13.

Ruvolo, M. (1993). Mitochondrial COII sequences and modern human origins. *Molecular Biology* and *Evolution, 10*, 1115–1135.

Ruvolo, M. (1997). Molecular phylogeny of the hominoids: inferences from multiple independent dna sequence data sets. *Molecular Biology* and *Evolution, 14*, 248–265.

Ruvolo, M. (2004). Comparative primate genomics: the year of the chimpanzee. *Current Opinion in Genetics & Development, 14*, 650–656.

Ruvolo, M., Zehr, S., von Dornum, M., Pan, D., Chang, B., & Lin, J. (1993). Mitochondrial COII sequences and modern human origins. *Molecular Biology and Evolution, 10*, 1115–1135.

Sabater Pi, J. (1984). *El chimpancé y los orígenes de la cultura.* Barcelona: Anthropos.

Sabater Pi, J., Véa, J. J., & Serrallonga, J. (1997). Did the first hominids build nests? *Current Anthropology, 38*, 914–916.

Sala, N., Arsuaga, J. L., Pantoja-Pérez, A., Pablos, A., Martínez, I., Quam, R. M., . . . Carbonell, E. (2015). Lethal interpersonal violence in the Middle Pleistocene. *PLoS One, 10*(5), e0126589.

Sandrock, O. (1999). *Taphonomy and Paleoecology of the Malema Hominid Site, Northern Malawi.* (Ph.D. dissertation), University of Mainz.

Sang-Hee, L. (2005). Is variation in the cranial capacity of the Dmanisi sample too high to be from a single species? *American Journal of Physical Anthropology, 127*, 263–266.

Santa Luca, A. P. (1980). The Ngandong fossil hominids. *Yale Univ. Publ. Anthrop., 78*, 1–175.

Santonja, M. & Pérez-González, A. (2010). Mid-Pleistocene Acheulean industrial complex in the Iberian Peninsula. *Quaternary International, 223–224*(0), 154–161.

Sarich, V. & Wilson, A. C. (1967a). Immunological time scale for hominid evolution. *Science, 158*, 1200–1203.

Sarich, V. & Wilson, A. C. (1967b). Rates of albumin evolution in primates. *Proceedings of the National Academy of Sciences, 58*, 142–148.

Sarmiento, E. E. (1998). Generalized quadrupeds, committed bipeds, and the shift to open habitats: an evolutionary model of hominid divergence. *American Museum Novitates, 3250*, 1–78.

Sarmiento, E. E. & Marcus, L. F. (2000). The os navicular of humans, great apes, OH 8, hadar, and *Oreopithecus*: function, phylogeny, and multivariate analyses. *American Museum Novitates, 3288*, 1–38.

Sartono, S. (1971). Observations of a new skull of *Pithecanthropus erectus* (Pithecanthropus VIII) from Sangiran, Central Java. *Proceedings of the Academy of Science, Amst. B, 74*, 185–194.

Satta, Y., Takahata, N., Schönbach, C., Gutknecht, J., & Klein, J. (1991). Calibrating evolutionary rates at major histo-compatibility complex loci. In J. Klein & D. Klein (Eds.), *Molecular Evolution of the Major Histocompatibility Complex* (pp. 51–62). Heidelberg: Springer.

Schaaffhausen, H. (1880). Funde in der Sipkahöhle in Mahren. *Sonderbericht der niederrheinischen Gesellschaft für Natur- und Heilkunde*, 260–264.

Schaeffer, B., Hecht, M. K., & Eldredge, N. (1972). Phylogeny and paleontology. *Evolutionary Biology, 6*, 31–46.

Schick, K. D. (1994). The Movius Line reconsidered: perspectives on the Earlier Paleolithic of Eastern Asia. In R. Corruchini & R. Ciochon (Eds.), *Integrative Paths to the Past: Paleoanthropological Advances in Honor of F. Clark Howell* (pp. 569–594). NJ: Prentice-Hall.

Schick, K. D. & Toth, N. (1993). *Making Silent Stones Speak.* New York, NY: Simon & Schuster.

Schlanger, N. (1994). *Châine opératoire* for an archaeology of the mind. In C. Renfrew & E. B. W. Zubrow (Eds.), *The Ancient Mind. Elements of Cognitive Archaeology* (pp. 143–151). Cambridge: Cambridge University Press.

Schlebusch, C. M., Skoglund, P., Sjödin, P., Gattepaille, L. M., Hernandez, D., Jay, F., . . . Jakobsson, M. (2012). Genomic variation in seven Khoe-San groups reveals adaptation and complex African history. *Science, 338*, 374–379.

Schmid, P. (2002). The Gladysvale project. *Evolutionary Anthropology: Issues, News, and Reviews, 11*(S1), 45–48.

Schmincke, H. & Van den Bogaard, P. (1995). Die datierung des Mashavera-Basalt lavastroms. *Jahrb. RGZM, 42*, 75–76.

Schmitt, D. (2003). Insights into the evolution of human bipedalism from experimental studies of humans and other primates. *Journal of Experimental Biology, 206*, 1437–1448.

Schmitz, R. W. (2003). Interdisziplinäre Untersuchungen an den Neufunden aus dem Neandertal Johann Carl Fuhlrott (1803–1877) gewidmet. *Mitteilungen der Gesellschaft für Urgeschichte, 12*, 25–45.

Schmitz, R. W., Serre, D., Bonani, G., Feine, S., Hillgruber, F., Krainitzki, H., . . . Smith, F. H. (2002). The Neandertal type site revisited: interdisciplinary investigations of skeletal remains from the Neander Valley, Germany. *Proceedings of the National Academy of Sciences, 99*(20), 13342–13347.

Schoetensack, O. (1908). *Der Unterkiefer des Homo heidelbergensis aus den Sanden von Mauer bei Heidelberg.* Leipzig: Wilhelm Englemann.

Schrenk, F., Bromage, T. G., Betzler, C. G., Ring, U., & Juwayeyi, Y. M. (1993). Oldest *Homo* and Pliocene biogeography of the Malawi Rift. *Nature, 365*, 833–836.

Schrenk, F., Kullmer, O., Sandrock, O., & Bromage, T. G. (2002) Early hominid diversity, age and biogeography of the Malawi-Rift. *Human Evolution, 17*(1),113–122.

Schultz, A. H. (1930). The skeleton of the trunk and limbs of higher primates. *Human Biology, 2*, 303–348.

Schuster, M., Duringer, P., Ghienne, J.-F., Roquin, C., Sepulchre, P., Moussa, A., . . . Brunet, M. (2009). Chad Basin: paleoenvironments of the Sahara since the Late Miocene. *Comptes Rendus Geoscience, 341*(8–9), 603–611.

Schuster, S. C., Miller, W., Ratan, A., Tomsho, L. P., Giardine, B., Kasson, L. R., . . . Hayes, V. M. (2010). Complete Khoisan and Bantu genomes from southern Africa. *Nature, 463*(7283), 943–947.

Schwarcz, H. P., Grün, R., & Tobias, P. V. (1994). ESR dating studies of the australopithecine site of Sterkfontein, South Africa. *Journal of Human Evolution, 26,* 175–181.

Schwarcz, H. P., Simpson, J. J., & Stringer, C. B. (1998). Neanderthal skeleton from Tabun: U-series data by gamma-ray spectrometry. *Journal of Human Evolution, 35,* 635–645.

Schwartz, J. H. (2000). Taxonomy of the Dmanisi crania. *Science, 289*(5476), 55–56.

Schwartz, J. H. (1984). Hominoid evolution: a review and a reassessment. *Current Anthropology, 25,* 655–672.

Schwartz, J. H. (1999). Evolutionary provocations. Paul Sondaar, the evolution of the horse, and a new look at the origin of species. In J. W. F. Reumer & J. De Vos (Eds.), *Elephants Have a Snorkel! Papers in Honour of Paul Y. Sondaar* (Vol. 7, pp. 283–296). Rotterdam: Deinsea. Annual of the Natural History Museum.

Schwartz, J. H. (2004). Getting to know *Homo erectus*. *Science, 305*(5680), 53–54.

Schwartz, J. H. (2011). Developmental biology and human evolution. *Human Origins Research, 2*(1), e2.

Schwartz, J. H. & Tattersall, I. (1996a). Significance of some previously unrecognized apomorphies in the nasal region of *Homo neanderthalensis*. *Proceedings of the National Academy of Sciences, 93,* 10852–10854.

Schwartz, J. H. & Tattersall, I. (1996b). Whose Teeth? *Nature, 381,* 201–202.

Schwartz, J. H. & Tattersall, I. (2002). *The Human Fossil Record*. New York, NY: John Wiley & Sons.

Schwartz, J. H., Tattersall, I., & Eldredge, N. (1978). Phylogeny and classification of the primates revisited. *Yearbook of Physical Antropology, 21,* 95–133.

Sémah, F., Sémah, A. M., Djubiantono, T., & Simanjuntak, H. T. (1992). Did they also make stone tools? *Journal of Human Evolution, 23*(5), 439–446.

Semaw, S., Renne, P., Harris, J. W. K., Feibel, C. S., Bernor, R. L., Fesseha, N., & Mowbray, K. (1997). 2.5-million-year-old stone tools from Gona, Ethiopia. *Nature, 385,* 333–336.

Semaw, S., Rogers, M., & Stout, D. (2009). The Oldowan-Acheulian transition: is there a "Developed Oldowan" artifact tradition? In M. Camps & P. Chauhan (Eds.), *Sourcebook of Paleolithic Transitions* (pp. 173–193). New York: Springer.

Semaw, S., Simpson, S. W., Quade, J., Renne, P. R., Butler, R. F., McIntosh, W. C., . . . Rogers, M. J. (2005). Early Pliocene hominids from Gona, Ethiopia. *Nature, 433*(7023), 301–305.

Semendeferi, K. & Damasio, H. (2000). The brain and its main anatomical subdivisions in living hominoids using magnetic resonance imaging. *Journal of Human Evolution, 38,* 317–332.

Senut, B. (1991). Origine(s) de la bipédie humaine: approache paléontologique. In Y. Coppens & B. Senut (Eds.), *Origine(s) de la bipédie chez les hominidés* (pp. 245–257). Paris: Editions du CNRS.

Senut, B., Pickford, M., Gommery, D., Mein, P., Cheboi, K., & Coppens, Y. (2001). First hominid from the Miocene (Lukeino Formation, Kenya). *CR Acad. Sci., 332,* 137–144.

Serre, D., Langaney, D., Chech, M., Teschler-Nicola, M., Paunovicz, M., Mennecier, P., . . . Pääbo, S. (2004). No evidence of Neandertal mtDNA contribution to early modern humans. *PLoS Biology, 2,* 313–317.

Serre, D. & Pääbo, S. (2004). Evidence for gradients of human genetic diversity within and among continents. *Genome Research, 14,* 1679–1685.

Sesé, C. & Gil, E. (1987). Los micromamíferos del Pleistoceno Medio del complejo cárstico de Atapuerca (Burgos). In E. Aguirre, E. Carbonell, & J. M. B. d. Castro (Eds.), *El hombre fósil de Ibeas y el Pleistoceno de la Sierra de Atapuerca* (pp. 74–87). Valladolid: Junta de Castilla y León. Consejería de Cultura y Bienestar Social.

Shang, H. & Trinkaus, E. (2008). An ectocranial lesion on the Middle Pleistocene human cranium from Hulu Cave, Nanjing, China. *American Journal of Physical Anthropology, 135,* 431–437.

Shapiro, H. L. (1976). *Peking Man: The Discovery, Disappearance and Mystery of a Priceless Scientific Treasure*. London: Allen & Unwin Ltd.

Sherwood, R. J., Ward, S. C., & Hill, A. (2002). The taxonomic status of the Chemeron temporal (KNM-BC 1). *Journal of Human Evolution, 42*(1–2), 153–184.

Shreeve, J. (1994). 'Lucy', crucial early human ancestor, finally gets a head. *Science, 264,* 34–35.

Shreeve, J. (1995). Sexing fossils: a boy named Lucy? *Science, 270,* 1297–1298.

Shunkov, M. (2005). The characteristics of the Altai (Russia) Middle Paleolithic in regional context. *Indo Pacific Prihstory Association Bulletin, 25 (Taipei Papers, volume 3),* 69–77.

Sibley, C. G. & Ahlquist, J. E. (1984). The phylogeny of the hominoid primates, as indicated by DNA-DNA hybridization. *Journal of Molecular Evolution, 20,* 2–15.

Sibley, C. G., Comstock, J. A., & Ahlquist, J. E. (1990). DNA hybridization evidence of hominoid phylogeny: a reanalysis of the data. *Journal of Molecular Evolution, 30,* 202–236.

Siddall, M. (1998). The follies of ancestor worship. *Nature Debates, November, Nature debates, November, 18.*

Sillen, A. (1992). Strontium-calcium ratios (Sr/Ca) of Australopithecus robustus and associated fauna from Swartkrans. *Journal of Human Evolution, 23,* 495–516.

Sillen, A., Hall, G., & Armstrong, R. (1995). Strontium calcium ratios (Sr/Ca) and strontium isotopic ratios (87Sr/86Sr) of *Australopithecus robustus* and *Homo* sp. in Swartkrans. *Journal of Human Evolution, 28,* 277–285.

Sillen, A., Hall, G., Richardson, S., & Armstrong, R. (1998). 87Sr/86Sr ratios in modern and fossil food-webs of the Sterkfontein Valley: implications for early hominid habitat preference. *Geochimica et Cosmochimica Acta, 62,* 2463–2473.

Silverman, N., Richmond, B., & Wood, B. (2001). Testing the taxonomic integrity of *Paranthropus boisei* sensu stricto. *American Journal of Physical Anthropology, 115,* 167–178.

Simanjuntak, T., Sémah, F., & Gaillard, C. (2010). The palaeolithic in Indonesia: nature and chronology. *Quaternary International, 223–224*(0), 418–421.

Simons, E. L. & Pilbeam, D. (1965). Preliminary revision of the Dryopithecinae (Pongidae, Anthropoidea). *Folia Primatologica, 3*, 81–152.

Simpson, G. G. (1931). A new classification of mammals. *Bulletin of the American Museum of Natural History, 59*, 259–293.

Simpson, G. G. (1945). The principles of classification and a classification of mammals. *Bulletin of the American Museum of Natural History, 85*, 1–350.

Simpson, G. G. (1944). *Tempo and Mode in Evolution*. New York, NY: Columbia University Press.

Simpson, G. G. (1951). The species concept. *Evolution, 5*, 285–298.

Simpson, G. G. (1953). *The Major Features of Evolution*. New York, NY: Columbia University Press.

Simpson, S. W., Quade, J., Levin, N. E., Butler, R., Dupont-Nivet, G., Everett, M., & Semaw, S. (2008). A female Homo erectus pelvis from Gona, Ethiopia. *Science, 322*(5904), 1089–1092.

Singer, R. (1954). The saldanha skull from hopefield, South Africa. *American Journal of Physical Anthropology, 12*(3), 345–362.

Singer, R. & Wymer, J. (1982). *The Middle Stone Age of Klasies River Mouth in South Africa*. Chicago, IL: University of Chicago Press.

Siori, M. S. & Sala, B. (2007). The mammal fauna from the late Early Biharian site of Castagnone (Northern Monferrato, Piedmont, NW Italy). *Geobios, 40*(2), 207–217.

Sirakov, N., Guadelli, J. L., Ivanova, S., Sirakova, S., Boudadi-Maligne, M., Dimitrova, I., ... Tsanova, T. (2010). An ancient continuous human presence in the Balkans and the beginnings of human settlement in western Eurasia: a Lower Pleistocene example of the Lower Palaeolithic levels in Kozarnika cave (North-western Bulgaria). *Quaternary International, 223–224*(0), 94–106.

Skelton, R. R. & McHenry, H. M. (1992). Evolutionary relationships among early hominids. *Journal of Human Evolution, 23*, 309–349.

Skinner, M. M., Gordon, A. D., & Collard, N. J. (2006). Mandibular size and shape variation in the hominins at Dmanisi, Republic of Georgia. *Journal of Human Evolution, 51*, 36–49.

Skoglund, P. & Jakobsson, M. (2011). Archaic human ancestry in East Asia. *Proceedings of the National Academy of Sciences, 108*(45), 18301–18306.

Slatkin, M. & Hudson, R. R. (1991). Pairwise comparisons of mitochondrial DNA sequences in stable and exponentially growing populations. *Genetics, 129*, 555–562.

Smith, F. H. (1980). Sexual differences in European Neanderthal crania with special reference to the Krapina remains. *Journal of Human Evolution, 9*(5), 359–375.

Smith, H. F. & Grine, F. E. (2008). Cladistic analysis of early *Homo* crania from Swartkrans and Sterkfontein, South Africa. *Journal of Human Evolution, 54*(5), 684–704.

Smith, K. (2006). Homing in on the genes for humanity. *Nature, 442*(7104), 725–725.

Smith, T. M., Tafforeau, P., Reid, D. J., Pouech, J., Lazzari, V., Zermeno, J. P., ... Hublin, J.-J. (2010). Dental evidence for ontogenetic differences between modern humans and Neanderthals. *Proceedings of the National Academy of Sciences, 107*(49), 20923–20928.

Smith, T. M., Toussaint, M., Reid, D. J., Olejniczak, A. J., & Hublin, J.-J. (2007). Rapid dental development in a Middle Paleolithic Belgian Neanderthal. *Proceedings of the National Academy of Sciences, 104*(51), 20220–20225.

Smouse, P. E. & Li, W. H. (1987). Likelihood analysis of mitochondrial restriction-clavage patterns for the human-chimpanzee-gorilla trichotomy. *Evolution, 41*, 1162–1176.

Soares, P., Alshamali, F., Pereira, J. B., Fernandes, V., Silva, N. M., Afonso, C., ... Pereira, L. (2012). The expansion of mtDNA Haplogroup L3 within and out of Africa. *Molecular Biology and Evolution, 29*(3), 915–927.

Soergel, W. (1933). Die geologische Entwicklung der Neckarschlinge von Mauer: Ein Exkursionsbericht. *Paläontologische Zeitschrift, 15*, 322–341.

Solecki, R. (1953). The Shanidar Cave sounding, 1953 season. With notes concerning the discovery of the first Paleolithic skeleton in Iraq. *Surner, 9*, 229–232.

Solecki, R. (1971). *Shanidar: The First Flower People*. New York, NY: Knopf.

Solecki, R. (1975). Shanidar IV, a neanderthal Flower Burial in Northern Iraq. *Science, 190*, 880–881.

Sondaar, P. Y., Van den Bergh, G., Mubroto, B., Aziz, F., de Vos, J., & Batu, U. L. (1994). Middle Pleistocene faunal turnover and colonization of Flores (Lesser Sunda Islands, Indonesia) by *Homo erectus*. *C. R. Acad. Sciences, Paris, 319*, 1255–1262.

Soni, V. S. & Soni, A. S. (2010). Large size cleaver-like flakes and Hoabinhoidal elements from terminal Pleistocene to mid-Holocene epoch sites. *Quaternary International, 223–224*(0), 242–244.

Spencer, F. (1990). *Piltdown: A Scientific Forgery*. Oxford: Oxford University Press.

Spencer, M. & Demes, B. (1993). Biomechanical analysis of masticatory system configuration in neandertals and inuits. *American Journal of Physical Anthropology, 91*, 1–20.

Sponheimer, M. & Lee-Thorp, J. A. (1999). Isotopic evidence for the diet of an early hominid, *Australopithecus africanus*. *Science, 283*, 368–370.

Spoor, F., Gunz, P., Neubauer, S., Stelzer, S., Scott, N., Kwekason, A., & Dean, M. C. (2015). Reconstructed *Homo habilis* type OH 7 suggests deep-rooted species diversity in early *Homo*. *Nature, 519*(7541), 83–86

Spoor, F., Leakey, M. G., & O'Higgins, P. (2016). Middle Pliocene hominin diversity: Australopithecus deyiremeda and Kenyanthropus platyops. *Philosophical Transactions of the Royal Society B: Biological Sciences, 371*(1698). DOI: 10.1098/rstb.2015.0231.

Spoor, F., Leakey, M. G., Gathogo, P. N., Brown, F. H., Anton, S. C., McDougall, I., ... Leakey, L. N. (2007). Implications of new early Homo fossils from Ileret, east of Lake Tur-kana, Kenya. *Nature, 448*(7154), 688–691.

Spoor, F., Wood, B., & Zonneveld, F. (1994). Implications of early hominid labyrinthine morphology for evolution of human bipedal locomotion. *Nature, 369*, 645–648.

Sporns, O., Chialvo, D. R., Kaiser, M., & Hilgetag, C. C. (2004). Organization, development and function of complex brain networks. *Trends in Cognitive Sciences, 8*(9), 418–425.

Stebbins, G. G. (1950), *Variation and Evolution in Plants*. New York, NY: Columbia University Press.

Stedman, H. H., Kozya, B. W., Nelson, A., Thesier, D. M., Su, L. T., Low, D. W., . . . Mitchell, M. A. (2004). Myosin gene mutation correlates with anatomical changes in the human lineage. *Nature, 428*, 415–418.

Steele, J. (1999). Stone legacy of skilled hands. *Nature, 399*, 24–25.

Steele, J. & Gowlett, J. A. J. (1994). Communication networks and dispersal patterns in human evolution: a simple simulation model. *World Archaeology, 26*(2), 126–143.

Stern, J. T. & Susman, R. L. (1983). The locomotor anatomy of *Australopithecus afarensis*. *American Journal of Physical Anthropology, 60*, 279–317.

Stern, J. T. & Susman, R. L. (1991). 'Total morphological pattern' versus the 'magic trait': conflicting approaches to the study of early hominid bipedalism. In Y. Coppens & B. Senut (Eds.), *Origine(s) de la bipédie chez les hominidés* (pp. 99–111). Paris: Editions du CNRS.

Stewart, T. D. (1958). First views of the restored Shanidar I skull. *Sumer, 14*, 90–96.

Stoneking, M. & Delfin, F. (2010). The human genetic history of East Asia: weaving a complex tapestry. *Current Biology, 20*(4), R188–R193.

Stoneking, M., Jorde, L. B., Bhatika, K., & Wilson, A. C. (1990). Geographic variation of human mitochondrial DNA from Papua New Guinea. *Genetics, 124*, 717–733.

Strait, D. S. & Grine, F. E. (2004). Inferring hominoid and early hominid phylogeny using craniodental characters: the role of fossil taxa. *Journal of Human Evolution, 47*(6), 399–452.

Strait, D. S. & Wood, B. (1999). Early hominid biogeography. *Proceedings of the National Academy of Sciences, 96*(16), 9196–9200.

Strait, D. S., Grine, F. E., & Moniz, M. A. (1997). A reappraisal of early hominid phylogeny. *Journal of Human Evolution, 32*, 17–82.

Straus, W. L. (1957). Saldanha man and his culture. *Science, 125*(3255), 973–974.

Straus, L. G. (2001). Africa and Iberia in the Pleistocene. *Quaternary International, 75*, 91–102.

Street, M., Terberger, T., & Orschiedt, J. (2006). A critical review of the German Paleolithic hominin record. *Journal of Human Evolution, 51*(6), 551–579.

Stringer, C. B. (1983). Some further notes on the morphology and dating of the Petralona hominid. *Journal of Human Evolution, 12*(8), 731–742.

Stringer, C. B. (1984). The definition of *Homo erectus* and the existence of the species in Africa and Europe. In P. Andrews & J. L. Franzen (Eds.), *The Early Evolution of Man with Special Emphasis on Southeast Asia and Africa* (Vol. 69, pp. 131–143). Frankfurt: Cour. Forsch.-Inst. Senckenberg.

Stringer, C. B. (1985). Middle Pleistocene Hominid Variability and the Origin of Late Pleistocene Humans. In E. Delson (Ed.), *Ancestors: The Hard Evidence* (pp. 289–295). New York, NY: Alan R. Liss.

Stringer, C. B. (1986). The credibility of *Homo habilis*. In B. Wood, L. Martin, & P. Andrews (Eds.), *Major Topics in Primate and Human Evolution* (pp. 266–294). Cambridge: Cambridge University Press.

Stringer, C. B. (1987). A numerical cladistic analysis for the genus *Homo*. *Journal of Human Evolution, 16*, 135–146.

Stringer, C. B. (1993). Secrets of the pit of the bones. *Nature, 362*, 501–502.

Stringer, C. B. (2012). Evolution: what makes a modern human. *Nature, 485*(7396), 33–35.

Stringer, C. B. & Gamble, C. (1993). *In Search of the Neanderthals*. London: Thames and Hudson.

Stringer, C. B., Dean, M. C., & Martin, R. D. (1990). A comparative study of cranial and dental development with a recent British sample and among Neandertals. In C. J. DeRousseau (Ed.), *Primate Life History and Evolution* (pp. 115–152). New York, NY: Wiley-Liss.

Stringer, C. B., Grün, R., Schwarcz, H., & Goldberg, P. (1989). ESR dates for the hominid burial site of Es Skhul in Israel. *Nature, 338*, 756–758.

Stringer, C. B., Howell, F. C., & Melentis, J. K. (1979). The significance of the fossil hominid skull from Petralona, Greece. *Journal of Archaeological Science, 6*(3), 235–253.

Stringer, C. B., Hublin, J. J., & Vandermeersch, B. (1984). The origin of anatomically modern humans in Western Europe. In F. H. Smith & F. Spencer (Eds.), *The Origins of Modern Humans: A World Survey of the Fossil Evidence* (pp. 51–135). New York, NY: Alan R. Liss.

Susman, R. L. (2008). Evidence bearing on the status of *Homo habilis* at Olduvai Gorge. *American Journal of Physical Anthropology, 137*, 356–361.

Susman, R. L. & Stern, J. (1979). Telemetered electron myography of flexor digitum superficialis in *Pan troglodytes* and implications for interpretation of the OH 7 hand. *American Journal of Physical Anthropology, 50*, 565–574.

Susman, R. L. & Stern, J. T. (1991). Locomotor behavior of early hominids: epistemology and fossil evidence. In Y. Coppens & B. Senut (Eds.), *Origine(s) de la bipédie chez les hominidés* (pp. 121–131). Paris: Editions du CNRS.

Susman, R. L., de Ruiter, D., & Brain, C. K. (2001). Recently identified postcranial remains of *Paranthropus* and Early *Homo* from Swartkrans Cave, South Africa. *Journal of Human Evolution, 41*, 607–629.

Susman, R. L., Stern Jr., J. T., & Jungers, W. L. (1984). Arboreality and bipedality in the Hadar hominids. *Folia Primatologica, 43*, 113–156.

Sutton, M. B., Pickering, T. R., Pickering, R., Brain, C. K., Clarke, R. J., Heaton, J. L., & Kuman, K. (2009). Newly discovered fossil- and artifact-bearing deposits, uranium-series ages, and Plio-Pleistocene hominids at Swartkrans Cave, South Africa. *Journal of Human Evolution, 57*(6), 688–696.

Suwa, G. (1988). Evolution of the 'robust' australopithecines in the Omo succession: evidence from mandibular premolar morphology. In F. E. Grine (Ed.), *Evolutionary History of the "Robust" Australopithecines* (pp. 199–222). New York, NY: Aldine de Gruyter.

Suwa, G., Asfaw, B., Beyene, Y., White, T. D., Katoh, S., Nagaoka, S., . . . WoldeGabriel, G. (1997). The first skull of *Australopithecus boisei.* Nature, *389*, 489–492.

Suwa, G., Asfaw, B., Haile-Selassie, Y., White, T., Katoh, S., Woldegabriel, G., . . . Beyene, Y. (2007). Early Pleistocene *Homo erectus* fossils from Konso, southern Ethiopia. *Anthropological Science, 115*, 133–151.

Suwa, G., Asfaw, B., Kono, R. T., Kubo, D., Lovejoy, C. O., & White, T. D. (2009). The *Ardipithecus ramidus* skull and its implications for hominid origins. *Science, 326*(5949), 68–68e67.

Suwa, G., Kono, R. T., Simpson, S. W., Asfaw, B., Lovejoy, C. O., & White, T. D. (2009). Paleobiological implications of the *Ardipithecus ramidus* dentition. *Science, 326*(5949), 69–99.

Suwa, G., White, T. D., & Howell, F. C. (1996). Mandibular postcanine dentition from the Shungura Formation, Ethiopia: crown morphology, taxonomic allocations, and Plio-Pleistocene hominid evolution. *American Journal of Physical Anthropology, 101*, 247–282.

Suzuki, H. & Takai, F. (Eds.). (1970). *The Amud Man and his Cave Site.* Tokyo: Keigaku Publishing Co.

Swisher III, C. C., Curtis, G. H., Jacob, T., Getty, A. G., & Widiasmoro, A. S. (1994). Age of the Earliest Known Hominids in Java, Indonesia. *Science, 263*, 1118–1121.

Swisher III, C. C., Rink, W. J., Antón, S. C., Schwarcz, H. P., Curtis, G. H., Suprijo, A., & Widiasmoro, A. S. (1996). Latest *Homo erectus* of Java: potential contemporanity with *Homo sapiens* in Southeast Asia. *Science, 274*, 1870–1874.

Swisher, C. C., Rink, W. J., Schwarcz, H. P., & Antón, S. C. (1997). Dating the Ngandong humans. *Science, 276*, 1575–1576.

Sylvester, A. D., Mahfouz, M. R., & Kramer, P. A. (2011). The effective mechanical advantage of A.L. 129-1a for knee extension. *The Anatomical Record, 294*(9), 1486–1499.

Szalay, F. S. & Delson, E. (1979). *Evolutionary History of Primates.* New York, NY: Academic Press.

Szmidt, C. C., Normand, C., Burr, G. S., Hodgins, G. W. L., & LaMotta, S. (2010). AMS 14C dating the Protoaurignacian/Early Aurignacian of Isturitz, France. Implications for Neanderthal–modern human interaction and the timing of technical and cultural innovations in Europe. *Journal of Archaeological Science, 37*(4), 758–768.

Tague, R. G. & Lovejoy, C. O. (1986). The obstetrics pelvis of AL 288-1 (Lucy). *Journal of Human Evolution, 15*, 237–255.

Taieb, M. (1974). *Évolution Quaternaire du Bassin de l'Awash.* (Ph.D. dissertation), University of Paris VI.

Taieb, M., Johanson, D. C., Coppens, Y., & Aronson, J. L. (1976). Geological and paleontological background of Hadar hominid site, Afar, Ethiopia. *Nature, 260*, 289–293.

Tajima, F. (1983). Evolutionary relationship of DNA sequences in finite populations. *Genetics, 105*, 437–460.

Takahata, N. & Nei, M. (1985). Gene genealogy and variance of interpopulational nucleotide differences. *Genetics, 110*, 325–344.

Tangshan Archaeological Team & Archaeology Department of Peking University (1996). *Locality of the Nanjing Man Fossils.* Beijing: Cultural Relics Publishing House.

Tattersall, I. (1986). Species recognition in human paleontology. *Journal of Human Evolution, 15*, 165–175.

Tattersall, I. (1995). *The Fossil Trail.* Oxford: Oxford University Press.

Tattersall, I. (1998). Neanderthal genes: what do they mean? *Evolutionary Anthropology, 6*, 157–158.

Tattersall, I. & Schwartz, J. H. (1999). Hominids and hybrids: the place of Neanderthals in human evolution. *Proceedings of the National Academy of Sciences, 96*, 7117–7119.

Tauxe, L., Deino, A. L., Behrensmeyer, A. K., & Potts, R. (1992). Pinning down the Brunhes/Matuyama and upper Jaramillo Boundaries: a reconciliation of orbital and isotopic time scales. *Earth and Planetary Science Letters, 109*, 561–562.

Tavaré, S. (1984). Line of descent and genealogical process and their application in population genetics model. *Theoretical Population Biology, 26*, 119–164.

Tchernov, E. (1987). The age of the Ubeidiya Formation. *Israeli Journal of Earth Sciences, 36*, 3–36.

Tchernov, E. (1999). The earliest hominids in the southern Levant. *Proceedings of the International Conference of Human Palaeontology, Orce, Spain, 1995* (pp. 369–406). Orce: Museo de Prehistoria y Paleontologia.

Tchernov, E. (Ed.) (1986). *Les Mammifères du Pléistocene Inférieur de la Vallée du Jourdain a Oubeidiyeh* (Vol. 5). Paris: Association Paléorient.

Teaford, M. F. & Ungar, P. S. (2000). Diet and the evolution of the earliest human ancestors. *Proceedings of the National Academy of Sciences, USA, 97*, 13506–13511.

Terhune, C. E., Kimbel, W. H., & Lockwood, C. A. (2007). Variation and diversity in Homo erectus: a 3D geometric morphometric analysis of the temporal bone. *Journal of Human Evolution, 53*(1), 41–60.

Thackeray, J. F., Kirschvink, J., & Raub, T. D. (2002). Palaeomagnetic analyses of calcified deposits from the Plio-Pleistocene hominid site of Kromdraai, South Africa. *South African Journal of Science, 98*, 537–540.

The Chimpanzee Sequencing and Analysis Consortium (2005). Initial sequence of the chimpanzee genome and comparison with the human genome. *Nature, 437*(7055), 69–87.

Thoma, A. (1966). L'occipital de l'Homme mindélien de Vértesszöllös. *L'Anthropologie, 70*, 495–534.

Thompson, J. L. & Nelson, A. J. (2000). The place of Neandertals in the evolution of hominid patterns of growth and development. *Journal of Human Evolution, 38*(4), 475–495.

Thorne, A., Grün, R., Mortimer, G., Spooner, N. A., Simpson, J. J., McCulloch, M., . . . Curnoe, D. (1999). Australia's oldest human remains: age of the Lake Mungo 3 skeleton. *Journal of Human Evolution, 36*, 591–612.

Thrush, P. W. (1968). *A Dictionary of Mining, Mineral, and Related Terms.* Washington, DC: U.S. Bureau of Mines.

Tianyuan, L. & Etler, D. A. (1992). New Middle Pleistocene hominid crania from Yunxian in China. *Nature, 357*, 404–407.

Tiemei, C., Quan, Y., & En, W. (1994). Antiquity of *Homo sapiens* in China. *Nature, 368*, 55–56.

Tiemei, C. & Sixun, Y. (1988). Uranium-series dating of bones and teeth from Chinese paleolithic sites. *Archaeometry, 30*, 59–76.

Tiemei, C., Sixun, Y., & Shijun, G. (1984). Using the uranium method to investigate important Palaeolithic dates in northern China. *Acta Anthropol Sin, 3*, 259–269.

Tiemei, C. & Yinyun, Z. (1991). Paleolithic chronology and possible coexistence of *Homo erectus* and *Homo sapiens* in China. *World Archaeology, 23*, 147–154.

Tiercelin, J.-J., Schuster, M., Roche, H., Brugal, J.-P., Thuo, P., Prat, S., . . . Bohn, M. (2010). New considerations on the stratigraphy and environmental context of the oldest (2.34 Ma) Lokalalei archaeological site complex of the Nachukui Formation, West Turkana, northern Kenya Rift. *Journal of African Earth Sciences, 58*(2), 157–184.

Tobias, P. V. (1964). The Olduvai Bed I hominine with special reference to its cranial capacity. *Nature, 202*, 3–4.

Tobias, P. V. (1965). *Australopithecus, Homo habilis*, tool-using and tool-making. *South African Archaeological Bulletin, 20*, 167–192.

Tobias, P. V. (1966). The distinctiveness of *Homo habilis*. Nature, 209, 953–960.

Tobias, P. V. (1967a). *The Cranium and Maxillary Dentition of Australopithecus (Zinjanthropus) boisei. Olduvai Gorge, Vol. 2.* Cambridge: Cambridge University Press.

Tobias, P. V. (1967b). *Olduvai Gorge Vol 2: The Cranium and Maxillary Dentition of Australopithecus (Zinjanthropus) boisei.* Cambridge: Cambridge University Press.

Tobias, P. V. (1967c). Pleistocene deposits and new fossil localities in Kenya. *Nature, 215*, 479–480.

Tobias, P. V. (1980). 'Australopithecus afarensis' and A. africanus: critique and alternative hypothesis. *Palaeonto. Afr., 23*, 1–17.

Tobias, P. V. (1982a). The antiquity of man: human evolution. In B. Bonné-Tamir (Ed.), *Human Genetics, Part A: The Unfolding Genome* (pp. 195–214). New York, NY: Alan R. Liss.

Tobias, P. V. (1982b). Man the tottering biped: the evolution of his erect posture. In D. Garlick (Ed.), *Proprioception, Posture and Emotion* (pp. 1–13). Sydney: Committee in Postgraduate Medical Education, the University of New South Wales.

Tobias, P. V. (1985a). The conquest of the savannah and the attaining of erect bipedalism. In C. Peretto (Ed.), *Homo: Journey to the Origin of Man's History* (pp. 36–45). Venezia: Cataloghi Marsilio.

Tobias, P. V. (1985b). Single characters and the total morphological pattern redefined: the sorting effected by a selec-

tion of morphological features of the early hominids. In E. Delson (Ed.), *Ancestors: The Hard Evidence* (pp. 94–101). New York, NY: Alan R. Liss.

Tobias, P. V. (1986). Delineation and dating of some major phases in hominidization and hominization since the Middle Miocene. *South African Journal of Science, 82*, 92–94.

Tobias, P. V. (1987). The brain of *Homo habilis*: a new level of organization in cerebral evolution. *Journal of Human Evolution, 6*, 741–761.

Tobias, P. V. (1991a). The age at death of the Olduvai *Homo habilis* population and the dependence of demographic patterns on prevailing environmental conditions. In H. Thoes, J. Bourgeois, F. Vermeulen, P. Crombe, & K. Verlaeckt (Eds.), *Studia Archeologica: Liber Amicorum Jacques A.E. Nenquin* (pp. 57–65). Gent: Universiteit Gent.

Tobias, P. V. (1991b). The environmental background of hominid emergence and the appearence of the genus *Homo*. *Human Evolution, 6*, 129–142.

Tobias, P. V. (1991c). *Olduvai Gorge, Volume IV. The Skulls, Endocasts, and Teeth of Homo habilis.* Cambridge: Cambridge University Press.

Tobias, P. V. (1991d). Relationship between apes and humans. In A. Sahni & R. Gaur (Eds.), *Perspectives in Human Evolution* (pp. 1–19). Delhi: Renaissance Publishing House.

Tobias, P. V. (1992). The species *Homo habilis*: example of a premature discovery. *Ann. Zool. Fennici, 28*, 371–380.

Tobias, P. V. (1994). The evolution of early hominids. In T. Ingold (Ed.), *Companion Encyclopedia of Anthropology. Humanity, Culture and Social Life* (pp. 33–78). London: Routledge.

Tobias, P. V. (1997). El descubrimiento de Little Foot y la luz que proporciona sobre cómo los homínidos se volvieron bípedos. *Ludus Vitalis, 8*, 3–20.

Tobias, P. V. (2003). Encore Olduvai. *Science, 299*, 1193–1194.

Tobias, P. V. & Clarke, R. J. (1996). Faunal evidence and Sterkfontein Member 2 foot of early man: response to J.C. McKee. *Science, 271*, 1301–1302.

Tobias, P. V. & von Koenigswald, G. H. R. (1964). A comparison between the Olduvai hominines and those of Java and some implications for hominid phylogeny. *Nature, 204*, 515–518.

Tocheri, M. W., Orr, C. M., Larson, S. G., Sutikna, T., Jatmiko, Saptomo, E. W., . . . Jungers, W. L. (2007). The primitive wrist of *Homo floresiensis* and its implications for hominin evolution. *Science, 317*(5845), 1743–1745.

Toth, N. (1985a). Archaeological evidence for preferential right-handedness in the lower and middle Pleistocene, and its possible implication. *Journal of Human Evolution, 14*, 607–614.

Toth, N. (1985b). The oldowan reassessed: a close look at early stone artifact. *Journal of Archaeological Science, 12*, 101–120.

Toth, N. & Schick, K. D. (1993). Early stone industries and inferences regarding language and cognition. In K. R. Gibson & T. Ingold (Eds.), *Tools, Language and Cognition in Human Evolution* (pp. 346–362). Cambridge: Cambridge University Press.

Toussaint, M., Macho, G. A., Tobias, P., Partridge, T., & Hughes, A. R. (2003). The third partial skeleton of a late Pliocene hominin (Stw 431) from Sterkfontein, South Africa. *South African Journal of Science, 99*, 215–223.

Trauth, M. H., Maslin, M. A., Deino, A. L., Strecker, M. R., Bergner, A. G. N., & Dühnforth, M. (2007). High- and low-latitude forcing of Plio-Pleistocene East African climate and human evolution. *Journal of Human Evolution, 53*(5), 475–486.

Tribolo, C., Mercier, N., Valladas, H., Joron, J. L., Guibert, P., Lefrais, Y., . . . Lenoble, A. (2009). Thermoluminescence dating of a Stillbay–Howiesons Poort sequence at Diepkloof Rock Shelter (Western Cape, South Africa). *Journal of Archaeological Science, 36*(3), 730–739.

Trinkaus, E. (1981). Neanderthal limb proportions and cold adaptation. In C. B. Stringer (Ed.), *Aspects of Human Evolution* (pp. 187–224). London: Taylor & Francis.

Trinkaus, E. (1983). *The Shanidar Neandertals*. New York, NY: Academic Press.

Trinkaus, E. (1984). Neanderthal pubic morphology and gestation length. *Current Anthropology, 25*, 509–514.

Trinkaus, E. (1987). The Neandertal face: evolutionary and functional perspectives on a recent hominid face. *Journal of Human Evolution, 16*, 429–443.

Trinkaus, E. (1988). The evolutionary origins of the Neandertals, or, why were there Neandertals? In E. Trinkaus (Ed.), *L'Homme de Neandertal. Volume 3. L'anatomie* (pp. 11–29). Liège: Etudes et Recherches Archéologiques de l'Université de Liège.

Trinkaus, E. (2001). The Neandertal paradox. In C. Finlayson (Ed.), *Neanderthals and Modern Humans in Late Pleistocene Eurasia* (pp. 73–74). Gibraltar: The Gibraltar Museum.

Trinkaus, E. (2006). Modern human versus Neandertal evolutionary distinctiveness. *Current Anthropology, 47*, 597–620.

Trinkaus, E. (2009). The human tibia from Broken Hill, Kabwe, Zambia. *PaleoAnthropology, 2009*, 145–165.

Trinkaus, E. & Howells, W. W. (1980). Neandertales. *Investigación y ciencia, enero. En E. Aguirre (ed.), Paleontología humana, Barcelona, Prensa Científica*, pp. 15–26.

Trinkaus, E. & Shipman, P. (1993). *The Neandertals. Changing the Image of Mankind*. New York, NY: Alfred A. Knopf.

Trinkaus, E. & Smith, F. H. (1985). The fate of Neandertals. In E. Delson (Ed.), *Ancestors: The Hard Evidence* (pp. 325–333). New York, NY: Alan R. Liss.

Trinkaus, E. & Zilhao, J. (2002). Phylogenetic Implications. In J. Zilhao & E. Trinkaus (Eds.), *Portrait of the Artist as a Child. The Gravettian Human Skeleton from the Abrigo do Lagar Velho and its Archeological Context* (pp. 497–558). Lisbon: Portuguese Institute of Archaeology.

Tryon, C. (2006). Le concept Levallois en Afrique. *Fontation Fyssen – Annales, 20*, 132–145.

Turner, A. (1992). Large carnivores and earliest European hominids: changing determinants of resource availability during the Lower and Middle Pleistocene. *Journal of Human Evolution, 22*, 109–126.

Tuttle, R. H. (1981). Evolution of hominid bipedalism, and prensile capacities. *Philosophical Transactions of the Royal Society B: Biological Sciences, B292*, 89–94.

Tuttle, R. H. & Basmajian, J. V. (1974). Electromyography of brachial muscles in *Pan, Gorilla* and hominid evolution. *American Journal of Physical Anthropology, 41*, 71–90.

Tuttle, R. H., Webb, D. M., & Tuttle, N. I. (1991). Laetoli footprint trails and the evolution of hominid bipedalism. In Y. Coppens & B. Senut (Eds.), *Origine(s) de la bipédie chez les hominidés* (pp. 187–198). Paris: Editions du CNRS.

Tuttle, R. H., Webb, D., Tuttle, N. I., & Baksh, M. (1990). Further progress on the Laetoli trails. *Journal of Archaeological Science, 17*, 347–362.

Tzedakis, P. C., Hughen, K. A., Cacho, I., & Harvati, K. (2007). Placing late Neanderthals in a climatic context. *Nature, 449*(7159), 206–208.

Ungar, P. S. & Grine, F. (1991). Incisor size and wear in *Australopithecus africanus* and *Paranthropus robustus*. *Journal of Human Evolution, 20*, 313–340.

Ungar, P. S., Grine, F. E., Teaford, M. F., & El Zaatari, S. (2006). Dental microwear and diets of African early Homo. *Journal of Human Evolution, 50*(1), 78–95.

Ungar, P. S., Krueger, K. L., Blumenschine, R. J., Njau, J., & Scott, R. S. (2011). Dental microwear texture analysis of hominins recovered by the Olduvai Landscape Paleoanthropology Project, 1995–2007. *Journal of Human Evolution, 10.1016/j.jhevol.2011.04.006*, Jul 22. [Epub ahead of print].

Ungar, P. S., Scott, R. S., Grine, F. E., & Teaford, M. F. (2010). Molar microwear textures and the diets of *Australopithecus anamensis* and *Australopithecus afarensis*. *Philosophical Transactions of the Royal Society B: Biological Sciences, 365*(1556), 3345–3354.

Uno, K. T., Polissar, P. J., Jackson, K. E., & deMenocal, P. B. (2016). Neogene biomarker record of vegetation change in eastern Africa. *Proceedings of the National Academy of Sciences, USA, 113*(23), 6355–6363.

Valladas, H., Reyss, J. L., Arensburg, B., Belfer-Cohen, A., Goldberg, P., Laville, H., . . . Vandermeersch, B. (1987). Thermoluminiscence dates for the Neanderthal burial site at Kebara in Israel. *Nature, 330*, 159–160.

Valladas, H., Reyss, J. L., Joron, J. L., Valladas, G., Bar-Yosef, O., & Vandermeersch, B. (1988). Thermoluminiscence dating of Mousterian 'Proto-Cro-Magnon' remains from Israel and the origin of modern man. *Nature, 331*, 614–616.

Vallender, E. J. & Lahn, B. T. (2004). Effects of chromosomal rearrangements on human-chimpanzee molecular evolution. *Genomics, 84*, 757–761.

Vallois, H. (1935). Le *Javanthropus*. *Anthropologie, 45*, 71–84.

Vallois, H.V. (1960). L'Homme de Rabat. *Bulletin d'Archéologie Marocaine, 3*, 87–91.

Vallois, H. V. (1965). Des hommes mindeliens en Hongrie. *L'Anthropologie, 69*, 595–596.

Vallois, H. V. & Vandermeersch, B. (1975). The Mousterian skull of Qafzeh (Homo VI). An anthropological study. *Journal of Human Evolution, 4*, 445–455.

van den Bergh, G. D., Kaifu, Y., Kurniawan, I., Kono, R. T., Brumm, A., Setiyabudi, E., . . . Morwood, M. J. (2016). Homo floresiensis-like fossils from the early Middle Pleistocene of Flores. *Nature, 534(7606),* 245–248.

van den Bergh, G. D., Li, B., Brumm, A., Grün, R., Yurnaldi, D., Moore, M. W., . . . Morwood, M. J. (2016). Earliest hominin occupation of Sulawesi, Indonesia. *Nature, 529(7585),* 208–211.

van den Bergh, G. D., Meijer, H. J. M., Due Awe, R., Morwood, M. J., Szabó, K., van den Hoek Ostende, L. W., . . . Dobney, K. M. (2009). The Liang Bua faunal remains: a 95 kyr. sequence from Flores, East Indonesia. *Journal of Human Evolution, 57(5),* 527–537.

van der Made, J. (1992). Migrations and climate. *Courier Forschungsinstitut Senckenberg, 153,* 27–39.

van der Made, J. (2011). Biogeography and climatic change as a context to human dispersal out of Africa and within Eurasia. *Quaternary Science Reviews, 30(11–12),* 1353–1367.

van der Made, J. & Mateos, A. (2010). Longstanding biogeographic patterns and the dispersal of early *Homo* out of Africa and into Europe. *Quaternary International, 223–224(0),* 195–200.

van der Merwe, N. J., Francis Thackeray, J., Lee-Thorpa, J. A., & Luyta, J. (2003). The carbon isotope ecology and diet of *Australopithecus africanus* at Sterkfontein, South Africa. *Journal of Human Evolution, 44,* 581–597.

van der Plicht, J., van der Wijk, A., & Bartstra, G. J. (1989). Uranium and thorium in fossil bones: activity ratios and dating. *Applied Geochemistry, 4,* 339–344.

van Heekeren, H. R. (1952). Rock-paintings and other prehistoric discoveries near Maros (South West Celebes). *Laporan Tahunan Dinas Purbakala, 1950,* 22–35.

Vandermeersch, B. (1978). *Le crâne pré-würmien de Biache-St-Vaast.* Paper presented at the Les Origines Humaines et les Epoques de l'Intelligence, Paris.

Vandermeersch, B. (1981). *Les Hommes Fossiles de Qafzeh (Israël).* Paris: CNRS.

Vandermeersch, B. (1989). The evolution of modern humans: recent evidence from southwest asia. In P. Mellars & C. Stringer (Eds.), *The Human Revolution: Behavioural and Biological Perspectives on the Origins of Modern Humans* (pp. 155–164). Princeton, NJ: Princeton University Press.

Vandermeersch, B. (1996). New perspectives on the 'Proto-cromagnons' from Israel. *Human Evolution, 11,* 107–112.

Vandermeersch, B. & Garralda, M. D. (1994). El origen del hombre moderno en Europa. In C. Bernis, C. Varea, G. Robles, & A. González (Eds.), *Biología de poblaciones humanas: problemas metodológicos e interpretación ecológica* (pp. 26–33). Madrid: Ediciones de la Universidad Autónoma.

Vanhaeren, M., d'Errico, F., Stringer, C., James, S. L., Todd, J. A., & Mienis, H. K. (2006). Middle Paleolithic shell beads in Israel and Algeria. *Science, 312(5781),* 1785–1788.

Vaufrey, R. (1931). Les progrès de la Paléontologie humaine en Allemagne. *L'Anthropologie, 41,* 517–551.

Vekua, A., Lordkipanidze, D., Rightmire, G. P., Agusti, J., Ferring, R., Maisuradze, G., . . . Zollikofer, C. (2002). A new skull of early Homo from Dmanisi, Georgia. *Science, 297(5578),* 85–89.

Velasco, J. D. (2008). Species concepts should not conflict with evolutionary history, but often do. *Studies in History and Philosophy of Science Part C: Studies in History and Philosophy of Biological and Biomedical Sciences,* 39(4),407–414.

Venter, J. C., Adams, M. D., Myers, E. W., Li, P. W., Mural, R. J., Sutton, G. G., . . . Zhu, X. (2001). The sequence of the human genome. *Science, 291(5507),* 1304–1351.

Verhaegen, M. (1996). Morphological distance between australopithecine, human and ape Skulls. *Human Evolution, 11,* 35–41.

Verhoeven, T. (1958). Pleistozäne Funde in Flores. *Anthropos, 53,* 264–265.

Vermeersch, P. M., Paulissen, E., Stokes, S., Charlier, C., Van Peer, P., Stringer, C., & Lindsay, W. (1998). Middle Paleolithic burial of a modern human at Taramsa Hill, Egypt. *Antiquity, 72,* 475–484.

Vieillevigne, E., Bourguignon, L., Ortega, I., & Guibert, P. (2008). Analyse croisée des données chronologiques et des industries lithiques dans le grand sud-ouest de la France (OIS 10 à 3) *Paléo [En ligne], 20.*

Vigilant, L., Stoneking, M., Harpending, H., Hawkes, K., & Wilson, A. (1991). African populations and the evolution of human mitochondrial DNA. *Science, 253,* 1503–1507.

Vignaud, P., Duringer, P., Mackaye, H. T., Likius, A., Blondel, C., Boisserie, J.-R., . . . Brunet, M. (2002). Geology and palaeontology of the Upper Miocene Toros-Menalla hominid locality, Chad. *Nature, 418,* 152–155.

Villa, P. (2001). Early Italy and the colonization of western Europe. *Quaternary International, 75,* 113–130.

Villmoare, B. (2005). Metric and non-metric randomization methods, geographic variation, and the single-species hypothesis for Asian and African *Homo erectus. Journal of Human Evolution,* 49(6), 680–701.

Villmoare, B., Kimbel, W. H., Seyoum, C., Campisano, C. J., DiMaggio, E. N., Rowan, J., . . . Reed, K. E. (2015). Early Homo at 2.8 Ma from Ledi-Geraru, Afar, Ethiopia. *Science,* 347(6228), 1352–1355.

Vlcek, E. (1978). A new discovery of *Homo erectus* in central Europe. *Journal of Human Evolution, 7,* 239–251.

Vlcek, E. & Mania, D. (1977). In neuer fund von *Homo erectus* in Europa: Bilzingsleben (DDR). *Anthropologie, 15,* 159–169.

Vlcek, E., Mania, D., & Mania, U. (2000). A new find of a Middle Pleistocene mandible from Bilzingsleben, Germany. *Naturwissenschaften, 87,* 264–265.

Vogel, G. (1999). Chimps in the wild show stirrings of culture. *Science, 284,* 2070–2073.

Vogel, J. C. & Waterbolk, H. T. (1967). Groningen Radio Carbon dates VII. *Radio Carbon, 9,* 145.

von Koenigswald, G. H. R. (1936). Een Nieuwe *Pithecanthropus* Ontdekt. Een schedeltje van "slechts" driehonderd duizend jaar een merkwaardige vondst in de buurt van Soerabaia. *A.I.D. (Algemeen Indisch Dagblad) De Preangerbode, Mar. 28, Zaterdag, No. 88, Derde Blad,* 9. Bandung.

von Koenigswald, G. H. R. (1938). Ein neuer Pithecanthropus-Schädel. *Proceedings of the Academy of Science, Amst., 41*, 185–192.

von Koenigswald, G. H. R. (1949). Zur stratigraphie des javanischen Pleistocän. *Ing. Ned-Indië., I*, 185–201.

von Koenigswald, G. H. R. (1981). Davidson Black, Peking Man, and the Chinese Dragon. In B. A. Sigmon & J. S. Cybulski (Eds.), *Homo erectus: Papers in Honor of Davidson Black* (pp. 27–39). Toronto: University of Toronto Press.

von Koenigswald, G. H. R. & Weidenreich, F. (1939). The relationship between Pithecanthropus and Sinanthropus. *Nature, 144*, 926–929.

von Petzinger, G. & Nowell, A. (2014). A place in time: situating Chauvet within the long chronology of symbolic behavioral development. *Journal of Human Evolution, 74*(0), 37–54.

Vrba, E. S. (1974). Chronological and ecological implications of the fossil Bovidae at the Sterkfontein Australopithecine site. *Nature, 250*, 19–23.

Vrba, E. S. (1980). Evolution, species and fossils: how does life evolve? *South African Journal of Science, 76*, 61–84.

Vrba, E. S. (1982). *Biostratigraphy and chronology, based particularly on Bovidae of southern African hominid-associated assemblages: Makapansgat, Sterkfontein, Taung, Kromdraai, Swartranks; also Elandsfontein (Saldanha), Broken Hill (now Kabwe) and Cave of Hearths.* Paper presented at the 1er. Cong. Int. Paléont. Hum., Nice.

Vrba, E. S. (1984). Evolutionary patterns and process in the sister-group Alcelaphini-Aepycerotini (Mammalia: Bovidae). In N. Eldredge & S. M. Stanley (Eds.), *Living Fossils* (pp. 62–79). New York, NY: Springer-Verlag.

Vrba, E. S. (1985). Ecological and adaptive changes associated with early hominid evolution. In E. Delson (Ed.), *Ancestors: The Hard Evidence* (pp. 63–71). New York, NY: Alan R. Liss.

Wadley, L. (2010). Compound-adhesive manufacture as a behavioral proxy for complex cognition in the Middle Stone Age. *Current Anthropology, 51*(S1), S111–S119.

Wagner, G. A., Fezer, F., Hambach, U., von Koenigswald, W., & Zöller, L. (1997). Das Alter des *Homo heidelbergensis* von Mauer. In B. K. Wagner GA (Ed.), *Homo heidelbergensis von Mauer. Das Auftreten des Menschen in Europa.* (pp. 124–133). Heidelberg: Universitätsverlag Winter.

Wagner, G. A., Krbetschek, M., Degering, D., Bahain, J.-J., Shao, Q., Falguères, C., . . . Rightmire, G. P. (2010). Radiometric dating of the type-site for *Homo heidelbergensis* at Mauer, Germany. *Proceedings of the National Academy of Sciences, USA, 107*(46), 19726–19730.

Wagner, G. A., Maul, L. C., Löscher, M., & Schreiber, H. D. (2011). Mauer – the type site of *Homo heidelbergensis*: palaeoenvironment and age. *Quaternary Science Reviews, 30*(11–12), 1464–1473.

Walker, A. (1981). The Koobi Fora hominids and their bearing on the origins of the genus *Homo*. In B. A. Sigmon & J. S. Cybulski (Eds.), *Homo erectus: Papers in Honor of Davidson Black* (pp. 193–215). Toronto: University of Toronto Press.

Walker, A. (1993). Perspectives on the Nariokotome discovery. In A. Walker & R. E. F. Leakey (Eds.), *The Nariokotome Homo erectus Skeleton* (pp. 411–430). Berlin: Springer-Verlag.

Walker, A. & Leakey, R. E. F. (1978). The hominids of East Turkana. *Scientific American, 239*, 44–56.

Walker, A. & Leakey, R. E. F. E. (1988). The evolution of *Australopithecus boisei*. In F. E. Grine (Ed.), *Evolutionary History of the "Robust" Australopithecines* (pp. 247–258). New York, NY: Aldine de Gruyter.

Walker, A. & Leakey, R. E. F. (Eds.). (1993). *The Nariokotome Homo erectus skeleton.* Berlin: Springer Verlag.

Walker, A., Leakey, R. E. F., Harris, J. M., & Brown, F. H. (1986). 2.5 Myr *Australopithecus boisei* from West of Lake Turkana, Kenya. *Nature, 322*, 517–522.

Walker, J., Cliff, R. A., & Latham, A. G. (2006). U-Pb isotopic age of the StW 573 hominid from Sterkfontein, South Africa. *Science, 314*(5805), 1592–1594.

Wallace, J. A. (1975). Did La Ferrassie I use his teeth as a tool? *Current Anthropology, 16*, 393–401.

Walsh, J. E. (1996). *Unravelling Piltdown: The Science Fraud of the Century and Its Solution.* New York, NY: Random House.

Walter, R. C. & Aronson, J. L. (1982). Revisions of K/Ar ages for the Hadar hominid site, Ethiopia. *Nature, 296*, 122–127.

Walter, R. C. & Aronson, J. L. (1993). Age and source of the Sidi Hakoma Tuff, Hadar Formation, Ethiopia. *Journal of Human Evolution, 25*, 229–240.

Wang, W., Crompton, R. H., Carey, T. S., Günther, M. M., Li, Y., Savage, R., & Sellers, W. I. (2004). Comparison of inverse-dynamics musculo-skeletal models of AL 288-1 *Australopithecus afarensis* and KNM-WT 15000 *Homo ergaster* to modern humans, with implications for the evolution of bipedalism. *Journal of Human Evolution, 47*(6), 453–478.

Wanpo, H. (1960). Restudy of the CKT *Sinanthropus* deposits. *Vertebrata Palasiatica, 4*, 45–46.

Wanpo, H., Ciochon, R., Yumin, G., Larick, R., Qiren, F., Schwarcz, H., . . . Rink, W. (1995). Early *Homo* and associated artefacts from Asia. *Nature, 378*, 275–278.

Ward, C. V. (2002). Interpreting the posture and locomotion of *Australopithecus afarensis*: where do we stand? *Yearbook of Physical Anthropology, 85*, 185–215.

Ward, C. V., Kimbel, W. H., Harmon, E. H., & Johanson, D. C. (2012). New postcranial fossils of Australopithecus afarensis from Hadar, Ethiopia (1990–2007). *Journal of Human Evolution, 63*(1), 1–51.

Ward, C. V., Kimbel, W. H., & Johanson, D. C. (2011). Complete fourth metatarsal and arches in the foot of *Australopithecus afarensis*. *Science, 331*(6018), 750–753.

Ward, C. V., Leakey, M. G., & Walker, A. (2001). Morphology of Australopithecus anamensis from Kanapoi and Allia Bay, Kenya. *Journal of Human Evolution, 41*, 255–368.

Ward, C. V., Plavcan, J. M., & Manthi, F. K. (2010). Anterior dental evolution in the *Australopithecus anamensis—afarensis* lineage. *Philosophical Transactions of the Royal Society B: Biological Sciences, 365*(1556), 3333–3344.

Ward, R. & Stringer, C. (1997). A molecular handle on the Neanderthals. *Nature, 388*, 225–226.

Ward, T. B., Smith, S. M., & Finke, R. A. (1999). Creative cognition. In R. J. Sternberg (Ed.), *Handbook of Creativity* (pp. 189–212). New York: Cambridge University Press.

Washburn, S. L. (1957). Australopithecines; the hunters or the hunted? *American Anthropologist, 59*, 612–614.

Washburn, S. L. (1967). Behaviour and the origin of man. *Proc. Roy. Anthrop. Inst. Great Britain and Ireland, 1967*, 21–27.

Watanabe, H., Fujiyama, A., Hattori, M., Taylor, T. D., Toyoda, A., Kuroki, Y., . . . Sakaki, Y. (2004). DNA sequence and comparative analysis of chimpanzee chromosome 22. *Nature, 429*(6990), 382–388.

Watson, E., Esteal, S., & Penny, D. (2001). *Homo* genus: a review of the classification of humans and the great apes. In P. V. Tobias, A. A. Raath, J. Moggi-Cecchi, & G. A. Doyle (Eds.), *Humanity from African Naissance to coming Millennia: Colloquia in Human Biology and Palaeoanthropology* (pp. 307–318). Firenze: Firenze University Press.

Weaver, T. D. & Hublin, J.-J. (2009). Neandertal birth canal shape and the evolution of human childbirth. *Proceedings of the National Academy of Sciences, 106*(20), 8151–8156.

Webb, S., Cupper, M. L., & Robins, R. (2006). Pleistocene human footprints from the Willandra Lakes, southeastern Australia. *Journal of Human Evolution, 50*, 405–413.

Weber, J., Czarnetzki, A., & Pusch, C. M. (2005). Comment on "The Brain of LB1, Homo floresiensis". *Science, 310*(5746), 236b.

Weckler, J. E. (1954). The relationships between Neanderthal man and Homo sapiens. *American Anthropologist, 56*(6), 1003–1025.

Weidenreich, F. (1927). Der Schadel von Weimar-Ehringsdorf. *Verhandlungen der Gesellaschaft für physische Anthropologie, 2*, 34–41.

Weidenreich, F. (1933). Über pithekoide Merkmale bei Sinanthropus pekinensis und seine stammesgeschichtliche Beurteilung. *Z. Anat. Entw. Gesh., 99*, 212–253.

Weidenreich, F. (1936). The mandibles of *Sinanthropus pekinensis*: a comparative study. *Paleontol. sin., Ser. D., 7 III*, 1–163.

Weidenreich, F. (1940). Some problems dealing with ancient man. *Am. Anthropol., 42*, 375–383.

Weidenreich, F. (1941). The extremity bones of *Sinanthropus pekinensis*. *Paleont. sin., New Ser. D, 5*, 1–150.

Weidenreich, F. (1943). The skull of *Sinanthropus pekinensis*: a comparative study on a primitive hominid skull. *Palaeontol. Sin. New Ser., 10*, 108–113.

Weiner, J. & Stringer, C. (2003). *The Piltdown Forgery (50th Anniversary Edition)*. Oxford: Oxford University Press.

Weiner, S., Xu, Q., Goldberg, P., Liu, J., & Bar-Yosef, O. (1998). Evidence for the use of fire at Zhoukoudian, China. *Science, 281*(5374), 251–253.

Weinert, H. (1936). Der Urmenschensch~idel von Steinheim. *Zeitschriftfiir Morphologie und Anthropologie, 35*, 413–518.

Weinert, H. (1950). Über die Neuen Vor- und Frühmenschfunde aus Afrika, Java, China und Frankreich. *Zeit. Morph. Anthropol., 42*, 113–148.

Wendt, W. E. (1976). 'Art mobilier' from the Apollo 11 Cave, South West Africa': Africa's oldest dated works of art. *South African Archaeological Bulletin, 31*, 5–11.

Westaway, K. E., Roberts, R. G., Sutikna, T., Morwood, M. J., Drysdale, R., Zhao, J. x., & Chivas, A. R. (2009). The evolving landscape and climate of western Flores: an environmental context for the archaeological site of Liang Bua. *Journal of Human Evolution, 57*(5), 450–464.

Westaway, K. E., Sutikna, T., Saptomo, W. E., Jatmiko, Morwood, M. J., Roberts, R. G., & Hobbs, D. R. (2009). Reconstructing the geomorphic history of Liang Bua, Flores, Indonesia: a stratigraphic interpretation of the occupational environment. *Journal of Human Evolution, 57*(5), 465–483.

Weston, E. M. & Lister, A. M. (2009). Insular dwarfism in hippos and a model for brain size reduction in Homo floresiensis. *Nature, 459*(7243), 85–88.

White, M., Scott, B., & Ashton, N. (2006). The Early Middle Palaeolithic in Britain: archaeology, settlement history and human behaviour. *Journal of Quaternary Science, 21*(5), 525–541.

White, M. J. & Schreve, D. C. (2001). Island Britain-Peninsula Britain: palaeogeography, colonisation and the Lower Palaeolithic settlement of the British Isles. *Proceedings of the Prehistoric Society, 66*, 1–28.

White, R. (2001). Personal ornaments from the Grotte du Renne at Arcy-sur-Cure. *Athena Review, 2*(4), 41–46.

White, T. D. (1980a). Additional fossil hominids from Laetoli, Tanzania: 1976–1979 specimens. *American Journal of Physical Anthropology, 53*, 487–504.

White, T. D. (1980b). Evolutionary Implications of Pliocene hominid footprints. *Science, 208*, 175–176.

White, T. D. (1986). *Australopithecus afarensis* and the Lothagam mandible. *Anthropos, 23*, 73–90.

White, T. D. (1994). Ape and hominid limb lenght. *Nature, 369*, 194.

White, T. D. (2003). Early hominids–diversity or distortion? *Science, 299*(5615), 1994–1997.

White, T. D. & Suwa, G. (1987). Hominid footprints at Laetoli: facts and interpretations. *American Journal of Physical Anthropology, 72*, 485–514.

White, T. D., Ambrose, S. H., Suwa, G., Su, D. F., DeGusta, D., Bernor, R. L., . . . Vrba, E. (2009). Macrovertebrate paleontology and the Pliocene habitat of *Ardipithecus ramidus*. *Science, 326*(5949), 67–93.

White, T. D., Asfaw, B., Beyene, Y., Haile-Selassie, Y., Lovejoy, C. O., Suwa, G., & WoldeGabriel, G. (2009). *Ardipithecus ramidus* and the paleobiology of early hominids. *Science, 326*(5949), 29–40.

White, T. D., Asfaw, B., DeGusta, D., Gilbert, H., Richards, G. D., Suwa, G., & Clark Howell, F. (2003). Pleistocene *Homo sapiens* from Middle Awash, Ethiopia. *Nature, 423*(6941), 742–747.

White, T. D., Johanson, D. C., & Kimbel, W. H. (1981). *Australopithecus afarensis*: its phyletic position reconsidered. *South African Journal of Science, 77*, 445–470.

White, T. D., Lovejoy, C. O., Asfaw, B., Carlson, J. P., & Suwa, G. (2015). Neither chimpanzee nor human, Ardipithecus reveals the surprising ancestry of both. *Proceedings of the National Academy of Sciences, 112*(16), 4877–4884.

White, T. D., Suwa, G., & Asfaw, B. (1994). *Australopithecus ramidus*, a new species of early hominid from Aramis, Ethiopia. *Nature, 371*, 306–312.

White, T. D., Suwa, G., & Asfaw, B. (1995). Corrigendum. *Australopithecus ramidus*, a new species of early hominid from Aramis, Ethiopia. *Nature, 375*, 88.

White, T. D., Suwa, G., Hart, W. K., Walter, R. C., WoldeGabriel, G., Heinzelin, J., . . . Vrba, E. (1993). New discoveries of *Australopithecus* at Maka in Ethiopia. *Nature, 366*, 261–265.

White, T. D., Suwa, G., Simpson, S., & Asfaw, B. (2000). Jaws and teeth of *Australopithecus afarensis* from Maka, Middle Awash, Ethiopia. *American Journal of Physical Anthropology, 111*, 45–68.

White, T. D., WoldeGabriel, G., Asfaw, B., Ambrose, S., Beyene, Y., Bernor, R. L., . . . Suwa, G. (2006). Asa Issie, Aramis and the origin of *Australopithecus. Nature, 440*(7086), 883–889.

Whiten, A., Goodall, J., McGrew, W. C., Nishida, T., Reynolds, V., Sugiyama, Y., . . . Boesch, C. (1999). Cultures in chimpanzees. *Nature, 399*, 682–685.

Whiten, A., Horner, V., & de Waal, F. B. M. (2005). Conformity to cultural norms of tool use in chimpanzees. *Nature, 437*(7059), 737–740.

Whitworth, T. (1966). A fossil hominid from Rudolf. *South African Archaeological Bulletin, 21*, 138–150.

Widianto, H. & Zeitoun, V. (2003). Morphological description, biometry and phylogenetic position of the skull of Ngawi 1 (East Java, Indonesia). *International Journal Osteoarchaeology, 13*, 339–551.

Wiley, E. O. (1978). The evolutionary species concept reconsidered. *Systematic Biology, 27*(1), 17–26.

Wilkins, J. (2013). *Technological Change in the Early Middle Pleistocene: The Onset of the Middle Stone Age at Kathu Pan 1, Northern Cape, South Africa.* (Ph.D. dissertation), University of Toronto.

Will, M., Bader, G. D., & Conard, N. J. (2014). Characterizing the Late Pleistocene MSA lithic technology of Sibudu, KwaZulu-Natal, South Africa. *PLoS One, 9*(5), e98359.

Williams, F. L. & Patterson, J. W. (2010). Reconstructing the paleoecology of Taung, South Africa from low magnification of dental microwear features in fossil primates. *Palaios, 25*, 439–448.

Williams, M. A. J., Dunkerley, D. L., De Deckker, P., Kershaw, A. P., & Stokes, T. (1993). *Quaternary Environments.* London: Edward Arnold.

Williams, S. A. (2010). Morphological integration and the evolution of knuckle-walking. *Journal of Human Evolution, 58*(5), 432–440.

Wills, C. (1995). When did Eve live? An evolutionary detective story. *Evolution, 49*, 593–607.

Wilson, E. O. (1975). *Sociobiology: The New Synthesis* (Hay ed. castellana ed.). Cambridge, MA: Harvard University Press.

Wintle, A. G. & Jacobs, J. A. (1982). A critical review of the dating evidence for Petralona Cave. *Journal of Archaeological Science, 9*, 39–47.

Wise, S., Boussaoud, D., Johnson, P. B., & Caminiti, R. (1997). Premotor and parietal cortex: corticocortical connectivity and combinatorial computations. *Annual Review of Neuroscience, 20*, 25–42.

Withfield, L. S., Sulston, J. E., & Goodfellow, P. N. (1995). Sequence variation of the human Y chromosome. *Nature, 378*, 379–380.

WoldeGabriel, G., Ambrose, S. H., Barboni, D., Bonnefille, R., Bremond, L., Currie, B., . . . White, T. D. (2009). The geological, isotopic, botanical, invertebrate, and lower vertebrate surroundings of *Ardipithecus ramidus*. Science, *326*, 41–45.

WoldeGabriel, G., Haile-Selassie, Y., Renne, P. R., Hart, W. K., Ambrose, S. H., Asfaw, B., . . . White, T. (2001). Geology and palaeontology of the Late Miocene Middle Awash valley, Afar rift, Ethiopia. *Nature, 412*(6843), 175–178.

WoldeGabriel, G., White, T. D., Suwa, G., Renne, P., Heinzelin, J., Hart, W. K., & Heiken, G. (1994). Ecological and temporal placement of early Pliocene hominids at Aramis, Ethiopia. *Nature, 371*, 330–333.

Wolpoff, M. H. (1971a). Competitive exclusion among Lower Pleistocene hominids: the single species hypothesis. *Man, 6*, 601–614.

Wolpoff, M. H. (1971b). The evidence for two australopithecine lineages in South Africa. *American Journal of Physical Anthropology, 39*, 375–394.

Wolpoff, M. H. (1971c). Is Vértesszöllöss an occipital of *Homo erectus*? Nature, *232*, 867–878.

Wolpoff, M. H. (1977). Some notes on the Vértesszöllös occipital. *American Journal of Physical Anthropology, 47*, 357–363.

Wolpoff, M. H. (1980). *Paleoanthropology*. New York, NY: Knopf.

Wolpoff, M. H. (1981). Cranial capacity estimates for Olduvai Hominid 7. *American Journal of Human Anthopology, 56*, 297–304.

Wolpoff, M. H. (1982). *Ramapithecus* and hominid origins. *Current Anthropology, 23*, 501–510.

Wolpoff, M. H. (1984). Evolution in *Homo erectus*: the question of stasis. *Paleobiology, 10*, 389–406.

Wolpoff, M. H. (1991). Levantines and Londoners. *Science, 255*, 142.

Wolpoff, M. H. (1999). The systematics of *Homo*. Science, *284*, 1773.

Wolpoff, M. H. (2002). The species of humans at Dmanisi. *American Journal of Physical Anthropology [Suppl], 34*, 167.

Wolpoff, M. H. & Caspari, R. (2000). The many species of humanity. *Anthropological Review, 63*, 3–17.

Wolpoff, M. H., Hawks, J., & Caspari, R. (2000). Multiregional, not multiple origins. *American Journal of Physical Anthropology, 112*, 129–136.

Wolpoff, M. H., Thorne, A. G., Selinek, J., & Yinyun, Z. (1994). The case for sinking *Homo erectus*: 100 years of *Pithecanthropus* is enough! *Cour. Forsch.-Inst. Senckenberg, 171*, 341–361.

Wolpoff, M. H. Spuhler, J., Smith, F., Radovcic, J., Pope, G., Frayer, D., . . . Clark, G. (1988). Modern human origins. *Science, 241*, 772–774.

Wood, B. (1984). The origin of *Homo erectus*. In P. Andrews & J. L. Franzen (Eds.), *The Early Evolution of Man with Special Emphasis on Southeast Asia and Africa* (Vol. 69, pp. 99–111). Frankfurt: Cour. Forsch.-Inst. Senckenberg.

Wood, B. (1985). Early *Homo* in Kenya, and its systematic relationships. In E. Delson (Ed.), *Ancestors: The Hard Evidence* (pp. 206–214). New York, NY: Alan R. Liss.

Wood, B. (1988). Are "robust" australopithecines a monophyletic group? In F. E. Grine (Ed.), *Evolutionary History of the "Robust" Australopithecines* (pp. 269–284). New York, NY: Aldine de Gruyter.

Wood, B. (1991). *Koobi Fora Research Project, vol. 4: Hominid Cranial Remains*. Oxford: Clarendon Press.

Wood, B. (1992a). Early hominid species and speciation. *Journal of Human Evolution, 22*, 351–365.

Wood, B. (1992b). Origin and evolution of the genus *Homo*. Nature, *355*, 783–790.

Wood, B. (1992c). Evolution of the Australopithecines. In S. Jones, R. Martin, & D. Pilbeam (Eds.), *The Cambridge Encyclopedia of Human Evolution* (pp. 325–334). Cambridge: Cambridge University Press.

Wood, B. (1997a). Mary Leakey 1913–1996. *Nature, 385*, 28.

Wood, B. (1997b). The oldest whodunnit in the world. *Nature, 385*, 292–293.

Wood, B. (2002). Hominid revelations from Chad. *Nature, 418*, 313–315.

Wood, B. (2005). *Human Evolution. A Very Short Introduction.* New York, NY: Oxford University Press.

Wood, B. (2011). Did early Homo migrate "out of" or "in to" Africa? *Proceedings of the National Academy of Sciences, 108*, 10375–10376.

Wood, B. & Chamberlain, A. T. (1987). The nature and affinities of the "robust" Australopithecines: a review. *Journal of Human Evolution, 16*, 625–641.

Wood, B. & Collard, M. (1999a). The changing face of the *Homo* genus. *Evolutionary Anthropology, 8*, 195–207.

Wood, B. & Collard, M. (1999b). The human genus. *Science, 284*, 65–71.

Wood, B. & Collard, M. (1999c). The systematics of *Homo* – response. *Science, 284*, 1773.

Wood, B. & Constantino, P. (2007). Paranthropus boisei: fifty years of evidence and analysis. *Yearbook of Physical Anthropology, 50*, 106–132.

Wood, B. & Harrison, T. (2011). The evolutionary context of the first hominins. *Nature, 470*(7334), 347–352.

Wood, B. & Lieberman, D. E. (2001). Craniodental variation in *Paranthropus boisei*: a developmental and functional perspective. *American Journal of Physical Anthropology, 116*, 13–25.

Wood, B. & Lonergan, N. (2008). The hominin fossil record: taxa, grades and clades. *Journal of Anatomy, 212*, 354–376.

Wood, B. & Richmond, B. G. (2000). Human evolution: taxonomy and paleobiology. *Journal of Anatomy, 196*, 19–60.

Wood, B. & Strait, D. (2004). Patterns of resource use in early *Homo* and *Paranthropus*. *Journal of Human Evolution, 46*, 119–162.

Wood, B. & Turner, A. (1995). Out of Africa and into Asia. *Nature, 378*, 239–240.

Wood, B., Li, Y., & Willoughby, C. (1991). Intraspecific variation and sexual dimorphism in cranial and dental variables among higher primates and their bearing on the hominid fossil record. *Journal of Anatomy, 174*, 185–205.

Wood, B., Wood, C., & Konigsberg, L. (1994). *Paranthropus boisei*: an example of evolutionary stasis? *American Journal of Physical Anthropology, 95*, 117–136.

Wood, B. A. (1974). Olduvai Bed I post-cranial fossils: a reassessment. *Journal of Human Evolution, 3*, 373–378.

Wood, B. A. (1993). Early *Homo*: how many species? In W. H. Kimbel & L. B. Martin (Eds.), *Species, Species Concepts, and Primate Evolution* (pp. 485–522). New York, NY: Plenum Press.

Woodward, A. S. (1921). A new cave man from Rodhesia, South Africa. *Nature, 108*, 371–372.

World Health Organization (1992). World malaria situation in 1990. *Bulletin of the World Health Organization, 70*(6), 801–805.

Wu, L., XianZhu, W., YiYin, L., ChengLong, D., XiuJIe, W., & Shuwen, P. (2009). Evidence of fire use of late Pleistocene humans from the Huanglong Cave, Hubei Province, China. *Chinese Science Bulletin, 54*, 256–264.

Wu, R. (1985). New Chinese *Homo erectus* and recent work at Zhoukoudian. In E. Delson (Ed.), *Ancestors: The Hard Evidence* (pp. 245–248). New York, NY: Alan R. Liss.

Wu, X., Holloway, R. L., Schepartz, L. A., & Xing, S. (2011). A new brain endocast of *Homo erectus* from Hulu Cave, Nanjing, China. *American Journal of Physical Anthropology, 145*, 452–460.

Wu, X., Schepartz, L. A., & Liu, W. (2010). A new Homo erectus (Zhoukoudian V) brain endocast from China. *Proceedings of the Royal Society B: Biological Sciences, 277*(1679).

Wu, X., Schepartz, L. A, & Norton, C. J. (2010). Morphological and morphometric analysis of variation in the Zhoukoudian *Homo erectus* brain endocasts. *Quaternary International, 211*, 4–13.

Wurz, S. (1999). The Howiesons Poort backed artefacts from Klasies River: an argument for symbolic behaviour. *South African Archaeological Bulletin, 54*(169), 38–50.

Wurz, S., le Roux, N. J., Gardner, S., & Deacon, H. J. (2003). Discriminating between the end products of the earlier Middle Stone Age sub-stages at Klasies River using biplot methodology. *Journal of Archaeological Science, 30*, 1107–1126.

Wymer, J. (1955). A further fragment of the Swanscombe skull. *Nature, 176*, 426–427.

Wymer, J. (1958). Further work at Swanscombe, Kent. *The Archaeological News Letter, 6*, 190–191.

Wynn, T. (1989). *The Evolution of Spatial Competence*. Urbana, IL: University of Illinois Press.

Xinzhi, W. (1981). A well-preserved cranium of an archaic type of early *Homo sapiens* from Dali, China. *Sci. Sin., 241*, 530–539.

Xinzhi, W. (1999). Investigating the possible use of fire at Zhoukoudian, China. *Science, 283*, 299.

Xinzhi, W. (2004). On the origin of modern humans in China. *Quaternary International, 117*(1), 131–140.

Yellen, J. E., Brooks, A. S., Cornelissen, E., Mehlman, M. J., & Stewart, K. (1995). A Middle Stone Age worked bone industry from Katanda, Upper Semliki Valley, Zaire. *Science, 268*, 553–556.

Yinyun, Z. (1998). Fossil human crania from Yunxian, China: morphological comparison with *Homo erectus* crania from Zhokoudian. *Human Evolution, 13*, 45–48.

Yokoyama, Y., Falguères, C., Sémah, F., Jacob, T., & Grün, R. (2008). Gamma-ray spectrometric dating of late Homo erectus skulls from Ngandong and Sambungmacan, Central Java, Indonesia. *Journal of Human Evolution, 55*(2), 274–277.

Yongyan, W., Xiangxu, X., Leping, Y., Jufa, Z., & Shungtang, L. (1979). Discovery of Dali fossil man and its preliminary study. *Sci. Sin., 24*, 303–306.

Youngeun, C. & Mango, S. E. (2014). Hunting for Darwin's gemmules and Lamarck's fluid: transgenerational signaling and histone methylation. *Biochimica et Biophysica Acta (BBA) – Gene Regulatory Mechanisms*.

Yuan, Z. X., Lin, Y. P., Zhou, G. X., Zhang, X. Y., Wen, B. H., Jiang, C., . . . Wei, H. K. (1984). Field report of an excavation at Yuanmou Man's site. In G. X. Zhou & X. Y. Zhang (Eds.), *Yuanmou Man* (pp. 12–22). Kunming, China: Yunnan People's Press.

Zaim, Y., Ciochon, R. L., Polanski, J. M., Grine, F. E., Bettis Iii, E. A., Rizal, Y., . . . Marsh, H. E. (2011). New 1.5 million-year-old Homo erectus maxilla from Sangiran (Central Java, Indonesia). *Journal of Human Evolution, 61*(4), 363–376.

Zaitsev, A. N., Wenzel, T., Spratt, J., Williams, T. C., Strekopytov, S., Sharygin, V. V., . . . Markl, G. (2011). Was Sadiman volcano a source for the Laetoli Footprint Tuff? *Journal of Human Evolution, 61*(1), 121–124.

Zechun, L. (1985). Sequence of sediments at Locality I in Zhoukoudian and correlation with loess stratigraphy in northern China and with the chronology of deep-sea cores. *Quaternary Research, 23*, 139–153.

Zeitoun, V., Détroit, F., Grimaud-Hervé, D., & Widianto, H. (2010). Solo man in question: convergent views to split Indonesian *Homo erectus* in two categories. *Quaternary International, 223–224*(0), 281–292.

Zelditch, M. L., Swiderski, D. L., Sheets, H. D., & Fink, W. L. (2004). *Geometric Morphometrics for Biologists: A Primer*. London: Elsevier Academic Press.

Zeuner, F. E. (1945). *The Pleistocene Period: Its Climate, Chronology and Faunal Succession*. London: The Ray Society.

Zhang, D. D. & Li, S. H. (2002). Optical dating of Tibetan human hand- and footprints: an implication for the palaeoenvironment of the last glaciation of the Tibetan Plateau. *Geophysical Research Letters, 29*, 1072–1074.

Zhang, P., Huang, W., & Wang, W. (2010). Acheulean handaxes from Fengshudao, Bose sites of South China. *Quaternary International, 223–224*(0), 440–443.

Zhang, S. (1985). The Early Palaeolithic of China. In W. U. Rukang & J. W. Olsen (Eds.), *Palaeoanthropology and Palaeolithic Archaeology in the People's Republic of China* (pp. 147–186). London: Academic Press.

Zhu, R., An, Z., Potts, R., & Hoffman, K. A. (2003). Magnetostratigraphic dating of early humans in China. *Earth-Science Reviews, 61*, 341–359.

Zhu, R., Potts, R., Pan, Y. X., Yao, H. T., Lu, L. Q., Zhao, X., . . . Deng, C. L. (2008). Early evidence of the genus *Homo* in East Asia. *Journal of Human Evolution, 55*(6), 1075–1085.

Zhu, R. X., Potts, R., Xie, F., Hoffman, K. A., Deng, C. L., Shi, C. D., . . . Wu, N. Q. (2004). New evidence on the earliest human presence at high northern latitudes in northeast Asia. *Nature, 431*(7008), 559–562.

Ziaei, M., Schwarcz, H. P., Hall, C. M., & Grün, R. (1990). Radiometric dating of the Mousterian site at Quneitra. In N. Goren-Inbar (Ed.), *Quneitra: A Mousterian Site on the Golan Heights* (pp. 232–235). Jerusalén: Monographs of the Institute of Archaeology, Hebrew University, 31.

Zilhão, J. (2006). Chronostratigraphy of the Middle-to-Upper Paleolithic Transition in the Iberian Peninsula. *Pyrenae. Revista de Prehistòria i Antigüitat de la Mediterrània Occidental, 37*, 7–84.

Zilhão, J. & Trinkaus, E. (2002). Hisorical implications. In J. Zilhao & E. Trinkaus (Eds.), *Portrait of the Artist as a Child* (pp. 545–558). Lisboa: Trabalhos de Arqueologia 22.

Zilhão, J., d'Errico, F., Bordes, J. G., Lenoble, A., Texier, J. P., & Rigaud, J. P. (2006). Analysis of Aurignacian interstratification at the Chatelperronian-type site and implications for the behavioral modernity of Neandertals. *Proceedings of the National Academy of Sciences, 103*(33), 12643–12648.

Zipfel, B., DeSilva, J. M., Kidd, R. S., Carlson, K. J., Churchill, S. E., & Berger, L. R. (2011). The foot and ankle of Australopithecus sediba. *Science, 333*(6048), 1417–1420.

Index

Note: page numbers in *italics* refer to Figures, Tables and Boxes, whilst those in **bold** refer to Glossary

A

Abbate, Ernesto 297, *307*, 311, 314
Abell, Paul 198
Aborigines *14*, *416*
accretion model 397, *398*
accumulation model *397*
Acheulean (Mode 2) culture 229,
    241–3, 329, 330, **495**
  in Asia 249
  and cultural dispersion 243–6
  temporal distribution of *272*
  transition from Oldowan culture
      (Mode 1) to 239–49
  transition to Mousterian
      (Mode 3) 253–61
Adam *420*, 424
adapiform **495**
adaptation **495**
  and bipedalism 71–7
adaptive radiation **495**
adaptive value **495**
Adcock, Gregory 445, 446
adenine *8*, *25*
Adgantole (Ethiopia) 152
ADU-VP-1/3 cranium 365, *366*
Afar (Ethiopia) 86, *87*, 360
Africa 78
  ancestry 431
  archaic *Homo sapiens* 357
  cercopithecid zones *118*
  climatic zones *89*
  dispersal of ancient hominins
      in 225–8
  effects of glacial cycle in *310*
  first hominin dispersals 269–3
  grade II transitional
      specimens 361–68
  grade III transitional specimens 368
  Hominini tribe 83
  hominins from the Mindel-Riss
      interglacial period 357–68
  Homo ergaster specimens 278–82
  hybridization in 461–3
  isolation of African and Asian
      populations 272

main areas where early hominins
    have appeared *85*
main sites with hominin
    occupation in the corridors
    crossing the Sahara *313*
Middle Stone Age (MSA) 261–5,
    360, *433*, *474*
migration corridors out of *269*
P. boisei sites *178*
possible exit routes to Europe
    from *312*
sites with Pliocene and Pleistocene
    hominins *108*
sub-Saharan 83
temporal distribution of Oldowan
    (Mode 1) and Acheulean
    (Mode 2) in *272*
transition from to Stone Age 261–5
African apes 230
  clades *52*
  hands *62*
  ilium of *61*
  infraspinous region 149
African great apes *see* African apes
Africans, introgressive variants in
    sub-Saharan Africans *462*
Agassiz, Louis *307*
age(s)
  of Chiwondo Beds 199
  at death indicated by Olduvai
      fossils *194*
  and hypodigm of ARA-
      VP-6/500 *134*
  and their dating *44*
Aguirre, Emiliano 163, 269, 318, 341,
    346, 353
Agustí, Jordi 78, 79
Ahern, James 203
Ahern, J. C. N. *402*
Aiello, Leslie 136, 145, *184*, 187
Ain Hanech (Algeria) 312
Akazawa, Takeru 478
A.L. 137-50 ulna 145
A.L. 199-1 maxilla 143
A.L. 200-1a palate 143

A.L. 288-1 specimen (Lucy) *142,
    143–4*, 145, 146, *149*, 224
A.L. 333-45 specimen 145
A.L. 333-105 specimen 145
A.L. 417-1d specimen 145
A.L. 438–1 specimen 145, 146
A.L. 444–2 cranium *144*, 145, 147
A.L. 666-1 maxilla 200–2
ALAVP-2/10 mandible 132, *133*
ALA-VP-2/11 hand phalanx 132, *133*
ALA-VP-2/120 ulna and humerus
    shaft *133*
Alba, Davis 139
Aubert, Maxine 491
Alberts, Bruce *134*
Alcelaphine 366
Alcelaphini *51*, 89
Alemseged, Zeresenay 149
Alexeev, Valerii 204
Allahendu, Maumin 200
alleles *12*, *13*, **495**
Allia Bay (Kenya) 150
allometry **495**
allopatric **495**
allopatry 33
allotaxa *302*
Almécija, Sergio 139
Alouatta palliata, dentition *157*
Altai (Siberia) *448*, *459*
altruism 29, **495**
  reciprocal altruism and group
      selection 31
Alu gene 50
aluminum isotopes 52
Am-Ado (Ethiopia) 92, *93*, 152
America 431
AME-VP-1/71 pedal phalanx
    132, *133*
amino acids **495**
Amud (Amud Valley, Israel),
    Neanderthal specimen 408, 412
anagenesis 23, **495**
anagenetic speciation 33, 37
analogy 23, 31–2, **495**
Anatolia (Turkey) *269*

anatomically modern human **495**
  see also modern humans
ancesters 38
  number of *426*
ancestral Adam 424
ancestral characters 46
ancestral human populations
  divergence estimates for human
    genomic sequences *456*
  size of 421–4
ancestral Neanderthal populations,
    divergence estimates for
    Neanderthal genomic
    sequences *458*
Andersson, Johan Gunnar *291*
Andreae, Achilles *347*
Andrews, Peter 78, 79, 80, *132*
animal proteins 235
antelopes 109
anthropocentrism 52
anthropoids (Anthropoidea) 45, **495**
Anthropopithecus erectus 268
antibiotic resistance 12
antibodies 46, **495**
antigens **495**
Antilopini tribe 89
Antón, Susan 84, 270, 278, 297, *302*,
    370, 372, *406*
anvils **495**
apes 52, 54, **495**
Apollo cave (Namibia) 485
apomorphic characters
    (apomorphies) 36, **495**
Arabian Peninsula *315, 316*
Aralee Issie 92, *93*, 152
Arambourg, Camille 96
Aramis (Middle Awash,
    Ethiopia) 152
  age of *130*
  location of *131*
  skeleton 133–4
ARA-VP-1/129 deciduous molar 131
ARA-VP-1/401 mandible *157*
ARA-VP-6/500 skeleton *61*, 133, 134,
    136, 138, 154
ARA-VP-7/2 arm *130, 137, 138, 139*
ARA-VP-14/1 maxilla with
    dentition *152*
arboreal behavior *79*
arboreal palmigrade 138
archaic *Homo sapiens* 358, 376,
    377, **495**
archaic introgression *462*
Arcy-sur-Cure artifacts 489, 490
Ardipithecus 53, 54, 131, 132, 139, 140,
    154, 157, **495**
  –*Australopithecus* dispersal 227
  craniodental evolution 155

hands and feet features 35
hypodigm of *134*
locomotion 155
phylogenetic relationship of
    Australopithecus and *158*
species and time of emergence 125
Ardipithecus kadabba 133, *153*, 154, 157
Ardipithecus ramidus 73, 81, 131, 132,
    133, *134*, 137, 138, *139*, 140, 141,
    145, 153, 154, 155, 156, 187, 233
  ARA-VP-6/500 133
  evolutionary relationships
    between *A. anamensis* and
    A. afarensis 157–59
  last common ancestor
    locomotion 137
  left hand of *137*
  locomotion of *139*
  mandibles *156*
  pelvis and posterior limbs *139*
  primitive traits 136
  temporal range of *155*
Ardipithecus ramidus kadabba 132, 133
Ardu beds (Aduma, Ethiopia) lithic
    artifacts 365
argon-argon 41, **495**
Argue, Debbie *385*, 387
Armelagos, George 277
Aronson, James 91, 144
Arsuaga, José Luis 353
Arsuaga, Juan Luis 77, 318, 340
arthropods **495**
articulation **495**
Asa Duma site (Gona, Ethiopia) 131
Asa Issie site (Ethiopia) *131*, 152
Asa Koma specimens 94, 133
Ascenzi, Antonio 324, 339, *340*
Asfaw, Alemayehu 360
Asfaw, Berhane 130, 158, *180*, 220,
    240, 297, 335, *337*, 364
ash deposits 66, *251*
Asia 85
  ancient Mode 2 in 249
  archaic *Homo sapiens* 368–77
  Homo erectus specimens from
    282–304
  hybridization in 463–5
  isolation of African and Asian
    populations 272
  occupation of 243
  relationships between climatic
    cycles and hominin dispersal
    to Asia 273
  temporal distribution of Oldowan
    (Mode 1) and Acheulean
    (Mode 2) in 272
ASI-VP-2/334 right maxillary
    dentition *152*

ASI-VP-2/334 teeth 152
ASI-VP-2 specimens 152
ASI-VP-5/154 femur 152
ASK-VP-3/78 specimen *133*
ASKVP-3/160 specimen *133*
ASS (Aurora Stratigraphic Set) 322
assemblages **495**
Assimilation Hypothesis (AH) 466
assortative (selective) mating 24
astronomical polarity time scale
    (APTS) *141*
Atapuerca (Spain) 318, *319*, *354*
  evolutionary transition indicated
    by 341–4
  transition to the Neanderthals
    340–1
ATD6-5 mandible 322, 323
ATD6-14 left maxillar fragment *326*
ATD6-15 frontal bone 319, 321, *326*
ATD6-58 specimen 322
ATD6-69 partial face 322
ATD6-96 mandible 236, 322, 323, *326*,
    327, 341
ATD6-112 child mandibular
    fragment 322
ATD6-113 mandible 323, *327*
ATE9-1 mandible 318, 322,
    342, 343
Atlanthropus mauritanicus 281
Aurignacian culture (Mode 4) 229,
    257, 261, 485, **495**
Australians, genotypic and allelic
    frequencies for the M-N blood
    groups in 24
Australopithecinae 53, 173
Australopithecines 45
  allopatric separation of ancestors of
    chimpanzee, gorillas and 83
  in Chad 166–7
  evolutionary scheme for the gracile
    and robust australopithecines
    in the Middle and Upper
    Pliocene 182–91
  numbers of "robust" specimens
    of 175
  phylogenetic role of the South
    African "gracile" 189–90
  in South Africa 161–6
Australopithecus 37, 53, 54, 73, 81, 92,
    107, 140–2, 157, 172, 173, 338,
    382, **495**
  apomorphies of 141–2
  brain size 18
  defining the genus 160
  diet *171*
  dispersal of genus 160
  diversification of 167–69
  ear semicircular ducts 75

monophyly 183–5
origin of *228*
–*Paranthropus* dispersal 228
passage of fetus through birth
    canal in female *76*
phylogenetic relationship of
    Ardipithecus and *158*
from Pliocene site 12 KT 122
species of *141, 161*
traits characteristic of 169
Australopithecus aethiopicus 160, *181*
Australopithecus afarensis 40, 63, 74,
        106, 126, 132, 142–51, 157, 160,
        161, *181,* 185, 187, 190, 234, *394*
    as ancestor of East Africa
        specimens 185
    bipedalism *144*
    characterization of 147
    dentition *157*
    derived traits of 144
    Dikika specimens of 147–49
    enamel 155
    evolutionary relationships
        between *Ar. ramidus* and
        A. anamensis 157–59
    extension of the temporal range
        of 144–5
    forelimb 146
    in the Hadar sample 145–6
    Laetoli fossil footsteps (Tanzania)
        and 65
    Laetoli specimens of 146–7
    locomotion of 72–3, *148*
    locomotor patterns *63*
    Maka specimens of 149
    mandibles *156*
    metatarsal of *144*
    pelvis 77
    postcranial elements of 145
    primitive features shared with
        KNM-WT 17000 *190*
    sexual dimorphism in *144*
    temporal range of *155*
    West Turkana and South Turkwel
        specimens of 150–1
    Woranso-Mille specimens of 147
Australopithecus africanus 116, 140,
        146, 147, 160, 161–2, *163, 166,*
        169, 178, 181, 184, 189, 190,
        194, 233
    age at death of fossils *194*
    as ancestor of South African
        "robust" specimens 185
    diet 170, 171
    in the genus *Homo* 203
    locomotion of 74
    pelvis 76
    phylogenetic placement of *180*

Australopithecus africanus
    aethiopicus 180
Australopithecus africanus habilis 194
Australopithecus africanus
    transvaalensis 147
Australopithecus anamensis 100, 140,
        147, 148, 150–3, 155, 156, 158,
        160, 184, 189, 233
    dentition *157*
    derived traits of 153–4
    enamel 157
    evolutionary relationships
        between *Ar. ramidus* and
        A. afarensis 157–59
    hypodigm and age of 150–2
    mandibles *156*
    temporal range of *155*
Australopithecus bahrelghazali 166, 167,
        183, 187, 189, *312*
Australopithecus boisei 160
Australopithecus cf. *afarensis* 124
Australopithecus cf. *africanus 126*
Australopithecus garhi 160, *168,* 169,
        183, 189, 220–1
Australopithecus habilis 194
Australopithecus (Kenyanthropus)
    platyops 184
Australopithecus platyops 187, 188, 189
Australopithecus ramidus 130, 132
Australopithecus robustus 160, 194
Australopithecus sediba 141, 165–6,
        183, 189, 222
Australopiths 233
    diet of *170*
    gracile and robust *170*
    pelvis of *60*
    specimens 64
autapomorphies 36, 77, 269,
        404, **495**
    Homo erectus 266, 267, *268,* 278
    Homo sapiens 403, 416
    Neanderthals 403
autosomes **495**
Ayala, Don Lope Sánchez de *420*

B

Baab, Karen 299, 300, 302, 338, *385*
Baba, Hisao 286, 372
Backwell, Lucinda 232
Bader, Gregor 263
Bae, Christopher 376, 377
Baena, Javier 244
Bailey, Geoffrey 314
Balkans, Aurignacian sequences of
    the *487*
Balme, Jane 477
Banks, William 485
Bantu ethnic group 428

Bapang formation (Kabuh, Java)
    282, 285
BAR 1002 00 130
Barbetti, Michael 250
Bardon, J. 399
Baringo District (Kenya) 101
Barnfield Pit (Swanscombe, Kent,
    England) 349
Bartstra, Gert-Jan 370
Bar-Yosef, Ofer 249, 271, 410, 415,
    478, 479, 483
Basell, Laura 262
base pairs **495**
Basmajian, John *79*
Bastir, Markus 472
Bate, Dorothy 412
bats 39
    wings 31
beads 262, 263, 477, 484, 489
Beagle, HMS 1, 22
Beaulieu, Christophe 187
Bednarik, Robert *395,* 479–80,
    481, 484
beds *40,* **495**
bees, stingless 19–21
Begun, D.R. 78
Behrensmeyer, Anna 69, 87, 277
Behrensmeyer, Kay 96
Belfer-Cohen, A. 271
Belmaker, Miriam 244, 249
BEL-VP-1/1 specimen 150
Bennett, Matthew 277
Berckhemer, Fritz 349
Berger, Lee 108, 174, 183, 213, 221
Berger, Lee R. 165
Bergson, Henri 2
beryllium isotopes 42
Bettis, Arthur 296
Betzler, Christian 107
Bidditu, Italo 324
bifaces 236, 237, 239, 240, 241, *242,*
    243, 246, 249, 259, **495**
Bilzingsleben site (Germany) 350–1,
    481, 483
Binford, Lewis 234
biochronology 41
biogeography 6, **495**
    and climate 86–7
biological evolution 6
biological species' concept (BSC) 32,
    33, 34, 338
biometricians 4
biostratigraphy **495**
biota **495**
bipedalism 52, 56, 57, 63, 134, 138,
    268, 416, **496**
    and adaptation 71–7
    in animals *58*

bipedalism (*continued*)
  Australopithecus afarensis 142
  of Australopiths 72
  biomechanical analysis *63*
  changes related to 59–61
  energy efficiency of 72–3
  experimental studies of *67*
  fast-speed 75
  forms of 58–9
  genetics of *61*
  hypothetical models of the origin
    of *72*
  incipient 71
  indicated by Laetoli footprints
    67–8
  knuckle-walking 71
  and limb morphology 62
  locomotion 127
  partial vs complete 73–4
  phylogeny of 222
  postures in human lineage 75
  slow-speed 74–5
  speed and efficiency of 72
birds
  bipedalism in *58*
  wings 31, 32
birth canal, in modern women 77
birth, recreating Neanderthal 473,
    474
Bizat Ruhama (Israel) 245
Black, Davidson 253, 290, 294
bladelets **496**
blades **496**
Blake, C. Carter *396*
Blombos cave (South Africa) 435, 474
  chrono-cultural sequences at *264*
  cultural changes 485
  engravings *475*, 484
Blumenschine, Robert 196
Boaz, Noel 199, 253
Bobe, René 87
Bodo (Ethiopia) 358, 360
Bodo man *360*, 440
body weight *270*
Bohlin, Birger 290
bones, forelimb 31
bone tools 231, *232*
Bon, François 485
Bonmati, Alejandro 355
bonobos 133
Booth, Laura 207, 208
bootstrapping 27
Bordes, François 257
Bose basin (Southern China) 249
bottlenecks **496**
Boule, Pierre Marcellin 399
Bouri formation (Ethiopia) 94, 169,
    238, 297, 337, 363, 364

BOU-VP-2/66 cranium 297
BOU-VP-12/130 cranial
    fragments 169
BOU-VP-16/1 cranium 364, 365
Bouyssonie, L. 399
bovids 87, 89
Bpg 2001.04 maxilla 286, *286*
Brace, Loring 194, 399, 407
brachiation 55, 79, 134, 135, *135*, 138,
    496
brain case **496**
Brain, Charles 116, 252
Brain, C. K. 174
brain expansion 468–71
  beginning with *Homo erectus 373*
brain size *56*
brain, the 55–7
Bräuer, Günther 215, 358, 361, 379,
    440
Brazil 431
breccia *110*, **496**
Breuil, Abbe 485
Briggs, Adrian 453, *453*
Brillanceau, Abel 166
Broadfield, Douglas 371
Brock, Andrew 115
Broecker, Wallace 305
Broken Hill specimens (Zambia)
    358, 359
Bromage, Timothy 107, 199
Bronn, Heinrich Georg *347*
Brooks, Alison 358, 360
Broom, Robert 115, 160, 169, 173, 281
Brose, David *257*
Brown, Barbara 124, *179*
Brown, Francis 100, 205, 279, *361*
Brown, Frank *179*
Brown, Peter 382, *383*, *385*, 386
brow ridge **496**
BRT-VP-2/73 foot *93*, 159
Bruckner, Eduard 305
Brumm, Adam 389, 392
Bruner, Emiliano 298, 468, 470
Brunet, Michel 126, 127, 166, 187
BSNN49/P27 pelvis 275–6
Buia village (Eritrea) 280
Bulbeck, David 476
burials 478–9
burin **496**
"burnt earth cave" 254
Burtele (Ethiopia) 92, 159
Bushman Rock Shelter, jaw 475
Busk, George 2
Butzer, Karl 116

C
calcaneus **496**
calcareous **496**

calotte **496**
Campbell, B. 193
Campbell, Bernard 349
Campogrande area (Italy) *328*
Candida 25, 26
canines **496**
  size reduction of 56, *57*
Cann, Howard *429*
cannibalism 176, 236, 253, 348
Cape of Good Hope 435
Caramelli, David *446*
carbon 14 (^{14}C) 42, **496**
Carbonell, Eudald 231, 318, 321, 329,
    341, 353
carbon isotope analysis (^{13}C/^{12}C) 171
Carlson, Kristian 166
carnivores 25, 121
Carotenuto, Francesco *385*
carpal bone **496**
Carroll, Sean 416
carrying capacity 86
Caspari, Rachel 345
catarrhines (African, European, and
    Asian monkeys) 45, **496**
categories **496**
cats (*Dinofelis*) 121
Caune de l'Arago (France) 351–3
Cavalli-Sforza, Luigi Luca 421
Cavallo, Grotta del (Italy) 489
cave art 261, *493*
ceboids **496**
Cenozoic Era 44, **496**
Central Africa 83
central base foraging model *86*
Central Europe, Aurignacian
    sequences of *487*
CEPH Diversity Panel *429*, 430
Ceprano site 324–8, *329*,
    331, 339–40
cercopithecoids (Old World
    monkeys) 45, *112*, 121, **496**
cerebral cortex **496**
Chad 83, 187, 226, *228*, 311, 312
  Australopithecines found in 166–7
  TM266 vertebrates *122*
châine opératoire ("working
    sequence") 241, 242, 249, 259,
    333, 485
Chaline, Jean 81
Chamberlain, Andrew *174*, 198
Chaplin, George 431
characters **496**
Châtelperronian culture *257*, 489,
    490, **496**
Chauhan, Parth 249
Chauvet cave (France) 261, 485
Chemoigut formation (Kenya) 179
Chenchiawo (China) 288

Chesowanja (Kenya), paleomagnetic studies 252
chicken *25, 26*
childbirth 75–6, 77
chimpanzees 45, *46*, 48, 52, 53, *79*, 133
  allopatric separation of gorilla, australopithecine ancestors and *83*
  area of occupation of 84
  DNA sequence genome 49
  enamel 79
  genetic distance between humans and 35, *36*
  knuckle-walking 79
  last common ancestor of humans and *51*
  mitochondrial DNA genetic diversity of *447*
  passage of fetus through birth canal in female *76*
  pelvis of *61*
  phenotypic differences between humans and 50
  population size 55
  teeth fossils 167
  use of stones and sticks to get food 230–1
China 288
  colonization of 288–90
  cultural modes in 246–49
  evolutionary meaning of the most advanced specimens of *Homo erectus* in 376–7
  major sites with hominin fossils and stone artifacts of Lower and Middle Pleistocene in *248*
  morphological proximity of *Homo erectus-Homo sapiens* in 375
  paleomagnetic age of Chinese sites with the oldest hominin evidence *290*
  permanence of *Homo erectus* in 373–5
  sites with the oldest human presence in *288, 289*
  southeast sites related with *Homo erectus* 287
  transition in 373–7
  use of fire in 252–3
Chinese names *290*
chins **496**
  origin of the human *402*
Chiwondo Beds, Karonga district (Malawi) 106–7, 180, 199
choppers *333*, **496**
chopping-tool cultures, distribution of *245*
Chordates phylum 30

Christianity *2*
chromosome 2 *61*
chromosomes 8, **496**
  mutations of 12, *13*
chromosome X *426*
chromosome Y 424, *426*
chronospecies 33, 37–8, **496**
  replacement of *448*
chronostratigraphy **496**
chrons 43, 44
Churchill, Steven 407
Ciochon, Russell 124, 294
Clactonian industry 259, *351*
clades **496**
cladistic event *see* cladogenesis
cladistics 23, 25, *33*, 36–7, 39, 53, 183, **496**
  phylogenies *36*
  problems of 37–8
  traditional *38*
  transformed *38*
cladogenesis **496**
cladograms 36, 38, *39, 137, 184, 188, 337, 389*, **496**
Clarke, Ron *112, 162, 162, 163, 164, 174, 203*, 368
  reconstruction of the Ceprano cranium 324, *331*
Clarke, Ronald *195*
Clarke, Ron J. 203
Clark, Geoffrey 259, 452
Clark, Geoffrey A. 479
Clark, Grahame 229, *230*
Clark, J. Desmond 240, *252*, 364
Clark, Le Gros 193, *268*, 349
classes **496**
clavicle **496**
cleavers 240, **496**
Cliff, Robert 111, 163
climate, and biogeography 86–7
climate change, Upper-Miocene 81
climatic cycles 271–3
Clochon, Russell 288
closed feedback loops 56
cluster analysis 24
Clyde, William 80, 81
coalescence theory 418, *419*, 424, **496**
codons 9, **496**
cognitive capacity **496**
Cohen, Mark 277
Collard, Mark 136, 154, 187, 194, 222, 268, 299, *379*
Collard, Nicole 218
colonization, of North Europe 332
competitive exclusion principle 205
Conard, Nicholas 263
Condemi, Silvana 397
Conroy, Glenn 198, 360, 376, 397

consequences of insularity (to check) 385–6
conspecific **496**
continental drift **496**
convergence 23, **496**
convergent adaptation 47
Cooke, Basil *102*, 115
Cooke, H.B.S. 173
Cooper, Alan *447*
Coppens, Yves 81, 97, 146, 167, 175
core **496**
cosmogenic nuclides 42
C/P3 complex 128, 133, 149, 155
cranial index *208*, 209
cranial variability, in Dmanisi hominin crania *219*
cranial volume estimation *208*
cranial volumetric continuity *268*
cranial volumetric measurements *394*
cranium 55, **496**
Creator, the *2*
Cretaceous **496**
Crick, Francis 5
Crocodylus 106
Crocuta (hyenas) *89*
Cro-Magnon 260, 445, 446, 460, **496**
Cropper, Margaret 278
Crow, Tim *467*
Crummett, Tracy 289
cuboid **496**
cultural traditions
  in India *246*
  names of *230*
culture, taphonomic indications of 233–4
cuneiform **496**
curation *480*
Curnoe, Darren *116*, 201, 210, 463, 464
Curtiss, Garniss *102*
cytochrome-c 23, 25, 27
  rate of evolution *30*
  substitution of nucleotides in gene coding of 28
cytosine 8, *25*
Czarnetzki, Alfred 328, 383

D

D211 mandible 123, 217
D2280 calotte 217
D2282 cranium 217, *218*
D2600 mandible 218, *219*
D2700 cranium 218, *220*
D2735 mandible *218*
D3444 cranium 218, *220*
D3900 mandible 218, *219, 220*
D3901 specimen 219
D4167 specimen 219

D4507 humerus 219
Daam Aatu Basaltic Tuff (DABT) 130
Dali cranium 374, 376
Danakil cranium 280, 281
Danube River valley *307*
Dar-es-Soltan (Morocco) 439
Dart, Raymond 116, 160, 161, *163*,
 189, 232, 233, 234, *438*
Darwin, Charles 1, *2, 7*, 22, 55–6,
 84, 125
 canine size reduction
 hypothesis 57
 natural selection 2
Darwin, Erasmus *2*
Darwinian fitness 7
dating methods 6, **496**
 radiocarbon **500**
 U-series disequilibrium *493*
daughter species 36
Davidson, Ian 478, *479*
Dawson, Charles 57
Day, Michael 102, 109, *131*, 274, 357,
 362, 363, 399, 478
Deacon, Hilary 434, 435, 437
Deacon, Terrence 474
Dean, David 217, 397
Dear Boy (OH5) 175, 176, *177*
de Castro, José María Bermúdez 318,
 322, 329, 330, 340, 342, 343, 344,
 353, 355, 454, 472, *473*
de Castro, José M. Bermúdez 215
deciduous dentition **496**
Dederiyeh cave (Syria) 478
De'e (Guangxi Province, China) 463
Deino, Alan *107*, 199
de la Torre, Ignacio 241
de León, Marcia Ponce *473*
deletion of chromosomes *13*, **496**
Delson, Eric 116, 117, 186, 191, 217,
 371, 400
de Lumley, Henry 351, 352
de Lumley, Marie-Antoinette 324,
 351, 352
deMenocal, Peter 89
Denen Dora (Ethiopia) 91, *92*,
 141, 144
Denisovans 453–4, **496–7**
 genome 459
 phalanx *454*
Dennell, Robin 330, 342, 454
dental arcade **497**
dental development 472–3
dental traits 78
dental wear 172, 400
dentition
 changes in *157*
 deciduous **496**

locomotion and 79–81
 permanent **500**
Denton, George *307*
deoxyribonucleic acid *see* DNA
derived characters 36
derived traits **497**
d'Errico, Francesco 232, 485, 489
de Ruiter, Darryl 109, 174
de Ruiter, Darryl J. 172
descendents 38
*Descent of Man and Selection in Relation
 to Sex, The* (Darwin) 1, 55,
 396, 429
DeSilva, Jeremy *148*
de Vries, Hugo 4
diastema **497**
DID-VP-1/80 hand phalanx *133*
Die Kelders 1 (DK1) *437*
Die Kelders 1 (DK1) (South
 Africa) 435
Diepkloof Rock Shelter (South
 Africa) 263, *264*
diet
 of *A. africanus* 170
 and enamel 169–72
 of first hominins 156
 Neanderthals *407*
 of *Paranthropus* 170
diffusion coefficient 84
digits **497**
DIK-1-1 *142*, 147, 149
DIK-2-1 mandible 148
Dikika (Ethiopia) 231, 410
 specimens of *A. afarensis* 147–49
Dinaledi Chamber, Rising Star cave
 system 211–4
diploid **497**
Dirks, Paul 214
dispersal *243*
distances, in current humans 299
diversity 1, 33
Djebel Irhoud (Morocco) 367, 439,
 463
Dmanisi (Georgia) *122*, 123, 215, *216*,
 217, *243*, 244, 269
 cranial variability in *219*
 cultural changes in *244*
 new taxon for 217–20
 tools *244*
DN 7 cranium 174, 175, *177*
DNA 5, 7–10, **497**
 ancient 441, 442, *443*
 cloning 5
 double helix *9*
 eukaryotic organisms 8
 high-throughput DNA
 sequencing 452–3

methylation 15
 sequencing 5–6
 from Sima de los Huesos *461*
DNA–DNA hybridization 46, 49
Dobzhansky, Theodosius 4, *5*, 32
dogs *25, 26*
Dolni Vestonice (Czech Republic) 479
dolomitic caves 107, *108*
dolphins 39
domestic cats, evolution of 25
dominant characters **497**
Domínguez-Rodrigo, Manuel *226*, 361
donkeys *25, 26*
Douglas, Rodney *368*
Douka, Katerina 489
"dragon teeth" and "dragon
 bones" *291*
Drapeau, Michelle 144, 146
DRB1 genes 422
 genealogical tree of *422, 423, 425*
 lineages *426*
 as a molecular clock *424*
 nucleotide differences among *422*
Dreyer, Thomas 367
Drimolen (South Africa) 118–9
 bone tools 232, *233*
 stratigraphic relationships *118*
Drummond, Alexei 447
Dryopithecus 139
 enamel 78
 locomotion and enamel of *79*
Dubois, Eugène 57, *58*, 266, *268*
ducks *25, 26*
duplication of chromosomes **497**
Durband, Arthur 298, 371, 372
Dusseldorp, Gerrit 475
Duyfjes, Johan 282, *283*
dwarfism 382–3, *384*, 385, 389
dwellings, of the first hominins 233

E
Early Aurignacian 486
Early Pleistocene *267*
Early Stone Age 261, *262*
East Africa 87
 changes in vegetation in *172*
East Rudolf (Lake Turkana,
 Kenya) 296
ecology 6
edentulous maxilla 182, *183*, 219
edentulous skull 271
effective population size **497**
Ehringsdorf (Germany) 397
electron spin resonance (ESR) 43, **497**
electrophoresis **497**
Elephas antiquus fossils 347
Elephas (elephants) 89

el-Wad caves, Mount Carmel
(Israel) 410
enamel 132, 155, 157, **497**
of *Ardipithesus* 132
and diet 169–72
and locomotion 79
strontium/calcium ratios in 171
encephalization **499**
endemic **499**
endemism, of species in remote
archipelagos 22
endocast/endocranial cast **497**
Engis Cave (Belgium) 395
infantile cranium 396
England, glacial and interglacial
periods with evidence of
human presence in 331
engravings 261, 262, 396, 481, 483
Ennouchi, Émile 366
En, Wu 376
enzymes **497**
Eocene **497**
EP 1000/98 tibia 182
EP 1500–1501 maxilla 182
epigenesis **499**
epigenetic processes 7, 15
epithelium **497**
epoch **497**
Equatorius africanus 139
Equus (modern horses) 37
era **497**
Erk-el-Ahmar formation (Israel)
244
Ethiopia 200
ethnic differentiations 431
Etler, Denis 289
Etler, Dennis 376
eukaryotes 8, **497**
Eumetopias jubata 19
Eurasia 431
Europe 86
"classic" glaciations in 309
colonization of 304–35
crossing the Mediterranean 314–18
dispersal routes of modern
humans in 486
effects of glacial cycle in 310
first hominins 318–28
glacial cycles 305–11
hominin species 339
Homo antecessor as terminal lineage
in the occupation of 343
mammal zones 334
occupation of 306, 307, 330
Paleolithic record before and after
c. 500,000 years 305
population encounters in
485–7

principal sites with lithic industry
after MIS 15 and before MIS 7
255
regional phases of European
Paleolithic cultures 257
Saharan barrier to migration
to 311–4
short chronology vs. ancient
occupation 305
sites with Mode 1 utensils dated
prior to c. 0.5 Ma 333
temporal distribution of Oldowan
(Mode 1) and Acheulean
(Mode 2) in 272
EV 9001 cranium 373
EV 9002 cranium 373
Evans, Alistair R. 472
Evernden, Jack 102
evolution
first scientific proposals on 2
Lamarck's theory of 31
molecular 5–6
molecular clock of 27–9
by natural selection 6–7
synthetic theory of 4
theory of 1, 2
evolutionary ethology 6
evolutionary history
of 20 species 26
reconstruction of 29
evolutionary species' concept
(ESC) 32, 33, 34, 338
evolutionary trees 23–7
distance methods 23–4
methods for constructing 23
statistical degree of confidence
of 27
exact randomization 35
exons 8, **497**
expensive tissue hypothesis 471
extant **497**
Eyasi Plateau (Tanzania) 106

F

Falk, Dean 383, 384, 394, 394
families 39, **497**
Far East, sites with "archaic"
sapiens 377
faults **497**
fauna 121, **497**
Olduvai (Tanzania) 104
faunistic comparison 41
Feldhofer cave (Germany) 396, 444
Feldhofer cranium (Neander,
Germany) 442
femoral head **497**
femoral neck **497**
femur **497**

Fengshudao site (China) 249
Ferring, Reid 217, 244
fibrinopeptides, rate of evolution 30
fibula 62, **497**
Finarelli, John 80, 81, 81
finches 22
Finlayson, Clive 400
fire, use of 250–3
Fisher, R.A. 4
fission of chromosomes 12
fission-track dating 42–3, **497**
fitness 16, **497**
of genotypes 17
flakes 235, 259, 333, **497**
Fleagle, John 363
flora **497**
Flores Island (Indonesia) 246, 317,
379, 380
carvings 394
industry of 389–92
utensils 393
Florisbad (South Africa) 367,
367, 368
Foley, Robert 262, 309, 387, 476
folivores **497**
foot, partial (BRT-VP-2/73) 159
Footprint Tuff 66
foragers **497**
foramen magnum 59, 131, **497**
Forbes Quarry (Gibraltar) 395, 399
specimens 401, 406
forelimb bones 32
formations 40, **497**
Formicola, Vincenzo 277
fossil footprints 65–71
conservation of 71
Laetoli (Tanzania) 65–7
Turkana footprints 68–71
fossilization 40
fossil populations 34
species identification of 35
fossil record, study of 40
fossils 43, 472, **497**
founder effect 426
founder principle 15
FOXP2 gene 49, 50
France 258
Franciscus, Robert 386, 406
frequency-dependent selection 17–8
Frida Leakey Korongo (FLK)
(Tanzania) 105
frugivores **497**
Fuhlrott, Johann 395
Fuller, Chad 361
Function, Phylogeny and Fossils
(Begun et al) 78
fusion of chromosomes 12
FwJj14E site (Kenya) 71, 277

G

Gàala Tuff Complex 130
Gabunia, Léo 217, *219*
Gadeb site (Ethiopia) 97, 240, 252
Gaillard, Claire 246
Galápagos Islands 11, *32*
Gamble, Clive 397, 413
gametes 497
gamma ray spectroscopy
    technique 53
Ganopolski, Andrey 87
Gantt, David *79*
Gargett, Robert 479
Garralda, María Dolores 275, *400*
Garrod, Dorothy 397, 410, 412, 414
Gatogo, Patrick 205
Gaudzinski, Sabine 328
Gause, Georgyi Frantsevitch *205*
Gee, Henry *132*
Geffen, Eli 403 ·
Gelasian Age 44
Geleijnse, Vera 434
gene control mechanisms 9
gene duplication 27
gene expression 50
gene flow 14
gene frequencies 14–5
gene migration 14
gene pool 12
genera 39
Generalized Procrustes Analysis 299
genes 8, 418, **497**
    paralogous 37
    positive selection of 466–8
genetic change 14–5
genetic code 9, *10*
genetic distances 35, *421*
genetic drift 7, 15, *447*, **497**
genetic monomorphism 19
genetic polymorphism 19
genetics
    of bipedalism *61*
    and family relationship *420*
Genetics and the Origin of Species
    (Dobzhansky) 14
genetic variation, in a population
    22–4
genomes 6, **497**
genome sequencing, coverage
    of *460*
genotypes **497**
    fitness of *17*
genotyping **497**
genus 53–4, **497**
    adaptive concept of 49–50
    concepts of 38–9
geographical grouping 429
geological stages 44

geomagnetic polarity time scale
    (GPTS) *141*
geometric symbols *477*
Georgia 215, 243
Gerbillus grandis 366
Germany 333
germ-plasm theory 3–4
Gesher Benot Ya'aqov (Israel)
    245, 252
Ghiselin, Michael *33*
Giannecchini, Monica 277
gibbons 45, *79*, 136
Gibbons, Ann *130*, *134*
Gibraltar Strait 316
Gilbert, W. Henry *298*, *337*
Gil, Enrique 319
Gilligan, Ian 403
Ginzburg, Avishag 403
glabella **497**
glacial periods **497**
glaciations 87
    "classic" *308*, *309*
    discovery of *307*
    in Europe 305–11
    measurement by oxygen
        isotopes *307*
Gladysvale specimens (South
    Africa) *118*
Glazko, Galina 124
globin genes, phylogeny of *38*
gluteus maximus 73, 74
Gómez-Olivencia, Asier 404
Gona culture (Ethiopia) *236*
Gona site (Ethiopia) 237
    pelvis 275–6, *276*
    tool assemblages *238*
Gongming, Yin 374
González-José, Rolando 185, 338
Goodall, Jane 230
Goodman, Morris 45
Goodwin, William 450
Gordon, Adam 218, *385*
Goren-Inbar, Naama 252, 482
Gorham cave (Gibraltar) *400*
Gorilla 81, **497**
    locomotion and enamel of *79*
    phylogenetic relation between *Pan*,
        and *Homo* and *48*
Gorilla gorilla 206, 267
gorillas 45, 46, 52, 53, *79*
    allopatric separation of chimpanzee,
        australopithecine ancestors
        and *81*
    enamel 79
    knuckle-walking 79
    locomotion of *60*
    mitochondrial DNA genetic
        diversity of *447*

population size 55
    use of tools by *232*
Gorillinae 53
Gowlett, John *86*
Gracia, Ana 355
gracile australopiths 169, 170, **497**
grades *47*, **497**
Grafenrain gravel pit (Germany)
    346, *347*
Grand Canyon (Bright Angel Trail,
    Nevada, United States) *51*
Gran Dolina (Atapuerca, Spain)
    318–24, 340, 472
    stragigraphy and corresponding
        paleomagnetic poles in *323*
    TD6 328
Graves, Ronda 275
Gray, Tom *142*
great apes 52, 53, **497**
Greece, examples of ancient homins
    from *350*
Green, David 149
Greenfield, Leonard 124
Green, Richard 429, 454, 456, 457,
    459
Grine, Frederick 160, 169, 170, 171,
    185, *186*, 188, 203, 210, 335, 437,
    438, 439
Grotta di Fumane (Lessini Mountains,
    North Italy) 488, 489
Grotte du Renne (Arcy-sur-Cure,
    France) 489, 490
Groves, Colin 124, 279, 397, 434
Grün, Rainer 109, 296, 370, 413
GSI (Geographic Information
    System) 482
guanine *8*, 25
Guanjun, Shen 294
Guipert, Gaspart 353
Gunz, Phillipp 473

H

Haberer, K.A. *291*
habitat **497**
Hadar culture (Ethiopia) *236*
Hadar (Ethiopia) *40*, 64, 73, *75*, 200
    Dikika 92
    localization of *141*
    number of *A. afarensis* species in
        sample at 145–6
    specimens 142–3
Haeusler, Martine 224
Hailemichael, Million 144
Haile-Selassie, Yohannes *130*, 147,
    153, 159, 364
Haldane, J.B.S. 4
half-life **497**
Hallgrimsson, Benedikt 225

Hammer, Michael 461
hammerstone **498**
hand and wrist, *Australopithecus sediba* 166
handaxes 235, *240*, *245*, **498**
hands 62
  H1 213
  of hominins 139
  left hand of *Ardipithecus ramidus* 137
Han, Fei 352
Hanson, Brooks *134*
haplodiploidy 20
haploid **498**
Happisburgh (Norfolk, UK) 404
Hardy–Weinberg law 12–3, 14, **498**
Harmand, Sonia 232
Harris, John 252
Harrison, Terry 138, 182, *186*
Harrod, James B. 480
Hartwig-Scherer, Sigrid 224
Harvati, Katerina 350, 400
Hathnora site (Central India) *246*
Haughton, S. H. 173
Häusler, Martin *145*
Hawaii 22
Hawks, John 345
Hay, Richard 102
Hb^S allele 16, *17*
HCRP RC 911 (maxilla) 180, 199
HCRP UR 501 (mandible) 199
Heaton, Jason *112*
Heim, Jean-Louis 404
hemoglobin 27
  rate of evolution 30
  sickle cell anemia 16, *17*
Hennig, Willi 36
Henshilwood, Christopher 475, 484
hereditary variation 1, 7
heredity 12
Herries, Andy 114, 163
Herrmann, Esther *232*
Herto BOU-VP-16/1 specimens *365*
heterogeneity 19
heterozygote **498**
Hexian (China), cranium 374
Higham, Thomas 489
high-throughput DNA sequencing *see* PEC (primer extension capture)
Hill, Andrew 83, 101, *107*, 124, 200, 201
Hinton, Martin 57
hip 61
hip bones, fossil specimens of *61*
hip shape 59
histone proteins 8, 15
Hofmeyr (Eastern Cape, South Africa)

cranium 438–9
skull 475
Hohle Fels cave (Germany) 486
Holliday, Trenton *386*
Holloway, Ralph 169, *208*, 397, 468
holotypes **498**
Holton, Nathan *406*
homeobox genes 9
home range *86*, 270
Hominidae family 30, 45, *51*, 53
Hominid Corridor Research Project 106
Hominina 53
Homininae 53
hominin evolution
  "bushy" nature of *184*
  climatic and phylogenetic events accompanying 90
  correlation between global climate changes, presence of large lakes in East Africa and 88
Hominini tribe 53–4, 83, **498**
  emergence of 124
  first phylogenetic changes in 154–7
  fossil record 125
  genera and species 84
  phylogenetic tree 54, 184
  taxa of 125
  taxonomic considerations of 45
hominin lineage 78
hominins **498**
  chimpanzee separation event 124
  cladogram of ancient 221
  climate and phylogenetic events accompanying the evolution of 273
  dispersal of ancient hominins in Africa 225–7
  diversity of Middle and Upper Pliocene 187–9
  dwellings of the first 233
  European specimens between MIS 15 and MIS 7 348
  evolutionary model for the Lower and Middle Pleistocene 335–44
  gracile 160
  mandibles 325
  Middle and Upper Pliocene 160–91
  from the Mindel-Riss interglacial period 357–68
  Miocene 124–34
  origin and dispersal of 228
  phylogenetic relationships of the Miocene and Lower Pliocene 157–9
  population dispersal 83–6

relationships between climatic cycles and hominin dispersal to Asia 273
robust 160
taxonomy during the Lower and Middle Pleistocene 296–7
time and place of the appearance of 124–5
Toros-Menalla 127
transition to Upper Plicoence 345–440
Tugen Hills region 128–30
Hominin tribe 30
hominization grades 301
Hominoidea **498**
hominoid lineages
  age of 45–8
  role of locomotion in the divergence of 134–40
hominoids (apes and humans) 45, **498**
  brains 56
  cladogram of current 135
  monophyletic groups 47
  paraphyletic groups 47
Homo 50–2, 53, 54, 135, 140, *168*, 190, 252, *498*
  changing views of 51
  emergence of the genus 192–228
  fossils, from Dmanisi 123
  including *A. africanus* in genus 203
  Linnaeus classification of 415
  locomotion and enamel of 79
  locomotion and monophyly of the first 222–5
  Myosin gene 50
  phylogenetic relation between *Pan*, and *Gorilla* and 48
  radiation during Middle and Lower Pleistocene 266–344
  South African transition 221–2
  species from the Plio-Pleistocene 267
  species of genus 161
  time of emergence of species of 125
  transition to 220–2
Homo antecessor 324, 329, 340, 341
  apomorphies of 326
  cranial traits of 326
  dental traits 326
  identified individuals of 340
  as LCA of Neanderthals and modern humans 341, 344
  mandibular traits 326
  phylogenetic tree 342
  as terminal lineage in the occupation of Europe 343

Homo erectus 15, 22, 35, 37, *127*, 175,
176, 192, 193, 198, 222, *268*, 345,
369, *379*
  Acheulean (Mode 2) culture 243
  in Asia and Africa 300–3
  biological reality of *269*
  bipedalism 75
  body mass 270
  brain expansion beginning
    with *373*
  from Ceprano site 324
  characterization of 273–8
  in China 373–5
  cranial features of *212–3, 297*
  dispersal process of 84
  ear semicircular ducts 75
  environment and adaptation in
    Asia 294
  European archaic 346–57
  evolutionary transition between
    Homo neanderthalensis and 397
  evolution to *Homo sapiens* 378
  hypotheses in the evolutionary
    transition from *Homo
    habilis* 341
  in Java 285–6
  in the Lower Pleistocene 266
  main sites of the entire sample
    of *267*
  Mauer mandible 346–8
  occupation of Flores Island by *317*
  permanence of 369–70
  size and body mass of 274–6
  Sm 3 (cast) *274*
  smaller size specimens 303–4
  species definition 266–9
  specimens of *336*
  study of cranial shape variation
    in *302*
  systematic of 337–9
  taxonomy of *269*
  variability of 298–300, 304
Homo erectus habilis 194
Homo erectus newyorkensis 371
Homo erectus pekinensis 290
Homo erectus petraloniensis 345
Homo erectus sensu lato *73, 269,* 296
Homo erectus sensu stricto *269,*
  296, 345
Homo erectus (unspecified
  subspecies) 194
Homo ergaster *180*, 198, 245, *269*, 270,
  296–8
  African specimens 278–82
  body mass 270
  cranial features of *296*
  Olduvai (Tanzania)
    specimens 278–9

transitional species between *Homo
  sapiens* and 377–79
Homo floresiensis 73, 270, 379–94
  older fossils of 384–5
  phylogenetic consideration
    of 386–89
Homo gautengensis 203, 210–211,
  335, 338
Homo georgicus 215–20, 223, 269, 337
Homo habilis 75, 102, *163*, 176,
  192–203, 229, 278, 296, *379, 379*
  age at death of fossils *194*
  apomorphies of 194–5
  body mass 270
  cranial and mandibular traits 194
  cranial capacity *207, 207*
  cranial features of *212–3*
  cranial size and coefficient of
    variation *208*
  dental traits 194
  discovery and proposal of 229
  foot 65
  grouping with *H. rudolfensis* 221
  hypodigm of *205*
  hypotheses in the evolutionary
    transition to *Homo erectus* 341
  locomotion of 222
  locomotor patterns *63*
  morphology of 208
  OH 62 and variability
    within 195–6
  Oldowan culture (Mode 1) 243
  Olduvai specimens of 195
  postcranial elements of 194–5
  sensu stricto and sensu lato 207
  sexual dimorphism of 205–7
  in South Africa 201–3
  taxon to encompass the whole
    sample of 207–10
  Turkana specimens of 196–201
Homo habilis sensu lato 204, 205–10
Homo habilis sensu stricto 204, 205,
  206, 207
Homo heidelbergensis 341, 345, 356,
  376, 378, 379, *397, 397*, 448
  Mauer mandible *346*
Homo helmei 345, 368
homoiology 32
Homo (Javanthropus) soloensis 345
Homo (javanthropus) soloensis 369
homologous **498**
homologous traits 27
homology 23, 27, 31–2
Homo modjokertensis 199
Homo naledi 211–214
Homo neanderthalensis 260, 328, 339,
  357, *379*, 395–415, *397*, 433
  "classic" Neanderthals 398–400

differences between *Homo sapiens*
  and 416–7, 442
evolutionary transition between
  Homo erectus and 397
the first Neanderthals 395–98
genetic distance between *Homo
  sapiens* and 441–68
Gibraltar I cranium *406*
hybridization traces with *Homo
  sapiens* 458–9
similarities between *Homo sapiens*
  and *442*
Homo paleohungaricus 351
Homo/Pan divergence 49–50
homoplasies **498**
Homo rhodesiensis 341, 345, *379*
Homo rudolfensis 187, 188, 199,
  203–10, 278
  grouping with *H. habilis* 221
  hypodigm of *205*
  postcranial elements of *204*
  specimens of *204*
  validity of taxon *204*
Homo saldanensis 361
Homo sapiens 15, 22, 30, 35, 54, 105,
  175, *268*, 345, 415–40
  African archaic 357
  apomorphies and climates related
    to *417*
  brain size 18
  cranial size and coefficient of
    variation *208*
  differences between *Homo
    neanderthalensis and* 416–7, *442*
  evolution from *Homo erectus* 378
  exit out of Africa for 476–7
  Founder effect 424–6
  genetic distance between *Homo
    neanderthalensis* and 441–68
  hybridization traces with *Homo
    neanderthalensis* 458–9
  Klasies River 433–8
  older African specimens of 439–40
  scope of the taxon *379*
  similarities between *Homo
    neanderthalensis* and *442*
  tool-making techniques 261
  transitional species between *Homo
    ergaster* and 377–79
  use of grades to explain the
    transition to *358*
Homo sapiens daliensis 376
Homo sapiens idaltu 363
Homo sapiens neanderthalensis 407,
  412
Homo sapiens rhodesiensis 361
Homo sapiens sapiens 407, 412
Homo sapiens steinheimensis 349

Homo soloensis 286, 372
Homo steinheimensis 349
Homo swanscombensis 345, 349
Homotherium (saber toothed
    tigers) *89*
homozygote **498**
Hooker, Joseph *2*
Hooton, Earnest 398
horses *24, 25, 26*
Höss, Matthias *444*
Hovers, Erella 478, 479, 483
Howell, Clark 193
Howell, F. Clark 199, 261, *347,* 348,
    400, 401, *402, 404,* 414
Howells, William 280, 416
Howells, W.W. 150, 298
Howieson's Poort (HP) traditions
    (South Africa) 262, 265,
    474, 484
HOX D gene *61*
Hrdlika, Ales 360
Hrvatsko Zagorje (Croatia) 490
Huang, Weiwen 249
Hublin, Jean-Jacques 77, 268, 361,
    366, 397, 416, 463
Huffman, O. Frank 282, *283*
Hughes, Alun 201
Hull, David *33*
Hulu Cave (China) 291, 341, 374
"human accelerated regions"
    (HARs) 49
human evolution, genera and
    species 83
Human Genome Project 49
human hemoglobin 16
human hunter-gatherer groups, area
    of occupation of 84
human language 55
Human Leukocyte Antigen
    (HLA) 421
human lineage
    bipedal postures in *75*
    traits of 54–65
humans *see* modern humans
Humbu formation (Tanzania) 240
humerus *139,* **498**
humidity 84
hunter-gatherer **498**
Huxley, Julian 5, 299
Huxley, T. H. *3*
hybridization 39, **498**
hybridization hypothesis 345, **498**
hybrids 22
hydroxyapatite, fossil teeth 53
hyenas, ancestral 233
Hylobates 79, 136, **498**
hymenopterans, haplodiploid
    reproduction of *30*

Hyodo, Masayuki 284
hyoid bone **498**
hypodigm **498**
hypotheses, falsifying *208*
Hypperion (horses) *89*
Hyracotherium 37

I

Ileret fossil footprints (Kenya) *70,* 71,
    277, 277
ilium, of African great apes *61*
immune systems, study of *46*
Inbar, Moshe 368
inbreeding **498**
incisors **498**
India, cultural traditions in *246*
industries, transitional 487–89
infant–mother mass ratio
    (IMMR) 76
infraorders 45
insectivores **498**
insects 17
interbreeding 21–2
interglacial period **498**
International Code of Zoological
    Nomenclature (ICZN) 53, *147*
International Commission on
    Stratigraphy 44
International Paleontological
    Research Expedition 363
International Stratigraphic Chart
    44, 308
introgression **498**
introns 8, **498**
inversion of chromosomes 12, **498**
Irhoud Djebel (Morocco) 366, 472
Isaac, Glynn *86, 238, 239,* 241
Ishida, Hidemi 124
isolation *33*
    reproductive 37
Isolation Hypothesis *385*
Italy 333

J

Jablonski, Nina 167, 431
Jacobs, Zenobia 262
Jacob, Teuku 370, 372, 382, 383
Jacovec Cave (or Jakovec)
    (Sterkfontein, South
    Africa) 109, *110,* 112, 162
    dates of samples obtained in *114*
    specimens from *163*
Jaeger, Jean-Jacques 361
Jalvo, Fernández 104
James, Steven 252, 253
Jankovic, Ivor 487
Java (Indonesia) 266, 282
    carving of tools in *246*

coexistence of *Homo erectus* and
    Homo sapiens in *373*
colonization of 282–3
cultural modes in 246–9
evolutionary meaning of the
    advanced specimens of *Homo
    erectus* in 372–3
fossil associations and tools in *284*
Homo erectus' variability in 285–6
morphological proximity *Homo
    erectur-Homo sapiens* in 370–2
permanence of *Homo erectus* 371–2
transition in 369–373
use of fire in 253
Jiménez-Arenas, Juan Manuel *208*
Jinniushan (China) *373,* 374
Johanson, Donald 115, *142, 143,* 146,
    195, 198
Jolly, Clifford *79*
Jolly, Keith 361
Jordan River 269
Jouffroy, Françoise 225
Judaism *2*
Jungers, William 210
Jungers, W.L. 202
junk DNA 8

K

Kabarnet Trachyte Formation
    (Kenya) 101
Kabua (Kenya) 440
Kabwe (Zambia) 359
Kada Hadar (Ethiopia) 91, *92, 141,*
    143, 144
Kaifu, Yousuke 285
Kaitio (Kenya) 100
Kalambo Falls (Olduvai,
    Tanzania) 240, 253
Kaletepe Deresi 3 (Turkey) *269*
Kalochoro (Kenya) 100
Kamikatsura subchron *324*
Kanapoi (Kenya) 150, *151*
kangaroos *25, 26*
Kappelman, John 277
Kapthurin (Baringo, Kenya) 167,
    280
Karavanic, Ivor 490
Karkanas, Panagiotis 350
K/Ar method *see* PEC (primer
    extension capture)
Kataboi (Kenya) 100
Kathu Pan 1 site (Northern Cape,
    South Africa) 262
Katoh, Shigegirho 280
Kattwinkel, Wilhelm 102
Kay Behrensmeyer Site (KBS) 98–9,
    101, 102, 179
Kay, Richard *79*

Kean, W.F. *251*
Kebara (Mount Carmel, Israel) 403, 408, 478
Keilmesser complex 339
Keith, Arthur 56, 134
Kennedy, Gail 267, 268, 273, 397
Kent, Dennis 100, *307*, 310
Kenyanthropus 172, 187, **498**
Kenyanthropus (*Australopithecus*) *platyops* 167–8
Kenyanthropus platyops 168, 187, 188, 189
Kenyanthropus rudolfensis 187
Kenyapithecus 78, 81
Kenyapithecus africanus 79
Kenyapithecus wickeri 79
Keyser, André 174
Keyser, Andrew 118
KGA10-1 mandible 280
KGA10-525 specimen 179, *180*, 186, 187
KGA10 specimen 179
Khoe-San people 427, 428, 434, 474
Khrameeva, Ekaterina 461
Kibish formation *361, 362*
Kidder, James 298
Kimbel, William 92, 147, 157, 182, 190, 198, 200, 201, 203, 299
Kimble, William 115
Kimeu, Kamoya 197, 279
kingdoms **498**
King, Geoffrey 314
Kingston, John 83
King, William 395, *396*
kin selection 19–21, **498**
Kirschvink, Joseph 116
Klasies River (South Africa)
  caves *436*
  chrono-cultural sequences at *264*
  Homo sapiens 433–39
  localities of *435*
  mandibles *437*
  members and sections of the main site of the *436*
  sample contained in the analysis of *437*
Klein, Richard 262
KNM-BC 1 temporal bone 201
KNM-CH1 cranium 179
KNM-ER 406 cranium 179, 279
KNM-ER 732 skull 179
KNMER 992 mandible 296
KNM-ER 999 femur 440
KNM-ER 1470 cranium 187, 196–7, *197*, 203, *204*, 205, 207, 210
  age of 205
  features of *206*
  sexual dimorphism 206

KNM-ER 1472 femur 196, *204*
KNM-ER 1475 femur 196
KNM-ER 1500 specimen 179
KNM-ER 1805 cranium and mandible 198
KNM-ER 1813 specimen 197–8, 203, *204*, 205, 207
  features of *206*
  sexual dimorphism 206
KNM-ER 3733 cranium 279
KNM-ER 3735 specimen 198, 224
KNM-ER 3883 cranium 279
KNM-ER 3884 cranium 440
KNM-ER 6000 mandible 210
KNM-ER 15000 specimen 244
KNM-ER 25520 mandible *179*
KNM-ER 42700 cranium 280, 304
KNM-ER 62000 mandible 199, 210
KNM-KP 271 humerus 150
KNM-KP 29281 mandible 150, *151*
KNM-KP 29283 maxilla 150
KNM-KP 29287 mandible 151, 152
KNM-KP 34725 specimen 152
KNM-LT 329 jawbone 124
KNM LU 335 molar 128, 129
KNM-OC 45500 specimen 304
KNM-OL 45500 specimen *303*, 335
KNMP-KP 29287 mandible. *153*
KNM-SH 8531 maxilla 124, *126*
KNM-TH 28860 specimen 139
KNM-WT 4000 specimen 167, *168*
KNM-WT 8556 mandible *150*, 168
KNM-WT 15000 specimen 274–5, 280
KNM-WT 16005 mandible 180, 181
KNM-WT 16006 mandible 150
KNM-WT 17000 cranium 180, *181*, 182, 185, 191
  affinities with different taxa *190*
  primitive features shared with *A. afarensis 190*
  species closed to *191*
KNM-WT 18600 mandible *179*
KNM-WT 22936 mandible 150
KNM-WT 38350 specimen 168
KNM-WT 40000 cranium 167, 168, 187, 188
KNM-WT 42718 specimen 201
knowledge, declarative and procedural *241*
knuckle-walking 135, 136, **498**
  as a derived trait 139, *140*
  evidence of *138*
  indicative plesiomorphies of 138
  as a primitive trait *140*
  wrist joint during the swing phase and support phase *136*
Kokiselei 4 (Kenya) 243
Konigsberg, Lyle 187

Konso formation (Ethiopia) 97, *98*, 179, *180*, 280
Konso-Gardula (Ethiopia) 240
Koobi Fora formation (Kenya) 69, 73, 97, 98–100, *101*, 178, 182, 280, 440
  footprints 276–8
  fossils of *100*
  Homo habiliso specimens 196
  Homo specimens 278
  paleomagnetic studies 250
  specimens of *Homo habilis* sensu stricto and *Homo rudolfensis* 205
Koppe, Thomas 406
Koro-Toro (Chad) 122
Korsi Dora (Ethiopia) 92, *93*, 147
Kozarnika Cave (Bulgaria) 332
  choppers *333*
  location of *334*
Kramer, Andrew 207
Kramer, Patricia 148
Kramers, Jan 114
Krause, Johannes 448, 454
Kretzoi, Miklós 351
Krings, Matthias 446, 447
Kromdraai site (South Africa) 89, 115–6, 233
Ksar Akil (Lebanon) 408
KSD-VP-1/1 hip bone *61*, 147, 159
KT 12/H1 mandible 166, 167
Kullmer, Ottmar 180
Kuman, Kathleen 203, 368
Kun, Ho Chuan 288
Kurashina, Hiro 240
Kurtén, Björn 291

L
L7A-125 mandible 181
La-Chapelle-aux-Saints (France) 398, 399
La Chapelle-aux-Saints (France) 470
La-Chapelle-aux-Saints (France), "Old Man" 399
Lacruz, Rodrigo 109
Ladina basaltic tuff (LABT) 133
Laetoli fossil area (Tanzania) 106
Laetoli fossil footsteps (Tanzania) 57, 65–7
  studies of *65*
  type of bipedalism indicated by 67–8
Laetoli Hominid 18, 361
Laetolil Layers 106
Laetoli (Tanzania) 142, 146, 361
  "archaic *sapiens*" crania 368
  localization of *141*
  stratigraphic column and radiometric dating of *183*

La Ferrassie (Dordogne, France) 399
 neanderthal *365*
Lagar Velho (Portugal) 450, *451*
Lahr, Marta Mirazón 262, 387, 476
Lake Baringo region (Kenya) 255
Lake Eyasi (Tanzania) 361
Lake Mungo (Australia) 445, *479*
lakes 86
Lake Turkana basin (Kenya) 87,
 97, *99*
 Kanapoi and Allia Bay 100
 Paranthropus from *179*
Lamarck, Jean-Baptist de Monet,
 chevalier de *2*
 evolutionary theory 31
Lam, Yin Man *437*
language, human 55
Lapedo specimen 450
Laporte, Léo 69, 70, 277
large-brained human species
 (LBH) *469*
large mammals, migration out of
 Africa 86
Larick, Roy 283
Larson, Susan *386*, 386
Lartet, Edouard 485
last common ancestor *51*
 of Neanderthals and modern
 humans 341
Late Pleistocene *267*
Late Stone Age (LSA) 262, 474
Latham, Alfred 111, 115, 163
Latimer, Bruce 134, 136, 147
Latina Valley (Italy) 324, 340
Lattman, Jeffrey *371*
LB1 hominin specimen 379, *384*
 brain 383–4
 cranium *382*, *383*
 dwarfism 385
 endocranium *394*
 facial asymmetries of *383*
 femur and tibia *382*
 femurs *386*
 humerus *386*
 mandibles *382*
 morphology 382
 pelvis *386*
 postcranial elements of 386
 reasons for dwarfism of 382–3
LD 350-1 mandible 201
Leakey, Louis 78, 103, *104*, 105, 175,
 176, 192, *195*, 229, 230, 239, 243
Leakey, Mary (nee Nichols) *104*, *105*,
 146, 150, 178, 192, *195*, 230, 241
Leakey, Mary (née Nichols) 102
Leakey, Mary (nee Nichols)
 cultural sequence at Olduvai *239*
 instruments at Olduvai 239

Laetoli Hominid 18 specimen 361
 zygomatic bone fragment 182
Leakey, Meave 150, 152, 153, 155,
 156, 167, 175, 199, 210
Leakey, Richard 97, *102*, 178, 182,
 196, 197, 198, 224, 274
Leang Jarie cave (Indonesia) 491
Leang Timpunseg cave
 (Indonesia) 491, *492*
Ledi-Geraru mandible 201
Lee-Thorp, Julia 171
Lee-Thorp, Julia A. *172*
Le Fur, Soizic 127
Le Gros Clark, Wilfrid 78
Le Moustier site (Dordogne,
 France) 253
Leonard, William 84, 270
leopards, evolution of 35
Lepre, Christopher 100, 243
Lesotho 384
lesser apes 51, **498**
Levallois points 260
Levallois technique 255, *256*, 259, **498**
Levantine Corridor 243, 314, **498**
Lewin, Roger 191
Lézignan 334
L.H.-4 specimen 143, 146, *147*
Liang Bua site (Flores Island,
 Indonesia) 379–82, 389
 sedimentary units *388*
 utensils 392–4
Lieberman, Daniel *186*, 187, 204, 206,
 225, 389
limb morphology 62–5
lineages 39
Linnaean taxonomy 38
Linnaeus, Carolus 29–30, 32, 415
lions *31*, 39
Lister, Adrian 385
lithic industry, of TD6 *325*
lithic traditions 229–65, **498** *see also*
 tools
Little Foot specimen 63, 64, 162–5
Liu, Wu 291, 375, 376
Liu, Xiling 468, 471
location *40*
Lockwood, Charles 299
locomotion
 of *Ar. ramidus 139*
 of *Australopithecus afarensis 147*
 and dentition 79–81
 and enamel *79*
 primitive and derived traits
 of 136–40
 role in divergence of hominoid
 lineages 134–40
locus **498**
Lokalelei culture *236*

Lokalelei stone artifacts *237*
Lombard, Marlize 475
Lomekwi basin (Kenya) 100
 specimens *168*
 tools 232, *238*
Lompoa Leang (Indonesia) 491
Lonergan, Nicholas *301*
Longgupo Cave (Wushan site,
 China) 247, 288, 289
Longlin specimen (China) 463,
 464, 465
Lonsdorf, Elizabeth 232
Lope, Don *420*
Lordkipanidze, David 217,
 218, *219*
Löscher, Manfred 348
Lothagam Hill (Kenya) 124, *126*
Louys, Julien 295
Lovejoy, Owen 73, 136
Lower Laetoli Beds (Tanzania) 106
Lower Pleistocene
 evolutionary model for the
 hominins of the 335–44
 lithic tools 404
 migratory path of *311*
Lower Pliocene 140–2
loxodonta (elephants) 121
LRAT gene 428
LSA *see* Late Stone Age (LSA)
Luca, A. P. Santa 372
Lucy *see* A.L. 288-1 specimen (Lucy)
Lukeino Formation (Tugen Hills,
 Kenya) *128*
Lyell, Sir Charles *2*

M

McBrearty, Sally 167, 262, 357, 360
Macchiarelli, Roberto 472
McClean, Richard *251*
McCown, Theodore 413
McDougall, Ian 100
McFadden, Phillip 115
McHenry, Henry 224
MacLatchy, L. 78
McNulty, Kieran 81, *385*
McPherron, Shannon 231
Madagascar, pygmy hippo 385
magnetostratigraphy **498**
Magori, Cassian 361, 362
Mahfouz, Mohamed *148*
Maillo-Fernández, José Manuel 262
Major Histocompatibility Complex
 (MHC) 421
Majuangou (China) 246–7
Majuangou, stratigraphic horizons
 with lithic tools of *247*
Maka (Ethiopia), specimens of
 A. afarensis 149

Makah Mera 152

Makah Mera (Ethiopia) 92, 93

Makapansgat (South Africa) 89, 115, 480
  classic section of *115*
  hominins *163*

MAK-VP-1/12 mandible 149, *149*, *156*

Malapa (South Africa) 119–20, 165

malaria 16, 49

Malawi 106–7

Malema, Chiwondo Beds (Malawi) 180

Malez, Mirko 455

Mallegni, Francesco 339

Maludong specimen (China) 463, *465*

mammals 30, 39

mammary glands 36–7

Mammuthus meridionalis 310

mandibles *437*, **498**

mandibular symphysis **498**

Manthi, Frederick 157

manuport *480*

Manzi, Giorgio 298, 339, 340, 344, 348

Marchal, François *397*

Margulis, Lynn 441

marine isotope stages (MIS) 255, *256*, 307

Marino, Bruno 83

Marston, Alvan 349

Martín, Fernando Díez 259

Martin, Lawrence 80, 170, *186*

Martinón-Torres, María 330, 342, 454

Martin, Robert 224, 383, 385

mass spectrometry *353*

Mata Menge (Flores Island, Indonesia) 384, 389
  fossils 387
  stone tools *392*

Mateos, Ana 83

mating
  assortative/selective 14
  random 13–4

Mauer mandible (Germany) 346–48

maxillas *183*, **498**

maximum likelihood methods 27

maximum parsimony methods 25

Mayr, Ernst 5, 32, 40, 53, 154, 266, *268*, 294

Mazza, Paul Peter Anthony *307*

Mediterranean Europe, Aurignacian sequences of *487*

Mediterranean, the 314–18

MEF2A (myocyte enhancer factor 2A) gene *468*

megadontia 140, 150, 157

Mehlman, Michael 361

Meiri, Shai *385*

Melanesia 431

melanin 11

melanoma 17

Mellars, Paul 476, 479

members 40, **498**

Mendel, Gregor 3–4

mentum **498**
  see also chins

Mercier, Norbert 490

Mesgid Dora (Ethiopia) 92, *93*, 152

messenger RNA (mRNA) 17, **498**

metacarpals **498**

metatarsals **498**

Metazoan kingdom 30

Mexico 431

Mezmaiskaya Cave (northern Caucasus) *444*
  child specimen 1 459
  Mousterian handaxes from *260*
  mtDNA comparison from Neanderthals and specimens of *444*–6
  specimen 443, *444*

MH1 (*A. sediba* specimen) 165

MHS-Bayasi (Tanzania) 241

microRNA (miRNA) 17–8

Middle Awash (Ethiopia) 94–6, 130–3, 141, 149, 169, 237
  Adu Asa formation 94
  Bouri formation 94
  Matabaetu formation 94
  paleomagnetic studies 252
  Sagantole formation 94
  Weahietu formation 96

Middle East, sites where "classic" and "progressive" Neanderthals are found *409*

Middle Miocene 78

Middle Paleolithic 259, 487

Middle Pleistocene 267, 366
  evolutionary model for the hominins of the 335–44
  MIS from *308*

Middle Pliocene 140, 160, 345
  consistency of the evolutionary scheme for the gracile and robust australopithecines in the 182–91
  hominin diversity of 187–89
  species of general *Australopithecus, Paranthropus* and *Homo* 161

Middle Stone Age (MSA), in Africa 261–5, 360, *433*, 474

migration 243

Milankovitch, Milutin *307*

Millard, Andrew 361

Millennium Man *129*

Miller, Joseph 208

Milton, Katherine *471*

Mindel-Riss interglacial period, African hominins from the 357–67

minor elements 102

Miocene 85, **498**
  hominins 124–34
  hominoid evolution 78–9
  human genera 153
  known remains of different types of Miocene hominin *154*
  monkeys and apes 79
  phylogenetic tree of hominoid genera *82*

Mirazón Lahr, Marta 309, 312, *313*

Miri (Kedung Cumpleng) 246

MIS 16 glaciation 344

Mishra, Sheila *241*, 246, 249

mitochondria 441, **498**, **499**

mitochondrial DNA 417–18, 424, 441, 442
  genealogical tree of 418–21
  Neanderthals 442–4
  and nuclear DNA 444
  phylogenetic tree of complete *455*
  phylogenies *420*
  populations and species 450–2
  protocol for retrieving fossil mtDNA *443*
  relationships among different mtDNA haplogroup lineages *427*
  retrieval of ancient 442

mitochondrial DNA (mtDNA) 45, **499**

mitochondrial Eve 418, *420*, 421

modern humans 46, 52, 53, 417–8
  adaptive phenotypic changes 431–3
  brain distance between Neanderthals and 468–74
  and chimpanzees last common ancestor 51
  chromosomes 12
  classification of *31*
  cognitive distance between Neanderthals and 474–94
  cytochrome-c amino acids of 24
  dispersal routes in Europe *486*
  encounters with Neanderthals 477–8
  evolutionary history 26
  evolutionary trees 23
  fossil record of 433
  genetic distance between chimpanzees and 35, *36*

genetic diversity 429–31
genetic material flow from
    Neanderthals *459*
hands 62, 139
imprint of the first 427–9
locomotion of *60*
minimum number of nucleotide
    differences in the genes coding
    for in *25*
mitochondrial DNA genetic
    diversity of *447*
mitochondrial DNA phylogeny
    of *420*
molecular evolution of *48*
nasal and frontal sinuses of *406*
Neanderthals and 441–94
in the Near East 414–6
passage of fetus through birth
    canal in female modern *76*
pelvis of modern *61*
phenotypic differences between
    chimpanzees and 50
phylogenetic tree *447*
relational models between
    Neanderthals and *408*
relationship between the cultural
    developments of Neanderthals
    and *257*
scapula 149
substitutions and suppressions of
    bases in genomes of *458*
suprageneric taxonomy of *53*
traces of first modern humans in
    Europe *476*
trait polarity in Neanderthals
    and *404*
traits that distinguish other
    primates from *55*
modern mind 261, 262, **499**
    emergence of 474–6
Moggi-Cecchi, Jacopo 203
"Mojokerto child" 282
molars **499**
molecular biology 5
molecular clock 46–7, **499**
    biases of *48*
    of evolution 27–9
Möller, Hugo *347*
Moncel, Marie-Hélène 481
Monferrato (Italy) 334
Moniz, Marc 160
monkeys *25, 26*
monopedalism *58*
monophyletic groups
    39, **499**
Monte Poggiolo (Italy) 332
Moore, Mark 392, 394

morphological similarity 52
morphometry, advanced
    techniques 34–5
Mortillet, Gabriel de 253–4
Morwood, Michael 389, *393*, 394
Morwood, Mike 379, 381, 382, 387
mosaic evolution *134*
Mosquera, Marina 231
mosquitos *31*
moths *25, 26*
Mounier, Aurélien *397*
Mousterian culture (Mode 3) 229,
    257, 259–61, 400, **499**
    handaxes from Mezmaiskaya Cave
        (Caucasus) *260*
    sites with tool speciemens of *401*
    transition from Acheulean
        (Mode 2) to 253–61
    transition to Aurignacian
        (Mode 4) 261
Movius, Hallam 245, 329, 479
Movius line *245*, 246, 249
Moyà-Solà, Salvador 139
MSD-VP-5/16 mandible 159
mtDNA *see* mitochondrial DNA
Mturi, Amini 240
Mughuret el-Kebara cave (Mount
    Carmel, Israel) 410
"Multiregional Evolution
    Hypothesis" 345
multiregional hypothesis (MH) 346,
    357, 368, 376, 395, 408, 412, 417,
    418, 441, 445, 458, **499**
mutants **499**
mutationism 4
mutations 4, 10–1, **499**
    of chromosomes 12, *13*
    deletions 12
    duplication 12
    effects of 11–2
    gene 10, 11
    point 10, *11*
    rates of 11–2
    translocations 12
Muttoni, Giovanni *307*, 310, 332, 339
myoglobin 27

N

Nachukui formation (Kenya) 97, *99*,
    100
naked ape *416*
Nanjing 1 specimen (China) 375–6
Napier, John 136, 192, 229
Nariokotome (Kenya) 100, 218, 219,
    274, 279, 360
nasal **499**
Nassarius krausianus 484

Natoo (Kenya) 100, *179*
natural remanent magnetism
    (NRM) 250, *251*
natural selection 2, *2*, 16–7, 33, 84, **499**
    directional 18
    diversifying 18
    evolution by 6–7
    impact of Darwin's theory of *4*
    kin 19–21
    modes of 17–9
    reciprocal altruism and group
        selection 21
    sexual 19
    stabilizing 18
    see also sexual selection
Navajo Indians, genotypic and allelic
    frequencies for the M-N blood
    groups in *14*
navicular **491**
"Neanderthal question" 407–8
Neanderthals 34, 57, 259, 260, 261,
    266, 395, **491**
    adaptation to colder climates 403–7
    as ancestral species to *H. sapiens* 407
    birth canal in *77*
    boundary of occupation by *405*
    brain distance between modern
        humans and 468–74
    burials 478–9
    childbirth 77
    classic *401, 402, 412*
    cognitive abilities 489
    cognitive distance between modern
        humans and 474–94
    diet *407*
    as different species from modern
        humans 407
    encounters with modern
        humans 477–8
    in Far East Asia *377*
    features of *399*
    genetic material flow to modern
        humans *459*
    genome 457, 459–61
    geographic extension of sites with
        specimens of *449*
    high-throughput DNA
        sequencing 452–3
    interpretive model 407
    mitochondrial DNA genetic
        diversity of *447*
    and modern humans 441–94
    morphology of 400–2
    mtDNA 442–4
    mtDNA comparison from
        Mezmaiskaya and specimens
        of 444–6

Neanderthals (*continued*)
  multiplication of Neanderthal mtDNA studies 446–8
  nose *406*
  ontogeny of *402*, 471–2, *473*
  origin and age of samples of *453*
  origin of apomorphies of *403*
  phylogenetic tree *447*
  phylogeny of the mtDNA of *454*
  presence in North Africa 366
  "progressive" *412*
  reconstruction of a newborn *473*
  relational models between modern humans and *410*
  relationship between the cultural developments of modern humans and *258*
  remote African origin for *453*
  retrieval of nuclear DNA 454–7
  in Siberia 448–50
  sites with specimens of *398, 401*
  as subspecies of *H. sapiens* 407
  substitutions and suppressions of bases in genomes of *458*
  trait polarity in modern humans and *404*
Neander Valley (Germany) 395
  cranium 57
  discovery of specimens at *396*
Near East
  ages of the caves in the 412–4
  chronological sequences of Upper Pleistocene sites in the *414*
  modern humans in the 414–5
  Neanderthals 397
  populations of the *412*
  sites where "classic" and "progressive" Neanderthals are found *409*
  species in the 411–2
neighbor-joining 24
Nei, Masatoshi 124
Nelson, Andrew 472
Nenguruk Hill (Tanzania) 182
neo-Lamarckism 3
Neolithic 277, **499**
neoteny **499**
neurocranial geometry analysis *463*
Neurospora 25, *26*
Neuville, René 410
Nevell, Lisa *385*
Ngandong site (Java) 281, 286, 322, 369
  age of 370
  Homo erectus 298, 345, 369
  Homo soloensis 286
Ngawi I cranium (Java) 371
Ngebung (Sangiran, Java) 246, *247*
Ngeneo, Bernard 196

niches 22, **491**
Nichols, Mary *see* Leakey, Mary (née Nichols)
Nicolás, María Elena 355
nitrogen bases *8*, 457, 458, **497**
Niujiangbao (China) 247, 289
NM-KP 29285 tibia 150
NMT-W64-160 mandible 177
Noah's Ark hypothesis 418
Noble, William 478, *479*
nodes 24, 27, 36, 37, 38, 39, 53, **499**
Nomade, Sébastien 340
nomads **499**
non-coding RNA (ncRNA) 15
non-utilitarian objects 479–81
Noonan, James 456
Northern Sahara 281–2
northern sea lion 19
North Europe, colonizations of *332*
North Pole 43
Norton, Christopher 294, 376
noses 348, 401, *406*, 407
nuclear DNA 444
nucleic acids 8, 463, **499**
nucleotides 8, 11, **499**
nucleus **499**
numerical taxonomy *33*
Nzube, Peter 152, 195

O

O-69 sample (Spain) **491**
O-87 sample (Spain) **491**
Oase Cro-Magnon 461
occipital torus **499**
Oceania 431
ocean navigation, earliest evidence of *393*
ocher pigment use 475, 476, 477, 484, 489, 494
ODVP-1/1 parietal fragment 360
OH 4 molar and mandibular fragment 192
OH 5 cranium 176, 182
OH 6 teeth and cranial fragments 62, 192
OH 7 (Jonny's Child) 192, *225*, 225
  cranial capacity *208*
  lower jaw and parietal *193*
OH 8 specimen 64–5, 192, *193*, 225
OH 9 specimen 278, 302
OH 10 big toe phalanx 192
OH 12 specimen 278
OH 13 mandible and cranial fragments 176, 192, *194*, 197, 203
OH 16 cranial fragments 192, 206
OH 22 mandible 278
OH 24 (Twiggy) 195
OH 28 femur 274, 278, 353
OH 35 tibia and fibula 192

OH 43 metatarsals 192
OH 48 clavicle 192
OH 62 specimen 195, 198, 224
  femur 225
  fossil fragments *196*
  and variability within *Homo habilis* 195–6
OH 65 specimen 195, 196, 210
OH 86 specimen *226*
Ohman, James 274
Okladnikov (Russia) 403, 448–50
"Old Man" (La Chapelle-aux-Saints, France) 399, 478
Oldowan culture (Mode 1) 229, 234–8, 330, **499**
  advanced *239*
  symbolic behavior *480*
  temporal distribution of *272*
  transition to Acheulean Mode 2 239–49
Olduvai (Tanzania) 42, 64, 73, 102–5, 160, 192, 234, 362
  age at death of fossil *194*
  cultural sequence at *239*
  Homo ergaster specimens 278–9
  Homo habilis sensu stricto specimens *205*
  Homo habilis specimens 195
  Homo rudolfensis specimens *205*
  human fossil specimens 105
  subchron 44
Old World monkeys **499**
Oligocene 78, 91, 422, **499**
Olorgesailie (Kenya) 240, 241
Olson, Todd 145
omnivores **499**
Omo-323-1976-896 specimen 179
Omo (Ethiopia) *95*, 96–7, 179, 199, 440
Omo II cranium 363
Omo I specimen 363
On the Origin of Species by Means of Natural Selection (Darwin) 1, 3
ontogenetic **499**
ontogeny **499**
Oppennoorth, W.F.F. 286
Oppennoorth, Willem 369
Optimal Body Size Theory (OST) 385
orangutans 45, 52, 53, *79*
  brachiation 79, 135, 136
  enamel 79
  population size 55
orbit **499**
orders 30, **499**
organelles **499**
Oron, Maya 482
Orrorin 53, *125*, 139, 154, **499**
  hominin status *130*
  hypodigm of *129*

Orrorin tugenensis 102, 128, *129*, 133, *154*
Orschiedt, Jörg 339
orthognathic **499**
orthograde **499**
orthologous genes or chromosomes **499**
orthologous traits 27
os centrale, fusion of *139*
Otte, Marcel *239*, 241, 249, 259, *317*
OTU (Operational Taxonomic Unit) *33*
Ouranopithecus 124
Out of Africa hypothesis *see* replacement hypothesis
Ovchinnikov, Igor 444
ovum **499**
Owen, Richard 31
oxygen isotopes *256*

P
Pääbo, Svante 430, 442, 453
palaeoart 481
palate **499**
paleoanthropology 32, 34
Paleocene **499**
Paleolithic **499**
paleomagnetic events, of the last 2.6 Ma *323*
paleomagnetic studies 250
paleomagnetism 43, 109, **499**
paleontology 32
Paleopithecus sivalensis 58
Palmqvist, Paul 208
Pan 81, 135, **499**
    locomotion and enamel of *79*
    phylogenetic relation between *Gorilla*, and *Homo* and *48*
Panagopoulou, Eleni 350
Pangea *6*
pangenesis *3*
Panini tribe 53, 124, 154
Pan troglodytes (common chimpanzee) *133*, *209*, *267*
Papio species 117
Papio ursinus 117
paralogous traits 27
Paranthropinae 173
Paranthropus 53, 54, *125*, 140, 145, 160, 169, 172–82, 192, 266, **499**
    characteristic traits of *174*
    diet *171*
    ear semicircular ducts *75*
    molars *186*
    monophyly 185–7
    phylogenetic relationships between species of 190–1
    rejection of genus *178*
    species of genus *161*

Paranthropus aethiopicus 106, 177, 179, 180–2
Paranthropus boisei 119, 175–80, *181*, 182, 185, 187, 191, 229, 238
    sister group with *P. robustus 186*
    sites *178*
    variability of *186*
Paranthropus crassidens 174
Paranthropus robustus 115, 116, 118, 160, 169, 173–5, 180, 185, 190, 191, 252
    diet 171, 172
    of Drimolen *177*
    sister group with *P. boisei 186*
paraphyletic groups 39, **499**
paratype **499**
Paraustralopithecus aethiopicus 181
Parés, Josep M. 318
Parfitt, Simon 404
parsimony **499**
Partridge, Timothy 109, *110*, 115, 116, 162
Passarello, Pietro 298
Patel, Biren 139
patella **499**
Paterson, James 117
Patterson, Bryan 150
PCR (polymerase chain reaction) 5, 442, **444**, **500**
Pearson, Karl 4
Pearson, Osbjorn *437*
PEC (primer extension capture) 452–3, **500**
Pei-Hua, Huang 293
pelvis 60–1, *61*, 74, 75, 76, **500**
    fossils 61
    Gona (Ethiopia) 275–6
Penck, Albrecht 305
penguins *25*, *26*
Peninj (Tanzania) 241
Pérez-Claros, Juan A. 208
Pérez-González, Alfredo 319–20
permanent dentition **500**
Pestera cu Oase (Romania) 460
Petraglia, Michel 476
Petralona cave (Greece) cranium 349–50
phalanges **500**
phenetics *33*, **500**
phenograms *33*
phenotype 15, **500**
Philip Tobias Korongo (PTK) (Tanzania) *226*
phyletic **500**
phyletic lineages, problems of 37–8
phylogenesis **500**
phylogenetic distances 47
phylogenetic trees 81, *82*, **500**
phylogenies 23, 35, 37, 39, **500**

phylum **500**
Pickering, Robyn 112, 200, 201, 203, 221
Pickering, Travis *112*, 236
Pickford, Martin 101, 128, *129*
Pierolapithecus 138
pigeons *25*, *26*
pigs *25*, *26*
Pike, Alistair 491
Pilbeam, D. 193
Pilbeam, David 206
Piltdown fossil (England) 56–7
Pirro Nord (Italy) *307*, 333, 334
Pithecanthropus 58, 285, 399, **500**
Pithecanthropus alalus 268
Pithecanthropus erectus 57, *58*, 192, 266, *268*
Piveteau, Jean 397
Plasmodium falciparum 16
plate tectonics *6*
platyrrhines (American monkeys) 45, **500**
Plavcan, Michael *145*, 157
Pleistocene 44, 51, 85, **500**
    evolutionary events of *267*
    maximum extension of ice in the northern hemisphere during *405*
plesiadapiform **500**
Plesianthropus 173
plesiomorphies 36, 140, **500**
Pliocene **500**
    climate change *89*
    environmental change 87
Pliocene hominins
    diversity of 159
    geographic and temporal distribution of *227*
    taxa *227*
Plio-Pleistocene 44, *267*
Poirier, Frank 376, 397
polymerase chain reaction (PCR) 5, 442, **444**, **500**
polymorphism **500**
polypeptides **500**
polyphyletic groups 39, **500**
Pongidae family 45, 53
Pongo 79, 136, **500**
Pontzer, Herman *225*, 471
Popper, Karl 451
population dispersal variables 270–1
populations **500**
population size 55, *86*, 426
positive selection *467*
postcranial **500**
potassium/argon (K/Ar) method 41, **500**
potassium isotopes 42
Potts, Richard 303

Poulianos, Aris 349, 350
Po Valley (Italy) 332, *336*
Praeanthropus africanus *147*, *168*
Prat, Sandrine 201
premolars **500**
Prentice, Michael *307*
previtamin D$_3$ 431
primates 30, **500**
  suborders 45
  traditional classification of *46*
primer extension capture
    (PEC) (high-throughput
    sequencing) 452–3, **500**
primitive characters 36
primitive traits **500**
Proconsul 78, *139*
  hominoid clades 80
prognathism 207, **500**
prosimians **500**
Prossinger, Hermann 348
proteins **500**
  evolution of *453*
Protoaurignacian 486, **500**
Proto-Cro-Magnons 414
Prüfer, Kay 459, 460
pseudogenes 27
purines 25
Pusch, Carsten 328, 383
pyrimidines 8, 25

Q

Qafzeh cave (Israel) *365*, 408, 410,
    413, *433*, 479
Qafzeh specimens (Israel) *400*, 411,
    412, 483
quadrupedalism **500**
Quan, Yang 376
Quaternary **500**
Quinqi, Xu 374
Quneitra fragment *482*, *482*
Quneitra (Golan Heights, between
    Israel and Syria) 482, 483
Quneitra/nl(Golan Heights, between
    Israel and Syria), artifact *483*

R

rabbits *25*, *26*
races 429
radioactive isotopes 41–2
radioactivity **500**
radiocarbon dating **500**
radiometric methods 41–3
radius
  and central joint complex
    (CJC) *139*
Rae, Todd 406
Rahmstorf, Stephan 87
Raia, Pasqual *385*

Rak, Yoel 180, 203, 403
Ramapithecus *132*
ramus **500**
random mating 13–4
ranks 39
rattlesnakes *25*
Raub, Thimothy 116
Ready, Elspeth *407*
recessive **500**
Reck, Hans 102, 105
recombinant DNA techniques 9
red ocher 475
Red Sea 314
Reed, Kaye 144
reflectance index *432*
relative fitness 7
Reno, Philip *145*
replacement hypothesis 345, *346*,
    357, 368, 408, 417, 418, 429, 441,
    461, **500**
reproductive isolation 22, 37
Revolution of Middle Pleistocene 87
rhesus monkeys 23
  cytochrome-c amino acids of *24*
ribonucleic acid (RNA) 7, **500**
ribonucleotide reductase M2
    polypeptide pseudogene 4
    (RRM2P4) *420*
Richards, Michael *407*
Richmond, Brian 136, 153, *181*,
    186, *204*
Riel-Salvatore, Julien 479
Rift Valley, Tugen Hills (Kenya) 81,
    83, 86, *89*, 91–107, 140
  Hadar region 91–2
  main fossiliferous areas of *91*
  specimens from 280–1
  WORMILPRP (Woranso-Mille
    Paleontological Research
    Project) 92
Rightmire, G. Philip 266, 358, 360,
    361, 435
Rightmire, G. Philippe 294
Rightmire, Philip 207, 358, *359*, 360,
    433, 434
Rightmire, Philip G. 377
Ring, Uwe 107
Rising Star cave system (South
    Africa) 214
RNA 7, **500**
Robertson, Marcia 84, 270
Robinson, John 115, 231, 281
robust
  australopiths 170
  hominins 169
  specimens **500**
Roche, Helene 236
Rodriguez, Xosé Pedro 231

Roebroeks, Wil 305
Rogers, Michael 230
Rolian, Campbell 62, 225
Ron, Hagai 244
ROR2 gene 428
Rosas, Antonio 215, *472*
Rosenberg, Karen 374
Rosenberg, Noah 429
Rozzi, Fernando Ramírez *472*, *473*
Rozzi, Ramirez *132*
RRM2P4 pseudogene *420*
Ruff, Christopher 224, 275,
    277, 374
Rukang, Wu 288, 374
Runnels, Curtis 350
running 72, 73
  and walking 74–5
RUNX2 gene 428, *429*, *467*
Ryonggok cave (North Korea) 376

S

Sabater Pi, Jordi 230
Saccharomyces *25*, *26*
sacrum 61, 162, **500**
sagittal crest *161*, **500**
Sagri, Mario *307*, 311, 314
Sahara, barrier to migration from
    Africa to Europe 311–4
Saharan Africa *228*
Sahelanthropus 122, *125*, *128*,
    154, **500**
Sahelanthropus tchadensis 127, *154*,
    166, 187, 312
St. Acheul site (France) 239
St. Césaire (Charente-Maritime,
    France) 400
Sala, Naomi 356
Salé (Morocco) specimen 281,
    361, 470
Sambungmacan (Java) 286
  Sm1 371
  Sm3 calotte *274*, 370, 371, *372*, 376
  Sm4 288, 372
  specimens 322, 370, 371, 372
Samburu Hills (Kenya) 124
San Bushmen 428, 476
Sang-Hee, Lee 218, *219*
Sangiran (Java) 199, 246, 282, 283–5
  map of *247*
  skullcaps *285*
Sarstedt (Germany) 328, 339
savannas 83, **500**
  emergence of *89*
  hypothesis of adaptation to the
    open 87–91
Savin, Samuel 144
Saylor, Beverly 153
Scandia, Giancarlo *307*

scapulas 149, **500**
Scardia, Giancarlo 310
Schaafhausen, Hermann *396*
Schepartz, Lynne 291, 294, 376
Schick, Kathy 240, 243
schizophrenia *467*
Schlanger, Nathan 241
Schlosser, Max *291*
Schmerling, Phillip Charles 395
Schmid, Peter *145*
Schoetensack, Otto *347*
Schrenk, Friedemann 107
Schreve, Danielle 330
Schultz, J. 203, 210
Schultz, Michael 215
Schwarcz, Henry 109, 414
Schwartz, Jeff 289
Schwartz, Jeffrey *219*, 335, 404, *451*
scrapers 260
screwworm flies *25, 26*
SDCCAG8 gene 428
sedimentary rocks **501**
selection *see* natural selection
selection coefficient 16
selective mating 14
selective pressure *62*
selective value *see* adaptive value
Sémah, François 246
Semaw, Sileshi 131, 237, 239
sensu lato **501**
sensu stricto **501**
Senut, Brigitte *79, 129*, 130, 136
Serengeti 172
Serre, David 430, 447, 448
Sesé, Carmen 319
sex cells *see* gametes
sex chromosome **501**
sexual dimorphism 206–7, *402*, **501**
sexual selection 19
Shang, Hong 375
Shanidar (Iraq) 397, 408, 411, 478
    cranium 412
    specimen 478
shared derived characters **501**
Shaw, John 114, 163
shells
    marine 489
    perforated *485*
Sherwood, Richard 201
Shipman, Pat *396*
Shungura culture *236*
Shungura formation (Ethiopia) *95*,
    96, 100, 179, 199
siamangs 79
Siberia, Neanderthals in 448–50
Sibudu cave (South Africa) 263, *264*
Sicilian Channel 314, *317*
sickle cell anemia 16

Sidi Hakoma (Ethiopia) 91, 92,
    141, 144
Silal Artum (Upper Ndolanya Beds,
    Tanzania) 182
Silberberg Grotto (South Africa) 109,
    111, 112, *113*, 114
Sillen, Andrew 116, 171, 252
Silverman, Nicole *181, 186*
Sima del Elefante (Spain) 318, *320–1*,
    340, 342, 343
    fauna and age of *321*
    stone tools 322
Sima de los Huesos (Spain) 214, 215,
    353–7, 481
    age at death of samples found
        at *355*
    crania 355, *356*
    DNA from *461*
Simanjuntak, Truman 246
simian **501**
Simons, E. 193
Simpson, George Gaylord *5*, 5, 134
Simpson, G. G. 45
Simpson, John 414
Simpson, Scott 275
Sinanthropus lantianensis 288
Sinanthropus pekinensis (Peking
    man) 290, *294*, 375
Singer, Ronald 361
single nucleotide polymorphisms
    (SNP) 427, *457*, 466, **501**
single-species hypothesis *204*, 205, **501**
Sirakov, Nikolay 332
sister groups **501**
sister species 38
sites *40*
sitting 59
Siwaliks (Pakistan) *58*
Sixun, Yuan 293
SK 52 facial fragment 174, *176*
SK 847 partial cranium 203, *205*, 210,
    211, *211*, 212–3, 335
Skhul cave (Israel) 408, 410, *433*
    age of 413
    cranium *410*
    specimens *400*, 411
Skinner, Matthew 218
skin pigmentation 432
Skuhl (Near East) 259, 260, 478,
    479, 483
skulls **501**
SKW 18 cranium 174, *176*
Smilodon 89
Smith, C.M. *437*
Smith, Fred *402*, 490
Smith, Grafton Elliot 56
Smith, Heather 203, 335
Smith, Tanya 472, 473

snakes *26*
SNP (single nucleotide
    polymorphisms) 427, *457*,
    466, **501**
Soa basin, Flores (Indonesia) *391*
Soanian-type traditions *246*
Social Darwinism 3
sociobiology 6
Soejono, R.P. 379
Soergel, Wolfgang *347*
soft-hammer technique 243, **501**
Solecki, Ralph 397, 411, 478
"Solo Man" (*Homo (javanthropus)
    soloensis*) 369
Song Terus Cave (Punung, East
    Java) 246
Sonia, Harmand *238*
South Africa 73, 83, 87
    emergence of the modern
        mind 474–6
    hominins 172
    Homo erectus fossils 281
    Homo habilis in 201–3
    sample 428
South African Bushmen 368
South Africa sites 107–20, *434*
    dating sediments 109
    Drimolen 118–9
    Gladysvale *118*
    Koro-Toro 122
    Kromdraai site 115–6
    Makapansgat 115
    Malapa 119–20
    north of the Rift Valley 120–3
    Sterkfontein cave 109–14
    structure and dating of 107–9
    Swartkrans Cave (South
        Africa) 115–6
    Taung 116–7
    use of fire in 252
    where SB and HP artifacts were
        found *263*
South America 1
South Pole 43
South Turkwel (Kenya), specimens of
    A. afarensis 150–1
Spain 333
speciation 37, **501**
    species and 21–2
species **501**
    concepts of 32, 38
    different types of 32–4
    evolutionary concept 32
    identification 35
    phenetic concept 32
    polyphyletic species concept 32
    and speciation 21–2
    trait variation among *338*

speleothems *108, 112*
Spencer, Herbert *3*
Sponheimer, Matt *172*
Spoor, Fred *304*
SPTLC1 gene *428*
STD-VP-2 teeth and partial
    clavicle *133*
Stebbins, George Ledyard *5*
Stebbins, G. Ledyard *5*
Steele, James *86, 236*
Stegodon *392, 393*
Steinheim cranium (Germany) *348–9*
Steininger, Christine *174*
Stekelis, Moshe *410*
stem species *36*
Sterkfontein cave (South Africa) *63,*
        89, 109–14, 161, 201, 204
    ages of members *114*
    bones *233*
    diversity of *162*
    fauna *112*
    main hominins from *163*
    map *110*
    paleomagnetic sequence *112*
    paleomagnetism analysis of *114*
Stern, Jack *225*
Stewart, T. Dale *411*
Still Bay (SB) (South Africa) *262, 265,*
        474, 484
stochastic clocks *28–9*
Stone Age *253*
stone tools *229, 322*
    cut marks *236*
    Mata Menge (Flores) *392*
Stout, Dietrich *239*
Strait, David *136, 147, 159, 188*
strata *40,* **501**
stratigraphy *40–1,* **501**
stratocladistics *81,* **501**
Straus, William *361*
Street, Martin *339*
Stringer, Chris *207, 268, 274,*
        294, 346, 397, 406, 413, 414,
        416, 478
Stringer, Christopher *346, 250*
Structure (program) *430, 431*
Sts 19 cranium *203*
StW 53 cranium *202, 203, 210, 211,*
        234
    cranial features of *212–3*
    fragments *211*
Stw 151 specimen *203*
Stw 183 specimen *162*
Stw 255 specimen *162*
StW 431 partial skeleton *162*
StW 505 cranium *161, 162*
StW 573 specimen (Little Foot) *63,*
        64, 162–5

suborders *45*
suids (swine) *102, 121*
Sulawesi (Indonesia) tools *491*
SULF2 gene *428*
Sungir (Russia) *479*
superfamilies *45*
supraorbital torus **501**
Susman, Randall *174, 225*
suspension *79, 138*
suspensory behavior *139,* **501**
Sutikna, Thomas *379, 381*
Sutton, Morris *116*
sutures **501**
Suwa, Gen *130, 157, 179, 186,*
        205, 296
Suzuki, Hishasi *410*
Swanscombe cranium (England) *349*
Swartkrans Cave (South Africa) *108,*
        109, 115–6
    bone tools *232*
    dating of *116*
    photograph of *117*
    stone industry *233*
    use of fire in *252*
Swisher III, Carl *370*
Sylvester, Adam *148*
symbolism *481,* **501**
Symbolism Mode 1 (SM1) *481*
Symbolism Mode 2 (SM2) *481–3*
Symbolism Mode 3 (SM3) *481, 484,*
        489–90
Symbolism Mode 4 (SM4) *481, 485,*
        490–4
symbols *479,* **501**
sympatric species *33, 205, 234,*
        302, **501**
synapomorphies *25, 36,* **501**
synthetic theory of evolution *4*
Systema Naturae (Linnaeus) *30*
systematics **501**

T

Tabun (Mount Carmel, Israel) *408,*
        410, 414
    age of *412*
    mandible specimen *410*
    specimens *411*
Taieb, Maurice *91, 237*
talus **501**
    Little Foot *63*
Talus Cone Deposit *116*
Tanzania *86*
taphonomy *451,* **501**
Taramsa (Egypt) *439*
tarsal bones **501**
Tattersall, Ian *289, 346, 371, 377,*
        404, 451
Taung Child *107, 109, 148, 161*

Taung (South Africa) *116–7, 160*
Tautavel Man *351, 352*
taxa *30,* **501**
taxonomy *29–40, 45–77,* **501**
    classification of living beings *29–31*
    immunological methods *45*
    nominalist *32*
Tchad basin (South Africa) *121*
TD6 specimens
    age *324*
    lithic industry of *325*
    schematic plan of *325*
    stratigraphy of *327*
Teaford, Mark *156*
Technique 454 (T454) *453*
teeth, wear patterns *170*
    see also dentition
Telanthropus *281*
temperature *84*
temporal bone **501**
temporomandibular and mastoid
        joints *59*
Terberger, Thomas *339*
Terhune, Claire *299*
Ternifine specimens (Algeria) *281*
Tertiary **501**
Teshik-Tash (Uzbekistan) *449, 479*
tetrapods **501**
Thackeray, Francis *116*
THADA gene *466*
thalamus **501**
thermal-ionization
        mass-spectroscopy (TIMS) *353*
thermoluminescence method *43,* **501**
thermo-remanent magnetism
        (TRM) *250*
Theropithecus gelada *157*
Thoma, Andor *351*
Thompson, Jennifer *472*
Thorne, Alan *370, 434, 445*
Throckmorton, Zack *148*
thymine *8, 25*
Tianyuan, Li *376*
tibia *62,* **501**
Tiemei, Chen *293, 374, 376*
Tiemei, Chen *374*
tigers, evolution of *25*
time *34, 37*
tissues, gene expression in *50*
TM 262–01–60–1 specimen *127*
TM266 vertebrates (Chad) *122*
TM 1517 *173*
TM 1517 specimen *115, 116, 173*
Tobias, Phillip *87, 109, 146, 169, 178,*
        180, 192, 195, 197, 202, 206,
        210, 229
Tocheri, Matthew *386*
tokogenetic relations *36*

tombs 478–9
tools 221
  carving of tools in Java *246*
  Dmanisi, Georgia *244*
  flakes *246*
  Levallois technique 259
  making of 229–65
  non-manufactured *235*
  Oldowan culture (Mode 1) 199, 234
  pre-cultural uses of 230–3
  for scavenging and hunting 192
  use in apes *232*
  uses of bones as 232
Toros-Menalla (Chad) 120–2, 126–8
Torres, Trinidad 353
tortoises 1
total energy expenditure (TEE) 471
Toth, Nicholas 235, 240, 243
Toussaint, Michel 162
trace elements 71, *102*
Trachypithecus cristatus, dentition *157*
traits, in mosaic evolution *134*
transfer RNA (tRNA) 7
transgenerational epigenetic
    inheritance (TEI)
    phenotypes 15
transitional industries *257*
transitions, nucleotide bases 25
translocations of chromosomes 12
transversions, nucleotide bases 25
Trauth, Martin 86, 87
Tree of Life *34*
Trinil (Java)
  femur 266, *268*
  fossil 57
  specimens *267*
Trinkaus, Eric 346, 375, 402, 478
Trinkaus, Erik 375, *396*, 402, 407, 411
triplets **501**
Tryon, Christian 262
tuff, volcanic **501**
Tugen Hills region (Kenya) 101–2
  chronostratigraphy of *104*
  hominins 128–30
tuna *25, 26*
Turkana footprints (Kenya) 68–71
Turkana (Rift Central, Kenya) 86, 87
  Homo habilis specimens 195–200
Turner, Alan 289, 295
turtles *25, 26*
Tuttle, Russell *79*
Tutu, Desmond, genome
    sequencing *428*
Twiggy (OH 24) 195
type specimen **501**
Tzedakis, P.T. *400*

U
UA 31 cranium 280, *281*
UB 335 incisor 244
Ubeidiya (Israel) 244
ulna *138*, **501**
Uluzzian technocomplexes 489
unfair competition, hypothesis of 330
Ungar, Peter 156, *157*, 170, 171
ungulates *89*
universal application species 32
U/Pb method 42
Upper Laetoli Beds (Tanzania) 106
Upper Miocene 83, 121, 124
Upper Ndolanya Beds (Tanzania)
    106, 182
Upper Paleolithic 487, 488
Upper Pleistocene 366
  chronological sequences of Upper
      Pleistocene sites in the Near
      East *414*
  hominid crania *366*
  MIS from *308*
  species of the 395–440
Upper Pliocene 51, 140, 161
  adaptations 170
  consistency of the evolutionary
      scheme for the gracile and
      robust australopithecines in
      the 183–92
  hominin diversity of 188–90
  hominin transition to 347–442
  species of general *Australopithecus,
      Parathropus* and *Homo* 162
Upper Semliki Valley (Zaire) 233
UR 501 mandible 179, 199, 200,
    201, 204
uranium series **501**
Ursus (bears) *89*
U-series disequilibrium dating *493*
UVMED dose *432*, 432
UV-radiation 17

V
V1519 incisors 288
Vallesian crisis 79
Vallois, Henry-Victor 351
van den Bergh, Gerrit D. 385
van der Made, Jan 83, 84, *89*, 271, *272*
Vandermeersch, Bernad 410
Vandermeersch, Bernard 275, *397,
    400*, 410, 414, 416, 478, 479, 483
van Donk, Jan 305
van Kolfschoten, Thijs 305
variation *4*
Vatte di Zambrana 470
Vaufrey, Raymond 348
vegetation, changes in East
    Africa *172*

Vekua, Abesalom 217, 218
Velika Pécina (Croatia) 490
Verhaegen, Marc 80
Verhoeven, Theodor 389
vertebrate **501**
Vertés, László 351
Vértesszöllös (Hungary) 350, 351
vertical climbing 137
Vi33.16 (previously Vi.80) 457, *458*
VI33.25 457
Vi33.26 457
Vi-80 (now Vi33.16) 455
Vieillevigne, Emmanuelle 257
Villafranchian fauna 87, *89*, 105, **501**
Villmoare, Brian 201, 298, 299, 300
Vindija (Croatia) *407*, 455, 457
vitamin D 11, 432
Vivien Evelyn (Fuchs) Korongo
    (VEK) *105*
volcanoes 41, 42
von Koegniswald, G. H. Ralph 284
von Koenigswald, Gustav
    Heinrich 246
von Koenigswald, Gustav Henry
    Ralph 282
Vrba, Elizabeth 87, 89, 109, 169

W
Wadi Amud (Israel) 410
Wagner, Günther *347*
Walanae Basin (South Sulawesi) *491*
Walker, Alan 150, 156, *179*, 180, *181,
    182*, 191, 198, 274, 275
Walker Circulation 87
Walker, Joanne 111, 163
walking 72
  and running 74–5
Wallace, Alfred Russell 2
Walter, Robert 91, 92
Wang, Wei 250
Wanpo, Huang 288, *288*, 291, 374
Ward, Carol *134, 144*, 146, 147, 151,
    156, 157
Ward, Stephen 124
Washburn, Sherwood 136, 233
Watson, James 5
Weaver, Timothy 77
Weber, Jochen 385
Weidenreich, Franz *268*, 345, 372
Weiner, Steve 253
Weinert, Hans *147*, 348
Weismann, August 3–4
Wernicke's area **501**
Westaway, Kira 379, 381
Weston, Eleanor 385
West Turkana (Kenya) 100, 182, 279–80
  Nachukui formation *103*
  specimens of *A. afarensis* 149–50

White, Mark 330
White North Americans, genotypic and allelic frequencies for the M-N blood groups in *14*
White, Randall 490
White, Tim 115, 130, 134, 136, 138, 146, 156, 158, 187, 198, *337*, 363, 365
   analysis of post-cranial remains of *Ar. ramidus* 137
   analysis of traits of KNM-WT 17000 190
   and classification of Lucy 143
Widianto, Harry 371
Wiley, Edward 32
Wilkins, Jayne 262, 474
Willerslev, Eske *447*
Williams, Frank 117
Williams, Scott 139
Will, Manuel 263
Wilson, Allan 442
Wilson, Edward 84
wings 32
Witti Mixed Magmatic Tuff (WMMT) 133
Wolpoff, Milford 52, *79*, 198, *257*, 289, 345, 346, 349, 351, 373, 378, *379*, *402*, 407, 412, 487
   cranial volumn estimation *208*
   gracile and robust forms of *Australopithecus* and *Paranthropus* 171
   single species hypothesis *205*
   and *S. tchadensis* 125
Wood, Bernard 59, 68, 154, *181*, 185, 187, 194, 198, 203, 204, 221, 237, 289, 294, 299, *346*, *379*, *385*, 387, 416
   characteristic traits of *Parathropus* 174

dentition 153
determining meaning of the genus *Homo* 223
evolutionary context of the first hominins *186*
evolution of bipedalism 75
Homo erectus and 268
hominization grades *301*
   morphology of KNM-ER 1470 and 1813 206
   obituary of Mary Leakey 102
   taxonomic classification 53
Wood, Christopher 187
Woodward, Arthur Smith 360
Woranso-Mille site 92, *93*, 152, 153
   postcranial elements 159
   specimens of *A. afarensis* 147–8
WORMILPRP (Woranso-Mille Paleontological Research Project) 92
Wright, Sewall 4
wrist-radius joint 138
Wu, Liu 253
Würm glaciation 249, 412, 414
Wu, Rukang 293
Wurz, Sarah 475
Wu, Xinzhi 375
Wu, Xiujie 294, 295, 375, 376
Wymer, John 349

X

Xingren, Dong 374
Xinzhi, Wu *253*, 375, 376

Y

Yangtze riverbed, Sichuan province 247
Yesuf, Ali 200
Yingkou (China) 274

Yinyun, Zhang 374, 375, 376
Yiron (Israel) 244, 269
Y molars **501**
Yokoyama, Yuji 370
Yuanmou (China) 246
Yunxian (China) 373
Yuzhu, You 374

Z

Zaim, Yahdi 285
Zaire, River 81, 167
Zdansky, Otto 290
Zechun, Liu 291
Zeitoun, Valery 286, 371, 372
ZFY genes 424, 426
Zhang, Pu 249
Zhang, Yinyun 375
Zhisheng, An 288
Zhoukoudian III (ZKD III) cranium 292
Zhoukoudian specimens 253, 285, 286, 289–95, 298
   absolute ages *296*
   evidence of use of fire 253
   Locality 1 *253*, 293, 294
Zhoukoudian V (ZKD V) cranium *295*, 375
Zhu, Rixiang 289
Ziegler, Reinhart 397
Zilhão, Joao 485, 486
Zinjanthropus 230
Zinjanthropus boisei 176, 192
Zipfel, Berhard 166
zircon crystals 42, 43
Zuné, Lü 374
Zwigelaar, Tom 359
zygomatic arch **501**
zygomatic bone **501**
zygotes **501**